15th Process Plant Safety Symposium 2013

(PPSS)

Topical Conference at the 2013 AIChE Spring Meeting and 9th Global Congress on Process Safety

San Antonio, Texas, USA
28 April - 2 May 2013

ISBN: 978-1-62748-445-9

Printed from e-media with permission by:

Curran Associates, Inc.
57 Morehouse Lane
Red Hook, NY 12571

Printed by Curran Associates, Inc. (2013)

AIChE
3 Park Avenue
New York, NY 10016-5991

Phone: (203) 702-7660
Fax: (203) 775-5177

www.aiche.org

Additional copies of this publication are available from:

Curran Associates, Inc.
57 Morehouse Lane
Red Hook, NY 12571 USA
Phone: 845-758-0400
Fax: 845-758-2634
Email: curran@proceedings.com
Web: www.proceedings.com

TABLE OF CONTENTS

Business and Regulatory Aspects of Effective Near-Miss Management Implementation

Ulku G. Oktem, PhD
Near-Miss Management LLC
Philadelphia, PA USA (Tel: 267-603-2378; e-mail: oktem@nearmissmgmt.com)

Wharton Risk Management and Decision Processing Center
University of Pennsylvania, Philadelphia, PA 19104 (Email: oktem@wharton.upenn.edu)

Iclal Atay, PhD
Chief Bureau of Release Prevention, Division of Environmental Safety and Health
Department of Environmental Protection, New Jersey, USA (Email: Iclal.Atay@dep.state.nj.us)

Prepared for Presentation at
American Institute of Chemical Engineers
2013 Spring Meeting
9th Global Congress on Process Safety
San Antonio, Texas
April 28 – May 1, 2013

Business and Regulatory Aspects of Effective Near-Miss Management Implementation

Ulku Oktem, PhD
Co-Founder and President, Near-Miss Management LLC
Philadelphia, PA USA (Tel: 267-603-2378; e-mail: oktem@nearmissmgmt.com)

Adjunct Professor and Senior Research Fellow, Risk Management Center
Wharton Business School, University of Pennsylvania (Email: oktem@wharton.upenn.edu)

Iclal Atay, PhD
Chief Bureau of Release Prevention, Division of Environmental Safety and Health
Department of Environmental Protection, New Jersey, USA (Email: Iclal.Atay@dep.state.nj.us)

Abstract

Accidents typically have a history of warning signals that are leading indicators or precursors of major events causing business disruptions. Collectively these signals are called near-misses. The definition of a near-miss was debated until about a decade ago, but now the concept is much better understood by most corporate safety, health and environmental groups. Yet, its effective implementation still lags behind standard safety practices.

In this paper, principles of an effective near-miss management system are discussed and both regulatory and business aspects of near-miss management are addressed. The regulatory review targets various near-miss related regulations and the challenges faced in their implementation. The business aspect focuses on cost effective and results oriented methodologies to meet regulatory obligations. Examples are presented to facilitate implementation process.

1. Introduction

Every year chemical process industry experiences significant financial losses due to accidents and unplanned shutdowns. This number for chemical and petrochemical industry in US alone is greater than $10 billion per year excluding major accidents, such as 2010 BP Deepwater Horizon accident at Gulp of Mexico and 2005 Texas City refinery accident. A quick look at Chemical Safety Board's (www.csb.gov) website will be sufficient to be reminded of major accidents – between January 2010 and February 2013, there are 14 completed investigations whose reports are released and 7 ongoing investigations. All these accidents are tip of the pyramid and almost all cases have numerous near-misses associated with it.

The near-miss concept has long been understood in chemical industry but integration of near-miss management into the culture and day to day operations in a manner that makes everyone in the organization both responsible and accountable from near-miss management is still in its infancy in many organizations. This is one area where regulations are not very effective beyond encouraging management.

In this paper, we will first present the regulatory space mainly focusing on New Jersey. Next, we will talk about key components of integrating near-miss management into the business fabric and give some examples, and finally, we will bring up some ideas on connection (or disconnection) between the two systems.

2. Regulatory Landscape

Federal Regulations

Under the authority of section 112 (r) of the Clean Air Act, the United States Environmental Protection Agency (USEPA) promulgated the Chemical Accident Prevention (CAP) rules in 1998.

The CAP rules are built upon existing industry codes and standards and require facilities that use certain flammable and toxic substances at or above specified threshold quantities to develop a Risk Management Program (RMP). The RMP must include the following: hazard assessment including an evaluation of worst-case and alternative accidental releases that details the potential effects of an accidental release, an accident history of the last five years; registration information for the facility including identification of its processes, regulated substances handled, and amounts; prevention program that includes safety precautions and maintenance, monitoring, and employee training measures; and emergency response program that spells out emergency health care, employee training measures and procedures for informing the public and response agencies (e.g. the fire department) should an accident occur.

An important element of the prevention portion of the RMP is to have procedures for investigating incidents. This RMP requirement is promulgated at 40 CFR 68.81. This provision requires the owner or operator of a covered facility to investigate each incident which resulted in or could have reasonably resulted in a catastrophic release of a regulated substance. 40 CFR 68.81 does not have a definition for the term "incident."

New Jersey's Regulations

Following the 1984 catastrophic release of methyl isocyanate at Bhopal India, New Jersey enacted the Toxic Catastrophe Prevention Act (TCPA) in 1986 and the implementing regulations in 1988 (see N.J.A.C. 7:31). Following the promulgation of the CAP rules, New Jersey amended its TCPA rules and obtained delegation from USEPA. In lieu of the undefined term "incident," New Jersey proposed its rules using the term "Extraordinarily Hazardous Substance (EHS) accident" and provided the following definition: "EHS accident means an unplanned, unforeseen, or unintended incident, situation, condition, or set of circumstances which directly or indirectly results in an EHS release." The definition of EHS release must also be looked at to understand the meaning of EHS accident. The TCPA rules provide the following definition: "EHS release means a discharge or emission of an EHS from a piece of EHS equipment in which it is contained, excluding discharges or emissions occurring pursuant to and in compliance with the conditions of any State permit or regulation". In summary, as proposed New jersey rules would have required investigation of incidents which resulted in an unpermitted release of a regulated hazardous substance regardless how small the quantity of the release. To delegate the implementation of the CAP rules, USEPA required New Jersey to also require investigation of the incidents that did not result in any release and the incidents that resulted in a permitted but unplanned release. In response, New Jersey adopted amendments to the TCPA rules to require investigation of "EHS accidents" and "Potential catastrophic events." The rules provided the following definition: "Potential catastrophic event" means an incident that could have reasonably resulted in a catastrophic release of an EHS."

"Near-Miss" defined as "Potential Catastrophic Event"

New Jersey clarified in the adoption documents of the TCPA rules published in the July 20, 1998 and March 16, 2009 New Jersey Registers that the term potential catastrophic event is equivalent to the term "near miss" by using both terms interchangeably (see 30 N.J.R. 2733 Response to comment 53; and 30 N.J.R. 2740 Summary of Agency-Initiated Changes, items 2., 4., and 6.) New Jersey further clarified that the term near miss has the same meaning as the term near miss used in the publications by the Center for Chemical Process Safety of the American Institute of Chemical Engineers (AIChE-CCPS) (see 41 N.J.R. 1218 and 12 19, response to comments 65 and 66).

The AIChE-CCPS provides definitions and examples of process safety near misses in two of its publications: the CCPS Book, Guidelines for Investigating Chemical Process Incidents, Second Edition, March 2003, page 61; and Process Safety Leading and Lagging Metrics – You Don't Improve What You Don't Measure, January 2011, page 36. Some of the examples include the following:

- Excursions of process parameters beyond pre-established critical control limits.
- Activation of layers of protection such as relief valves, interlocks, rupture discs, blowdown systems, halon systems, vapor release alarms, and fixed vapor spray systems.
- Activation of emergency shutdowns.

USEPA found the approach of New Jersey outlined above to be at least equivalent to CAP rules requiring investigation of incidents.

Summary of New Jersey EHS Reports

Regulated facilities located in New Jersey are required to submit annual reports to the Department of Environmental Protection. Prior to 2010 some facilities were submitting triennial reports. The reports include a description of significant changes to the management system, a process hazard analysis report if one was completed in the past year, an inherently safer technology report if one was completed in the past year, the compliance audit report, and a summary of EHS accidents and potential catastrophic events. The summary of incidents includes the name and quantity of EHS released if any, the date and time of the incident, the equipment involved, and the basic contributory causes.

Table 1 below summarizes the incidents reported to the New Jersey Department of Environmental Protection between 2002 and 2011. Each year less than fifty percent of the facilities reported that they had one or more incidents. Most of the incidents reported involved the release of an EHS. The number of near misses reported average less than 1 per facility. This may not be an accurate representation of the near misses that actually occur. The cause for not reporting near misses could be a misunderstanding of the reporting requirements or a company culture that does not encourage reporting of near misses. It may also be a concern with perception because the term used for near misses is "potential catastrophic events." A list of potential catastrophic events in an annual report may be misleading to people who are not directly involved in the process.

Table 1. EHS Incidents Reported in TCPA Annual/Triennial Reports

Calendar Year	Total # of Incidents Reported	# of Near Misses (no release)	# of Facilities that Reported	Ave. # of Incidents per Facility	Max. # of Incidents Reported by a Facility	# of Facilities that Reported an Incident	Max. Quantity Released (lbs.) *See Note 1
2002	169	13	90	1.9	30	34	42,717
2003	209	12	79	2.6	58	34	8,900
2004	199	5	87	2.3	53	24	75,000
2005	267	25	76	3.5	82	24	51,666
2006	298	37	72	4.1	93	28	63,704
2007	327	39	80	4.1	97	28	7,219
2008	244	18	67	3.6	67	25	122,000
2009	209	12	68	3.1	54	24	7,600
2010	220	20	77	2.9	68	21	54,800
2011	228	9	79	2.9	89	22	17,955

*Note 1: Each year, the maximum quantity of an EHS released was a Flammable Substance or Flammable Substance Mixture.

3. Business Approach to Near-Miss Management

Fundamentals of Near-Miss Assessment

Near-misses have three fundamental characteristics: (a) frequency, (b) actual damage, and (c) maximum potential damage of related worst-case scenarios that could result from specific categories of near-misses. All of these should be recorded, monitored, and used to track progress in safety, operability, and reliability of any manufacturing process. These three fundamental characteristics form the base of leading indicators. From a process control perspective, the use of the process performance indicators as a basis for prediction and control has been a central research area since the work of Pontryagin and Kalman in the 1950's and 1960's, leading to the revolution in real-time control.[1]

"Near-Miss Management" concept is based on identification of early stress signals and taking the proper action to eliminate the stressor. There are numerous examples of major accidents in manufacturing industry whose earlier signals were ignored. After listing several major accidents in chemical industry Michael Dolan in his key note speech at 2012 AIChE Spring Meeting reiterated the important fact: "Each and every one was preventable with good engineering practice and attention to detail. And beneath each of these is a pyramid of near misses in many plants that could have been equally disastrous".

There are numerous examples of major accidents in manufacturing industry whose earlier signals were ignored. In this section, a methodology is presented for integration of near-miss management into the culture and day-to-day operations in a manner that underlines the critical connections between near-misses (both personal and process related) and behavior (both employees and management). This was also emphasized by Michael Dolan in his same speech as "the need to make the safety and security of our operations even more central to our profession than we already do". This methodology, which resulted from observations of successful near-miss management in various industries and research on the principles of risk management, is based on three principles, all of which are equally essential for successful reduction in accidents and significant improvements in operational reliability:

a. Organic integration of near-miss management into the operational fabric

Effective utilization of near-misses requires its components to be organically integrated into the operational fabric. That is, every employee and manager has an active role and participates fully in the near-miss management process (such as the eight-step process described in the next section).

b. Categorization of near-misses

To be able to track and monitor in a manner that helps with decision making, the overall risk has to be divided into meaningful, track-able categories, such as technology, facility, personnel

c. Tracking and monitoring of near-misses in each category

[1] For a concise summary of the history of optimal control theory and its applications to business problems, see Sethi and Thomson (2000).

As mentioned above, the tracking of near-misses in each of the categories should be done based on their three characteristics:

a) *Frequency* refers to the number of events per unit period.

b) *Actual damage* is the observed damage and applies only to those cases where an incident has occurred.

c) *Potential damage* is associated with each and every near-miss regardless of the presence of actual damage.

In a recent article, J.B. Titus poses the question: "Which factor has the highest priority for assessing hazard risk levels: possibility, severity, or frequency?" We think to measure and monitor the overall performance, if it will be helpful – not required, one can assign different weights to various categories and calculate their weighted sums. But, it's most important that the underlying details of the summed value be transparent, that is, directly addressable, since they point to the key problem areas.

The above categorization, combined with the track-able characterization of near-misses, would make a powerful tool to manage near-misses so as to reduce accidents. The most accurate near-miss information is available at the basic worker level. Operators, who are responsible for day-to-day operations of a process, are the best people who know the disturbances, causal factors of these disturbances, and the difficulties in overcoming such disruptions.

Eight-Step Near-Miss Process Management

An eight-step near-miss management process, which was developed as part of the Wharton Risk Center study (Phimister et al., 2003; Oktem U., 2003; Muermann and Oktem, 2002), incorporates all the key factors necessary to address near-misses in a facility to reduce accidents and catastrophes. Below is brief explanation of each step:

1. Identification

Identification is the first step of the process where an individual recognizes an incident or a condition as a "near-miss". To execute this step successfully each facility should have a clear definition of a near-miss which is communicated to each employee effectively, at all times.

2. Disclosure (Reporting)

Once a near-miss is identified it must be disclosed, preferably in a written form. This can be done either by the person who identified the near-miss or by a supervisor to whom a near-miss is reported verbally. Having a clear, simple, and possibly multi-channel, procedure for reporting would encourage this process and would increase the probability of reporting most near-miss observations.

3. Prioritization

Once an incident is reported, it needs to be prioritized. This very critical step determines the path to be followed in the subsequent steps; the level of attention that will be given to the incident, the depth of analysis that will be performed in finding causes, the resources that will be dedicated to finding and implementing solutions, and the extent to which the information about this incident will be disseminated. It is important for each corporation to set their own criteria for prioritization, based on the above mentioned factors: (a) frequency, (b) actual damage, and (c) maximum potential damage of related worst-case scenarios.

4. Distribution

Based on the priority and the nature of a near-miss, the information is distributed to the people who would be analyzing the causes of the incidents. The distribution mechanism for various priority levels must be determined during system development by the management team.

5. Identification of Causes (Causal Analysis)

This step includes identification of both direct- and root-causes of a near-miss. The implementation of this step can be as simple as the reporter inputting his/her ideas to identify the causes. On the other hand, for the highest priority near-misses, a committee may be formed to undergo an extensive root-cause analysis.

6. Solution Identification

The most important feature of this step is looking for a solution for each identified cause. Sometimes, several causes can be corrected with a single solution. In other cases, there may not be a feasible, effective solution, and hence, a less than ideal corrective action may be taken. All solution decisions, even "no solution for the time being", should be noted. In the end, each cause must have been addressed.

7. Dissemination

Once solutions are identified, they should be communicated to the people who will execute these solutions, assuming they have not participated in the solution identification process. This step also includes an important intermediate function, which, if overlooked, can stall the system; i.e., obtaining permission from the manager with resources to implement the solutions. Another function of the dissemination step is to inform all possible interested parties of the particular near-miss. This may extend well beyond the site or corporate functions and may include customers, contractors, suppliers, etc.

8. Resolution (Tracking)

Once solutions are identified and implementers are informed, it is important to track all suggested changes to ensure they are properly executed. Also, when all of the changes are completed, to encourage future participation, the reporters of near-misses should be informed of the results from their identifications of near-misses.

How Effective Near-Miss Management in Reducing Accidents:

A well managed near-miss system is expected to help identify problems at the early stages and eliminate causal factors, reducing the probability of having major accidents/releases. Below are three examples which demonstrate that at facility level paying attention to safety issues does reduce accidents.

Example 1: Norsk Hydro

Figure 1 shows the results of a study published by Jones et al. (1999) on Norsk Hydro data. This is an offshore drilling and on-shore refining company in Norway. They started to implement a near-miss program in 1985 at their off-shore drilling facility where they employed 3500 people. As we can see from the data on this chart, between 1985 and 1997, their lost time injuries decreased by tenfold. During the same period, their Near-Miss reporting rate went from zero to 0.5 near-miss per person per year.

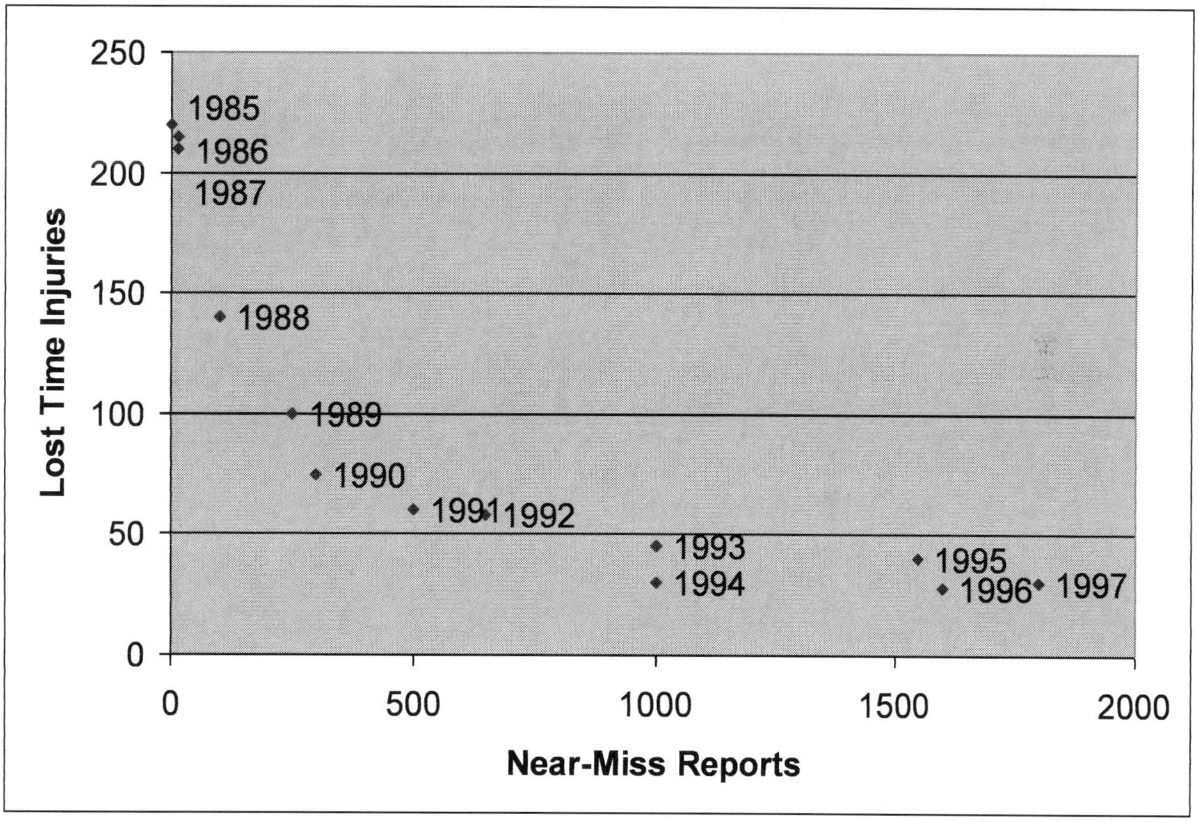

Figure 1: Norsk Hydro data (Jones et al., 1999)

Example 2: Dow Chemical

A recent study (Nesmith et al., 2012) reported by Dow Chemical show that a pioneering site has the same trend as Norsk Hydro, that is, as the number of reported process safety near-misses increased the number of incidents decreased – confirming that a well-managed near-miss program can help reduce the number of accidents. In this case, a strong and personal leadership support for PSNM reporting, including creation of a network for sharing experiences, accelerated implementation of the program.

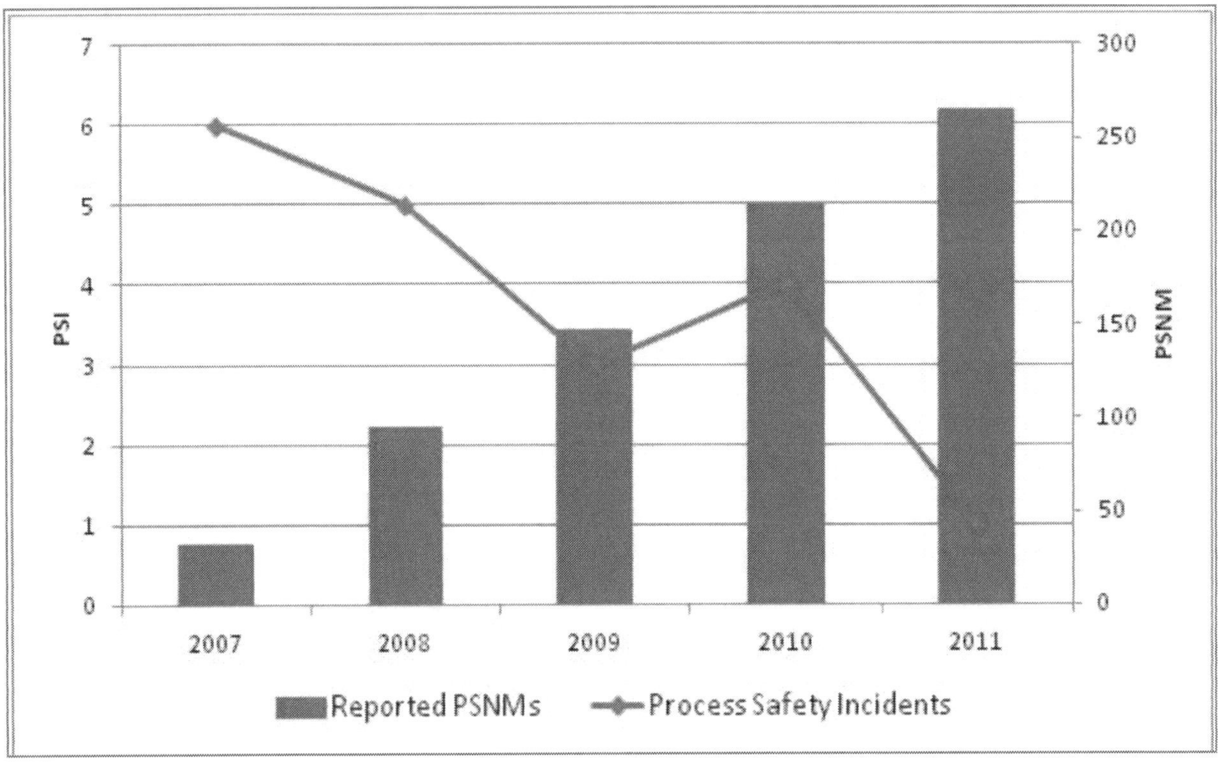

Figure 2: One of Dow Chemical's North American Sites (Nesmith et al., 2012)

Example 3: DuPont

In this example for DuPont data, even though there is no specific number of near-miss reporting presented, however, if one assumes that the increase in the audit scores has some relevance to paying attention to process safety issues, bringing them to forward and addressing them such as the near-miss program again, a similar trend on reduction of incidents is observed.

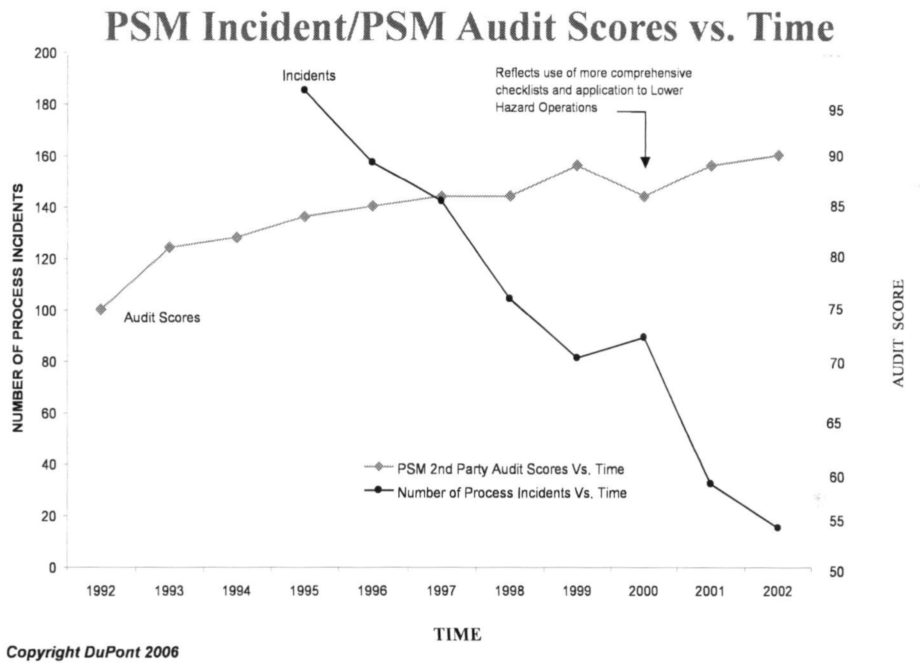

Figure 3: DuPont data (Kleindorfer, 2006)

.

4. Can Facilities' Success be Repeated at the State Level?

Looking at New Jersey DEP's data , it is difficult to draw conclusions. There are many variables, and data that is reported is very limited.

New Jersey has facilities in the following sectors: chemical manufacturing, refineries, food processing, power generation, water and waste water, and miscellaneous other types. Refineries are the most complex facilities with many regulated processes. Some chemical manufacturing facilities are also very complex with multiple regulated processes. Warehouses are the least complex regulated entities.

Every year more than 50 % of the regulated facilities report that they had no near misses or accidents. The reasons for not reporting any accidents or near misses may vary from facility to facility. They may not be doing a good job at identifying and addressing incidents and near misses, or they may be doing a good job but electing not to report for concerns of public perception or other reasons. Hence, there is not enough information about incidents to draw any conclusions.

Of the facilities that report any accidents and near misses, the largest number of accidents and near misses are reported by complex multi process chemical manufacturing facilities and refineries. This is expected.

What incidents are considered significant?

In this case the largest releases are reported by the facilities that also reported the largest number accidents/near misses. Hence, a straight forward extrapolation of results obtained by individual facilities cannot be confirmed at the state level. To encourage facilities run a more rigorous near-miss management program, identifying and reporting higher number of near-misses, state may evaluate the reasons why large percentage of facilities are not reporting near misses and take actions to address those reasons.

Table 2. Quantity of the extraordinarily hazardous chemical releases reported to New Jersey DEP from 2007 through 2011 (There is no reported significant damage to equipment or facilities.)

Quantity of Extraordinarily Hazardous Substance (Chemical) Releases (Q in lbs.)	**Number of Facilities Reported Releases**
100,000 < Q	1 (from a refinery)
10,000 < Q < 100,000	5 (all from refineries)
1,000<Q <10,000	25 (21 from refineries)
100 < Q < 1,000	60 (50 from refineries)

5. Conclusions

Often, near-miss management has played an ex post forensic role in risk management rather than an alerting one Above examples show that an effective near-miss management program, which identifies leading indicators and precursors of hazardous conditions can help improve safety and environmental releases significantly. This requires strong commitment and integration of the concept into the culture and day to day operations in a manner that underlines the critical connections between near-misses and behavior.

While this paper provides some roadmap for an effective near-miss management at the facility level it points out that regulatory landscape appears to not provide a very conducive environment to help flourish the concept. A solution to this issue may be achieved by forming a public private partnership group to facilitate companies to report near-misses and get support in improving their safety and environmental releases.
The Center for Chemical Process Safety (CCPS) developed a membership based Process Safety Incident Database (PSID). The purpose of the CCPS PSID is to pool process safety incident experience among participating companies so they can learn from the experiences of others without suffering the consequences of failures, while minimizing corporate liability. PSID includes process safety incidents with a potentially important lesson to be learned from incidents that did or could have resulted in a fire, explosion, fatality, multiple injuries, significant release of hazardous materials, or any other unique process safety incident (including near-misses). Currently a fraction of the CCPS member entities are members of the PSID data base. The PSID database currently has information about more than 700 incidents.

6. References

Jones S., Kirchsteiger C., Bjerke, W. (1999), "The importance of near miss reporting to further improve safety performance." *J. Loss PreVention Process Ind., 12* (1), 59.

Kleindorfer P.R. (2006), "Workplace Safety at Alcoa Case Discussion", Presentation at ARKEMA Executive Program, INSEAD.

Muermann A., U. G. Oktem, (2002). "The Near-Miss Management of Operational Risk", *Journal of Risk Finance*, Fall, 25-36.

Nesmith G., Keating J. T., and Zacharias L. A. (2012). "Process Safety Near Misses in Dow Chemical: A Proactive Program to Improve Process Safety Performance," 2012 AIChE Spring Meeting, April 1-4.

Oktem, U. G. (2003). "Near-miss: a tool for integrated safety, health, environmental and security management". 37th Annual AIChE Loss Prevention Symposium. March 30–April 3.

Phimister J. R., U. G. Oktem, P. Kleindorfer, and H. Kunrether (2003). "Near-Miss Incident Management in the Chemical Process Industry", *Risk Analysis*, **23** (3), 445-459.

Sethi, S. P. and G. L. Thompson (2000). *Optimal Control Theory: Applications to Management Science and Economics*, Second Edition, *Kluwer Academic Publishers*, Boston.

Titus J.B., "Risk level assessment priority: Possibility, severity, or frequency?," *Control Engineering*, January 2013, p25.

Process Near misses Identification, Investigation and Challenges

Pawan Mittal / Parminder Sawhney
Reliance Industries Limited
Jamnagar
pawan.mittal@ril.com/parminder.sawhney@ril.com

Prepared for Presentation at
American Institute of Chemical Engineers
2013 Spring Meeting
9th Global Congress on Process Safety
San Antonio, Texas
April 28 – May 1, 2013

UNPUBLISHED

Process Near misses Identification, Investigation and Challenges

Pawan Mittal / Parminder Sawhney
Reliance Industries Limited
Jamnagar
pawan.mittal@ril.com / parminder.sawhney@ril.com

Keywords: Process Safety near miss, reporting, investigations

Abstract

The benefits of reporting and investigating Near misses have long been established in the safety management systems, as they result in prevention of actual incidents from taking place. At Reliance in line with this belief, reporting and investigation of Process safety Near miss is given top priority by the management.

The paper presents the systems and benefits of process safety near miss reporting and investigation at Reliance. Few of the key elements of it are: i) IT tool for reporting of Near miss, ii) categorization of near miss based on the risk rating, iii) investigation based on risk rating using the same procedure which is being used for Process Safety Incidents, iv) review of investigations results by the top management, v) reporting of near miss one of the key parameters in team KRAs (Key Result Appraisal).

Also the paper discusses few near misses which have helped in identifying opportunities for strengthening our Process Safety management systems. The incidents range from challenges to Equipment integrity, errors in operating procedures, faults in instrumented systems etc.

1. Introduction

Process Safety near miss incidents are the most important indicators of safety performance. Process safety incidents are actual encounter of the consequences whereas Process Safety near miss incidents are considered "weak signals", some of which contain the signature of more serious and adverse effects. Therefore, addressing Process Safety near miss incidents in a timely and disciplined manner reduces the frequency and consequences of larger problems. Hence, reporting and analyzing them, especially those with high potential, is a pre requisite for maintaining and sustaining strong Safety Management Systems within any organization.

2. Definition:

2.1 Process Safety Incidents:

The Process Safety Incidents are those incidents that represent failures in a line of defense in the overall HSE program but *with* consequence.

2.2 Process Safety Nearmiss:

The Process Safety Nearmiss are those incidents that represent failures in a line of defense in the overall HSE program but *without* consequence.

2.3 Examples of some Process Safety Nearmiss:

A safety interlock that discovered during a routine interlock check (say during a turnaround) was not function properly at the correct set point. During the normal operations, it would have caused a Process Safety Incident.

A PSM critical valve (either a relief valve or an emergency shut-off valve) is found to be defective during a routine maintenance inspection; non-functioning of that valve during normal operations could lead to serious consequences.

The interlock set point for a safety interlock is found in the field to not be as is documented in the control copy of the unit's Process Technology. It could have triggered a Process upset and if interlock set point is at a higher set point than the actual, it could cause a Process Safety Incident also.

A conservation vent or an overflow line on a tank is found to be plugged and would not work properly if the tank experienced a change in pressure or level. Failure of pass flow leads to low flow on tripping of circulating pump. It would cause over pressurization/rupture of tank.

Any time a safety instrumented system malfunction and operates when demand is not placed on the system (i.e. nuisance failure) resulting into process upset.

Failure to open of a rupture disk, a pressure control valve to flare or atmospheric release, or a pressure safety valve when the system conditions reach or exceed the prescribed trigger point could lead to over pressurization of equipment and can result in explosion/fire.

System pressure exceeded the Integrity Operating Window[1] (IOW) during normal operations can not only cause immediate incident, it can lead to undesirable effect on equipment and process (corrosion, stress etc.)

3. Nearmiss Reporting & Investigation

1. **Identifying of Process Safety Nearmiss & Reporting:** Once the nearmiss is identified, it should immediately be reported / logged. There should be well defined system /procedure in place in order to report near miss incidents. At Reliance, an online portal system is there wherein any employee can report a near miss and subsequently, the near miss is validated and approved by the Plant-HOD (Head of Department) and the HSEF (Health, Safety, Environment & Fire) head. Even if a near miss is not approved and is purely an unsafe condition, necessary actions are initiated for corrective/preventive actions.

2. **Formation of Investigation team: -** Once a Process Safety near miss incident, is identified, a committee shall be constituted to investigate the Process Safety near miss incident. This team should be a cross functional team with a `Root Cause Analysis facilitator who is trained on the subject. The team consists of a chairman, coordinator and two-three members based on criticality of the near miss.

3. **Corrective /Preventive action:-**Take immediate corrective action, (if cause seems apparent) even prior to the formation of formal team for investigation and without waiting for their recommendation.
   For Citation, during loop checking it was found that the PSV installed in the loop was of lower rated than the design. Same had been replaced immediately with the required design rated capacity. To avoid reoccurrence, investigation team was formed to find out the root cause.

---

[1] As per API RP 584

4. **Communication of the Incident: -** Discuss the Process Safety near miss incident in daily meetings at plant level, interplant level and site level so that they are well communicated across the site to Operations & Maintenance group.

5. **Verify the gaps:-**All Operation & Maintenance leads to verify, whether similar problem / gap exists in their area through sharing of Process Nearmiss from other areas. At Reliance, we have following system to achieve this-
   a. Discussion on process safety near miss at different levels such as plant level meeting, interplant level meeting and site level meeting, so that concerned persons of the plant and other plants are aware of the near miss and review the system in their area also.
   b. After investigation, LFI (Learning from Incident) is being circulated to all plant managers and maintenance leads for review of recommendation applicable to their system and follow up. For this we have well developed LRTS system (Learning from incidents and Recommendations Tracking System) – The identified actions for the process safety near miss are entered in Corrective Action Tracking (CAT) system for ensuring closure of the recommendations.

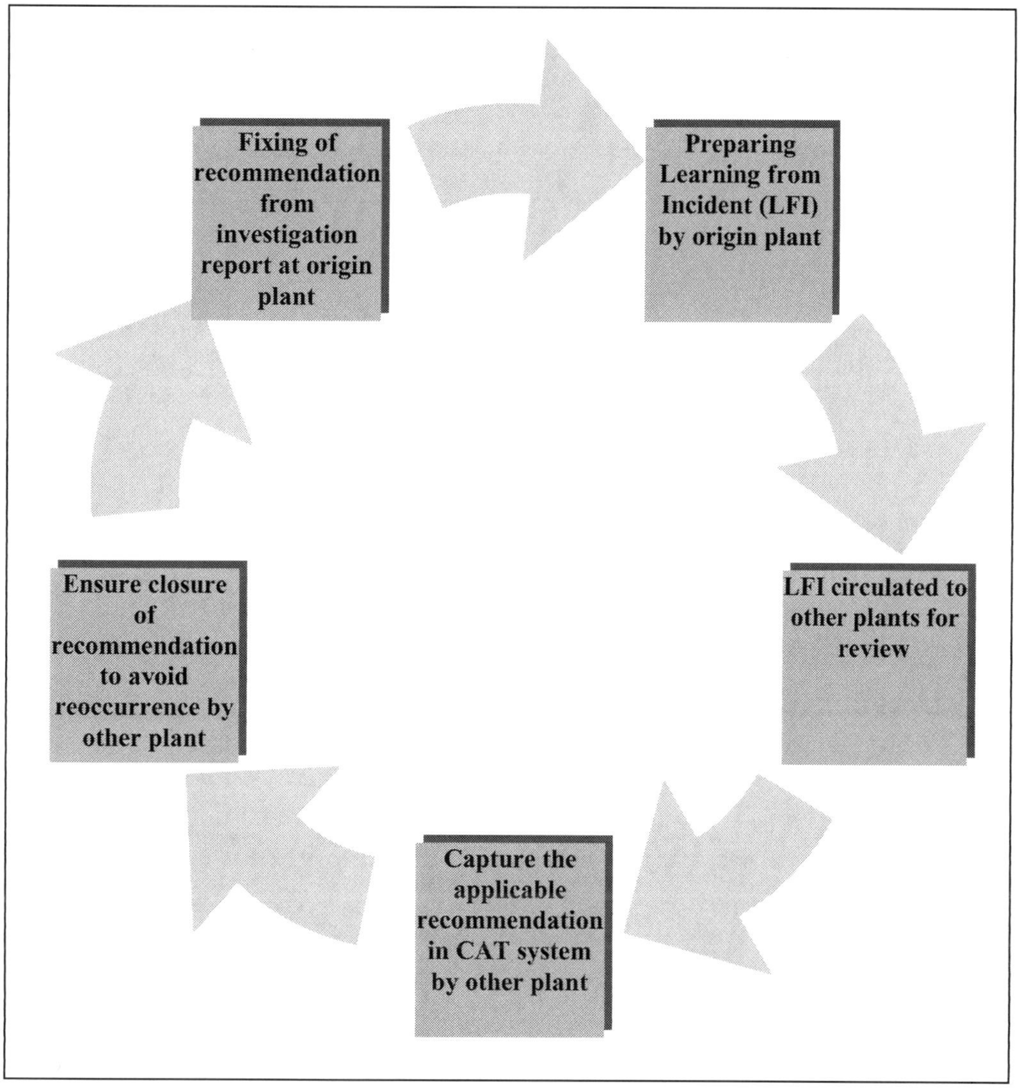

6. **Feedback about the actions: -** Concerned Operation & Maintenance leads should confirm to the site management that either such gap doesn't exist has been rectified or is being identified in their respective areas.

7. **Inputs to Investigation team: -** It is to ensure that the corrective actions taken are shared with the investigation team. This will facilitate investigation team to fix the key factors and do a thorough and fair investigation. At Reliance, we have well designed and defined web based reporting portal (iRIL- Incident Reporting, Investigation and Learning). The corrective actions taken are captured while reporting near miss in web based portal. Same will be visible to investigation team while they initiate the investigation process.

8. **Summary of Root causes & Corrective action:-** Maintain summary of the corrective action taken and identified root causes for future reference & analysis.

At Reliance, the identified failed PSM[2] and OD[3] (Operational Discipline) element for the process safety near miss is being captured and categorized in web based portal. The summary is being fetched at six monthly bases and analyzed and report is generated. Based on this report action plan is developed for implementation and strengthening the process safety management and operational discipline.

9. **Preparation of LFI (Learning from Incident):-**It is ensure that timely preparation of Learning from incident (LFI) and publish the same for sharing in other plants for identifying and correcting similar issues.

Various benchmarking methods are followed at Reliance for Nearmiss Reporting - like Nearmiss Reported/Employee, Maximum number of Nearmiss reported in an area etc. to promote Nearmiss reporting and infuse a healthy reporting culture. The selected Nearmiss reporters are also awarded to recognize their effort and timely action.

The employees who report Nearmiss have to be convinced by the actual actions that their reported Nearmiss are being tracked and actions initiated for closure to prevent re-occurrence. It not only motivates employees for continuous reporting but it also brings a positive Safety Culture within an organization.

At Reliance, Process Nearmiss are investigated (based on criteria) like any other incident, root cause analysis is done to identify root causes (physical, human & system) and actions taken. The whole idea is to prevent re-occurrence as Nearmiss could easily convert into an incident. The difference between a near miss and an accident often is a fraction of a second or an inch.

---

[2] PSM elements: Process Technology, Process Hazard Analysis, Operating Procedure and Safe Practices, Management of Change, Quality Assurance, Pre Startup Safety review, Mechanical Integrity, Training and Performance, Contractor Safety performance, Incident investigation, Emergency Planning and Response, Auditing

[3] OD element: Leadership by Example, Common Shared Values, Strong Team Work, Pride in the Organization, Sufficient and Capable Resources, Employee Involvement, Active lines of Communication, Up to date documentation, Practice Consistent with Procedures, Absence of Short cuts, Excellent Housekeeping

4. Typical Business Flow to identify the Process Safety Nearmiss incidents & incidents

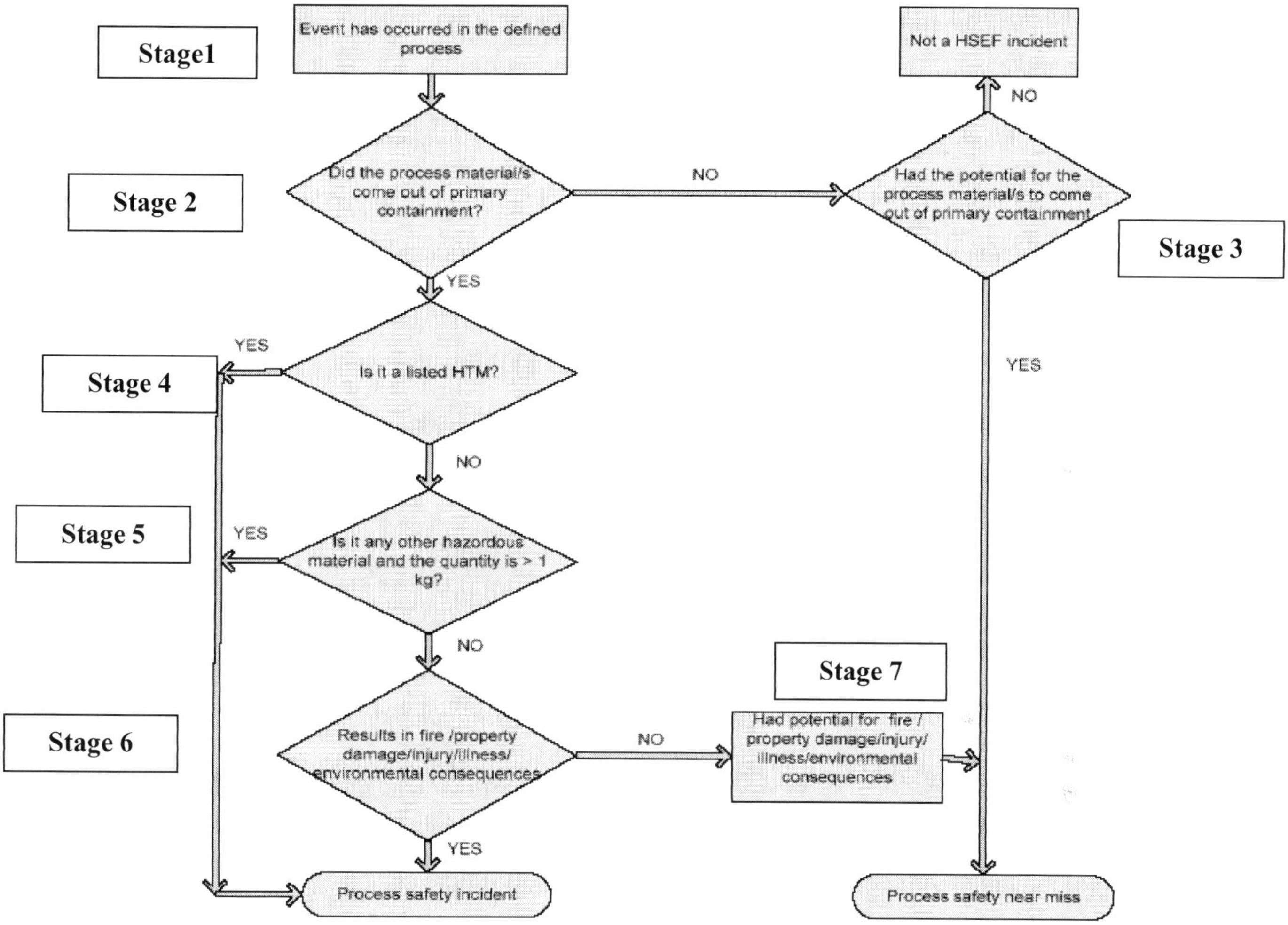

Stage-1: Event has occurred in the defined process

The incident or event that has occurred must be in the defined process to qualify as a process incident or near miss. For example an electrical tube light fell from height in the warehouse is not considered as process event whereas over pressurization of process vessel due to malfunction of instrument is considered as process event.

Stage-2: Did the process material/s come out of primary containment

The intention of this question is to check whether the material involved in the event had come out from primary containment or not.

Stage-3: Had the potential for the process material/s to come out of primary containment

If the decision of stage-2 is "No", then check whether the event had a potential for the process material to come out of primary containment. If it is no then it should not be a considered as a HSEF incident and if it is yes then it will be considered as Process Safety Nearmiss.

Stage-4: Is it a listed HTM (Highly Toxic Material)

If the decision of stage-2 is "Yes", then check whether the material involved in the event is listed as highly toxic material or not. At Reliance, we have identified 14 chemicals as "Highly Toxic Material"[4]. If the decision of stage-4 is "Yes" then the event will be reported as Process Safety Incident irrespective of the quantity of material that came out from primary containment.

Stage-5: Is it any other hazardous material and the quantity is > 1 kg.

If the decision of stage-4 is "No", then check whether the released material is listed in any other hazardous material list and whether the release quantity is greater than 1 kg. If the decision of stage-5 is "Yes" then the event shall be reported as Process Safety Incident. At Reliance, each plant has identified the list of hazardous materials in the Process Technology package.

Stage-6: Result in the fire/ property damage/ injury/ illness/ environment consequences

If the decision of stage-5 is "No", then check whether the consequence of event is fire/ property damage/ injury/ illness/ environmental impact. If it is yes then the event will be reported as Process Safety Incident.

Stage-7: Had potential for fire/ property damage/ injury/ illness/ environment consequences

If the decision of stage-6 is "No" then check whether the event had a potential for fire/ property damage/ injury/ illness/ environmental impact. If it is yes then it will be considered as Process Safety Nearmiss.

---

[4] Highly Toxic Materials: Ammonia (anhydrous), Chlorine, Dimethyl amine, Di Methyl Di Sulphide (DMDS), Ethylene Oxide, HF (anhydrous), HCl Anhydrous, Hydrogen Cyanide, Hydrogen Sulphide (anhydrous liquid), Methyl Mercaptan, Monomethylamine, Sulphurdioxide (anhydrous), Titanium Tetra Chloride, Trimethylamine

5. Challenges

5.1 Reporting Recriminations

Employee may be reluctant to report near misses due to potential recriminations such as –

5.1.1 Peer pressure
5.1.2 Investigation Style
5.1.3 Direct Disciplinary Action
5.1.4 Unintended Disciplinary Action

These issues are addressed by management in the following ways-

- Rights of near miss reporting to all employees.
- Demonstrate leadership by example by reporting near miss and involvement in investigation.
- Discussion of each and every process safety near miss at appropriate management level.
- Take immediate corrective actions and inform to employee/ contractor who had reported the near miss.
- Conduct training program for effective investigation.
- Not to write name of any person in the investigation report.
- Drill down to system factors in the investigation and not limit the investigation to physical or human factor only.

5.2 Lack of Incentives: Those employees and contractors should be awarded suitably who reports nearmiss incidents specifically high potential work place or process safety neramiss incidents. An organisation should have a structured programme / scheme to recognize such employees / contractors. Actually, these types of schemes facilitate in higher reporting and later this translates into a cultural change wherein people start reporting even without any awards / incentives. There are various ways to give awards or incentives to employees or contractors who reported the process safety near miss.

- Appreciation to employees/ contractors at appropriate level such as plant level meeting, interplant level meeting and site level meeting.
- Reporting of near miss should be considered during year end performance.
- Display of name & photo of employee/ contractor who reported the best process safety near miss.
- Incentives to employee/ contractor

5.3 Individual Confusion: There are occasions, when employees and contractor personnel may be having some confusion in identifying incidents and nearmiss incidents; hence there should be clearly defined guidelines for better understanding the concept of near miss and incidents. Awareness campaign and awareness session should be arranged especially for new entrants. In fact, employees should be motivated to observe and report any unsafe act/condition even if it is not deemed fit as a `Nearmiss'.

Number of activities can be carried out during awareness campaign on identification and reporting of process safety near miss such as Special training programs on guidelines of identification and reporting of process safety near miss; Special tool box talk by senior management to motivate the employees; Display of posters and banners; Screening of films; Quiz contest; Appreciation or awards distribution to employees who reported process safety near miss; etc.

5.4 Lack of Management Commitment: Commitment of the management plays a vital role in the process of reporting near miss and incidents as well. Therefore, leaders from all disciplines should demonstrate their commitment for strengthening the reporting process. Motivation and encouragement of employees /contractors by the leaders of the company certainly plays a vital role in enhanced nearmiss reporting. If necessary corrective actions are not implemented in a timely manner, it may create a sense of demotivation in the employees and they feel that no action is being taken on their reported observations. It may actually hamper the overall Nearmiss reporting. This signifies the role of monitoring and tracking actions generated from the investigation of Nearmiss.

Repeat Nearmiss is not a very healthy sign and it somewhere indicates that the System is not robust enough for sustenance and needs to improve.

Management commitment on near miss reporting is demonstrated by the followings-

- Rewarding the employees/ contractors for near miss reporting.
- Showing sense of commitment in fixing corrective and preventive actions.
- Provide necessary resources for addressing the corrective and preventive actions appropriately.
- Widely communicate and discuss the learning's from other plant in appropriate forums.
- Ensure framing of appropriate action plan based on analysis report of near miss reported to prevent reoccurrence.

6. Maximizing the Near Miss Reporting

Near miss reporting provides valuable data for improving the process safety management systems at a facility. The following processes can maximize the benefits from a process safety near miss program.

1. Use process safety lagging indicator, process safety near miss, and management system leading indicators to build a process safety performance pyramid.

2. When evaluating process safety near misses, consider the potential adverse impacts. The level of response to a near miss (i.e. investigation, analysis, and follow-up) should be determined using the potential as well as the actual consequences of the event.

3. Tie the near miss data to the deficient management system in order to drive system improvements from near misses as well as from actual incidents.

4. Place value upon reporting near misses. Consider reward / recognition for reporting near misses as well as rewards for bottom line performance.

7. Conclusion

Article proposes a near miss management framework that adds operational and strategic value to corporate environmental, health and safety practice. To improve process safety near miss management system one must focus on the details of the process by which near miss are identified, reported and analyzed. Definitely there are some challenges such as peer pressure, disciplinary actions, lack of incentives, lack of knowledge, etc. but management commitment towards reporting and investigation of near miss, leadership by example, follow up of investigation and recommendations, incentive scheme, appreciation to employee/ contractor at appropriate level are some of the ways to override these challenges.

“Learning from near misses is much cheaper than learning from accidents.”

Near Misses – Private or Public Concern?

Jay Brabson, P.E.
Accidental Release Prevention Program – State of Delaware
715 Grantham Lane
New Castle, DE 19720
Jay.Brabson@state.de.us

Prepared for Presentation at
American Institute of Chemical Engineers
2013 Spring Meeting
9th Global Congress on Process Safety
San Antonio, Texas
April 28 – May 1, 2013

UNPUBLISHED

Near Misses – Private or Public Concern?

Jay Brabson, P.E.
Accidental Release Prevention Program – State of Delaware
715 Grantham Lane
New Castle, DE 19720
Jay.Brabson@state.de.us

Keywords: Near Miss, RMP, Incidents, Regulation, Regulator, PSM

Abstract

Near miss incidents include those releases and/or fires that do not result in significant employee injury or costs. In many cases there are federal reporting requirements that include reports to the National Response Center (NRC). Loss of containment releases are often near misses for catastrophic incidents and provide evidence of process safety program weaknesses. Facilities have an interest in conducting incident investigations to find root causes, lessons learned, and make safety improvements. Small sites often do not have the expertise to conduct good investigations while some large sites may tend to blame operator error rather than admit management system errors or process design issues [9].

This paper explains 2002 amendments of the Delaware requirement to report releases. The amendment adds the reporting of flammable gas releases and expands the existing written follow up report requirements to explain the 'facts and circumstances leading to the release and the measures proposed to prevent the future releases and remedy any deficiencies in the prevention program.' Having this 'public' examination of near miss investigations puts pressure on both industry and our regulatory program to seek root causes and make practical improvements in site process safety programs – without adding any substantial regulatory burden to industry.

1. Introduction

There is a positive story concerning improvements to industry process safety management (PSM) programs and adoption of better safety cultures at large chemical companies, small manufacturing sites and even oil refineries. Yes, there are still near miss accidental and environmental releases occurring and the Chemical Safety Board (CSB) has their hands full investigating continuing catastrophic accidents. As the CSB can attest, many of these releases occur at small sites, often not even involving extremely hazardous substances (EHS) [1]. And unfortunately, there are still many releases occurring at refineries and other 'PSM covered' sites. But these releases should not hide the fact that better process safety programs and safety management cultures have improved process safety at most of the facilities that are regulated under the Occupational Safety and Health Administration (OSHA) PSM program and the Risk Management

Program (RMP) under the Environmental Protection Agency (EPA). Some facilities, especially large corporations, are adopting process safety program elements even when there is no regulatory requirement – it just makes good business sense. But, with accidental releases still occurring and attracting media attention or environmental concerns, there continues to be discussion of potential new regulatory initiatives or new regulations.

In the real world of resource constraints, there are not enough regulatory resources and expertise to adequately cover the universe of stakeholders subject to regulatory requirements now, let alone the many sites covered under only 'general duty' requirements due to low quantity or no EHS onsite. This story is about working smarter as regulators to deploy regulatory resources into process safety issues that need attention. How does this relate to near misses?

Most accidental releases reported under EPA federal rule 40 CFR Part 355 – *Emergency Planning and Notification* are near misses involving no injuries and no significant equipment damage. There is also a recognition that if catastrophic releases are at the top of the pyramid, serious near misses include minor injuries as well as minor equipment damage[2][3]. There are numerous papers and studies explaining the importance of doing good incident investigations on accidents and near misses as a means for improving process safety programs. This paper explores the concept of applying regulatory resources to accidental releases, most of which are near misses, in order to improve the overall process safety of all manufacturing facilities including those not directly covered by process safety regulations.

2. Finding an appropriate regulatory 'tool'

The EPA's federal rule 40 CFR Part 355 is not a useful tool for tracking and preventing accidental releases [4] due to its emphasis on trigger reporting and emergency notification. Reporting chemical substances released above a certain quantity may be of some value to various community right-to-know entities, but this information alone, reported to the NRC without further investigation has limited utility as a means to determine a facility's ongoing process safety performance. We are fortunate in Delaware to have a state regulation entitled *Report of Discharge of a Pollutant or an Air Contaminant*, promulgated in 1991, that mirrors and expands upon these federal reporting regulations [5]. In 2002 the regulation was amended to update reporting quantities. At that time we made several additional changes that make this Delaware regulation a useful tool for investigating release incidents by our Accidental Release Prevention program. Amendments included:

- Adding RMP flammables and Delaware explosive substances (reporting quantity of 1000 pound release within 15 minutes to an hour).
- Adding a general duty to provide a written follow-up report, regardless of the substance or quantity as requested if there is concern for public health or safety or environmental welfare.
- Adding, beyond the 40 CFR Part 355 written follow-up report requirements, two report requirements; 'facts and circumstances leading to the release and the

measures proposed to prevent the future releases and remedy any deficiencies in the prevention program'.

While this may sound like a lot, these amendments only build on federal requirements already in place. Assuming facilities conduct incident investigations after accidental releases anyway, Delaware requirements do not add to the existing federal regulatory burden. We knew from experience that accidental releases occur at manufacturing sites using a variety of hazardous substances and we simply wanted to ensure that good quality incident investigations with practical recommendations for improvement were being conducted.

This idea is not anything new under the sun [6]. The Center for Chemical Process Safety (CCPS) of the American Institute of Chemical Engineers (AICHE) has a fee based private and confidential project; the Process Safety Incident Database (PSID) for 'sharing incident information among members with a focus on lessons learned from the incident'[7]. While many industry leaders recognize the value of these incident investigations, there are also many manufacturers that are not members of CCPS [8]. This paper suggests there is a place for regulators representing interests of workers, the public and the environment, to review a subset of incident investigations for all facilities, after 'reportable' releases of hazardous substances.

3. Results

These written follow-up reports submitted after reportable incidents and accidental releases allow the ARP program (and potentially air and water environmental programs) to get a better idea of the quantity released, what happened, and the incident investigation recommendations. This includes facilities both large and small that might have a release involving any of over 1100 hazardous substances. With this Delaware regulatory reporting regulation in place for over 10 years we have found a reduction in releases, improvements to process equipment and design and improvements to process safety programs. Admittedly, improvement in safety programs can be hard to measure and antidotal based on inspection reports, compliance audits, and recommendations. Measurable reductions in reportable releases include reductions to 4 or 5 per year in 2003-2005 down to 2 per year by 2010-2012. (This ignores oil refinery flaring reports which can number over 100 per year.) Would this have been the result even if there were no regulatory hazardous substance release reporting program? Maybe, as noted earlier, many facilities see the value in improving their process safety programs and safety culture. However, common sense suggests that having facility incident investigation reports and recommendations open to review by regulators is a strong motivator for facilities to conduct quality incident investigations and allocate dollars for equipment improvements in the face of tight facility budgets. Anecdotally, we know from experience that just having a regulator show up to review an incident and the accompanying investigation recommendations catches upper management's attention. And, where appropriate, we have added to a facility's recommendations providing technical recommendations for further improvements to process safety programs, equipment and management control systems. We have even had some cases where facilities, particularly smaller sites, use this written follow-up report as sent to regulators to increase leverage for having recommendations implemented and resources deployed by corporate executive management [9].

It has been my experience as a process safety regulator in a small state that many of the covered facilities we inspect have well documented process safety programs. During full RMP inspections it can take days of reviewing documentation to find even minor recommendations for improvements. Near miss incidents involving accidental releases of EHS are opportunities for both the facility and our program to uncover weaknesses and holes in process safety programs. We have also helped smaller sites find best process and equipment practices particularly when they have no real engineering staff. As regulators we see a variety of processes and can share non-confidential 'best practices' we observe in our inspections. We have found this reporting tool to be a win-win for both industry and the regulators representing the public and environment.

4. Observations

Clearly, complex manufacturing processes increase the probability of accidental releases, and there is evidence that chemical companies and petro-chemical companies with more complex processes lead to a greater risk of accidental releases. Upon reviewing the EPA nationwide database for risk management plan submissions on their Chemical Data Exchange (CDX) webpage, simpler manufacturing processes involving chlorine (most often used for water treatment) and ammonia (most often used in refrigeration or fertilizers) have approximately seven percent of facilities with reportable accidents. They make up 70% of the total RMP submissions and not surprisingly, the total report of accidents is also about 7 percent for all facilities. The chart below shows that the much smaller universe of chemical and petro-chemical facilities report higher accident rates (20 and 38 percent respectively when I checked in 2012) [10].

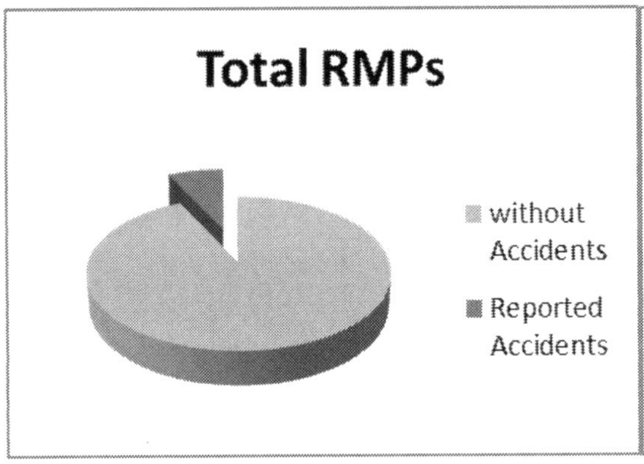

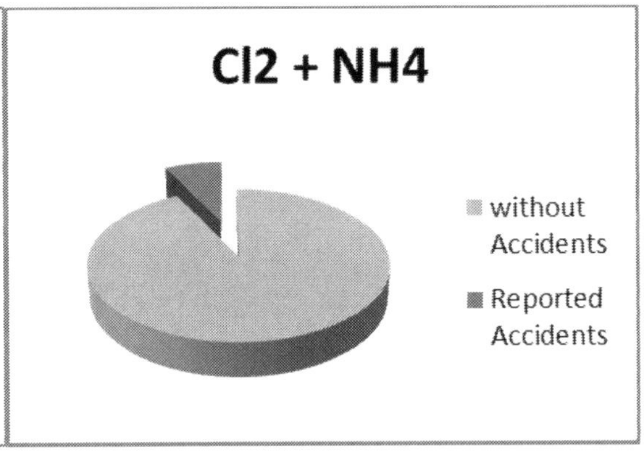

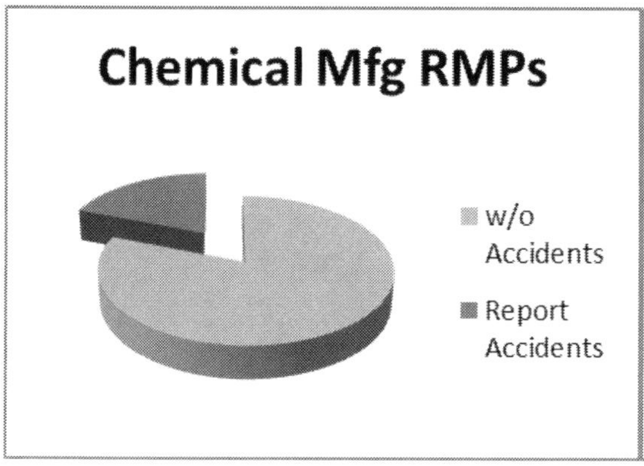

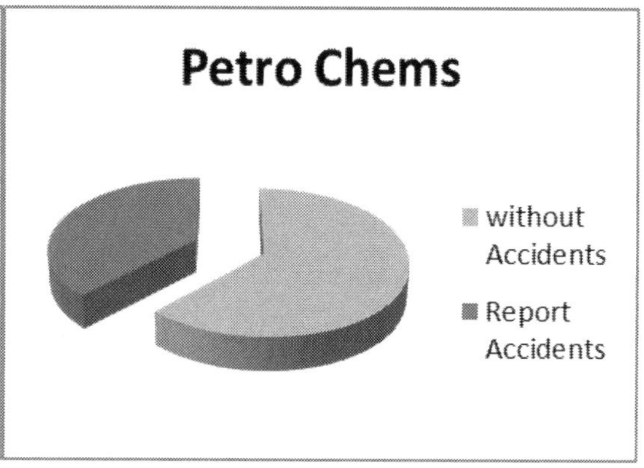

Also, as evidenced by past CSB accident data, many small (and some larger) sites may not recognize certain risk and safety issues when dealing with substances not categorized as EHS including dust, reactive substances, and flammables.

And finally, we find that no matter how good the process safety program, outdated process designs and older equipment can contribute to accidental releases. At this time in 2013, most refineries have improved process safety programs. But while equipment has been replaced and updated, much of the piping and associated supports and some equipment and process designs are over 30 years old with associated 'lean' maintenance crews. Refineries may have good pressure vessel and pipe testing programs. They may relentlessly look for root causes after incidents and provide good recommendations. But like an old house or an old car, there is always something ready to fail due to age (wear and tear), corrosion or vibration.

In making this point in the past, people asked me for concrete examples. I want to share an interesting example of a robust process safety program at a Delaware facility that was still having accidental releases/near misses. It involved a recent ammonia refrigeration inspection. A review of all Delaware ammonia releases reported to the NRC from 2005 through 2012 indicated a total of 10 'reportable' ammonia releases. Of those 10, six came from just one particular site. There are 15 separate sites in Delaware using ammonia, and 13 of those sites have ammonia refrigeration. Six out of 10 releases from one site would certainly not be expected. This particular ammonia refrigeration facility has two different sites in Delaware. We generated a facility 'report card' through a review of the Delaware required incident reports database which indicated that of those 6 releases, 5 came from just one of the four separate ammonia systems at this one site. Not surprisingly, these five releases came from the oldest of the four refrigeration systems.

(See graph below)

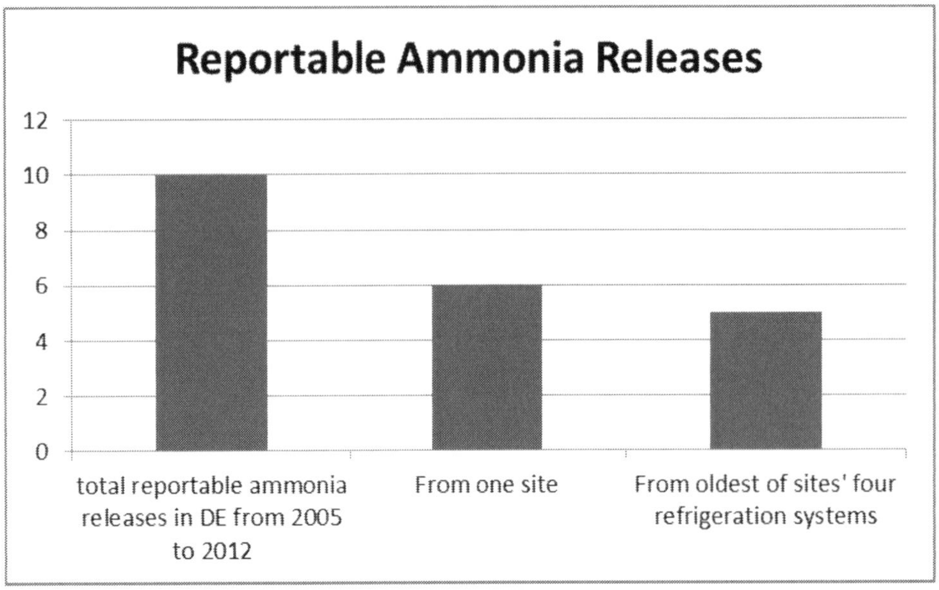

During our inspection we found an organized process safety program led by competent facility and corporate PSM experts with good management support. They appeared to have a good mechanical integrity program with well trained and knowledgeable refrigeration engineers. We gave them this 'report card' on near miss accidental releases to review and recommended their PSM team 'discuss releases and incidents associated with this system and explore options for improvements'. After review, the PSM committee and corporate managers budgeted over 600k for process safety improvements. This case is further evidence that older process equipment can lead to accidental releases despite good process safety programs. This is an interesting local case study that could have national applicability – the idea that there is outdated equipment and process designs out there that could potentially be preemptively targeted for inspection based on a history of accidental releases reported to the NRC.

The above example represents the core message of this paper; promoting industry process safety improvements through the use of near miss reportable accidental releases to measure a site's performance and the use of incident investigations to improve process safety programs. Continuous process safety improvement is not only a public and environmental mandate; it is a sound corporate investment. Delaware's ARP program works with industry to reduce accidental releases in the spirit of partnership, performance, and efficiency.

Federal regulators continue to be focused on regulatory compliance and enforcement with limited interest in partnerships and limited technical or practical processing knowledge. Their inspection programs are governed by high profile accidents, 'emphasis' programs, and worst case scenarios as reported in RMP submissions.

Therefore, based on our experience, state and local governments have an important role to play in this process safety improvement story. Being 'closer to the action' so to speak, local government can play an important role in developing constructive safety partnerships with industry. However, states embarking on such a path need to be sure to hire regulators with sufficient industry experience to genuinely make government-industry partnerships work successfully. The CSB would appear to agree that local jurisdictions have a better track record. In their August 28, 2008 investigation report on the Bayer CropScience, LP release in Institute, West Virginia they make an eloquent two page recommendation (recommendation 8.3) to the Kanawha-Charleston Health Department to establish a 'Hazardous Chemical Release Prevention Program' [11].

5. Conclusion

There are still accidental releases occurring, both many near misses and occasional catastrophic accidents. There is room for continuous improvement and near miss incident investigations are a powerful tool for process safety improvement whether used by regulators or facilities. Based on the EPA risk management plan accident data presented earlier, some releases and accidents appear to be due to process complexity; an issue not directly addressed as part of a typical Process Hazard Analysis (PHA). This is an issue that might yield additional insight in further study. Why does risk analysis on all the parts of processes not result in an overall elimination of releases? Are there subtle interactions or intersecting variables in complex processes that are not adequately covered in a typical risk analysis? Are mechanical integrity program staffing and resource allocations legitimate questions for PHA? Or, are there just black swan accidents that defy risk analysis [12]?

Facilities with older equipment and outdated process designs also have more releases. There is evidence this is the case in spite of high-quality PSM programs. Facilities need to step back and take an honest look at where their near miss accidents are occurring over time. Sometimes improving the process safety program isn't enough – engineer's need to point out the need for capital improvements based on real performance data.

Finally, in a world of limited resources and technical knowledge, regulatory programs representing the public should be identifying ways to utilize resources that encourage sustainable process safety improvements. Statistically, there will always be those facilities at lowest end of the bell curve that need some level of enforcement to encourage compliance with regulations. But the good news is that most facilities have taken process safety seriously and developed quality PSM and RMP programs. Regulators should wake up to the fact there is a bell curve of facility performance and consider using near miss accidental releases to measure process safety program performance. Reviewing near miss incident investigations for reportable releases is much less technically and resource oriented than full scale multi-day, week or month inspections. Taking into account the limited manufacturing experience of many process safety regulators, this represents a better use of resources and technical abilities.

6. References

[1] John W. Herber, "Smaller Companies Struggle with Process Safety", Process Safety Progress (2012), 346-349

[2] James R. Phimister, Ulku Oktem, Paul Rl Kleindorfer, and Howard Kunreuther, "Near-Miss Management Systems in the Chemical Process Industry," Risk Management and Decision Processes Center Operations and Information Management the Wharton School of Management, University of Pennsylvania, 2003

[3] Heinrich, H.W., "Industrial Accident Prevention", New York, McGraw-Hill, 1931.

[4] Environmental Protection Agency (EPA), 40 CFR Parts 300 and 355 extremely hazardous substances list and threshold planning quantities; emergency planning and release notification requirements, first published Federal Register 52:77, pp. 133378-13410, 22 April, 1987

[5] State of Delaware, Administrative Code 1203 Effective date April 1, 1991, Revised December 11, 2002 and July 11, 2004

[6] Riwik U., Kuwait National Petroleum Co., "Risk-Based Approach to near miss," Hydrocarbon Processing October 2002.

[7] CCPS Process Safety Incident Data Base (PSID) Project http://www3.aiche.org/PSID/PDF/PSID.pdf (accessed December 2012)

[8] "Process Safety Performance Indicators for the Refinery and Petrochemical Industries," ANSI/API Recommended Practice 754 First Edition, April 2010

[9] Jay Brabson, P.E., "Regulatory Initiative for Improving Industry Process Safety Performance," Process Safety Progress (2010), 305-307

[10] EPA Central Data Exchange https://cdx.epa.gov/ (Accessed January 2012)

[11] U.S. Chemical Safety and Hazardous Investigation Board, Report No. 2008-080WV, 'Pesticide Chemical Runaway Reaction Pressure Vessel Explosion', January 2008

[12] John W. Murphy and Jim Conner, "Beware of the Black Swan: The Limitations of Risk Analysis for Predicting Extreme Impact of Rare Process Safety Incidents," Process Safety Progress (2012), 330-333.

9TH GLOBAL CONGRESS ON PROCESS SAFETY

Are You Prepared for the Coming Economic Recovery?

John W. Herber
Process Hazard Management Services, LLC.
11 Moonlight Bay
Stillwater, MN 55082
john@phms.us.com

Prepared for Presentation at
American Institute of Chemical Engineers
2013 Spring Meeting
9th Global Congress on Process Safety
San Antonio, Texas
April 28 – May 1, 2013

UNPUBLISHED

Are You Prepared for the Coming Economic Recovery?

John W. Herber
Process Hazard Management Services, LLC.
11 Moonlight Bay
Stillwater, MN 55082
john@phms.us.com

Keywords: Economic recovery, Restarting equipment, PSSR, Training, Preparation

Abstract

The world economy has been depressed for several years. During that time, changes have occurred; these include reductions in funding, headcount reduction, reduced production, equipment temporarily out of service, and reduced equipment integrity inspections to name a few. Some of these changes may have been reviewed, but many especially those involving human resources may not have been thoroughly analyzed or were only reviewed to reassign duties until a replacement can be named. Restarting mothballed equipment and a return to "full" production will make these gaps more obvious.

Gaps may be the availability of qualified workers, training resources for new employees as production returns to pre-recession level. If these gaps are not recognized, the economic upswing will challenge the existing resources & systems and potentially lead to unsafe or even catastrophic situations. Also, there may be shortages of raw materials if suppliers are not prepared to meet your production levels.

This paper serves to alert the petrochemical industry to the impending problems that may be encountered during the economic recovery, to encourage planning to address resource gaps, and to review existing PS systems to verify their ability to handle the influx of changes that will occur.

1. Introduction

"It is worth discussing radical changes, not in the expectation that they will be adopted promptly, but for two other reasons: One is to construct an ideal goal so that incremental changes can be judged by whether they move the institutional structure toward or away from that ideal. The other reason is very different. It is so that if a crisis requiring or facilitating radical change does arise, alternatives will be available which have been carefully developed and fully explored." Milton Friedman, U.S. economist. [1]

This describes the situation the U.S. chemical manufacturing industry is in today. It is teetering on the edge of a recovery, although when it will tip into more consistent improvement is unknown. This paper will explore the various management and economic factors prevalent in the U.S. economy and how they impact key decisions that affect process safety.

1.1 history

Over my 35 + year chemical engineering career, the U.S. and many other economies have experienced various downturns. They may have affected only a small sector of the economy or a single region of a nation's economy. I was the beneficiary of the upturn following a 2+ year economic recession in the mid-1970s. When I was hired, the plant was recovering from staff reductions across many areas – production, maintenance, and support functions. Many remaining employees had been placed in positions that were not their primary skill area. At that time, as during the present "recession", 'a job was a job'. However as the economy recovered the plant began to return to pre-downturn operations, they needed to hire and train many new workers. In addition, as other more suitable positions opened up in other areas, people moved onto new job responsibilities. By the end of my first year, 25-35% of the operators had less than 1 year of experience.

Today, the general economy and the petrochemical industry in general are in a similar position. When they are pushed for production, they must pull in more resources. This may mean hiring temporary workers or utilizing overtime to stretch the existing resources. Companies are reluctant to hire permanent resources due to the uncertain future which may result in yet another layoff. The "on again/off again" nature of a recovery can also place suppliers of material and equipment in that same operating mode.

1.2 economic situation

It is at best difficult to project future economic trends. The current economic recession has baffled economists since its beginning in 2007, so this article will not attempt to predict when the current economy will improve. It may be doubtful that things will even return to what was 'normal' prior to the recession. This paper cannot define what the second decade of this millennium holds. However, with the number of changes in economic and management practices it is not difficult to foresee problems and pitfalls that may present themselves as the economy improves.

A recent article in Chemical Engineering Progress [2] predicted growth for the U.S. economy and the chemical industry in general. Those predictions are summarized in Figure 1 below. Another consistent prediction from economists is that this recovery will be slow and long in duration. On the surface this may seem to be the best situation, a steady 3-5% growth rate. It is relatively easy to forecast single-digit growth and to adapt business plans production schedules and capital forecasts to meet this pace. However, what if the economists are wrong? What if the most aggressive rate of improvement is what really happens? Further consider that compounding a 3 to 5% growth rate over the course of five years becomes a 16-28% percent increase. Somewhere in that period, resources and capacity are going to be overtaxed.

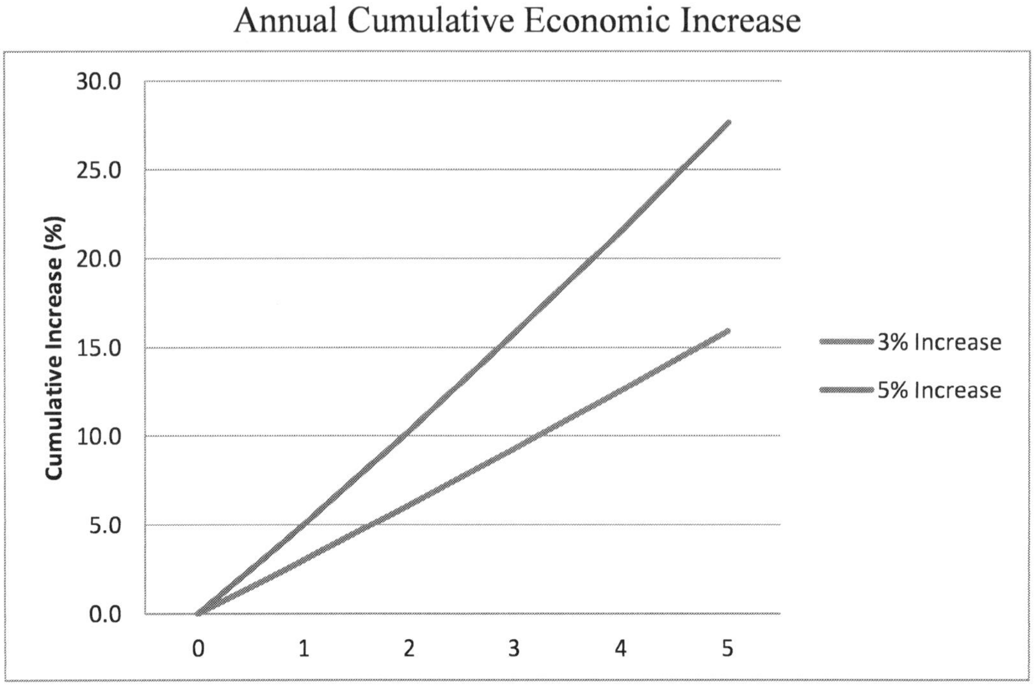

Figure 1.

The chemical industry is capital intensive requiring highly specialized equipment. There is no inventory of such equipment sitting on a show room floor waiting to be installed in any petrochemical plant. Lead time for equipment, particularly large specialized equipment can be over two years. This means that the chemical industry must be looking ahead at least four or five years to implement what growth they anticipate today. For large continuous operations such as refineries single-digit gains in productivity may be achieved by planned efficiency or de-bottlenecking efforts. For smaller specialized or batch operations, this may mean installing another reactor system very similar to those that exist today. These two scenarios are easy to predict, estimate and have approved by management. However when the smaller incremental improvements no longer meet the projected demand, a major capital project and even perhaps an entirely new site may be required. Projects of this magnitude require many years to design, acquire land and permits, install, check-out and eventually operate. That is why it is necessary to prepare for economic recovery before existing equipment and resources are overtaxed.

This paper is intended to spark thinking and discussion around the topic of the possibility that the economy may improve more quickly. If the common theory that the U.S. stock market leads the general economy by six to nine months, the time for planning was the last quarter!

"It is generally true that, the more preparations one has for an event, the more inconveniently fast the event will occur." (Monica Fairview, Darcy Cousins, 2010) [3]

What if the economic recovery comes later than planned? Consider yourself lucky! A good plan for restarting process can always be improved and some pre-start-up activities can be performed well in advance of start-up. A good start-up plan is developed by a cross-functional team that knows the process and equipment well. The start-up plan requires procedures written with sufficient detail to allow operators who have not performed those tasks in many years to complete them safely. Test methods and results need to be documented in case there is a delay or change of assignment before start-up.

2. General Approach

The difficulty in presenting this topic is that the operation of any manufacturing plant is like a spider web. Any perturbation in one area affects most others to some degree. For instance, a change to operating schedules not only impacts operations, but also the entire supply chain, maintenance, and other staff functions. Other factors at work in today's economy are its global nature, "just-in-time" inventories, and a higher reliance on contractors and outsourced workers to name a few.

Also, to understand how manage this situation, we need to understand what has happened over the last two to five years that has made this issue much different from other economic recessions. To best dissect this, the paper will sort the issues by personnel, equipment, and materials factors. As noted above, there may not be clear delineation among these areas, but this is one way to cover the important points.

3. Personnel Issues

Many of the observations made here derive from personal experience, contact with petrochemical companies, and general articles in newspapers and magazines.
For at least 10 years, the U.S. economy and perhaps others have been using a headcount mentality and have made generalizations about the "correct" number of people a given operation requires. Unfortunately, this may be based on not science, but on loosely related ratios of the number of people required to some production metric. It may be salespeople per sales dollar or the ratio of engineers to the capital budget. This very short-sighted thought process ignores many key risk based process safety principles such as Process Safety Knowledge, Process Management Systems, Management of Changes and Process Safety Competency. Consider the very ends of operation: When production is zero, i.e., during a turnaround or shutdown, are all employees laid off? Actually the maximum number of people working a given area probably occurs immediately before operations actually begin. This principle also does not account for the need for highly specialized skills in many areas of the chemical industry plus a significant amount of experience.

Compounding this management trend is yet another aspect of the U.S. economy. Over the past five years, many companies have encouraged more experienced employees to consider retirement and have provided incentives for early retirement. This strategy would not have significant impacts if the company were entirely exiting a particular business and would no longer need that experience or expertise in the foreseeable future. Many companies have mothballed equipment or reduced production rates and therefore reassigned the people from particular production area. Some companies do not apply management of change to personnel issues. Some may review differences in production shift schedules, but loss of technical expertise across several years may not have been evaluated for its impact and probably not considered a threat to future production.

Looking forward, demographics indicate that the "graying of America" is a sobering reality. Many companies are stocked with people ready to retire as soon as their investments seem sufficiently secure. Even if the workforce's average age is under 55, a major turnover of personnel may occur as workers migrate to new positions opened by other career opportunities within or outside their current company.

The unemployment rate in the U.S. has ranged between 8 and 10 percent over the last two years, and stood at 7.8% at the end of 2012 [4] dropping about one percentage point per year. The current shortage of highly skilled workers will worsen. When an influx of new workers is required, a company may struggle to find sufficient workers who meet their basic selection criteria for open positions. One strategy to address this problem is to use contract resources, outsource the work to a different region, in the country or offshore. If the economy in general is improving, there will be competition for skilled laborers and the ability of contractors to supply skilled labor will also decline. Additionally, transferring operations overseas would require even more planning and near-flawless implementation to meet a sudden increase in demand.

3.1 personnel change backlog

Over the past three to five years, job mobility has been limited by the number of openings in companies. As a result, when recovery begins that there may be a number of employees who are looking for new opportunities either inside their company or at a different company. This is yet another risk factor when looking at staffing in an organization. Expansion may create new positions and employees may be promoted. This generates more demand for entry-level positions as experienced workers move to new or different positions. There may be also a loss of experienced employees who take positions in other companies. These factors alone are difficult to predict but are obvious threats to steady operation.

3.2 recommended personnel change actions

1. Review the seniority and job posting policies to ensure they are robust enough to handle the expected shuffle of job openings.
2. Renew/revise the policies for promotions, job bidding, etc. in the changing employment environment. This may require discussions with labor representatives.
3. Communicate the policies to personnel at the first sign of increased job change activity so that all understand the system's function.
4. Utilize Management of Change when modifying staffing level, either up or down and when assigning new personnel to a process area.
5. Conduct an open discussion with technical staff to spotlight their internal career options before they look elsewhere to meet their career goals.
6. Acknowledge the value of their expertise and loyalty shown during earlier economic difficulties.

3.3 recruiting new personnel

3.3.1 operations & craft skilled employees

In a perfect world, there would be a ready supply of highly trained and skilled people, just waiting to be hired. If there were this pool of candidates, how can you ensure they are skilled, qualified, or willing to work for you? The local labor representative may manage the seniority pool and also provide classes to keep those in the pool current on their training and certifications. However, in many markets, there is a shortage of skilled workers, or the only available people lack acceptable work histories.

3.3.2 technical employees

All comments noted above for operations and craft-skilled personnel, also apply to technical employees. They fall into two categories – experienced and new grads; each has its pluses and minuses. Experienced technical staff come with a practiced set of skills and proven work habits. They arrive with installed hazard awareness and risk acceptance that may be different from your company's. On the other hand, those skills may not be state-of-the-art. More experienced workers may be more set in their ways and less adaptable to your systems and methods.

New grads are the flip-side of the experienced engineer with less hazard/risk awareness, but more adaptable and probably have more current technical skills especially in computerized areas.

3.3.3 recommended personnel recruiting actions:

1. Consult with the local labor representatives to determine what sources of labor exist and compare this to foreseeable labor needs.
2. Look beyond the local labor pool; if possible to find other workers who can fill hourly or technical resource needs. These resources may require a different orientation or training plan.
3. Use existing employees to recruit acquaintances who may be underemployed or looking for new opportunities.
4. Do not assume that new grads or experienced workers understand your processes and hazards. They must all receive training that enables them to perform safely even if it means some redundant training.

3.4 personnel training

New employees will require basic training. However, if the minimum worker proficiency cannot be found, new employees may require far more training than in the past. Secondly, training resources in a company may have included some of those people who retired early or transferred to other areas. That jeopardizes the ability to provide adequate training to a significant number of new employees.

3.4.1 recommended training actions:

1. Inventory existing training requirements to ensure all the required training materials are still available, current, and include new training requirements that were developed. This is also an excellent time to cull out any training that has exceeded its effectiveness.
2. Once the necessary training 'courses' have been defined, assess if sufficient resources exist inside or outside the organization to provide the courses on the required schedule.

3. Using the gaps identified in 1 & 2 above, develop a more complete training strategy that utilizes internal and external resources to meet the training needs.
4. Use Chemical Safety Board (CSB) videos [5] and other incident-based examples to increase employees' sense of vulnerability of serious events.
5. Set a long-term goal of establishing an e-learning library. Develop new training materials for this format in mind.

4. Equipment

Of the areas to analyze in preparation for economic recovery, the equipment issue may be the easiest. Equipment, unlike people has a nameplate capacity and can be factually evaluated for its fitness for duty. When needed, a preparation plan can be developed based on past experience and a PreStart-up Safety Review (PSSR).

There are a number of industry practices that can be employed to inspect and prepare equipment. However, they are only effective if they are performed well and necessary corrective actions are taken. Some inspections are based on runtime, but others, such as corrosion under insulation, have been at work even while equipment has been idle. Despite pressures to restart equipment, these inspections must all be performed to avoid a catastrophic event or merely a nuisance shutdown.

Other key areas for review are the instruments and PLC logic. All field instruments need to be tested to verify they still function as expected. Similarly, the control logic should not have changed over time, but how do you know? A vice president that I worked with had an excellent quotation for this, "In God we trust. All others bring data." [6] All control logic needs a functional review that may include revisiting the initial check-out and start-up plans. Likewise, mechanical equipment also needs to be checked out. Even basic attributes such as proper rotational direction are important. Since it is very difficult to determine what is critical, the most practical approach may be to check-out everything in detail. Remember that shipping containers may also be delinquent in their required inspections. Containers, from intermediate bulk containers (IBCs) through tank trailers to railcars, need regular inspection to protect products in transit. These inspections may not have occurred while the shipping containers were inactive.

4.1 recommended equipment actions

1. Develop a check-out and restart plan with the assistance of workers experienced in that area, even if it means moving some people back to old positions temporarily.
2. Equipment that has been inactive for months or years doesn't magically repair itself. Review repairs and inspections that may have been deferred and address them as part of the re-start/start-up plan.
3. Cost-cutting measure over previous years may have led to "cannibalization" of equipment, spare parts or components. Include this potential in the inspection procedures.
4. Revisit PHAs as part of the re-start efforts and use hazard identification techniques to evaluate restart activities.

5. Never assume an erroneous reading is 'just a bad instrument'. Restarting equipment takes the process into territory it has not experienced recently or ever. Reasonable estimates of process parameters must be made to provide the operators with ranges that can be expected until steady-state is attained. Start-up procedures need to provide remedial action for when the expected range is exceeded.
6. Renew process hazard training for equipment that has been inactive for longer than a few months.
7. Include mitigative safeguards and equipment in the restart plan since there is a higher risk of leaks, spills or off-spec material.

6. Materials

During the years since full scale production, materials in storage have aged, and former sources of supply may have disappeared. As a rookie process engineer, I was tasked with starting up a batch fine chemicals reactor. The planned check-out of the equipment went as expected; some minor repairs were needed. The first batch produced an unexpectedly high yield. The second batch was more normal even though all the raw materials came from the same sources. What was probably the fourth batch had a drastically low yield. What happened?? The raw material tank had stratified over the three years the system was inactive, so the first batch had a higher concentration of reactant and the fourth was deficient. This was proven by lab analysis after the fact. Wouldn't it have been smarter to sample the tank before the first batch and recognize the need to mix the material even though mixing had never been required in the past?

If tanks or pipelines have been used for other material over a shutdown period, residual contaminants or cleaning materials may still be present. Flushing and testing of the lines may seem like overkill, but in reality it is cheap insurance against quality issues or a major process upset.

Sources of supply that provided key raw materials may have gone out of business or altered their range of products. That key material once readily available may require the supplier to restart a process and mothballed equipment. For certain materials, the manufacturing location may have changed, and subtle differences may exist even though the containers are still labeled the same way with the same company logo.

When new sources of raw materials are required, they need particular evaluation. One of the best evaluation methods is to progress from lab scale through pilot scale evaluations before making a full scale material change. Leaks, spills and off-spec material may cause new and/or different waste streams. Like other suppliers, waste handling contractors will need to be ready and capable of handling the influx of wastes.

1. Do not assume that materials which have been in inventory maintain the same potency as when initially purchased; test and confirm. Lower potency may not correlate to lower reactivity; inhibitors and stabilizers also lose effectiveness over time.
2. Verify past suppliers are still viable and are prepared to meet your renewed demand.
3. Where necessary, use the company's practices to search for new suppliers and to verify the quality of their products. This may take several iterations to successfully complete.
4. Any material changes should receive a thorough Management of Change (MOC) review.

5. Be prepared to handle new and different waste streams during cleaning, checkout and start-up. This includes collection, storage, and disposal.
6. When inventorying raw materials, double check that spill response chemicals and equipment are also at proper levels.

6. Conclusions

1. "Better to prepare and prevent than repair & repent" – Ezra Taft Benson U.S. Secretary of Agriculture
2. "Before anything else, preparation is the key to success". Alexander Graham Bell U.S. (Scottish-born) inventor (1847 - 1922)
3. The ability to react quickly and correctly to the future depends on the degree of anticipation and preparation that can be made in advance.
4. To be prepared requires a reasonably accurate vision of the future and a plan on how your business will fit there.
5. After years of being denied funding, expect more, positive responses to funding proposals. Therefore, be careful what you ask for; you just might get it.
6. Well-thought-out proposals will be those that get approved and implemented.
7. Your suppliers may also struggle with the same restart and resource issues. Use extra diligence to avoid problems that been noted in.

7. Acknowledgements

The author would like to acknowledge the reviews by Ms. Kathleen Clemmer, whose editing and constructive comments definitely improved this paper.

8. References

1. Milton Friedman, Capitalism & Freedom 2nd ed. (1982) pg ix
2. Chemical Engineering Progress January 2013 pg 26. " Shale Gas Brightens Outlook for U.S. Chemicals" T. Kevin Swift & Martha Gilchrist Moore
3. Monica Fairview, Darcy Cousins, (2010)
4. Bureau of Labor Statistics (BLS) website. (http://www.bls.gov/bls/unemployment.htm) last accessed 3/3/2013
5. Chemical Safety Board website (http://www.csb.gov/videoroom/detail.aspx?) last accessed 3/3/2013
6. Chuck Oesterlein, Vic President, 3M Company (retired) (unpublished)

Additional references not cited:

Guidelines for Risk Based Process Safety, CCPS (2007) Wiley

Recognizing Catastrophic Incident Warning Signs in the Process Industry, CCPS (2012) Wiley

Guidelines for Hazard Evaluation Procedures, 3rd ed. CCPS (2008) Wiley

Guidelines for Mechanical Integrity Systems, CCPS (2006) Wiley

Guidelines for Performing Effective Pre-Startup Safety Reviews, CCPS (2007) Wiley

9. Author's Biography

John is a Staff Consultant for the Center for Chemical Process Safety (CCPS) and principal consultant for Process Hazard Management Services LLC (PHMS).
John concluded a 33 year career with 3M Company in 2009 that included engineering, production operations and corporate safety. In Corporate Safety Services, John developed programs including PSM metrics, Process Hazard Management policy and implementation guidance. John's diverse experience has shaped his personal passion for process safety.

John has made presentations at 2 CCPS global congresses, participated on writing committees for several CCPS Guidance books and chairs the Process Safety Beacon and Process Safety Messages committees, and is a Process Safety Boot Camp instructor. He was named a CCPS Fellow in 2009.

John has a BS in Chemical Engineering from Purdue University. He resides in Stillwater Minnesota and is married with 2 married children and 2 grandchildren.

How to Create a Sustainable Safety Culture
Moving from Pound and Blame to Partnering

Richard N Knowles, Ph.D.
The Safety Sage
Richard N. Knowles and Associates, Inc.
6083 Bahia Del Mar Circle
Unit 564
Saint Petersburg, FL 33715
RNKnowles@aol.com
www.SafetyExcellenceForBusiness.com

Prepared for Presentation at
American Institute of Chemical Engineers
2013 Spring Meeting
9th Global Congress on Process Safety
San Antonio, Texas
April 28 – May 1, 2013

UNPUBLISHED

How to Create a Safety Culture
Moving from Pound and Blame to Partnering

Richard N. Knowles, Ph. D.
The Safety Sage
Richard N. Knowles and Associates, Inc.
6083 Bahia Del Mar Circle
Unit 564
Saint Petersburg, FL 33715
RNKnowles@aol.com

Keywords: Safety in the Workplace, Employee Engagement, Leadership, Safety Culture

Abstract

Safety cultures do not spontaneously occur. They are created and nourished by honest, authentic operating managers and leaders. These people must insist on high standards and foster trust and interdependence within their organization. This requires open, freely flowing information that people at all levels in the organization can access, analyze and use to keep their workplace, the hazardous materials and energies under control. This work is focused and relentless, requiring discipline on the part of the whole organization, including the top level managers, front line supervisors, engineers and hands-on employees. This paper presents an operating manager's experience on how he evolved his "pound and blame" managerial culture into a "partnering" culture and successfully improved the organization's safety performance on both the plant's personal safety and process safety metrics. This effort contributes to the well-known "business case" argument for a robust process safety culture, as measured by the significant increase in the plant's operating profit and corresponding process safety-related improvements.

1. Introduction

Personal and process safety are critical factors in safely operating a chemical plant and must be conducted at the highest levels of competence and integrity. As a plant manager I had total responsibility and accountability for the operation of our facility; this was an awesome experience. Any major mistake could hurt people on the plant, the neighbors and cause a major financial impact on the company.

Total safety performance, and process safety in particular, are core technologies for those operating a chemical plant using hazardous materials. Sloppy operations with injuries, leaks and explosions are totally unacceptable. We have no right to be careless, hurting our people, the neighbors and the environment.

In managing a plant where highly hazardous materials, like phosgene, anhydrous ammonia, chlorine, oleum, hydrogen cyanide, were used, I placed high value on personal and process safety management. I also respected the engineers who did the analyses, the managers who had to make tough decisions, the operating people and mechanics who worked safely and did the operations, inspections, maintenance and all the follow-up. It takes a focused operating team to thoroughly address process hazards and to ensure the safety, health and welfare of everyone inside and outside the plant.

I had to help to create a culture of operating discipline, openness, honesty, respect, integrity, listening, sharing, supporting each other and doing the things we agreed needed to be done. Excellence in personal safety flows into and supports excellence in process safety management and visa-versa; it is truly a team effort so that we can operate with the highest levels of safety and process integrity and standards.

This is an effort that also reaches out into the communities surrounding our plants. Our neighbors need to have the confidence that we are operating in a way that will protect them from the potential problems we could face. We have to lower both injury rates, and the probabilities of an incident or release to the very lowest levels. This is an issue of our freedom to operate and avoid excessive regulations. These are issues of survival for our industry.

The Challenge:

As an example of the success of shifting the culture to excellence in both total safety and in process safety management, I want to share the story of our work in the Kanawha Valley in West Virginia. When I was assigned as the Plant Manager of the DuPont Belle Plant in April, 1987, the initial challenge was to restore safety and operating discipline and performance.

When the safety gets right, everything gets right. The results that the Belle Plant people achieved were extraordinary. Total Recordable Injury Rates dropped by over 95%, emissions dropped by 88%, productivity rose by 45% and earnings rose 300%. In the first 2-3 years we achieved significant reductions in injuries and incidents. Then we had to move beyond the initial successes putting systems and processes into place over the next several years to sustain this work

2. Our Story

Safety Leadership is the Key to Sustainable Safety Excellence

Everything that happens in an organization happens through people. We depend on people doing everything within their capabilities to do their best. We need to have open, honest, knowledgeable, competent, people of high integrity doing the safety and process hazards work and in carrying out the various analyses, installations, inspections and all the routine follow up work like relief valve inspections, pressure testing and thickness measurements as scheduled. We can have the most sophisticated analyses possible, but if people don't take the initiative to follow through, then all is for nothing. Cutting corners, ignoring the use of PPE, covering up, not doing the inspections, keeping inadequate records, not following up, bullying and lying are unacceptable.

My mantra was "I don't have a right to make my living where it is okay for you to get hurt. Now we have to get on top of the safety stuff and build a strong business."

Figure 1 illustrates these ideas of how the technology and the people are both critical to build a sustainable safety culture. Most organizations put most of their emphasis on the technological side of things doing excellent analyses and studies. They develop recommendations for the best practices in managing the processes and in minimizing the risks. This is often called "the hard side" and is critical to do well.

A few organizations combine this work with the people side. All the technological work and recommendations must flow through people. The engineers and designers, those doing the installations, the supervisors, the operators, the mechanics and the safety people who must train and monitor things all need to be involved and committed to things being done right. The processes put into place to do the proper follow-up inspections and checks must be rigorous all the way through the design, installation and onward for the life of the equipment. The best technical work will be wasted if these things are not done. The people side is often called the "soft side". I've heard it said "the hard side is the easy side and the soft side is the hard side".

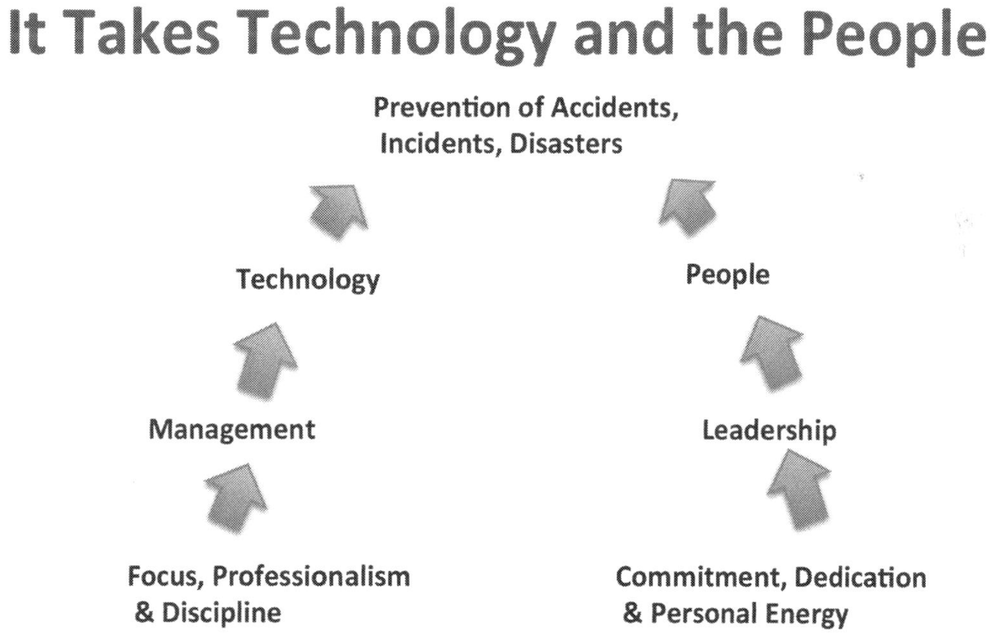

Figure 1

The shift in the safety culture is focused, relentless work, requiring discipline on the part of the whole organization.

One of the most important tasks for the operating manager is to create a culture where:

- information is freely flowing so people can find and use it in their work and analyses,
- there is a climate of trust and interdependence,
- people can and do tell the truth as they see it,
- we listen to each other seeking understanding,
- people, at any level, including first line supervisors and hands-on employees, can take appropriate leadership action to fix a problem that they encounter,
- People act on the information so that the plant can be the safest it can be

Many people know critical information but it is often not shared or discussed. This can lead to disasters. Incidents like the Challenger Disaster, where the engineers knew the low temperatures could cause the O-Rings to fail but their managers rejected or ignored their cautions, are unacceptable. Similar cultural conditions were present in the Deep Water Horizon disaster.

When I was first assigned as manager of the DuPont plant in Belle, WV in 1987 there were about 1,300 people working there and the plant's injury performance was very poor. This is quite apparent in Figure 2. The standards were terrible, things were being covered up, people would not tell the truth about incidents, various groups did not get along with other groups, management didn't get along with the hourly people and visa versa, everyone blamed everyone, else, the bullies blocked improvements etc. I was expected to get the performance much better before we killed someone. I went on the warpath and raised a lot of fuss. I pounded, verbally, on people and pushed hard. I talked with as many as I could find, but most tried to stay away from me. I had lots of arguments with the managers reporting directly to me. I chastised some people publically, which was poor on my part. I spent hours and hours each day in the plant and many evenings and nights as well. I walked everywhere. I wore glove so I could pick up trash. When people refused to work more safety or were bullying their work mates, they went into the discipline system. We terminated those who just didn't get the message…working safety was important!

In a climate like this everyone was under a lot of pressure, Healthy communications were non-existent. I was not happy about the plant or myself, and neither were most of the others. But our injury performance, safety standards and housekeeping got a lot better.

In two years the injury rate had dropped to 0.74 and leveled out. This injury rate was close to the DuPont rate at the time and that was pretty good. But we could not sustain this without a lot of effort and pressure. We were all sick of the fighting.

As things stabilized we began to talk together about how to shift our culture to one where the performance could become sustainable and better. In 1991 we shifted to teams across the entire plant. You can see in Figure 2 that the injury rate improved, dropping to 0.26. We learned to talk together more effectively, we learned together more effectively and the sustainability in our performance showed up as the years rolled by. The depth of our total safety effort significantly improved so process safety was better, uptime improved, earnings increased. Figure 2 shows that this level of performance between about 0.3 and 0.0 sustained for many years after I was reassigned in 1995. My role shifted from being a tough taskmaster, driving and pounding all the time to become a cheerleader. We were becoming partners. This was a lot more fun.

In all this work we shared information abundantly, trust and interdependence built, we talked a lot and listened even more. Good ideas emerged from all over the place about how we could improve customer service, improve our environmental performance, shorten the turn-around time between product campaigns from 3-4 weeks to just 3-4 days through better planning and cooperation. Resistance to change almost disappeared as people came up with good ideas and implemented them. In my first few years at the plant we struggled with change making only 3-4 major changes. As we all learned together creating the new ideas, resistance to change dropped to where we were making 2-3 changes a month and I was slowing them down so that we would not out-run our technology. Management of change is critical in process safety management.

Figure 2

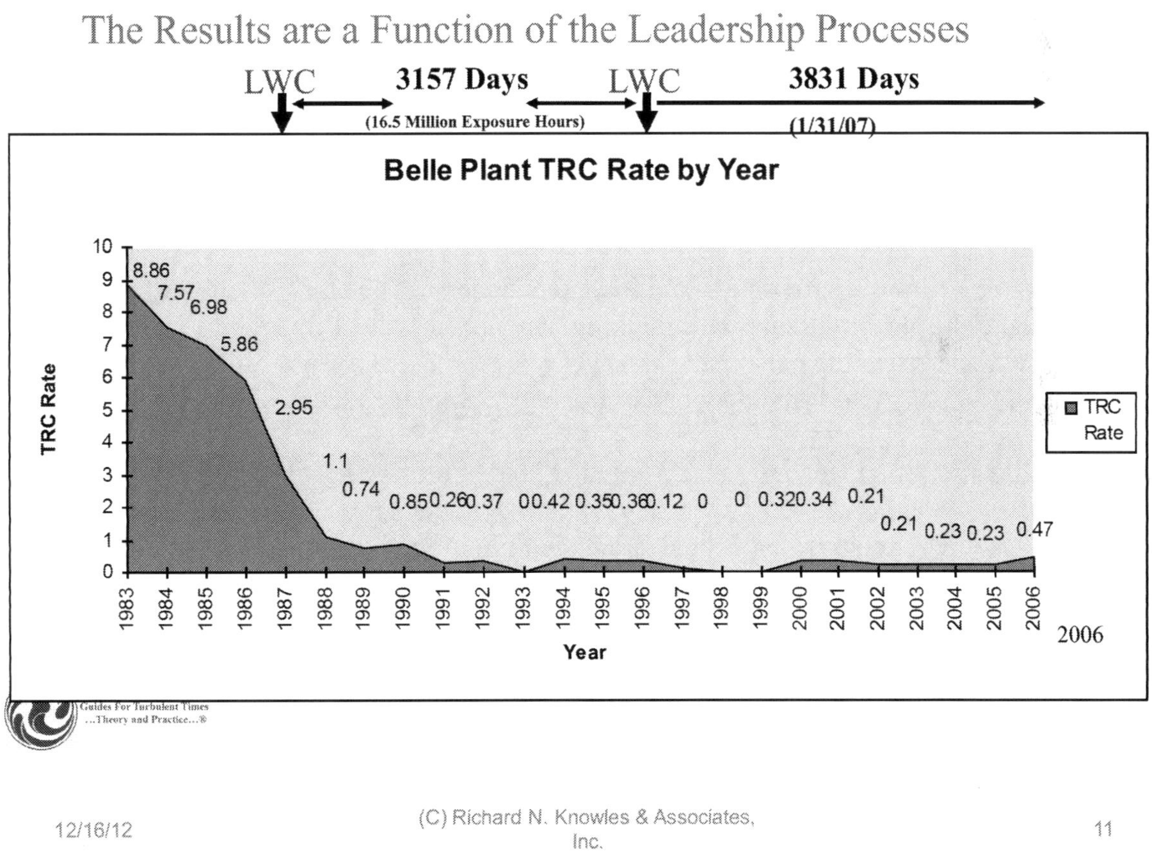

Note: LWC means lost workday case and TRC means total OSHA recordable injury rate.

In reflecting on all this work to improve safety and our over-all performance, and in the consulting I've done around the world since retiring 17 years ago, I have been able to develop a simple, 4-step process that summarizes our work at Belle and works quickly in every organization in which we've worked. In working this way the injury rates drop quickly, the communications open up, people develop more energy and are more creative. They experience a cultural change like we did at Belle. This shift does not require any capital investment.

This cultural shift begins with a 4-step Safety Leadership Process™.

1. The top leaders have to talk together, openly and honestly, about what they really want to have happen and develop the commitment to do it. They need to be very clear and coherent about what they are really going to do to move towards sustainable safety excellence. Are they really going to live and do the critical work? The Process Enneagram© tool is used to hold the focused, open conversations. This tool is the only tool that I know of that helps people solve complex problems, make the social connections they need to get the work done and releases the emotional energy and commitment to do the work quickly…all at the same time. My book, The Leadership Dance, provides a lot of information on the creation and development of this tool [1].

   This first workshop takes about a day to develop their Safety Strategic Plan. During the course of this first step of the work, the fierce conversations are held and trust and interdependence build. The top leaders develop their Principles and Standards, and ground rules as to how they will work together and with the others in the organization. They agree to share information and to hold each other accountable to live up to their Principles and Standards. They agree to "walk the talk".

2. Having co-created a Safety Strategic Plan, the leadership goes into their workplaces to talk about the plan and share their thinking, listen, and to look for improvements in the Plan with the operating personnel. This work is visible and open for all to see, and with everyone sharing and listening together trust and interdependence build.

3. Each day involves listening, talking together, and exploring new and better ways to do the safety work more effectively. They learn to conduct simple, effective Safe Acts Audits, actually seeing what people do and learning together and acting on what they are learning to prevent incidents and injuries. Most of the problems that emerge are related to systems and communications problems.

4. Then they help create and support the environment where the people can make changes that they have discovered together.

The Safety Leadership Process includes listening, talking, sharing, and co-creating the future together. This is done over and over as the organization and its environment evolve. This Safety Leadership Process is developed more fully in my new book, "Partnering for Safety and Business Excellence" [2]. In this process we partner with the people to achieve excellent, sustainable performance.

Once the leaders are clear and unified in their thinking about safety, they must step out of their offices, step into their workplaces and step up to the challenges and ensure that problems are promptly fixed. During the course of this work, as everyone is learning, the organization becomes "leaderful" - that is, when anyone sees a problem, they step up to take the initiative to get it fixed.

An example of the sort of environment we were able to create occurred one morning when I was walking around the plant. A methylamines operator came up to me to tell me he'd shut down the amines plant, on his own initiative, because he'd found a small, weeping type of leak at the base of a relief valve on top of the 150-foot distillation column. He wanted to prevent a more serious leak so he took the initiative to shut-down to fix it. I felt that this was a great contribution when an operator can take this sort of initiative and prevent a problem.

Just look at the way our safety performance changed and improved over 23 years. Our safety leadership process before 1987 was one of talking about safety but doing little to make it happen. When someone did get hurt, there was a lot of blame and shame for the operators or mechanics. When I came into the plant (1987-1990), our process was a typical, hard, top-down process that I applied to everyone. We drove things hard, argued a lot and reduced the injury rates to about 0.7, but could not sustain this level of performance. As soon as the pressure was released, things grew worse. While we were at this level of compliance, the system was full of holes and people kept getting hurt. When we learned to partner with the operating personnel in 1991, look at how well they performed and sustained this excellent level of performance for years after I was reassigned in 1995.

As we re-established safety and operating discipline our process safety improved as well. Sharing information, building trust and interdependence, and helping everyone see the importance of their work and how it all fit together spread to other aspects of our work. The inspections were done as scheduled, and things got repaired in a timely way. More and more people at the operating level got involved as the story of the amines operator illustrated. The number of releases went down and the severity was significantly reduced. The engineers were doing the analyses with the appropriate technology, thoroughly and with the highest integrity. When they would review their work with me, I would often ask, "Is your analysis strong enough that you would be okay if your family lived at our fence line?" We had families living next to our fence line and we were responsible for their safety. If the engineers hesitated in their answer, they went back to do the analyses again.

Having learned to shift the safety culture to one of sustainability and excellence the positive impacts showed up everywhere. When the safety gets right, everything gets right. The results that the Belle Plant people achieved were extraordinary. Total Recordable Injury Rates dropped by over 95%, emissions dropped by 88%, productivity rose by 45% and earnings rose 300%.

3. Conclusion

The top operating managers have to create the desired culture, spend time in the plant to know the people, see what is happening, pick up any clues and be sure that people are doing their best. They have to listen openly to each other working with the highest level of respect and integrity.

Sometimes tough judgments have to be made about just what has to happen to improve the safety. There may be disagreements between the engineers, operators, mechanics, supervisors and operating managers about the level of hazard and what needs to be done. Since we can never get away from this tension, everyone can live in it providing we listen for understanding with integrity, openly discuss everything and struggle to come to the best agreements we can make. We need to co-create our future. Safety cultures do not spontaneously occur. They must be created and nourished by operating managers. This includes partnering with all the people with high operating discipline, and fostering openness, the free flow of information, and working to build trust and interdependence within their organization.

4. References

[1] Knowles, Richard N., *The Leadership Dance, Pathways to Extraordinary Organizational Effectiveness.* Niagara Falls, N.Y. The Center for Self-Organizing Leadership. (2002)

[2] Knowles, Richard N., *Partnering for Safety and Business Excellence.* St. Petersburg, FL. The Center for Self-Organizing Leadership. (2013)

9TH GLOBAL CONGRESS ON PROCESS SAFETY

Process Safety: At the Heart of Operations

Chung Duc Nguyen
Gulf of Suez Petroleum Company (GUPCO)
Palestine Street, Maadi, 11511, P.O. Box 2400, Egypt

Prepared for Presentation at
American Institute of Chemical Engineers
2013 Spring Meeting
9th Global Congress on Process Safety
San Antonio, Texas
April 28 – May 1, 2013

UNPUBLISHED

Process Safety: At the Heart of Operations

Chung Duc Nguyen
Gulf of Suez Petroleum Company (GUPCO),
Palestine Street, Maadi, 11511, P.O. Box 2400, Egypt

Keywords: Process Safety, KPI, Operations, Management, Implementation

Abstract

Process safety excellence can only be achieved when it is placed at the heart of operations. The majority of oil and gas companies have well-defined process safety frameworks and process safety resources at the corporate level. However to make the program work at a site, additional factors are required. This paper will address aspects that are critical to a successful site process safety program, focusing on the key success factors, aspects of process safety which need the most attention and resources from operating managers, and methods to effectively communicate the most important aspects of process safety to the site team

1. Introduction

Site implementation will determine success of any process safety program. All frameworks, processes and procedures set at the corporate level will be implemented at a site by front line people. Having processes in place and disciplined operations are pre-requisite conditions for success. Additional factors are also required to make process safety truly owned by operations and driven by operations

2. Key success factors for a well-functioning process safety program

2.1 Operations discipline

Operations discipline is probably the most important factor for a successful process safety program. First and foremost, a just culture should be established with compliance being the expected behavior. The site team should be very well aware of the boundary between accepted and non-accepted operational behavior. Secondly, there should be a sense of ownership and pride for disciplined operations amongst the site team. Operations discipline has to become a personal thing for everyone at the site, something that people want to do, not something that they have to do. A good practice is to have the site team get involved in the development of operating procedures, assurance processes, and key performance indicators. Site leaders need to instill the belief within their team that they can achieve "world class operations" or "best in class process safety performance". They could also have the site team agree upon and sign a "How we operate" document which clearly defines expectations for safety and operations behaviors, and display it in the form of big posters in the control room, mess hall and public locations on-site.

2.2 Site leader is a passionate process safety role model

This could be a real challenge for companies that have never experienced significant losses caused by process safety incidents or those that are very much production-oriented. Passion for process safety has to come from a deep understanding about process safety, about the fact that safety is a long and continuous journey that never ends, and from genuine concerns for the safety of frontline people. This is reflected in a strong commitment for disciplined operation, process safety performance, and in giving process safety an equal place in the operations agenda with production priorities. Site leaders need to go to great lengths to understand site operations and not take things at face value. A passionate site leader will always ask questions to probe for potential issues, i.e. challenge control and mitigation measures for risk assessment, and question the reason for overdue critical maintenance or deferral requests.

One of the quickest ways to destroy commitment from the workforce for a process safety program is seeing leaders do not practice what they preach. Site leaders should never let production prevail if there is a conflict between production and safety. For example, if a site leader is willing to carry out a closure test on a critical pipeline isolation valve even though the test may lead to production loss if the valve fails to open after the test, the leader clearly demonstrates the priority of safety. With a strong role model, the site leader's people are much more likely to feel empowered to make critical decisions such as shutting the plant down in case of an emergency. If a leader constantly looks for opportunities to improve process safety performance, the whole organization will react in the same way. The whole site team looks to the site leader as the representative of the company, therefore it is important that site leaders walk the talk and set a role model for process safety

2.3 Process safety is part of the performance contract and reward program

If expectations are set for process safety but do not tie in with some form of annual performance contract and reward program, the workforce will have fewer incentives to improve process safety performance. Policies are set at the corporate level, however having a program that ties individual process safety contribution with performance evaluation will support the site leader in driving process safety performance. Site leaders need to ensure contribution to process safety is a significant item in the performance contract for site personnel. Department managers need to commit to carrying out a certain number of process safety audits, providing training to ensure competency of the workforce, and improving process safety key performance indicators (KPIs). Offshore installation managers (OIMs) and superintendents need ensure that all critical registers are up to date, procedures are complied with, and process safety performance is communicated to the crew. These are just a few examples of process safety priorities in performance contracts.

2.4 Process safety capability of the organization

A process safety adviser role should be established to support the site leader to deliver process safety programs such as providing process safety training for the site team, setting up reporting guidelines, determining KPIs and providing support for specific process safety issues.
The whole site team also needs to have a certain level of understanding about process safety and about their role and responsibility for safe and reliable operations. For example, an OIM should

be able to look at the process safety scorecard and explain the risks on his platform and areas that need focus of his crew. A shift supervisor knows that an effective shift handover should be carried out in both spoken and written formats and the handover should include critical process safety information. An operator should understand that operating and work procedures need to be complied with and that any anomalies found during the shift should be reported. All of this can be done through training and by reinforcing disciplined operations.

2.5 Process safety assurance processes and reporting tools

2.5.1 Processes and reporting tools

Key process safety assurance processes should be established and made available to site personnel, such as management of change, prestart-up assurance, deviation from normal operations, incident investigation, risk assessment, inspection and maintenance procedures. The site leader could request a gap analysis against the company's process safety framework or a well-established management system (OSHA Process Safety Management, for example) to identify if any important processes are missing and develop a plan to close the gap. Reporting tools should also be available and accessible for the frontline people. Common process safety reporting tools are incident and near miss reporting, process upset reporting, and anomalies reporting. The site leader needs to ensure that reports are followed up, actioned, and information is then fed back to the organization either in the form of lessons learned or follow up actions.

2.5.2 Process safety Key Performance Indicators (KPIs)

A fit for purpose set of KPIs provides a snapshot of site process safety performance and helps site leaders to identify areas that require intervention. Common mistakes when setting up KPIs are: defining too many KPIs such that they are not really "key" anymore, adopting KPIs from other sources that are not representative for site operations, or choosing KPIs without considering the organization's reporting capability. KPIs for a refinery may be different from KPIs for a modern platform and different from those for aging facilities. A process safety adviser who really understands processes, facilities, and the site team's capability would be able to help the site leader to develop a set of KPIs that are a true reflection of site operations.

2.5.3 Communication

Communication is very often the missing link in the process safety loop. Decisions and interventions are made at the site management level but very little information may be passed on to the front line workers who interact with the plant every day and play an important role in keeping the plant in a safe operating condition. A communication tool that transfers the most critical information to front line people should be in place. For example, top risks at site, long-term overrides, high risk wells, overdue safety actions, and temporary changes should be clearly communicated. When developing such a communication tool, one has to bear in mind that the information needs to be presented in a way that it is understandable for front line personnel, that and is concise and visual. An example of such a process safety communication tool is described in Section 5.

3. Process safety - important aspects for site leader

As mentioned earlier, a thorough understanding about all element of the process safety management framework is necessary for site leaders. Below are aspects of process safety that seem to be systematic issues at operating sites and need further attention.

Operating and work procedures: Site leaders need to ensure all procedures are up to date, accessible and followed. This is easier said than done. In the real world of operations, people often take shortcuts or violate procedures with the very best intention to get things done. This results in "procedures not up to date" or "procedure not followed" being common immediate causes in major incident investigation reports. It is a good practice to get operations teams involved in the development of operating procedures to ensure that the procedures are fit for purpose and increase ownership of procedures for operating personnel.

Hot work and breaking containment: Sources of ignition and fuel are two main elements of a recipe for disaster. Hot work and breaking containment jobs should have rigorous attention when it comes to writing procedures, assigning crews and performing work at the site. Compliance should be a strict requirement. A good practice is to identify critical hold points in the procedure that need to be signed off before moving on to the next steps. Site leaders should request compliance audits to determine if procedures are being followed.

Risk assessment: Two common pitfalls associated with risk assessment that the site leader should be aware of are: risk normalization and confirmation bias. Confirmation bias means the purpose of the risk assessment is to justify continued operations instead of to evaluate the risk. This leads to a number of issues: a poor quality of risk assessment, lower risk ranking of potential hazards, and inadequate control and mitigation measures. The site leader, therefore, needs to set a clear expectation for the risk assessment team and challenge the risk assessment before signing off.
Risk normalization tends to happen in organizations that have to manage risk on a more frequent basis, such as aging facilities. The level of risk that the team is willing to take increases gradually and accepting greater risks becomes the norm. The site leader needs to set a clear limit for the acceptable risk level, i.e. only a certain number of deviations from normal operations on a plant are allowed, any additional deviation will lead to plant shutdown. Examples of deviation from normal operations are running the plant outside of the operating envelope or with known integrity issues (for example pipeline thickness) and impaired critical safety systems. Another example is to set a time limit for continued operations with deviation, and shut the plant down when the limit expires.

Pre-startup safety review: Start-up of a plant creates significant risks compared to normal operations. Site leaders need to ensure that a start-up assurance process will be utilized to support safe and reliable start-up of plant and equipment. A set of fit for purpose prestart-up checklists for different start up scenarios may be required.

Shift handover: Shift handover needs to be carried out in a structured way, in both verbal and written formats. Site leaders need to ensure critical process safety information is recorded in the handover note and discussed, for example, impaired safety instrumentation or equipment, and operational status for the process. Audits of the shift handover process could be carried out to ensure compliance.

Incident investigation and learning: Learning is always the difficult part of the incident investigation process. Looking back at major process safety incidents, it is not uncommon to find the same root causes for multiple incidents i.e. overriding critical instrumentation, not following procedures, incompetent personnel, and inadequate engineering. Site leaders need to make sure that lessons from incidents that have happened on site and lessons from external sources that are relevant to the site are really internalized, not just by increasing awareness during safety stand downs, but through specific actions such as reviewing engineering design, updating procedures, and reinforcing operational discipline.

Management of change (MoC): One of the biggest challenges for process safety in operations is to manage changes. MoC has been covered in numerous process safety discussions; however, it is worth mentioning here again with a few key points that site leaders should consider. Firstly, a comprehensive MoC procedure that provides guidance for different types of changes should be in place. Secondly, the site leader needs to ensure all stakeholders in the MoC process receive adequate MoC training and are competent. Finally, ensure the process is followed, including: adequate engineering review and risk evaluation, personnel are trained on changes that occur, and procedures and engineering documents are updated.

4. How to know if your process safety program is working

4.1. KPIs trending in the right direction

The easiest way to see if a process safety program is working is to look at trends: improvement in leading indicators and reduction in lagging indicators. If there is no improvement or the trend is going in the reverse direction, site leaders need to dig deeper to find out what is happening and intervene if required, by: providing additional resources to clear safety critical maintenance backlogs, speeding up technician assessment programs to improve competency KPIs, and carrying out more emergency drills.

4.2 Level of engagement of the organization

This is all about a positive attitude towards process safety and a commitment for continuous improvement. This soft part of process safety is hard to measure, however there are different ways that a leader can read the level of engagement of the organization. For example, the practice of reporting all incidents, near misses, and process upsets in a timely manner becomes the expected behavior. Management of change meetings have the right level of attendance from the relevant departments. The workforce shows a strong compliance with work procedures. Department managers see monthly process safety meetings as a forum to discuss issues and an opportunity to drive process safety performance.

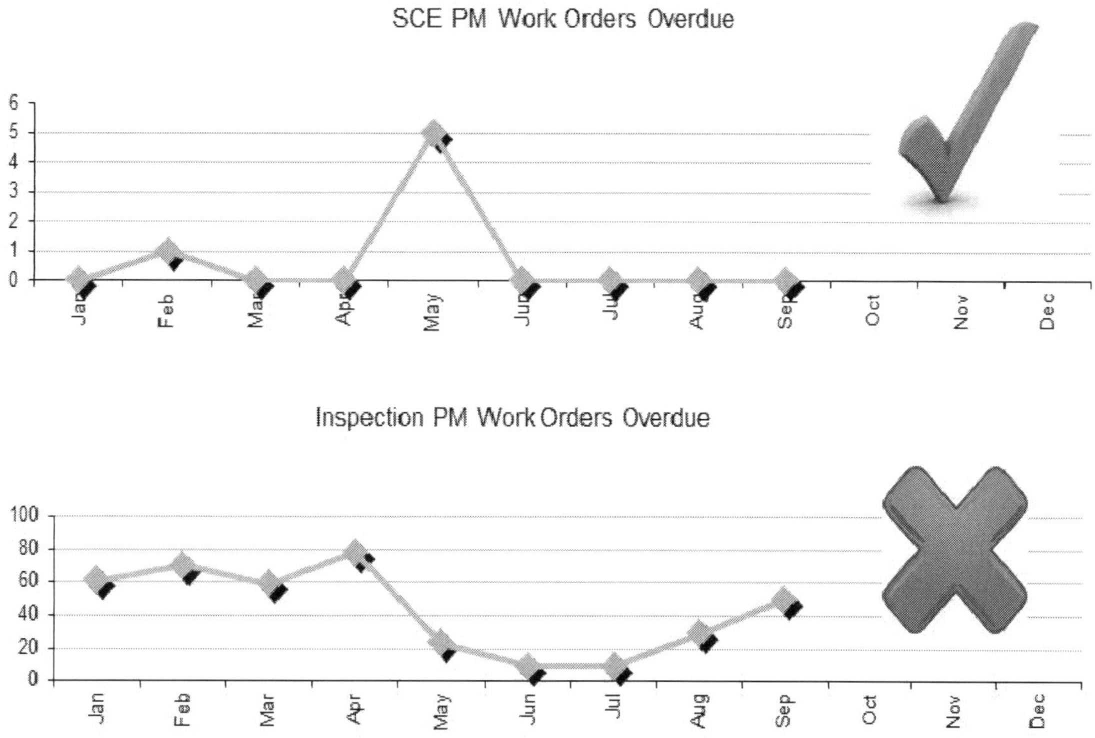

Figure 4.1.1 Examples of KPI trending for safety critical equipment and inspection preventive maintenance (PM)

4.3 Process safety audits

Process safety audits could be carried out to evaluate effectiveness of the process safety program at the site. A three year site audit against the corporate process safety management system is the practice in a number of organizations. The quality of the audit team is important to ensure that the audit will reveal any hidden issues and provide practical resolution to address them. At the site level, a site leader can request compliance audits on certain elements of the process safety framework on a more frequent basis. For example, compliance with the permit to work system, compliance with the management of change process, work procedures, and risk assessment can all be audited if deficiencies are suspected or to determine the general health of these management systems. A team led by the process safety adviser with participants from other disciplines and supporting functions should be able to deliver this.

5. Case study: Process safety scorecard in GUPCO

Process safety "at the heart of operations" means it is really owned by operations and driven by operations. In GUPCO's operations, process safety scorecards have been effectively used to facilitate the process (see example Figure 5.1). The site process safety adviser set up various sections with leaders, managers and site team to go through the process safety framework, layers of protection concept, and discuss key performance indicators for each department. This top-down and bottom-up enrolment of staff ensures clarity regarding KPIs and ownership by

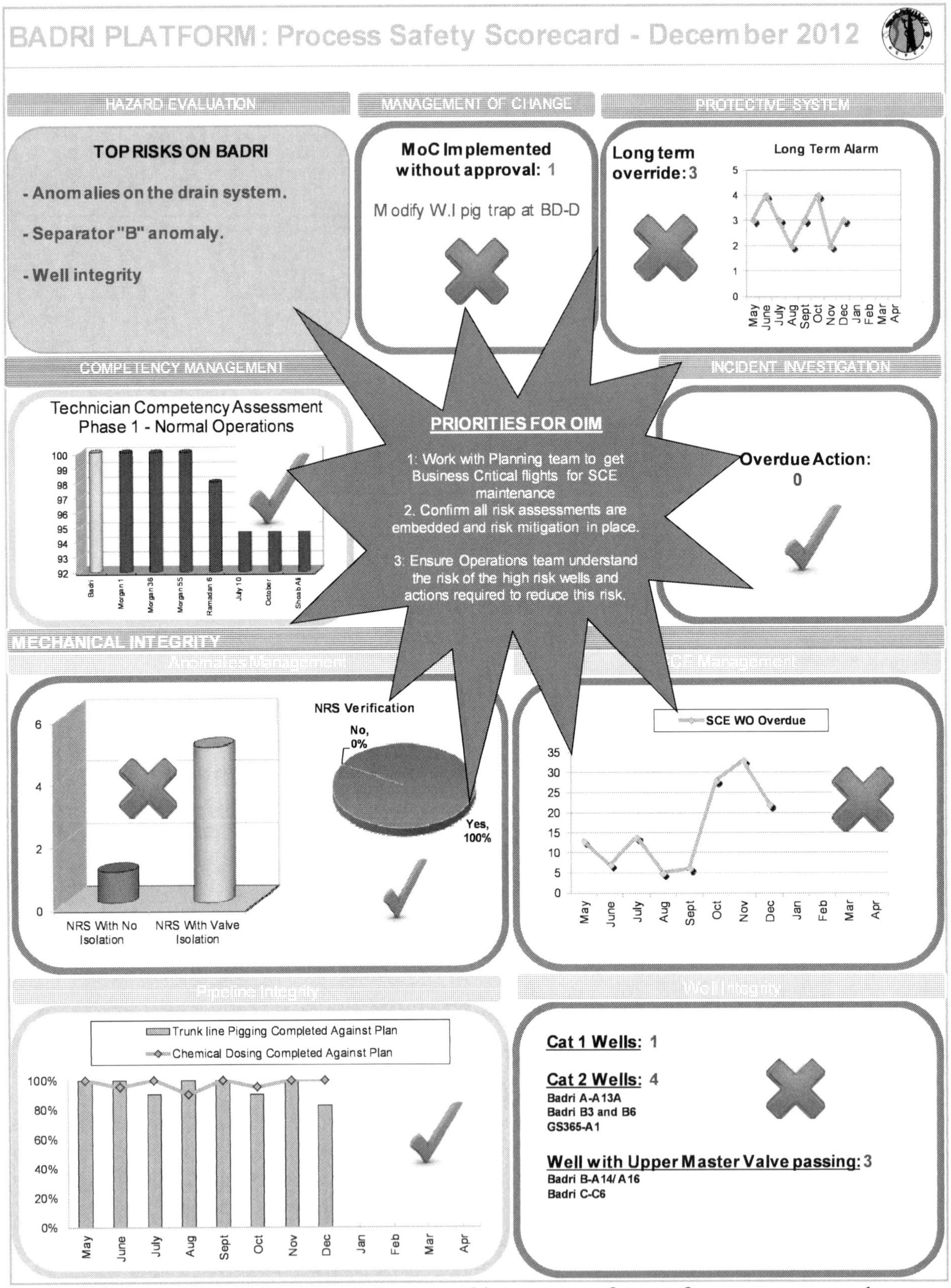

Figure 5.1 Example of Operations monthly process safety performance scorecard

functional departments who play a key role in "driving" these KPIs. A central database is created for functional departments to update their monthly key performance indicators. Operations managers then work with offshore installation managers (OIMs) to develop process safety performance scorecards using information from this central database. Elements and sub-elements of the process safety management framework are color-coded in the scorecard to highlight process safety issues and areas that need intervention and priorities. Monthly process safety meetings are established for OIMs and supervisors to communicate the scorecard to the team and for the crew to provide feedback for issues identified. The site leader and functional department managers also participate in the meetings. The scorecard is posted in the control room at each operations location, serving as a communication tool for the OIM to the crew, and as a reminder of risks and priorities. It is also used as a tool to assist site management for process safety audits.

6. Conclusion

Process safety is always at the heart of safe and reliable operations. The way to process safety excellence is a long journey with a disciplined operation being the pre-requisite condition. In addition to that, process safety assurance processes, leader commitments, a reward program that connects with process safety performance, and site process safety capability are required to ensure a successful implementation of process safety programs at site.

7. References

[1] Hopkins, Andrew, (2009) *Failure to Learn: The BP Texas City Refinery Disaster, Page 44.* Sydney Australia: CCH Australia Limited

[2] Reason, James, (2008) *Managing the risks of organizational accidents, Page 195.* England: Ashgate Publishing Limited

30, 60, 90 Day Plan for Operations Managers to Assess Process Safety and Prevent the Next Serious Incident

Brian D. Wilson, P.E.
Arkema, Inc
9502B Bayport Blvd
Pasadena, TX 77507
brian.wilson@arkema.com

Prepared for Presentation at
American Institute of Chemical Engineers
2013 Spring Meeting
9th Global Congress on Process Safety
San Antonio, Texas
April 28 – May 1, 2013

UNPUBLISHED

30, 60, 90 Day Plan for Operations Managers to Assess Process Safety and Prevent the Next Serious Incident

Brian D. Wilson
Arkema, Inc
Arkema, Inc
9502B Bayport Blvd
Pasadena, TX 77507
brian.wilson@arkema.com

Keywords: process safety, assessment, action plan, incident prevention, leadership, roadmap

Abstract

Operations managers today are faced with numerous challenges as they begin a new role and during their entire tenure in the role. Key job areas include: EH&S (Environmental, Health, and Safety) – especially process safety, operating cost, raw material budgets, logistics, reliable product supply, product quality, capital planning, and organizational stewardship. During the time the operations manager is coming up to speed in a new role, the organization is at higher risk as a key organization leader's focus can be diluted and distracted. Also, senior managers can be faced with similar "job overload" and lose focus on key priorities. It is absolutely critical that operations managers avoid these pitfalls and make process and personal safety a cornerstone of their day-to-day focus. This paper discusses why process safety must be a top priority for operations leaders and suggests key milestones over 30, 60, and 90 days to assess the site process safety culture and layers of protection. Practical tips are given on what and how to assess and what to do with gaps to prevent the next serious incident. These efforts might save a life, prevent harm to the community, or save the enterprise.

1. Operations managers – new leader challenges and no roadmap

Operations managers can have a very varied background. Some may have over 30 years of experience in multiple businesses, sites, regions, and technologies. For others, it can be their first assignment in a lead operations role. Regardless, this position is generally designed as a "linchpin" to integrate the process safety, EH&S, business, manufacturing, and supply chain to meet corporate goals. Even though this role is extremely critical and paramount to managing site and enterprise risk, training for the new role can be "ad hoc" and poorly defined. Many times the new operations manger must take the lead in "learning what they don't know," which can leave much to chance.

This paper outlines key process safety objectives and milestones as a new operations leader comes up to speed on the organization (people), technology, equipment, work processes, and management systems. This road map is intended to assist in prioritizing learning objectives and minimizing PSM (Process Safety Management) [1] oversight gaps during the leadership transition.

2. T-minus 60 days: due diligence of the new role – are you the right fit?

2.1 Due diligence

Preparation for the new operations role should not begin the day you start the new role. It should start as you begin your research for the new position. A due diligence review of a new role should include the familiar items: a site visit and tour, business review, discussion of key challenges of the new role, travel involved, location, compensation package, what success looks like, and your personal/family goals. The items below focus on the process safety challenges that should be investigated.

2.1.1 Preparing for the job interview

As you setup your site visit, you should compile as much information as possible on the health of the site's PSM systems and culture. The new position may be an area, site, region, or global role; however, many of the questions to answer will be the same.

Key questions to answer might include:

- How and who conducts your PHAs (Process Hazard Analysis) [2]? Are they current? Consider that the experience and skill level of PHA teams can vary greatly. If the initial PHAs were completed by a consulting group, they may or may not be available for discussion. Also, these reviews can often be "rubber stamped" by a fairly inexperienced team during the required 5-year renewals. Of course, the best case is to have a senior risk analyst professional with a strong background in the technology, past events, and RCIs (Root Cause Analysis) serve as the PHA facilitator and be a risk steward at the site or for the business.
- What is the status of the PHA action items? Who is driving closure?
- When was the last PSM audit [3] completed? Who did it, and what were the auditor's qualifications? What is the status of actions?
- What are the worst case scenarios in the RMP (Risk Management Plan) [4]? How many people on the site leadership team know what they are? Do the Emergency Response Team commanders know these scenarios?
- Is the mechanical integrity program functioning [5]? Are there established equipment inspection plans, inspection reports, notification of deficiencies, and high level management approvals for deferrals?
- How disciplined is the SWP (Safe Work Permit) process [6]? Does a good training, certification and audit process exist? What metrics are reported?
- Did the past leader and current staff understand and drive process safety improvements?
- What do the PSM leading and lagging metrics [7] look like at the site. What are the frequency and severity of process safety events? What type of leading indicators are being tracked?

It is best that you do not show up with a long list of questions and no answers. Site information can generally be found online, site contacts, calls to friends, using online RMP information, and through the OSHA (Occupational Safety and Health Administration) website. Therefore, do your homework and due diligence. This will make the interview process more productive and you will appear much more informed.

2.1.2 Understanding the needs of the position

It is critical that you perform an honest assessment of the new role, your knowledge and job skills, and your fit. Unfortunately, you will likely have a very incomplete picture of the PSM quality and health during your due diligence review. However, you should be able to define the process safety risk at the site and your process safety skills. A 3 x 3 matrix can help you identify the job risk vs. your skills (see Figure 1).

Figure 1. Site PSM hazards vs. your process safety/EH&S skills matrix

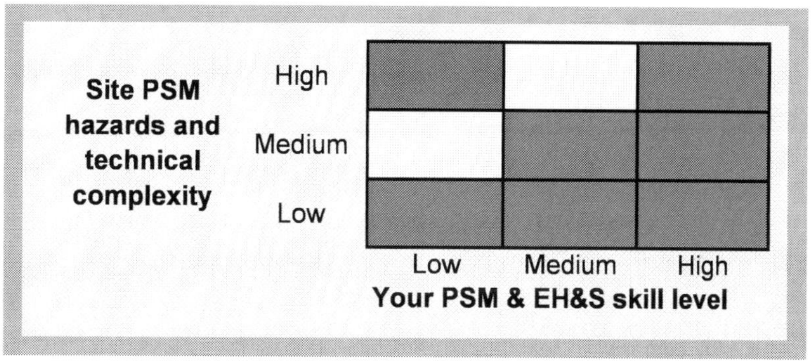

This table provides a warning (upper left quadrant) that a person with weak PSM/EH&S skills should not move to a role in a highly hazardous, complex plant or business. For some job sites, senior managers understand the process safety job risks are high and decline this high risk assignment. This sometimes results in a junior person being appointed to the role that may not fully understand the risk. Bottom line, go in with your eyes open, do your due diligence, and be willing to turn the job down if you are not a good fit for the organization's needs.

2.1.3 Managerial courage to ask the tough questions and set boundaries with the new boss

Following your due diligence review, you need to ask some direct questions of your potential new boss. Specifically, when have they had to make tough decisions between running a unit with high process safety risk vs. having to shutdown the unit? How did they deal with the business pressure to run the unit? What are the expectations of you to lead with process safety and EH&S as top priorities versus a "pounds first" mentality?

You need to be direct during the interview and communicate that you will not run an "unsafe" unit or plant. Some past examples where you have made tough decisions or took an unpopular stance to "do the right thing" are good to discuss. This may or may not help you get the job.

However, you should be willing to live with not getting a job working for a boss or business that does not support process safety.

3. T-minus 30 days: you've got the job – now what?

You asked the tough questions during the interview and you were offered the job. You feel based on the process safety risk analysis and your background, you are a good fit and you accept the job. Now what?

3.1. Wrap up the old job - quickly

It is critical to wrap up the old job as quickly as possible. The first step should be to document a summary of your key job tasks and DOA (Delegation of Authority) responsibilities. Generally, a table listing job task, frequency, and comments is a good place to start. Second, get the personnel MOC (Management of Change) process moving with your replacement on the old job. Begin bringing your replacement "up to speed" as quickly as possible.

3.2 Begin learning the new organization and processes, and prepare for the personnel MOC review (new leader review)

Learning the names of people in the organization is critical. Learning their names quickly shows them that they are important to you. It also allows you to begin the process of determining the organization's strengths and weaknesses.

If the organization has a defined "New Leader Review" process, ask to receive the study documents BEFORE you start the job. A new leader review typically contains items such as [8]:

- Organization overview and years of service in key roles
- Process overview
- Reactive chemistry hazards
- Reactive chemistry cardinal rules
- Past process safety incidents at the site and in the industry
- Status of key audits and action items (PSM, RMP, PHA)

3.3 Get clear vision of what success looks like

Have several discussions with the new leadership team on what they feel success would look like over the next year and then three years. This will help you begin formulating a roadmap to move the organization forward. This roadmap should include all key operations areas – but keep process safety and EH&S first on the list. Ensure the key elements of PSM are identified to allow you to make deliberate plans in each key area.

4. T+1-5 days: getting the first week right – everyone's watching

You have found a house in the new location, moved your family and home goods, and it is your first day at work. Do not underestimate what perception people take away based on your first few actions and words on the new job. If you say, "I'm here to return us to profitability", that will be their take away. The best leaders I have seen gather the leadership team together relatively quickly and give them an idea of the direction you will ask them to move in. The challenges of "low cost, reliable supply" are obviously a high priority for operating mangers. However, the leader makes it clear that process safety and EH&S performance are the absolute top priority. Personal examples of severe incidents the leader has seen can strengthen this call to champion process safety and general EH&S performance.

4.1 What you do speaks so loudly I can't hear what you are saying

Be careful of your actions – especially during the first few weeks. If all you do is mull over operating cost, the S&OP (Sales and Operations Planning) forecast, and business projections, folks will quickly feel your words on process safety and EH&S priority are hollow and not genuine. Therefore, spend time the first week doing field audits, speaking with process safety risk analysts, process safety and EH&S managers, SWP trainers, operators, and mechanics. Success in these conversations might give the impression that you are a bit of a zealot or fanatic when it comes to PSM and EH&S. There may even be a few jokes that begin with the new operations manager is the new process safety and EH&S manager. If that is what it takes to ensure the organization knows that safety is priority #1, then "so be it"!

4.2 Establishing a DOA timeline and communication

As you come on board, discuss with your new boss and who you are replacing the expectation of when you will receive the DOA to begin authorizing critical MOCs and process safety related escalations for approval. The organization needs to be clear that you should not be authorizing SWP escalations, deviations from normal work process, or SIS (Safety Instrumented System) deactivations until you are properly trained and properly understand the hazards.

4.3 Meetings with key leaders to get you up to speed

The following meetings should be set-up in the first 30 days to begin the learning process. These include:

Administrative leader in charge of on-boarding new employees: Review the new employee checklist – generally part of the on-boarding process that includes all administrative details and high level functional orientation.

EH&S manager. Questions should center on 1) health of the MOC process [9], 2) PSM concerns, 3) critical training you need, 4) key priorities over next 18 months, and 5) a review of the "cardinal rules" for safety, environmental operations, and reactive chemistry. These cardinal rules generally include:

> *Safety Cardinal Rules* – following MOC, SWP (especially the critical to life SWP standards), SIS deactivation, and procedure use policies

Critical to Life SWP standards generally cover LOTO (lock out/tag out), L&EO (line and equipment opening), confined space entry, hot work, and work at elevation

Environmental Cardinal Rules – following MOC policy, not bypassing critical environmental controls without proper authorization, ensuring all spills or releases are reported

Reactive Chemistry Cardinal Rules – following strict operating discipline to ensure chemicals are not inadvertently mixed, safe operating limits are not exceeded, or only proper materials of construction are used in highly hazardous service

Distributed Control System (DCS)/Safety Instrumented System (SIS) Control Engineer: Meet with the control engineer to determine if a clear SIS list has been created. Do testing procedures exist to test the systems, are the instruments used to test properly calibrated, how many cycles of testing have been completed over the past 5 years, and what have the findings been (as found/as left)? Do any test frequencies need to be adjusted based on the findings?

Ensure you understand how any SISs can be bypassed or deactivated. Who has the authority to bypass these devices, and what is the process used to ensure risks are managed? How are bypasses documented (when and how bypassed, when returned to service)?

Maintenance manager and the senior mechanical technicians (fixed, rotating, and Instrument and Electrical (I&E)): The goal of the meeting is to determine the health of the Mechanical Integrity (MI) system. Have PSM covered equipment been identified? Have equipment inspection plans been completed based on risks? Does a computerized maintenance management system (CMMS) exist to track the inspection preventive maintenance (PMs) tasks? Have all initial internal inspections been completed? Is there a mechanical integrity deficiency review process? How many deficiencies exist and who has approved continued operation of this equipment?

Senior foremen and technicians: These meetings are to ensure you fully understand the SWP procedures, especially for the critical to life SWPs (line and equipment opening, confined space entry, hot work, work at elevation, and lock-out/tag-out). You also need to understand the "escalation policy" for special hazards (when lines cannot be fully cleared, work in atmospheres above STEL (Short Term Exposure Limit)/IDLH (Immediately Dangerous to Life and Health), use of chemical hoses, etc.).

Training coordinator: Discuss how operators, maintenance personnel, and contractors are trained and tested for proficiency. It is critical that rigorous training and testing be completed.

Contractor manager [6]: Meet with the contractor manager and key contractor leaders to see what orientation and training contractors receive. Who do they go to with questions? Can they report hazards without being penalized? Does the contactor have a zero tolerance policy for major safety infractions?

5. T+ 5-30 days: understanding the hazards

5.1 Research the highly hazardous chemicals (HHCs) handled at the site

As you begin weeks 2-6, you need to understand the major hazards of the process. One potential place to start is to create a spreadsheet with all key chemicals, physical properties, chemical formula, and exposure limits (TWA – time weighted average, STEL, IDLH, ERPG – emergency response planning guideline). For senior managers, it is often good to add in a few familiar components for comparison to better ground yourself on the hazards of the new chemicals versus ones you have worked with in the past. This can also assist others in understanding how hazardous certain chemicals may be. For instance, the IDLH for hydrogen cyanide is 50 ppm. No one is surprised by the low number. However, the IDLH for more common materials such as acetic acid (50 ppm), ammonia (300 ppm), and for chlorine (10 ppm) are also quite low. The key is not to become complacent with "everyday" chemicals. They can be surprisingly toxic relative to other highly hazardous chemicals.

5.2 What do the RMP, air dispersion modeling, and quantitative risk analysis (QRA) identify as major hazards?

Now is the time to review past process hazard analyses (PHAs). This might include Hazard and Operability studies (HAZOPs), the RMP, air modeling results, facility siting studies, and any QRAs [10] and LOPAs (Layer of Protection Analysis) [11] that exist. It is unlikely you will be able to read through all of the documents. However, you should certainly read the management summaries and understand any gaps identified versus OSHA 1910.119 and your corporate standards.

5.2.1 Fatality versus frequency curves

For any completed QRAs, you should review the frequency vs. number of fatalities (FN) curves as compared to your corporate safety risk criteria [12]. Does your site meet this criteria? How do the FN curves fit with published industry guidelines? Who in your organization really understands how this risk compares to other process safety risk in the company.

5.2.2 ERPG hazard zone boundaries

Ensure you understand the key "worst case scenarios", evacuation distances, and how to alert people to evacuate. This should be part of the plant emergency response capabilities. If you have a corporate role, ensure you understand how corporate emergency response can support the corporation's and sites' emergency response plans [13 & 14].

5.3 Emergency response training and capabilities

Spend a day touring the facility focusing on site emergency response equipment. This includes field equipment (deluge systems, water monitors, foam systems, short stop systems, and SCBA (self contained breathing apparatus). It also includes what mobile fire fighting capability the site might have and the training of emergency response personnel. Training may include materials for dealing with fires, spills, and releases, as well as first aid and paramedic capabilities.

6. T+ 31-60 days: auditing the systems

Now that you have a better grasp of the key hazards, risks, work processes, and personnel at the site, it is time to begin auditing for compliance. The next section discusses what you can do to identify gaps and begin setting priorities for closure.

6.1 Key reviews: RMP, PSM audit, PHA summaries and action item status

Get ready to pull a few late nighters and weekends to review the PHA work. You need to review key summaries of PHAs, the RMP report, and the latest PSM audit. Understanding the key findings, gaps, and action items is critical to establishing a site PSM roadmap and setting goals for the next 12-18 months.

6.2 MOC discipline: MOC, PSSR (Pre-startup safety review), and RBS (required before start-up) documentation

Ensure you are invited and attend any MOCs that are occurring at your site. Once you understand the work process fully and the capabilities of the personnel, you may choose to delegate. The key is, don't delegate until you know who and what you are delegating.

We all understand an MOC review should be conducted for non-like-for-like changes and personnel reviews. However, how well-disciplined is this process at the site? Also, are the MOC leaders good at identifying the hazards and associated levels of risk for changes and how these risks will be mitigated? Do "first time" and infrequent activities receive a critical review? Are the right people involved? Are all actions documented?

For the PSSR, do the right people attend and can they identify any remaining hazards or risks? Are RBS actions clearly documented and does the management system check to ensure these are complete prior to start-up?

A strong MOC system is likely the single best defense for preventing a serious incident.

6.3 Safe work permit : how good are you on critical to life standards (line and equipment opening, lock-out/tag-out, confined space entry, hot work, work at elevation)?

Set up standing meetings with your SWP experts. This might be the SWP safety technician or a senior foreman or operator. Complete at least one audit per month on different SWPs (especially LOTO, CSE, hot work, and L&EO). In the author's experience, sites generally over estimate the accuracy and discipline used to issue SWP permits. At one site, the author found less than 70% of SWP permits are completed fully and accurately as measured by an independent safe work permit coach.

Several "watch out" areas include: (1) quality of the maintenance work package – how clear is the package on the scope of work and how it should be completed; (2) how clear are line-break points identified to prevent inadvertent opening of the wrong line; and (3) how well-trained is the maintenance team? Many times the maintenance team is oriented in a 4-8 hour session listing scores of chemicals and hazards. This can lead to poor retention of the chemical hazards involved and even the names of operating areas.

Field auditing will also help you see if the persons accepting safe work permits actually follow all of the instructions and restrictions on the permit (e.g. PPE (Personnel Protective Equipment) required, scope of work allowed, tools allowed, and when to shutdown work).

6.4 Procedure use policy and discipline

As boring as it may sound, use of procedures is absolutely critical to safe operations of a site. At least once per month, you should review if procedures that require each step to be signed-off are actually being properly authorized and if any step deviations are being approved at the right management level. Without oversight and proper discipline, the procedure use policy can quickly become "use when it is convenient". Don't let this happen at your facility!

6.5 Fatigue Management

Ensuring plant personnel are fit for duty is important and has been a challenge in industry. The most demanding area is ensuring personnel do not become overly fatigued by long work schedules. As the operations manager, you should set clear expectations on maximum allowable work hours per day and maximum days that can be worked co-currently [15, 16].

7. T+ 60-90 days: formal training sign-off by process safety, EH&S and management leadership and development of roadmap

In the 60–90 day period, your assessment and training should be completed. All formal sign-off should be completed at that time verifying that you are ready to properly manage the hazards and risks at the site.

You should also begin completing a 12-18 month roadmap to drive the key objectives at the site. Process safety and EH&S should be at the top of the list.

8. Evolution or revolution?

You now have the first 90 days behind you. They pass quickly. You should know the people, process, and systems. Along the way, you should have corrected any short term gaps you found in the key PSM systems. Now it's time to make some longer term decisions and implement the action plan. After firming up the action plan, you need to decide what should be completed in-house and where you might need outside help.

8.1 Evolution - driving continuous improvement

If you are fortunate enough to be working at a site that has a strong PSM system in place, that is great. Your focus and action plan then will be evolutionary. Specifically, how do you take this mature process safety management system and continue to improve on it? This is generally accomplished by continuing training, driving the right process safety metrics, and continuing to improve technology (especially inherently safer systems).

8.2 Revolution – when drastic change is needed

As you determine the health of the key systems (MOC, SWP, PHA, MI, SIS, process safety metrics), you may find serious gaps which cannot wait on a "continuous improvement" approach. A decision needs to be made on what can be completed "in house" and what may need to be outsourced. Where possible, staffing from within is generally best so you can retain the knowledge at the site. The next best choice is to get corporate resources to assist. If these are not available, you need to look outside. The key point is to get any critical gaps closed ASAP (as soon as possible) and be willing to make a strong case for the resources.

The chemical and petroleum industries have been able to find significant efficiencies via use of system automation, economies of scale, and energy integration. Continued efficiencies and leaner staffs are expected and can be implemented safely. However, these efficiencies must come after good PSM and reliability systems are in place. Ensure you do not blindly follow a "cost cutting" program. It may result in an extremely high price if the proper systems, organization, and technology are not in place.

8.3 Everyone needs a coach

Now that you know the organization, you should have a good feel for who would be a good mentor. This is a person to contact from time to time to discuss priorities, time schedules, and ideas. This can also be a person who can give you feedback on your performance and "what they are hearing" about your performance. Enlist the help of a mentor to help you on your journey to manufacturing and process safety excellence.

9. Conclusion: it ain't easy, but we have to get it right

The transition of an operations leader into a new role has many pitfalls. This can include organizational behavior missteps, poor emotional intelligence, bad financial decisions, and lack of supply to customers. However, no mistake or oversight can have the disastrous consequences of poor process safety management. Operational manager "on–boarding" should be structured and not left to chance. Ideally, the new organization has a detailed "on-boarding" program. However, the author's experience, in over a dozen operational roles, has seen most programs are lacking. A new operations leader will likely be well served by generating an on-boarding timeline with key milestones. This will help force the right prioritization based on the numerous demands on an operational leader's time. The clock is ticking and a new leader has to get it right.

10. Acknowledgements

Thanks to Karen Study, Phil Myers, and Don Rhea for their review and expert comments.

11. References

[1] *Process Safety of Highly Hazardous Chemicals (29 CFR 1910.119)*, U.S. Occupational Safety and Health Administration, May 1991

[2] *Hazard Evaluation Procedures*, New York: Center for Chemical Process Safety, American Institute of Chemical Engineers, P. 115, 2008

[3] *Guidelines for Auditing Process Safety Management Systems,* New York: Center for Chemical Process Safety, American Institute of Chemical Engineers, 1993

[4] *General Guidance on Risk Management Programs for Chemical Accident Prevention* (40 CFR Part 68), U.S. Environmental Protection Agency, 1994

[5] *Guidelines for Mechanical Integrity Systems,* New York: Center for Chemical Process Safety, American Institute of Chemical Engineers, 2006

[6] *Guidelines for Safe Process Operations and Maintenance,* New York: Center for Chemical Process Safety, American Institute of Chemical Engineers, P. 226, 239 1995

[7] *Guidelines for Process Safety Metrics,* New York: Center for Chemical Process Safety, American Institute of Chemical Engineers, 2010

[8] Study, Karen A., "Training Management of Change Approvers", American Institute of Chemical Engineers, 6th Global Congress on Process Safety, 2010

[9] *Guidelines for Management of Change for Process Safety,* New York: Center for Chemical Process Safety, American Institute of Chemical Engineers, 2008

[10] *Guidelines for Chemical Process Quantitative Risk Analysis,,* New York: Center for Chemical Process Safety, American Institute of Chemical Engineers, P. 395, 2000

[11] *Layer of Protection Analysis,* New York: American Institute of Chemical Engineers, 2001

[12] *Guidelines for Developing Quantitative Safety Risk Criteria,* New York: Center for Chemical Process Safety, American Institute of Chemical Engineers, 2009

[13] *Corporate Crisis Management*, New York: Center for Chemical Process Safety, American Institute of Chemical Engineers, 2005

[14] *Guidelines for Technical Planning for On-Site Emergencies*, New York: Center for Chemical Process Safety, American Institute of Chemical Engineers, 1995

[15] Baker, James, et al. *The Report of the BP U.S. Refineries Independent Safety Review Panel*, P. 87, 2007

[16] *Standard RP-755 - Fatigue Prevention Guidelines for the Refining and Petrochemical Industries*, American Petroleum Institute, Washington D.C., 2010.

Empowering Operating Managers with Decision Support System to Prevent Process Safety Incidents

Principal Author: Mr. Deepak Phanse

Centre for HSE Excellence
Reliance Industries Ltd.
Reliance Corporate IT Park,
Thane Belapur Road, Ghansoli,
Navi Mumbai, Maharashtra
India-410 007

deepak.phanse@ril.com

Other Authors: Ms. Prashanthi Bhupathi, Ms. Manasa Godavarthy, Mr.Rohan Vardekar

Centre for HSE Excellence
Reliance Industries Ltd.
Reliance Corporate IT Park,
Thane Belapur Road, Ghansoli,
Navi Mumbai, Maharashtra
India-410 007

prashanthi.bhupathi@ril.com, manasa.godavarthy@ril.com, rohan.vardekar@ril.com

Prepared for Presentation at
American Institute of Chemical Engineers
2013 Spring Meeting
9th Global Congress on Process Safety
San Antonio, Texas
April 28 – May 1, 2013

UNPUBLISHED

Empowering Operating Managers with Decision Support System to Prevent Process Safety Incidents

Principal Author: Mr. Deepak Phanse
Centre for HSE Excellence
Reliance Industries Ltd.
Reliance Corporate IT Park,
Thane Belapur Road, Ghansoli,
Navi Mumbai, Maharashtra
India-410 007

deepak.phanse@ril.com

Other Authors: Ms. Prashanthi Bhupathi, Ms. Manasa Godavarthy, Mr. Rohan Vardekar
Centre for HSE Excellence
Reliance Industries Ltd.
Reliance Corporate IT Park,
Thane Belapur Road, Ghansoli,
Navi Mumbai, Maharashtra
India-410 007

prashanthi.bhupathi@ril.com, manasa.godavarthy@ril.com,
rohan.vardekar@ril.com

Keywords: Operation engineer, incident investigation, near-miss, metrics, alarm rationalization, IT enabled, risk

Abstract

The operating manager can be considered as a "forward" or "finisher" as in a football (US soccer) team. He is an essential link between the strategic and tactical level of management and the shop floor personnel who are in close proximity to the process and its hazards.

Having a robust Safety Management System is essential for an organization. However, effective methods to get the operating managers to understand their role, the parts of the system that they need to deeply understand, the way to evaluate their plant's progress or lack of it, and their ability to get the operating team to do the right things is a crucial link in the delivery of good process safety.

This paper broadly outlines Reliance Industries Limited's (RIL's) endeavor to sensitize and facilitate the operating managers in their efforts to effectively manage process safety. Some of the programs initiated by RIL include:

- Establishing a key list of leading and lagging metrics to be monitored on a regular basis
- Bringing clarity on the need to achieve production targets while being within the stipulated safe limits
- Using a Job Cycle Check to improve compliance to operating procedures
- Developing and monitoring ways to share and implement learning from internal and external incidents.
- Using a Safety Observation program to influence safe behaviours at the workplace.
- Establishing non-negotiable actions such as completion of 2nd party audits, timely incident investigations, and adherence to Life Protection rules

This paper explains the usefulness of these efforts and the challenges faced.

1. Introduction

Reliance Industries Limited's (RIL) commitment to excellence in Health, Safety and Environment (HSE) performance is one of the Group's core values. The company is unwavering in the policy of "Safety of persons overrides all production targets", which drives all employees to work towards continual improvement in the safety management systems for the benefit of our people, communities and protection of property in which we operate. In bringing this vision to reality, the operating manager is a vital link between the strategic, the tactical levels of management and the shop floor personnel who are in close proximity to the process and its hazards.

The HSE vision drove the organization to define its strategies to achieve excellence through establishing standards and procedures in line with the global practices, involving all levels of employees with an accountability to ensure increased involvement, measuring performance at various stages, dedicating sufficient resources, increasing HSE competency, establishing a governance system and continual improvement tracked by various metrics. The initial stages of instituting the process safety program involved extensive emphasis on the development of various processes and procedures for the safe operations of the plants involving cross functional teams to establish a standardized program across all RIL operations.

To enable the operating manager to take decisions for safety enhancement while he/she is confronted with conflicting priorities, it is essential to sensitize them to the pertinent aspects of process safety.

The foundation of the RIL HSE Management Framework is "Health, Safety and Environment is the responsibility of the Line Organization". RIL believes that the only proven way to excel in Health, Safety and Environment management is for all members of the line organization to accept responsibility for their personal safety and for the safety of their team.

At Reliance Industries Limited, the Process Safety Management program stands as a stepping stone for the behavioral and the cultural shift towards a safer environment at the workplace. The involvement of the line management in the implementation of PSM is done by adopting the subcommittee approach at the sites. This also includes other initiatives like inculcating behavioral changes through safety observation programs. This has proven to be successful in the few years of our PSM journey.

The establishment of a robust process safety management program at Reliance follows a top-down approach. This is being demonstrated by the active involvement of all levels of employees, from the top management to the shop floor employees in all initiatives like the safety observation programs to the participation in the subcommittees.

2. Initiatives for enhancing employee involvement

Analyses of internal and external incidents over the years reveal that 'compliance to procedures' is one of the major contributing factors. Such incidents cannot be attributed to merely non-compliance by operating personnel, but the root causes could be systemic factors like unclear instructions through the laid down procedures.

These analyses prompted RIL to launch the 'Project Health Check' initiative. This involved identification of critical procedures across the sites by a team consisting of operating managers and shop floor personnel. These critical procedures were then studied for completeness and clarity of instruction.

A Job Cycle check program was introduced to validate the practice against procedures, thereby identifying a scope of improvement in the existing standard operating procedures. These checks prompt continual improvement in the procedures, making them safer and user-friendly. Job cycle check created a sense of ownership among the operating personnel about the procedures. This has now emerged as a tool for continual improvements of the procedures as a cyclic process. As a part of the Job Cycle Check, evaluation of the preparatory measures taken for the job and execution of the job as per the procedures is done, covering aspects like differences between procedures and practice, unsafe conditions and human factors like accessibility, clear identification of the equipment etc. This enables the operating personnel to capture major improvement opportunities and also enhance the compliance to procedures. Any revisions to the procedures based on the feedback from a job cycle check are made by a team of trained and competent operating personnel.

This time targeted initiative helped sensitize the operating managers to the importance of not only compliance to procedures but the importance of adequacy of procedures. While excellence in this area still remains a challenge in the journey, a step in the right direction was taken.

3. Establishing data systems to monitor and analyse leading and lagging metrics for improving process safety performance.

The performance of the Safety Management System is monitored by a set of leading and lagging indicators. RIL has set up monitoring, reporting and tracking systems for deviations and other performance parameters critical to process safety that are derived from global best practices advocated by CCPS and the local regulations. These leading and lagging metrics enable the operating management to measure, improve and sustain reliable execution of process safety management systems.

The data for some of the leading indicators used at RIL for performance assessment are from the safety observation process, near miss incidents, metrics that showcase progress in various PSM elements, with the lagging indicators being the incident and injury metrics.

The Process, Fire, Distribution, Environment and Work place incidents are reported and logged into a home-grown Information Technology (IT)-enabled system across all businesses. The cycle from the reporting of an incident, investigation, generation of recommendations, to action tracking is being done through this system. The output of the investigation generates a report stating various contributing factors (system, physical and human factors), the failed PSM elements and the operational discipline characteristics. This data is available for the operating manager, which enables him/her to identify the system deficiencies in the plant or area and take corrective actions to prevent recurrence of such incidents.

A formal process of learning from incidents is also enabled through this IT system where internal and external incidents are analyzed for learnings that may be applied at similar processes or plants and shared across the organization. Learnings are shared through various means, including displays at strategic locations, through e-communications, discussed at the beginning of meetings, etc. The recommendations arising from learnings from incidents are tracked through an IT-enabled system, where in many cases; the operating manager is the action party.

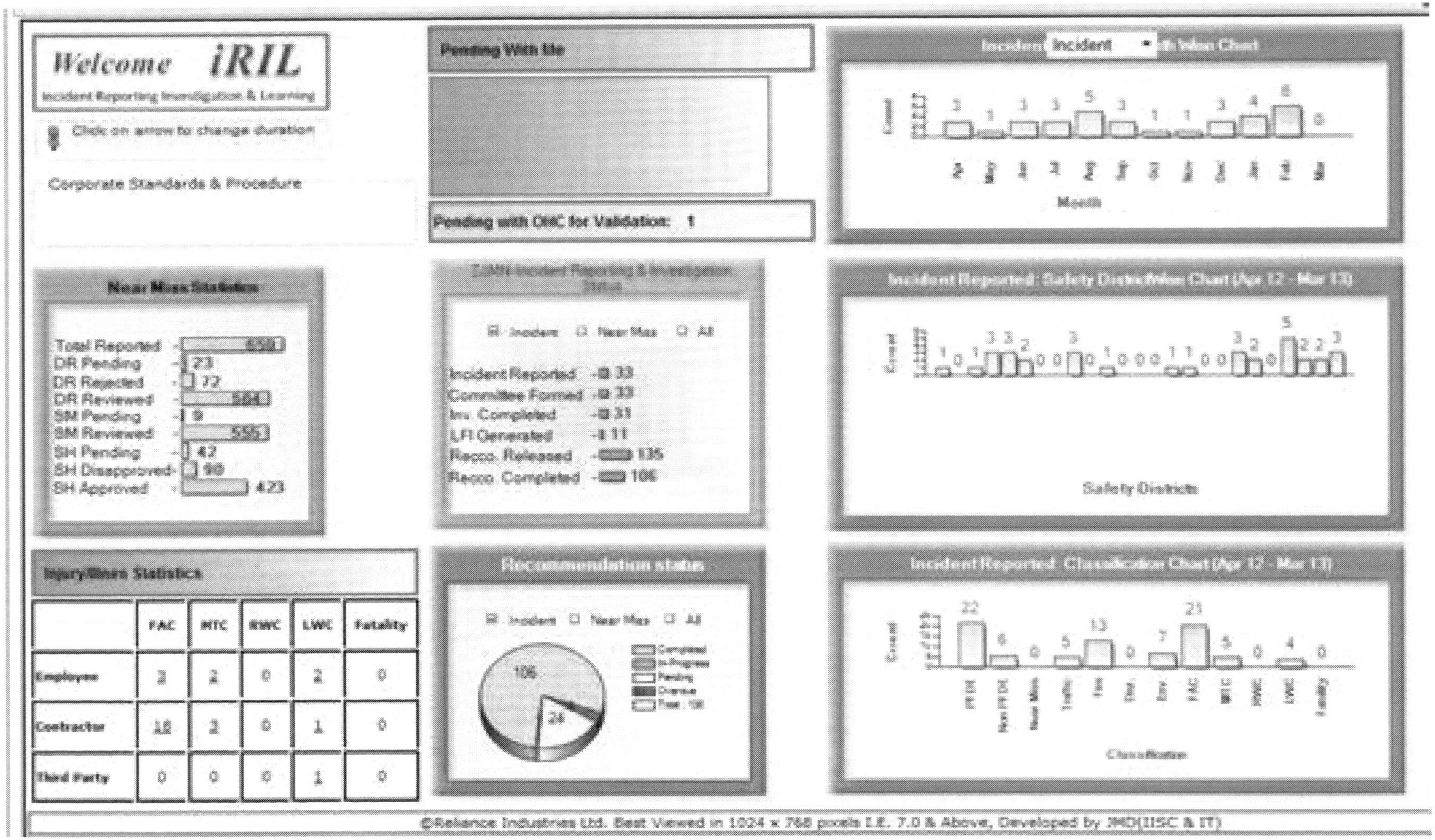

Fig1 : Snapshot of the Incident reporting system

While it is recognized that IT solutions alone cannot deliver high safety performance, RIL believes that the power of IT coupled with a strong subcommittee structure comprising members from line management, can deliver a superior functioning safety system in the long run for a large multi location organization.

4. Safety observation programs to influence behaviors at the workplace

RIL has taken an approach towards behavioral safety through a 'Safety Observation program'. The methodology of a safety observation program practiced at RIL follows the global approach of ‘observe’, ‘discuss’, ‘comment’, ‘agree’, ‘discuss other safety issues’ and ‘acknowledge’.

The objective of the Reliance Safety Observation program is to reduce errors in working and manage defenses. By observing various unsafe acts and conditions, the operating manager, by virtue of his process knowledge, is able to create a situational awareness of what can go wrong in the surroundings, and by initiating a conversation with the employee at work and engaging the employee in the observation process, a behavioral change is initiated and reinforced. The situational awareness is a critical step practiced at RIL which is different from the conventional approach of a safety observation program. Findings of the Safety Observation Program are recorded electronically and analyzed as part of leading metrics.

While most organizations use the observation process primarily for workplace activities and behavior change, RIL is making efforts to dovetail PSM elements in the safety observation process to get maximum impact on the process safety related activities.

5. Procedures underpinned with technology

In pursuit of operating and safety excellence, RIL has evolved in synchronization with the global trend with respect to IT solutions. With technology, information inundation is quite possible. Therefore, to empower the operating managers with better decision making tools, support is taken from different software solutions.

Excessive alarms are a common issue faced by process operators with modern Distributed Control Systems (DCSs). This is a contributing factor for diminished capacity for the operator who has to attend to both critical and spurious alarms. These factors increase the likelihood that a critical alarm may be overlooked making it difficult to respond to a plant upset, and raising the chance of an unplanned shutdown or an incident.

To avoid such a situation, the alarm rationalization process was initiated to evaluate all potential or existing alarms against principles established through an alarm philosophy. This system minimizes spurious alarms and enables the operator to focus on the critical alarms to capture rationale such as cause, consequence and corrective action which can be used to guide operator response.

RIL is in the process of implementing an IT solution which will help the operating managers to more easily input as well as use all necessary information to arrive at informed decisions in the field quickly. Such application will also enable information sharing across all levels of the organization. The operations logbook feature of this application maintains an electronic log that provides shift summary reports, tracks daily operations and improves process and plant reliability. Operations monitoring helps users track operating performance against targets, systematically carry out data monitoring and highlight problem areas.

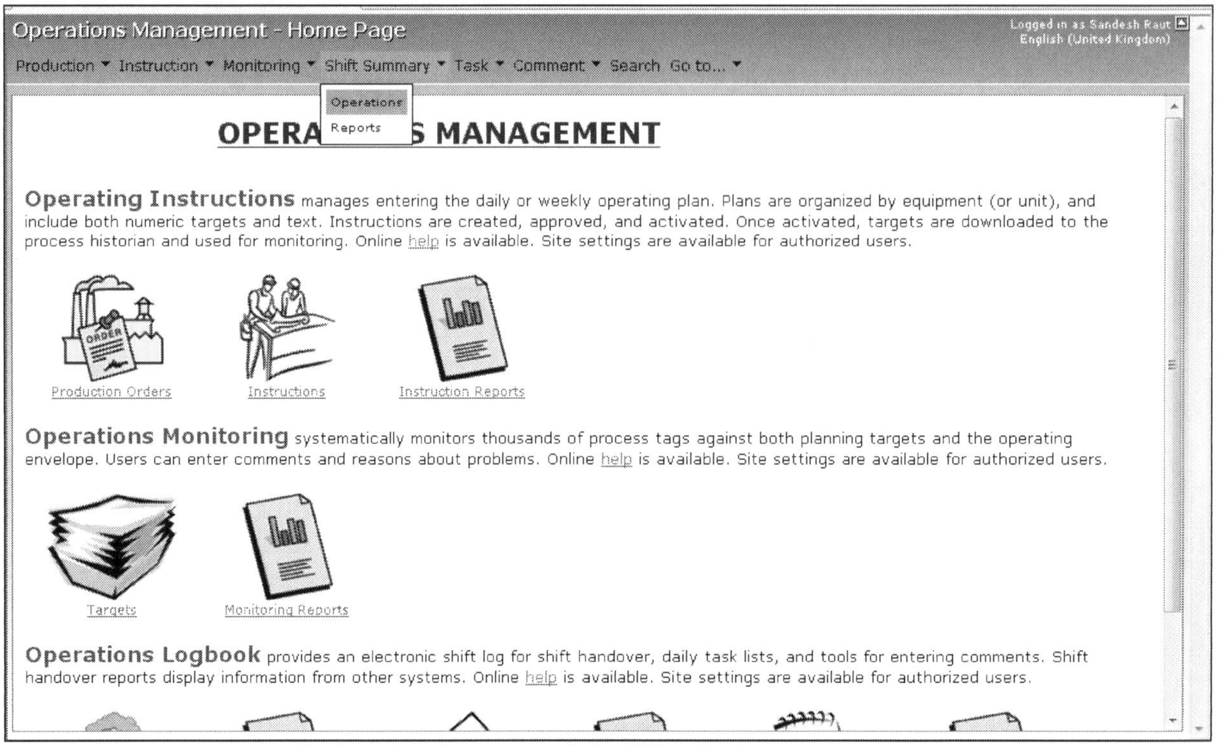

Fig 2 : Key features of the IT solution

Key advantages of the software solution are

- Integrated database to ensure that all functions (maintenance, operations, technology, safety) work on the same database
- Monitoring of deviations with respect to planning targets and standard operating conditions
- Knowledge repository of reasons for deviations
- Consistent and comprehensive e-enabled Interlock Bypass Management
- Complete and consistent record of hand over and take over activities carried out during the shift

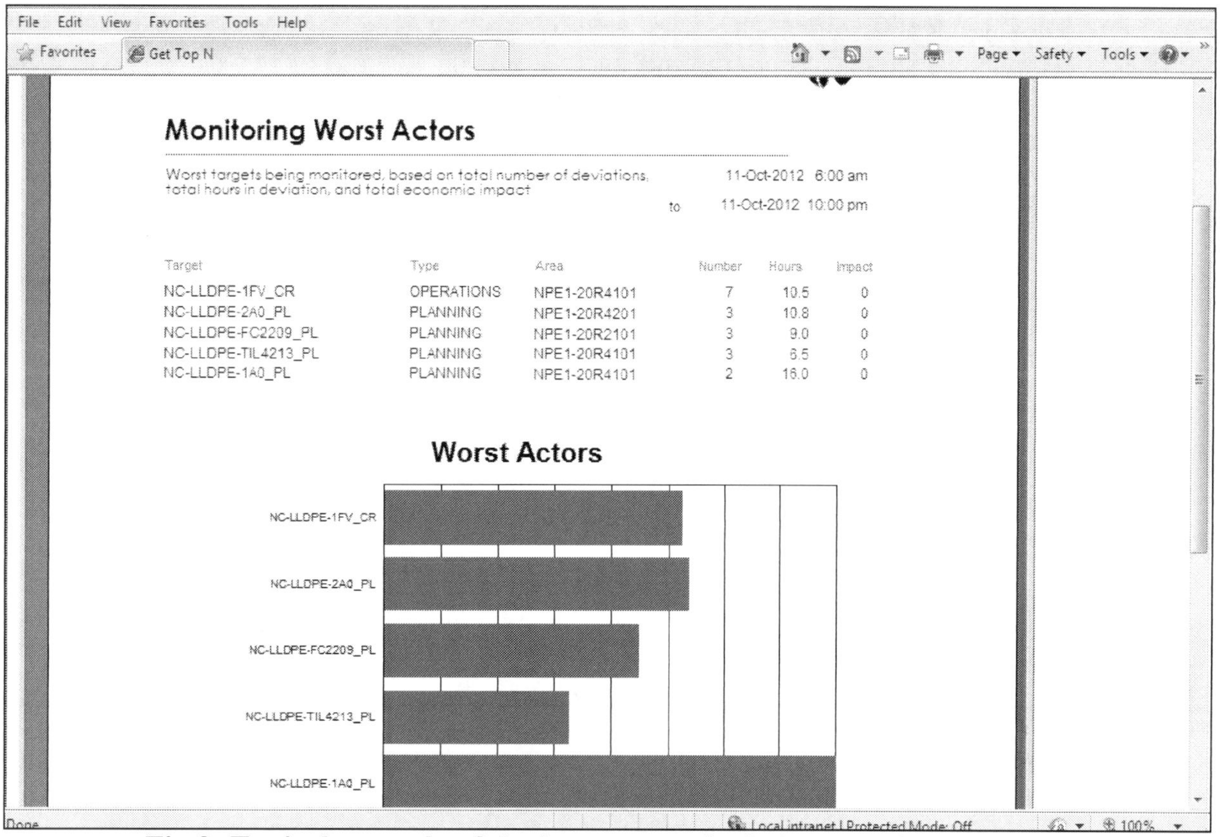

Fig 3: Typical example of the key process deviations monitored

This solution shall also enable the operating managers to establish risk-based inspection procedures, clarity on achieving production targets within the stipulated safe limits and reliability centered maintenance activities thereby resulting in increased health of the equipment and safe operations.

6. Developing understanding of risk-based approach during decision-making in case of emergencies

Risk is managed by identifying hazards, assessing consequences and probabilities, and evaluating and implementing preventive and mitigative measures. Periodic risk assessments are performed by qualified operating and maintenance personnel. Risk assessments are updated at specified intervals and as changes occur.

Risk perception has increased significantly among the operating personnel through their involvement in risk studies such as Process Hazard Analysis, Layer of Protection Analysis, and Quantitative Risk Analysis conducted for all the operating units as a part of the RIL PSM program. There has been an increased awareness among the operating group about the hazards that could be manifested in case of barrier failure. An important outcome of these risk studies is identification of barriers critical to process safety and maintenance of them. While preventive barriers prevent loss of primary containment and require periodic upkeep, mitigative barriers act as a last line of defence in case of an incident.

Having identified the maximum credible and worst case scenarios through risk assessment studies, operating managers are enabled to plan for an emergency by assessing the potentiality of damage and the resources required for handling an emergency. For clearly defined responsibilities and accountability during such conditions, to enable effective response and decision making during emergencies, systems like pre-incident planning and mock drills have been established at RIL.

7. Internal certification processes for increased competency of several levels of employees

A tiered competency certification method for process safety management has been designed taking in to account the level of competency an employee requires for his application during operations. Development of "Reliance Certified PSM Belts" has been initiated with the objective of bringing in the right level of PSM awareness for all levels of employees.

The target group for each level of certification may vary depending on their role in operations and the extent of the process safety applications they have to execute in their daily routine operations. The minimum required competency of the technical, functional skills of operating personnel also additionally defines the level of competency required that would be appropriate to the process safety demands for the job or role. The four levels of certification depend on the process safety demands on the functions of an employee, classified as the: awareness level, knowledge level, skill level, and mastery level of process safety knowledge. This certification process is also extended to all the employees – the maintenance managers, technical services, commercial team, and all levels of the management.

The first-level certified group of employees may not need to work extensively for process safety but need to have a basic understanding of the principles. The groups of personnel identified for the second belt certification are those who may have to apply PSM principles in their daily operational activities and contribute to increase the PSM trickle. Over a period of time it is expected that the third and fourth levels of certification will produce a group of PSM experts who will contribute to the continuous growth of PSM establishment in the organization by taking up coaching, and directing projects focusing on PSM and similar initiatives. The certification process is defined to complete with a formal validation in the system.

8. Conclusions

The journey to process safety excellence is a long one. Organizations all over the world are converging to some basic requirements to establish and improve process safety. However, best practices continue to evolve and proactive organizations continue to improve through internal experiences. The initiatives described above may in the overall sense not be very different from other organizations. However the finer differences are the ones which decide the extent and depth of success that each program can have in sensitizing as well as empowering the operating personnel in improving PSM. RiL continues to evolve systems in its continuous journey towards PSM excellence.

References

[1] RIL Group Process Safety Management Standard,- internal document

[2] *Guidelines for Process Safety Fundamentals in General Plan Operations,* New York: Center for Chemical Process Safety, American Institute of Chemical Engineers, 1993

[3] *Guidelines for Implementing Process Safety Management Systems,* New York: Center for Chemical Process Safety, American Institute of Chemical Engineers, 1994

[4] CCPS, "*Process Safety Leading and Lagging Metrics*", AICHE, New York, 2011. http://www.aiche.org/ccps/resources/overview/process-safety-metrics/recommended-process-safety-metrics

9TH GLOBAL CONGRESS ON PROCESS SAFETY

Barrier Diagram (Bow Tie) Quality Issues for Operating Managers

Dr Robin Pitblado
DNV (USA) Inc.
1400 Ravello Drive, Katy TX 77449
Robin.pitblado@dnv.com

Pieter Weijand
Governors
Netherlands
Pieter.Weijand@gmail.com

Prepared for Presentation at
American Institute of Chemical Engineers
2013 Spring Meeting
9th Global Congress on Process Safety
San Antonio, Texas
April 28 – May 1, 2013

UNPUBLISHED

Barrier Diagram (Bow Tie) Quality Issues for Operating Managers

Dr Robin Pitblado
DNV (USA) Inc.
1400 Ravello Drive, Katy TX 77449
Robin.Pitblado@dnv.com

Pieter Weijand
Governors
Netherlands
Pieter.Weijand@gmail.com

Keywords: Barrier diagrams, Bow Ties, Quality, Operations

Abstract

The Oil & Gas (O&G) and petrochemical industries have been much more successful at reducing occupational safety events as compared to major process accidents over the past 20 years. A novel approach based on barrier diagrams, called Bow Ties, directly addresses major accidents and the focus on barriers and maintaining barrier functions has been shown to be effective in O&G operations in the North Sea. While not mandated in OSHA 1910.119, many companies are starting to apply barrier methods to improve operations and reduce major accidents. The Bow Tie approach is deceptively simple and many errors are made in Bow Tie diagrams. This is due to a lack of a formal standard or accepted methodology for the approach. This paper describes many common errors that appear in Bow Ties used for operational safety and how these might be rectified.

Since Operating Managers will rely on these risk diagrams as a key part of their process safety program, it is essential that they have good quality and not contain structural errors.

1. Introduction

The Oil & Gas and petrochemical industries have been very successful at enhancing occupational safety performance, achieving close to a factor of 10 reduction in OSHA lost time or recordable events, as reported in company annual sustainability reports. Conversely, major accident performance has hardly improved over the period, shown both by anecdotal evidence of major accident headlines, Chemical Safety Board lists of events, or Marsh data on 100 largest losses [1]. The relative performance between these two domains is shown in Figure 1.

While OSHA 1910 and EU Safety Cases have driven part of the improvement in occupational safety, they have been less successful for major accidents as shown in the figure. The authors believe this is due to a lack of focus on major hazards during operations. OSHA 1910 focuses more on design safety and typically the required PHA (Process Hazard Analysis) documents are not routinely used during operations. Conversely in the UK, the Piper Alpha accident in 1988 led to a detailed public inquiry chaired by Lord Cullen. That recommended new offshore safety case regulations that include a focus on barriers addressing major accidents. These were called Safety Critical Elements, and each required a Performance Standard (what it is expected to do, its reliability and its survivability) and a Written Scheme (the required inspection, maintenance and competence programs to keep the element functioning at its performance standard for the facility lifetime). This focus on barriers has led to close to a factor of 10 reduction in major leaks (which have the potential to lead to total platform loss) based on UK HSE (Health and Safety executive) statistics. Norway is broadly similar in barrier requirements. Statistically, there were 2 total loss events in the North Sea from 1980-1988, and none since then in the intervening 25 years. This supports a factor of 10 reduction – although the period is short.

The focus on barriers has led many companies to embrace a barrier approach to managing major accidents. The theory under-pinning this was developed by James Reason and Patrick Hudson, and is crystalized in the well-known Swiss cheese model of accident causation. The Bow Tie diagram was developed in the early 1990s, simultaneously in Australia and the Netherlands. This approach assists operations activities to consider major accidents and is considered the most effective of the various barrier methods as it provides both a clear summary of barriers deployed and a visual format that is easy to communicate. Unfortunately, there is no accepted international standard for Bow Tie diagrams and this leads to quality issues with many Bow Ties having fundamental errors that weaken the value of the method and hence the potential to reduce major accidents.

Figure 1. Comparison of performance - occupational vs major accident safety

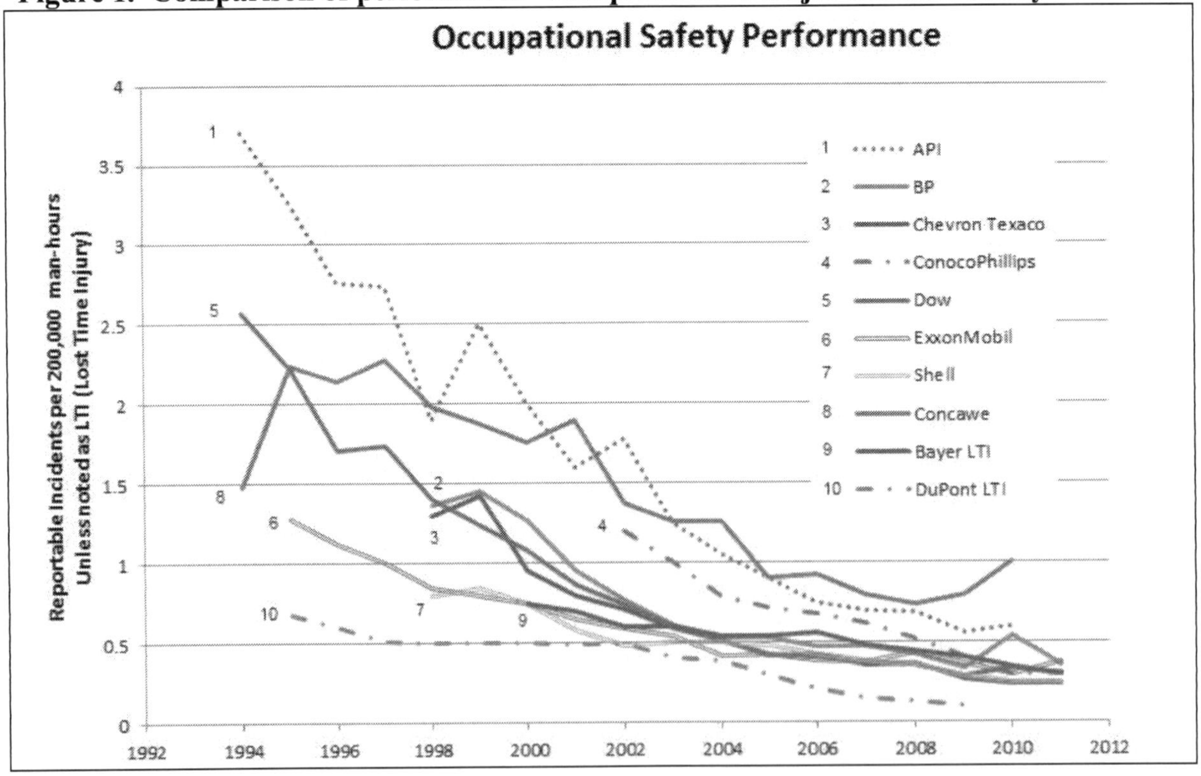

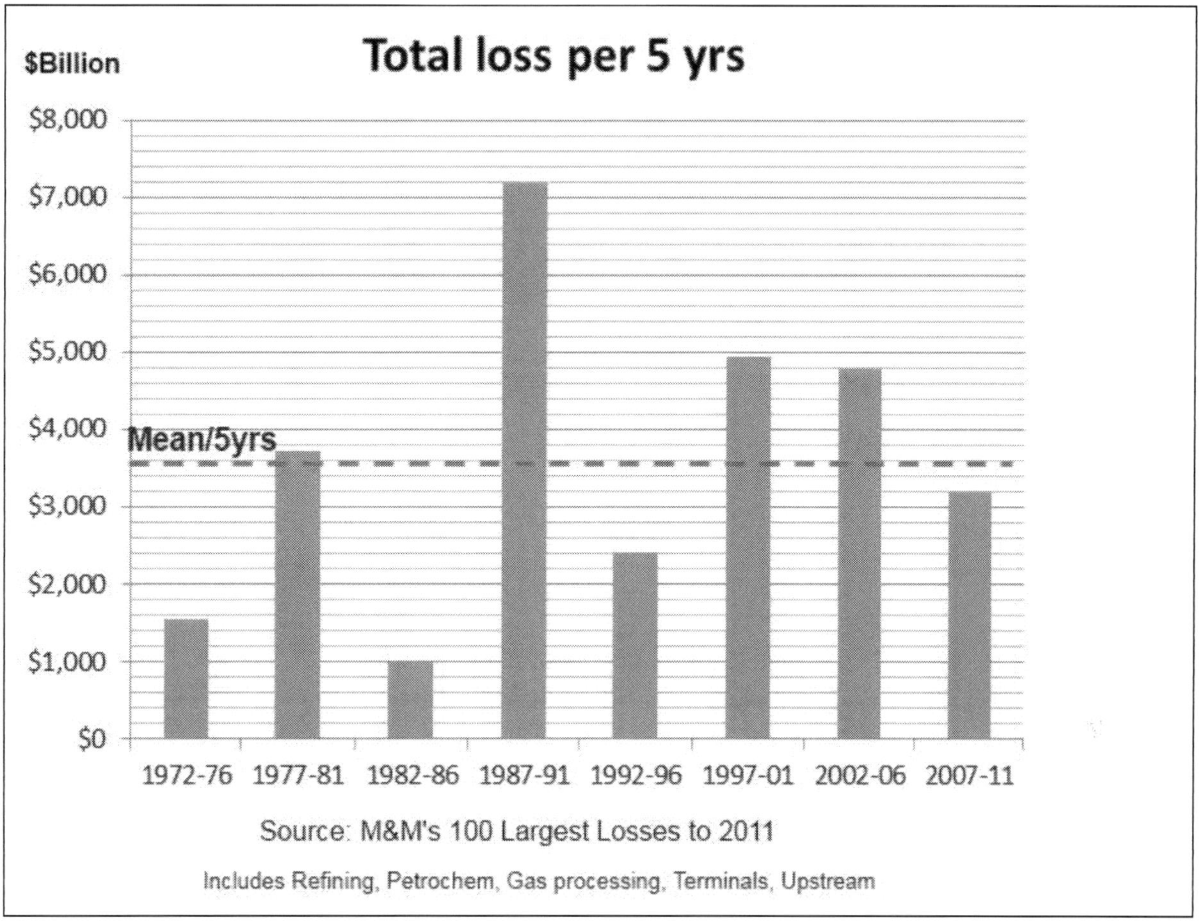

2. Bow Tie barrier diagram methodology

2.1 Method Overview

There are several descriptions of the Bow Tie method. These include a paper by Shell to the EU (Zuijderduijn [2]), the EU ARAMIS research project (Salvi and Debray [3]), the Norway BORA Project [4], and various company standards. The authors have worked with BP and Shell standards. While these are all good documents, they are not mutually consistent and none is accepted as a universal standard for method or terminology.

The methodology for Bow Tie diagrams appears straightforward and a typical diagram is shown in Figure 2.

Figure 2. Example Bow Tie diagram demonstrating method

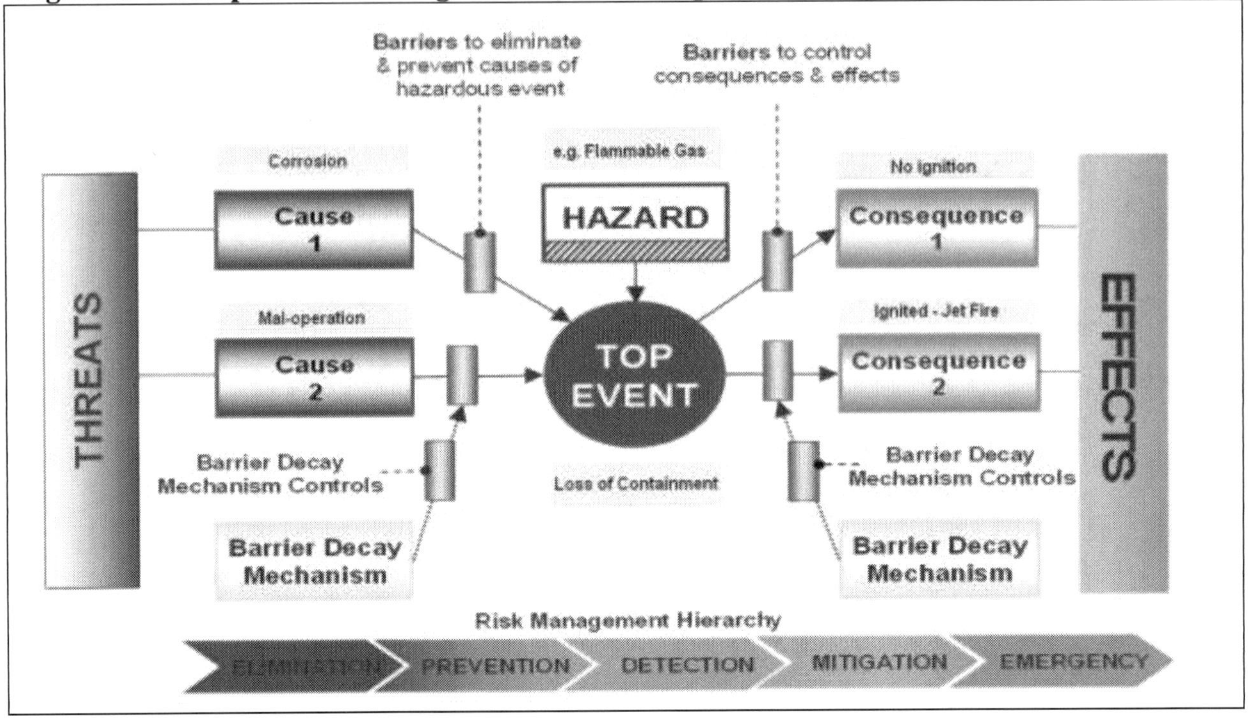

While Figure 2 is a good overview of the method, real Bow Ties can be extensive. A typical Bow Tie might have 8-10 threat causes and 3-5 outcome consequences. Each arm might have 3-5 barriers, but more are possible. Also several barriers will have Barrier Decay Mechanisms built-out.

A misconception by some is that the barriers, both prevention and mitigation, are simply marks on a threat pathway showing the existence of a barrier. This allows for good communications to showing all the barriers deployed and who is responsible for each. The specification of the barrier may be loosely defined and potentially inconsistent. However if the only intent is to list all the barriers on a figure, then this is acceptable and is simply a variant of providing a list.

While this is an important use of Bow Ties and enhances staff and contractor communications and understanding, it is only a partial use of the tool, and may contain structural errors that defeat more extensive and value-adding uses. The key is to understand the scientific basis for Bow Ties. The ARAMIS project provided a graphic (Figure 3) showing the Bow Tie as a combination of Fault Tree Analysis (FTA) on the left hand prevention side, and Event Tree Analysis (ETA) on the right hand mitigation side.

Figure 3. Bow Tie under-pinnings

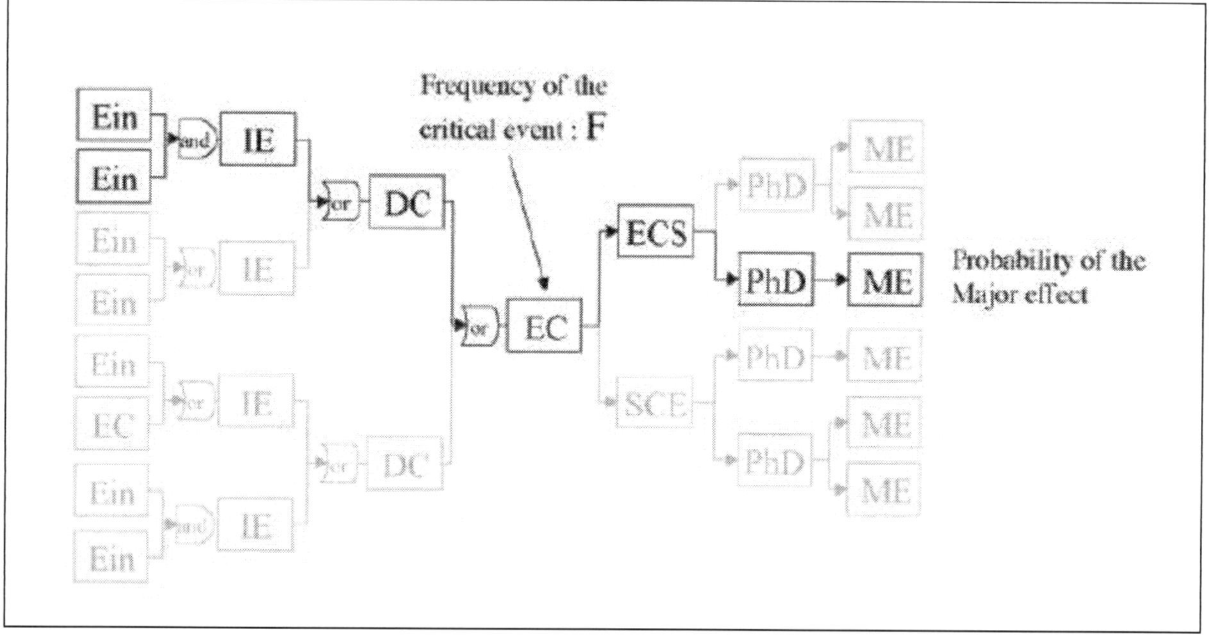

2.2 *Fault Tree Basis*

In order that the Bow Tie barrier method can be exploited for its maximum potential contribution to reducing major accidents, it is vital to link it conceptually to theoretically sound and more mature methodologies, which have formal rules for structure and for arrangement of events. A Bow Tie barrier is actually a fault tree AND-gate construct with a "demand" combined with "barrier fails" as shown in Figure 4.

Figure 4. Barrier is Fault Tree AND-gate

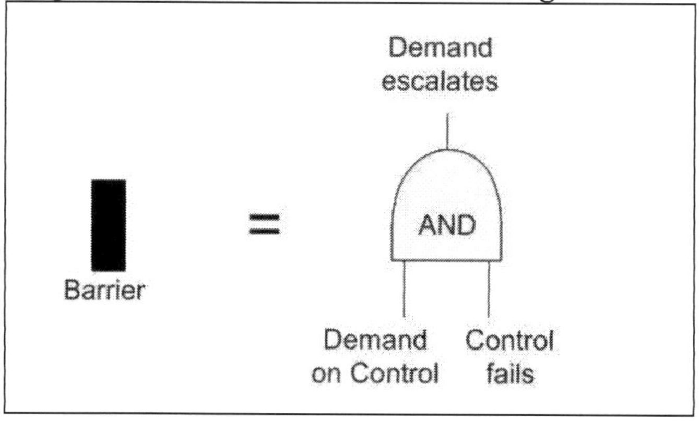

The different Threat arms coming into the Top Event are combined through a Fault tree OR-gate construct.

The Mitigation-side can be thought of as Event Tree Analysis as suggested in the ARAMIS project, but instead of the prevention side AND-gate format, the barrier on the mitigation side is an Event Tree branch with barrier fails and barrier succeeds as two outcomes. Since "barrier succeeds" terminates the event, that arm is discarded and the Bow Tie only builds-out the sequence of barrier failures until the ultimate consequence is reached. This explanation works at a first level, but it fails when Barrier Decay Mechanisms (escalation factors) are introduced. A Barrier Decay Mechanism is a fault tree build-out of how the main pathway barrier can degrade and what additional barriers are deployed to prevent this (Figure 5). This build-out is used equally on both sides and mitigation barriers can degrade just as prevention ones may. Also some barriers conceptually work after a consequence is realized, and the temporal order can be distorted for the sake of a 'clear' Bow_Tie diagram shape.

Figure 5. Barrier Decay Mechanism

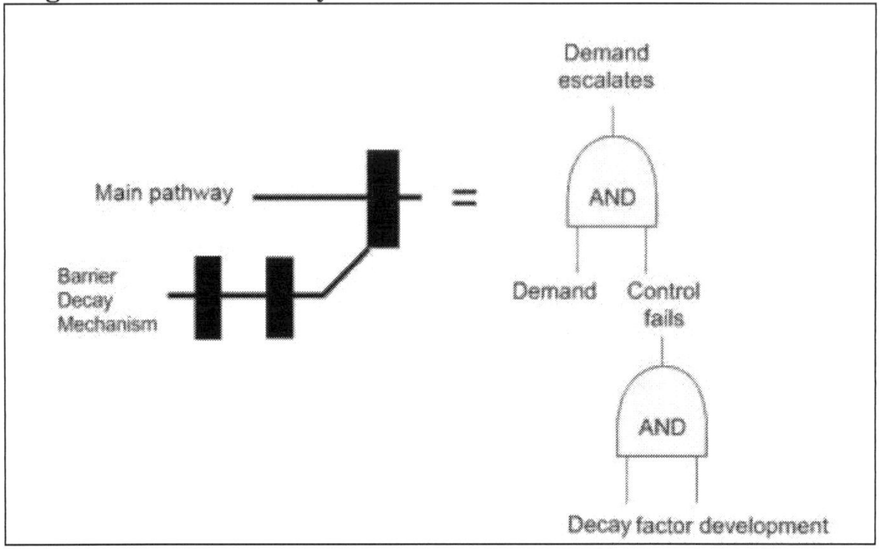

This build-out only applies to FTA, there is no equivalent form in ETA. Thus a Bow Tie uses fault tree methods on both sides, not just the prevention side. This is important as the aim is to reinforce the theoretical basis for Bow Ties, so that formal rules for construction and combination can apply, and fuller deliveries are possible from the Bow Tie. Examples of these include correct barrier counts between the Threat and the Top Event (Barrier Decay Mechanism barriers below the main pathway do not count in that total), a clarity in construction that places barriers in their correct locations, not simply plotted on a "useful graphic", an ability to draw correct conclusions knowing that the figure complies with Fault Tree systematics and that decisions taken on the basis of the Fault Tree are valid. A good overview of fault tree systematics and rule sets is provided by NASA [5].

2.3 Basic Definitions

The Bow Tie example shown in Figure 2 has several key elements. These can be defined as follows:

- The Hazard – this is the potential (material or activity) to lead to a Top Event
- Top Event – this is the event that can lead to undesired outcomes (usually loss of control or loss of containment) – it is named based on Fault Tree terminology
- Causes – these are the categories of challenges to safety systems that if not terminated will lead to the Top Event
- Outcomes or Consequences – these are the categories of undesired outcomes that can occur if the event starting from the Top Event is not terminated.

- Prevention barriers – these are the barriers / controls / safeguards that can terminate (or reduce the likelihood) of the cause before it reaches the Top Event
- Mitigation barriers – these are the barriers / controls / safeguards that can mitigate an event sequence before it reaches the undesired consequence
- Barrier Decay Mechanism or Escalation Factor – build-outs from the main pathway barrier for reasons it might degrade and the barriers deployed to prevent this.

There are several definitions for barriers; a good one is from Sklet [6]: "Safety barriers are physical and/or non-physical means planned to prevent, control, or mitigate undesired events or accidents". This definition allows for both hardware barriers (e.g. Emergency Shutdown Systems) and for non-physical barriers (e.g. inspection programs, training) and this allows the holistic deployment of barriers in a comprehensive major accident protection system to be properly mapped.

3. Problems observed in Bow Tie analysis

Given this understanding of the fundamental under-pinning of Bow Tie analysis by scientifically valid process safety methodologies, it is now possible to identify potential problems in Bow Ties produced without reference to such methods.

3.1 Hazard

The authors have observed several deficiencies with Hazards defined in Bow Tie analysis. A Hazard is a material, energy source or activity with a potential for an undesired outcome typically if there is a loss of containment or other form of loss of control. Some example valid Hazards and other incorrect hazards are shown in Table 1. The test for validity is that the Hazard should identify the source of (or damage to / disruption of) the problem and lead naturally to the Top Event, but should not be the top event itself. Note Hazards can have multiple Top Events, but there can only be one Bow Tie for each Top Event. There can be hazards which are correct for one Top Event but incorrect for another. Thus Collision may be a valid Hazard for a Top Event of "hull punctured", but not valid for a Top Event of "loss of steering" (where Collision would be a consequence).

Table 1. Valid and incorrect hazards

Valid Hazards	Incorrect Hazards
• Cyanide • A 100t sphere of pressurized propane • Driving a heavy vehicle at night • Excess overtime being worked	• A leak of propane • Ignition • Control system malfunctions

The significance of an error in defining the Hazard is that it might lead to a faulty Top Event and getting this right is core to success. For example if the Hazard is incorrectly defined as the Top Event (e.g. Leak of Propane in Table 1) then the resulting Top Event might be "large flammable cloud" and that would be incorrect. This is closer to an undesired outcome and this could lead to generation of incorrect barriers.

3.2 Top Event

The Top Event can be thought of as a point in time where there is some change in state that causes a loss of control over the hazard. Most often for major accidents this might be loss of containment or loss of control. Top Events are not the outcome; usually there are multiple barriers that can be deployed between the Top Event and the final consequence(s). Thus "explosion" would generally make a poor Top Event – it would be better to define this as a leak of hazardous material as this flows more naturally from the hazard. Then show how barriers can be deployed before a large flammable cloud forms that can explode: e.g. gas detection systems, emergency shutdown systems, ignition controls, and management of occupied spaces, etc. Some examples of good and poor Top Events are shown in Table 2. The Top Event should indicate how control is lost and / or give an indication of scale. The definition of a Top Event also depends on perspective of the analysis, for instance a Top Event defined from one perspective could be inappropriate from another perspective. For example for the Hazard: Reactive Chemicals a Top Event: Loss of Control of Reaction would make sense, but not a Top Event Incorrect Mixture. In this context Incorrect Mixture is a Threat, but it could be a valid Top Event in another context.

Table 2. Examples of good and poor Top Events

Good examples:	Poor or Incorrect Top Events

• Large leak of pressurized propane
• Small spill (<2kg) of eco-toxic material
• High speed turbine disintegration
• Control of exothermic reaction is lost
• Error in control of work
• Structural failure of equipment item
• Loss of intended separation
	• Explosion of propane cloud
	  o Outcome not Top Event
	• Ignition of flammable cloud
	  o Intermediate event in mitigation sequence
	• Crane dropped object
	  o Outcome, not loss of control
	• Corrosion
	  o Threat, not Top Event
	• Eco-toxic leak to environment
	  o Consequence, not Top Event

3.3 Cause and Effect

In the Bow Tie diagram it is important that there be a clear cause and effect between the various elements:

- Hazard should lead logically to the Top Event
- threats (which might have multiple separate causes) should lead to the Top Event
- the Top Event should lead logically to one or some set of Consequences
- Barrier Decay Mechanisms should lead logically to main pathway barrier failure

If the cause and effect linkage is absent or unclear, then it is likely that the Bow Tie will be incorrect, and this means that barriers may be missed or that new barriers are introduced that have no effect on major hazards.

3.4 Threats

Threats lead to the Top Event and are normally collected as a sequence of causes. These are often developed as a brainstorm of all possible causes. This might be 20+ causes. However, many of these causes (e.g. galvanic corrosion, thinning corrosion) may have similar controls and there is no value building out two pathways barriers – a single pathway would show all important barriers. Typical errors we see are too many causes identified, or conversely too few with important causes not enumerated. The suitability of barriers deployed against a major accident can only be assessed if all important threat causes are listed.

Not all threat causes have the same likelihood. For example, for a Top Event: "Large Leak of Propane", the threat corrosion is continuously present, whereas others like operator error or dropped object are high and low probability threats respectively. All should be treated similarly in terms of barrier build-out but there can be a different target for the number of barriers needed to manage the threat. The number of barriers required is discussed later. While it is a matter of judgment, some threats can be excluded (e.g. plane crash, earthquake, flood) although the proximity to airports, local earthquake category and flood susceptibility may change this assessment. In the last case – there is also not a direct link between flood and loss of containment.

Some general rules for defining threats are:

- Threats should address all significant risks leading to that Top Event (otherwise the Bow Tie would lead to misleading over-confidence in control)
- threats need to be direct (e.g. factors affecting human performance – such as drug and alcohol abuse – are not a direct cause but combine with other factors to affect Fitness for Duty, so Human Error might be the threat and alcohol and drug abuse would be a Barrier Decay Mechanism under Fitness for Duty)
- threats can be combined where the barriers deployed are the same
- all threats should lead logically to the Top Event and thereafter to every Consequence
- threats should be independent – two threats should not need to combine to lead to the Top Event (e.g. wet weather threat and dust on conductors threat leading to Top Event: short circuit – better to define a threat as Exposed Dirty Conductors)
- barrier failures are not threats (e.g. maintenance is a barrier, and failed maintenance is the barrier failing in a threat pathway and not a threat itself – referring back to the Fault Tree model: Threat AND Maintenance fails – so the Threat here will be corrosion or mechanical wear, etc.)

A good example of threats developed in the BORA Project [5] is provided in Table 3. This separates several threat classes from technical degradation of the system to external events. When building a Bow Tie the team needs to assess all these events to see if any can be combined or whether any can be eliminated as too unlikely.

Table 3. Example threats for offshore process safety barrier diagrams (from [4])

Initiating Event Type	Initiating Events
A. Technical degradation of system	1. Degradation of valve sealing 2. Degradation of flange gasket 3. Loss of bolt tensioning 4. Fatigue 5. Internal corrosion 6. External corrosion 7. Erosion 8. Other causes
B. Human intervention introduction latent error	1. Incorrect blinding/isolation 2. Incorrect fitting of flanges or bolts during maintenance 3. Valve(s) in incorrect position after maintenance 4. Erroneous choice or installations of sealing device 5. Maloperation of valve(s) during manual operation* 6. Maloperation of temporary hoses.
C. Human intervention causing immediate release	1. Break-down of isolation system during maintenance. 2. Maloperation of valve(s) during manual operation* 3. Work on wrong equipment, not known to be pressurised
D. Process disturbance	1. Overpressure 2. Overflow / overfilling
E. Inherent design errors	1. Design related failures
F. External events	1. Impact from falling object 2. Impact from bumping/collision

* This may lead to either introduction of a latent error or an immediate release

3.5 Consequences

Consequences are the logical outcome of all the mitigation barriers failing. They are usually grouped into 3-4 broad categories reflecting the range of mitigation measures deployed. Typical categories may include:

- Safety impact (Fatalities/injuries)
- environmental impact
- asset damage
- reputation impact

Consequences are events themselves and can be developed further. This aspect is most relevant if an actual event has occurred and the barrier diagram is being used for accident investigation (as with the BSCAT investigation approach [7] – Barrier-based systematic cause analysis technique). Thus a consequence can be a Top Event leading to further consequences. For example if the consequence is Environmental Impact, this can be converted to a Top Event leading to Consequences of short term Fish death, long term ecosystem damage, water treatment plant closure, etc. There will be prevention and mitigation barriers for this sequence as well.

For this reason consequences in the Bow Tie may not be the actual loss. Other factors may lead to larger or smaller losses.

3.6 Barriers

As noted earlier, a barrier as plotted in a Bow Tie is an AND-gate combining demand and barrier fails. If this happens then the next barrier in the sequence will be tested and so on until the Top Event is reached, and a similar process occurs on the Mitigation side until the Consequence is reached.

It may be seen that if there are 3 effective barriers on each side and each contributes a factor of 10 or more reduction (i.e. 90% effective) then 6 orders of magnitude reduction will occur from the original threat frequency. This is often very hard to achieve. However some Bow Tie diagrams show 10 or more barriers on both sides – and this is almost always incorrect as it implies a 20 order of magnitude reduction in frequency.

Some rule sets for barriers – these apply to both the Prevention and Mitigation side

- Barriers can be physical or non-physical
- barriers are real things that be tested or measured, non-physical should be systems
- barriers can be passive – pressure containment equipment, dikes
- barriers can be active – sprinklers, emergency shutdown systems
- barriers can be critical tasks – procedural systems, inspection programs
- barriers should be independent – this is based on fault tree rule sets
- sociotechnical aspect(s), barriers are mixtures of the above (frequently)

Some rule sets suggest that a barrier must be capable of completely terminating an event pathway. The authors believe this is too stringent. The basic concept being implemented here is the Swiss cheese model and this acknowledges that no barrier is 100% effective. This can be shown by way of example. There are important barriers such as water spray systems for toxic releases of ammonia. These have several issues affecting their effectiveness – their design basis (ammonia leak rate and water discharge rate and droplet size), their leak detection and actuation speed, and the spray deployment. It is common only to erect such sprays in the downwind direction towards significant population exposures. Thus such arrays, regardless of their technical efficiency can never protect leak events drifting in other directions and according to the rule that they have the potential to completely terminate the event – they would be excluded. This would not be correct – even partially effective barriers should be included. There is no real difference between partial effectiveness and reliability issues. Thus we would add a further rule

- Ideally barriers can terminate an accident pathway, but partially effective barriers should be included if these can make a material difference to the outcome

It is in the area of barriers where most errors are found in Bow Tie construction. Typical errors include:

- Too many similar barriers (Procedure 1 + Procedure 2 + Procedure 3, instead of a single barrier Procedural Control)
- inclusion of trivial or hardly effective barriers (e.g. sunglasses as a barrier to vehicle collision) *(Linked to scale of Top Event)*

- confusion of accident event sequence with barriers (e.g. Ignition – the correct barrier would be Ignition Controls, and if these fail then ignition may occur)
- inclusion of system element barriers on a main pathway rather than the system itself (e.g. escape and evacuation, escape lighting, muster area, etc., rather than Emergency Response System – the system details belong in the Barrier Decay Mechanism build-out as, if they fail, the Emergency Response System is degraded)

3.7 Barrier Decay Mechanism / Escalation Factor

This aspect of Bow Ties is poorly understood and many errors are seen here. The title that is most often used (Escalation Factor) contributes to this as there is no escalation of the event but there is an increase in failure likelihood. For this reason, the title Barrier Decay Mechanism is preferred as this clearly indicates the function of this pathway. The function is easily understood in fault tree terms (Figure 5). The main pathway barrier should be holistic procedural or administrative system or a complete hardware system. These systems have component elements and failures in these degrade the overall system and, thus, should properly be collected in the Barrier Decay mechanism.

Consider a critical shutdown procedure that is to be invoked if certain process parameters are exceeded. Key elements of such a procedure might be:

- Procedure correct and up-to-date
- staff all trained in the procedure
- staff competent to recognize the event and confident to act in a timely manner
- staff fit for duty
- staff supervision

It would be incorrect to include all of these on the main pathway. If simple order of magnitude reductions for each barrier are applied (i.e. 90% effective) this would imply that the elements of this procedure contribute to a 6 orders of magnitude risk reduction. This is wrong. This difference is more clearly shown in Figure 7 – located at end of paper.

A superficial inspection of part a) of the figure might satisfy the reader that the threat is very well protected – with 7 barriers. But many of these are dependent barriers and the correct manner to draw the barrier diagram is shown in part b) with a Barrier Decay Mechanism. This is titled Shutdown fails and it transposes all the main pathway shutdown barriers from part a) into the Barrier Decay Mechanism and leaves one overall Shutdown procedure. The diagram now correctly shows only 2 barriers and this is generally insufficient for an exothermic reactor hazard.

Some conclusions from this are that a Procedure Document is not a main pathway barrier – a piece of paper does not reduce risk. Similarly Training should not be a main pathway barrier. Training supports the real barrier, the Procedural Control.

3.8 Scoring Barriers

A key benefit of barrier diagrams is that they make explicit the critical safety barriers. An obvious benefit is to judge whether there are sufficient barriers in place for each threat. While there is no formally agreed method, Shell [3] outlined its approach that identifies needed barrier counts:

- For high risk threats – 3 barriers on both sides of the Bow Tie
- for medium risk threats – 2 barriers on each side
- for low risk threats – 1 barrier on each side

A surplus of barriers on one side cannot make up for a deficiency on the other side. That target still must be achieved. Also, since there are multiple threats, all with different likelihoods, the number of barriers on the Mitigation side will be determined by the highest risk threat.

Barriers can only be counted on a Bow Tie diagram if the structure is correct. This means that the rules identified above need to be implemented, and especially that Barrier Decay Mechanism barriers do not appear in the main pathway. When drawn correctly, it is often very difficult to achieve 3 barriers on the Prevention side (left side), although somewhat easier on the Mitigation side (right side).

3.9 Overall Complexity

Some Bow Tie diagrams can become very complex. One example is shown in Figure 6 – this is deliberately made small to show the complexity – it is not intended to be read. This may be correct, but it is too complex for good communication. However, without detailed examination, there appear to be some issues with this figure – on one Threat arm there are 9 barriers and on one Consequence arm there are 11 barriers. If this were true, then these pathways would have infinitesimal likelihood. More likely, this may mean that Barrier Decay Mechanism barriers have been located onto the main pathway. If the intent is purely to log barriers this is acceptable, but if the aim is to judge whether there are sufficient barriers for each threat – then this diagram may mislead and lead to incorrect safety expenditures.

Figure 6. Overly complex Bow Tie diagram

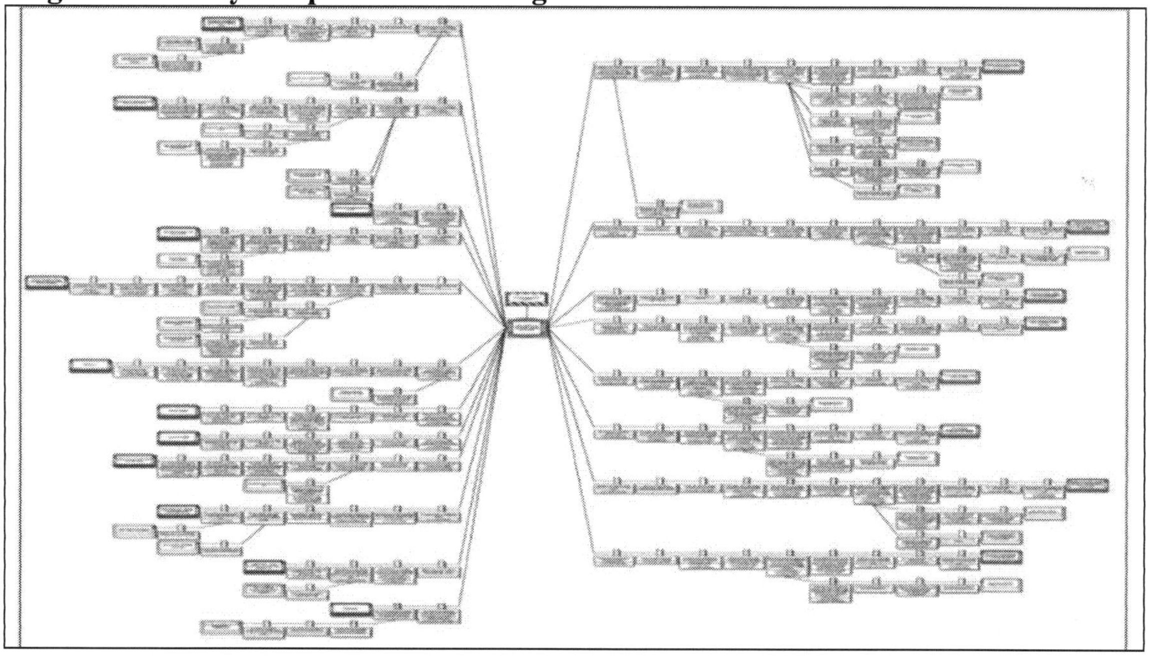

There are conflicting goals with Bow Tie diagrams. For facility managers and safety personnel – they need to know about all barriers. For regulators they may only need to know about main pathway barriers, not Barrier Decay Mechanisms. Individual operational or maintenance staff need clarity and simplicity as they might be confused if all detail for all staff is shown. They need an overview of the system of controls and all of the details for their contribution. Typically, software for Bow Tie diagrams provides for these different needs and diagrams can be customized from the same dataset to show only what is needed.

4. Conclusions

The Bow Tie barrier diagram method is growing rapidly in popularity – partly due to its sharp focus on major accidents, its compact format, and its ability to communicate critical safety barriers to staff and contractors. While several published documents exist describing the method, there is no globally accepted standard. This has led to many quality issues that detract from its use in an operational environment.

This paper has shown that the Bow Tie method should be carefully linked to its fault tree under-pinnings. FTA is the well proven and theoretically sound method that gives the Bow Tie diagram a real basis in safety science, much beyond being a convenient picture and aid for communication. One of the most important of these benefits is an ability to judge whether there are sufficient barriers for each identified threat.

The authors encourage the development of a standard document that would aid the industry in developing better quality diagrams without structural errors. Software can be used to increase or decrease the amount of detail shown to aid communication.

5. References

[1] Marsh (2011) 100 Largest Losses (1972–2011), Large Property Damage Losses In The Hydrocarbon Industry (22nd Ed), published by Marsh and McLennan Group of Companies.

[2] Zuijderduijn C., RISK MANAGEMENT BY SHELL REFINERY/CHEMICALS AT PERNIS, THE NETHERLANDS Implementation of SEVESO-II based on build up experiences, using a Hazards &Effects Management Process., EU Safety Conference Athens (2000)

[3] Salvi O., Debray B. (2006), A global view on ARAMIS, a risk assessment methodology for industries in the framework of the SEVESO II Directive, Journal of Hazardous Materials 130, 187–199

[4] BORA Handbook (2007) Barrier and Operational Risk Assessment, Rev 0, Norway Preventer Report for NFR / HSE / OLF.

[5] NASA (2002) Fault Tree Handbook with Aerospace Applications (Ver 1.1) Lead author William Vesely, available from www.nasa.gov

[6] Sklet S (2006) Safety barriers: Definition, classification, and performance. Journal of Loss Prevention in the Process Industries 19, 494–506.

[7] Pitblado R.M. and Fisher M., (2010), Novel Investigation Approach Linking Management System and Barrier Failure, European Refining Technical Conference (ERTC), Barcelona.

Figure 7. Comparison of incorrect and corrected barrier layouts

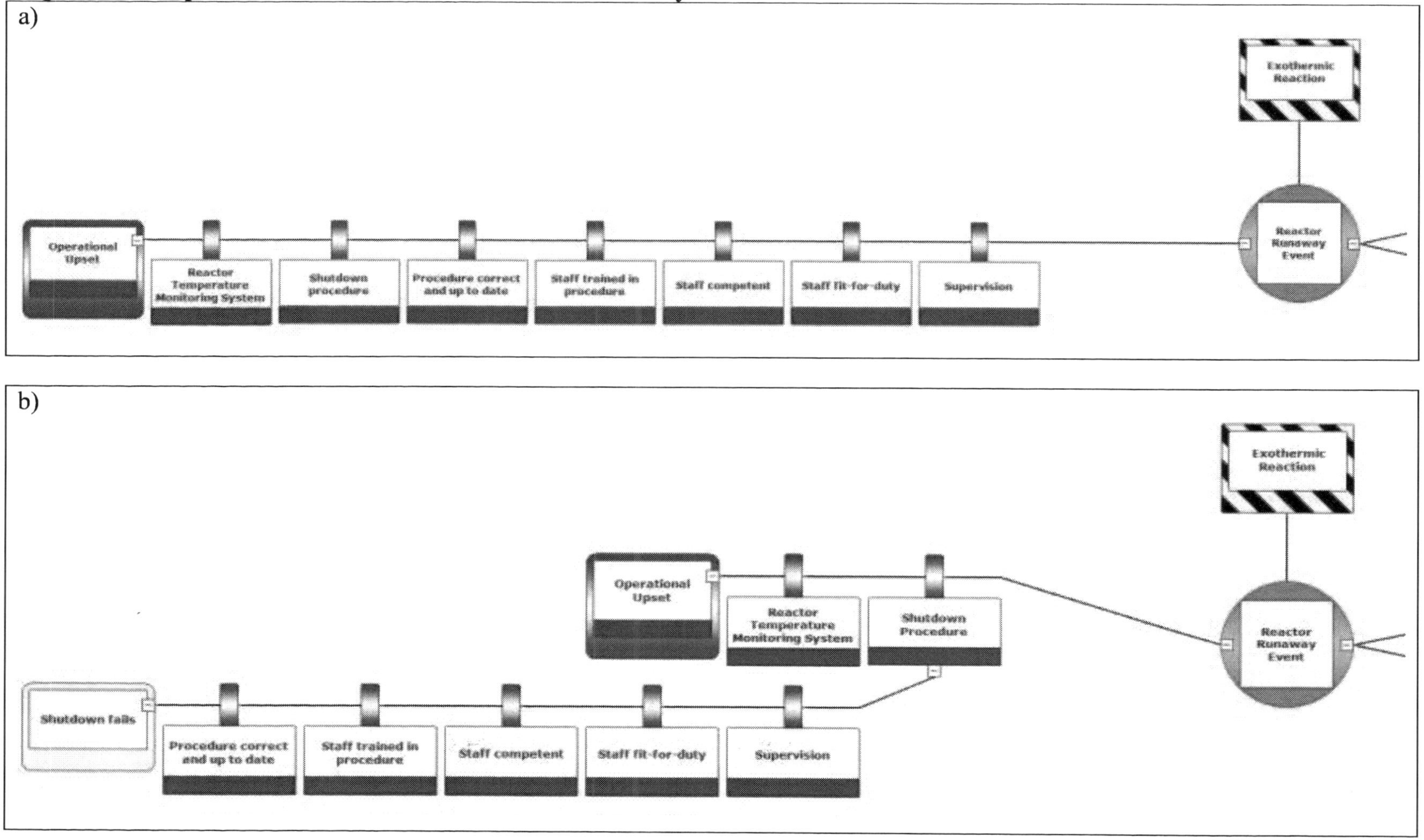

(43a) Learning From High Potential Near Misses in the Refining Sector

Ramesh VM, Research Scholar, Bharathiar University, 641046, India and Dr. Sakthivel R, Management Studies, Govt Arts College, 641018

Abstract: Safety is the first priority for any industry. Though steps are taken to ensure an accident-free environment, "Near Miss" is one phenomenon whose possibility of occurrence cannot be ruled out completely. What matters here is the way in which such instances are tackled preventing accidents. Moreover, these incidents can be used as an effective tool in education and training. Each near miss thus offers a free learning tool and a whole new experience with a scope to preach and practice efficient safety mechanisms. Near misses can be classified into two types- Low potential near miss and High potential near miss. All near misses whether it is of Low or high potential require in-depth analysis leading to identification of the root cause so as to avoid the recurrence of such incidents in future. The outcome of every near miss is to reach down the line so as to spread the useful information to all working in the industry. Many workers do not report "near misses" due to fear of negative actions from the management that can deprive them of their future progressions in the Industry. Steps need to be taken to overcome this fear as capturing of all near misses will definitely help the industry in bringing out innovative ideas to ensure safety. The exercise of accident prevention will bear fruits only if the effort to spread the message of usefulness of near miss reporting is spread down the line. Some of the near misses that occurred in the refinery sector are identified to be of High potential and major mishaps were averted only because of the timely action in the second of need. The ability of the concerned to do the right job at the right hour can be attributed to the training given to every employee on fire fighting which includes the analysis of near miss as well as the analysis of various other accidents happened in the other industries.I wish to share few high potential near misses which can be a learning to all working the refinery sector.

Felt Leadership:
An Essential Ingredient in Successful PSM Implementation

Brian D. Rains
DuPont Sustainable Solutions
974 Centre Road; CRP-730
Brian.D.Rains@DuPont.com

Prepared for Presentation at
American Institute of Chemical Engineers
2013 Spring Meeting
9th Global Congress on Process Safety
San Antonio, Texas
April 28 – May 1, 2013

UNPUBLISHED

Felt Leadership:
An Essential Ingredient in Successful PSM Implementation

Brian D. Rains
DuPont Sustainable Solutions
974 Centre Road; CRP-730
Brian.D.Rains@DuPont.com

Keywords: Felt Leadership, PSM, High Performing Work Organizations

Abstract

"Felt Leadership" is a term that is frequently mentioned in today's PSM conferences and informal conversations as a quality that must be demonstrated by management if PSM implementation is going to be successful. Interestingly, these two words were first introduced by the DuPont Company over 20 years ago and the context was very different then and not even directly related to safety. Felt Leadership was one of the six attributes of a "High Performing Work System." As one of these 6 attributes, it was deemed essential to achieve outstanding operations and business results through people. This paper will not only describe "felt leadership" in meaningful, practical terms but will also deliver insights as to why it is both appropriate and necessary for effective PSM implementation as well as high performing work organizations.

1. Introduction

One of the clear lessons learned from the Baker Panel Report of the BP-Texas City incident was the role of leadership in successfully implementing and sustaining excellent PSM performance. Of the Panel's ten recommendations, five spoke directly to changes that leadership needed to make. One of the panel members, Dr. Paul V. Tebo, former DuPont Vice President of Safety, Health, and Environment, has actually stated that in his opinion, all ten recommendations speak to leadership. He went on to say that the fundamental, underlying issue at BP-Texas City was "leadership, leadership, leadership."

- **RECOMMENDATION # 1 – PROCESS SAFETY LEADERSHIP ****
- **RECOMMENDATION #2 – INTEGRATED AND COMPREHENSIVE PROCESS SAFETY MANAGEMENT SYSTEM**
- **RECOMMENDATION #3 – PROCESS SAFETY KNOWLEDGE AND EXPERTISE**
- **RECOMMENDATION #4 – PROCESS SAFETY CULTURE ****

- **RECOMMENDATION #5 – CLEARLY DEFINED EXPECTATIONS AND ACCOUNTABILITY FOR PROCESS SAFETY ****

- **RECOMMENDATION #6 – SUPPORT FOR LINE MANAGEMENT ****

- **RECOMMENDATION #7 – LEADING AND LAGGING PERFORMANCE**
- **NDICATORS FOR PROCESS SAFETY**

- **RECOMMENDATION #8 – PROCESS SAFETY AUDITING**

- **RECOMMENDATION #9 – BOARD MONITORING ****

- **RECOMMENDATION #10 – INDUSTRY LEADER**

**** Direct Leadership Responsibility**

As helpful and appropriate as these corporate recommendations were at the time and continue to be, the subject of PSM leadership can, at times, need to be brought down to a very individual and even personal level. In recent presentations at PSM Conferences, the term "felt leadership" has been used to attempt to describe a brand or style of leadership that is also necessary in complex chemical processes and operations. It has been used to describe a quality of leadership that can be sensed or "felt" rather than just heard or seen. It suggests leadership with passion and authenticity and that is genuine and even humble.

2. Origin of Felt Leadership in DuPont

The term "felt leadership" was first used in the DuPont Company in a systemic way in 1991. And the context was very different. In fact, it was not used in connection with safety in any direct way.

A group of Plant Managers, of which I was one, had been formed into a body called the "Plant Managers Board." We had a very specific task and that was to find and identify best practices related to "High Performing Work Systems" or HPWS for short. High Performing Work Systems were described then as "high involvement systems of accomplishing work in which all employees have developed the capability to connect with and drive the quality of business results to be world class."

This board was made up of approximately 10 Plant Managers from all regions of the world. Some were quite young whereas others were more experienced. Some managed very large plants whereas others managed small plants. (I was personally leading a site organization of 20 DuPont employees.) Some spoke English as a second (or third) language whereas others only spoke English. Some were engineers whereas others had educational backgrounds spanning the disciplines. We came together face-to-face at least quarterly for roughly two years. And we almost always visited different operating sites to observe and analyze. Sometimes the facilities were DuPont's whereas in other cases, we visited such diverse facilities as the Honda car assembly plant in Marysville, Ohio, USA.

After a visit to a DuPont facility in rural Illinois, this group met together in a hotel conference room and attempted to synthesize our collective observations into something usable and concise; something that we could attempt to package and leverage across DuPont Operations. We filled tear sheet after tear sheet of paper until we literally covered the four walls in the room. We had seen and observed much, even too much as there was so much data.

Then, the key moment arrived. One of our members, who later rose to become DuPont's Sr. Vice President of Operations (and interestingly spoke English as a second language) took the floor and articulated in simple terms what became to be known as the "Six Attributes of HPWS." The first attribute was "Vision, Mission and Strategic Intent Known by All." The second was "Committed and Dedicated Leadership 'Felt' within the Organization", or simply "Felt Leadership."

Six Attributes of High Performing Work Systems in DuPont

1. Vision, mission, strategic intent known by all.
2. Committed and dedicated leadership "felt" within the organization.
3. Every person knows the paying customer and how their work impacts the customer.
4. Teamwork and people working in teams is part of the structure.
5. Everyone learns, and teaches. "Learn-Teach-Learn".
6. Principle driven.

Phrases that were used at the time to further clarify the meaning of Felt Leadership included:

- Creates vision and energy
- Stretches goals
- Invites positive push back
- Results and output oriented
- "Felt" sense of purpose

Further insight into what was the intended meaning of Felt Leadership was captured in a "Six Attributes of HPWS Self-Assessment Tool." The questions asked were:

- Do the leaders have a clear vision for change based on needs of all stakeholders?
- Are the leaders visible, knowledgeable and open with people?
- Do the leaders value upgrades from anywhere?
- Do the leaders continually set objectives that demand that people develop and grow skills needed to move toward the vision?
- Do the leaders engage people in a process that builds accountability, willingness, and confidence to act in an empowered way?

3. Felt Leadership and PSM

With all this as a backdrop, what really does "Felt Leadership" have to do with Process Safety Management?

A more descriptive term for what process safety is all about might be "Operational Risk Management." Operational Risk Management implies that:

- It is the operation or the materials processing activities that must be controlled.
- It is the reduction of risk to an acceptable level with the ultimate elimination of significant process incidents that is the goal.
- It is the management of these risks that is one of the organization's key responsibilities and challenges.

The degree of success that is achieved in the management of operational risks can be equated with the quality of decision making by and within the operations organization. And leadership is all about influencing and improving the quality of decision making. Thus, Felt Leadership can best be described in the context of PSM as the approach that leaders follow to insure the highest quality of decision making by the organization in the management of the existing operational risks. Decision making links Felt Leadership to PSM Performance. Felt Leadership becomes a necessary ingredient in successful PSM implementation and execution.

4. Demonstrating Felt Leadership

Within the organization of one of our global clients, inspired leaders captured the vision of their roles with respect to PSM and embraced the concept of providing Felt Leadership. But they did not know how to act differently or more effectively in the management of their significant operational risks. They understood the concept of Felt Leadership as it related to PSM but with no role models, they struggled to visualize or project themselves effectively into their/environment or culture in a uniquely additive way. They needed help and they asked for help. They wondered what supportive but probing questions they should even ask to strengthen a culture of PSM Excellence through Felt Leadership. We provided the following as initial guidance:

1. While visiting a control room and speaking to panel board operator, the focus of the discussion could be on safe operating limits for the process:
    - Does he/she know the safe operating limits, and the consequences of operation outside those limits?
    - Is he/she currently operating with any interlocks by-passed? If so, was the by-pass procedure appropriately followed?
    - How long has this situation (interlock bypassed) existed and how long is the by-pass condition expected to last?
    - How many panel alarms are there per shift? Is there any program to reduce the number of panel alarms?

- Is he/she aware of any work going on in the plant, so that he can alert people to any process emergency?

2. While visiting a plant, question the operations supervisor to see what work permits have been prepared for the day. Take time and review them, looking for the quality of the risk assessment and the precautions that have been put in place to mitigate the identified risks.

3. While visiting a plant, a review of a broad range of PSM elements can occur, such as asking operations personnel about:
   - Any incidents that have occurred in other similar operations.
   - The frequency and nature of emergency and disaster mock drills.
   - The number and nature of near misses that have occurred recently and the results of investigations into these and any other incidents that have occurred in the plant.
   - The results of recent PSM audits and what weaknesses were identified.
   - Whether there are any tests or process changes active in the plant.

4. In conversation with operators in the field, discuss the procedure governing the activity underway and how well it is written and followed.

The coaching of these leaders was actually very similar to what had occurred previously inside DuPont as we strived to strengthen our PSM Performance and the quality of Felt Leadership. Simple "Felt Leadership Cards" were printed to help Operations Leaders consider how best to demonstrate their commitment and passion for PSM.

These cards are still in use today and one is duplicated on the next two pages.

Personal "Felt Leadership" – PSM

What is my spoken and unspoken message on PSM expectations to the organization – integral to core values and business success or a "necessary evil"?

What have I done this month to review PSM metrics and performance indicators for the site?

What am I doing to reinforce Operational Discipline?

Is my organization providing adequate "rewards and recognition" for PSM accomplishments?

How is my organization performing versus annual PSM COT's and objectives?

Where do we need to enhance short and long term PSM resources and organizational capabilities?

What is the status of PSM integration improvements at recently acquired sites or businesses?

What are my direct reports doing to reinforce the value, expectations and accountabilities for PSM on an ongoing and consistent basis?

Are we providing enough support and resources to maintain site equipment and infrastructure?

What have I done personally to engage new leaders in my organization that plays key PSM roles to ensure effective MOC- Personnel?

Have I established clear measures on overall PSM program performance (not just a focus on incidents)?

The Goal is Zero

Ten Key PSM Questions for the Line Organization

1) What trends are your PSM tracking systems showing – do you have any overdue PSM action items within the work group?

2) What recent improvements or accomplishments in PSM does your organization feel good about?

3) What is your biggest concern about the area's PSM program right now and what are you doing about it?

4) What additional focus or support do you need, if any, to improve and maintain PSM?

5) What are your first party PSM auditing and incident trend analysis showing across operations?

6) What are you doing to improve your people's knowledge and capability in PSM to eliminate hazards and incidents?

7) How are you driving a consistent "no leak" and "no alarm" operating philosophy in the operating area?

8) What are you personally doing to demonstrate "felt leadership" in PSM on a routine basis and is it effective?

9) How do you ensure all your people training and written operating procedures up to date and complete?

10) Can you show the status of mechanical integrity tests and PSM work orders in the tracking system and are they up to date?

Other words have been used to describe how Felt Leadership applies to PSM Performance. The following ten basic characteristics have been drafted to summarize this thinking.

Ten Characteristics of PSM Felt Leadership

1. Be visible to the organization.
2. Be relentless about time with your people.
3. Recognize your role as a teacher/trainer.
4. Develop your own safety functioning skills and pass them along to the organization.
5. Behave and lead as you desire others to do.
6. Maintain a self-safety focus.
7. Confirm and reconfirm safety as the Number 1 VALUE.
8. Continuous emphasis and clarity around SHE expectations.
9. Show a passion for ZERO injuries, illnesses, incidents.
10. Celebrate and recognize "ZERO" successes.

Some additional guidance for each of these ten characteristics is included in the appendix.

Additional specific actions that management can take to demonstrate Felt Leadership and their commitment to safety include:

- Leading by example in following all safety rules and procedures;
- Showing a passion for safety & a genuine care for individuals;
- Leading safety meetings with direct reports;
- Including safety in all routine meetings and discussions;
- Participating in safety observations regularly;
- Reviewing/critiquing all incident investigations and personally getting involved in high potential and serious investigations;
- Meeting with your organization to express concern and reinforce the learnings from incident investigations; and
- Using company events to recognize/celebrate good safety performance by individuals and groups.

Other attributes of Felt Leadership can be demonstrated by always following this guidance:

- I always practice what I preach with reference to safety.
- I know all potential hazards in my area.
- I audit for hazards.
- I do not permit unsafe shortcuts in order to get a job completed.
- I check on the use of PPE.
- I correct unsafe practices without delay.
- I make sure workers have proper tools.
- I make sure tools are in good condition.

- I encourage reports on unsafe conditions.
- I promptly respond to workers' safety concerns.
- I investigate incidents promptly and fully.
- I encourage safe suggestions.
- I encourage housekeeping for safety.
- I consider my own health and safety.
- I regularly attend safety meetings.
- I set high standards and try to be the example for everyone else.
- I ensure that all workers are properly trained in safety.
- I stop work if necessary to correct unsafe practices.
- I remove unsafe workers from the job.
- I encourage safety communications and follow-up on corrective actions.

5. Conclusion

The successful implementation of robust Process Safety Managing Systems in complex chemical operations requires a high degree of order and high quality decision making. Leadership is fundamentally charged with creating a culture where all members of the organization respond to these needs and perform at a very high level. Although the term Felt Leadership was initially used in DuPont in the context of an essential ingredient of a High Performing Work System, its use as a descriptor of the necessary leadership qualities in Process Safety Management is both appropriate and accurate.

Felt Leadership is essential to the successful implementation and execution of Process Safety Management.

6. Appendix

Ten Characteristics of PSM Felt Leadership

1. **Be visible to the organization:**
   - "Walk the Talk".
   - Provide a positive example to your employees.

2. **Be relentless about time with your people:**
   - Find ways to spend time with employees/contractors.
   - This will encourage you people to do the same.

3. **Recognize your role as a teacher/trainer:**
   - Teach safety through your:
     - Own behavior
     - Interactions with your employees/contractors
   - You are always 'on stage'.

4. **Develop your own safety functioning skills and pass them along to the organization:**
   - Continuously develop and hone skills such as safety observations, incident investigations, improving leadership skills (Felt Leadership), etc.
   - Make sure your people have the time and resources to develop their safety skills.
   - Reinforce and support what they learn when they return from training.

5. **Behave and lead as you desire others to do:**
   - What people see you do is what they believe.
   - When you notice an unsafe situation and do something about it, you are behaving as you want all of your employees to behave.

6. **Maintain a self-safety focus:**
   - Assess yourself in terms of how 'good' or 'bad' a job you are doing.
   - Stop and think about – 'How did that go?'; 'What might I have done differently to get an even better result?'

7. **Confirm and reconfirm safety as the Number 1 VALUE:**
   - Make comments about safety as you walk throughout the work area.
   - Identify and reinforce what is being done right.
   - Identify unsafe conditions and behaviors and correct them before they cause injuries.
   - Learn what safety issues exist on the shop floor.
   - Learn how the issues are being handled.

- Determine how well employees understand and apply the core principles in their work.

8. **Continuous emphasis and clarity around SHE expectations:**
    - Establish and publicize SHE expectations.
    - Include expected safety behavior.
    - Continuously emphasize their importance by:
        - Including them in communications.
        - Weaving them into most, if not all, conversations you have with your organization.

9. **Show a passion for ZERO injuries, illnesses, incidents:**
    - Make clear your expectation that 'zero' is the goal.
    - Put safety upfront on all interactions.
    - Ensure that your safety performance expectations of your contractors is the same as your own employees.
    - Require that everyone work on a steadily improving path in our journey to ZERO.
    - Periodically focus special attention on an aspect of needed safety improvement; be specific.

10. **Celebrate and recognize "ZERO" successes:**
    - Live safety, not just preach it.
    - Recognize and reward improvement and success.
    - Recognize the achievement at all levels and make sure leadership recognizes and shows appreciation.
    - Recognize specific aspects of safety that have improved as a result of targeted efforts.

9TH GLOBAL CONGRESS ON PROCESS SAFETY

Localization of control rooms using a MINLP approach, based on accidental explosion scenarios

Jaime E. Cadena
Chemical Engineering Department, Universidad de los Andes
Cra 1 Este No 19A – 40, Bogotá, Colombia
je.cadena21@uniandes.edu.co

Lina M. Sierra, Sergio E. Rodríguez, Manuel A. Ramos, Jorge M. Gómez, Felipe Muñoz
Chemical Engineering Department, Universidad de los Andes
Cra 1 Este No 19A – 40, Bogotá, Colombia
lm.sierra443@uniandes.edu.co, se.rodriguez68@uniandes.edu.co, ma.ramos43@uniandes.edu.co, jorgomez@uniandes.edu.co, fmunoz@uniandes.edu.co

Prepared for Presentation at
American Institute of Chemical Engineers
2013 Spring Meeting
9th Global Congress on Process Safety
San Antonio, Texas
April 28 – May 1, 2013

UNPUBLISHED

Localization of control rooms using a MINLP approach, based on accidental explosion scenarios

Jaime E. Cadena
Chemical Engineering Department, Universidad de los Andes
Cra 1 Este No 19A – 40, Bogotá, Colombia
je.cadena21@uniandes.edu.co

Lina M. Sierra, Sergio E. Rodríguez, Manuel A. Ramos, Jorge M. Gómez, Felipe Muñoz
Chemical Engineering Department, Universidad de los Andes

Keywords: facility layout, optimization problem, uncertainty reduction, Monte Carlo simulation, UVCE modeling

Abstract

Facility layout for the process industry plants is an important element in the basic engineering stage, which is fundamental in the design of the process, influencing significantly the future safety performance of the operation and its related costs. A control room serves as an operations centre where the facility can be monitored and controlled, therefore, its location and design is a priority for the safety of the process unit operation. Given that the control room is the 'core' of control and monitoring of the process, any significant damage to it resulting from a process incident might jeopardize the whole process and generate subsequent incidents. This work presents a novel methodology intended to serve as an additional tool for aiding the decision-making process regarding the location of control rooms, taking into account the risk of unconfined vapor cloud explosion (UVCE) and several passive barriers against it. The methodology is based on a mathematical model that bonds the design and localization of a control room into a computational tool, which has been developed taking into account key characteristics of a process or plant layout for explosion scenarios. The tool combines stochastic estimation of atmospheric parameters through raw Monte Carlo simulations with a mixed integer nonlinear programing (MINLP) optimization to determine the control room location. This process also helps to choose the best construction strategy for the control room among a portfolio of strategies fed to the model by the user. The final results of the methodology are the optimal location for a control room in an existing facility and the minimum costs related to the risk level to which it is exposed.

1. Introduction

A process incident such as a fire, explosion or dispersion can cause injuries and/or fatalities, but also a disruption of operations and undesired consequences, business-continuity wise. This makes determining an optimal facility layout and the location of the control room, one of the priorities for safe operations within a production plant. Given that facility layout is one of the elements determined in the early plant design stages, it also represents one of the first safety barriers in a hazard-incident representation such the one presented in Figure 1.

Figure 1: Hazard-Incident representation and location of facility layout as a safety barrier (FLB) and additional safety barriers (SB) [1].

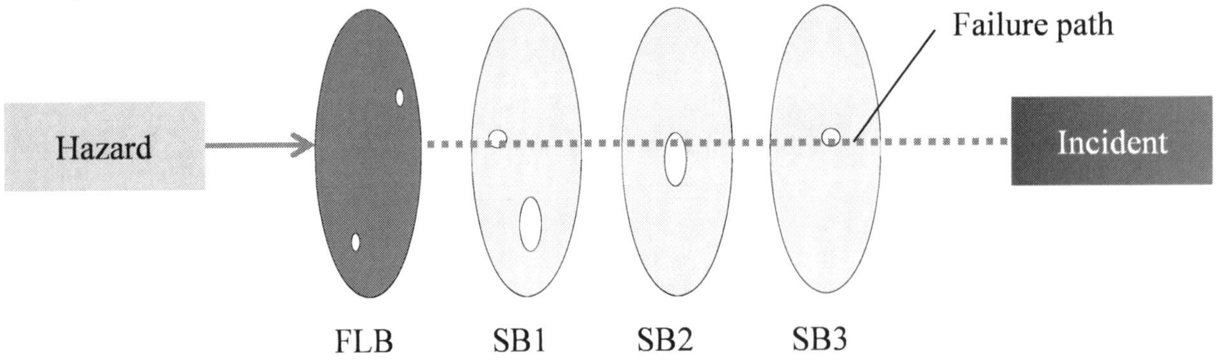

Figure 2 represents the desired function of facility layout as a safety barrier, including the adequate location of the control room. However, this desired representation is not always taken into account by the design philosophy implemented, which leads to process incidents with low probability of occurrence and high consequences, such as Flixborough (1974), Texas City (2005), Buncefield (2005), T2 Labs (2007), Puerto Rico (2009) and Amuay (2012) [2]. Although these incidents involve facility layout issues, Flixborough and Texas City involve serious damage to control rooms and its consequences has made them examples of scenarios and failures that should always be taken into account when determining the final facility layout.

Traditionally the facility layout problem has been approached with heuristic and technical recommendation published by several organizations such as the Center for Chemical Process Safety (CCPS) [3]. However, the integration of such approaches with current edge-cutting engineering tools such as non-linear optimization algorithms and solving platforms allow presenting novel approaches according to characteristics of the problem. In general, three types of problem can be identified: formulation, space treatment, unknown parameters estimation. For each type of problem a different approach is needed and this is summarized in Table 1.

Table 1. Approaches used depending on type of problem

Type of problem	Solution approach	Advantages	Disadvantages
Formulation	Linear [4]	The solution of the problem is usually simple and assures a global optimum under linearity assumptions	Does not allow the modeling of non-linear physic-chemical behavior and/or requires simplifications and strong assumptions
	Non-linear [3][5-7]	Applicable approach to complex systems with a high degree of non-linearity	Obtaining a solution requires more iterations than with a linear approach
Space treatment	Continuous [3][5-7]	Allows determining the exact location of buildings and equipment. Use of different shapes to represent footprints.	Complicated mathematical formulation. Obtaining a solution requires high computational times.
	Discrete [4]	Simple formulation and the solution	Exact locations are not possible. Shapes for

Type of problem	Solution approach	Advantages	Disadvantages
		can be obtained after a few iterations.	footprints are limited to squares o rectangles.
Unknown parameters estimation	Stochastic [3][5-7]	Better representation of real behavior.	Complex procedure that demands large amounts of input data.
	Deterministic [6]	Simple procedure based on worst-case scenario approach.	Derived distribution show few applications.

This work defines describes a methodology which identifies the sets of parameters, variables, functions and restrictions that represent the control room's location problem. Furthermore, the methodology implements a strategy that allows reducing the atmospheric conditions uncertainty when modeling UVCE and dispersion scenarios. The output of the modeling (given by overpressure) allows determining risk values, with consequences on people and infrastructure. Finally a case study is presented and the results analyzed.

2. Methodology

Although no exact and unique solution exists for a highly complex engineering problem such as facility layout, the proposed methodology (Figure 2) intends to support the decision making process in which final layouts are determined and the protection strategies are selected. The methodology is configured as a non-linear optimization of discrete control room's location, having an underlying process of stochastic parameters (atmospheric, risk) estimation. This configuration of the problem requires the following assumptions:

- A square process plant area is selected
- The process plant has only one hazardous process unit, hence only one hazards source
- The process plant area is divided into a uniform computational grid
- A control room can occupy one and only one node of the computational grid
- Historic atmospheric data must be available

The methodology begins by characterizing the hazardous process unit in terms of the possible Loss of Primary Containment (LOPC) scenarios, allowing an estimation of the liberation quantities of hazardous materials (HazMat). From the LOPC, failure trees can be depicted, allowing a classical calculation of the UVCE occurrence probability. Here the liberation of HazMat stops driving the analysis and focus is shifted towards the next driven force: atmospheric conditions. In order to take into account real atmospheric data of the determined location of the plant, it is necessary to acquire the historical atmospheric data, process it, organize it and finally fit it to probability distributions. These distributions are the input for the next step, the Monte Carlo simulation. Since the final aim is to obtain the resulting consequences of the UVCE, both dispersion and an explosion models are implemented within the Monte Carlo simulation that has a fixed number of iterations, *N*. The output of the Monte Carlo simulation is a probability distribution of overpressure-impulse measurements for each node of the grid. This constitutes a risk map, which combined with the calculation of the damages that a determined level of overpressure can cause to humans or infrastructure (i.e. control room) at a specific node, leads to the consequences calculation. The previous steps and costs information on materials, construction and installation of the control room, feed the non-linear optimization algorithm with the required data to run it and obtain an optimum control room's location.

Figure 2: Detailed steps of the proposed methodology

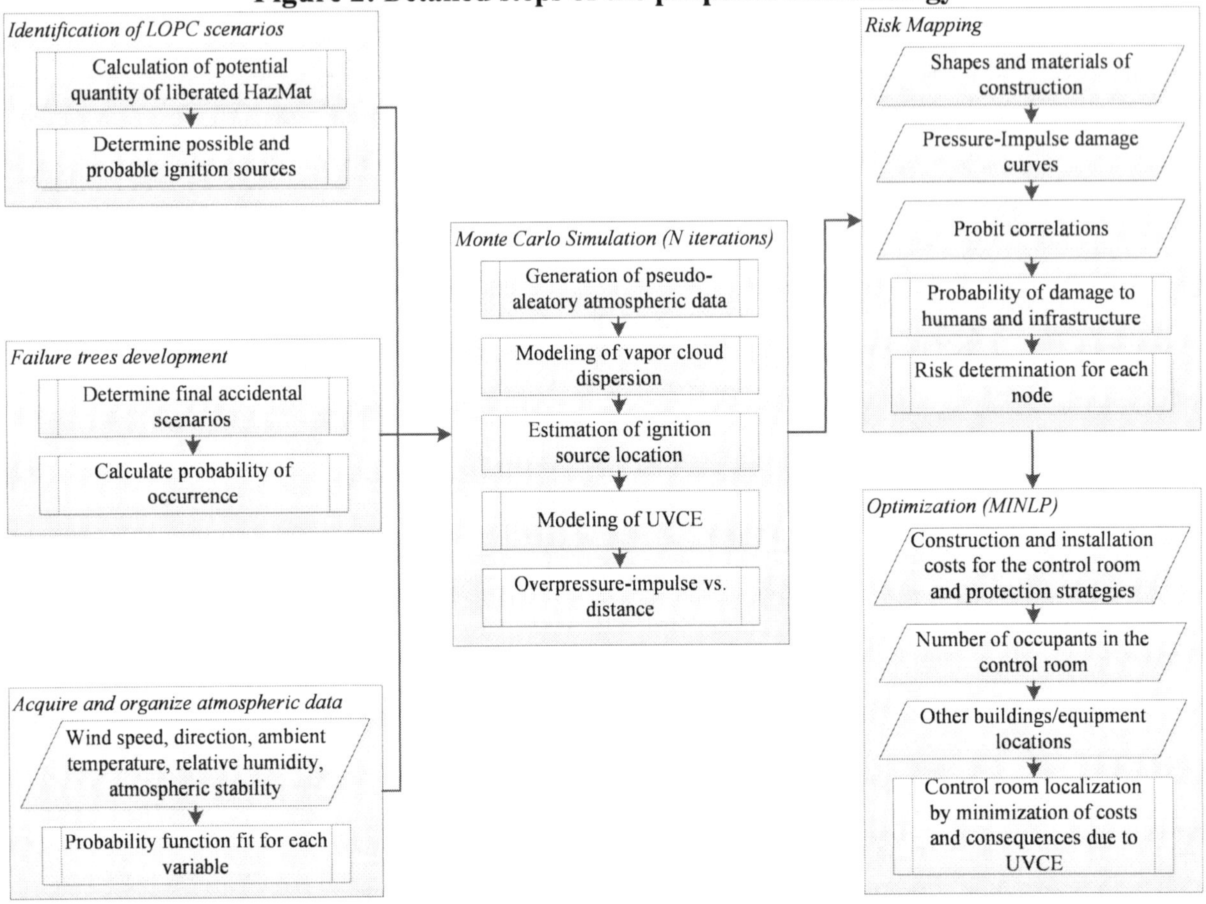

Following the general description of the methodology, it is implemented on a case study and the results are analyzed to establish advantages, disadvantages and improvement opportunities to be implemented in future work.

3. Case study: Hexane-Heptane separation unit

In order to implement the proposed methodology, a facility at Corpus Christi, TX, USA is selected as case study [13]. The hazardous process unit consists of a distillation unit and the following assumptions apply to the subsequent analysis of the system:

- The HazMat consists of 28 tons of flammable material
- The operating temperature of the distillation tower is 130°C and 160°C for top and bottoms, respectively
- The operating pressure is 4 barg
- The vapor transport pipeline has a 0.5 internal diameter and a 25 m length
- The liquid transport pipeline has a 0.15 internal diameter and a 25 m length
- The process plant area is 100 m in length and 100 m wide

- The grid consists of uniform nodes of 10 m by 10 m (see Figure 3)

The previous information and assumptions allow implementing the methodology in 7 sequential steps. The following is a summary of the most important developments in each step.

Figure 3: Representation of the grid and the nodes (labeled) in the process plant area

k01	k02	k03	k04	k05	k06	k07	k08	k09	k10
k11	k12	k13	k14	k15	k16	k17	k18	k19	k20
k21	k22	k23	k24	k25	k26	k27	k28	k29	k30
k31	k32	k33	k34	k35	k36	k37	k38	k39	k40
k41	k42	k43	k44	Hazardous process unit		k47	k48	k49	k50
k51	k52	k53	k54			k57	k58	k59	k60
k61	k62	k63	k64	k65	k66	k67	k68	k69	k70
k71	k72	k73	k74	k75	k76	k77	k78	k79	k80
k81	k82	k83	k84	k85	k86	k87	k88	k89	k90
k91	k92	k93	k94	k95	k96	k97	k98	k99	k100

3.1 Hazards identification

This step aims at identifying a representative group of scenarios that can result in a UVCE accidental event due to the LOPC of the involved HazMat. The found applicable scenarios are catastrophic failure of a vessel within the unit (column, condenser, reboiler), catastrophic failure of the vapor transport pipeline and catastrophic failure of the liquid transport pipeline.

The use of proper source models allow calculating that for both pipeline failures, the total HazMat available would be released in less than 120 seconds in vapor state at averaged atmospheric conditions. Therefore, it is assumed that any of the three mentioned scenarios has a leak/liberation time of 120 seconds. After the characterization of the LOPC, a event tree is developed to obtain the frequency of occurrence of the possible accidental events, using annual occurrence frequencies and event probabilities within the tree, obtaining the results presented in Table 2 [13].

Table 2. Annual occurrence frequencies for the identified accidental events [11]

Scenario	Probability	Frequency (year^{-1})
UVCE	0.34	6.5 x 10^{-6}
Flash Fire	0.34	5.1 x 10^{-6}
BLEVE	0.25	5.1 x 10^{-6}

3.2 Atmospheric data

A 10 year period (2002-2012) of atmospheric data with entries by the hour for the Corpus Christi area was obtained from the National Climatic Data Center (NCDC-NOAA) [19]. Wind speed & direction, ambient temperature, cloud coverage and relative humidity were extracted from the whole data set. This filtered data set allowed determining the Pasquill stability class each entry [8], using the relationship between solar elevation angle (determined an NCDC tool [17]) and cloud coverage as shown in Table 3. The final results of this step allow obtaining 109000 data registries, input for the next step.

Table 3. Relationship between solar elevation angle and cloud coverage

Cloud coverage	**Solar elevation angle α**		
	60°< α	**35°< α<60°**	**15°< α<35°**
Less than 4/8	Strong	Moderate	Minor
Medium altitude clouds	Moderate	Minor	Minor
Low altitude clouds	Minor	Minor	Minor

3.3 Dispersion and explosion models

The SLAB model negative floatability gases, is selected to model the dispersions due to its fit to the case study conditions. Such conditions involve a HazMat (hexane-heptane mixture) with a density higher than that of the air by a 2.5 factor. This model calculates the atmospheric dispersion solving a variant for mass, momentum, energy and species spatial and temporal averages [22]. Additionally, the SLAB model is considered a 'top hat' model, presenting the great advantage of its repetitive use within the Monte Carlo simulation. Once the model is applied, a concentration-distance profile is generated for each data set being evaluated and the ignition point is located as shown in Figure 4. An analysis of the model and the results obtained show that the accuracy of the SLAB model in the near-field is poor [22].
The TNT model is selected for the UVCE intensities calculation (overpressure and impulse) for each node of the computational grid imposed on the process plant area. Even though this model can be considered as the simplest one, its easy implementation in line with the Monte Carlo simulation and its well-known limitations [23] make it an attractive option. This model relies on an explosion efficiency factor (η) that is usually assumed to be between 1% and 10%, however a fixed 1% value is assumed since higher values presented unrealistic results in the near field when compared to typical UVCE overpressure values [4][8][25] .

Figure 4: Assumptions regarding ignition source location

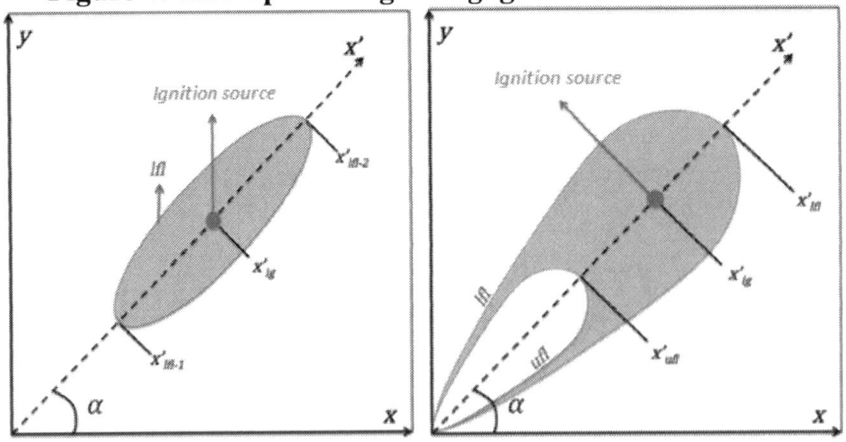

Figure 5: Example of a concentration-distance profile with its ignition source located

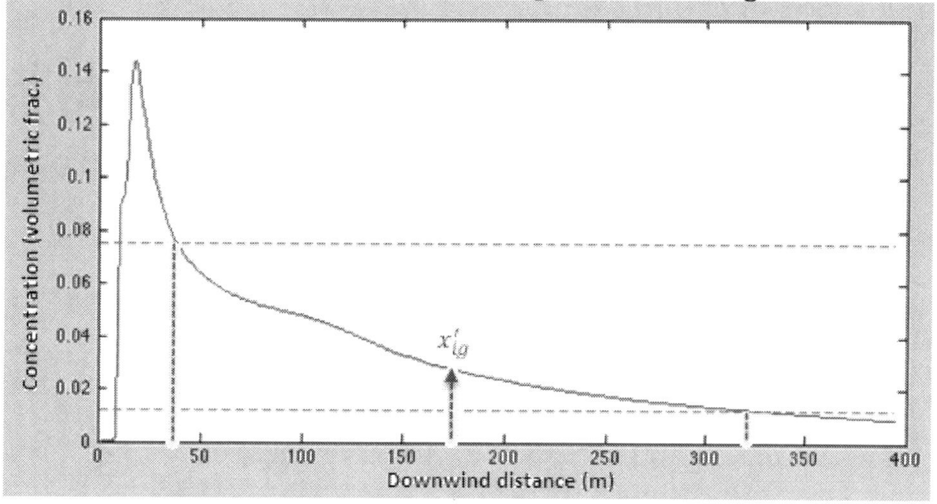

3.4 Monte Carlo simulation and risk mapping

Steps 3.1 and 3.2 are needed inputs for the Monte Carlo simulation and its output is given by the probability distributions of the fitted results of the *N* iterations performed over the dispersion and explosion models. For this case study the number of iterations was taken as 1000 as an initial guess, later proving to be a reasonable number of iterations when comparing computational time against the results' mean behavior. The results (profiles of concentration and overpressure against distance) are organized in form of histograms. From these results the expected values of overpressure (Figure 5) and impulse can be calculated for each node, necessary input to develop the risk maps. It must be noted that the expected value of the wind direction coincides with the non-symmetrical trend in the intensities maps such as Figure 5.

Risk mapping consist on assigning risk values or scores to each node of the grid taking into account structural damage to infrastructures and human affectation. The risk values for infrastructure damage depend on the type of protection strategy selected for the control room, having four different strategies as follows:

1. Steel reinforcement, walls (panels), roof and frame
2. No reinforcement, concrete reinforced walls, concrete roof and steel frame
3. Prefabricated concrete coating, concrete reinforced roof and walls, and steel frame
4. Reinforced masonry walls, light steel roof and no frame

For each construction configuration for the control room, there is an associated risk map calculated from Pressure-Impulse curves (PI) and Probit approach for human affectations. Once this risk maps are established assuming a Risk = Probability x Consequence relationship, the only step left previous to the optimization is the determination of additional parameters such as related costs of installation, construction and the location of buildings and equipment in the process plant area[4][5][12][18].

Figure 5: Overpressure map on the process plant area

35.01	47.41	29.74	29.24	28.53	26.52	24.48	22.61	21.11	20.09
30.49	30.84	30.17	31.36	32.44	31.44	28.44	25.06	22.39	20.70
28.73	29.89	33.48	40.49	48.05	47.33	38.31	29.73	24.43	21.57
28.45	31.39	40.55	64.92	109.75	109.16	63.14	37.52	26.97	22.47
28.60	32.77	48.20	109.78	Hazardous process unit		108.27	45.59	28.86	22.93
28.77	32.17	47.52	109.14			107.84	45.21	28.51	22.59
28.02	29.42	38.43	62.97	108.03	107.67	61.84	36.37	25.94	21.48
27.98	26.05	29.62	37.07	45.07	44.77	36.12	27.84	22.77	20.04
25.63	22.86	23.86	26.08	27.93	27.70	25.36	22.47	20.20	18.80
21.74	20.23	20.29	20.92	21.42	21.29	20.50	19.47	18.66	18.41

3.5 Optimization of the control room's location

The output of the 3.4 step along with some additional parameters, represent the needed input for to carry out the control room's location optimization. This step is performed through a MINLP formulation [24] of the optimization problem and a Simple Branch and Bound (SBB) solving algorithm, implemented on the GAMS software platform [25]. The optimization problem is structured as follows:

$$\begin{aligned} \min: \quad & z = f(x,y) \\ \text{s.t.} \quad & gj(x,y) \leq 0 \qquad j \in J \\ & x \in X \subseteq R^n \\ & y \in \{0,1\}^m \end{aligned}$$

The formulation can be divided into: parameters, sets, variables, constraints and objective function. A brief description of each one is presented as follows:

1. Parameters: the required parameters to solve the optimization problem are comprised of the protection strategies, PI curves, Probit parameters for human affectation, human occupancy, risk maps, construction/interconnection/installation costs, minimum separation distances, existing buildings/equipment & related location and the coordinates of each node within the computational grid (x_k, y_k).

2. Sets: the problem is divided in three sets. The first one (*F*) represents the existing buildings/equipment, the second one (*K*) the node set and the third one (*R*) the protection strategies.
3. Variables: the optimization problem solves two different kind of variables, integer and continuous. The former are given by binary variables that define the location of the control room at a specific node (*k*) and the active protection strategy (*r*). The continuous variables are given by the coordinates of the control room's location (x_c, y_c), but this is merely a formulation requirement since the possible value of these variables is given by the values of the node coordinates (x_k, y_k) specified as parameters.
4. Constraints: the following constraints assure that the obtained solution in the optimization is within physical and logical bounds:
    a. Control room's coordinates: the sum of the node-coordinate by binary variable of location product, of each node allows to determine the location of the control room.
    b. Unique location: the sum of the binary variable of location for each node must equal one.
    c. Protection strategy: the sum of all the binary variables of active protection strategy must equal one.
    d. Separation distances: the control room must be located taking into account the minimum separation distances established. This is a Pythagoras distances' formulation and finds its origins in [26].
5. Objective function: this function comprises the total cost related to the control room's obtained location and the protection strategy selected. Being a cost function, the optimization problem seeks to minimize it.

Once the elements of the optimization problem are defined, it is executed. The results show that the optimum control room's location for the case study is the *k87* node (using the 3rd protection strategy). It is important to note that this location does not correspond to the node with the lowest risk score in the grid, on the contrary, its risk score is 'medium'. The selection of this node balances the cost components involved in the control room's construction, installation and other key parameters evaluated in the objective function.

Regarding the optimum protection strategy, it is found to be the one that resist the highest levels of overpressure even when it also represents the most expensive one. This variable (protection strategy) and its influence demonstrates that it directly affects the decision making process and has a direct relationship with the plant's safety and business continuity.

4. Conclusions

This work proposes a methodology that integrates economic and safety analysis in order to determine the optimum location of a control room through the use of risk scores (risk mapping) and MINLP formulation of an optimization problem. This methodology seeks to aid the decision making process of facility layout.

The proposed methodology was applied to a previously documented case study and the results of it showed that it can effectively locate the control room within a discretized process plant area, while taking into account different protection strategies related to the control room's construction.

Nonetheless, the proposed methodology has diverse and important improvement opportunities in the treatment of the process plant space and the selection of dispersion and explosion models. These models do not allow the modeling of chemical species mixtures and do not take into account the interaction with objects, opposed to a Computational Fluid Dynamics (CFD) approach. The assumption that a building occupies a whole node tends to overestimate the real size of the facilities, inducing uncertainty to the results.

5. Perspectives

Future work should be focused in incorporating risk acceptability within the optimization formulation, as well as the interaction of the dispersion and the blast wave with the facility's buildings and equipment, e.g. through a CFD approach.

A spatially continuous approach should be developed in order to capture in the highest detail the consequences to humans due to structural damages, hence increasing the models sensibility to the number of occupants inside the control room.

Finally, it would be a considerable step forward to include dynamic risk scores related to the location of diverse hazardous process units and the risk represented by these (even if considered fixed).

6. References

[1] Sklet, S. 2006. "Safety barriers: Definition, classification, and performance". Journal of Loss Prevention in the Process Industries, 19, 494-506, http://dx.doi.org/10.1016/j.jlp.2005.12.004.

[2] Vyas, K. & Minaya, E. 2012. "Explosion at Amuay refinery in Venezuela kills 41; fire still burning". Hydrocarbon Processing. Punto Fijo, Venezuela.

[3] C. O. Diaz, R. Vázquez-Román, S. Jung, and M. S. Mannan, "A Comparison of Deterministic and Stochastic Approaches to Solve the Facility Layout Problem with Toxic Releases," in 19th European Symposium on Computer Aided Process Engineering, vol. Volume 26, Elsevier, 2009, pp. 93–98.

[4] S. Jung, D. Ng, C. D. Laird, and M. S. Mannan, "A new approach for facility siting using mapping risks on a plant grid area and optimization," Journal of Loss Prevention in the Process Industries, vol. 23, no. 6, pp. 824–830, Nov. 2010.

[5] S. Jung, D. Ng, J.-H. Lee, R. Vazquez-Roman, and M. S. Mannan, "An approach for risk reduction (methodology) based on optimizing the facility layout and siting in toxic gas release scenarios," Journal of Loss Prevention in the Process Industries, vol. 23, no. 1, pp. 139–148, Jan. 2010.

[6] C. Díaz-Ovalle, R. Vázquez-Román, and M. Sam Mannan, "An approach to solve the facility layout problem based on the worst-case scenario," Journal of Loss Prevention in the Process Industries, vol. 23, no. 3, pp. 385–392, May 2010.

[7] R. Vázquez-Román, J.-H. Lee, S. Jung, and M. S. Mannan, "Optimal facility layout under toxic release in process facilities: A stochastic approach," Computers & Chemical Engineering, vol. 34, no. 1, pp. 122–133, Jan. 2010.

[8] D. A. Crowl and J. F. Louvar, Chemical Process Safety: Fundamentals with Applications, 2nd ed. Prentice Hall, 2001.

[9] C. Cannalire, R. Clarke, B. Mazzarotta, and R. Nicoletti, "Optimizing safety of control rooms in case of flammable releases," Journal of Loss Prevention in the Process Industries, vol. 6, no. 5, pp. 313–317, Oct. 1993.

[10] Q. Baker, Explosion Risk ans Structural Damage Assestment Code (ERASDAC). Arlington, VA: 30th DoD Explosive Safety Seminar, Department of Defense Explosive Safety Board, 2002.

[11] DDESB (Department of Defense Explosive Safety Board), "Technical Paper 14." Approved Methods and Algorithms for DoD Risk-Based Explosives Siting, 2009.

[12] API, "Risk Based Inspection." Base Resource Document. API Publication 581, 2000.

[13] Center for Chemical Process Safety (CCPS), Guidelines for Chemical Process Quantitative Risk Analysis, 2nd ed. Wiley-AIChE, 1999.

[14] S. Mannan, Ed., Lees' Loss Prevention in the Process Industries, Third Edition: Hazard Identification, Assessment and Control, 3rd ed. Butterworth-Heinemann, 2005.

[15] F. Díaz Alonso, E. González Ferradás, J. Francisco Sánchez Pérez, A. Miñana Aznar, J. Ruiz Gimeno, and J. Martínez Alonso, "Characteristic overpressure–impulse–distance curves for the detonation of explosives, pyrotechnics or unstable substances," Journal of Loss Prevention in the Process Industries, vol. 19, no. 6, pp. 724–728, Nov. 2006.

[16] M. Swisdak, Simplified Kingery Airblast Calculations. New Hampshire: Naval Surface Warfare Center, 1980.

[17] Q. Baker and C. Oswald, "Vulnerability model for occupants of Blast Damaged Buildings." 34th Annual Loss Prevention Symposium, Session 3, 2000

[18] Center for Chemical Process Safety (CCPS), Guidelines for Facility Siting and Layout, 1st ed. Wiley-AIChE, 2003.

[19] National Oceanic and Atmospheric Administration, "Climate data online: map and application search," National Climatic Data Center, 24-Oct-2012. [Online]. Available: http://www.ncdc.noaa.gov/cdo-web/. [Accessed: 24-Oct-2012].

[20] National Oceanic and Atmospheric Administration, "Solar Calculation Details," National Climatic Data Center, 2009. [Online]. Available: http://www.esrl.noaa.gov/gmd/grad/solcalc/calcdetails.html. [Accessed: 02-Oct-2012].

[21] Center for Chemical Process Safety (CCPS), Guidelines for Use of Vapor Cloud Dispersion Models, 2nd ed. Wiley-AIChE, 1996.

[22] D. Enmak, User's Manual for SLAB: An Atmospheric Dispersion Model for Denser-than-Air Releases. California: University of California, 1990.

[23] Van der Bosch, C. J. H., Weterings, R. A. P. M. & Stoffen, N. C. P. v. R. d. G. 1997. Methods for the Calculation of Physical Effects: Due to Releases of Hazardous Materials (liquids and Gases): Yellow Book, Directorate-General for Social Affairs and Employment, Committee for the Prevention of Disasters.

[24] J.A. Caballero, I.E. Grossmann. "Una revisión del estado del arte en optimización, Revista Iberoamericana de Automática e Informática Industrial". RIAI, 4 (2007) 5-23.

[25] A. Brooke, D. Kendrick, A. Meeraus, R. Raman. "GAMS User's Guide". Washington DC, USA, 1998.

[26] J.E. Cadena, M.A. Ramos, J.M. Gómez, F. Munoz, "Facility layout optimization detailed research through MILP and MINLP", in AIChE - 2012 AIChE Spring Meeting and 8th Global Congress on Process Safety, Conference Proceedings , 2012.

9TH GLOBAL CONGRESS ON PROCESS SAFETY

Risk analysis for Overpressure from Pressure Vessel Burst

Seungho Jung
Air Products and Chemicals, Inc.
7201 Hamilton Blvd.
Allentown, PA 18195
Jungs2@airproducts.com

Elizabeth M. Lutostansky (lutostem@airproducts.com)
Joan M. Schork (shorkjm@airproducts.com)
Air Products and Chemicals, Inc

Prepared for Presentation at
American Institute of Chemical Engineers
2013 Spring Meeting
9th Global Congress on Process Safety
San Antonio, Texas
April 28 – May 1, 2013

UNPUBLISHED

Risk analysis on overpressure of Pressure Vessel Burst

Seungho Jung
Air Products and Chemicals, Inc.
Allentown, PA 18195
Jungs2@airproducts.com

Elizabeth M. Lutostansky, Joan M. Schork
Air Products and Chemicals, Inc

Keywords: Pressure Vessel Burst (PVB), explosion, Blast curves, geometry factor

Abstract

A pressure vessel burst (PVB) is a type of explosion that involves the burst of a pressure vessel containing gas at elevated pressure. Any vessel that is pressured or has the capacity to build high pressure before bursting can be a candidate for generating a PVB. At burst, a blast wave propagates outward from the source by the sudden expansion of a compressed gas. Fragment hazards are also present when the shell of the vessel along with attached external accessories is thrown. PVBs are possible with vessels containing any gas or mixture including inerts and flammables.

Historically, there have been incidents related to PVBs. Work that has been done to predict the overpressure impacts is covered in Guidelines for Vapor Cloud Explosion, Pressure Vessel Burst, BLEVE and Flash Fire Hazards, 2nd Edition by CCPS [1]. Moreover, as Facility Siting has received significant attention recently, Air Products has included PVBs in Facility Siting studies to safely locate occupied building.

This paper focuses on how to select an appropriate PVB model and correctly use it in calculating vulnerability inside an occupied building. It provides an overview of a simplified method for estimating the characteristics of pressure vessel bursting for practicing engineers.

1. Introduction

Explosions can be classified into two types, oxidations such as Vapor Cloud Explosions (VCE), dust explosions and condensed phase explosion and sudden expansions such as Pressure Vessel Bursting (PVB), Boiling liquid expanding vapor explosions (BLEVE), and rapid phase transitions (RPT). Pressure vessel burst (PVB) is involves a rupture of a vessel containing gas at elevated pressure. The source of the energy released in the blast is the compressed gas. A BLEVE is caused by the rupture of a vessel containing a pressurized liquid above its normal boiling point.

With a BLEVE, the liquid in the vessel undergoes a phase transition from liquid to gas. This flashing contributes to the energy released. Many engineers become confused by the difference between PVBs and BLEVEs because of their similarities. Both liquid and gas may be present in vessels undergoing PVB or BLEVE and blast effect calculations look similar. It is important to remember that in the case of PVB, liquids present in the vessel do not undergo phase transition while for BLEVEs, the primary source of energy is the phase transition. .

The scope of this paper is limited to PVBs. It presents a simplified method for obtaining PVB overpressure results for a building siting study.

2. Methods for predicting blast effects from pressure vessel bursts

A blast curve method is highly recommended because it is more easily implemented than doing numerical simulations. There is an outstanding set of blast curves based on a systematic one-dimensional numerical study by Tang et al. [2]. The Baker-Tang curves assume all of the energy is employed to make a blast effect rather than to be used into the fragments as kinetic energy. These curves are for spherical, free air burst which do not include ground reflection.

Figure 1. Positive overpressure curves for various vessel pressures (courtesy copied from Guidelines for Vapor Cloud Explosion, Pressure Vessel Burst, BLEVE and Flash Fire Hazards, 2nd Edition by CCPS [1])

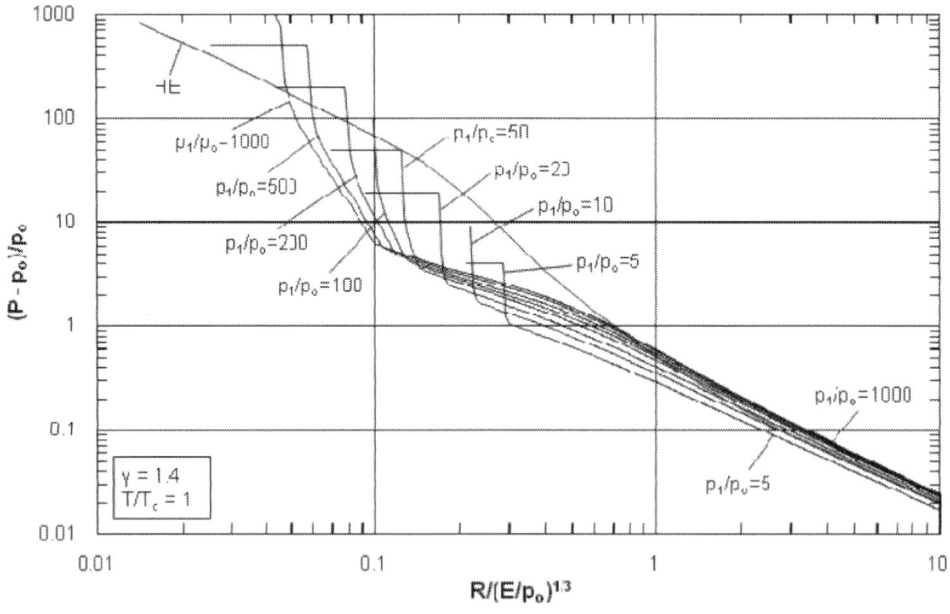

There were not much experimental data for PV burst but Table 1 shows the set of experimental data at relatively lower operating pressure. After early 1980s, there appeared many experimental studies that reported PV burst overpressure measurements but they cover different vessel shapes in different locations such as ground and air [3]. Therefore only data points with spherical free air burst are chosen to compare and validate against the Baker-Tang curves.

Table 1. Summary of literature data

Author	Year	Shape of vessel	Type of burst	Substances	Operating pressure ratio to atmospheric pressure (p1/p0)	Overpressure range
BAKER	1977	Spherical	Free air burst	Air, Argon	6	2.7~50 kPa
BAKER	1977	Spherical	Free air Burst	Freon 12	18 ~ 22	0.73~7.3 kPa

Figure 2 shows good validations against two experimental data sets qualitatively. As the peak overpressure line with the ratio, p1/p0=20 is higher than that with the ratio, p1/p0=5, the experiment data points with Air, Argon are distributed well above the data point with Freon 12.

Figure 2. Validation of Tang's curves on lower pressure cases against experimental data

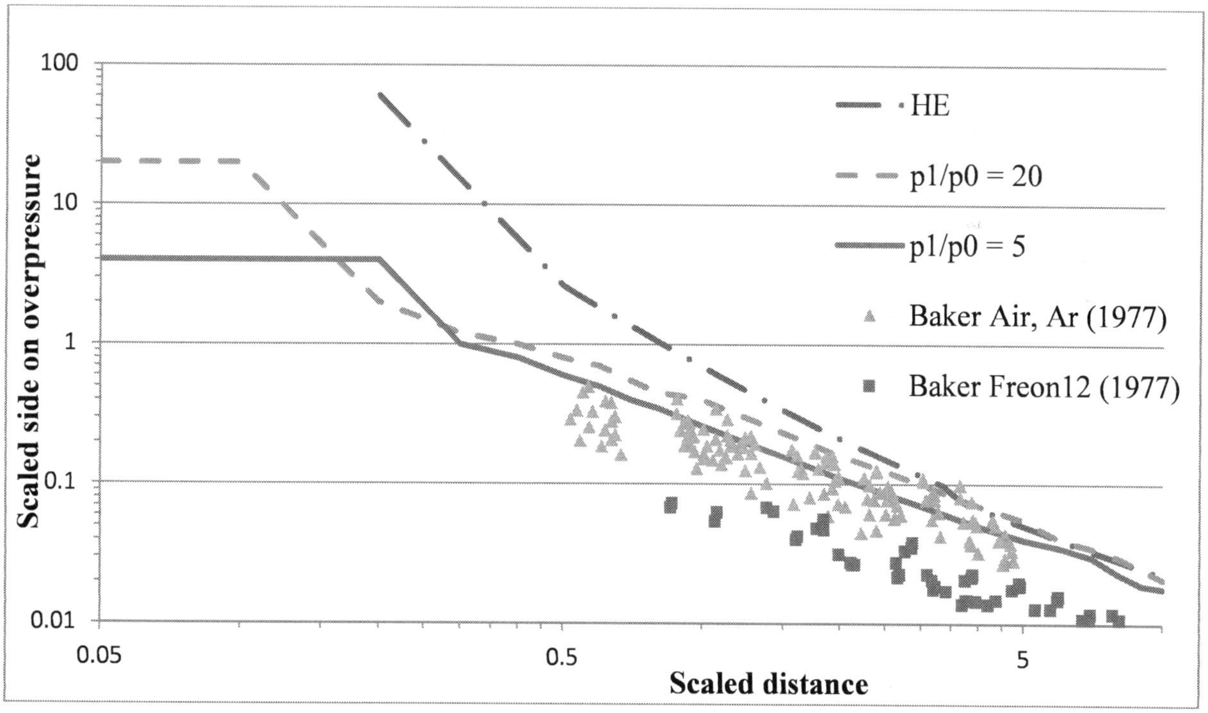

In the far-field, the HE line is close enough to all the other pressure ratio curves as seen in Figure 1. Though a recommendation was made to find the scaled side on overpressure based on p1/p0 ratio, it is quite helpful and faster to use the HE for the far-field further than scaled distance ($\bar{R}$) > 1. 6th order polynomial equation can better fit the HE curve in Figure 1 and suggested as

$$log\bar{P_s} = c_0 + c_1 log\bar{R} + c_2(log\bar{R})^2 + c_3(log\bar{R})^3 + c_4(log\bar{R})^4 + c_5(log\bar{R})^5 + c_6(log\bar{R})^6 \quad \text{[Eq. 1]}$$

where:

$\{c_0, c_1, c_2, c_3, c_4, c_5, c_6\} = \{-0.4102, -1.8282, 0.3661, 0.1122, -0.0830, -0.0042, 0.0046\}$

$\bar{P_s}$ is the dimensionless overpressure, or the overpressure scaled by ambient pressure (Multiply by 14.696 to convert to psig).

3. General calculating procedure for Pressure Vessel Bursts

3.1 Calculate PV energy

The internal energy U of a system is the sum of the kinetic, potential and intermolecular energies of all the molecules in the system. For an ideal gas it is

$$U = \frac{pV}{\gamma - 1} \quad \text{[Eq. 2]}$$

where p is the absolute pressure of the gas, V is the volume of the gas, and γ is the ratio of specific heats of the gas in the system, i.e. C_p/C_v. Therefore, the Brode's PV burst energy is

$$E_{ex,brode} = \frac{(p_1 - p_0)V_1}{\gamma_1 - 1} \quad \text{[Eq. 3]}$$

where the subscript 1 refers to initial state and the subscript 0 refers to the ambient conditions. For ground burst, the effective energy is doubled to account for the ground reflection as

$$E_{eff} = \frac{2(p_1 - p_0)V_1}{\gamma_1 - 1} \quad \text{[Eq. 4]}$$

3.2 Calculate scaled distance, $\bar{R}$

Once the explosion energy of a PV burst is estimated, the scaled distance, $\bar{R}$ can be obtained from

$$\bar{R} = R\left(\frac{P_0}{E_{eff}}\right)^{1/3} \quad \text{[Eq. 5]}$$

where R is the distance from the blast at which blast parameters are to be determined.

3.3 Determine scaled positive side on overpressure, $\overline{P_s}$

To use this graphical method, select the overpressure curve in Figure 1 that corresponds to the vessel burst pressure ratio p1/p0. Read the $\overline{P_s}$ = (p1-p0)/p0 value. For simplicity, if $\bar{R}$ is larger than 1, the equation correlated with the Highly Explosive (HE) line can be used

3.4 Adjust scaled positive side on overpressure, $\overline{P_s}$

There are two adjustment factors; one is for temperature, the other is for geometry. Higher temperature increases the overpressure near the vessel but the effect decreases with distance away from the vessel. If the target is far enough from the vessel, this adjustment effect is negligible and the use of ambient temperature in Figure 1 can be adequate.
Geometry adjustment factors are used with of cylindrical vessels and slightly elevated spherical vessels. The multiplier factors shown in Table 2, for$\overline{P_s}$ initially suggested by Baker et al. [4], are widely used.

Table 2. Factors for geometrical effect

Multiplication factors on $\overline{P_s}$ for slightly elevated spherical vessel	
$\bar{R} < 1$	2
$\bar{R} > 1$	1.1
Multiplication factors on $\overline{P_s}$ for cylindrical vessel	
$\bar{R} < 0.3$	4
$0.3 < \bar{R} < 1.6$	1.6
$1.6 < \bar{R} < 3.5$	1.6
$3.5 < \bar{R}$	1.4

However, these factors were made based upon high explosives experiments, and it was found by Geng et al. and written in the CCPS book [1] as the factors are not directly applicable to PVBs. It was suggested using a group of adjustment factor specifically for PVBs with bursting pressure ranges from 10 to 100 for both the cylindrical and elevated spherical PVBs. Note the factors vary continuously with scaled distance, $\bar{R}$. For simplicity, it is recommended that a factor of 1.8 be used for cylindrical vessels and a factor of 2.0 for slightly elevated spherical vessels as these are the maximum values given by Geng.

3.5 Calculate Vulnerability

Given the adjusted $\overline{P_s}$ the vulnerability of people inside a building is next determined. A vulnerability curve can be generated as Figure 3 using information from Guidance for the Location and Design of Occupied Buildings on Chemical Manufacturing Sites, published by the UK's Chemical Industries Association (CIA). The overpressure-vulnerability relationship for type 3, typical domestic building, is presented in Figure 3.

Figure 3. Vulnerability curve against overpressure (courtesy coped from Marx, et al. [5])

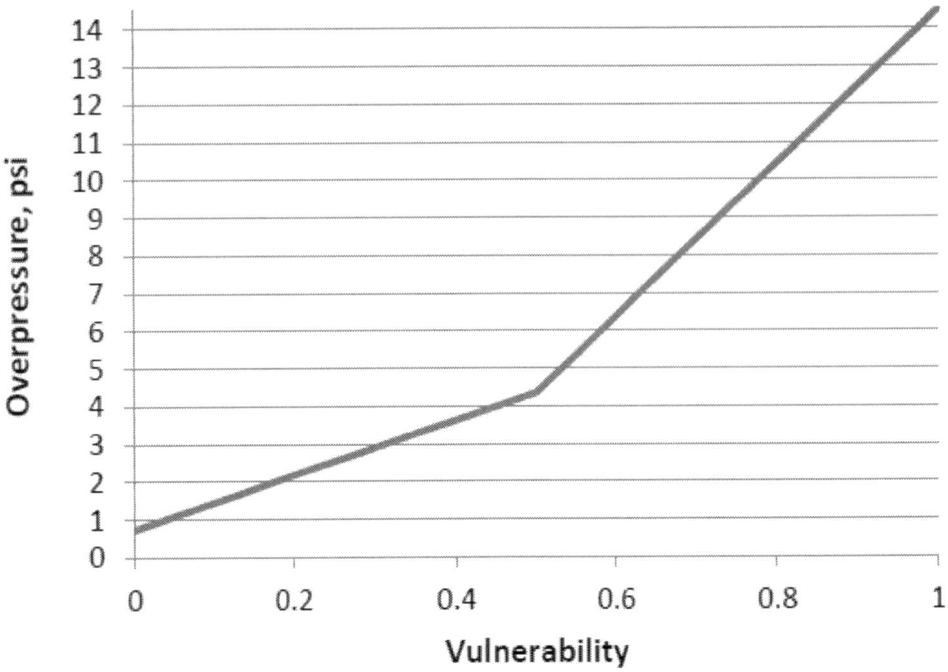

4. Example calculation case study

A $100m^3$-cylindrical pressure vessel at grade was under pressure test with nitrogen gas at 20 barg and it ruptured. There was a control room located 50 me from the vessel. What would be the side on overpressure at the control room? And what would be the vulnerability inside the control room if the building type is CIA type 3.
Table 3 presents the details of PVB calculation and recommendations.

Table 3. PVB calculation flowchart

Procedure	Details of calculation	Note
Collect data	Internal pressure is 21 bara The ratio of specific heats of nitrogen is 1.40 Ground burst, ambient pressure is 1 bara	
Calculate PV burst energy	$E_{ex,brode} = \frac{(p_1 - p_0)V_1}{\gamma_1 - 1} = \frac{(21-1)100}{1.4-1} = 500\ MJ$	
Calculate effective energy	$E_{eff} = \frac{2(p_1 - p_0)V_1}{\gamma_1 - 1} = 1\ GJ$	Multiply 2 if it is ground level explosion
Calculate scaled distance, $\bar{R}$	$\bar{R} = R\left(\frac{P_0}{E_{eff}}\right)^{1/3} = 50\left(\frac{10^5\ Pa}{10^9\ J}\right)^{1/3} = 2.32$	R = 50 m

Read or calculate scaled positive side on overpressure, $\overline{P_s}$	$\overline{P_s} = 0.13$	$\bar{R}$ is larger than 1, equation (1) is proper to use
Adjust scaled positive side on overpressure, $\overline{P_s}$	Multiply 1.8 for a cylindrical vessel geometry factor $\overline{P_s} = 1.8 \times 0.13 = 0.23\ barg$	
Vulnerability inside building	40 % from Figure 3.	CIA type 3 building
Recommendation	If the consequences are judged to be unacceptable, the blast-pressure effects shall be mitigated by the location of the equipment within suitable shields, walls, or barriers. Or the personnel in the control room should be evacuated when the test is ongoing.	

5. Conclusion

Young engineers are often told that BLEVEs are one of the highest impact phenomena in the process industry, even more so than PVBs. PVBs tend to have lower energy releases involved compared to BLEVEs since PVBs do not involve energy from liquid flashing. However, PVBs cannot be ignored since there can be many pressured gas vessels and pressure tests on a gas industry site. This paper provides an overview of method for estimating the characteristics of pressure vessel bursting for practicing engineers. It has been presented to include advanced modeling technology on PVBs and how to properly apply a simplification in adapting blast curves or equation and geometry effects.

6. References

[1] "Guidelines for Vapor Cloud Explosion, Pressure Vessel Burst, BLEVE and Flash Fire Hazards", 2nd ed., CCPS

[2] Tang, M. J., Cao, C. Y., Baker, Q. A., "Calculation of Blast Effects from Bursting Vessels", 1996 AIChE National Meeting

[3] Chen, Fei, Lutostansky, Elizabeth M., Schork, Joan M., "Improvement of BLEVE Overpressure Method Using Historical Data", 2011 CCPS conference

[4] W.E. Baker, J.J. Kulesz, R.E. Richer, R.L. Bessey, P.S. Westine, V.B. Parr, G.A. Oldham, "Workbook for predicting pressure wave and fragment effects of exploding propellant tanks and gas storage vessels", NASA CR-134906, NASA Scientific and Technical Information Office, Washington, 1975 and 1977

[5] Marx, Jeffery D., "API RP 752 Building Siting Using F-N curves", 2013 MKOPSC international symposium

Effects of Process Variables on Peak Relief Rates Estimated by Dynamic Simulation for a Multiple Distillation Column System

John Wilkins
Staff Process Consultant
Smith & Burgess LLC
5700 Northwest Central Drive Ste. 301
Houston, Texas 77092
John.Wilkins@smithburgess.com

Dustin Smith
Vice President of Operations
Smith & Burgess LLC
Dustin.Smith@smithburgess.com

Prepared for Presentation at
American Institute of Chemical Engineers
2013 Spring Meeting
9th Global Congress on Process Safety
San Antonio, Texas
April 28 – May 1, 2013

UNPUBLISHED

Effects of Process Variables on Peak Relief Rates Estimated by Dynamic Simulation for a Multiple Distillation Column System

John Wilkins
Staff Process Consultant
Smith & Burgess LLC
5700 Northwest Central Drive Ste. 301
Houston, Texas 77092
John.Wilkins@smithburgess.com

Dustin Smith
Vice President of Operations
Smith & Burgess LLC
Dustin.Smith@smithburgess.com

Keywords: dynamic simulation, VMG Sim, distillation columns, sensitivity analysis, pressure relief valves

Abstract

Process engineers and designers have recently turned to using dynamic analysis as a more realistic method for sizing relief systems. The primary goal of using dynamic analysis is to ensure that modifications are truly required instead of making unnecessary and costly upgrades of relief systems based on overly conservative calculations. In work previously presented by the authors, it has been shown that, for certain overpressure scenarios, the dynamic relief rates predicted by dynamic simulation are significantly affected by certain operating conditions [1].

This paper shows the effects of process variables on the relief loads estimated by dynamic simulation for multiple columns. This paper will also show how these effects would impact the flare load for a unit with multiple columns. The multiple column system is based on a system the authors have encountered at a refinery. The sensitivity analysis shows that changes in the column liquid levels significantly affect the initial relief times, relief loads, and relief durations.

1. Introduction

To stay in compliance with industry, regulatory, and typically corporate guidelines, it is important for a plant to conduct and document analyses of their relief systems. It also allows for plant engineers and designers to see the effects of their process on existing or proposed relief systems. Relief system analyses traditionally use steady-state equations to determine the relief rate used to size relief devices. For complex systems, these equations are based on conservative simplifying assumptions. This usually results in a larger relief rate that requires a larger relief device. The drawback to using traditional methods, when analyzing an existing system, is that they can overstate the required relief rate. Sometimes this results in the recommendation of costly relief systems upgrades where only minor modifications are needed.

Dynamic analysis of complex systems involves creating a simulation that calculates changes to the system over time. This analysis can be used to better estimate the relief requirements to protect from overpressure for a specific relief scenario. The primary goal of dynamic relief system analysis is to better understand how the system will respond to the upset scenario. An additional benefit is that the detailed analysis can be used to support making limited changes in lieu of more extensive modifications. The fundamental differences between these two methods are important to understand in order to make sure they are both being used to determine relief rates that are conservative.

Equation 1 shows a simplistic steady-state equation that can be used to determine the relief rate through a relief valve installed on top of a distillation column that has lost cooling, feed, and reflux, but has continued heat input (a typical total power failure scenario).

$$\dot{M}_{relief} = \frac{\dot{Q}_{reb}}{\Delta H_{vap}} \qquad \text{[Eq. 1]}$$

In order to use the above equation, the right hand side of the equation would be evaluated at relief conditions, which may vary from normal operating conditions [2]. Assumptions such as the composition of the fluid on the process side of the reboiler can have effects such as a lower heat of vaporization or a larger reboiler duty that will increase the predicted relief rate.

Equation 2 is an example of how the above equation would be represented in a dynamic environment. It shows the same equation as functions of time as well as initial process conditions such as pressure, temperature, etc.

$$\dot{M}_{relief}(t, P_i, T_i, etc) = \frac{\dot{Q}_{reb}(t, P_i, T_i, etc)}{\Delta H_{vap}(t, P_i, T_i, etc)} \qquad \text{[Eq. 2]}$$

Dynamic simulations calculate the changes in the system based on incremental time steps from an initial starting point. Therefore, the selection of initial conditions is important and will affect the results [1]. These initial conditions are typically based on operating conditions that represent the normal operations of the facility. Operating conditions are controlled in ranges, and depending on the specific item, there may be no such thing as a normal condition. Liquid levels, temperatures, pressures, and reflux rates for distillation columns can vary during operation and can change due to upsets elsewhere in the plant. This is important to consider when global relief scenarios like power failure and cooling failure affect the entire plant, as these variables are all essentially assumptions.

API 521 Section 5.22[3] states that if dynamic simulations are used for column relief system design, the model must be conservative with respect to the maximum relief load. It goes on to state that the assumptions used for the simulation *shall* be checked by sensitivity analyses to assess the impact on the column relief rate. A sensitivity analysis is the study of how the uncertainty of the input variables affects the output of a mathematical model. Therefore, the input variables of a dynamic simulation must be checked in order for the designer to ensure that the simulation is conservative.

2. Methodology

2.1 Setting up the dynamics file

A steady-state model containing 2 debutanizers, a depropanizer, a gasoline splitter, and 12 PSVs was created using VMG Sim. The flow diagram of this system is displayed in Figure 1.

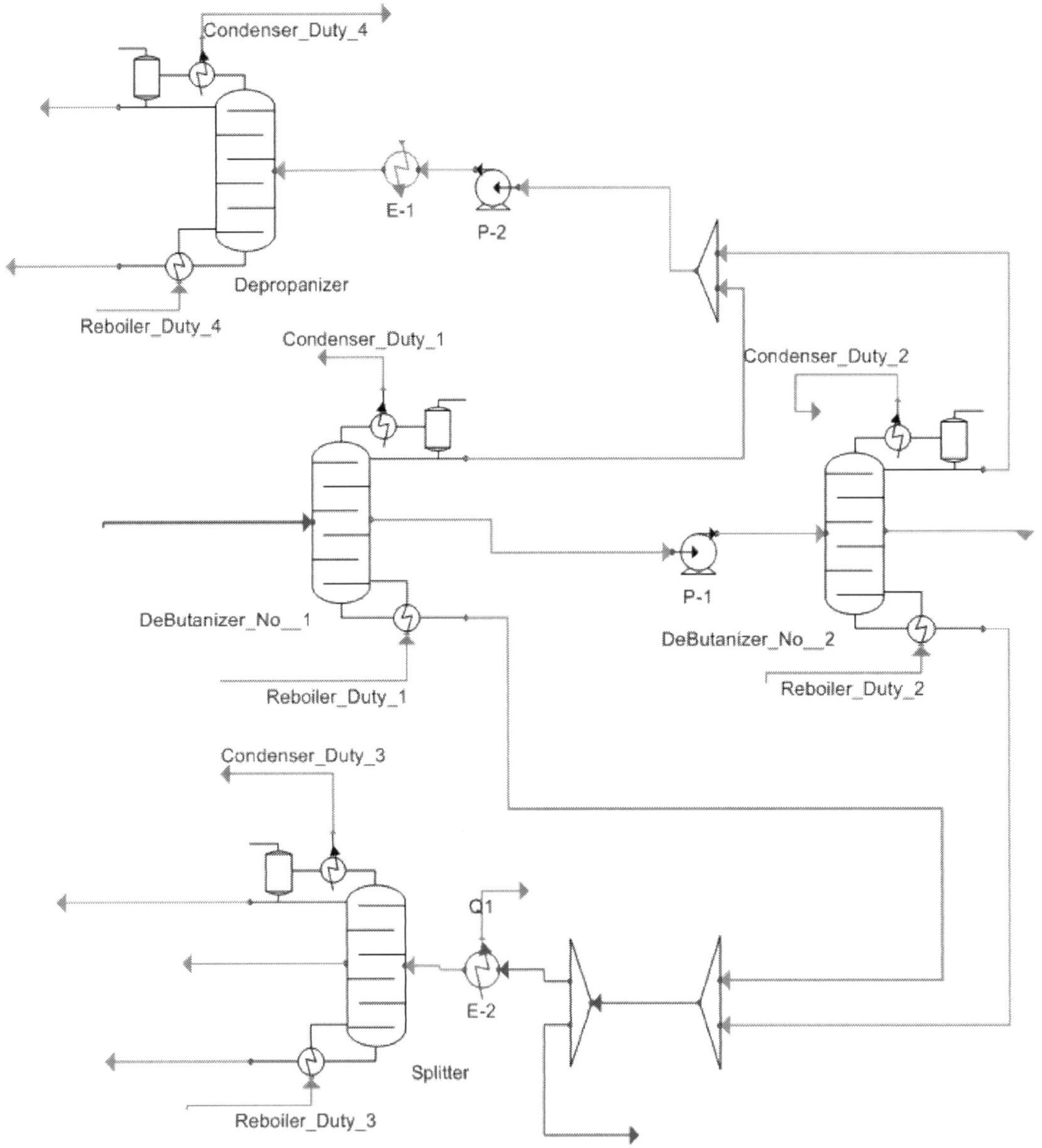

Figure 1 : Process Flow Diagram of a Multi-Column System

The steady-state model was used to create a dynamic model that uses 21 controllers set to match the operating parameters of the equipment. Controllers are used in dynamic simulations to achieve specified flowrates, pressures, temperature, and other physical conditions. The controller attempts to achieve the user's specification by opening or closing a valve located on a material or energy stream. This is different from a steady-state model where flowrates, duties, and physical properties are typically specified or calculated. The action of the controller is based on the tuning parameters provided by the user.

Twelve PSVs were added to the dynamic model to simulate relief. Table 1 displays information on these relief valves.

Table 1: Summary of PSV Information

Name	Location	PRV Size	Set Pressure (psig)
PSV-1	DeButanizer No. 1 vapor overhead	6Q8	184
PSV-2	DeButanizer No. 1 vapor overhead	6Q8	190
PSV-3	DeButanizer No. 1 accumulator	3L4	180
PSV-4	DeButanizer No. 2 vapor overhead	6Q8	190
PSV-5	DeButanizer No. 2 accumulator	1.5H3	190
PSV-6	Splitter vapor overhead	6R8	30
PSV-7	Splitter vapor overhead	6R8	30
PSV-8	Splitter vapor overhead	6R8	30
PSV-9	Splitter vapor overhead	6R8	30
PSV-10	DePropanizer vapor overhead	6Q8	290
PSV-11	DePropanizer accumulator	3K4	274
PSV-12	DePropanizer accumulator	3K4	274

In order to perform the sensitivity analysis, different column liquid levels were used as operating conditions. Setpoints of the respective level controllers for the columns were adjusted to achieve the desired liquid levels. Three simulations were created for a low liquid level of 20%, a medium liquid level of 40%, and a high liquid level of 60%. These percentages represent the fraction of liquid height in the lower section of the column that does not contain trays.

2.2 *Initiating relief*

Cooling water failure was chosen as the overpressure scenario to review as it affects all of the four columns simultaneously. Total loss of cooling water is expected to result in the loss of overhead condenser duties, which may lead to overpressure in the column due to excess vaporization. This was accomplished by specifying duties of zero for all condensers. It was assumed that all of the condensers lost cooling at the same time.

All controllers in the dynamics model were turned off (set to manual) to avoid taking credit for positive controller action. This was done in accordance with API 521 Section 4.2.4[3]. Changing the controller mode from "automatic" to "off" causes the controlled valve to hold its position during relief.

For each liquid level, the integrator (essentially an electronic strip chart recorder in VMG Sim) was turned on and recorded the pressure in the columns as it slowly increased. The overhead PSVs opened at their respective set pressures and flow data was collected. The sum of the mass flow rates relieved by the PSVs was recorded in order to view the total load into the flare.

3. Results

Graphs of dynamic relief for the three liquid levels are displayed below. PSV-5 was not graphed as it does not contribute a significant relief load. The initial time (t = 0), is the time at which the cooling to the tower was stopped and the overpressure scenario began. Note that the change in liquid levels shifts the magnitude of the relief rate and the peak point in time for some of the towers.

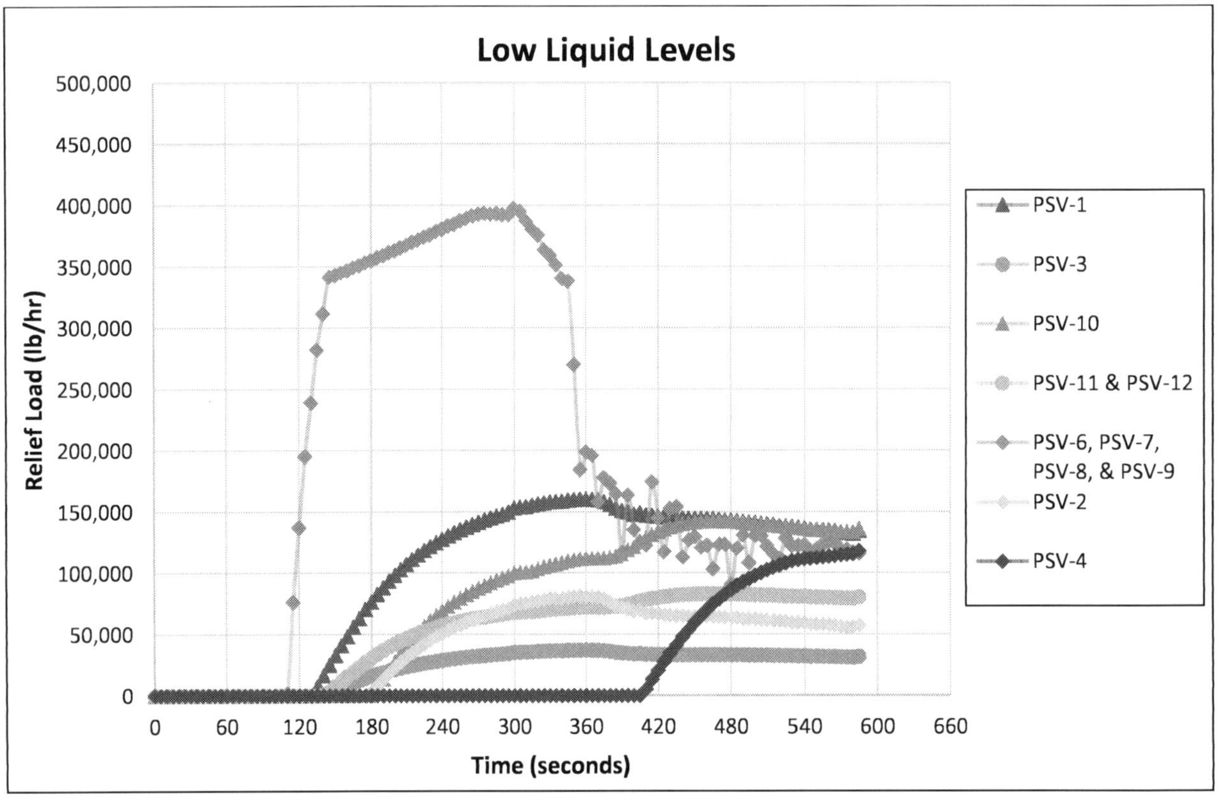

Figure 2: Cooling failure relief rates with low liquid levels

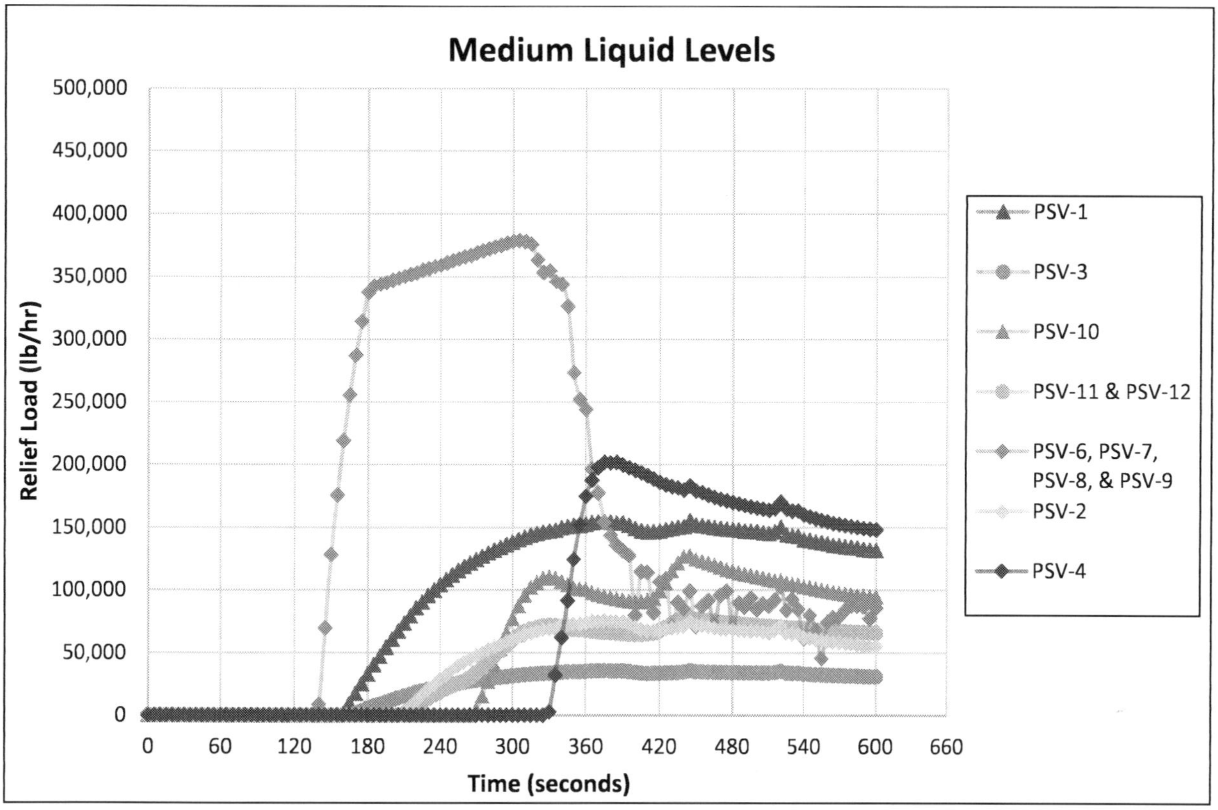

Figure 3: Cooling failure relief rates with medium liquid levels

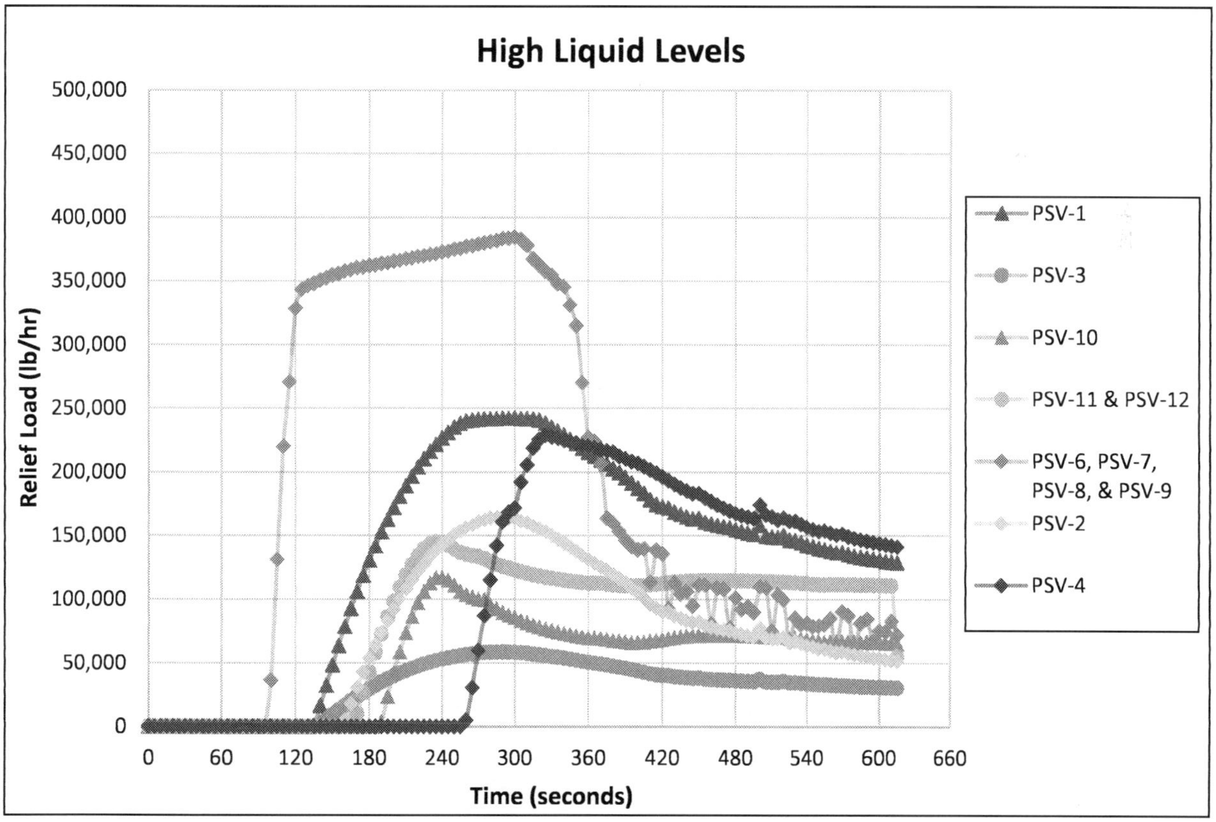

Figure 4: Cooling failure relief rates with high liquid levels

Figure 5 shows the combined flare loads for the three liquid levels. The curves were made by adding the relief loads from the previous graphs over the same time frame. The combined load represents the load that would enter the flare header. The horizontal line represents the steady-state relief load for the four columns estimated by traditional methods.

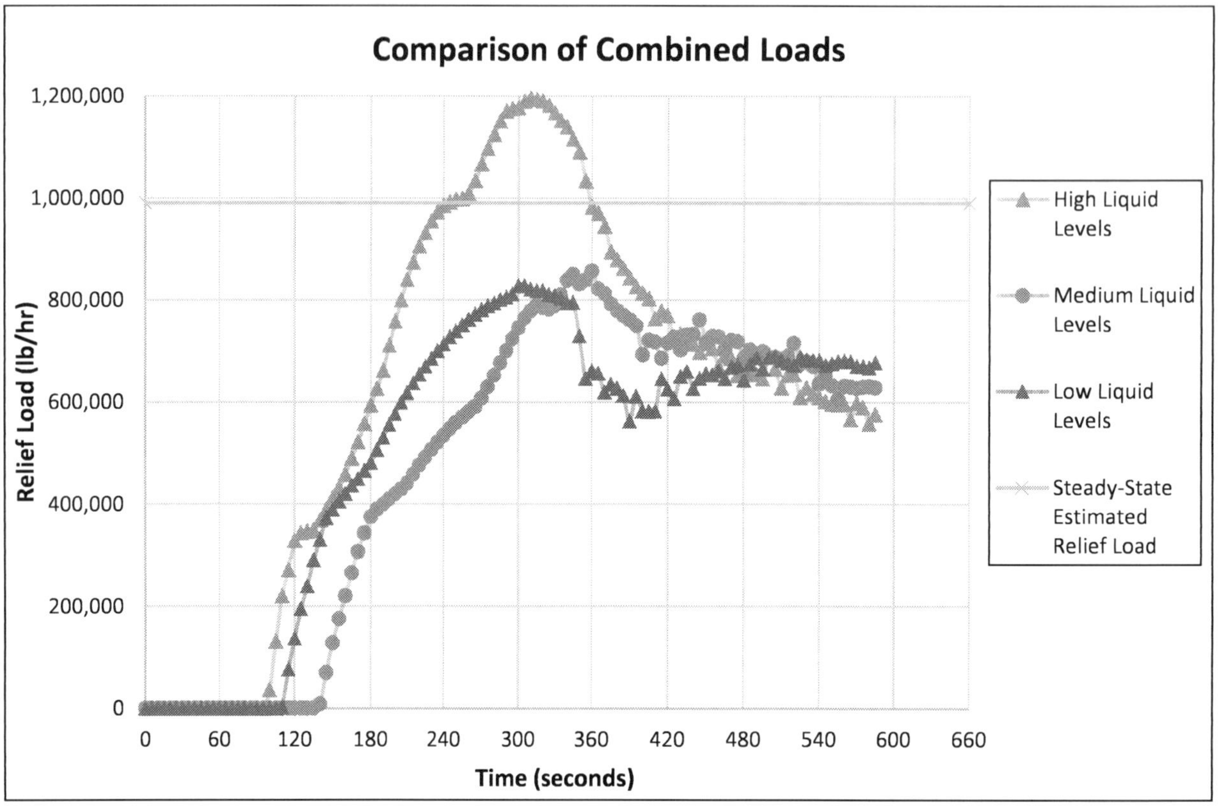

Figure 5: Combined cooling failure loads for different column liquid levels

The steady-state estimation of the combined relief load of the columns assumes that the reboilers have a reduced capacity due to the reduction in the log mean temperature difference (see Reference 1 for an additional explanation). The dynamic model was based on a constant heat input, and thus may over-predict the duties of the reboilers under upset conditions. A further improvement could be made by modeling the reboilers based on their individual UA characteristics.

The three curves in Figure 5 have roughly the same shape as well as time of peak flow. The noticeable difference is the peak relief load for the high liquid level curve. There is a 43% increase in the peak of the combined relief load for all PSVs when the liquid levels are increased by 50%. Lowering the liquid levels by 50% had little effect on the combined peak load.

The increase in the combined peak load appears to be caused by two factors. The peak flows of PSV-1 and PSV-2 located on the 1st debutanizer increased by 57% from normal liquid levels to high liquid levels. The peak time for PSV-4 located on the 2nd debutanizer lowered when the liquid level increased, which resulted in PSV-4 contributing significantly to the combined load peak. Changing the liquid levels in the columns affected the time to initial relief, the time to reach peak load, and the height of the peak for these valves. This behavior is present to a lesser extent in the other relief valves.

4. Conclusions

Dynamics simulation can be a useful tool to show how a system reacts to sudden changes and how the system uses control valves to stabilize itself. Dynamic analysis is very different from hand-calculated methods which are traditionally used to size relief systems. Although dynamic analysis can be used to size relief systems, it must be done knowing how process variations affect the relief rates.

The results show that variations in the liquid levels affect relief times and peaks for a system of multiple columns. Increasing the liquid levels of four columns in a system by 50% increased the combined peak load into the flare by 43%. This was due to increases in the peak flowrate of two PSVs and the shifting of a large peak towards the combined load peak.

This result shows that changes in process variables can have a large impact on the relief rate used to size a flare system. Consideration must be given to the effect process variables can have on the time of initial relief, the time to reach the peak load, and the magnitude of the peak flow.

Because there are multiple variables that can impact the peak relief rate for a flare, a full sensitivity analysis of a system with multiple columns would be time consuming and costly. However, if dynamic analysis is used to size a relief system, a sensitivity analysis must be performed to ensure that the analysis will lead to a safe design.

5. References

[1] Wilkins, John & Dustin Smith, P.E. *Effects of Process Variables on Dynamic Relief Load Estimates for a Depropanizer and Debutanizer*. Oct. 2012. Electronic poster presented at the AIChE Regional Process Technology Conference, Texas City, TX.

[2] Cristea, Nicholas and Dustin Smith, P.E. *Making Relief Load Estimates Match Reality*. 2013

[3] ANSI/API Standard 521 5th Edition, ISO 23251, May 2008

A Hybrid Model for Characterizing the Source of Hazardous Material Release

Shahryar Khajehnajafi, PhD
Anjana Meel, PhD
SAFER Systems LLC
5284 Adolfo Road, Suite 100
Camarillo, CA 93012
shah@safersystem.com

Prepared for Presentation at
American Institute of Chemical Engineers
2013 Spring Meeting
9th Global Congress on Process Safety
San Antonio, Texas
April 28 – May 1, 2013

UNPUBLISHED

A Hybrid Model for Characterizing the Source of Hazardous Material Release

Shahryar Khajehnajafi, PhD
Anjana Meel, PhD
SAFER Systems LLC
5284 Adolfo Road, Suite 100
Camarillo, CA 93012
shah@safersystem.com

Keywords: hazardous chemicals, chemical spill, source of release, source of hazardous material, plume model, Source Area Locator (SAL), source characterization, Gaussian plume model, Bayesian method, Bayesian inference, hybrid model, odor source, source location estimation, burro

Abstract

On Sept 10, 2012, Los Angeles Fire department and 911 operators received many complaints about a strong foul smell. The complaint covered an area of 10,000 square miles. There were many speculations for the source of the release; sewage, chemical spill, pipeline carrying natural gas. However, after many hours of investigation by South Coast Air Quality Management it was discovered that the source of the foul odor was dead fish at the Salton Sea churned up by a thunderstorm that had occurred that evening. The Salton Sea is a 376-square-mile saltwater lake about 150 miles southwest of Los Angeles. The odor remained strong for up to 150 miles downwind of the source, and quite unusual. Fortunately the above case did not pose any health effects.

It is common for industrial sites carrying hazardous chemicals to receive calls from surrounding communities complaining of a chemical presence in the ambient air. Following a complaint, industrial sites must work promptly check and verify if the odor source is related to a chemical release on their property, especially when there are other industrial sites in the vicinity, for mitigating and reducing the extent of the possible hazard. Locating the source and amount of release for predicting the geographical extent of the exposure in a timely manner is, therefore, essential to take an effective response action for evacuation or shelter-in-place.

With only basic knowledge of meteorological information; wind speed, wind direction, and stability along with two or more gas concentration measurements, it is possible to quickly frame the zone of chemical release. This is an inverse problem to the usual forward dispersion calculation in which the source location and release rate is known and the concentrations at sensors positions are predicted. There are many publications in the area of source characterization; location, and amount of chemical being released. The two most cited methods are: Bayesian inference using Markov Chain Monte Carlo (MCMC) sampling methodology, and adjoint advection- diffusion equation. The Bayesian methodology fits well for application to atmospheric dispersion where ambient and gas sensor data are stochastic in nature, however, it takes a long time to converge.

In this paper, we briefly describe a simple least-square based method for source characterization using Gaussian plume model. We apply the Bayesian method to address the same problem and compare the results of the two methodologies for few selected release scenarios as well as a set of field data. Subsequently, we will present a hybrid model using Bayesian inference which utilizes the result of the simple least- square model as a seed for narrowing the search domain and reducing the time for locating source of chemical release. We present the improvement in the run time using the hybrid model which makes it suitable for emergency response and provides a reasonably accurate prediction of source characterization.

1 Introduction

In this paper, the focus is on estimating the source area location of a hazardous release using the sensor measurements obtained from sensors located throughout a facility. If parameters of a system are known, the outcomes of measurements can be predicted. That is known as the 'forward problem'. On the contrary, estimation of the values of the parameters of the system using the results of the measurements is called a 'backward problem'. Identifying source parameters for location, quantity of release using sensor measurements is a backward problem. Tarantola (2005) discussed the inverse problem theory and methods for model parameter estimation in great detail. Usually a forward problem has a unique solution, but the inverse problem can have multiple solutions. Therefore, in an inverse problem, any apriori information should be utilized explicitly and uncertainties in the data should be handled carefully. When nonlinearities are available, it is important to describe uncertainty and apriori information using probability distribution. Bayesian theory based methods can be used to describe uncertainty. When systems are not nonlinear, least-square methods are considered very efficient. Both least-square methods and Bayesian Theory based methods have their own advantages and disadvantages. Least-square method usually minimizes the error between measurements and prediction to calculate the system unknowns while Bayesian theory based methods updates the prior probability distribution of unknown parameters using the sensor measurement data to obtain a posterior distribution of the unknown parameters, which provides the updated information of the system.

Thus the source area location and release rate is indeed a very challenging and nonlinear problem, even for identifying the source area from a single sensor concentration measurement. Even for a single sensor measurement, there can be multiple solutions for release location all of which may lie on a line/curve with varying distance and release rates. For example, a large release rate at a further distance from a sensor may result in the same concentration values observed by a fixed sensor as a small release rate at a location closer to the sensor. When more than one sensor is involved, the problem becomes even more complicated.

The Bayesian theory has been used by several researchers for estimating the source area location. Yee et al. (2007, 2008) developed a Bayesian probabilistic inferential solution for source reconstruction (both location and emission rate) from a limited noise concentrated data and the methods were tested for JU2003 field study in Oklahoma City (urban dispersion) and the ETEX (long-range dispersion on continental scales). Kim et al. (2007) developed a source reconstruction algorithm using dynamic Bayesian inference and applied it on Yonggwang atmospheric tracer experiment in Korea to show that the posterior modes of the release point and the released source rate for this experiment converged to their true values. Keats (2009) developed Bayesian inference technique for source determination in atmospheric environment. Adjoint (backward) dispersion models were used to reduce the computational effort required from calculating one [forward] plume per possible source configuration to calculating one [backward] plume per detector. Markov chain Monte Carlo (MCMC) is used to efficiently sample from the posterior distribution for the source parameters; both the Metropolis- Hastings and hybrid Hamiltonian algorithms are used. These algorithms were applied on a full-scale field experiment (Joint Urban 2003) in Oklahoma City and the results from the algorithms were also compared with the US Environmental Protection Agency's standard CMB model using a test case involving PM data from Fresno, California. Neuman et al. (2005) used an urban puff model [the Defense Science Technology Laboratory's (Dstl) Urban Dispersion Model (UDM)] for prediction plume evolution in complex urban geometries that enabled rapid urban dispersion simulations by combining traditional Gaussian puff modeling with empirically deduced mixing and entrainment approximations. In their work, a preliminary reconstruction of an atmospheric release event was demonstrated using stochastic sampling algorithms and Bayesian inference together with the rapid UDM urban puff model based on point measurements of concentration. This model was applied to a prototype isolated building and for observations and flow conditions taken during the Joint URBAN 2003 field campaign at Oklahoma City. Ideally, the urban puff model is used to approximate the source location and strength. The more accurate CFD model can then be used to refine the solution. Zheng et al. (2011) reviewed inverse calculation approaches for source determination and classified them into probability modeling methods and optimization modeling methods. According to them, probability modeling methods are demonstrated to be insufficient during emergency responses due to their lacking of enough prior information of unknown parameters, while optimization modeling methods are efficient and become a new trend in source determination. Sharan et al. (2011) applied least square data assimilation for identification of point source emissions from limited number of atmospheric concentration measurements. Monache et al. (2008) applied Bayesian inference and Markov Chain Monte Carlo sampling to reconstruct a contaminant source on a continental scale. By running the Bayesian MCMC algorithm on a large parallel cluster the inversion results could be obtained in few hours as required for emergency response to continental-scale releases. Senocak (2008) developed a method for stochastic event reconstruction of atmospheric contaminant dispersion using Bayesian Inference. Cowles and Carlin (1996) wrote a review article describing the markov chain monte carlo convergence diagnostics. Convergence criteria to the right solution and optimum step size for chain movement are important topics in the application of Markov Chain Monte Carlo method. Haario et al. (2001) described the importance of choice of proposal distribution for MCMC methods for convergence of the algorithm. An adaptive metropolis algorithm with Gaussian proposal distribution was introduced.

As shown in the above references, a number of researchers have worked in source location estimation using Bayesian probabilities methods. Since probability based methods alone cannot be sufficient during emergency response due to long convergence time periods, in the approach proposed in this paper, first an initial solution is quickly estimated using a least-square based method based on backward wind corridor. Subsequently, to obtain a more rigorous solution using Bayesian probabilistic method (referred as MCMC), the initial solution based on least square method is provided as a seed solution along with a search area/boundary. Thus, this two-stage method decreases the overall convergence time for probabilistic method by providing an intelligent apriori information. Using this approach, the emergency responders can quickly start responding based on the initial solution and more accurate information using a probabilistic method can be provided to them in due course of time. Thus, a combined least square-Bayesian approach ensures that even a probabilistic method provides a solution in much smaller time frame.

2 Approach

In this section, our hybrid approach for estimation of source area location is described. At first, source area location is estimated using a simple and intelligent technique developed using backward wind corridor and least-square based method (Khajehnajafi and Burla, 2008). The least-square method is an optimization based approach and several key features of this approach are listed in Table. 1. This method is very efficient and gives a good estimate for source location for most of the wind stability conditions, except when the wind is too stable because the slope of wind corridor is very narrow.

Another approach used for source location estimation is a probability based method using Bayesian theory where a probabilistic distribution is assumed for source area location and the sensor measurements and markov chain monte carlo (MCMC) simulations are used to estimate the source area location. Since this is a statistical approach, it takes significantly more time to converge to a value. Key features of this approach are listed in Table 1.

A Hybrid approach is a combination of the two approaches as listed in Table 1. Deterministic approach assumes that the parameters of the system are known completely without any uncertainty while probabilistic approach accounts for uncertainty in the model at the expense of longer time for convergence. Therefore, having a method that can narrow down the initial search zone for the Bayesian approach could be very useful in decreasing the convergence time. Thus, a hybrid approach is suggested in this paper where at first a Source Area Locator (SAL) model is used to narrow down the area and then Bayesian approach (referred as MCMC) is used to zero down on the actual source area location. The idea is that the combination of these two is a great method that accounts for weaknesses of both the models and makes it a very efficient model to narrow down upon a location very close to the actual source for all wind and stability conditions. Each approach on its own can provide reasonable results with some limitation either in terms of a further away solution for certain atmospheric conditions, or really long convergence time. Thus, the hybrid approach is a very valuable approach to estimate the source area location. Next, specific details of each of the methods are provided.

Table 1: Deterministic, Stochastic, and Hybrid Approach to Source Area Location

Deterministic (Least-square based)	Stochastic (Probabilistic)
• Assumes that all the system parameters are known perfectly • Usually faster runtime • Good for emergency response situations • Referred as SAL – Deterministic approach based on least-square based method to narrow down the search area	• Assumes uncertainty in system parameters • Usually longer run-time • Probabilistic approach • Based on Bayesian - Markov Chain Monte Carlo method (MCMC)
Hybrid Approach	
• Initially solved using deterministic approach based on least –square method SAL • Solution of deterministic approach is used to narrow down the search area for Bayesian - Markov Chain Monte Carlo method (MCMC). Thus, reducing the run time and obtaining a more accurate prediction of source parameters.	

2.1 SAL (Least Square Based Model)

The least square based method is referred to as SAL (Source Area Locator) in this paper. In this technique, it was demonstrated that with only the basic knowledge of meteorological information; wind speed, direction, and stability along with two or more gas concentration measurements, it is possible to quickly frame a region for the source of chemical (Khajehnajafi, and Burla 2008). A technique called 'reverse corridor' or 'back trajectory' is used to narrow the search domain for release location as seen in Figure 1 (a). This simple methodology uses the horizontal dispersion obtained from surface data at receptor locations. The construction of the corridor is key to reduce search area, therefore, reducing the computational cost. In Figure 1(a) the yellow region is the search domain if only sensor 1 and 2 were impacted. Introducing sensor 3 value, the search area is further reduced to the meshed zone. Adding additional sensors may narrow the search area; however, it will significantly increase the simulation time and increase the nonlinearity of the problem.

For the simple example shown in Figure 1(a) initially, sensitivity of the model to main parameters for dispersion, e.g. wind meander, sensor concentration fluctuations, and atmospheric stability was studied. A Gaussian model was used for that study. The result of SAL model for different stability classes A through F (1 through 6) showed that the accuracy of the model prediction for source area decreased with the increasing in stability class especially for higher stability classes 5 and 6 and at very low release rates. The results for the source location and the release rates, even when the model was simulated with the noisy sensor data, viz. 40% and 80%, matched very well with the true data when very fine mesh size was use (except for highly stable conditions at low release rates) indicating that the back trajectory method coupled with a simple Gaussian plume model is in fact a very simple yet plausible technique for locating source of pollutant.

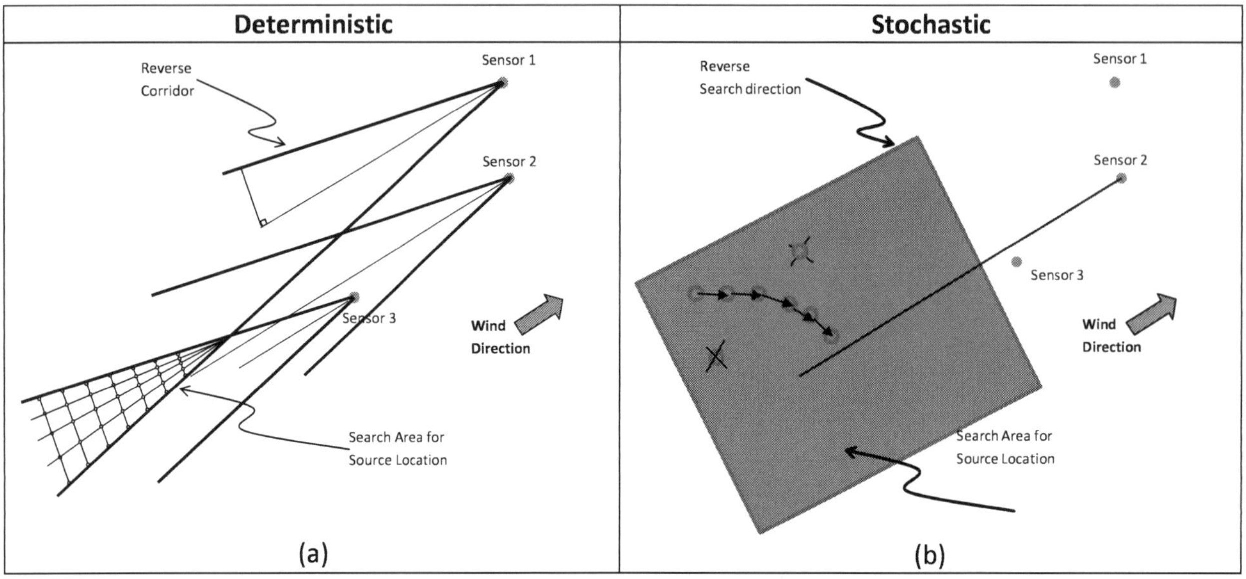

Figure 1: Pictorial view of deterministic and stochastic approach to source location of hazardous material release

2.2 Bayesian Model (Markov Chain Monte Carlo)

As mentioned in the introduction section, MCMC alone takes a lot of time to converge. Herein, based on the solution from the SAL model, an area was outlined as a narrowed search domain for the Bayesian probabilistic method. If the initial guess is chosen at random, the outcome from MCMC may take much longer time to converge and may even fails to converge sometimes because of broader search zone. In Bayesian method, a prior distribution is defined for parameters of interest (For example the location and release rate in this case). The likelihood function is estimated using the predicted concentration and measured sensor concentration. Then both prior distribution and likelihood distribution are used to estimate the posterior distribution. The posterior distribution gives us the latest knowledge of the original parameters of interest. The Bayes's theorem for the source reconstruction can be written as (Monache et al. 2008).

$$p(x,y,q|\,Sensor\ Data) = \frac{p(Sensor\ Data|x,y,q)p(x,y,q)}{p(Sensor\ Data)} \propto p(Sensor\ Data|x,y,q)p(x,y,q) \quad (1)$$

Where p( ) is the probability distribution. The x, y, q are the source coordinates and the release rate. *Sensor data* is the measurements from the different sensors. Bayes theorem relates the posterior distribution, $p(x,y,q|\,Sensor\ Data)$, to the product of the probability of measurements given the source parameters, $p(Sensor\ Data|x,y,q)$, also called the likelihood function, and the probability of the source parameters prior to any knowledge of the measurements, $p(x,y,q)$, also called the prior. For this study, x, y, q are assumed to be unknown parameters. One can include other parameters in the unknown list.

2.2.1 Prior Distribution

Let *m* be the model parameters i.e. all the parameters for which prior need to be defined. x, y, q are the prior parameters, x coordinate of the release location, y coordinate of the release location and the release rate, respectively. The prior distribution can be written as:

$$p(m) = p(x)p(y)p(q) \tag{2}$$

2.2.2 Likelihood Function and Acceptance Condition

The likelihood distribution is function of CP, predicted concentration for a sensor, and CM, measured concentration by an actual sensor. Here, the likelihood function is assumed to be of the following form (Monache et al. 2008):

$$\ln[(CM|S)] = \ln[L(\sigma, CP, CM)] = -\frac{\sum_i^N [log_{10}(CP_i) - log_{10}(CM_i)]^2}{2\sigma^2} \tag{3}$$

Where L is the likelihood function, CP_i is the predicted concentration values at the sensor location *i*, CM_i is the sensor measurements for sensor *i*, σ is an error parameter chosen accordingly to expected errors in the observations and predictions. *N* is the number of sensors. *S* is the state vector formed by all the unknown parameters (*x*, *y*, *q*). If the measured concentration values spread over several order of magnitudes, taking the logarithm of both predicted and measured values prevents large concentrations from dominating in the likelihood computation. The above likelihood is relatively simple, where the *N* prediction errors are assumed independent and expected to be proportional in size to the predicted concentration (as a result of log transforming both the observed and predicted concentration). If large errors are expected in the prediction/or measurements, large values of sigma should be assumed resulting in a broader-flatter posterior distribution. Here a sigma value of 0.8 is assumed. For a more complex scenario, it can also be treated as an unknown parameter and assigned a prior distribution.

The posterior distribution is sampled with an MCMC procedure via a Metropolis-Hasting algorithm.

2.2.3 Concentration Prediction Model

The concentration prediction model is based on a Gaussian plume dispersion model as shown below (Kim et al. 2011).

$$CP_i(x, y, z) = \frac{q}{2\pi U \sigma_y \sigma_z} \exp\left(-\frac{y^2}{2\sigma_y^2}\right) \tag{4}$$

Where, $CP_i(x, y, z)$ is predicted concentration at a particular location (x, y, z). q is the emission rate, U is mean wind speed, H is height of the release, x is distance along the wind, y is distance along the horizontal crosswind direction, z is the distance along the vertical axis, σ_y, σ_z are the standard deviations in the horizontal crosswind and vertical directions that are also called as Gaussian plume dispersion parameters and are defined empirically for different stability conditions. σ_y, σ_z are defined below:

$$\sigma_y = \zeta_1 x \,(1 + 0.0004x)^{-0.5}, \qquad \sigma_z = \zeta_2\, x \tag{5}$$

2.2.4 Sampling Procedure and Algorithm

Various algorithms exist for the sampling with MCMC. We have assumed uniform prior for the prior parameters. For the MCMC method, x is in the range -10,000 to 10,000 and y is the range 10,000 to 10,000 with (0,0) as the release location. q is uniform prior between 0.1 to 10 kg/s. U is the value given for the experiment of interest. H is assumed at the ground level. Ideally, parameter ranges can be assumed for wind direction, wind speed, ζ_1, ζ_2. However, for this study we have kept these constant. Of course, ζ_1 and ζ_2 will vary based on the stability class.
For the hybrid approach, the release location is first determined using the SAL approach. Limits for x, y are assumed to be ($-1000 + x_{sal}$, $1000 + x_{sal}$) and ($-1000 + y_{sal}$, $1000+ y_{sal}$) where x_{sal} and y_{sal} are release location coordinates obtained using the least-square method (SAL). Range for q is 0.001 to 10 Kg/s. The new value of x, y, q is obtained using adaptive sampling based on the current value and a step size.

2.2.5 Algorithm

An initial value for each parameter is sampled from its prior distribution. System parameters are assumed to be uniform prior with range as described above. The prior distributions are not completely uncertain as a good initial range is provided using the SAL method. Using the likelihood function, these sampled values from the prior distribution are selected to be part of the posterior distribution based on some criteria. A number of iterations involving source term sampling and dispersion simulation are executed until convergence to the posterior distribution is reached. Accepted parameter values collected during all the iterations constitute a Markov Chain which is defined as a set of consecutive values of parameter in which the probability of a value depends only upon the previous sampled value._Each iteration of the solution that becomes part of Markov Chain consists of following steps (Monache et al. 2008):

1. New value for source term is proposed (x_{prop}) according to $x_{prop} = x + dx$. Where x is the current value and dx is displacement from the current value modeled as a random walk sampled from a Gaussian distribution with zero mean and a standard deviation specified from the current step size. The Prior distribution $p(m)$ is used as a target distribution to estimate a prior likelihood of x_{prop}. If the prior likelihood of x_{prop} exceeds that of x, the proposed value x_{prop} replaces x. If not, a random (Bernoulli) "coin flip" determines whether the new proposal, even with its lower prior likelihood will be accepted. This ensures that sampling doesn't get stuck at a random point.

2. Repeat step 1 for the y and q parameters considered in the prior distribution. The assumption is that independent sampling of each parameter ensures that they reflect their likelihood better than sampling all of them at once and testing their likelihood
3. The predicted concentration for sensors is simulated using the current values of all the prior parameters using equation 4.
4. Likelihood of the current values of all the parameters which forms the proposed state is calculated using the predicted concentration and observed concentration by sensors. If the proposed likelihood is higher than the likelihood of previous state, it is accepted. If not, a random (Bernoulli) coin flip is used to ensure that the search reconstruction procedures continues to search the complete space of proposed source term parameter states, preventing the procedure from remaining indefinitely within the neighborhood of a local extreme.

After the proposed state is calculated, it is accepted if

$$\ln[L_{new}] - \ln[L] \geq \ln[rnd(0,1)] \quad (6)$$

Where L_{new} is the likelihood value of the proposed state and L is previous likelihood value and rnd(0,1) is a random number generated from a uniform distribution in the internal [0,1].

At a particular time instant, likelihood test for proposed state using measured data from all sensors at a particular time instant determines the acceptance or rejection of this state. Steps 1-4 are repeated to create a new proposed state with respect to all the parameters until the Markov Chain converges for a particular time instant. Note that in case of time varying information; step 1 to 4 can be further repeated for each time instant.

3 Case Studies: Comparison with the Field Data

3.1 BURRO EXPERIMENTS

BURRO experiments from the 1980 LNG (Liquefied Natural Gas) spill tests conducted by Lawrence Livermore National Laboratory (LLNL) and Naval Weapons Center (NWC) were used to compare the predictions from the model with the experiments. BURRO series experiments are one of the most comprehensive experimental analyses available in the literature for sensor measurement data of a chemical release (Koopman et al. 1980, Goldwire et al. 1980). In total nine experiments were conducted. Five experiments from the 1980 LNG (Liquefied Natural Gas) spill tests were used for this analysis. It is very difficult to get access to experimental data in this field and BURRO data are indeed very useful to compare modeling predictions with the experimental data for such a complex set of sensors.

This particular experimental setup employs close to 20-30 sensors. Not all sensors will see concentration values all the time so only the sensor with non-zero concentration reading are analyzed in this situation. This in itself is debatable as non-zero sensors may provide some information about the location which otherwise can be missed in case of uncertainty. Ideally, non-zero sensors should be included as well but for simplicity reasons we have not included those here and can be part of a further study. The next level of complication arises from the dynamic nature of the measured sensor values due to plume meander caused by horizontal turbulent fluctuation in the atmosphere. In ideal situation one expects the sensors to measure concentrations reliably based on the release rates. However, the uncertainty in wind direction, wind speed, measurement errors due to hardware and software and variability in atmospheric stability condition can contribute to a lot of uncertainty in the sensor data. For example, see the BURRO experiments report for concentration values measured by sensors for few experiments. Because of these uncertainties in the values and the simplified nature of our modeling approach, careful and intelligent decisions need to be made about the measured sensor readings that are used for this analysis as only a single measured value is used for each sensor in this analysis. Nonetheless, a more complicated model can be created which can include few/all aforementioned uncertainties at the expense of higher simulation time and resources. These may be included in the future studies of the project after carefully analyzing the tradeoffs involved in various approaches.

In our approach, the spill scenarios conditions and measured sensors values were used to obtain solution from SAL and MCMC approaches and then predictions for source area locations from the two models were compared against the actual experiment release location. In particular, five different spill scenarios from BURRO experimental series i.e. BURRO 2, BURRO 3, BURRO 7, BURRO 8, and BURRO 9 covering a range of atmosphere stability conditions are reported in this study (Table 2).

The release location was denoted as (0, 0) position in a new coordinate system, which is about 45 degree from the true north. The wind conditions were measured along the true north conditions. Therefore, wind directions for the new coordinate system are calculated by adding 45 degrees to wind direction values reported in the BURRO experiments. A plot in the new coordinate system containing actual release location and sensor locations is shown in Figure 2. The sensors were mainly distributed in four different arches as shown in Figure 2. Detailed location of these sensors and other towers can be found in the BURRO report part 1. The wind fields for different experiments are reported in the BURRO report part 2. To run SAL, a constant wind field (i.e. constant wind speed and wind direction) is assumed. For these short duration experiments, wind field does not change a lot so this is a reasonable assumption. Undoubtedly including variability in the wind field would result in a better prediction at the expense of more computational time.

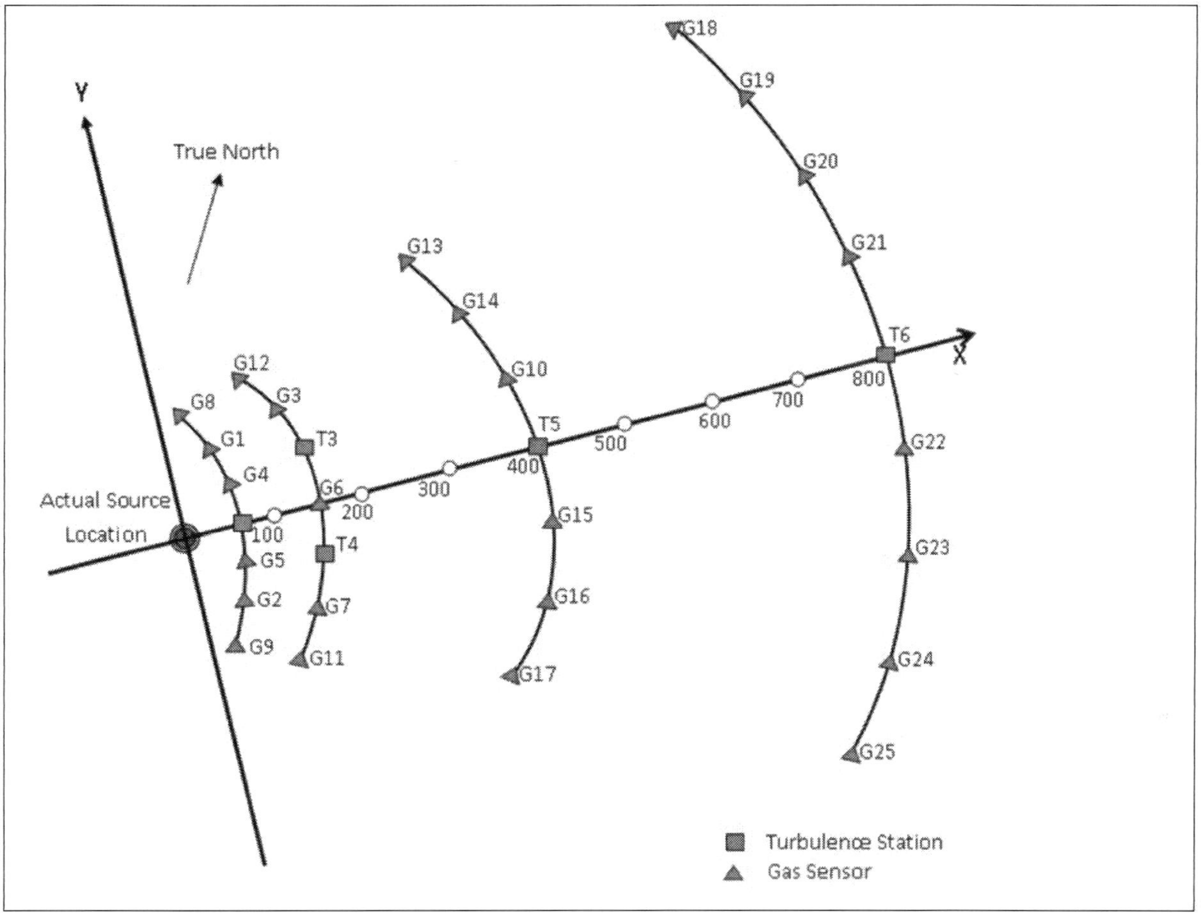

Figure 2: **Burro experiment setup of sensors.**

Table 2: Spill conditions for BURRO 3, BURRO 7, BURRO 8, and BURRO 9

Experiment Name	Chemical	Wind direction	Stability	Wind speed (m/s)	Spill Rate (m3/min)	Spill duration
BURRO 2	LNG	266	2	5.4	11.9	173
BURRO 3	LNG	270	2	5.4	12.2	166
BURRO 7	LNG	253	3	8.4	13.6	174
BURRO 8	LNG	280	5	1.8	16	107
BURRO 9	LNG	277	4	5.7	18.4	79

3.2 SAL Sensitivity to Different Cost Functions

A reasonably good approximation for source area location using SAL is our first target. To accomplish that goal we had to identify the best cost function. Three cost functions were identified; A simple concentration difference objective function **abs**(c_m-c_p), referred as CF 1, Normalized concentration based objective function [**abs**(c_m-c_p)/c_m], referred as CF 2; Distance and concentration normalization based objective function [abs (c_m-c_p)/c_m *$(x_{dist}^2+y_{dist}^2)^{0.5}$], referred as CF 3. Where c_m is the measured concentration and c_p is the predicted concentration.

Only a single measurement value per sensor is used for this analysis. Since it is a dynamic release scenario, various values can be selected from the sensor measurements. A sample of sensors output for Burro 3 is shown in Figure 3. Two set of data was used for SAL study. One was based on the single best sensor value for the entire time zone which reflects the best/average sensor values and the other is based on the maximum value for the entire time. All the sensors with non-zero sensor values were included in the analysis. For SAL simulation, thirteen different values of release rates ranging from 0.01 to 10 are used for source location. A local source point was identified which was average of these thirteen values and was used as an initial point for the next level search. First a coarse grid is used to narrow the search zone. Then a finer mesh grid search was carried out around that zone for various ranges of release rates to identify the source area location.

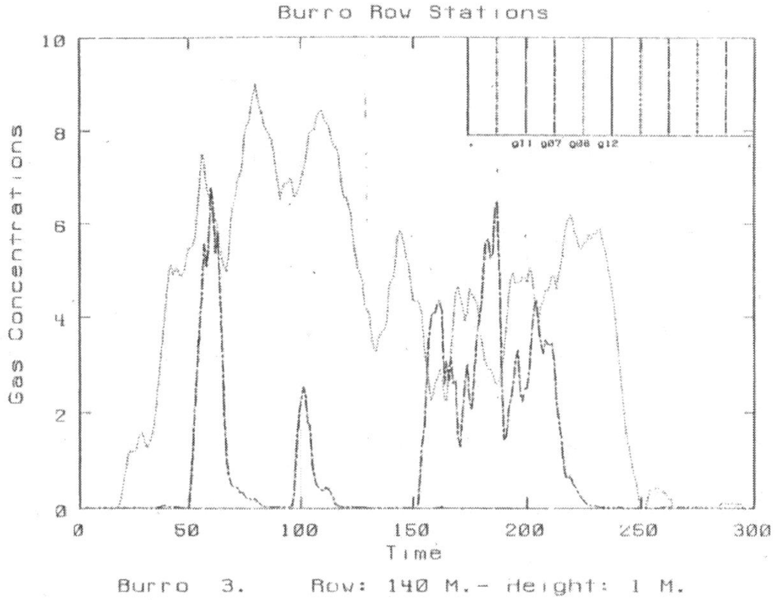

Figure 3: Gas sensors concentration measurement at 140 m arc.

The following tables represent the single sensor value obtained from each sensor positioned in the experimental grids for this study along with the source location predicted by SAL using three different cost functions.

3.2.1 Burro 2

Table 3: Sensor's value from BURRO 2 experiment

Sensors	G2	G5	G4	G1	G8	G3	G12	G13
ppm	8	5	1.2	1	2	5	4	0.02

Table 4: Source area location for BURRO 2 using sensor values

WD=266	CF1	CF2	CF3
Stability	2	2	2
X location	-65.1	-57.5	-56.6
Y location	9.2	7.5	8.8

3.2.2 Burro 3

Table 5: Sensor's value from BURRO 3 experiment

Sensors	G2	G5	G4	G1	G7	G6
ppm	9.0	4.0	17	15	4	5

Table 6: Source area location for BURRO 3 using sensor values

WD=270	CF 1	CF 2	CF 3
Stability	2	2	2
X location	-73.4	-117.2	-42.5
Y location	-14.8	40.2	16.3

3.2.3 Burro 7

Table 7: Sensor's value from BURRO 7 experiment

Sensors	G12	G3	T3	G13	G14	G18	G19
ppm	9.0	0.78	0.34	2.8	2.6	0.4	0.05

Table 8: Source area location for BURRO 7 using sensor values

WD = 253	CF1	CF2	CF3
Stability	3	3	3
X location	-111.6	69.5	191.6
Y location	-49.6	59.0	148.6

3.2.4 Burro 8

Table 9: Sensor's value from BURRO 8 experiment

Sensors	G9	G2	T2	G4	G1	G8	G11	G7	T4	G6	T3	G3
ppm	15	6	8	11	5	8	17	12.5	3	6	5	4.8
Sensors	**G12**	**G10**	**G16**	**G15**	**T5**	**G17**	**T6**					
ppm	2	0.02	0.7	0.7	2	3.8	0.25					

Table 10: Source area location for BURRO 8 using sensor values

WD = 280	CF1	CF2	CF3
Stability	5	5	5
X location	-633.9	-285.4	-355.2
Y location	119.8	70.6	77.9

3.2.5 Burro 9

Table 11: Sensor's value from BURRO 9 experiment

Sensors	G2	T2	G4	G7	T4	G6	T5	G16	T3	T6
ppm	6	1	7	1.2	2.2	6	0.25	0.2	4.5	0.2

Table 12: Source area location for BURRO 9 using sensor values

WD=277	CF1	CF2	CF3
Stability	4	4	4
X location	-70	63.5	66.3
Y location	8.8	-3.9	-5.3

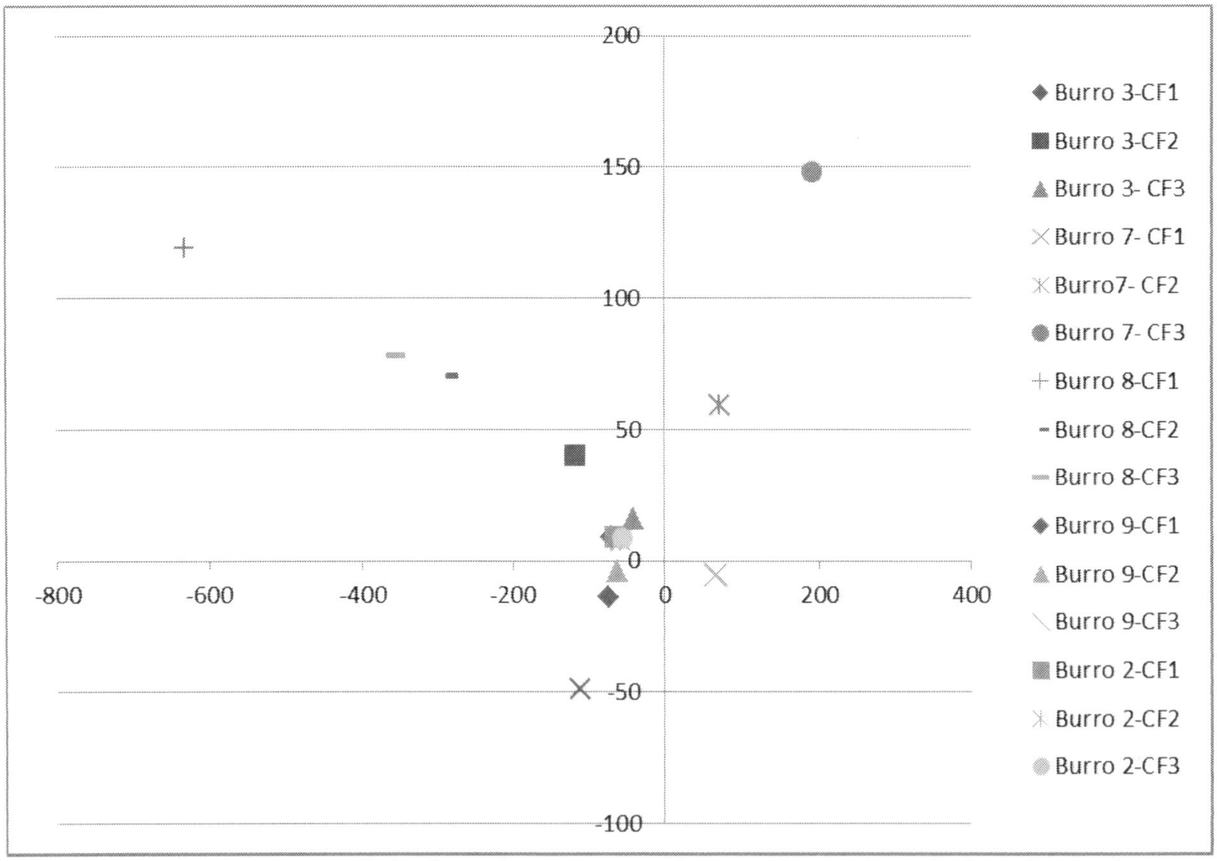

Figure 4: **Source location prediction for various Burro experiments using SAL (actual release is at 0,0).**

From this analysis we can see that the CF2 give better results out of all three objective functions. It was found that other than stability case 5, in most cases the source area was located within (+/- 70, +/- 70) range of the actual release location. SAL took only few seconds to run for all cases.

3.3 Hybrid Model

Subsequently, the solution from the hybrid approach with an initial guess from SAL to narrow down the search domain using Bayesian theory was obtained. BURRO MCMC refer to the results obtained using the hybrid approach in Figure 5. BURRO SAL refers to results obtained using the least-square based methodology. As evident from Figure 5, BURRO MCMC provides good solution for all stability conditions. Improvements in source location were seen for almost all the cases using the hybrid approach. These simulations were done for a wide variety of stability ranges including unstable to stable atmospheric conditions. Improvements seen in unstable cases using hybrid MCMC approach are more profound than stable cases as shown later in this report. The hybrid model improved the run time by a factor of two to three times.

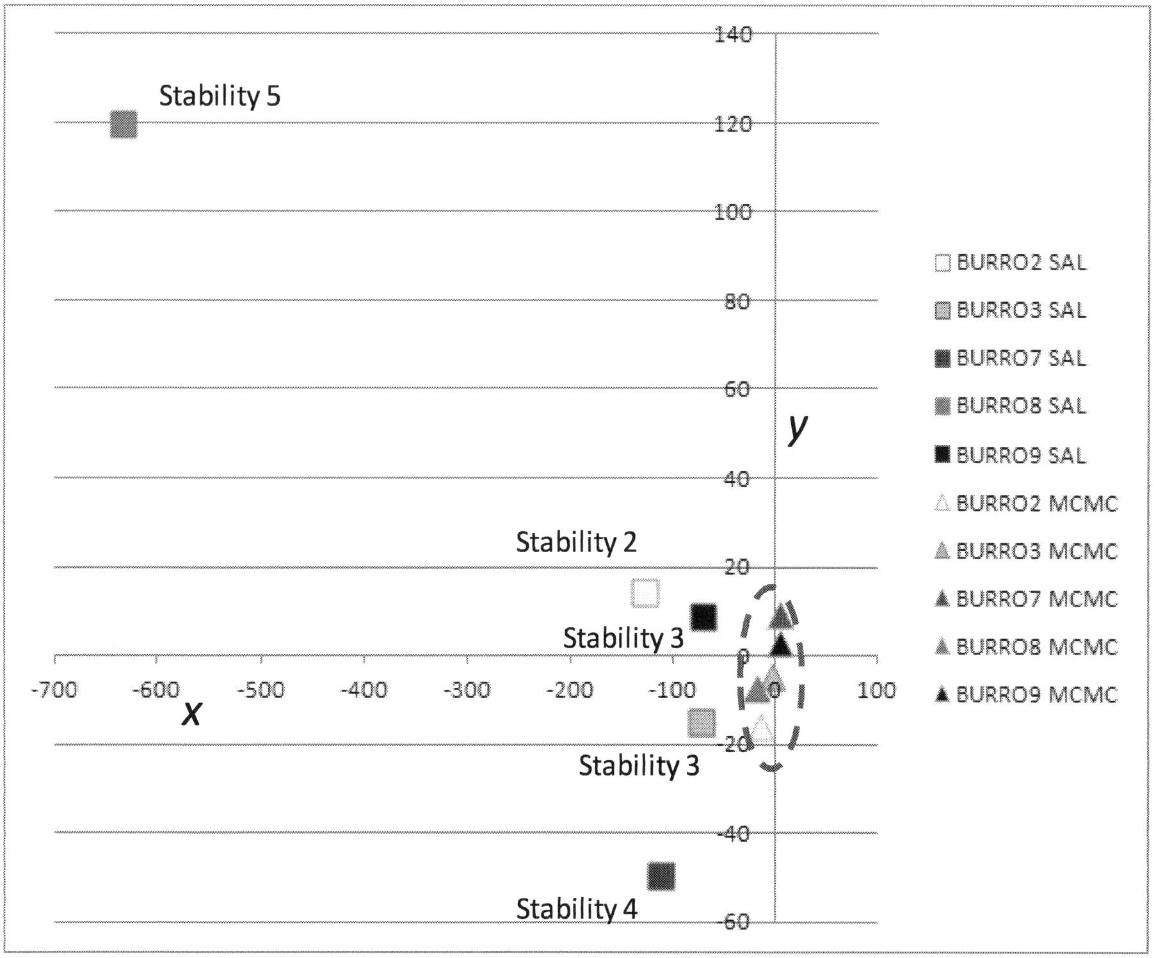

Figure 5: Burro experiment result from SAL and hybrid MCMC approach

Time-varying wind field, uncertainty in wind direction and speed, dynamic sensor readings, release duration, release time may be included in prior parameters for the hybrid approach where the main focus will be on including uncertainties and then identifying the location of the release.

If sensor data is available in dynamic fashion, these hybrid models can be extended to accommodate the dynamic Bayesian model (Johannesson et al. 2004) techniques.

4 Conclusion

The focus of this paper is mainly on estimating release location using the sensor measurements and providing an estimate for the release rate. An important method to estimate source area location using a hybrid approach based on least-square method (SAL) and bayeisan theory (MCMC) was provided. It was shown that the method provides a good solution for BURRO experiment data series for all stability conditions.

5 References

[1] Tarantola A., " Inverse Problem Theory and methods for model parameter estimation", Society for industrial and applied mathematics 2005, Philadelphia.

[2] Johannesson G., Hanley B., Nitao J., Dynamic Bayesian Models via Monte Carlo – An Introduction with Examples -, UCRL-TR-207173, Lawrence Livermore National Laboratory, October 2004.

[3] Cowles M. K. and Carlin B.P. ,Markov Chain Monte Carlo Convergence Diagnostics: A Comparative Review, Journal of American statistical association, vol 91 (434), 1996, pp. 883.

[4] Yee E., Lien F-S, Keats A., D'Amoursc R. , Bayesian inversion of concentration data: Source reconstruction in the adjoint representation of atmospheric diffusion, Journal of Wind Engineering and Industrial Aerodynamics, Vol 96, 2008, pp. 1805–1816.

[5] Yee E., Bayesian Probabilistic Approach for Inverse Source Determination From Limited and Noisy Chemical or Biological Sensor Concentration Measurements, Chemical and Biological Sensing VIII, Proc. of SPIE, Vol. 6554, 65540W, 2007.

[6] Kim J.Y., Jang H.K., and Lee J.K, Source Reconstruction of Unknown Model Parameters in Atmospheric Dispersion Using Dynamic Bayesian Inference, Progress in Nuclear Science and Technology, Vol. 1, 2011, pp. 460-463.

[7] Keats W.A. , Bayesian inference for source determination in the atmospheric environment, PhD thesis, University of Waterloo, Waterloo, Ontario, Canada, 2009.

[8] Neuman S., Glascoe L., Kosovi´c B., Dyer K., Hanley W., Nitao J., Event Reconstruction for atmospheric releases employing urban puff model UDM with stochastic inversion methodology, American Meteorological Society Conference, 2005, Atlanta GA.

[9] Zheng X., Chen Z., Inverse calculation approaches for source determination in hazardous chemical releases, Journal of Loss Prevention in the Process Industries 24 (2011) 293-301.

[10] Sharan M., Singh S. K., and Issartel J.P, Least Square Data Assimilation for Identification of the Point Source Emissions, Pure and Applied Geophysics, August 2011.

[11] Monache L.D. et al., Bayesian Inference and Markov Chain Monte Carlo Sampling to Reconstruct a Contaminant Source on a Continental Scale, Journal of Applied Meteorology and Climatology, pp 2600-2613, vol 47, 2008.

[12] Senocak I., Hengartner N. W., Short M.B., Daniel W.B., Stochastic Event Reconstruction of Atmospheric Contaminant Dispersion using Bayesian Inference, Atmospheric environment, Vol 42 (33), 2008, pp7718-7727.

[13] Koopman R. P., Baker J., Cederwall R. T., Goldwire H. C., Hogan W. J. LLNL/NWC (Lawrence Livermore National Laboratory/Naval Weapons Center) 1980 LNG (Liquefied Natural Gas) Spill Tests. Burro Series Data Report.

[14] Goldiwre H. C., Baker J., Koopman R. P., Cederwall R. T., Hogan W. J. LLNL/NWC (Lawrence Livermore National Laboratory/Naval Weapons Center) 1980 LNG (Liquefied Natural Gas) Spill Tests. Burro Series Data Report: The Appendices.

[15] S. Khajehnajafi, S. Burla, Prediction of Source Location and Rate of a Chemical Release for Emergency Response: Back Trajectory Approach, AIChE Spring Meeting, 2008

[16] Hogan W.R., Cooper G.F., Wallstrom G.L., Wagner M. M. and Depinay J-M, The Bayesian aerosol release detector: An algorithm for detecting and characterizing outbreaks caused by an atmospheric release of *Bacillus anthracis, Statist. Med. 2007;* **26**:5225–5252.

[17] Haario H., Saksman E., Tamminen J., An adaptive metropolis algorithm, Bernoulli, Vol 7 (2), 2011, pp. 223-242.

(43ak) Guidelines for Developing Site-Specific Human Error Rates and Human IPLs for LOPA Using a Safety Climate Approach

Farheen Khan, Human Factors Practice, Exponent, Inc., Chicago, IL, Sunil D. Lakhiani, Human Factors, Exponent, Failure Analysis Associates, Wood Dale, IL and Delmar Morrison, Thermal Sciences, Exponent, Inc., Warrenville, IL

Layer of Protection Analysis (LOPA) is a simple order of magnitude quantitative risk analysis technique commonly applied in the process industries. This powerful tool is especially useful for refining specific hazard scenarios during a process hazard analysis or a risk analysis. In LOPA, a hazard scenario is decomposed into a series of events, which may include the frequency or probability of equipment failures, personnel actions, and physical processes. By one estimate, 80 percent of incidents involve human error; thus, most risk analyses consider the likelihood and effect of human factors on potential scenarios. For LOPA, human error includes events initiated by the human in the work system and events where the human element failed to act as an independent layer of protection against an adverse event.

Although the LOPA method is widely used, there are few guidance sources for refining the human error component. The currently available standard LOPA estimates for the occurrence of human error assume optimal human factors conditions and that work systems are static environments unchanging over time. Work systems are, in fact, dynamic environments, which evolve over time through planned changes, as well as unplanned adaptations. Thus, no work system is completely free of error at any given time. Current LOPA estimates, therefore, fail to account for the dynamic nature of work systems where change from original conditions is not only possible but even necessary from time to time. This paper will examine the deficiencies in the LOPA human error probability estimates and propose a strategy that utilizes a safety climate approach to understanding error causation and incorporating it as a fundamental building block to human error estimation.

Benchmarking – An Important Milestone in the Journey towards PSM Excellence

Prerna Jain

Reliance Industries Limited

India

prerna.jain@ril.com

[1]CindiePridy, [2]Ravi Ramaswamy, [3]Robert Cravens

[1,3]Phillip Townsend Associates, Texas

[2]Reliance Industries Limited, India

cpridy@ptai.com

ravi.ramaswamy@ril.com

rcravens@ptai.com

Prepared for Presentation at

American Institute of Chemical Engineers

2013 Spring Meeting

9th Global Congress on Process Safety

San Antonio, Texas

April 28 – May 1, 2013

Benchmarking – An Important Milestone in the Journey towards PSM Excellence

Prerna Jain

Reliance Industries Limited

India

prerna.jain@ril.com

[1]CindiePridy, [2]Ravi Ramaswamy, [3]Robert Cravens

[1,3]Phillip Townsend Associates

[2]Reliance Industries Limited

Keywords: benchmarking, gap analysis, metric, performance gap, process safety management system, risk based process safety element, survey

Abstract

Reliance Industries Limited (RIL) has implemented a Process Safety Management (PSM) system through a structured and systematic approach over the last 6 years. In its PSM journey, RIL realized that there is a continuing need to improve and sustain PSM performance and to develop an understanding of the current status of RIL's existing PSM system, a benchmarking process of PSM practices, with reference to industry peers, is required. This was essential in order to identify and address the gaps to best performers within the peer group, define specific action items to obtain significant performance improvements, and ultimately achieve the best systems in the world in field of Process Safety Management.

To accomplish this goal, RIL conducted a PSM Benchmark Program Study designed by Phillip Townsend Associates (PTAI) and CCPS® at its six largest manufacturing locations in India. By helping to understand the relationship between process safety outcomes and management practices; this benchmarking program provided several key approaches for improving overall PSM performance. The program covered benchmarking of six elements related to a world class process safety management system. They are: process safety culture; compliance with standards, codes, regulations, and laws; hazard identification and risk analysis; asset integrity and reliability; management of change; and measurement and metrics.

The results of the study provided element specific quartile performance of each of the six manufacturing sites and Reliance as a whole in comparison to the industry peers who participated in the PTAI study. The gap analysis report presented by PTAI to the six sites

PSM team and senior management of RIL upon completion of the benchmarking study identified key areas of improvement for implementing and evolving effective Process Safety Management System at RIL.

This paper, presented by RIL and PTAI, discusses the benchmark process, methodology followed for the program, the value proposition of Process Safety Benchmarking, benefits of the program, key applications and lessons learned by RIL and new directions for improving PSM. The paper specifically focuses on the journey to success by realizing the gaps to top performers in the peer group, enhancing PSM knowledge and spreading PSM best practices.

1. Introduction

Chemical production and the associated processing industries dealing with hazardous and toxic materials have a significant commercial and societal responsibility to eliminate and/or mitigate the loss of primary containment or fires related to process materials while also ensuring that adequate safety systems are in place. Process Safety Management systems differ from personnel safety management in that process safety focuses directly on the processes, equipment and technology involved in causing an incident. The broader application of process safety management decision making and leadership behavior needs to address loss of containment events of greater and lesser consequences, safety system challenges and failures that could have led to failure, and operating discipline and management behaviors that ensure adequate layers of protection are being maintained. The focus shifts from the lagging indicators that track incidents to leading metrics that are used in a proactive manner to prevent incidents and continuously improve as shown in Figure 1: Process Safety Indicator Pyramid (API RP 754, 2010).

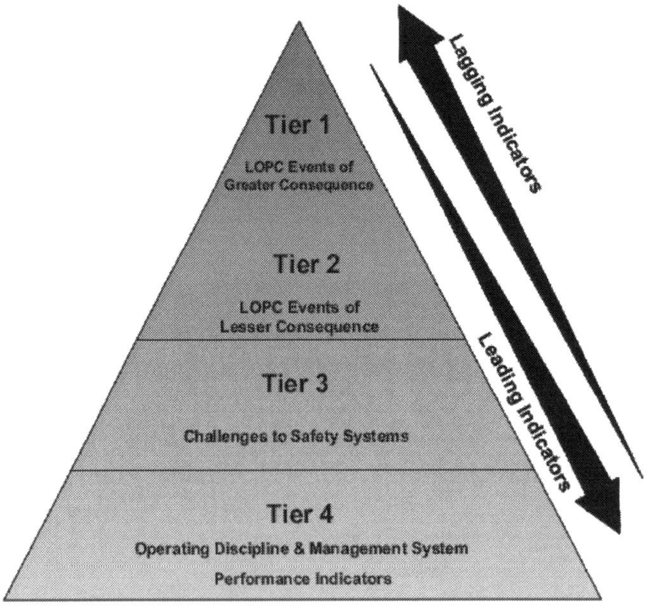

Figure 1: Process Safety Indicator Pyramid

The vision and recommendation of industry-wide process safety originated with the setting up of Centre for Chemical Process Safety (CCPS) by AIChE as a result of the December 3, 1984 Bhopal incident killing more than 3000 people due to toxic Methyl Iso Cyanate gas release. The need for continued focus on process safety got reinforced with the 2005 explosion at the BP refinery in Texas City, Texas. Since then, the Center for Chemical Process Safety (CCPS®) has identified 20 process safety elements to be used as guidelines for a risk based safety process (CCPS®, 2007). In 2009, the CCPS® and Phillip Townsend Associates, Inc. (PTAI) entered into an agreement to develop a benchmarking program for Process Safety Management Systems (PSM).

Viewed from a risk based perspective, the process safety elements can be grouped into four categories: Commitment to Process Safety, Understanding Hazards and Risk, Managing Risk, and Systematic Learning from Experience. PTAI and CCPS® designed the PSM benchmark selecting six of the 20 elements: Process Safety Culture, Compliance with Standards, Hazard Identification and Risk Analysis, Asset Integrity and Reliability, Management of Change, and Measurement and Metrics (CCPS®, 2007). The key goal was to seek an understanding of the relationship between process safety outcomes and management practices.

This paper discusses the details of the PSM benchmark, the value proposition to be gained, and specifically the experience of Reliance Industries Limited (RIL) in developing an understanding of the current status of RIL's existing PSM system, and the practices that will improve and sustain PSM performance. The objective of RIL is to take PSM systems and practices to the top quartile performance through a benchmarking process of industry peers relative to PSM practices. Specifically, the paper focuses on the most important decisions and behavior changes that leadership can make to spread process safety awareness throughout the organization, close the gap to the top performers, develop central and site specific action plans for improvement, and raise the internal level of all organizational units to the that of the top internal performer.

2. Methodology

Benchmarking identifies how well a company is doing compared to its peers in the industry, what the industry top performance is, how large the gap is to the top performers within the industry, and create a stimulus for continuous improvement. Benchmarking programs seek to understand the relationship between outcomes and management practices. The PSM benchmarking program is a systematic study of process safety between companies that will allow sites/companies to compare their management systems with the goal of improving the design and implementation of management practices that impact process safety.

2.1 Value Proposition of Process Safety Benchmarking

PSM benchmarking can elevate a company's process safety performance through four distinct steps:

2.1.1 Identify performance gaps

The identification of a company's performance in relation to Performance Quartiles will highlight key drivers for performance enhancement versus its peers. These quantifiable results replace perceptions with a data-based reality check. In addition, these results identify if the enterprise policies, practices, and procedures are operating as intended and are achieving the desired results.

2.1.2 Create performance enhancement goals

Data driven information provides a measure of the enterprise effectiveness – emphasizing strengths, weaknesses and potential for improvement over time. This helps focus attention on the critical aspects of an enterprise that create risk. It also underscores what improvements should be made – providing quantifiable performance enhancement goals.

2.1.3 Measure the effectiveness of improvement programs

PSM benchmarking can be a starting point for goal setting and performance initiatives aimed at boosting step change improvements in process safety. The clear outcome is that this will add significant shareholder value through reduced risks of catastrophic incidents.

2.1.4 Maintain continuous improvement

The PSM benchmark helps maintain a stimulus for continuous improvement. It provides the necessary motivation to ensure consistent top performance results.

2.2 Benchmarking Process

A peer group is established by PTAI of sites/companies agreeing to compare their process safety programs and performance in an effort to identify best practices and drive improvement. Ensuring consistency and protection of each participant's interests are critical to a successful benchmark. A confidentiality agreement was completed between PTAI and RIL along with execution of a participation agreement. The Data Collection Instrument (DCI), providing clear and consistent definitions for all participating sites/companies was made available. The use of a secure site over the internet ensured security of data and ease of use by participants as they are able to see all the survey questions and select their answers. A kickoff discussion to align key stakeholders at each of the sites is critical. In RIL case, this was done through a series of webinars between PTAI consultants, RIL's Centre for HSE Excellence and stakeholders at each of the participating sites.

RIL chose to benchmark PSM in six of its large Refining and Petrochemical manufacturing sites in India and a team of site focal points, corporate coordinator and PTAI worked in a seamless manner to plan and execute the benchmark program. Following submission of the data, PTAI validates the data on a site and company basis and provides analysis of the results. Validation was enhanced with RIL by individual site visits to each of the six sites by PTAI. This provided the opportunity for discussion and verification of the results reported to the on-line survey. Participation by key RIL personnel during this process added to the effectiveness and accuracy of data reported and provided a seamless voice to the sites between RIL's Center for HSE Excellence and PTAI.

In particular, statistical analysis of the data was performed by PTAI to evaluate various drivers of process safety using a Gap Analysis comparing RIL's performance to the top quartile of the peer group. The quantitative and qualitative results were distributed to participants with numerical scores for each element along with an overall site/company specific score. Strengths, as well as the main opportunities for improvement to close the gap towards the top performers were identified. A discussion of the study results and gap opportunities was held with the executive management in an executive management summary with specific individual gap analysis presentations to the top Line and HSE management of each of the six RIL sites benchmarked.

2.3Benefits of PSM Benchmarking

Management review is an integral part of moving from lessons learned to systematic improvement of PSM. Reviews of the performance gaps can identify potential problems and address them before an incident occurs; as well as, identify opportunities for improvement to move from current levels to the level of top performers. PSM benchmarking helps an organization prioritize the allocation of process safety resources, both financial and human, to focus on the highest identified risk areas.

Among the benefits of PSM benchmarking, there are certain ones that can have a big impact on an organization. There should be a marked increase in staff awareness of process safety issues. This awareness will expand the issues from the Health & Safety group to Engineering to Operations. Based on the organizational level of results discussion, it will potentially increase awareness from site management to impact those that are responsible for setting corporate strategy. A key learning for RIL was the differences between the individual site scores resulting in a changed perception by the Center for HSE Excellence and executive management. PSM benchmarking creates a common understanding and language on process safety matters. Benchmarking can also reduce risks by potentially providing an incident early warning signs.

Sometimes the opportunities for improvement can be very different from the perceived gaps in an organization. Identifying and gaining insight on perceived gaps versus actual gaps is greatly enhanced by an independent benchmark which effectively creates consistency across multiple sites within a company and with the others in the peer group.

3. Key Applications and Lessons Learned

Traditional process safety surveys have identified a list of both quantitative and qualitative issues. Translating the qualitative issues into measurable, actionable steps is a key issue for senior management decision making and allocation of resources. Benchmarking results must be measurable and use a scoring methodology that will not only identify the performance gaps, but also quantify the impact of selected actions. The

PSM benchmarking process designed by CCPS® and PTAI includes a scoring methodology which enables participants to have measurable action plans. Benchmarking is a systematic way to make site-to-site company comparisons and view a company's performance level in relation to that of top performers.

The application of the benchmarking results to identify performance gaps provides meaningful insight for organizations as they develop goals and the associated improvement plans to reach those goals. Repeated benchmarking monitors the rate of improvement. One example of a lesson learned that applies to numerous organizations relates to Process Safety Culture. The findings from the Process Safety Culture section indicate the largest range of results compared to the other sections. The underlying cause of the low scores was a direct outcome of not monitoring the process safety culture through the use of an employee survey. Without a way to assess their work force's values, attitudes and behaviors pertaining to process safety, an organization will have no meaningful basis for making and measuring improvements.

The global process safety incidents have produced tremendous need for the use of PSM benchmark assessment to measure and improve process safety performance. Initially focused on fatality rates, manufacturers quickly learned that having strong PSM system and effective layers of protection were important to prevent any loss of containment scenarios.

At RIL, PSM Benchmark assessment and results helped in identifying following three main interrelated but distinct benefits or observations: (a) Internal Benchmark of various RIL sites, (b) learning through process of understanding the benchmark questionnaire, and (d) need for a detailed analysis of existing systems implementation. We briefly discuss and illustrate examples of each benefit or observation, highlighting how the results could be used.

(a) Internal Benchmark of RIL multiple sites:

PSM Benchmark program conducted for six RIL sites brought to the surface best practices followed at different sites like: Number of indicators tracked and analysis of metrics trend, Quality control reviews of Process Hazard Analysis reports, action plan clearly defined from the Operational Discipline survey results, employee involvement through various PSM campaigns and some other actions related to operational discipline. These practices were beyond the existing standard and system requirements and became good learning for other sites. Also, these best practices will be additional inputs to the group level company documents on PSM. For example, the delta between the site with the highest score and the site with the lowest score in the Compliance with Standards section was 17%; offering a significant opportunity to analyze internal best practices and improve the level of performance to the highest score.

(b) Learning through process of understanding the questionnaire

The benchmarking questionnaire itself generated opportunities for learning new knowledge. Understanding the "why" part of questions called for discussions among various functions at each of the manufacturing sites and reinforced the understanding of PSM. A self-assessment process got started on the compliance to PSM requirement. Following table gives certain examples of learning and new knowledge for the questionnaire itself:

Table 1. Examples of learning and new knowledge from Benchmark questionnaire

Sr. no.	PSM Element	Question	Learning
1	Hazard Identification and Risk Analysis	How are overdue HIRA action items escalated in your organization?	Tracking system to be further strengthened for overdue HIRA action items.
2	Asset integrity and reliability	How is the metal thickness data from the periodic inspections analyzed for trends?	Thickness measurement trends should be analyzed, remaining life projections should be calculated, and adjustments to be made to inspection intervals.
3	Process safety culture	If your organization has utilized a survey tool to assess Process Safety Culture, how many questions were on the survey?	PSM Culture Survey to be established for the organization.

The validation visits from PTAI enhanced the quality of responses and the participation of various functions in the PSM benchmarking program.

(c) Need for a detailed analysis of existing systems implementation

The PSM benchmark results identified areas for further analysis to enhance the systems implementation for improved performance. At RIL, the trend of Process Safety incidents is varying for the last three years and there is no clear trend of reduction in loss of containment events from the metrics, inspite of implementing PSM systems for the last four years. For this, RIL has initiated a group level six sigma project on reduction in loss of containment events with support of an inter-site team. The project follows Define, Measure, Analyze, Improve and Control

(DMAIC) methodology of six sigma. Following is a brief of activities initiated for the above:

- A multisite team has been formed with a six sigma Black Belt guide.
- Literature on Process Safety and Six sigma circulated to team members.
- Project charter was prepared and circulated to team members and other stakeholders for inputs/suggestions and firmed up.
- Cause-Effect Matrix Format developed and Process mapping completed for PSM elements.
- Audit Report Analysis.
- Incident investigation report analysis.

4. New Directions and Next Steps

The PSM benchmark program results gave new insights to the RIL team on the status of PSM within RIL and in relation to the other participants. The gap analysis was converted to actionable recommendations at Group Level and individual site level. Next steps in the PSM excellence journey for RIL have been identified and these are

- Develop a formal system of Process Safety Culture Survey.
- Review of existing PSM metrics and make necessary amendments.
- Review Human Reliability/Human Factors/ Human-Machine Interface requirements, develop best practice requirements and make necessary amendments in Standards.
- Develop program for assessment of compliance with standards/codes/practices for existing equipment.
- Establish system to manage deviations from the industry consensus standards.
- Roll out the Management of Change –Personnel process formally.
- Actual implementation of Standards/Procedures/Practices in field:
    - Meeting target inspection/testing intervals, backlog of overdue inspections
    - Closure of recommendations of PHA, QRA studies
    - HIRA team members to meet qualification requirements
    - Closure of Audit recommendations
    - Implementation of temporary Management of Change
- Review RIL Group PSM Standards and Procedures for gaps and make necessary amendments.

A key benefit of the program for RIL is an Internal Benchmarking among its major manufacturing sites and sharing the best practices within the group. All the sites have developed action plans to close their own gaps in PSM practices.

The collaboration between PTAI and RIL gave new directions for RIL as well as for the Benchmark program itself. A development of two distinct Plan Do Check Act Assess (PDCAA) cycles, one for RIL and the other for PTAI has been the result of this collaboration.

4.1 Overview of PDCAA cycle

PDCAA is a 5-step methodology for process improvement. The stages are shown in Table 2.

Table 2. PDCAA stages

Plan	Establish objectives and how they will be achieved
Do	Put the plan into effect
Check	Verify that the process achieved the desired results
Act	Analyze any differences and their causes. Take action to improve things.
Assess	Analyze and reflect on evidence to make informed and consistent judgments on an ongoing basis for continual improvement.

The PDCAA Cycle for RIL is addressed through the following:

- Planning upgrade of RIL PSM standards to industry best levels and establishing systems to manage deviations, and developing appropriate leading and lagging metrics.
- The do part of the cycle will address the actual implementation of standards and practices in the field at different sites through a process of sharing the best practices by institutionalizing the inter-site network teams for the PSM elements.
- Check part will address strengthen the current PSM metrics tracking and review systems, IT enablement of dashboards for Process Safety at different levels of organization, and reflecting the updated requirements of compliance in the PSM audit protocols.
- A robust system of closure of recommendation and actions arising out of various studies and audits through a system of IT enabled universal work list and linking it to individual performance management system is under development which form the Act step of the cycle.

- Periodic reviews and assessment of PSM will complete the whole cycle of improvement.

4.2 Plan-Do-Check-Act-Assess- RIL Example: (Hazard Identification and Risk Assessment)

Plan

In this step, process needs are analyzed by examining a range of data available from the Benchmark assessment gap analysis relating questions in program to performance results. Data can be analyzed at the individual sites level and the company level. In this process, aspects on HIRA management system, including HIRA methodologies, team composition, criteria for HIRA team leaders/facilitators, level of training, techniques, risk criteria, recommendations and their status were covered.

Do

This step asks what group standards require and how they will be assessed. Forward action plan is made by looking at what is lagging in actual implementation in field ; such as, the timely closure of HIRA recommendations and minimum criteria/qualification for team members. Sites develop individual goals to improve their performance.

Check

In the "check" step, formal and informal assessments take place continually, to strengthen the performance of Metrics on HIRA elements. Sites can make use of comparative data to identify trends to study; for example, overdue HIRA action items.

Act

In this example the "act" step calls for standardization or having a robust system to ensure closure of overdue action items. Results from this cycle become input for the "assess" phase.

Assess

In this step, process of periodic reviews and monitoring of HIRA action items is done and tracked to ensure continual improvement.

4.3 Plan-Do-Check-Act-Assess- PTAI Example: (The benchmarking process and Process Safety Culture)

The PDCAA Cycle has also provided the opportunity for continuous improvement for PTAI and the PMS benchmarking survey.

Plan

Through the validation process the iterative nature of conference calls and site visits proved to be of significant value in providing accurate and consistent understanding of the survey questions. This resulted in comparable and meaningful results among the sites and RIL total scores.

Do

Reviewing the current benchmarking process and seeking feedback from RIL identified that enhanced validation and the creation of a more iterative process could be incorporated into the benchmarking process. A key area of needed improvement supporting the benchmarking process in the Process Safety Culture section is the measurement of Process Safety Culture. Both RIL and PTAI determined the need for a formal system for a Process Safety Culture Survey that provides more consistent measurement and comparable results of the safety culture that can be utilized for RIL internal site to site comparisons and RIL's sites to the external peer group. RIL and PTAI have discussed future collaboration on this project and the possible implementation to all of the RIL sites for internal benchmarking and review.

Check

Informal and formal process assessments are the key to continuous improvement of the PSM Benchmarking Survey and process.

Act

The act step calls for standardization of a formal process assessment tool to measure the benchmarking efforts required by participants and the actual benchmarking survey tool.

Feedback from participants needs to be evaluated and appropriate improvements to the program parameters and the benchmarking process implemented.

Assess

This is the process of continuous improvement and an environment of continuous learning. PTAI will periodically review the process and enhancements with CCPS to ensure an evolving process that grows in value-added elements.

5. Key success factors

The success of the PSM benchmarking survey hinged on four key factors; management support of a systematic approach, the strength of the validation process, the ability to report results with a detailed gap analysis that could be translated into specific action plans at a central and site level, and the seamless voice of RIL's Center for HSE Excellence and PTAI through strong communication and a systematic effort.

The management presentation of results was opened with comments from RIL's President, HSE, emphasizing the strategic importance and focus of PSM as a systematic process and not simply a process safety program. His comments were directed and focused on the areas he felt most important to taking RIL to the next level. This sent a very clear message of the strategic focus placed on PSM at RIL.

The continued dialogue and collaboration between RIL and PTAI has enhanced the double loop, PDCAA cycle. As a critical success factor this creates ongoing and new opportunities for both.

6. Conclusion

Process safety is a mature, but continuously evolving concept with practices and procedures varying widely among the refining, chemical and chemical processing industry. Moving from a compliance model to a risk-based model requires a comprehensive approach to managing process safety. Substantial improvement opportunities exist for even the current top performers. Specifically, no site has made a perfect score on any of the risk-based elements of the PSM benchmark.

The PSM benchmarking program is a unique methodology to allow companies to identify how large the gap is to top performers and to "learn" how to prioritize resource allocations based on the greatest impact opportunity to improve their process safety management system. This shifts the understanding of current practices to the lessons learned that are measurable and actionable. The benchmarking process identifies top quartile management practices while highlighting not only a company's strengths, but also potential development opportunities. The program's scoring system provides a marker to track improvements and measure the gaps within the company sites and against external top performers.

Using a systematic Process Safety Management System benchmark drives lessons learned to strategic and organizational learning.RIL considers periodical assessment of its PSM through repeat benchmarking at regular intervals as a good value proposition to monitor trends of performance and to achieve pacesetter performance compared to industry peers.

7. References

[1] American Petroleum Institute, API RP 754, "Process Safety Performance Indicators for the Refining and Petrochemical Industries, First Edition," 2010.

[2] Center for Chemical Process Safety CCPS®, "Guidelines for Risk Based Process

Safety," John Wiley & Sons, Inc., Hoboken, NJ, ISBN 978-0-470-16569-0: liii-liv, 2007.

9TH GLOBAL CONGRESS ON PROCESS SAFETY

"Process Safety Management - A Practical View"

German Luna-Mejias, P. Eng.
ACM Facility Safety
Suite 700 940 6th Ave SW, Calgary, AB, Canada
gluna@acm.ca
guatire2@gmail.com

Prepared for Presentation at
American Institute of Chemical Engineers
2013 Spring Meeting
9th Global Congress on Process Safety
San Antonio, Texas
April 28 – May 1, 2013

UNPUBLISHED

"Process Safety Management - A Practical View"

German Luna-Mejias, P. Eng
ACM Facility Safety
Suite 700 940 6th Ave SW, Calgary, AB, Canada
gluna@acm.ca
guatire2@gmail.com

Keywords: Process, Safety, Management, Practical

Abstract

Process Safety Management (PSM) was developed over 20 years ago to reduce hazards and risk for the Oil Industry. In the last 3 years serious accidents have happened in the Oil Industry that requires us to re-evaluate PSM programs. It has been observed in many HAZOP studies, that the HAZOP (part of a PSM program) does not reflect a real commitment to safety, but the "need" to meet regulations. This paper's initiative is to call organizations/companies to think and be safe instead of just fulfilling project or government rules. Full involvement of the organization, starting with managers, is required for the PSM program to work. In this paper, practical examples will be drawn from the successful implementation of a PSM program from several years ago in a refinery. Additionally, the reasons for this implementation to be considered successful will be discussed at length.

1. Introduction

1.1 History

On June 1st, 1974, a Cyclo-hexane vapor cloud explosion killed 28 workers and injured 53 others; 1800 houses were damaged in the nearby area in Flixborough, England. [1]

On Dec 3, 1984, an uncontrolled release of methyl-isocyanate (MIC) gas from a pesticide plant in Bhopal, India, resulted in more than 3,800 fatalities and approximately 11,000 injuries with different degrees of severity. [2]

In 1985, the American Institute of Chemical Engineers (AIChE) formed the CCPS (Chemical Center for Process Safety) quickly set out to publish its first process safety guideline book: Guidelines for Hazard Evaluation Procedures.

On July 6, 1988, the offshore platform Piper Alpha, located in the North Sea and owned by Occidental Petroleum, suffered a series of explosions and fires that resulted in the deaths of 165 crew members and the total loss of the platform. [3]

On Oct 23, 1989, an explosion in the Phillips refinery in Pasadena, Texas killed 23 people and injured 314 more. Impact of the explosion was registered as a 3.5 on the Richter scale. [4].

In 1990, the American Petroleum Institute issues the recommended practice 750: "Management of Process Hazards" which can be considered the first step onto the Process Safety Management programs. [5].

In 1992, OSHA formally recognized the vital importance for managing highly hazardous chemical processes within the US. A company that is involved in highly hazardous chemical processes or activities is mandated to implement a process safety management system. [9]

In 1994, CCPS modified the elements of the PSM program recommended in API RP-750 [5] to issue a new "standard" for PSM that includes more elements: "Guidelines for implementing Process Safety Management System"

After a period of "relative calm", big accidents in the oil and gas industry occurred again:

- BP Texas city, 2005 [6]
- BP Macondo, Gulf of Mexico 2010 [7]
- San Bruno, California 2011 [8]

Many disasters have impacted the process industry, from Flixborough to Macondo and elsewhere many more. Through the ashes of these tragedies, strong change has been advocated by the public and their government's regulators to hold companies accountable for their negligence.

1.2 Process Safety Management

Process Safety Management (PSM) is a methodology that was developed after the lessons from several major accidents showed that such events can be difficult to prevent using traditional occupational health and safety approach, which tends to focus on the interface between the individual operator and the equipment or process.

The 3 keywords of this philosophy are:

- PROCESS: related to process issues. i.e. fires, explosions, etc
- SAFETY: Safety regulation, safety concern
- MANAGEMENT: anyone is a manager in some degree, requires employment participation.

PSM is the application of management principles and systems to the identification, understanding and control of process hazards to prevent the release of highly hazardous chemicals such as toxic, reactive, flammable and/or explosive substances (process-related events), that may lead to injuries and incidents.

The goal is to minimize process incidents by evaluating the whole process. The PSM standard (OSHA 29- Part 1910.119) [9] is the regulatory framework for process industries in the US and is quickly becoming the industry "best practices" standard for responsible companies worldwide.

The PSM system originally suggested by the CCPS (Guidelines for implementing Process Safety Management System) consists of 12 elements:

1. Accountability: objectives and goals
2. Process knowledge and documentation
3. Capital project review and design procedures
4. Process risk management
5. Management of change
6. Process and equipment integrity
7. Human factors
8. Training and performance
9. Incident investigation
10. Company standards, codes and regulations
11. Audits and corrective actions
12. Enhancement of process safety knowledge.

The intent of this paper is to present some of the elements of PSM that were considered the most important for a specific company; not to debate the number of elements or which one is more important than the other.

Regulations should not be the sole driving force for the process industry to "think safe". Applying the PSM program accurately and effectively requires more than just the requirement to fulfill government regulation but a whole company culture to be involved in making the workplace safe. This paper's initiative is to encourage organizations/companies to think and be safe instead of simply fulfilling a project or complying with government regulations.

The examples given in this paper correspond to a company where PSM was not mandatory by law, but it was part of the company's standard as a way to make sure the company complied with the International O&G industry standards.

2. The practical Approach

In this section, examples of the approach taken by an O&G company to implement PSM are given for what was considered the most relevant part of the program. Not all elements are discussed as the key elements described below will support all other elements. The practical approach also includes unfortunate situations that happened before the implementation of the program to emphasize the need for it.

Elements analyzed in this document are:

- Management commitment (Element 1)

- Process Safety Information (Element 2)
- Process Hazard Analysis (Elements 4 & 5)
- Operating Procedures (Elements 6,7 & 8)
- Hot work Permit (Element 6)
- Training (Element 8)
- Pre-start up Safety Review (Element 9)
- Emergency Planning and Response (Elements 4 & 12)
- Program Audits (Element 11).

2.1 Management Commitment

Management commitment is fundamental to the PSM program. Exercising and eating well is something that people need to do for their own good and not because it is a mandate or a command to obey. Exactly the same happens with the PSM program: if people are not convinced of the importance of safety, the PSM program won't last. The first safety best practice identified, and one which is essential to any good safety program, is a management commitment to safety.

No company can have a successful PSM program if the management of that company is not committed to the concept of safety. Every company has an obligation to ensure, as much as possible, the safety of its employees. Not only does the company have a moral obligation to ensure the safety of their workforce, it is also good business to do so. Keeping the number of incidents down will also help the company to maintain few lost time days, low work injury indexes and increased profitability.

In a particular refinery, the General Manager wanted to develop a PSM program and designated a top manager to be in charge of the PSM program. This designated manager's first action was to convince all top managers of the importance of the PSM program. Convincing the other managers was not based on: "I said so, and I have been appointed by the GM" but with education and conviction.

The training of managers and cascading this knowledge to all personnel with supervisory responsibilities started with the video forum about the Piper-Alpha accident (one of the objectives of the forum was to recognize that this scenario may happen to that facility). The Piper-Alpha video, in conjunction with API 750 video, was presented to all personnel.

"Safety oriented Management behavior" was a key, as the employees know who the managers are, and they observed their commitment to safety. All personnel understood that they were "living by example" instead of "giving lip service to their own safety rules". The exposure of management risk and demonstration of Safety Concept by action prevented the scenario of: "Well, management doesn't need to operate safely, neither do we."

Additionally, financial incentives to personnel were also created during the earlier phases of the program to promote participation within it. After a period of time, people became conscious about safety and they started to behave safely even when "nobody is looking"; not because of a reward but they were convinced of PSM as a way of life.

Part of the management commitment to safety was the implementation of safety related training programs with the same level of importance as technical training. The program was successful because there was support from the entire management team, and management was involved in the formulation, communication and enforcement of its safety program. All levels of management were visibly demonstrating their commitment to safety.

2.2 Process Safety Information.

Process Safety information (PSI) includes design information, mechanical design and working knowledge. These 3 elements are linked to good quality P&IDS, which was the driving reason the company decided to move in the direction to update the P&IDs as the first step under the PSI element.

This was critical to the facility as it was a 60 year old refinery. In several cases the installation of jump-overs, minor changes and similar modifications were done in the field without any analysis or documentation on the P&IDs. Some minor incidents (spills, spray, etc.) were found to be a consequence of inadequate information on P&IDs.

As a result, a team composed with very experienced operators and young engineers was formed to update all the P&IDs of the facility, with two main purposes: pair young engineers with very experienced personnel and teach them how to update P&IDs and to comply with PSM.

A few years later, when most of the 30+ year operators were retired, these updated P&IDs were successfully used to train the new group of operators in the facility.

Another group, supported by Technical Services, began work to update and maintain the MSDS (Material Safety Data Sheets) for the facility. This action proved to be very effective for the firefighters as they were properly trained in the materials they may have to deal with in case of emergency. Before that they were relying on an operator telling them what to do during an emergency scenario.

Nowadays, this situation (outdated PSI) is a concern for many facilities: In a study performed in 2010 by OSHA IN 2010 [10]; 22.2% of the deviations (not compliance with PSM) were associated to non-updated PSI.

2.3 Process Hazard Analysis (PHA)

The purpose of this analysis is to minimize the likelihood of the occurrence and the consequences of a dangerous substance release by identifying, evaluating, and controlling the events that could lead to releases.

The management team and all personnel were asked to answer the following questions:

- Is my facility safe to operate?

- In the case of an event happening, am I protected?
- Is my facility protected?

This is how the HAZOP concept was introduced to the facility. A group of people were trained in HAZOP facilitation and they were in charge of implementing the PHA program.

Implementation of PHA, as with any other type of new procedure, was a challenge. At that time only the HAZOP was considered for implementation; later, additional PHA methods were added to the portfolio. Manual recording of PHA scenarios was a challenge, but the use of PHAs (manual or automated) continues to be vital to gauge whether a facility is protected or not.

In the earlier stages of the implementation, PHA-HAZOP was used as part of Management of Change element, but later was included in the evaluation cycle of all projects.

2.4 Operating Procedures

In the old facility, procedures were transmitted by oral tradition and there were very few written procedures, but even those written were not clear for non-experienced individuals. They were not developed with consideration for new personnel in that corresponding area.

In several opportunities, lack of written procedures led to accidents when one of the "guru's" was not there to execute a particular procedure or when a step was missed in a complicated procedure.

To prevent this from happening repeatedly, a team was built consisting of very experienced operators with experts in writing procedures and process engineers to start writing the procedures for each unit. Managers defined the most critical and most urgent ones to be done first. It was a medium-term effort that produced great results when a few years later the operations crews were modified to include people with technical background but almost zero operational experience.

2.5 Hot Work Permit

In traditional operation, the Operations Supervisor is also in charge of supervising the maintenance activities in his area. About 25 years ago, a request for a work permit was made to the Operations Supervisor; he erroneously determined that the risk was not too high, as he was busy with the day-to-day operation. He did not check the actual situation in the field. Unfortunately, the line where the work was going to be performed was not checked before and was filled with light naphtha (not properly drained) and this led to a fatality when welding.

After this accident and as a part of PSM program, a new work permit procedure was established:

- A dedicated Operations Supervisor for maintenance activities, being the only authority to sign-off work permits.
- A procedure was developed to include documentation and proper sign-off
- Training was developed for all involved parties about how to perform using the new Hot Working Permit, including contractors.

2.6 Training

The training of all personnel is a key element of Process Safety. The training must be on a continuous basis and at all levels of the organization. In this particular company, the kick-off of the PSM was to present the Piper-Alpha accident video to all levels of the organization from the top Management to the bottom of the organization. Also the API-750 Video was presented to all personnel, the objective: understand the risk, study the failures and realize that this situation was very likely to happen to that facility.

The importance of training is that it must to be continuous, refreshed on regular basis. Training must be developed for new employees and all contractors, with a rule of "nobody in" without safety training. The safety training included an exam; and those who do not pass the exam cannot get into the facility. To make sure PSM works, training is a key element and dedicated personnel to fulfill this target is recommended.

2.7 Pre start-up Safety Review

According with API RP 750 [5] the PSSR (Pre Start-up Safety Review) should be performed for new and modified facilities that are subject to this recommended practice to confirm that:

- Construction is according with specifications
- Procedures are in place (including safety procedures)
- PHA recommendations are closed out.
- Training for all operators is provided

It is fairly common to see confusion between PSSR for commissioning and PSSR during normal S/D for maintenance; as API 750 indicates, it is needed under both scenarios.

Before the implementation of PSSR, the responsibility of checking the facility prior to start up was only for operations. Normally this was done by an Operations Supervisor with other duties in the area and sometimes relying on another operator or on a Maintenance Supervisor. In this facility the need for PSSR was identified in an unfortunate incident:

An Operations Supervisor on his last night shift was doing the plant checking for start-up after he was informed by maintenance about completion of work. He was the only person in charge of checking the facility and provided the team with the "go ahead" for start-up 30 minutes before the end of his shift. The heating of the plant started and about 2 hours later a hot naphtha leak spilled over a hot heat exchanger producing a large fire that required 2 months of additional plant shutdown to repair damages. A good example of what "not to do".

A PSSR exercise takes a longer time to perform and needs a multidisciplinary team but will prevent incidents or accidents.

2.8 Emergency Planning & Response

Firefighter training is very important. In 1984, 100 municipal firefighters died when they were called to fight a fire in a heavy diesel tank in another location [11]. They handled it as a regular fire and high quantities of water were directed to the roof of the tank. No effort was made to cool off the walls of the tank and they were not aware of the presence of water in the bottom of the tank. When the diesel in the bottom of the tank reached the appropriated temperature, a boil over situation occurred.

One of the key elements in an Emergency Response plan is the knowledge of the hazards of a system that may create an emergency. As part of a PSM program, an intensive training on Process for non-technical personnel was given to firefighters (permanent and volunteers). Prior to the implementation of the PSM program, in the event of an emergency, the Firefighting Supervisor communicated directly with the Operations Supervisor on site, to obtain information about the potential hazards to people and facility. The Firefighting Supervisor had only a very limited knowledge of the facilities and associated hazards.

Emergency dry-runs were also implemented weekly according with a program developed on the most critical areas. This practice involved regular firefighters and volunteer firefighters.

Another step that was taken as a part of keeping the firefighters engaged was the annual recognition to all firefighters (rewards, celebrations, etc.).

2.9 Program Auditing

All the PSM elements should be audited on a regular basis to ensure the program effectiveness and performance. Programs like PSM require a lot of involvement from the Auditing team, the PSM leading team will define which parts of the program are most critical and it will be the start-up point for the auditing.

Audit process will always be needed, even when every person in the organization is "in the safety mindset".

3. Conclusions

- Implementation of the PSM program is a time consuming effort.
- Dedicated personnel are required for leading the change.
- Management actively involved in the PSM program is a must.

3.1 Tangible benefits from the implementation of PSM

- In a facility with a record of 4 fires per year, they were able to reduce the number of fires down to 1/year.
- Training of operators improved the operations of the facility reaching production records and reliability records.
- The Safety concept was built-in (be safe when nobody is looking)
- Ongoing, non-prescriptive, performance-based, participation at all levels.
- Facility received a Safety Award as "Safest in class" facility.

4. Recommendations

- PSM is a MUST for all O&G facilities
- It is a way of living! I encourage all O&G companies to follow this "way of living".

5. References

[1] Health and Safety Executive, 'The Flixborough Disaster: Report of the Court of Inquiry', HMSO, ISBN 0113610750, (1975)

[2] Chouhan TR "The Unfolding of Bhopal Disaster". Journal of Loss Prevention in the process industry **18** (4–6): 205–208. (2005).

[3] PatC-Cornell, M. E. "Learning from the Piper Alpha Accident: A Postmortem Analysis of Technical and Organizational Factors". Risk Analysis, Vol. 13, No. 2, 1993

[4] Bethea, R.M. "Explosion and Fire at the Phillips Company – Houston Chemical Complex, Pasadena, TX" Chemical Engineering Department, Texas Tech University, Lubbock, TX 79409-3121 (2007)

[5] API Recommended Practice API RP-750. "Management of Process Hazards". American Petroleum Institute, 1990.

[6] CSB. Investigation Report: Refinery Explosion and Fire (15 killed, 180 injured). Report # 2005-04-I-TX; Chemical Safety Board, Washington, DC. 2007

[7] British Petroleum Oil Company (BP). "Deepwater Horizon Accident Investigation Report" (2010)

[8] NTSB "Pacific Gas And Electric Company Natural Gas Transmission Pipeline Rupture and Fire" San Bruno, California. NTSB Report # PAR-11-01 (2011).

[9] OSHA. "Process safety management of highly hazardous chemicals". OSHA 29- Part 1910.119 (1992)

[10] Jordan Barab., "Learning from Industry Mistakes" presented at NPRA National Safety Conference, May 2010

[11] MARAVEN S.A. "Investigacion del Accidente de Tacoa" (Tacoa's Accident Investigation, In Spanish). Caracas, 1985.

(43an) Auditing the Process Hazard Analysis - Key Characteristics to Scrutinize in a Good-Quality HAZOP Study and Supporting Effective HAZOP Study Updates

Steven T. Maher[1], Michael A. Nour[1] and Kristin D. Norton[2], (1)Process Safety, Risk Management Professionals, Irvine, CA, (2)Process Safety, Risk Management Professionals, Inc., Irvine, CA

Although Process Safety Management (PSM), requiring good-quality Process Hazard Analyses, was promulgated in the United States in 1992, and HAZOP Study techniques date back to the mid-1970's, the HAZOP Study technique is still not uniformly applied in a high quality fashion. The objective of this paper is to provide experience-based tips on Process Hazard Analysis auditing, and some key characteristics that the auditor can scrutinize to assure quality implementation. Analogously, a similar set of characteristics that can facilitate future HAZOP Study updates can also contribute to a quality HAZOP Study that supports the spirit of Safety Management System objectives for the identification of hazards potentially impacting plant personnel, the community, and the environment. This paper will also review tips for structuring the HAZOP Study to facilitate later updates and revalidation.

9TH GLOBAL CONGRESS ON PROCESS SAFETY

Vapor Cloud Explosion Test of Air-Inflated Fabric Structure

Brad Horn
Baker Engineering and Risk Consultants, Inc.
3330 Oakwell Court, Suite 100
San Antonio, TX 78218-3024
BHorn@BakerRisk.com

Matthew T. Edel, P.E.
Baker Engineering and Risk Consultants, Inc.
3330 Oakwell Court, Suite 100
San Antonio, TX 78218-3024
MEdel@BakerRisk.com

Prepared for Presentation at
American Institute of Chemical Engineers
2013 Spring Meeting
9th Global Congress on Process Safety
San Antonio, Texas
April 28 – May 1, 2013

UNPUBLISHED

Vapor Cloud Explosion Test of Air-Inflated Fabric Structure

Brad Horn
Baker Engineering and Risk Consultants, Inc.
3330 Oakwell Court, Suite 100
San Antonio, TX 78218-3024
BHorn@BakerRisk.com

Matthew T. Edel, P.E.
Baker Engineering and Risk Consultants, Inc.
3330 Oakwell Court, Suite 100
San Antonio, TX 78218-3024
MEdel@BakerRisk.com

Keywords: vapor cloud explosion, explosion test, air-inflated structure, fabric structure, blast load, thermal load, structural response

Abstract

Portable buildings are gaining widespread use at refining and chemical processing facilities, particularly for facility siting requirements. Although trailers and blast resistant modules (BRMs) can be used for many modular applications, in certain situations fabric structures may provide a viable alternative. However, the blast response of fabric structures is neither clearly defined nor well understood throughout the blast design community.

One such modular fabric structure is an air-inflated structure. A manufacturer of this type of structure recently funded the testing of one of their specimens for the effects of a vapor cloud explosion (VCE). The VCE for the test was initiated using propane and an internal matrix of congestion inside a three-walled cubicle, referred to as the Deflagration Load Generator (DLG) test rig. The test involved a deflagration event that imparted a thermal flux and a maximum free-field pressure of 2.9 psi with an average duration of 31 ms at the front wall of the air-inflated structure.

The reduced blast loads and thermal effects inside the structure were compared to those outside. In addition, the attenuation of blast loads behind the structure was also examined, as well as shielding effects. The structural response of the test specimen was captured with high-speed video, and observations on the motion of the structure are provided.

1. Introduction

Baker Engineering and Risk Consultants, Inc. (BakerRisk) has developed a Deflagration Load Generator (DLG) test rig that can supply VCE blast loads to a variety of structures in order to provide structural response test data to manufacturers (i.e., product proof testing). The DLG rig produces a VCE blast load with a blast wave shape, pressure, and duration similar to the blast loads that are generated by accidental VCEs at chemical processing facilities. Dynamic Air Shelters, Ltd. (DAS) contracted a test of its manufactured air-inflated shelter.

The test involved imparting a VCE blast load on a DAS air-inflated shelter. The shelter was erected to the manufacturer's specifications and then subjected to a VCE blast load using the DLG test rig. The target blast loading for this test was a centrally peaked blast wave with a peak free-field blast pressure of 3.5 psi and 50 ms duration at the shelter wall closest to the DLG rig. Data recorded include pressure histories, thermocouple and heat flux measurements, and video.

2. Test Setup

This section describes BakerRisk's DLG test rig, the DAS air beam shelter, and the instrumentation used to measure and record data during the test.

2.1 DLG Test Rig

The DLG test rig is shown in Figure 1. The test rig is a steel box with one open side that is initially covered with a thin plastic sheet to allow a fuel-air mixture to be developed within the test rig. During testing, the test rig is filled with a propane-air mixture at a desired concentration, which is then ignited at the middle of the back wall. An obstacle array within the test rig causes the flame to accelerate and generate a blast wave. The resulting blast wave propagates out toward the target structure.

2.2 Air Beam Shelter

DAS constructed the air-inflated shelter shown in Figure 1, which was located directly in front of the DLG test rig during the test. The shelter system consisted of a ballistic-resistant wall and an air-inflated shelter with overall dimensions of approximately 46 ft. (14 m) wide by 56 ft. (17 m) long and 16 ft. (4.9 m) tall. The blast wall was made of PVC trapezoidal bags that were filled with material, in this case sand, and then hung between supporting wall frames spaced at approximately 6.7 ft. (2.0 m).

Figure 1. Air-Inflated Shelter (on left), Blast Wall, and DLG Test Rig (on right)

The air-inflated members were continuously pressurized and created the structural support for the shelter. Doors were located on two walls parallel to the direction of the blast propagation. The interior air beams are shown in Figure 2. Except for the instrumentation, the interior of the shelter was vacant during the test.

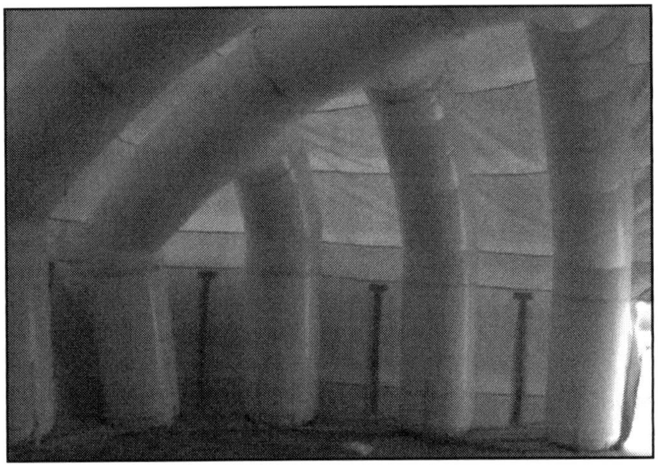

Figure 2. Air-Beam Columns

A schematic showing the location of the test rig, shelter and instrumentation is provided in Figure 3. The standoff distance between the test rig and the shelter was 21 ft. (6.4 m).

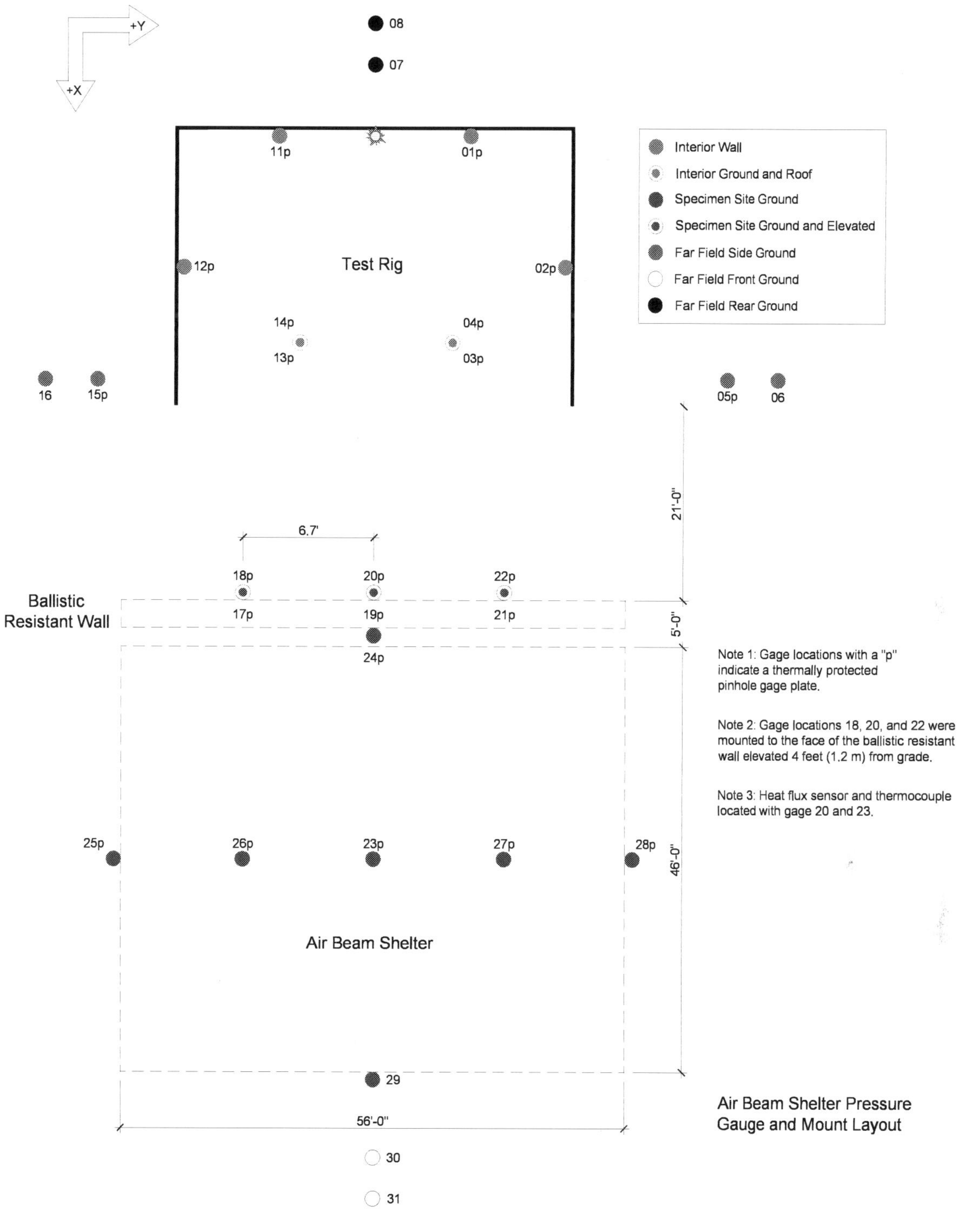

Figure 3. Instrumentation Layout Map

2.3 Instrumentation

During the test, the pressure, heat flux, and temperature histories were measured at locations of interest. Both regular and high speed video of the shelter structural response were captured.

A total of 29 dynamic pressure gauges sampled at 10,000 samples per second were used to measure the blast loads. Pressure gauge locations are shown in Figure 3; the gauge locations relative to the front of the test rig are provided in Figure 5. A total of 13 pressure gauges were positioned to measure the blast load at points of interest on the shelter and its components. Six gauges measured blast loads at the ballistic resistant wall, with three on the ground at the base of the wall and three mounted on the wall 4 ft. (1.2 m) above the ground gauges in order to measure the reflected load. One pressure gauge was located on the ground between the ballistic resistant wall and the shelter wall, and one pressure gauge was located at each shelter doorway. Three pressure gauges were located within the shelter interior, and one pressure gauge was located just outside the rear wall of the shelter.

Thermocouple and heat flux sensors were placed at the ballistic resistant wall and inside the shelter, coincident with pressure gauge locations 20 and 23. Each sensor was elevated 4 ft. (1.2 m) above grade and oriented toward the DLG test rig. Each thermocouple was sampled at 20,000 samples per second.

Video recording devices were used to document the test and record the shelter structural response to the blast loading. The high speed video (HSV) camera was positioned in the same plane as the leading edge of the ballistic resistant wall, while two high definition (HD) video cameras, two exterior spy cameras, and one interior spy camera recorded other angles.

3. Test Results

Figure 4 shows the structural response recorded by the high definition video camera, capturing at 30 frames per second.

Figure 4. VCE Loading Applied to Shelter

3.1 Blast Loads

A peak pressure measurement map with pressure values listed in green font is shown in Figure 5. Table 1 provides the blast load summary which includes the pressure gauge location, standoff distance from the test rig, measured peak pressure, positive impulse, and both the measured and calculated durations. The calculated duration assumes a perfectly triangular blast pressure history (i.e., duration is twice the impulse divided by the peak pressure). Table 2 provides a summary of the peak reflected blast pressures measured at the ballistic resistant wall and the corresponding calculated peak free-field (i.e., side-on, incident) blast pressures.

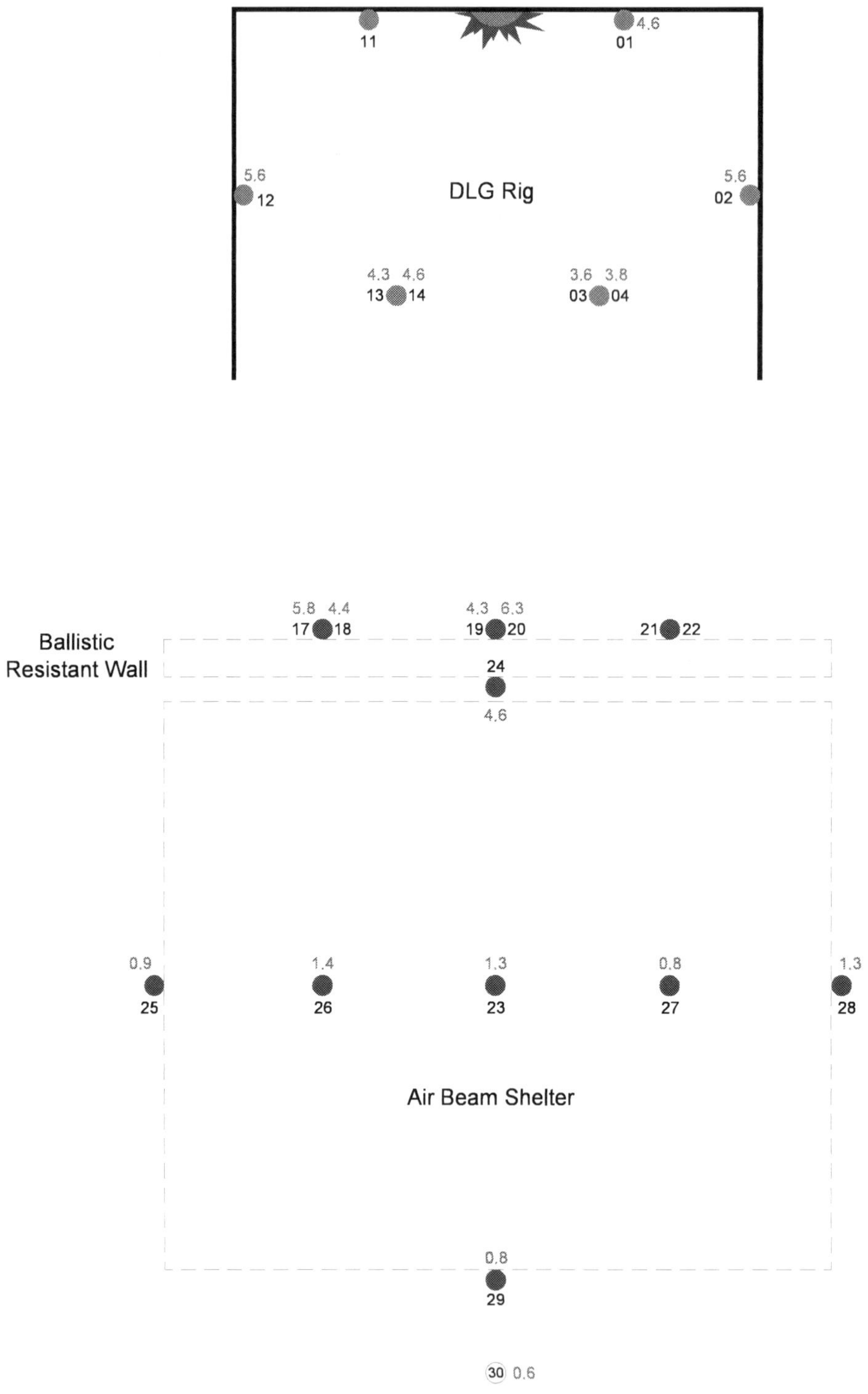

Note: Pressure values directly in front and behind the ballistic-resistant wall are reported as the adjusted peak values.

Figure 5. Peak Pressure Measurement Map (psig)

Table 1. Blast Load Summary

Location	Gauge	Distance From Test Rig (ft.)	Measured Peak Pressure (psig)	Measured Positive Impulse[1] (psi-ms)	Measured Duration (ms)	Calculated Duration (ms)
Rig Interior	2	0	5.6	206	120	74
Ballistic Resistant Wall[1]	17	21	5.4	90	48	31
	18	21	3.9	67	42	31
	19	21	3.9	70	51	32
	20	21	5.8	93	48	30
Front Shelter	24	24	4.6	95	57	41
Shelter Doorway	25	48	0.9	19	57	41
	28	48	1.3	28	57	41
Shelter Interior	23	48	1.3	29	53	45
	26	48	1.4	29	53	41
	27	48	0.8	16	53	41
Rear Shelter	29	71	0.8	16	45	41
Far Field	30	76	0.6	10	30	33

Table 2. Ballistic Resistant Wall Blast Pressure Summary

Location	Gauge	Distance From Test Rig (ft.)	Measured Peak Pressure (psig)	Adjusted Peak Pressure[1] (psig)	Calculated Free Field Pressure[2] (psig)
Ballistic Resistant Wall[1]	17	21	5.4	5.8	2.7
	18	21	3.9	4.4	2.0
	19	21	3.9	4.3	2.0
	20	21	5.8	6.3	2.9

[1] *Pressure values adjusted for pre-event signal ringing*
[2] *Calculated from reflection factor*

The maximum peak pressure in the DLG test rig was 5.6 psig (0.39 bar). Gauges at the ballistic resistant wall were positioned to measure the reflected load on the face of the wall or shelter.

The peak pressures and impulse at the wall were adjusted to account for the signal ringing prior to the arrival of the blast wave and are shown in Figure 7. The maximum adjusted peak pressure was 6.3 psig (0.43 bar). The pressure signals along the centerline of the shelter in the direction of the blast are shown in Figure 6. Two of the six gauges were invalid and were therefore discarded.

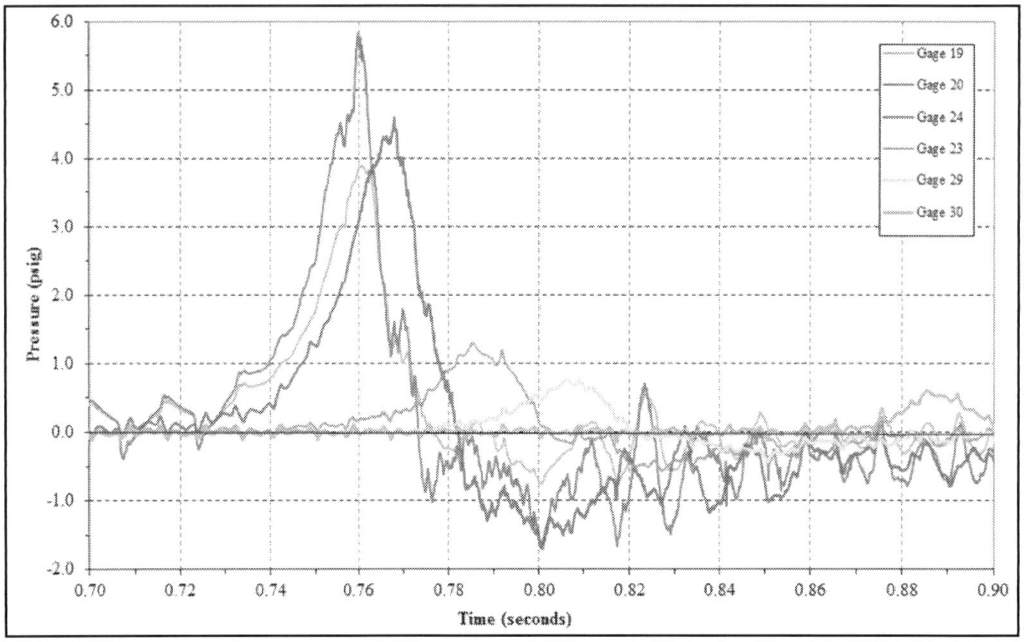

Figure 6. Pressure Histories along the Front Center Gauge Lane

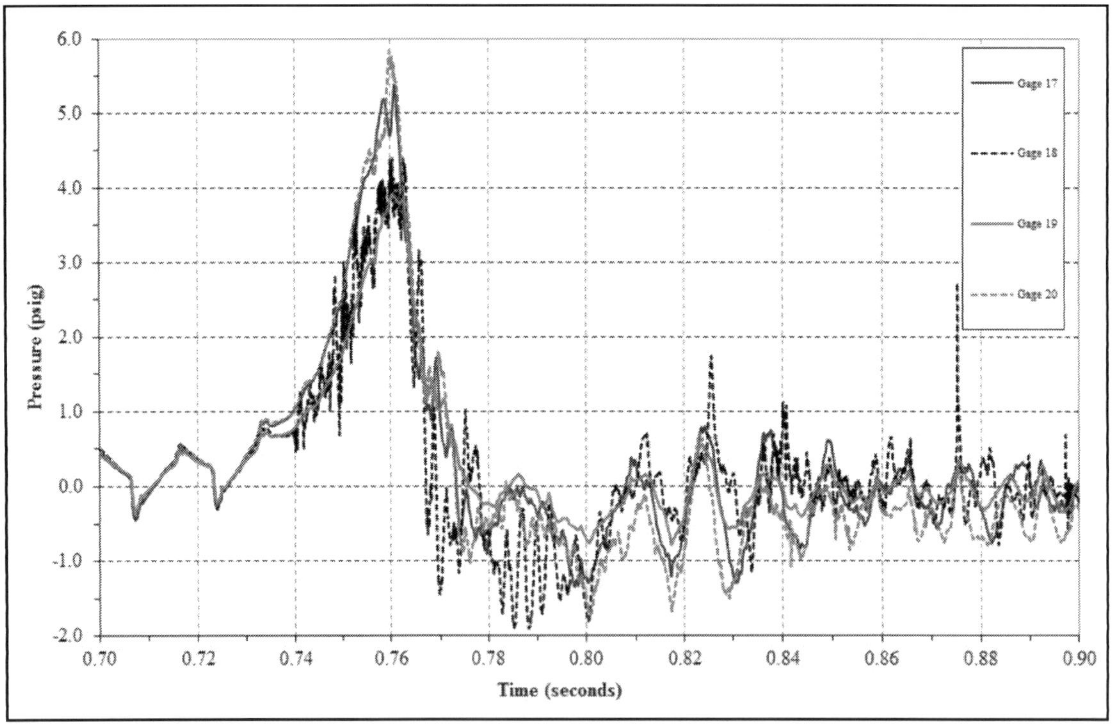

Figure 7. Pressure Histories at the Ballistic-Resistant Wall

The measured peak pressure and impulse between the ballistic resistant wall and air shelter were 4.6 psig (1.4 bar) and 95 psi-ms (6.6 bar-ms), yielding a calculated duration of 41 ms. Pressure histories were measured along the mid-span of shelter. The pressure histories at these mid-span locations are shown in Figure 8. For the doorway exterior locations, the largest measured peak pressure was 1.3 psig (0.09 bar) with an impulse of 28 psi-ms (1.9 bar-ms) and a calculated duration of 41 ms. The largest measured peak pressure among the interior gauges was 1.4 psig (0.1 bar) with an impulse of 29 psi-ms (2.0 bar-ms) and a calculated duration of 41 ms. Behind the shelter, the measured peak pressure was 0.8 psig (0.06 bar) with an impulse of 16 psi-ms (1.1 bar-ms) and a calculated duration of 41 ms.

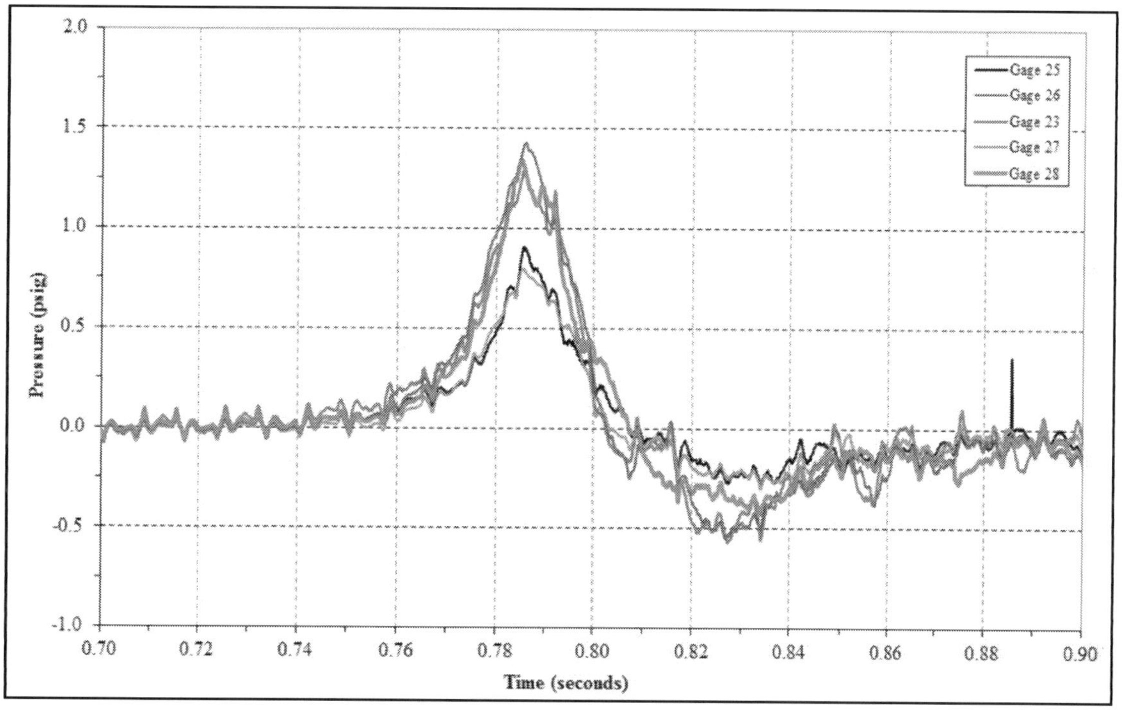

Figure 8. Pressure Histories at Shelter Mid-Span Locations

3.2 Thermal Loads

Figure 9 shows the measured temperature just past the exterior face of the ballistic resistant wall. This peak signal exceeded the range of the signal conditioner and was "clipped" at approximately 1100 °F (590 °C). The estimated actual maximum temperature at this location was just under 1500 °F (815 °C) based on an extrapolation of the signal. The measured signal inside the shelter indicates that the temperature had increased less than 5 °F (3 °C), as shown in Figure 10. That is, while the temperature just outside the shelter increased dramatically due to exposure to the fireball, the temperature within the shelter barely rose.

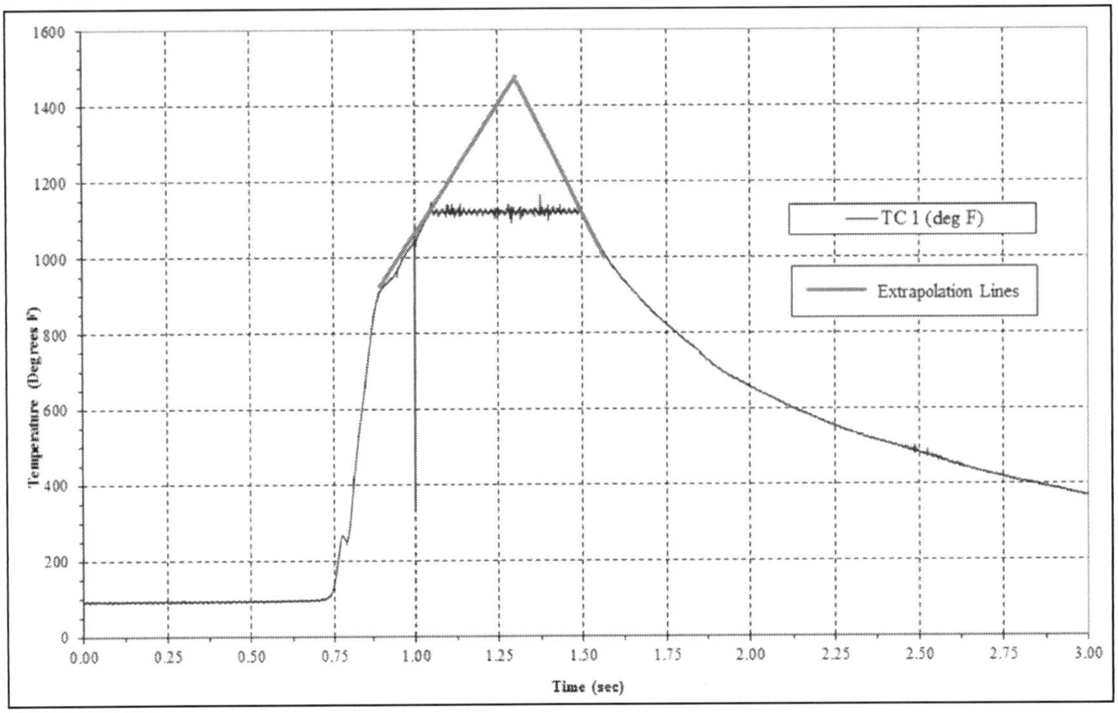

Figure 9. Exterior Temperature History

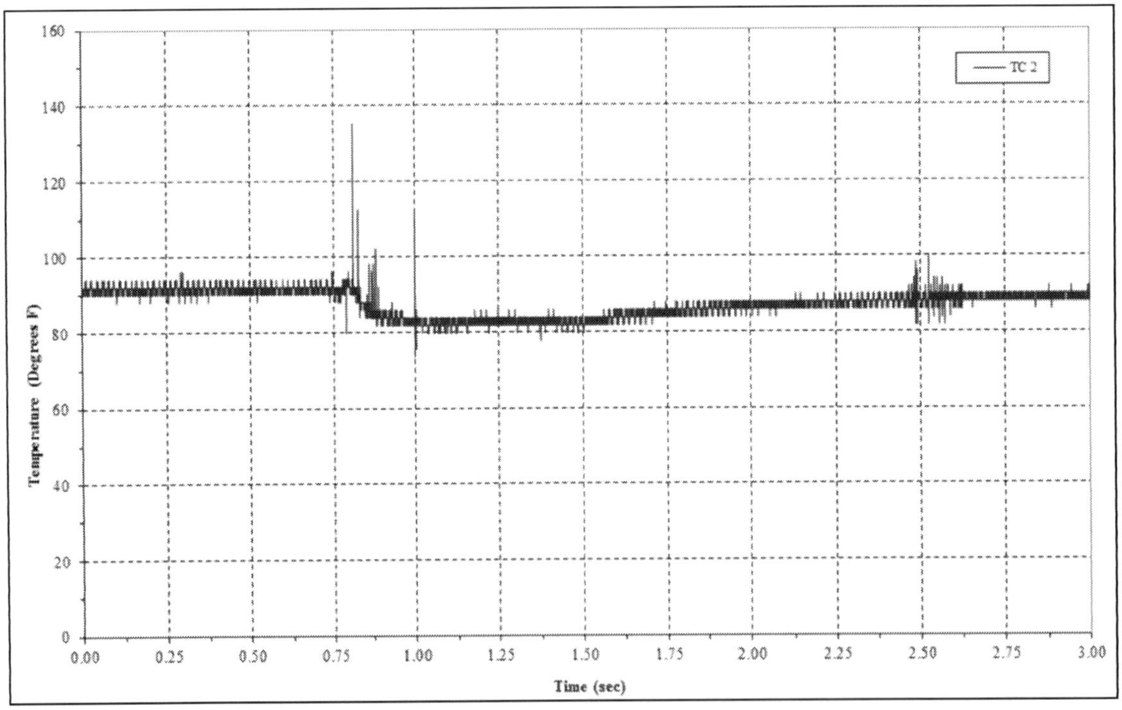

Figure 10. Interior Temperature History

The exterior heat flux signal is shown in Figure 11. This signal was filtered to provide a smooth curve with a maximum heat flux of 6 kW/m^2. The interior heat flux signal was invalid due to the low signal to noise ratio. The virtually zero interior heat flux is consistent with the measured temperature at this location.

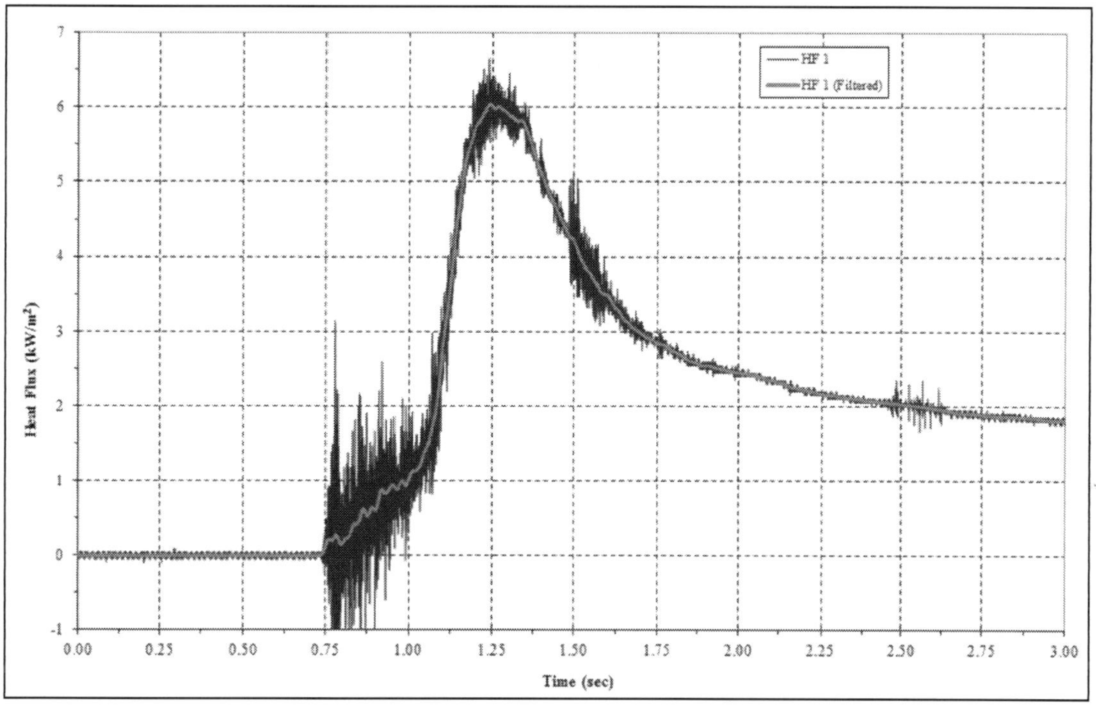

Figure 11. Exterior Heat Flux History

3.3 Structural Response

The air-inflated shelter sustained only minor damage from the VCE blast loading. The air-beams that create the arch section of the shelter experienced rapid temporary deflection as the blast wave propagated across the shelter. The ballistic resistant wall remained functional following the test, allowing safe egress from the shelter after the VCE. Figure 12 shows the wall deflection of about 1 ft. (0.3m) at the top. The yellow grid lines are approximately 1 ft. (0.3 m) apart. The yellow and blue markers indicate the roof deflected approximately 1.5 ft. (0.5 m) downward during the test and returned to the initial position within 2 seconds.

Figure 12. Shelter Deflection Comparison

4. Conclusions

A Dynamic Air Shelters, Ltd. air-inflated structure was subjected to a VCE having a blast load with a peak free-field pressure of 2.9 psi and a 31 ms duration at the wall nearest the DLG rig. The test resulted in a deflagration-type load with a centrally-peaked blast shape. The blast loads inside the structure were reduced from the levels recorded at the exterior ballistic resistant wall by approximately 75% but were comparable to the exterior pressures measured outside of the two doors. Shielding effects of the blast loads were anticipated directly behind the structure, but none were observed at this location relative to the loads recorded closer and further from the rig.

Significant thermal loads were observed on the front face of the structure due to its close proximity to the fireball emanating from the test rig. Yet even at nearly 1500 °F (815 °C) and 6 kW/m^2 at the exterior front face, the interior temperature increased less than 5 °F (3 °C) with virtually no heat flux. The air-inflated structure provided almost complete shielding from the thermal loads with negligible observable thermal damage.

The high speed and high definition video show a maximum of 1 ft. deflection of the ballistic resistant wall and less than 2 ft. deflection of the air-inflated shelter. The structure returned to its initial position within 2 seconds and had no permanent deflection. Internally, the structure had no debris hazards and remained inflated, allowing safe egress from the air beam structure.

Revisiting the Conditions for a CO2 Tank Explosion

Yu Zhang
Air Products and Chemicals Inc.
7201 Hamilton Blvd, Allentown, PA 18195-1501
zhangy15@airproducts.com

Joan Schork
Keith Ludwig
Air Products and Chemicals Inc.
7201 Hamilton Blvd, Allentown, PA 18195-1501
schorkjm@airproducts.com
ludwigka@airproducts.com

Prepared for Presentation at
American Institute of Chemical Engineers
2013 Spring Meeting
9th Global Congress on Process Safety
San Antonio, Texas
April 28-May 1, 2013

UNPUBLISHED

Revisiting the Conditions for a CO2 Tank Explosion

Yu Zhang
Air Products and Chemicals, Inc.
7201 Hamilton Blvd, Allentown, PA 18195-1501
zhangy15@airproducts.com

Joan Schork
Keith Ludwig
Air Products and Chemicals, Inc.
7201 Hamilton Blvd, Allentown, PA 18195-1501
schorkjm@airproducts.com
ludwigka@airproducts.com

Keywords: Boiling Liquid Expanding Vapor Explosion (BLEVE), Carbon dioxide, Overpressure

Abstract

Carbon dioxide is broadly used in a number of applications including food freezing, beverage production, enhanced oil recovery, supercritical extraction and welding. It is commonly stored as a pressurized liquid (e.g.: -18 °C/0 °F and 2 MPa/300 psia) in carbon steel tanks.

A number of CO2 tank explosions have been reported in the literature over the past 50 years. In order to design appropriate safety systems for liquid CO2 storage tanks, it is important to understand the worst case scenario. Should the energy release be modeled as a rapid vapor phase expansion (PV) or a Boiling Liquid Expanding Vapor Explosion (BLEVE)? The question arises because of a common definition of a BLEVE as an event resulting from the sudden loss of containment of a liquid that is at or above its atmospheric superheat limit temperature. Under the conditions listed above, CO2 is above its atmospheric boiling point but below its superheat limit. In this temperature range, will the vaporization rate be rapid enough to contribute to a blast overpressure in the event of a tank failure?

To answer this question, the authors made a critical review of information available in the literature including laboratory data and incident reports. This paper will describe the analysis conducted and present the results indicating that the potential for rapid vaporization should be included in evaluating the consequences of a CO2 tank rupture below the superheat limit.

1. Introduction

Carbon dioxide (CO2) is broadly used in a number of applications including food freezing, beverage production, enhanced oil recovery, supercritical extraction and welding. Since CO2 at its bubble point becomes a solid at a pressure below 5 bars, CO2 is commonly stored as a pressurized liquid at an elevated pressure of 20 bars absolute in carbon steel tanks.

When a pressurized liquid in an equilibrium state is depressurized instantaneously, it enters a superheated state (a meta-stable liquid above its boiling point) from which the liquid could vaporize in an explosive manner resulting in a blast wave and risk of flying fragments of the vessel. This is called a Boiling Liquid Expanding Vapor Explosion (BLEVE) [1,2]. When the degree of superheat of the liquid is small, the vaporization rate is not large enough to develop a significant blast effect. According to the Superheat Limit Theory developed by Kim and Reid [3], an explosive evaporation occurs when the depressurized liquid enters the spinodal decomposition temperature-pressure range. Based on thermodynamics principles and equation of state of fluids, Kim and Reid [3] predict a so-called BLEVE envelope (Figure. 1) which indicates the range of initial temperatures and pressures which will lead to a BLEVE following instantaneous depressurization. The superheat limit theory is cited in some industrial hazard evaluations as a criterion to determine whether or not a BLEVE should be considered [4,5].

The mechanism of BLEVE based on Kim and Reid's theory is briefly described here. When the boiling point of a liquid is reached, vapor bubbles start to form at active sites such as vessel walls. Nucleation at these sites is called *heterogeneous nucleation.* If bubble nuclei are large (approaching millimeter sizes), a small degree of superheat (defined as heating beyond the boiling point) is needed to make the bubble grow. If the bubble nuclei are small, for example when the vessel is very clean and smooth, they need a large superheat to grow. When there are not many active sites, the bubble nucleation must take place in the bulk of the liquid. Boiling in this case generally takes place at submicron nucleation sites at suspended particles, impurities, or ions. This requires a large degree of superheat to make the bubble grow. This means the liquid normal boiling point can be exceeded without boiling, and the liquid can be significantly superheated. There is, however, a temperature limit at a given pressure above which a phase transition from liquid to vapor has to happen, even though the bubble nulei are small. This superheat limit temperature (SLT) and pressure correspond to the conditions at which a spinodal decomposition of liquid happens. This can produce a very violent phase transition with boiling caused by *homogeneous nucleation* in the bulk of liquid. The superheated liquid state could also be reached by sudden pressure drop. For CO2, the experimental value of SLT at atmospheric pressure is 267 K (-6.2°C) [6].

Although the theory is commonly accepted, some authors question its applicability to a real case. For example, Birk and co-works [7] report that BLEVEs have been observed with propane when the pressure-liquefied gas was at ambient temperature (20 °C), well below its atmospheric superheat limit temperature of 53 °C. In the 2010 edition of the Center for Chemical Process Safety (CCPS) Guidelines to evaluate BLEVE [2], it says "There is no experimental evidence with a real pressure vessel where the phase change was by homogeneous nucleation. Homogeneous nucleation is extremely unlikely in industrial use due to temperature gradients, impurities, surface roughness and other factors that cause localized nucleation."

2. CO2 BLEVE

The end state of liquid CO_2 depressurized to 1 atmosphere is a solid and vapor mixture, which makes its depressurization process unique. CO_2 may evaporate explosively even when its temperature is below the STL, because CO_2 crystallizes at ambient pressure, thus presenting enough nucleation sites for liquid to flash [1].

Several fundamental questions are raised when considering CO_2 system. How does the formation of dry ice affect the blast energy? During the decompression process, does the vapor expand isothermally or isentropically? This initial report will only touch on two questions based on various incident reports on large scale CO_2 tank energy releases:

1. *Could a CO_2 BLEVE happen when the storage conditions (temperature and pressure) are outside the so-called BLEVE envelope predicted by the superheat limit theory?*
2. *If it goes BLEVE, could we use the algorithm presented in the CCPS guide to perform the consequence analysis?*

3. Historical experiences

Table 1 summarizes major CO_2 tank incidents we found with documentation in the open literature. The Worms, Haltern, and Repcelak incidents are summarized in CCPS's guide to BLEVE [1-2]. Information about the incident at Yuhang is collected from news reports in Chinese [8-9], and information about Fukushima's incident is translated from reports in Japanese [10-12].

Table 1. Some major CO_2 incidents

Date	Location	Cause	Initial Pressure [bar]	Initial Temperature [°C]
13 Nov. 2008	**Yuhang, China***	Brittle fracture	23	-15
21 Nov. 1988	Worms, Germany	Overpressure	20	-20
2 Sept. 1976	Haltern, Germany	Brittle fracture	23	-15
1 Mar. 1969	Fukushima, Japan	Overpressure	20	-20
2 Jan. 1969	**Repcelak, Hungary***	Overfilling	15	-30

* Overpressure tags are extracted from these incidents

Out of the five incidents, only two overpressure damage data points (tags) are obtained, which are used to validate the algorithm employed in the CCPS guidelines. There were many tags on projectiles, but we find these less reliable than the overpressure tags. One data point is from the Repcelak incident: a warehouse about 15-30 meters away from the explosion center was torn down due to heavy damage. The other data point is from the Yuhang incident: Window glass shattered within a 400 meter radius of the explosion center. Criteria on overpressure damage to buildings are obtained from CCPS's guide to BLEVE [1].

3.1 Large scale CO_2 tank energy release

Yuhang, Hangzhou, China, 13 November 2008

A transport ship carrying 130 cu. meter (95% full) of CO_2 exploded on November 13, 2008 in Yuhang, Zhejiang, China [8-9]. The tank's content was stored under a pressure of 23 bar at a temperature of -15 °C. At the moment of explosion, the ship was in the dock belonging to Rongsheng Chemicals. It was not in operation and there were few workers around as it was before regular business hours. The explosion completely destroyed the CO2 ship and sank H2SO4 and H2O2 carrying ships nearby. Two workers on the CO2 ship lost their lives instantly and 3 were injured due to projectiles.

Pieces of the ship were scattered over a 500 meter area around the explosion. The weight of these fragments varied from a few grams to several hundred kilograms. The shrapnel caused damage in several adjacent residential buildings. All the windows of a metal machining factory located around 500 meters away shattered. In the Yujiadouzhen, a village 500 meters away from the explosion, windows shattered in many residential buildings.

The cause of this accident is believed to be overloading and brittle failure of the CO2 tank. The tank was designed for on-shore use, but was modified by Rongsheng Chemicals for ship transportation. To lower the transportation cost, the company modified the level indicator, locked the relief valve, and overloaded the tank (95% filling level). The tank material, 16 MnR steel, has a working temperature from -20 °C to 475 °C. It was not suited for use under low temperature conditions.

Worms, Germany, 21 November 1988

This is one of the most catastrophic incidents involving a CO2 tank explosion. The detailed description of Clayton and Griffin is summarized here [13]. The liquid CO2 was stored in a 30 tons capacity tank at 20bar and -20°C. The vessel was shattered into a number of pieces in the explosion. Most of the tank was propelled into the Rhein River about 300 meters away from the epicenter. Only 20% of the tank was recovered originally. The Worms's investigation concluded that the vessel failed catastrophically, most likely due to overpressure from overheating with a frozen relief valve. The failure pressure was estimated to be between 31 and 51 bara. It was speculated that the failure caused a CO2 BLEVE [2].

Haltern, Germany, 2 September 1976

A rail car carrying 231,000 kg of CO2 exploded in Haltern, Germany [1]. The tank's contents were at 7 bar and -15°C. At the moment of the explosion, the car was passing through a railroad shunting yard. As it passed checkpoints, the car was observed to be releasing plumes of CO2 from the safety valve. It then exploded 15 m in front of a group of other rail cars. Parts of the tank were thrown up to 360 m. One person was killed in the explosion. Analysis of a recovered piece of the tank car showed that failure was due to brittle fracture. Brittle metal fracture is consistent with the near instantaneous overpressure created by a BLEVE event.

Fukushima, Japan, 1 March 1969

A CO2 storage tank in a steel mill exploded [10-12]. A slate-roofed factory within a 50 m radius was completely destroyed, and only its pillars remained. Windows of houses located within a 500 m radius were damaged. Before the explosion, the main safety valves were closed during repairs. Liquid remained in the tank (50% filling level) and the heat source of the external evaporator was left on. The vapor phase temperature and pressure in the tank kept rising, and eventually the tank exploded when the pressure exceeded the burst pressure of the tank. Based on heating element power, CO2 usage, and heating time, it is estimated that almost all liquid CO2 was turned into vapor before the explosion. Thus, this incident is determined to be a PV release.

Repcelak, Hungary, 2 January 1969
Two vessels containing liquid CO2 (15 bar, -30°C) in a CO2 production and filling plant exploded in rapid succession [1, 14]. The explosion completely destroyed the tank yard consisting of four liquid CO2 storage vessels and the equipment of the technological laboratory. Large fragments were scattered in a circle of approximately 400 m radius. A 2800 kg shell landed at a distance of 150 m and a fragment weighing 1000 kg landed 250 m away. During filling, the first vessel exploded. Some minutes later, another close-by vessel exploded. The second vessel probably failed as a result of impact from a fragment from the first vessel. The explosions tore the two vessels into pieces, and tore the third vessel off its foundation bolts. The third vessel was shot into the process laboratory like a rocket due to liquid CO2 rapid expansion through a hole on the bottom. The probable cause of the accident was overfilling the first tank due to a level indicator failure. The explosion of the second vessel was speculated to be caused by a CO2 BLEVE.

3.2 Small scale experimental investigation

Most experimental work on BLEVE has focused on flammable substances like LPG and propane [1,2]. Recently, there were several small scale lab tests performed to study CO2 BLEVE [15-18].

Kim and Reid, 1978
An important reason for considering the possible occurrence of BLEVE with CO2 is this paper by Kim and Reid which presents experimental and modeling work to investigate the phenomena of rapid depressurization of hot, high pressure liquids or supercritical fluids (CO2) [3].

The authors applied a thermodynamic model based on the spinodal fluid state, which represents the limit-of-stability for the liquid phase during expansion. For a BLEVE to occur, the substance has to find itself on the spinodal curve between 1 bar and the critical point where the curve ends. When a catastrophic failure occurs there is not time for heat to pass into the system and so the path during failure is adiabatic. In thermodynamic terms the entropy of the system is constant during failure and the conditions for BLEVE to occur means that the initial entropy has to be higher that the entropy of the intersection of the spinodal curve and the 1 bar isobar. As a result, a region of initial conditions that could give a spinodal state can be identified (Figure 1). Kim and Reid performed some simple experiments on depressurization of saturated liquid and supercritical CO2. None led to explosions regardless of the initial conditions relative to the envelope in Figure 1. The authors suggested the reasons for disagreement between theory and experiments were associated with factors like vapor bubbles on the wall of the vessel (sites for heterogeneous nucleation) and shock waves or disturbances from the mechanism that ruptured the burst disk of the test vessel.

Pettersen, 2002
Pettersen performed an experimental study aimed at clarifying the possible occurrence of BLEVE in CO2 systems [17]. A test vessel was built and instrumented in order to measure transient pressures during initial depressurization, and to record possible subsequent overpressure peak(s). The study concluded that none of the experiments, which were conducted over a wide range of initial conditions (3.5, 4.5, 6.0 and 10.0 MPa initial pressures with varying

liquid fill level and discharge opening), showed signs of BLEVE, which would be characterized by a sudden repressurisation to a level significantly above the initial pressure.

Wei, 2009

This master thesis presents experimental demonstration of CO2 BLEVE which happened below the superheat limit temperature (outside the BLEVE envelop in Figure 1) [18].

The list below highlights (taken from Wei's thesis) the most significant findings:

- The 'Superheat Limit Temperature' theory for predicting occurrence of a BLEVE was not supported with experiments in this work. A CO2 BLEVE can also occur when the initial temperature is below the superheat limit temperature of CO2 (-13.8 °C). Nevertheless, considering influencing factors including CO2 filling level, potential gas leaking and CO2 mixture with different phase compositions, the theory may still be acceptable.
- Two possible approaches for a safer CO2 storage include using an initial storage pressure of around 2 MPa and developing a safety valve that can further slow down pressure drop when an unexpected vessel opening and depressurization occurs.

3.3 Consequence analysis

An established procedure for calculation of the consequence of a BLEVE event can be found in CCPS's guide to BLEVEs [1,2].

1. Calculate the explosion energy. Ongoing debates on the calculation of the explosion energy from a BLEVE event involve discussions about the BLEVE mechanism and are outside the scope of this study. In this work, we consider three options where the explosion energy is
    a. Internal energy difference between the initial and final states of an isentropic expansion process [1],
    b. Internal energy difference between the initial and final states of an irreversible adiabatic expansion [19],
    c. The PV energy (energy from the vapor phase, excluding that from flashing liquid).

   Please note that the explosion energy will need to be multiplied by a factor of 2 if the burst is a ground level, to obtain the effective working explosion energy.
2. Once the explosion energy of a BLEVE is estimated, overpressure can be determined by employing the overpressure- scaled distance curve by Baker.
3. Once overpressure is calculated, a multiplier has to be applied to account for vessel shape effect.

According to CCPS, the method presented above is based on the similarity of the blast waves of pressure vessel bursts and high explosives. This similarity holds only at some distance from the explosion. In the near field (scaled distance < 2), the peak overpressure and impulse from a pressure vessel burst differ greatly from those of a detonation of a high explosive, except when the pressure vessel is filled with a very hot high-pressure gas. Baker et al. developed a method which is used to solely predict blast pressures in the near field. However, it is based on results of numerical simulations and involves solving an implicit equation. The details of this refinement can be found in CCPS's Guidelines [1]. Note that this refinement for the near field cannot be made for non-ideal gases.

4. Results

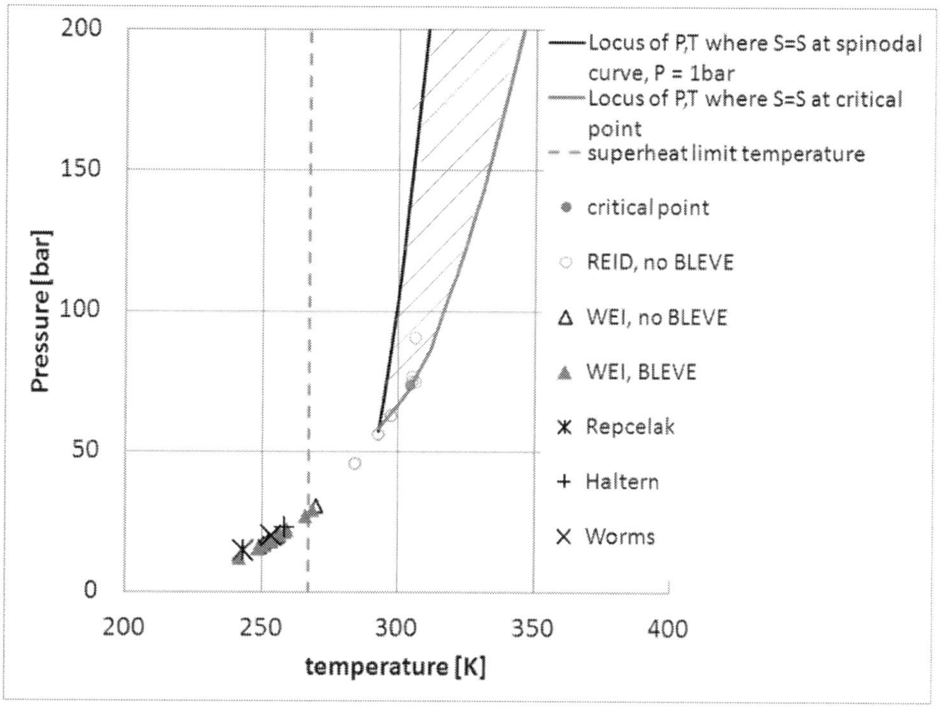

Figure 1. Summary of conditions involving CO2 BLEVE. Empty symbols are experiments with no BLEVE, and solid ones and crosses are experiments/events with BLEVE.

We summarize the literature data discussed in Section 3 in Figure 1. From Figure 1, we see that there is no clear dividing line between a BLEVE and a non-BLEVE event based on the superheat limit temperature, as indicated by experimental results and historical experiences. This might be due to the various processes involved in such an event, including nucleation of bubbles at the vessel wall. It is conclusive that the superheat limit temperature has no special significance to the BLEVE event observed in either past accidents or controlled experiments. Furthermore, Kim and Reid's experimental data along with others failed to produce a BLEVE in the laboratory when initial conditions were within the envelope predicted by the superheat limit theory.

We obtained two overpressure tags, one from Repcelak and the other fromYuhang:

1. The Repcelak: "light structure warehouse was so seriously damaged that it had to be pulled down." [14] The warehouse was 15 meters from the epicenter, and the overpressure was about 0.15-0.25 bar[1];
2. Yuhang: The windows of a metal machining factory located around 500 meters away, all shattered. In the Yujiadouzhen, a village 500 meters away from the explosion, windows shattered in many residential buildings [9]. The overpressure was about 0.035-0.07 bar [1].

Three different explosion energies, as discussed in Section 3.3, for each event are summarized in Table 2. The actual overpressure data *vs.* calculated overpressures using the CCPS's recommended method is shown in Figure 2. If the correct energy is chosen, the points should fall on or near the $y = x$ curve. Clearly, these results suggest that the explosion energy released from the incidents is more consistent with BLEVE type of explosion-energy where energy from the liquid phase contributes significantly.

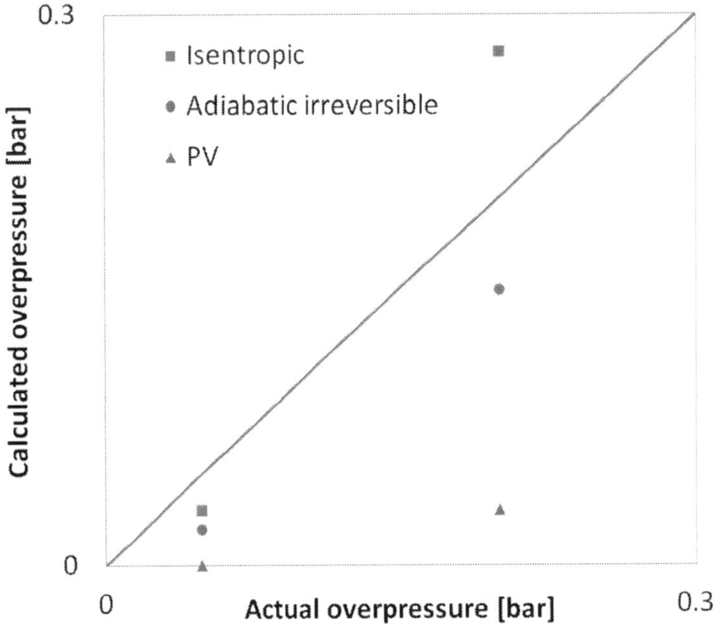

Figure 2. Actual overpressure versus calculated overpressure using different explosion energies for two events.

Table 2. Explosion energy, calculated overpressure (COP), and actual overpressure (AOP) tag for Repcelak and Yuhang incidents

Location	Initial filling level	Isentropic expansion [J]/COP[bar]	Adiabatic irreversible [J]/COP[bar]	PV [J]/COP[bar]	OP tag distance to epicenter R_tag [m]	AOP at R_tag [bar]
Repcelak	80%	3.5E9/0.28	9.9E8/0.15	3.0E7/0.03	15±5	0.20±0.05
Yuhang	95%	1.3E10/0.03	4.9E9/0.02	3.6E7/0.003	400±100	0.05±0.02

5. Conclusion

Through a critical review and analysis of historical and experimental reports of CO2 tank explosion, we conclude that a CO_2 tank rupture could lead to a BLEVE event even when the initial storage temperature is below the superheat temperature limit of CO_2. The consequence of a BLEVE event should be considered when assessing the risk associated with CO2 storage.

6. References

[1] "Guidelines for Evaluating the Characteristics of Vapor Cloud Explosions, Flash Fires, and BLEVEs, Center for Chemical Process Safety", *American Institute of Chemical Engineers*, New York, 1994

[2] "Guidelines for Vapor Cloud Explosions, Pressure Vessel Burst, BLEVE and Flash Fire Hazards, Center for Chemical Process Safety", *American Institute of Chemical Engineers*, New York, 2010

[3] Kim, M.E., Reid, R.C., "The rapid depressurization of hot, high pressure liquids or supercritical fluids", *Chemical Engineering at Supercritical Fluid Conditions*, Ann Arbor Science, 1983

[4] Calculation of BLEVE conditions for carbon dioxide, www.criticalprocesses.com/BLEVE.htm (accessed Jan. 10, 2013)

[5] "Final public report safety study for liquid logistics shipping concept (DNV)", http://cdn.globalccsinstitute.com/sites/default/files/publications/19006/final-public-report-safety-study-liquid-logistics-shipping-concept.pdf (accessed Jan. 10, 2013)

[6] Abbasi, T., Abbasi, S.A., "Accidental risk of superheated liquids and a framework for predicting the superheat limit," *Journal of Loss Prevention in the Process Industries,* 20 (2007) 165-181

[7] Birk, A.M., Cunningham, M.H., "The boiling liquid expanding vapor explosion", *J. Loss Prev. Process Ind.,* 7 (1994) 474–480

[8] Li, W., Di, G., Wang, R., "The cause analysis of a liquid CO2 tank explosion on a ship", *Materials Science Forum,* 689(2011)461-467

[9] Yuhang CO2 cargo ships explosion, http://zjnews.zjol.com.cn/05zjnews/system/2008/11/14/014973390.shtml (in Chinese, accessed Jan. 10, 2013)

[10] "High Pressure Gas Safety Inst. of Japan. Rupture of liquefied carbon dioxide in a low-temperature storage tank", *Accident investigation report (outline),* 1969 (in Japanese)

[11] "High Pressure Gas Safety Inst. of Japan", *Examples of accidents, Collection of examples of high-pressure gas accidents,* 1982 (in Japanese)

[12] Kitagawa, T., "Vapor explosion in liquid carbon dioxide tank", *Analysis of explosion hazard.* 1980 (in Japanese)

[13] Clayton, W.E., Griffin, M.L., "Catastrophic failure of a liquid carbon dioxide storage vessel", *Process Safety Progress* 13 (1994) 202-209.

[14] Voros, M., Honti, Gy, "Explosion of a liquid CO2 storage vessel in a carbon dioxide plant", *Loss Prevention and Safety Promotion in the Process Industries*, 1974

[15] van der Voort, M.M, van den Berg, A.C., Roekaerts, D.J.E.M., Xie, M., de Bruijin, P.C.J., "Blast from explosive evaporation of carbon dioxide: experiment, modeling and physics", *Shock Waves* 22 (2012) 129-140

[16] Bjerketvedt, D., Egeberg, K., Ke, W., Gaathaug, A., Vaagsaether, K., Nilsen, S.H., "Boiling liquid expanding vapour explosion in CO2 small scale experiments", *Energy Procedia* 4 (2011) 2285-2292

[17] Pettersen, J, "Experimental study on boiling liquid expansion in a CO2 vessel", 5th IIR Gustav Lorentzen Conference on Natural Working Fluids, Guangzhou, China, September 2002

[18] Ke, W., "CO2 BLEVE (Boiling Liquid Expanding Vapor Explosion)", Master Thesis, Telemark University College, Porsgrunn, Norway, 2009

[19] Chen, F, Schork, J.M., "Improvement of BLEVE Overpressure Method Using Historical Data. Conference Proceedings", 7th Global Congress on Process Safety. Chicago, IL, 2011.

9TH GLOBAL CONGRESS ON PROCESS SAFETY

New Ignition Source "Exploding Wire" for the Determination of Explosion Characteristics of Combustible Dusts in the 20-L-Sphere

Marc Scheid
Federal Institute for Materials Research and Testing (BAM),
Department 2 "Chemical Safety Engineering"
D-12200 Berlin, Germany
marc.scheid@bam.de

Alexander Klippel, Rico Tschirschwitz, Volkmar Schröder, Stefan Zirker, Christian Kusche
Federal Institute for Materials Research and Testing (BAM), Department 2 "Chemical Safety Engineering"
D-12200 Berlin, Germany

Prepared for Presentation at
American Institute of Chemical Engineers
2013 Spring Meeting
9th Global Congress on Process Safety
San Antonio, Texas
April 28 – May 1, 2013

UNPUBLISHED

New Ignition Source "Exploding Wire" for the Determination of Explosion Characteristics of Combustible Dusts in the 20-L-Sphere

Marc Scheid
Federal Institute for Materials Research and Testing (BAM),
Department 2 "Chemical Safety Engineering"
D-12200 Berlin, Germany
marc.scheid@bam.de

Alexander Klippel, Rico Tschirschwitz, Volkmar Schröder, Stefan Zirker, Christian Kusche
Federal Institute for Materials Research and Testing (BAM),
Department 2 "Chemical Safety Engineering"
D-12200 Berlin, Germany

Keywords: dust explosions, dust explosion testing, exploding wire, explosion characteristics, fuse wire, ignition source

Abstract

According to international standards several safety characteristics of dusts are determined in the 20-L-sphere (also known as SIWEK-Chamber). Dust cloud ignition is carried out using pyrotechnical igniters. Due to various disadvantages of pyrotechnical igniters the need for alternative ignition sources arises again and again. An alternative ignition source could be the so-called "exploding wire" or "fuse wire". The paper presents test results of a comparative study between both ignition sources for the determination of the safety characteristics Maximum Explosion Pressure and Maximum Rate of Explosion Pressure Rise of selected dusts in the 20-L-sphere. In addition to that the flame/arc propagation of both ignition sources was analysed by high speed and IR camera recordings. Tests in a windowed autoclave were performed to get information on the influence of turbulence generated due to dispersion of the dust on the shape of the generated flame and its propagation. Turbulence measurements in the windowed autoclave as well as in the 20-L-sphere with a LDA system allowed investigation if the igniter led to turbulence increase at the moment of ignition in comparison to tests without ignition.

1. Introduction

Safety characteristics are essential for the determination of explosion hazards during handling of combustible dusts and for the design of safety measures. The safety characteristics Maximum Explosion Pressure p_{max}, Maximum Rate of Explosion Pressure Rise $(dp/dt)_{max}$ and Lower Explosion Limit LEL are determined in closed vessels such as the 20-L-sphere (also known as SIWEK-Chamber). Tests are preformed according to international standards, for example EN 14034 part 1 and part 2 [2,3] or ASTM E1226 [1]. In order to determine the safety characteristics the dust samples are dispersed in air and ignited using two pyrotechnical igniters with energy contents of 1000 J (LEL) or 5000 J (p_{max}, $(dp/dt)_{max}$). Since the igniters could have an influence on the test results they are defined in the standards.

Due to various disadvantages of pyrotechnical igniters such as high costs, legal requirements concerning its storage and use as well as high energy input in comparison to most ignition sources in practice, the need for alternative ignition sources arises again and again; see also Krietsch [6]. Such an ignition source should be less expensive, readily available and the operator should be able to use it without a certificate of competence. In addition to that, the ignition energy should be adjustable over a wide range and if possible allow energy amounts less than 1000 J. Compared to the ignition energies described in the standards such energy amounts are more realistic for ignition sources found in practice, such as electrostatic discharges, mechanical sparks or hot surfaces.

An alternative ignition source which fulfils these requirements is the so-called "exploding wire" or "fuse wire". This type of ignition source is used for the determination of explosion limits of gases and is described in the standard EN 1839 [4].

This paper presents test results of a comparative study in order to find out, if the exploding wire is suitable as an alternative ignition source for the determination of explosion characteristics of dusts. For the study Maximum Explosion Pressure and Maximum Rate of Explosion Pressure Rise values from different dusts were determined with both ignition sources in the range of 100 J to 2000 J. The dusts were selected such that different combustion mechanisms were considered. High speed and IR camera recordings allowed comparison of flame front and electrical arc generated from the igniters. Calorimetric and electric measurements obtained information on the ignition energy of the igniters delivered to the mixture. With additional tests in a windowed autoclave it was possible to get information on the influence of pre-ignition turbulence on the shape of the generated flame/arc and its propagation as well as information on the turbulence generated from the ignition source. Turbulence measurements in the windowed autoclave were realised with a LDA system. The same system was used for tests in the 20-L-sphere to get information on the turbulence at the moment of ignition.

2. Test assembly and test settings

2.1 20-L-sphere

Main component of the test assembly was the 20-L-sphere (Cesana and Siwek [5]; EN 14034 1-2 [2,3]). In order to reach almost homogeneous dust dispersion into the test chamber the so called "rebound nozzle" was used.

Tests with pyrotechnical igniters as an ignition source were performed using the standard control unit of the 20-L-sphere. For tests with exploding wires an external ignition unit was necessary. Depending on the ignition energy different systems with isolating transformer, or capacitors were used. An isolation amplifier allowed the determination of the ignition energy by measuring current and voltage.

2.2 Windowed autoclave

A windowed autoclave with cylindrical shape was used for testing in order to get information on flame/arc propagation in a closed vessel as well as on the shape of flame/arc generated from both ignition sources. In this autoclave it was also possible to determine the influence of turbulence on the generated flames. The volume of the autoclave was approximately 11 L. Its diameter was 220 mm. Optical accessibility via the window allowed high speed recordings at different positions and turbulence measurements. A fan which was placed on the bottom of the autoclave enabled average turbulent velocity fluctuations (RMS-value) up to 1 m/s. In addition to that the system allowed measurements of turbulence generated from the ignition source.

2.3 Ignition sources

One of the fundamental differences of the two ignition sources pyrotechnical igniter and exploding wire is the ignition mechanism. While for the exploding wire an electrical arc and plasma is generated the pyrotechnical igniter emits flames and burning solids.

The ignition sources were placed in the middle of the vessel and connected to the lid of the vessel, such as it is described in the test standards. According to the European standards for the determination of the explosion characteristics Maximum Explosion Pressure and Maximum Rate of Explosion Pressure Rise two pyrotechnical igniters with energy of 5000 J shall be used, firing horizontally in opposite directions (EN 14034 1-2).

Instead of the ignition energy described in the standards tests were performed with ignition energies of 100 J, 500 J, 1000 J and 2000 J. In the first tests only a single ignition source with ignition energies of 100 J, 500 J and 1000 J was used. Further tests with a modified lid allowed tests with two igniters with ignition energies of 500 J and 1000 J each. For all tests with one ignition source the igniter was mounted such that it produced a flame or an electrical arc downwards. Electrodes and connected igniter for tests with a single ignition source are shown in Figure 1. In tests with two igniters the ignition source was installed such that the flame propagated in nearly horizontal direction such as it is described in the standard.

<u>Pyrotechnical igniter</u>

All pyrotechnical igniters used for this work have the same composition as the standard igniters used for determination of explosion characteristics in the 20-L-sphere. They contained 40 % by weight zirconium metal, 30 % by weight barium nitrate and 30 % by weight barium peroxide.

<u>Exploding wire</u>

The principle of the exploding wire is the evaporation and ionization of metal particles from a wire due to a sufficient current flow within milliseconds. During that process a metal vapour is created in which an electrical arc is generated between the electrodes because of the electrical conductivity of the plasma.

For all tests a nickeline wire with a diameter of 0.12 mm was used. In tests with a single ignition source the wire was tensed between the ends of two auxiliary electrodes which were connected two the main electrodes of the 20-L-sphere, see Figure 1. For tests with two exploding wires the lid of the 20-L-sphere had to be modified. It was designed such that it contained four electrodes.

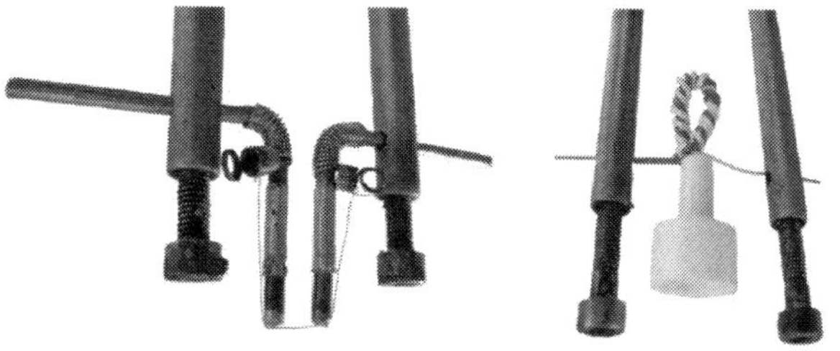

Figure 1. Ignition source for tests with a single igniter: Exploding wire with electrodes (left side) and pyrotechnical igniter (right side)

<u>Ignition unit with isolating transformer</u>

First tests with the exploding wire were performed with ignition energy of 100 J. In these tests the electrical power which is needed for melting the wire and generating the arc was supplied by an isolating transformer similar to that used for the determination of explosion characteristics of gases and vapors according to EN 1839 [4]. This system and its disadvantages such as not reproducible ignition delay are described by Scheid [7].

Ignition unit with capacitors
In tests with ignition energies of 500 J, 1000 J and 2000 J an ignition unit with capacitors was used instead of the isolating transformer. The main advantages of this new type of ignition unit are the higher ignition energies which the unit can produce and that it is able to ensure an exact ignition delay. A choke enabled elongation of the electrical arc during each test.

2.4 Recordings of flame front and electrical arc

In order to record the propagation of flame/arc of both igniters a high speed camera system was used. It allowed recordings of 5000 frames per second. So it was possible to determine maximum flame volume and length as well as the time delay for activating the ignition source and its burning time. In addition to that a fast acting IR camera system was used to estimate the maximum temperature level generated from the igniters. The IR camera allowed recordings with 850 frames per second.

2.5 Measurements of ignition energy (delivered to the mixture)

A combustion calorimeter working with the dry method allowed measurements of the ignition energy of both ignition sources. Dry method means, that the brass bomb of the calorimeter is covered with an insulating styrofoam and not with a water bath. This allows a more sensitive direct measurement of the temperature changes in the metal. Additionally an isolation amplifier was used to measure the burning current in the exploding wire as well as the voltage and enabled the determination of ignition energy generated in the electric arc during each test.

2.6 Turbulence measurements

A one-dimensional laser-Doppler-system (LDA) was used for turbulence measurements in the 20-L-sphere as well as in the windowed autoclave. Due to the big window of the autoclave the turbulence was measured at three different positions around the ignition source. Since the optical accessibility in the 20-L-sphere could only be realized via a windowed flange with a diameter of 40 mm it was only possible to measure the turbulence at one position next to the ignition source.

2.7 Tested dusts

Dust explosions generally show complex reaction mechanism, which depend e.g. on the chemical composition of the dust, the dust concentrations and the flow conditions. Van der Wel [8] differs between different reaction mechanisms depending on whether the reaction takes place in the gas phase due to evaporation or at the solid surface in form of gaseous products, solid or liquid material. The dusts used in this work were chosen such that they represent the different chemical classes and mechanisms as well as the dust explosion classes found in practice. As a result dusts from lignite, maize starch, niacin and anthraquinone were chosen. In addition to that a steel dust which was generated during a shot blasting process was tested.

3. Test results

3.1. Ignition energy measurements

Measurements with the isolation amplifier lead to different test results depending on the boundary conditions of the measurements. Table 1 shows the test results for measurements of ignition energy outside the test vessel without dust and flow, inside the test vessel with flow but without dust and inside the test vessel with flow and with a dust concentration of approximately 1000 g/m³ - 1250 g/m³ maize starch. Ignition unit was the isolating transformer.

Table 1. Ignition energy of igniter determined with the isolation amplifier; ignition source isolating transformer; ignition energy 100 J

Experiment-No.	Determined ignition energies		
	Outside the apparatus [J]	Inside the apparatus without dust [J]	Inside the apparatus with dust [J]
1	93.61	98.42	98.88
2	86.84	94.19	84.14
3	90.92	89.27	88.24
4	89.80	101.59	48.85
5	91.62	82.96	94.13
6	84.46	98.58	83.89
7	90.40	94.74	100.63
8	94.41	91.37	74.36
Ø	**90.26**	**93.89**	**84.14**
SD	**3.30**	**5.97**	**16.67**

Table 1 shows that the ignition energy values for tests without dust (inside and outside the test vessel) were almost similar. As a trend the values with flow (tests inside the vessel) and their standard deviation seem to be slightly higher. Tests with dust led to slightly lower ignition energies and a significant higher standard deviation of the values. Most of the tests led to ignition energies of more than 90 % of the adjusted value (set point adjustment). However, some tests with dust led to lower values. In one test less than 50 % of the adjusted value was measured. As a result the dust particles seem to influence the test results in some tests.

Figure 2 shows the arithmetic mean of the determined ignition energies as a function of dust cloud concentration for maize starch and lignite for adjusted ignition energy of 100 J. From the course of the mean values follows, that there is only a minor influence of the dust concentration on the ignition energy up to concentrations of approximately 1250 g/m³. For higher concentrations the influence is significantly higher.

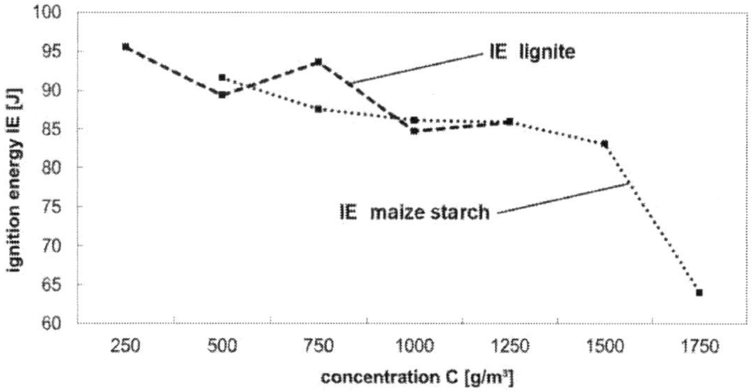

Figure 2. Arithmetic mean of ignition energy of igniter as function of dust cloud concentration for maize starch and lignite; adjusted ignition energy 100 J

Comparable test results were determined for ignition energies of 500 J, 1000 J and 2000 J using an ignition unit with capacitors. But as a trend the influence of the dust cloud on the ignition energy seems to be lower. For all tests the determined ignition energy was at least 80 % of the set point adjustment. Energy measurements with the combustion calorimeter confirmed the test results.

3.2 Propagation of flame/arc of ignition source

Tests in the windowed autoclave with maximum fan speed were performed in order to get information if pre-ignition turbulence has an influence on the flame/arc propagation of a 1000 J pyrotechnical igniter or an exploding wire with the same energy. However, for both ignition sources no visible influence on flame/arc propagation and maximum flame/arc volume was found.
Figure 3 shows the flame propagation for a 1000 J pyrotechnical igniter in comparison to an exploding wire with the same energy for tests in the windowed autoclave (no reaction of dust). The flame recordings show, that the flame of the pyrotechnical igniter propagates 3-4 times faster than the flame of the exploding wire and reaches a much higher maximum volume.

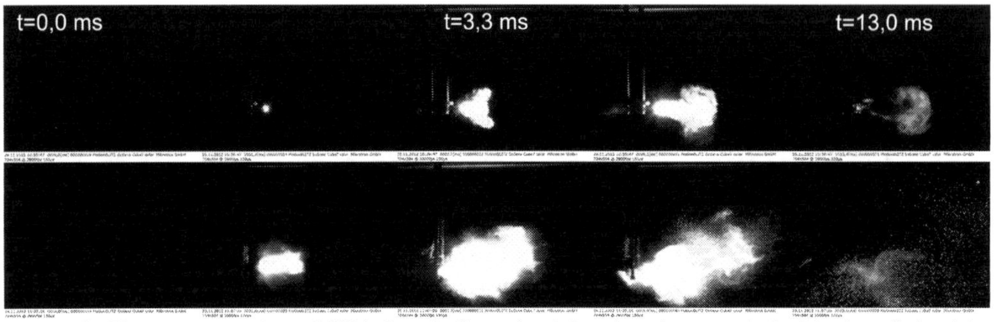

Figure 3. Flame propagation of 1000 J exploding wire (above) and pyrotechnical igniter (below) in a windowed autoclave with a volume of 11 L recorded with high speed camera

Recordings with a fast IR camera allowed an estimation of the maximum temperature level in the generated flame/arc and the maximum volume of flame or arc. Figure 4 shows four sequences of the flame/arc propagation of exploding wire (first and third sequence) and pyrotechnical igniter (second and fourth sequence) recorded with the IR camera. The sequences show a period of approx. 5 ms within the first 15 ms after triggering the igniter. Temperature range in the first two sequences was adjusted such that temperatures between 200 °C and 650 °C were visualized.

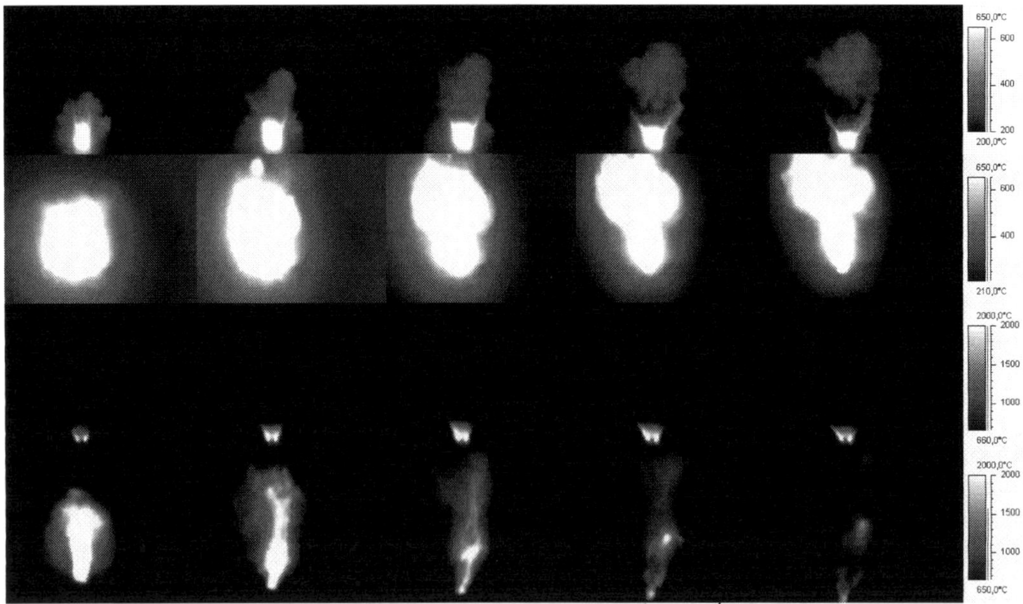

Figure 4. Maximum flame/arc volume of 1000 J exploding wire (1st and 3rd row) and pyrotechnical igniter (2nd and 4th row) recorded with a fast IR camera system

A comparison between sequence one and sequence two shows that the volume with the highest temperatures generated from the pyrotechnical igniter is much larger than the volume with such temperatures generated from the exploding wire. While for the exploding wire the volume with the highest temperatures does not increase with the propagation of the flame, it increases for the pyrotechnical igniter.

Sequence three and four visualize the temperature range between 650 °C and 2000 °C. For the exploding wire the highest temperatures were determined close to the wire. The volume in which such temperatures were determined stays almost constant in the whole sequence. In contrast to that the pyrotechnical igniter seems to generate a large volume with such high temperatures in from of a jet immediately after it was triggered. However, the volume with the highest temperatures seems to decrease significantly within 2 ms.

3.3 Turbulence measurements in 20-L-sphere and windowed autoclave

In tests in the windowed autoclave with operating fan (generated RMS-value of approx. 1 m/s) no differences in the measured RMS-value could be determined between tests without ignition and with a 1000 J exploding wire. Tests with the pyrotechnical igniter resulted in RMS-values of 2.5 m/s - 4.5 m/s depending on the measuring position. In tests in which the fan was turned off the exploding wire led to RMS-values of 0.5 m/s while the pyrotechnical igniter led to values of 2 m/s - 3 m/s. As a result turbulence measurements in the windowed autoclave showed that the turbulence generated from a 1000 J pyrotechnical igniter is stronger than the value generated from an exploding wire with the same ignition energy.

LDA-measurements in the 20-L-sphere showed, that the dispersion of dust with the rebound nozzle generated a flow with a maximum velocity of more than 40 m/s. This flow velocity decreased rapidly with decreasing pressure in the dust storage chamber and with closing the inlet valve. At the moment of ignition the flow velocity reached a value of approximately 10 m/s. The measured RMS-value at the moment of ignition was approximately 15 m/s.

In comparison to the turbulence generated in the windowed autoclave the value determined in the 20-L-sphere is much higher. Hence the test results from the windowed autoclave could not be transferred to the 20-L-sphere. As a result LDA-measurements in the 20-L-sphere were necessary. These measurements showed for tests with 1000 J ignition sources that at the moment of ignition for none of the igniters an increase of the RMS-value could be measured. However, the pyrotechnical igniter showed a not negligible velocity increase in one direction in comparison to the velocity generated due to the dispersion of dust.

3.4 Explosion behavior

Figure 5 presents the explosion pressure of maize starch as a function of the dust cloud concentration determined with both ignition sources with ignition energy of two times 1000 J. Both curves show a comparable maximum of approximately 8.2 bar for the exploding wire and 8.4 bar for the pyrotechnical igniter. For both igniters the maximum was determined for a concentration of 1000 g/m³. Deviations in explosion pressure between the different series were very low for dust explosion testing.

The determined rate of explosion pressure rise values as a function of the dust concentration are found in Figure 6. All test results show a good conformity concerning maximum rate of explosion pressure rise values which were almost similar. However, the reproducibility was not as good as for the explosion pressure. As a result the test values of the different test series differed more for both ignition sources. But this is very often found in dust explosion testing. Comparable test results were found for all tested dusts and ignition energies.

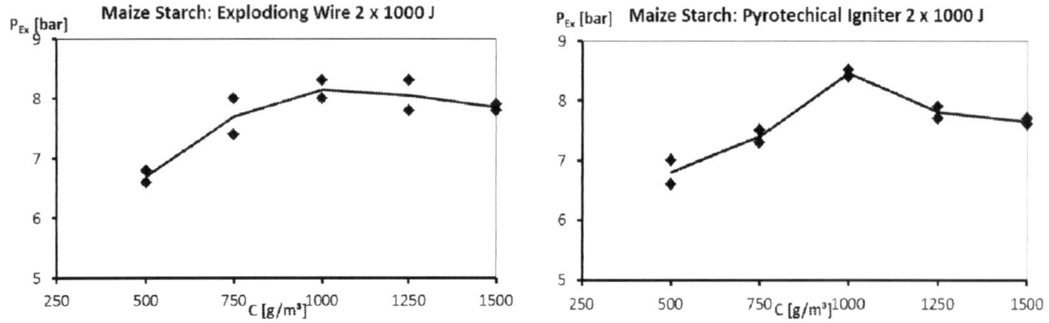

Figure 5. Explosion pressure of maize starch as function of dust cloud concentration determined with pyrotechnical igniters and exploding wires with ignition energies of 2 x 1000 J

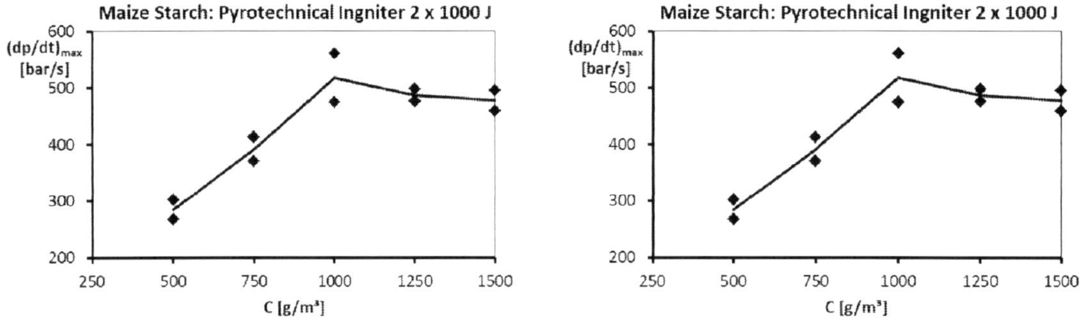

Figure 6. Rate of Explosion pressure rise of maize starch as function of dust cloud concentration determined with pyrotechnical igniters and exploding wires with ignition energies of 2 x 1000 J

Table 2 gives an overview of all determined p_{max} and $(dp/dt)_{max}$ values as well as the K_{St} values which were calculated from the $(dp/dt)_{max}$ values.

A comparison of the results determined with two 500 J ignition sources with one 1000 J ignition source in order to find out if the use of two ignition sources leads to different test results did not come to a clear result. While for lignite tests with two ignition sources led to slightly higher values for the Maximum Explosion Pressure, for maize starch one ignition source led to slightly higher values. The determined rate of explosion pressure rise values were for both dusts slightly higher in tests with two ignition sources. However, for both tested dusts the deviations in explosion pressure and explosion pressure rise between the different test series seem to be slightly lower for tests with two ignition sources.

Table 2 also includes test results which have been determined with two 5000 J igniters, which is the ignition energy described in the test standard for determining p_{max} and K_{St}. The tests showed that the Maximum Explosion Pressures values determined with ignition energies of 100 J, 500 J, 1000 J and 2000 J are less than 10 % lower than those with 10000 J. For steel dust and niacin the values determined with 10000 J were slightly lower than the values determined with the other ignition energies. However, the determined $(dp/dt)_{max}$ values and therefore the K_{St} values differed a lot from the values determined according to the standard procedure, especially for tests with comparatively low ignition energies. Tests with 10000 J led to values which were up to 30 % higher than the values determined with low ignition energies. Only tests with 2000 J ignition sources led to reasonable test results.

Table 2. p_{max}, $(dp/dt)_{max}$ and K_{St} values for all tested dusts determined with exploding wires and pyrotechnical igniters with ignition energies of 1 x100 J, 1 x 500 J, 2 x 500 J, 1x 1000 J 2 x 1000 J and 2 x 5000 J.

Dust		1 x 100 J		1 x 500 J		2 x 500 J		1 x 1000 J		2 x 1000 J		2 x 5 kJ
		Ex. Wire	Pyro. Igniter	Ex. Wire	Pyro. Igniter	Ex. Wire	Pyro. Igniter	Ex. Wire	Pyro. Igniter	Ex. Wire	Pyro. Igniter	Pyro. Igniter
Lignite median 37 µm	p_{max}	8.4	8.1	8.1	7.6	8.1	8.5	7.9	7.7	8.3	7.9	8.4
	$(dp/dt)_{max}$	601	545	681	630	711	619	671	608	723	705	784
	K_{St}	163	148	185	171	193	168	182	165	196	191	213
Maize starch median 14 µm	p_{max}	8.4	8.5	8.4	8.6	8.0	8.2	8.2	8.4	8.2	8.4	8.7
	$(dp/dt)_{max}$	408	470	528	480	500	527	487	468	504	530	614
	K_{St}	111	127	143	130	136	143	132	127	137	144	167
Niacin median 14 µm	p_{max}	-	-	8.3	8.2	-	-	8.2	8.0	-	-	7.9
	$(dp/dt)_{max}$	-	-	953	942	-	-	920	874	-	-	1051
	K_{St}	-	-	259	256	-	-	250	237	-	-	285
Anthraquinone median 3 µm	p_{max}	-	-	-	-	-	-	8.1	8.0	-	-	8.4
	$(dp/dt)_{max}$	-	-	-	-	-	-	1113	999	-	-	1379
	K_{St}	-	-	-	-	-	-	307	271	-	-	374
Steel dust median 24 µm	p_{max}	-	-	-	-	-	-	4.1	4.1	-	-	3.8
	$(dp/dt)_{max}$	-	-	-	-	-	-	398	394	-	-	432
	K_{St}	-	-	-	-	-	-	108	107	-	-	117

4. Conclusion

Measurements of ignition energy delivered to the mixture of the ignition source exploding wire using two different methods showed that the ignition energy generated with the exploding wire was reproducible. For ignition energies of 500 J and higher, the dust cloud did not influence the ignition energy significantly. However, some tests with ignition energy of 100 J and dust concentrations with more than 1000 g/m³ showed that the dust cloud was able to influence the ignition process such that the electric arc extinguished earlier. This resulted in lower ignition energy.

High speed recordings showed that the flame generated from the pyrotechnical igniter propagated faster than the flame/arc generated from the exploding wire and reached a larger volume. Recordings with a fast IR camera system confirmed these test results. In addition to that the recordings showed that as well the volume with a temperature level comparable to the Minimum Ignition Energy of most dusts as the volume in which very high temperatures were measured was much larger for the pyrotechnical igniter. In contrast to that in tests with the pyrotechnical igniter the temperature decreased earlier as in tests with the exploding wire. As a result of high speed and IR recordings as well as energy measurements for the 100 J exploding wire the igniter should have a significant influence on the determined p_{max} and $(dp/dt)_{max}$ values. However, such an effect was not determined during the measurements of p_{max} and $(dp/dt)_{max}$ of five different dusts. All tests resulted in comparable values of p_{max} and $(dp/dt)_{max}$ for both ignition sources.

A comparison of p_{max} values determined with two 500 J ignition sources with one 1000 J ignition source in order to find out, whether the use of two ignition sources will lead to different test results, did not come to a clear result. The determined pressure rise values were for both dusts slightly higher for tests with two ignition sources. However, for both tested dusts deviations in explosion pressure and rate of explosion pressure rise seem to be slightly lower for tests with two ignitions sources.

The p_{max} values determined with ignition energies of 100 J, 500 J, 1000 J and 2000 J were less than 10 % lower than values determined according to the test standard with two igniters of 5000 J each. As a result the influence of the ignition energy of the igniter on the Maximum Explosion Pressure seems to be almost negligible.

In contrast to that $(dp/dt)_{max}$ values determined according to the test standard led to 30 % higher values, especially in tests with ignition energies of 100 J and 500 J. Only tests with 2000 J ignition sources led to reasonable test results. The reason for that could not be solved totally. It is assumed that a combination of different effects has to be considered such as turbulence generated from the ignition source and the flame/arc volume of the ignition source. However, the effect of the flame/arc volume should result in different test results for exploding wire and pyrotechnical igniter. However, this could not be determined. Probably the turbulence generated from the igniters has a strong influence especially on the $(dp/dt)_{max}$ values. Tests in the 20-L-sphere showed for all ignition sources with energies up to 2000 J that the turbulence generated from the igniters seems not to have a significant influence on the turbulence field generated due to dispersion of the dust (pre-ignition turbulence). This is probably different for higher ignition energies.

In a next step tests with ignition energies up to 10000 J with two pyrotechnical igniters and two exploding wires will be performed. In addition to that a validation of the exploding wire with a greater number of dusts is planned.

5. References

[1] ASTM E1226-12, "Standard Test Method for Explosibility of Dust Clouds," ASTM International, US, 2012

[2] EN 14034–1, "Determination of explosion characteristics of dust clouds - Part 1: Determination of the maximum explosion pressure p_{max} of dust clouds," Beuth-Verlag, Berlin, Germany, 2011.

[3] EN 14034–2, "Determination of explosion characteristics of dust clouds - Part 2: Determination of the maximum rate of explosion pressure rise $(dp/dt)_{max}$ of dust clouds," Beuth-Verlag, Berlin, Germany, 2011.

[4] EN 1839, "Determination of explosion limits of gases and vapours," Beuth-Verlag, Berlin, Germany, 2003.

[5] Cesana C., Siwek R., "Manual KSEP 7.0," Kühner AG, <kuhnersafety.com/tl_files/kuhner/product/safety/PDF/B000_071.pdf>, Accessed 2013-01-04

[6] Krietsch A., Scheid M., Lasry M., Suter G., Teske H.-J., Stahmer K. W., Arnold A., Nemcak O., Cesana C., Siwek R., "Validation of alternative igniters for the determination of explosion characteristics of dust clouds in 20-l-sphere and 1m³-vessel," Proceedings 13th Int. Symp. on Loss Prevention and Safety Promotion in the Process Industries 6th – 9th June 2010, Brugge, 2010

[7] Scheid M., Kusche C., Schröder V., Barth U., "Tests on Suitability of the Ignition Source "Exploding Wire" for the Determination of Explosion Characteristics of Combustible Dusts in the 20-L-Sphere," Chemical Engineering Transactions, vol. 31, 2013 (in print)

[8] Van der Wel P. G. J., "Ignition and Propagation of Dust Explosions," Dissertation, Delft University Press, the Netherlands, 1993

Operational Discipline

Improving Process Safety Management Implementation through Behavior and Cultural Change

Charles A. Soczek
Principal Consultant
DuPont Sustainable Solutions
E.I. DuPont De Nemours and Company

Abstract

In 2007, US OSHA initiated a Petroleum Refinery Process Safety Management National Emphasis Program. This was followed in 2009 with a similar program focused on Chemical operations. These programs developed a series of findings, many of which were results of failure to implement OSHA 1910.119 PSM regulatory requirements. Many facilities had well documented programs, but there was poor implementation. In DuPont it was recognized that implementation is an Operational Discipline issue.

Operational Discipline (OD) as defined by DuPont is "the deeply rooted dedication and commitment by every member of an organization to carry out each task, the right way, and each time". OD is recognized as the key to successful implementation of a Process Safety Management program, turning a paper program into reality. Typically, companies have addressed OD improvements through behavioral modifications, but while the gains from programs to modify behavior can be substantial, they are difficult to sustain. Cultural change, while far more difficult to achieve, can produce more sustainable results. When done together, substantial improvements in PSM performance can be both rapid and sustained. This presentation will discuss the key attributes of operational discipline as well as provide an outline of the process to address behavior and culture change is provided.

Introduction

Operational Discipline (OD) as defined by DuPont is "the deeply rooted dedication and commitment by every member of an organization to carry out each task, the right way, and each time". Per the Business Dictionary, OD is defined as assurance that specific activities necessary for successful daily operation of a company's operations are running as they should be. This can include specific goals, missions, and guidelines. Whichever definition is chosen, OD is important to success of all aspects of an operating facility.

Process Safety Management has been in place for over twenty years. Many companies have good PSM systems, but based on the CSM Special Emphasis studies, implementation is poor. It is being increasingly recognized that Operating Discipline is a key factor that drives PSM implementation. Operating Discipline helps turn a paper PSM program into a reality

Improving Operating Discipline

Advances in Social/Psychological theory are showing that operating discipline improvements are influences by two factors, Behavior Change and Culture change. The Dejoy model helps explain this.

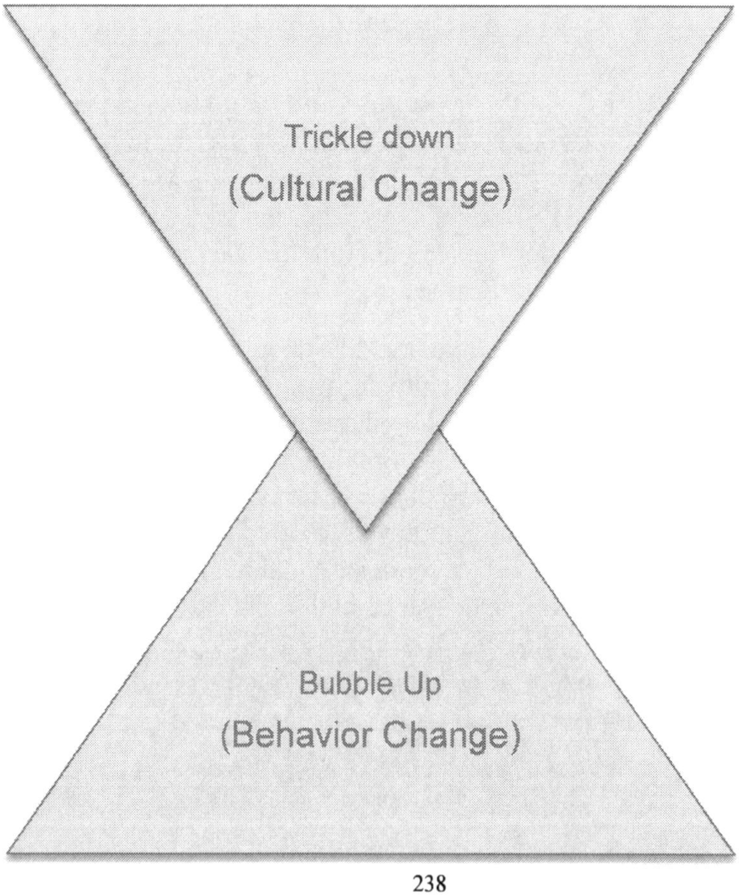

The basis for the Dejoy[1] model is that behavior change can be made quickly, but improvements are not sustainable without a large continuing effort, and that even with this, results will plateau (Bradley Curve). Safety culture change takes much longer, but once in place, superior performance can be both achieved and sustained when coupled with behavior improvement. In the Dejoy model, cultural change is driven by leadership whereas behavior change is driven by supervisors. This is an important distinction and helps focus the role of leadership and the role of supervision.

In a study of some DuPont sites, it has been found that multi-faceted Operating Discipline programs are in place. These programs address both behavior change and cultural change. Programs vary, and can consist of consists of a large number of small efforts or fewer larger programs involving all members of the organization. Either can be successful.

Dejoy[1] did a comparison of a cultural based program and a behavior based program.

Characteristic	Behavior -Based changes approach to safety	Cultural Based changes approach to safety
Background/origin	Operant psychology/ behavior modification	Organizational behavior/ anthropology
Key aspects	Bottom up approach	Top-down
	Analytic/data driven	Intuitive/ethnographic
	Setting Specific	Setting Specific
	continuous process	Self-sustaining
Typical Implementation	Identify and define critical behaviors	Assess aspects of culture (values, beliefs, assumptions, etc.)
	Set performance goals	Devise alternative vision
	Observe/sample behavior	Work with leadership and employees to implement change
	Provide contingent feedback and reinforcement	
Principal Strengths	Specific technology	Emphasizes organizational change
	Objective/empirical	Focuses on basic causes
	Shop floor focus	participatory (often)
	Participatory (usually)	Comprehensive
	Positive	
Principal Weaknesses	Victim-blaming	Diffuse technology
	Minimizes environment	Subjective/intuitive
	Focuses on immediate cause	Indirect

The above chart has highlighted a few of the key differences between the behavioral and cultural approach. In the bottom up, the emphasis is on the operators, maintenance workers, and contractors to follow procedures and rules. This requires good processes and standards and includes items as good communications, housekeeping, documentation, following procedures, and adequate resources. The top-down cultural improvement process is driven by leadership and is reflected in efforts such as leadership by example, building pride in the organization, expression of values, risk sensitivity, and being a learning organization that is innovative, agile, and continuously improves..

Some Definitions

- Operant conditioning psychology (sometimes referred to as instrumental conditioning) is a method of learning that occurs through rewards and punishments for behavior
- Psychological anthropology is an interdisciplinary subfield of anthropology that studies the interaction of cultural and mental processes. The subfield tends to focus on ways in which humans' development and enculturation within a particular cultural group—with its own history, language, practices, and conceptual categories—shape processes of human cognition, emotion, perception, motivation, and mental health. It also examines how the understanding of cognition, emotion, motivation, and similar psychological processes inform or constrain our models of cultural and social processes. Each school within psychological anthropology has its own approach
- Ethnography is a qualitative method aimed to learn and understand cultural phenomena which reflect the knowledge and system of meanings guiding the life of a cultural group

In behavior based approach, performance goals are set, whereas in a cultural based approach leadership works to build a vision of a better place (zero incidents).

Behavior programs are focused on the shop floor, whereas cultural change is focused on everyone in the organization.

Operating Discipline Characteristics

Organizations that have a high level of operating discipline exhibit an identifiable set of characteristics. DuPont has defined these characteristics as follows:

OD Characteristics

» Leadership by example
» Sufficient & capable resources
» Employee involvement
» Active lines of communication
» Strong teamwork
» Shared values
» Up-to-date documentation
» Practice consistent with procedures
» Absence of shortcuts
» Excellent housekeeping
» Pride in the organization

» Risk sensitivity
» Innovation, agility, and continuous improvement
» Justice and fairness

It is recognized that behavior change of people and cultural change are very complex. It is not proposed that the set of characteristics above is comprehensive, and it is expected that over time additional characteristics will be recognized and added to this list.

Implementation of an Operating Discipline Improvement Program

The process developed to implement an operating discipline program starts with an assessment based on 6 Sigma methodology. In the assessment, each characteristic is defined by a set of CTQs (Critical to Quality) and CTXs that define the CTQ.

These steps are

1. Pre-work

- Define scope of the assessment
- Form client OD team
- Develop assessment schedule
    - Capture input from 50% or more of the population
    - Capture current relevant data (see appendix A)
- Assessment team orientation
- Develop current state hypothesis

2. Field data collection (reach 50% of the workforce)

- Focus Groups – operators, mechanics, first line (2 hours)
- One-on-One- managers and leaders (60 minutes)
- Field Tours- employees actively performing work

3. Analysis

- Review and summarize key findings
- Apply a rating to each CTQ per rating guidance

4. Recommendations

- Each low rate CTQ has a recommendation(s) based on best practices. These best practices come from DuPont sites.
- The completed Assessment Model is used as the report to management

5. Action Plan

- Owned by line management based on findings

6. Continuous Improvement

    - Conduct reviews of the assessment and periodically update the action plan with additional "best practices"
    - Conduct re-assessments no more frequently than once per year

Examples of the Assessment Questions

Below is an example of the assessment tool developed that is used to understand each of the fourteen OD characteristics.

OD Characteristic: Leadership by Example

Leaders are demonstrating unrelenting passion for SHE and continuously model the behavior they expect of others

CTQ-1 Visible leadership presence and involvement across the organization

CTX – Build value in leaders that this is a critical part of their role and hold them accountable with specific expectations

- Is visible leadership evaluated as part of the overall personnel review process for line leaders, and if so, how ? Do Leaders at all levels have COTs relating to Operational Discipline?
- Do line leaders consistently take specific corrective action steps for non-conformance to SHE requirements?
- Line management is actively involved with planning, leading safety meetings & communications, and emphasizing broad participation
- Is visible leadership included in new supervisory/manager training and if so, is it effective?
- Is there consistency in approach and focus over time as leaders change?
- Does leadership review action plans to assure alignment with goals and objectives?
- Does leadership allocate adequate resources to support the safety program?
- Does leadership provide a personal example of following standards and procedures? (Wearing required personal protective equipment when in the field; using handrails, etc.)

CTX – Define and communicate what is "Visible Leadership" throughout the organization

- Does the site have formal/informal guidelines for leadership time spent in the field? How does the site measure this and is it effective
- Has the site clearly defined opportunities for visible leadership and personal leadership participation at all levels?
- Is leadership at all levels routinely in work areas conducting Safety Observations audits or other field evaluations?
- Has site leadership clearly articulated a vision and goals for Operational Discipline?
- Does leadership conduct periodic reviews of activities with all safety related teams?
- Does leadership critique audit and reports and follow-up to assure implementation of recommendations?

This characteristic has three CTQs and six CTXs. This is similar to the other characteristics.

The assessment process generates a large volume of information. The assessment team takes this information and develops action items. These items must be prioritized. Below are some examples of action items for leaders taken from an operating facility that could be generated from the process described above.

1. Increase field time and double safety audits
2. Devote 50% of PM meeting to safety
3. Off hours audits and safety coaching
4. Continue small things matter theme
5. Establish designated time in the field for leadership on safety
6. Revamp weekly safety audit meeting
7. Focus team on elimination of slips trips and falls
8. Review all letters of commendations

Summary

Sustainable improvements in Operating Discipline take time, and require facilities to continuously work on improving the safety culture and the behavior of people. This continuing focus on Operating Discipline is a common characteristic of high performing facilities.

References

(1) "Behavior change versus culture change: Diverging approached to managing workplace safety" David M. Dejoy Safety Science 43 2005.

(43g) Risk Analysis for Pressure Relief Devices – Managing Plant Safety and Reliability

Jose R. Aguilar-Otero Sr., Technology Application, COMIMSA, Saltillo, Mexico

Risk Analysis for Pressure relief devices – Managing Plant Safety and Reliability

Pressure relief devices are one of the most critical assets for plant risk and safety, therefore it is essential to be able to ensure PSV's operation, reliability and availability through the implementation of risk mitigation or prevention actions, either, corrective, preventive or predictive; these actions will allow plant owners to manage the risk associated with the device operation (or failure). This paper is intended to show a risk assessment using Weibull analysis focused on the identification for risk reduction opportunities and the selection for the best chance and dates to implement risk mitigation actions. Risk assessment methodology is based on API RP 581, Risk Based Technology.

Historically, maintenance planning for a PSV was carried out, based on pressure tests and engineering judgment; a pressure (POP) test is usually conducted to evaluate device's availability and functional condition and its frequency is linked to the protected equipment's maintenance program, regardless whatever the failure mode. Nevertheless, now a days, more quantitative and rigorous methodologies are required not just for maintenance planning but for developing a risk management strategy.

The methodology is a risk assessment. We evaluate the probability of failure (POF) times the consequence of failure (COF), for both, a valve failure and a protected equipment failure. Two failure modes are evaluated: fail to open on demand (FOD) and leak (LEAK). POF is calculated analyzing three main issues:

1. Historical inspection records and time in service, using a two parameters Weibull distribution, as follows:

$$F(t) = P_{fod}^{PSV} = 1 - \exp\left[-\left(\frac{t}{\eta}\right)^{\beta}\right] \quad (1)$$

2. Probability of failure from the protected assets,

$$P_f(t) = ffg * D_f(t) * F_{SA} \quad (2)$$

3. Probability of failure from process over pressure scenarios.

$$P_{f,j} = P_f(t) + \left(\frac{1-ffg}{3}\right)\left(\frac{P_{o,j}}{MAWP} - 1\right) \quad (3)$$

Consequences of failure (FOD, LEAK) are evaluated as a function of consequence area and as a function of its economic impact, for a valve failure and for a protected equipment failure.

FOD consequences are evaluated for four leak scenarios (1/4", 1", 4" and catastrophic rupture) for each protected equipment and using an overpressure as the discharge pressure. This overpressure is a higher pressure than the normal operating conditions. In this approach impact zones are used to evaluate fire, explosion and toxicity effects, in the same way we do it for a traditional quantitative consequence analysis. Meanwhile, for the leak failure mode, consequences are evaluated using a financial approach, no impact area is considered. Consequences categories for the leak mode are as follows:

- Loss of production costs
- Maintenance cost (repair costs)

- Product costs

The leak mode includes both, a small leak and a moderate leak. Moderate leak means valve stuck open, while small leak means leakage past valve (No seal).

Therefore, the global risk is calculated using the two risks approaches,

$$Riesgo^{PSV} = Riesgo^{PSV}_{FOD} + Riesgo^{PSV}_{LEAK} \quad (4)$$

Risk calculation involves a straightforward review of the process operating conditions using a layer of protection analysis (LOPA), it also requires analyzing the process conditions and initiating events than can challenge or demand the PSV and the layers affecting the initiating event frequency. The methodology also involves a process for evaluate as many overpressure scenarios as possible; this scenarios will also challenge the pressure relief device.

This risk analysis methodology has the following benefits:

a) Perform a What if...? Scenarios analysis, and evaluate risk mitigation actions

b) Update analysis modeling, using new data

c) Understanding risk drivers

d) Understanding event sequence for risk contributors

e) Set up a risk management strategy

The risk assessment was performed for 100 PSV, located on several offshore facilities. Results show risk contribution for every failure mode, risk ranking for the whole set of PSVs, new maintenance dates as well as reliability and availability estimations. The use of this approach is very useful, because we are able to deal with large data samples and also fits well with small data samples. Results can be expressed using failure mode information or time to failure information.

(43h) Enterprise Risk Management--Adopting Risk-Based Decision Making in a Chemical Enterprise From the Board Room On Down Through All Functions and All Levels

Jerry L. Jones, PE, Consultant, Menlo Park, CA

Enterprise risk management (ERM) practices have been in place for decades in many businesses. However, following a number of failures of large financial, energy and telecommunications ventures, the interest in broader adoption of effective ERM systems has increased dramatically. During the last decade in response to recommendations and guidance by leading professional societies concerned with financial accounting and corporate governance practices, many more publicly owned non-manufacturing and manufacturing businesses have begun implementing very structured and rigorous ERM systems.

While risk has historically been defined in relation to hazards (including natural perils) as a function of the probability and consequences (harm and financial loss) of an undesired exposure event, under various recent ERM guidelines risk has been more broadly defined in relation to business objectives:

o **Risk isthe effect of uncertainty on meeting objectives** (ISO 31000-2009: Risk Management Principles and Guidelines adopted in the US as ANSI/ASSE Z690.2-2011)

o **Risk isthe possibility that an event will occur and adversely affect the achievement of objectives** (COSO Framework for Integrated Risk Management, (Committee of Sponsoring Organizations for the Treadway Commission, 2004)

A key argument made for adopting ERM principles by both public and private companies and using risk-based decision making processes throughout the enterprise is the opportunity the approach offers to more effectively and efficiently protect current value (preventing losses, the downside of risk) and to create value (the upside of some risk exposures) for owners and other stakeholders. A further incentive is the potential to establish long-term strategic advantage by being able to more effectively and efficiently manage risks than competitors.

Both of the ERM system guidelines cited above recommend rigorous and comprehensive oversight of risk management practices by the Board of Directors and active engagement of the executive leadership team in managing the key risks. Such practices are exactly what have been recommended in relation to chemical process safety by a number of investigation panels or boards in the US and abroad following major chemical process industry accidents during the last decade.

In this presentation the wide range of risk exposures that must typically be managed by a chemical enterprise (showing that process safety is one of many requiring leadership attention) will be briefly reviewed with some specific examples provided for relevant consensus standards or guidelines for managing those risks. The following list of objectives that an ERM should achieve will be discussed along with some the challenges they present.

The ERM system should assure that:

o The company's significant existing risk exposures are identified, assessed and managed in a consistent manner by the various functional organizations and the results of such risk management activities regularly reviewed by the executive leadership team and the Board of Directors,

o The "right" risks are being taken based on consideration of risk/reward tradeoffs and the "risk appetite" of the organization, and resources for risk management are allocated commensurate with the risk exposures and risk control needs,

o Individual at all levels are actively engaged in risk-based decision making and risk management so as to leverage off the knowledge and experience of the entire organization,

o Individual responsibilities and accountabilities for ongoing risk control and monitoring are clearly defined in all functional organizations at all levels, including risk management related reporting requirements with metrics (both leading and lagging indicators) that go to the executive leadership team and to the Board of Directors,

o Internal auditing and self-assessment processes are in place to regularly evaluate the effectiveness of risk management processes and controls so as to keep risk exposures below the agreed upon risk tolerance limits.

A four-step business process for risk identification, analysis, risk-based decision making, and risk controls monitoring/reporting will be described that is consistent with ERM consensus guidelines and others including those for risk-based process safety. This generic business process can be readily adopted by various functional organizations within companies of all sizes to manage a wide range of types of risk exposures and used to support other business processes requiring risk-based decision making (e.g. process hazard analysis, product safety analysis, product quality assurance, management of change, incident investigations, supply chain security, mechanical reliability analysis, and business continuity analysis).

(43i) The Chemical Safety Board and the Prevention of Combustible Dust Hazards

Manuel R. Gomez, Chemical Safety Board, Washington , DC and Christina Morgan, Office of Recommendations, US Chemical Safety Board, Washington, DC

THE CHEMICAL SAFETY BOARD AND THE PREVENTION OF COMBUSTIBLE DUST HAZARDS

This presentation will:

- Briefly summarize the CSB's rationale for recommending an OSHA general industry combustible dust standard;
- Review OSHA actions up to date;
- Comment on the difficulties OSHA may be facing in the conduct of combustible dust rulemaking; and,
- Describe the CSB's ongoing and planned efforts to advocate successful rulemaking.

9TH GLOBAL CONGRESS ON PROCESS SAFETY

Performing a HAZOP on Start-up/Shutdown Procedures

Fareed Ebrahim
ACM Automation Inc.
#700, 940 – 6th Avenue SW
Calgary, Alberta
febrahim@acm.ca

German Luna-Mejias, P. Eng
ACM Automation Inc.
gluna@acm.ca

Prepared for Presentation at
American Institute of Chemical Engineers
2013 Spring Meeting
9th Global Congress on Process Safety
San Antonio, Texas
April 28 – May 1, 2013

UNPUBLISHED

Performing a HAZOP on Start-up/Shutdown Procedures

Fareed Ebrahim
ACM Automation Inc.
#700, 940 – 6th Avenue SW
Calgary, Alberta
febrahim@acm.ca

German Luna-Mejias, P. Eng
ACM Automation Inc.

Keywords: Procedure HAZOP, Start-up/Shutdown

Abstract

Statistics show that approximately 65% of process safety related incidents occur during the start-up/shutdown operations of the plant. Yet, the bulk of the risk assessments performed today focus on the normal operation of the plant. A major contributor to the lack of start-up/shutdown risk assessments is the lack of a strongly documented methodology. One method of performing this assessment is through the HAZOP of start-up/shutdown procedures. This HAZOP involves a step by step review of the procedure with consideration given to the sequencing, delay, or missing of steps involved in the procedure. By performing this review, required permissives are identified, contingencies are developed and essential safeguards are identified. This will help ensure the operations team is confident that they are performing the start-up/shutdown in a safe manner. This paper focuses on the framework for guidelines, procedures and templates that are necessary to perform this HAZOP. A practical approach with examples is presented.

1. Introduction

Currently, it is common practice to reserve a deviation in a continuous process HAZOP for start-up and shutdown issues. Unfortunately, during analysis, this deviation if often not explored to its full potential. This abrupt analysis is often due to the lack of available information and expertise, lack specificity within the deviation, and time constraints. Yet, statistics provide a compelling argument to improve the rigour associated with start-up/shutdown analysis. Approximately 65% of process safety related incidents occur during these modes of operations1. To improve the rigour associated with the assessment of start-up/shutdown, the HAZOP technique must be further extended to the procedures for these modes of operation. Many sources are available which explore the fundamentals of performing such a HAZOP1,2, and therefore, this paper will focus on developing a methodology by expanding upon the commonly used HAZOP technique for normal operating modes.

2. Preparing for the HAZOP

In order to ensure that a complete and accurate assessment is performed, compiling the correct documents and team members for the HAZOP is of paramount importance. Just as there are key documents required for a process HAZOP, specific documents are also required for the start-up/shutdown procedure HAZOP.

2.1 Required Documentation

Firstly, the procedure for start-up/shutdown must be available. Just as the P&IDs are the central document during a process HAZOP, the procedure will be the main focus in the start-up/shutdown analysis and therefore, should be the most up-to-date revisions available. This procedure can then be split into "nodes" or subsystems for analysis. Nodes or subsystems may be defined as each individual step in the procedure.

Depending on the complexity of each step, it may be required to split the step into simplified actions, which in turn become the node for analysis. Conversely, procedure steps may be combined to develop a node based on timing requirements and the requirement for actions needing to be performed simultaneously. Timing concerns associated with procedures will be further discussed with the HAZOP deviations. As the implications of each procedure step may not be immediately clear to the HAZOP facilitator, it is of utmost importance that nodes are developed in conjunction with the operations personnel.

Once the nodes have been identified, and the complexity of the procedure understood, a decision must be made on the degree of analysis for the HAZOP to be performed. Two methods of analysis can be employed for the HAZOP – a "What-If?" approach, or a guideword method. The "What-If?" approach follows the same method as that used for a continuous process. If only an extremely high level analysis is required, the "What-If?" methodology may be applied to the entire procedure instead of applying it to nodes. Conversely, the application of the guideword method provides a more structured approach. Generally, as the complexity of the procedure increases, more guidewords should be employed in order to ensure a thorough analysis is completed. PHA software3 contains pre-packaged guidewords to be used for analysis. Explanation and application of the guidewords will be explored in Section 3.

Supporting documents for the procedure HAZOP include the following:

- As-Built P&IDs
- Isolation markup drawings - showing spec blind placement and position as per start-up/shutdown modes
- Shutdown keys - including permissives
- Alarm lists
- Alarm/Trip bypass procedures - including contingency plans

It is important to note that the shutdown keys and alarm lists may not outline which systems may be bypassed during a start-up and shutdown. Therefore, it is critical to obtain the bypass information, as well as the operating window for all instrumentation that will be referenced during the start-up/shutdown operation.

2.2 Required Personnel

In addition to compiling the correct documents for the session, it is of equal importance to ensure the correct personnel are in the room during the HAZOP. Similar to a process HAZOP, representation from the process team, instrumentation, and operation is required. It is suggested to have extra representation from the operations group to provide a more diverse perspective for the brainstorming process. Furthermore, if the HAZOP is being performed at the commissioning phase, it is recommended for the construction manager or equivalent representative to be present.

3. Performing the HAZOP

3.1 Introducing Guidewords

Risk assessment software[3] often includes preloaded libraries of guidewords for the HAZOP of procedures. Table 1 below lists seven guidewords that can be used for the HAZOP. These guidewords are listed in a suggested order for use in the HAZOP. Guidewords may be changed, added or deleted based on the HAZOP team experience, process being analyzed, and corporate guidelines. Note that the performance of a start-up/shutdown procedure HAZOP does not negate the requirement to perform a pre-start-up safety review.

Table 1. Suggested Guidewords and Description for Procedure HAZOP

Guideword[3]	Description
Less	Not enough is done to adequately execute procedure action.
More	Action is performed to a further extent than specified.
No	Action is fully omitted.
Reverse	Actions in procedure are performed out of specified sequence.
Other than[3]	Action is performed that does not align with what is specified.
As well as[3]	The operator does another action, in addition to the specified step being done correctly.[2]
Sooner/Later[3]	Timing of action is incorrect.

All seven guidewords may not be used for each start-up/shutdown procedure being analyzed. Guidewords should be applied based on the complexity of the procedure. An example of a simplified method is the two-guideword approach[2] is contained in Table 2 below.

Table 2. Two-Guideword approach for Procedure HAZOP[2]

Guideword	Description
Omit	Step is not done or part of the step is not done.
Incorrect	Step is not performed as intended.

3.1 Application of Guidewords

In order to further illustrate the application of each guideword, a subsection of a mock, simplified procedure (listed below) for a Reformer will be used. Guidewords from the two word approach will not be further explored as they can be considered a subset of the initial seven. Note that nodes will be assumed to be the individual procedure steps for the purpose of this illustration. Consequences are not to be considered complete for use in industry HAZOPs but for the purpose of example only.

Simplified Excerpt of Reformer Procedure:

1. Once pilots are lit, start lighting main burners (two per cell).
2. After completing the first ten pilots, light another ten but evenly distributed per cell.
3. Wait for 30 minutes or until the heater outlet temperature is stable
4. Continue lighting burners to increase the heater temperature at a rate of 50°C/hour until a temperature of 380°C is reached.
5. Once temperature is stable and maintains at 380°C, start injecting HP Steam and continue lighting burners as required to prevent the heater temperature from dropping below 350°C.

3.2.1 Less

Example Deviation from Step 3: Adequate time is not allotted by operator for heater outlet temperature to stabilize in an attempt to speed up start-up.

Consequence: Potential for a sudden temperature increase in heater leading to refractory damage or flame impingement.

3.2.2 More

Example Deviation from Step 1: More than two burners are lit per cell in order to speed up start-up of unit.

Consequence: Potential for a sudden temperature increase in heater leading to refractory damage or flame impingement.

3.2.3 No

Example Deviation from Step 4: Additional burners are not lit by operator.

Consequence: If steam is injected into the heater when temperature is below 350°C there will be deactivation of catalyst within heater tubes.

3.2.4 Reverse

Example: Deviation from Step 3: Additional burners are lit prior to stabilization of heater outlet temperature.

Consequence: Potential for a sudden temperature increase in heater leading to refractory damage or flame impingement.

3.2.5 As well as

Example: Deviation from Step 2: Steam is injected by operator as well as lighting the additional burners.

Consequence: If steam is injected into the heater when temperature is below 350°C there will be deactivation of catalyst within heater tubes.

3.2.6 Other than

Example: Deviation from Step 5: Natural Gas is injected into the heater instead of Steam.

Consequence: Overheating of tubes in reactor leading to tube damage and potential explosion.

3.2.7 Sooner/Later

Example: Deviation from Step 5: Steam injected to heater prior to temperature reaching 350°C.

Consequence: If steam is injected into the heater when temperature is below 350°C there will be deactivation of catalyst within heater tubes.

3.3 Applying a Risk Ranking

As HAZOPs have developed over time, there has been a continued emphasis to add more semi-quantitative aspects to the study. A major element of this semi-quantitative aspect is the application of a risk matrix. Unfortunately, risk matrices are designed using a per year time frame. This per year concept is of course not applicable to a procedure. Therefore, when performing the procedure HAZOP, two options are available. One, is to avoid the risk ranking and perform a fully qualitative analysis. Second, is to apply the risk matrix by avoiding the yearly references contained in the likelihood, and setting a default likelihood of making an error in the procedure to correspond with highest likelihood available (ie. frequent) on the matrix. This is to quantify the fact that there is a possibility to make a mistake in executing the procedure at any given opportunity for execution. For example, if a procedure step is incorrectly worded, there is an opportunity for error every time that step is executed. The consequence levels outlined in the risk matrix are not time constrained and therefore are easily applied to a start-up/shutdown procedure HAZOP. Safeguards can then be applied in a fashion similar to process HAZOPs by dropping the likelihood by one level.

3.4 Adding Safeguards

When applying safeguards to a procedure, it is of course of utmost importance to adhere to the guidelines set out in IEC-61511 if they will be used as independent protection layers. In addition to this, it is extremely important to ensure that the safeguard will not be bypassed during the start-up or shutdown. In some cases, safeguards through a safety instrumented system may be disabled during start-up or shutdown. It is extremely important to identify these bypasses and ensure safeguards are not credited that will not be active. If a safeguard is bypassed, refer to contingency planning documents in order to identify an alternative. Furthermore, it is of utmost importance to ensure that the safeguards will in fact be within their operating range during start-up and shutdown. One control related safeguard that is far more predominant in start-up/shutdown procedure HAZOPs when compared to process HAZOPs are permissives. It is therefore important to have these permissives identified, optimally on the shutdown key. In addition to instrumentation safeguards, a double check to the operator performing the specified action also acts as a valid safeguard.

4. Conclusion

The application of the HAZOP technique to a start-up/shutdown procedure is an effective way to improve the quality of analysis in these modes of operation. Just as in a normal operating mode HAZOP, specific documentation and team members are required. Furthermore, the level of rigour of the study may be adjusted by using a "What-If?" approach or by varying the number of guidewords to be used. A risk ranking may be applied during a procedure HAZOP, but caution must be applied in order to avoid diluting the risk of a situation by ensuring the procedure is analyzed on a per opportunity (instead of a per year) basis. Safeguards must also be applied with caution due to the potential for bypasses or conditions being out of the operating window of certain instruments during start-up/shutdown. It is hoped that the application of the HAZOP methodology to start-up and shutdown procedures will greatly reduce the number of incidents that occur during these modes of operation.

5. References

[1] Bridges, Bill. "Paper: How to Efficiently Perform the Hazard Evaluation Required for Non-Routine Modes of Operation (Start-up, Shutdown, Online Maintenance)." Proc. of 2011 Global Congress on Process Safety., 2011

[2] CCPS. "9. Extensions and Special Applications." *Guidelines for Hazard Evaluation Procedures.* New York: Center for Chemical Process Safety of the American Institute of Chemical Engineers, 1992. 257-67.

[3] Dyadem International Ltd. *PHA-Pro.* Computer software. Vers. 8.2.2.0.

(43p) Chasing the Protection of Intermediate Bulk Containers, Year 10

John A. Leblanc, Engineering Standards, FM Global, Norwood, MA

The challenges of protecting composite intermediate bulk containers (IBC) storing liquids that burn have been under investigation by a number of groups for more than 10 years. The path to resolving this challenge chosen by FM Global looked at defining the fire hazard created by the liquids and then understanding how the containers reacted to fire. This effort has resulted in a strong understanding of the fire hazards and protection options for several liquid-IBC combinations:

Rack storage of Oil and alcohol in IBCs

Silicone fluids in palletized IBCs

Very high flash point liquids in palletized IBCs

Propylene glycol in palletized IBCs

Storage of Empty IBCs

Defining an FM Approved IBC

This paper will provide an overview of the fire hazards associated with storing liquids that can burn in IBCs and present the new protection options and their associated test programs.

Process Safety Challenges in the Hazardous Waste Industry

Hope A. Luebeck
Veolia Environmental Services, L.L.C.
700 East Butterfield Road, Suite 201
Lombard, IL 60148
Hope.Luebeck@Veoliaes.com

Prepared for Presentation at
American Institute of Chemical Engineers
2013 Spring Meeting
9th Global Congress on Process Safety
San Antonio, Texas
April 28 – May 1, 2013

UNPUBLISHED

Process Safety Challenges in the Hazardous Waste Industry

Hope A. Luebeck
Veolia Environmental Services
700 East Butterfield Road, Suite 201
Lombard, IL 60148
Hope.Luebeck@Veoliaes.com

Keywords: Process Safety, Hazardous Waste Industry, Recyclers' Safety Council

Abstract

The OSHA Process Safety Management (PSM) Standard, 29 CFR 1910.119, was promulgated in 1992 to prevent the release of highly hazardous chemicals (HHCs) following a series of tragic industrial accidents. During the last 20 years, the emphasis of this standard has primarily been focused on refineries, chemical plants, and similar processes. However, companies in the Hazardous Waste Industry (HWI) also handle HHCs and face a special set of challenges as they strive to be PSM-compliant in their operations.

Hazardous Waste Industry facilities are often smaller than typical petrochemical, chemical and pharmaceutical plants. They have fewer resources dedicated to their operations and, as a result, generally make slower progress in advancing their PSM programs. Reasons for these tendencies will be discussed, as will initiatives which are being taken to strengthen their programs.

1. Introduction

History has taught us that the most influential changes result from tragedies. The Process Safety Management (PSM) Standard 29 CFR 1910.119 [1] was established by the Occupational Safety and Health Administration (OSHA) in 1992 after a series of deadly industrial accidents including the methyl isocyanate release at a Union Carbide pesticide plant in Bhopal, India, in 1984 [2] and the vapor cloud explosion at the Phillips 66 chemical complex in Pasadena, TX in 1989 [3]. The 14-element safety standard includes specific requirements for managing risks associated with processes which utilize Highly Hazardous Chemicals (HHCs) such as were involved in the noted catastrophes. More recent events such as the BP Texas City, TX disaster in 2005 [4] have motivated OSHA to renew its scrutiny of process safety practices of PSM-covered organizations through a series of National Emphasis Programs [5].

There have been serious accidents in the Hazardous Waste Industry (HWI) as well, though these have not had the significant impact or notoriety as those in the chemical or petrochemical industries. Such recent HWI events include The Environmental Quality Company fire in 2006 [6] and the Veolia Environmental Services explosion and fire in 2009 [7] (See Table 1). As these sorts of incidents continue in the HWI as well as in other industries, companies must continue to further develop their PSM programs and prepare for potential future NEP inspections.

Table 1. Some Noteworthy Industrial Process Safety-Related Incidents

Industry	Company	Location	Year	Deaths	Injuries
Petrochemical	Union Carbide	Bhopal, India	1984	2000+	Unknown
	Phillips 66	Pasadena, Texas	1989	24	132
	BP	Texas City, TX	2005	15	180
Hazardous Waste	The Environmental Quality Company	Apex, NC	2006	0	30
	Veolia Environmental Services	West Carrollton, OH	2009	0	4

Many of the same issues that challenge the petrochemical industry in their quest to be PSM-compliant are shared by the HWI. However, the size and nature of their operations are often quite different; thus, their approach to PSM compliance can vary significantly.

This paper explains some unique aspects of PSM which apply to companies in the Hazardous Waste Industry (HWI). When the acronym HWI is mentioned, this is intended to apply broadly to all companies having business operations involving hazardous waste. A subset of HWI companies is Hazardous Waste Treatment (HMT) facilities. The HMT acronym is used to emphasize operations related to the treatment (e.g. purification or neutralization), recycling, or destruction of hazardous wastes.

2. Operational Differences by Industry

Companies in the refinery, chemical or pharmaceutical industries typically utilize the same types and quantities of raw materials on a daily basis. They tend to manufacture or store a limited number of products or families of products. Their manufacturing operations occur according to cycles or routine production runs. A given plant has one or more primary technologies which are key to their particular business. These facilities often have recognized that PSM was applicable to their operations long ago and thus have reasonably mature PSM programs.

Companies in the HWI have more variable operations. They utilize an assortment of unit operations to handle, process, recycle or destroy the chemical and biological wastes they receive on a regular or periodic basis. They combine, decant, recycle, fuel blend, distill, incinerate or otherwise dispose of the materials they receive from their customers. These are shipped to their facilities in a variety of container types and material phases. PSM challenges shift on a daily basis since the nature of what they need to process changes frequently and can fluctuate widely. PSM programs in the HWI must be adept at effectively managing these inconsistent hazards.

3. General PSM Challenges for the HWI

Companies in the HWI have to be prepared for a wider variety of process hazards due to the nature and kind of substances they accept from their customers. They refer to waste shipments of a given type they receive as a 'stream'. Even trace amounts of certain chemicals can have a tragic effect if these contaminate incoming streams without the HWI recipients' knowledge. HMT facilities need close ties with their customers to truly understand the nature of the wastes they receive and know how repeat streams may vary in chemical concentrations of major components and may contain different kinds of constituents.

The number and volume of these streams are always changing, even from the same customer. Customers of HWI companies don't intentionally generate waste regularly and the amounts of each are often related to either their production volumes or the number of process upsets they have experienced. So companies in the HWI can't always anticipate how much of what kinds of streams they will process in a business cycle. The generation of customer waste is extremely unpredictable. HWI companies have to be prepared for almost anything and have equipment which is designed for just that. Since waste can be generated at any time, HMT facilities need to be nimble and be able to respond quickly to new business or business for a newly-generated stream of waste from an existing customer. Compliance concerns which may result from new business need to be promptly addressed.

HWI companies are usually comprised of small facilities (with less than 100 employees per site) which have a multitude of regulations which they need to meet, including RCRA, RMP, and numerous others in addition to PSM. Personnel in these more-compact organizations tend to wear many hats and are distracted by the assorted duties cast upon them during the normal workday. They many even include the owner/operator of such a facility. Expediency in addressing PSM compliance issues may be affected by the amount of resources available in such organizations.

HMT facilities tend to have smaller budgets than comparable size facilities in larger companies or those in other industries. As such, monies to upgrade facilities, build infrastructure, or address hazard reduction projects can be more limited. Business cycles have a huge impact on what funding is available to address all compliance needs. And with the HWI downstream of the manufacturing industry, the HWI often goes as manufacturing does.

Companies and operating sites in the HWI have had multiple owners over the years, perhaps more than those in the chemical or petrochemical industries. This has led to mixed safety cultures both at individual sites and across organizations. Dealing with PSM issues may be controversial, since senior personnel or management might have experienced one or more safety cultures and younger personnel others. Companies must apply extra effort to address identified deficiencies since these cultural viewpoints may lead to disagreement regarding the extent of work necessary to effectively resolve compliance issues.

Organizations in the HWI industry previously treated their activities as simply waste-related operations, even if they were recycling or creating renewed end-products through processes such as distilling or fuel blending. But they really are another kind of chemical plant, albeit a more unique sort. These facilities need to be just as focused on maintaining their waste processing equipment as though they are generating a virgin product. Operations can have deadly consequences from improperly processing or having uncontrolled releases of highly hazardous chemicals, which is the reason for the PSM regulation in the first place. HWI companies have improved their performance in this arena in recent years, especially as OSHA continues its Chemical NEP activities [5].

4. Specific Challenges by PSM Element

Employee Participation
Effective Employee Participation can be limited by the number of personnel available at HMT facilities to participate in, lead and drive improvements and PSM-related activities. Companies must allot time and resources for their employees to participate in training, hazard analyses, safety committees, operating procedure reviews, incident investigations, emergency drills, etc. required for their PSM program to be successful. Site management may need to be resourceful in giving personnel time to meet PSM program obligations.

Process Safety Information (PSI)
Efforts must be ongoing to keep PSI current. Personnel must be encouraged to review and update P&IDs, equipment files, chemical and process hazard information, consequences of deviation in procedures and similar documents so they remain accurate and useful. One of the hardest challenges seems to be keeping good records of PSM-covered equipment in easy to access files. Older facilities have limited equipment design or code information available, which can prompt re-inspection or re-certification of equipment to verify these critical details. Sometimes design literature may need to be requested from a manufacturer if it becomes damaged or misplaced. If changes of service (say of storage tanks) have occurred but were poorly documented, records may need to be re-created and verified before inclusion in PSI databases.

Some HMT facilities have difficulty describing their 'process chemistry' per the requirements of the PSM standard. This is a dilemma unique to the HWI. True process chemistry is often not intended in typical HWI operations. If this applies to a given location, the site's process safety documentation must clearly explain this circumstance.

Process Hazard Analysis
A comprehensive Process Hazard Analysis (PHA) program is critical to understanding any site's hazards and ensuring that appropriate controls and acceptable levels of risk are assumed. Many PSM-covered sites in the HWI are older facilities which suffer from process designs which did not adequately cover facility siting concerns at the time of construction. If rebuilding such sites is not possible, locations instead have to focus on administrative and engineering controls to reduce or mitigate risk to their personnel and property from their operations. And facilities with multiple sites operated independently can face inconsistencies in risk tolerances when coming under a corporate PSM umbrella.

Some HWI organizations heavily rely on consultants to lead their PHAs. This can create a double-edge sword: a site gets support for a needed activity, but if they don't define the requirements for the PHA adequately nor hire a suitable consultant, the PHA results yielded may be substandard. The PHAs may not cover enough hazardous aspects of a process if their consultant is not properly informed and knowledgeable. PSI details and reports for these PHAs may be limited in their scope and may omit critical information (background or otherwise) relevant to the study. Sites must be vigilant to ensure that their PHA activities be thoroughly completed as required.

Operating Procedures

HMT facilities typically have very good basic operating procedures due to the nature of their operations. There would either be more accidents or more hazardous spills if these facilities were not being run safely. But, as is sometimes the case in any industry, the written procedures which exist may not be adequately comprehensive in their descriptions of consequences of deviation or emergency actions to take, including explaining what conditions warrant the shutdown of a process. Some procedures may direct the operator to consult with supervision during atypical or emergency operations. While this is not necessarily incorrect, what does an operator do when the process goes haywire on a weekend or off-shift when less supervision or technical support is available? Authors of such procedures need to clearly spell out common out-of-control responses and/or define when emergency shutdown should take place to provide the operators with the best ready guidance at all times.

On the maintenance side of operations, many more procedures often need to be composed to describe inspection, testing and preventive maintenance (ITPM) activities and how repairs and replacement of equipment ought to take place. A significant number of procedures should exist to define ITPM for PSM-critical equipment systems and controls. When predictive and reliability-centered maintenance (RCM) programs become utilized by more advanced organizations, the number of maintenance procedures should increase proportionally to define and address these specialized activities.

Training

Training can be executed in a variety of ways. Some companies depend upon computer-based training for much routine or refresher training, since employees can determine when they have time available to meet the required obligations and can fit in training modules while they are present during normal work hours. When facilities do not have this option available, they may need to be resourceful with training solutions. A familiar pattern is to schedule hourly workers for day long training sessions when they have a day off. Even this becomes an issue for companies who do not have excess staff to backfill work slots for other employees undergoing training. Companies can also organize special 'training day or days' when operations are intentionally curtailed so that particular training or orientation programs may be held for the entire facility at one time. Possibly other creative solutions may be sought to balance staffing with training needs and requirements as situations dictate.

Contractors

Managing contractors can be tricky, depending on the number and type of crafts or specialty groups on which a facility may rely. Overseeing the screening, performance and insurance requirements, among other factors, can require constant review. Staff dedicated to birddogging records and updates of same can help to keep these appropriately current. Ensuring that contractors have proper desired skills, certifications, and current site orientation may be another story. Some sites find it cumbersome to track down qualifications of contractors and have regular orientation sessions with them. That is why many smaller sites often partner with local/regional safety and health groups which specialize in OSHA and other common safety program training. In the greater Chicago area, one such group is the Three Rivers Manufacturing Association, which has a specialty Safety Center [8]. Similar such organizations exist in other parts of the country. For an annual membership fee, the safety groups can ensure their customers that a variety of contractors serving a region have certain basic safety and health skills, which then make it easier for member companies to quickly orient contractor personnel to their PSM-covered sites when jobs are required.

Sites in the HWI may also have their own employees serve as contractors for customer facilities when they work on special projects or are embedded at customer sites. These HWI personnel may need additional PSM-related training to be appropriately compliant both for their employer as well as for the customer. This is another unique aspect of the HWI business.

Pre-Start Up Safety Reviews (PSSR)
It is less common in the HWI for companies to debottleneck or re-design their processes. They don't generate as many cases where PSSRs are required. As a result, the PSSR process tends to be less well-known and less understood by site personnel. More time needs to be taken whenever PSSR opportunities exist to refresh personnel's memory of the purpose of the review and to ensure that a thorough and proper study is made of the new equipment or system before it is commissioned and HHCs are introduced to the process.

Mechanical Integrity
Maintaining the mechanical integrity of PSM-critical equipment is complicated for any organization. What makes this particularly difficult for smaller companies is establishing and staying on top of inspection, testing and preventive maintenance (ITPM) for what is often older equipment that had historically been run until it was broken. Emphasis is needed on implementing aspects of preventive and predictive maintenance to enhance the reliability and service life of their critical equipment, whether this is done in-house or via contract resources. Maintenance employees need appropriate training and/or certification according to the ITPM skills expected of them. Improving ITPM practices may even involve a type of 'culture change' which shifts focus from reactive to preventive/predictive maintenance practices. Management influence is essential to support this enhanced execution of maintenance activities.

Management of Change (MOC)
The MOC process is not only to be used for new equipment, capital projects, or changes in personnel with PSM responsibilities, but is also for new streams to be processed in an HMT facility. As waste compositions change from allowed norms, reviews need to occur to ensure that new streams or significant variations in existing ones will not result in any new hazards in the anticipated handling of the materials or the associated process operations. Changes in streams should likewise result in re-evaluating the materials of construction of the process equipment which may handle the new or different stream so that unanticipated failures of

equipment do not occur. Furthermore, since newer equipment is less common in the HWI, employees need to be diligent in understanding the difference between 'replacement in kind' (RIK) and non-RIK situations so they know when to apply MOC appropriately.

Hot Work
Hot Work operations in an HMT facility can be confusing to personnel. Some sites have difficulty defining hot work areas due to the fact that handling of waste streams may occur in many more locations around the site than at a standard manufacturing plant. Employees may need additional orientation and training to recognize Hot Work situations and zones where Hot Work operations may apply. As employees become more knowledgeable of their site's Hot Work requirements, proper preparation for and permitting of all Hot Work activities should occur as expected.

Incident Investigation
HMT facilities tend to have well-established incident investigation and emergency response planning. Many of their environmental permits already require these to be fairly robust. HWI organizations, like many other organizations, need to place more emphasis on leading indicators rather than lagging indicators. The new focus on standard API RP 754 *Process Safety Performance Indicators for the Refining and Petrochemical Industries* [9] is further emphasizing this trend. One leading indicator that is easy to track is near misses. Good PSM programs already have discipline in identifying incidents or spills to report. Now these need to identify events which may not have had an adverse result, but may have had the potential for the release of a highly hazardous chemical. Understanding the dangerous potential of these near miss situations can help sites to avoid tragic consequences from ignoring such warning signs.

Emergency Planning and Response
There needs to be diligence in conducting enough annual emergency response drills and in holding debriefings to help identify weaknesses in response programs and make necessary changes to improve. HWI organizations focus heavily on spills due to the nature of their environmental permits, but should equally consider drills on equipment failures or releases from pressure relief valves, for example. This 'balanced approach' to emergency drills should result in more comprehensive emergency response procedures and staffs which are better prepared to handle a greater range of adverse events.

Compliance Audits
Sites in the HWI are used to compliance audits from regulators on EPA and similar programs and tend to be well-versed in preparing for such audits. PSM auditing activities likewise need to be ongoing, including self-audits, to support PSM gap analyses which ought to be routinely occurring. Management should regularly review the results of these audits and put in place systems to successfully address identified deficiencies.

Trade Secrets
HMT sites do not tend to have many trade secrets, but, as with other organizations, they need to ensure that these are made available to personnel who need them to conduct PSM compliance activities. These personnel may need to enter into confidentiality agreements regarding access to this critical information to perform their normal duties.

5. Advancing PSM Programs at any PSM-Covered Facility

PSM programs are only as good as their corrective actions and improvement efforts. If you find problems in your systems but don't address them, it is as though you had no knowledge of them. Shortcomings identified in audits but which were not addressed have been contributing factors to major incidents, such as that which occurred in Texas City, TX in 2005 [4]. Management teams in every company need dedication to follow through on the recommendations from hazard analyses, incident investigations and compliance audits, among others, to address deficiencies and to correct issues. They must ensure continual improvement activities and projects exist and are led by appropriate champions. A variety of solutions may be needed to engage the work force to achieve better success with these aspects of PSM.

Sister plants can work together or company divisions can pool resources when there is an inadequate number of technical staff at a given location to lead PSM efforts. Today's internet-based applications and array of electronic gadgetry make these interactive communications easier than ever without excessive travel burdens.

Companies may consider utilizing process safety consultants to supplement their programs. This is a great way to get new ideas and access to industry best practices from trusted sources. Third-party auditors may also be employed to make unbiased suggestions for improvement. However, be sure your company actively partners with these groups for more successful PSM program advancement.

Organizations should get involved with trade organizations, professional societies and other groups that conduct seminars and training on process safety topics. Technical conferences are a great way to network with others in your own industry and learn about current PSM topics and newly identified best practices. The Global Congress on Process Safety (GCPS), which is a part of the Spring Meeting of the American Institute of Chemical Engineers (AIChE), is a fine example of an ongoing conference series dedicated to Process Safety education [10].

Companies in the HWI also have a new organization to supplement their PSM compliance efforts. The Recyclers' Safety Council (RSC) was created in 2011 by a consortium of industrial companies who recycle waste solvents and purify them for reuse. The intent of the RSC is to share best practices related to safety programs and compliance with the PSM standard. No similar group previously existed specifically for companies who recycle waste solvents. The RSC promotes collaboration among organizations who understand the unique set of challenges they face in achieving safety and process safety goals. My company, Veolia Environmental Services, is one of the founding members of this Council. Please contact me, the author of this article, for more information on participating in RSC activities, including an annual safety-themed conference.

6. Conclusions

HWI facilities, and HMTs in particular, require special diligence to effectively manage and progress their PSM programs. The size and nature of their operations require them to seek creative and possibly atypical measures to meet PSM compliance challenges. Though they are sometimes less advanced than their counterparts in other industries, they remain dedicated to addressing process safety gaps and continually improving their PSM systems. The recent formation of the RSC is an example of this commitment to process safety excellence.

PSM program success for any company is a function of the availability and dedication of the employees and the leadership of management. Limited resources, while a challenge, should not be an impediment to compliance. Be proactive and focused on your program and it will be possible to have a thriving process safety culture in your organization, regardless of size or the nature of your company's business.

7. References

[1] Process Safety Management of Highly Hazardous Chemicals, 29 CFR 1910.119, Occupational Health and Safety Administration , Occupational Safety and Health Standards.

[2] Jackson B. Browning, Jackson Browning Report, "Union Carbide Disaster at Bhopal (1993)" Available at www.bhopal.com/reports-and-studies/union-carbide-disaster-at-bhopal.

[3] "OSHA: Phillips 66 Company Houston Chemical Complex Explosion and Fire: A Report to the President, U.S. Department of Labor, Washington, D.C. (1990)".

[4] Baker, James et al 2007 "The Report of the BP U.S. Refineries Independent Safety Review Panel".

[5] OSHA National Emphasis Programs: Available at: http://www.osha.gov/SLTC/processsafetymanagement/index.html

[6] U.S. Chemical Safety Board Case Study 'Fire and Community Evacuation in Apex, N.C.', April 16, 2008. Available at: www.csb.gov/investigations .

[7] U.S. Chemical Safety Board Case Study 'Veolia Environmental Services Flammable Vapor Explosion and Fire', July 21, 2010. Available at: www.csb.gov/investigations .

[8] Three Rivers Manufacturing Association; Available at www.trma.org.

[9] API RP-754 Process Safety Performance Indicators for the Refining and Petrochemical Industries, 2010. Available at: www.api.org .

[10] Global Congress on Process Safety; Available at: www.aiche.org/resources/conferences.

A Novel Methodology for Maintaining Mechanical Integrity of High Pressure Relief Systems

Sandipan Laskar
Sr. Process Safety Manager
Clean Harbors Inc.
2027 Independence PKWY S
La Porte, TX 77571
laskar.sandipan@cleanharbors.com

Prepared for Presentation at
American Institute of Chemical Engineers
2013 Spring Meeting
9th Global Congress on Process Safety
San Antonio, Texas
April 28 – May 1, 2013

A Novel Methodology for Maintaining Mechanical Integrity of High Pressure Relief Systems

Sandipan Laskar
Sr. Process Safety Manager
Clean Harbors Inc.
2027 Independence PKWY S
La Porte, TX 77571
laskar.sandipan@cleanharbors.com

Keywords: Relief System, Mechanical Integrity (MI)

Abstract

Safe operation and maintenance of emergency systems are of primary importance to the process industries, including upstream oil and gas. The upstream production facilities are usually protected from overpressurization by high integrity pressure protection systems (HIPPS) which shuts in the well when local pressure levels exceed a certain threshold. The pressure relief system is often installed as a last safety barrier, when HIPPS fails. One of the critical elements of emergency system is proper design, documentation and to maintain high reliability of this system.

In particular, the pressure relief system is an integral part of upstream design and operation; prevention of an accidental release of toxic, flammable or noxious materials from a production facility requires correct design and construction, safe operation, and proper documentation of this system.

This paper proposes a basic change in the way relief systems are documented - a requirement in order to comply with OSHA 1910.119 and thereby describes a new methodology and a change in documentation to meet OSHA requirements and other industry standards (such as API 520/521, API-RP 75, 14J). This paper also describes the steps involved in a high pressure relief system evaluation and introduces concept of ever-greening to maintain a high reliability of this system. The paper illustrates how an up-to-date relief system leverages management of change (MOC) and helps in setting up the foundation for a strong overall process safety management (PSM) program. The paper will identify the drivers and the benefits of the proposed methodology for pressure relief documentation as well as explore the interdependency of the proposed program elements.

The paper will assist upstream facility personnel to utilize good engineering practice in design and documentation of safety relief systems, to meet both onshore and offshore process safety requirements, and to develop consistent company best practices for process safety management. In addition the paper will describe a comprehensive equipment based analysis protocol, preparation of documentation sufficient for design, operational, and audit purposes, and discuss maintenance both of the physical system and design data.

1. Introduction

Most oil and gas production, refining and petrochemical organizations have some kind safety management program that stems from the requirements outlined under OSHA's CFR 1910.119. Some have a very extensive program that covers most of the 14 elements of Process Safety Management (PSM) and is managed via an integrated platform while others follow an ad hoc approach that does not span all of the elements or is disjointed.

Maintaining accurate and current Process Safety Information (PSI) is a cumbersome and difficult task to accomplish due to the numerous sub-elements of Safety Management Program (SEMP). One of the key sub-elements of PSI maintenance is the management of the pressure relief documentation. Historically this is an element that has not been given the attention it deserves. However, recent incidents in the industry have elevated the importance of maintaining accurate and current pressure system documentation.

The ultimate goal is to allow all plant personnel to understand the basis of the current plant relief and flare header design and ensure that this current plant relief and flare header design provides adequate overpressure protection for all equipment. Thus, a new methodology for relief system documentation is created that is customized to fit the needs and resources of the operating company and had the ability to relate to the other Process Safety Information elements.

2. Pressure Relief Analysis

Pressure relief systems are applied in various installations in the oil and gas industry. The function of a pressure relieving system is to provide a positive means of protection for plant equipment from overpressure above limits permitted by the relevant codes and to dispose safely of any resultant relief of the process fluids.

The pressure relief analysis provides an efficient way to document the basis for relief system design basis. The major reasons for undertaking the pressure relief analysis by an operating company are listed below.

- Safety Initiatives
- Design/Capital Projects
- Audit/Compliance Regulations
- Corporate Standardization
- Acquisitions

Identification of the causes of overpressure, quantification of relieving rates required complying with codes, sizing of relief devices and the design of the relief disposal system is a complex task. Use of a systematic method for identification of contingencies, together with a set of project specific standards, provides the basis for sound relief system design. Quantification of contingencies using established calculation methods and process simulation is then used to size pressure relief devices and the associated relief header piping network. This approach provides an auditable safe design for a new plant and existing facilities. The different steps involved in a pressure relief analysis are illustrated below.

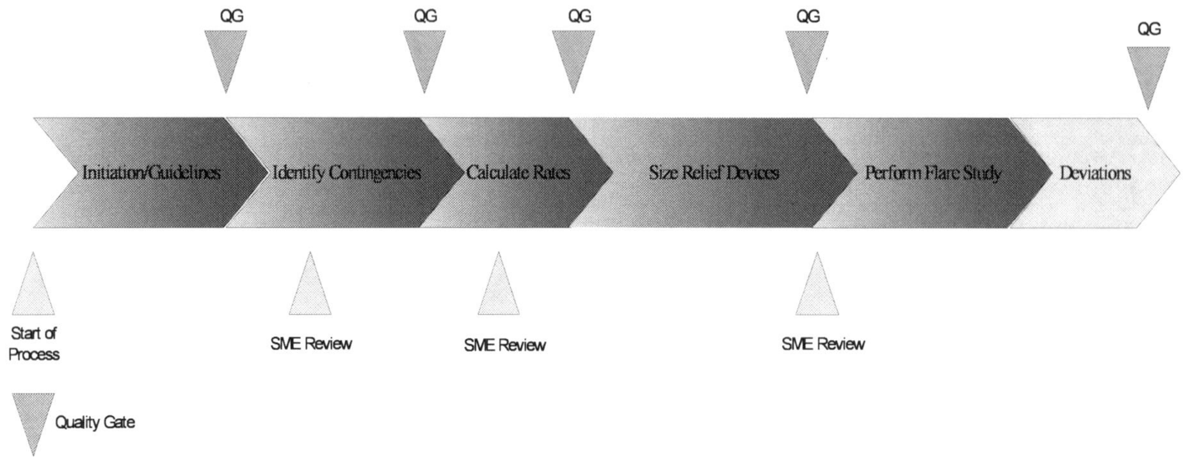

Figure 1 –Typical Phases of Pressure Relief Analysis

Mostly the pressure relief analysis is done as a one-time comprehensive project during the design phase or if there are any changes in the process. The major issues with one time comprehensive studies are listed below.

- Reactive instead of proactive approach
- Based on snap shot
- Underutilized valuable dormant data
- Missed training opportunities
- Out-of-sync with typical MOC programs
- Spend millions in 1-2 years for complete overhaul
- Increase risk of missing safety issues

3. Mechanical Integrity

Mechanical integrity is defined as the program or activities which ensure that equipment in a specified process is designed, constructed, installed and maintained to minimize the risk of accident release of high hazardous chemicals. Mechanical integrity can be achieved through the use of followings.

- Proper design to assure that equipment is suitable for its intended service
- Procedures to assure that we know how to properly operate, inspect and maintain equipment
- Training to assure everyone understands the procedures and the importance of doing things the right way
- Inspection to find flaws or deficiencies that may occur over a period of time or to confirm the status of equipment if nothing has changed
- Repair to maintain integrity
- Quality assurance to make sure that whenever something is done it is the done the right way.

Though there are many codes, standards and recommended practices that are available as a guide to ensure compliance with this element of PSM but still Mechanical Integrity (MI) presents the greatest implementation challenge. Mechanical integrity can be defined as an element of process safety management used to encompass activities that are designed to assure that mechanical equipment in an operating facility has the integrity to perform in the manner and service for which it was designed, locate, characterize, and rectify areas of deficiency that might lead to a condition that would be unsafe and endanger human life, health and/or property, assure that equipment is built and maintained in such a way as to perform it's desired function and ensure the ongoing durability and functionality of equipment achieved through a systematic approach. Although maintenance departments have successfully installed, repaired, and replaced plant equipment for decades, many of these same maintenance departments have struggled with OSHA's PSM requirements. One major challenge is prioritizing resources. Opportunities to improve will always exist; however, it is often difficult to effectively allocate money and manpower. Another challenge is simply getting organized. To meet the requirements of maintaining mechanical integrity, it is necessary to develop a program of testing safety devices and inspecting for both internal and external corrosion and erosion.

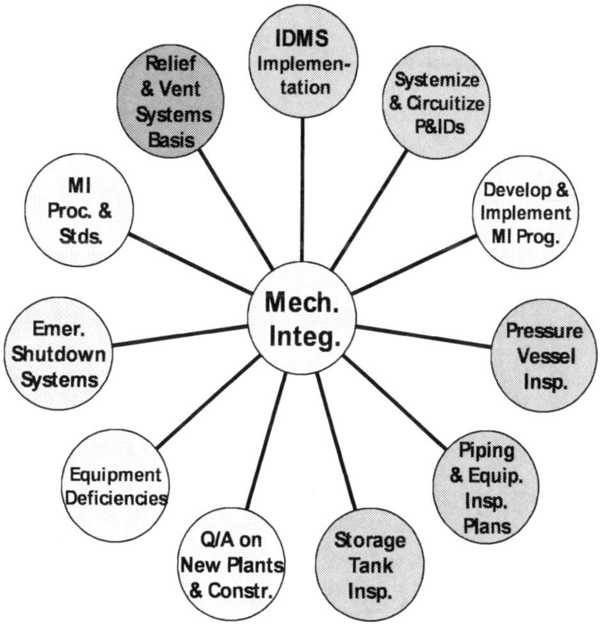

Figure 2 -Elements of Mechanical Integrity (OSHA's 29 CFR 1910.119, Para. j)

Mechanical integrity assures integrity and reliability of pressure equipment which in turn results in fewer safety incidents; reduction of mechanical failure (minimizes downtime and enables better planning), less process upsets and improves maintenance consistency and efficiency.

OSHA 1910.119 was written so that it would cover highly complex and hazardous processes handling a wide variety of hazardous chemicals. Production facilities are some of the simplest processes covered under this regulation. On the other hand, API RP 75 is written specifically for oil and gas production. It addresses offshore operations but much of the information is applicable to onshore operations as well. API helps in interpreting the application of OSHA's requirements to oil and gas facilities.

Many of the mechanical integrity activities are performed as part of a plant's maintenance program. Generally the OSHA guidelines rely on recognized and generally accepted good engineering practice (RAGAGEP), but allows companies to utilize alternate methods as long as they can prove that performance resulting from that method meets the standard. The specifications and techniques that form RAGAGEP are generally understood to mean recommended practices from the American Petroleum Institute (API), codes from the American Society of Mechanical Engineers (ASME), standards from the International Standards Organization (ISO), certification agency standards (DNV, Lloyd's, ABS), and other sources in regular use. The different elements of the PSM and SEMP program, is as shown below.

OSHA's PSM	SEMP
1. Employee Participation 2. Process Safety Information 3. Process Hazard Analysis 4. Operating Procedures 5. Training 6. Contractors 7. Pre-startup Safety Review 8. Mechanical Integrity 9. Hot Work 10. Management of Change 11. Incident Investigation 12. Emergency Planning and Response 13. Compliance Audits 14. Trade Secrets	1. Safety and Environmental Information 2. Hazard Analysis 3. Operating Procedures 4. Training 5. Pre-Startup Review 6. Assurance of Quality and Mechanical Integrity of Equipment 7. Safe Work Practices 8. Management of Change 9. Investigation of Incidents 10. Emergency Response and Control 11. Audit of Safety and Environmental Management Program Elements 12. Records and Documentation

The OSHA PSM guidelines are enforceable throughout the United States and specific US holdings for facilities having a threshold volume of hazardous chemicals, but are not applicable where another federal agency has occupational safety and health jurisdiction, including in federal waters controlled by Bureau of Ocean Management Regulation and Enforcement (BOEMRE, formerly Minerals Management Service (MMS)). Oil and gas drilling and servicing operations and normally unmanned remote facilities are specifically excluded from the standard, but typical facilities in the upstream industry such as production facilities, gas oil separation plants, and gas plants are covered if the threshold chemical volumes, typically hydrocarbons, are present.

In today's economic climate, we are operating with fewer people and demanding more from them than ever before. Therefore, it is imperative that the amount of documentation required be reduced to the minimum necessary to improve communication and increase safety. Excessive documentation places a time requirement that distracts people from their true and safe work practices. Documentation must be kept simple and concise so that it will be easy to use, with essential information readily at hand. The goal of the methodology presented in this paper is to improve safety without creating a large paperwork burden.

4. Methodology

The main goal of the proposed methodology is to develop a program that impacts the culture of the organization, thereby, drives and sustains accurate pressure relief information on a continual basis. The new program must leverage the overall, existing Safety Management Program with focus on the specific task at hand; keep perspective of risk in the current design. The new methodology eradicates the short comings of a one-time comprehensive pressure relief analysis and provides a standardized approach to ensure high reliability of the pressure relief system. The proposed methodology is further sub-divided into six different elements. These elements are Workflow, Resources and Processes, Tools, Training, Key Performance Indicators and Auditing. All these elements are interdependent for an overall success of the proposed methodology. The workflow process starts with a relief system review and generates the PRV Impact checklists, this checklist can be handed over to the MOC program. The actual workflow of the process is shown below.

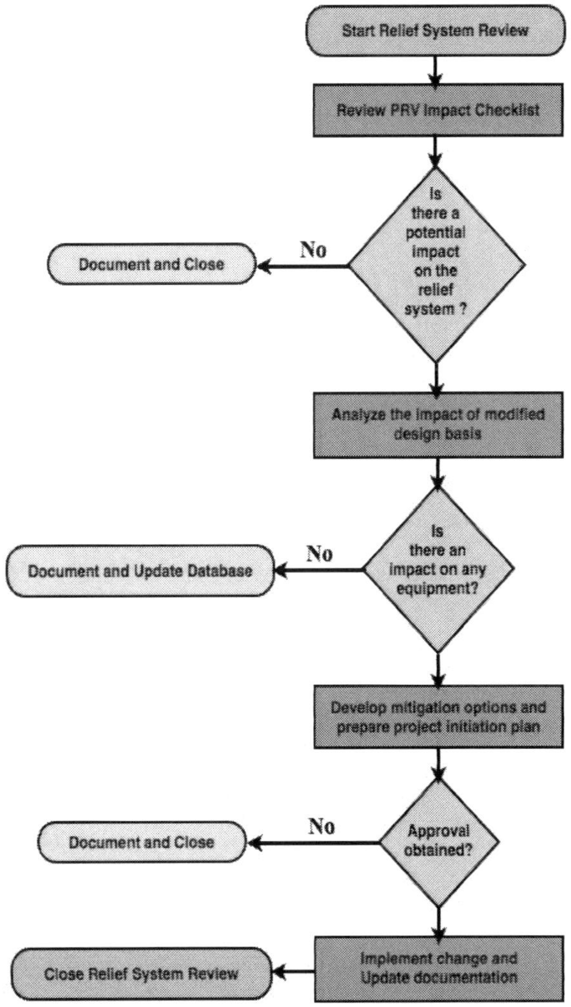

Figure 3 –Pressure Relief Valve Program Workflow

The steps involved in maintaining the data integrity and control of the proposed methodology are mentioned below.

- Develop a Relief System Impact checklist
- Incorporate checklist in the MOC program
- Control access to electronic databases
- Establish within MOC an exception check process for reconciling and validating data changes
- Develop a revalidation process for the plant-wide flare/relief system evaluation
- Complete a revalidation of each study within five years of the previous study
- Perform a complete redo of the study on a cyclical basis
- Manage the resolution of recommendations
- Establish a system of records for data
- Pool data into a central repository

The other elements of the program are to designate an owner for the program and appoint subject matter experts for engineering and information technology. An advisory group both internally at the site and externally at corporate level should be formed to support the program. There are readily available software in the market today to facilitate the actual calculations and sizing of pressure relief valves. Care should be taken to ensure that the tools used meet corporate standard, could be tailor-fitted to compliment internal company practices and also adapt to industry technology and technical practices.

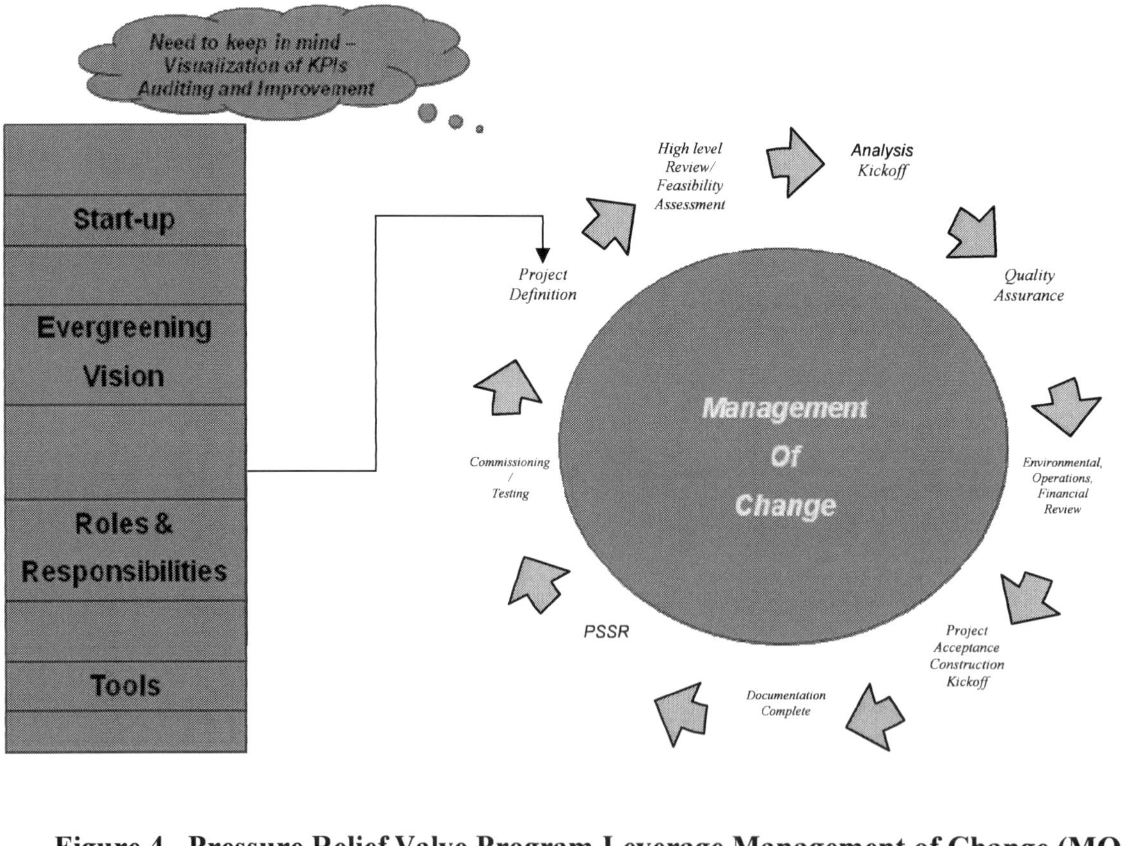

Figure 4 –Pressure Relief Valve Program Leverage Management of Change (MOC)

The above figure enunciates the importance of the proposed methodology in leveraging the existing management of change (MOC) program. The key performance indicators of the program are the data accuracy between different repositories, leverage existing information to measure performance and the program pull dormant data from relief system analysis to measure other PSM metrics. The major benefits of the proposed program are shown below.

- Standard approach
- Continual training
- Defined success factors
- Repeatable process
- Known roles and responsibilities
- Updated and accurate documentation

5. Conclusion

A good program work process starts from the projects to operations handover step on new construction or debottlenecks and runs all the way through the lifecycle of the plant encompassing modifications as a result of small and large projects. It provides a platform to communicate the PSI requirements through all disciplines at the operating facility in a concise manner and holds everyone at a high standard to meet these set requirements by continuously auditioning its effectiveness and providing training and improvement measures.

The cause of overpressure and the amount of fluid that is required to be removed from the system (vessel and piping) is unique for each installation. The proposed methodology ensures process safety without creating a large paperwork burden. The proposed methodology provides operating and maintenance staff, supervisors and foreman with the information, training and motivation they need to conduct their business in a safer manner. It is also intended that the information presented will help in proper documentation and thereby enable a faster response in case of an emergency.

The above mentioned methodology ensures most cost effective process for maintaining the mechanical integrity of high pressure relief systems while contributing up-to-date documentation of the emergency systems and thereby leveraging the existing MOC program. Particularly for high profile vertically integrated companies operating both onshore and offshore facilities, coordination of programs and practices provides an opportunity to streamline internal processes, comply with the new offshore regulatory environment, and operate more safely than ever before.

6. References

[1] "API RP 75, Recommended Practice for Development of a Safety and Environmental Management Program for Offshore Operations and Facilities", third edition. 2004. Washington, DC: API

[2] "Oil and Gas and Sulphur Operations in the Outer Continental Shelf--Safety and Environmental Management Systems", Bureau of Ocean Energy Management, Regulation and Enforcement Implementation (BOEMRE), *Final Rule*, 30 CFR Part 250, Volume 74, Number 199, (October 15, 2010)

[3] "Petroleum Refinery Process Safety Management National Emphasis Program", U.S. Occupational Safety and Health Administration (US OSHA), Directive CPL 03-00-004, (June 7, 2007)

[4] "PSM Covered Chemical Facilities National Emphasis Program", U.S. Occupational Safety and Health Administration (US OSHA), Directive CPL 09-06-002, (July 27, 2009)

[5] "Survey for API RP 75", Bureau of Ocean Energy Management, Regulation and Enforcement Implementation (BOEMRE), (September. 17, 2010)

[6] "API 520 Part 1, Standard for Sizing, Selection, and Installation of Pressure-Relieving Devices in Refineries, eighth edition. 2008. Washington, DC: API

A Framework for Managing Near-Miss Incidents

Presenter: GC Shah/Principal Author: GC Shah
Affiliation: Wood Group Mustang inc.
Address: 16401 Park Ten Place Blvd.; Houston, TX. 77084
email address: ghanshyam.shah@mustangeng.com

[

Prepared for Presentation at
American Institute of Chemical Engineers
2013 Spring Meeting
9th Global Congress on Process Safety
San Antonio, Texas
April 28 – May 1, 2013

UNPUBLISHED

A Framework for Managing Near-Miss Incidents

Presenter: GC Shah/Principal Author: GC Shah
Affiliation: Wood Group Mustang Inc.
Address: 16401 Park Ten Place Blvd.; Houston, TX. 77084
email address: ghanshyam.shah@mustangeng.com

Keywords:

ANSI/API RP-754-2010: Standard that creates "performance indicators" for process safety.
HAZID: Hazard Identification.
HAZOP: Hazard and Operability Study.
Lagging Indicators: Failures which have resulted in loss.
Leading Indicators: Indicators or evaluations that help predict the risk of unsafe events.
Near-miss incidents: Incidents that did not result in any loss, but have a potential for loss.
PHA: Process Hazard Analysis.

Abstract

Today, there is considerable emphasis being placed on learning from near-miss incidents and applying additional safeguards which would minimize possibility of escalating adverse impact of these events in the future. Not all near-miss incidents have potential for major adverse impact. It is important to focus on near-miss incidents which could result into incidents with major adverse impact. This paper applies a modified "Risk Assessment Matrix," that is commonly used for conducting PHA assessments such as HAZID and HAZOP, to estimate potential future risk indication conveyed by near-misses. In a strategic sense, near-miss management spans all work systems of an organization and its culture: IT system, operations, legal, and many others.

1. INTRODUCTION

In the hydrocarbon industry, as well as in many other industries for that matter, the "minor" events occur fairly often. While most of these events probably would end there and not result into major adverse impact, some of the events, if they recur perhaps under different circumstances, could have major impact on safety, environment, asset or a company's reputation. For instance, the event where a worker misses a rung on the ladder and falls and gets bruised may not develop into a major event. On the other hand, an event where there is a pipe leak which creates a cloud of flammable vapor that gets dissipated because of high wind, could under slightly different circumstances, result into a major event (potential fire and explosion). Even small leaks of flammable gases, if allowed to accumulate, could cause a fire and/ or an explosion.

OSHA defines near miss as "*incidents where no property was damaged and no personal injury sustained, but where, given a slight shift in time or position, damage and/or injury could have occurred.*" The definition seems to suggest that, the existing safeguards failed to work; but luckily the incident turned out to be a minor event. However, sooner or later luck may run out and a near miss can turn into a major adverse event. In terms of probabilities, one may say that statistically, there is a greater probability of occurrence of 'small impact" events and much smaller probability of high consequence or high impact events. From cost considerations, high impact events with high frequency simply can't be sustained financially.

Once near-miss events which have a potential to turn into "major" events have been identified, resources can be allocated to either strengthen the existing safeguards or add new safeguards in order to minimize the likelihood of occurrence of a major event.

2. IDENTIFYING POTENTIAL "TIME BOMBS."

Some experts suggest that near-miss is an accident waiting to get you into big trouble. The goal is to let it keep waiting!

How do we identify with a good degree of certainty that a near-miss will grow into a large-impact event? Experience with operations and maintenance could be a big help. You can develop checklists of incidents likely to develop into major events and incidents which are not likely to develop into large-scale events. Before we look at these lists, it is worth considering the meaning of the term "large-scale?" Stated broadly, the events that could cause one or more of the following may be regarded as the events that are likely to develop into "large-scale events."

OSHA recordables
Lost-time
Environmental RQ (Reportable Quantity) events
Environmental or safety violations
Events that could become a legal liability to a company
Public complaints

Some examples of specific events which could potentially turn into events with major adverse impact are:

Pin-hole leaks (e.g., pipes or flanges and fittings handling flammable, explosive or toxic materials; steam or condensate leaks on old boiler systems);
Leaking flanges, leaking mechanical seals;
Leaking tanks with flammables and/or toxic materials;
Spills under Reportable Quantity;
Lack of or poor lubrication of large pumps and/or compressors because pre-startup checks were not made or there was no checklist;
A worker narrowly escapes being asphyxiated in a confined space where nitrogen is being compressed;
Purge cycle on a furnace not working properly;

Welding sparks falling on hydrocarbon containing pipes;
Workers fail to understood a procedure (and therefore don't follow it) as it contained high-level technical jargon;
An isolation valve fails in open position;
Security infraction incidents (including cyber security);
Others.

It may be convenient to categorize the above events into the following groups:

- *Failure to follow procedures (OR poorly written or nonexistent procedures)*
- *Control system (and safety system) failures*
- *Old (older) equipment in challenging service*

Similar other systems of classification may be developed for your company.

Examples of events which will not likely develop into major events are:

Minor bruises or cuts;
Falls with a minor injury;
Potable water leaks;
A worker experiencing headache and taking non-prescription medication;
Other similar examples may be cited.

3. A FRAME WORK FOR NEAR-MISS MANAGEMENT

To be effective, the framework should be efficient. Workers should be encouraged to report ALL near-misses. Near-miss management start NOT after a near-miss, but way before such an event occurs. Guidance on leading indicators (as well as lagging indicators) is developed by API-754-2010. As Figure 1 shows near-miss is not a narrowly defined safety issue of simply reporting events; it spans through the entire organization and systems.

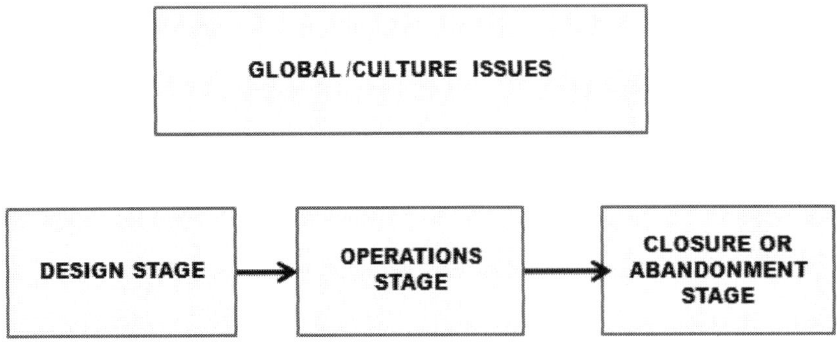

FIGURE 1: NEAR-MISS INCIDENT MANAGEMENT STAGES

3.1 Global and Cultural Issues

- Follow "good neighbors" policy. Establish good rapport with neighbors and regulators.
- Strategically important documents should receive thorough checks so as to avoid mis-interpretation.
- Encourage mutual trust between workers and management.
- Avoid bureaucratic systems in near-miss management.
- Encourage high-tech professionals (e.g., IT networks, advanced controls) to translate technical jargon into lay-person's language. The key is that persons, for whom the procedures (developed by technical professionals) are intended, should understand clearly.
- Ensure there are adequate resources to address near-miss incidents in reasonable time. Do not "dilute," safety.
- Acquisitions and divestitures also should (and typically they do) receive due diligence.

3.2 Design Stage:

- Ensure design is in accord with the applicable industrial standards and regulations.
- As far as practicable, follow the philosophy of fault-tolerant design. Stated differently, design should a mishap occur, there should be appropriate mitigative measures to constrain impact resulting from the mishap.
- Ensure documentation is easy to access and not ambiguous. (Workers referring to the documents in the future should get a clear understanding of the meaning and intent of the design).
- Perform periodic risk assessments; address findings thoroughly, and document.

3.3 Operations Stage:

- Encourage a collaborative rapport between your technical experts or engineers and operations and maintenance workers and contractors.
-
- Introduce New Technology only after a thorough risk assessment.
- Implement frequent training.
- Ensure the procedures and instructions are CLEAR to workers. Seek their input before implementing them.
- Conduct periodic audits of the facility. Some companies term these as "Loss Control Audits."
- Encourage workers to report ALL near-misses.
- Identify near-miss incidents which have potential to develop into major events. You may work out checklists workers can use to identify such incidents or consider the Risk Matrix shown later in this section. Of course checklists should be thorough so that workers won't miss out a near-miss that has a potential to grow into a major event.
- Provide resources to address near-misses within a reasonable time.

3.4 Closure and Abandonment

- Develop closure and abandonment policies. Apply Risk Management criteria to this phase of the organization.

3.5 Risk Matrix for Near Miss Incident Management

Figure 2 below shows a proposed Risk Matrix for near miss incident management. Risk is defined as a product of likelihood and consequence (severity of the outcome of an incident). Risk is grouped into three levels:

- High
- Medium
- Low

All near-miss incidents must be looked into. However, incidents identified as those having a potential for developing into "major" events require further investigation and possibly adding new safeguards or making existing safeguards more reliable or robust. Near miss management should be efficient. A small team of experienced workers can apply Risk Matrix (or checklists).

Figure 2: Risk Matrix for Near Miss Incident Management

		Risk Matrix for "Near-Miss" Incidents		
Consequence	**3**	**3**	**6**	**9**
	2	2	4	**6**
	1	**1**	**2**	**3**
		1	**2**	**3**
		Likelihood		

		Risk Matrix for "Near-Miss" Incidents		
		Likelihood		
		1	Relatively low likelihood of occurrence once in 10 yrs.	
		2	Relatively moderate likelihood of occurrence once every yr.	
		3	Relatively high likelihood of occurrence several times a year.	
		Consequence		
		1	Relatively "low" or "potentially low" level of impact: This events are NOT likely to develop into recordable or lost time or RQ events or higher.	
		2	Moderate level of impact or potential for medium level of impact such as OSHA recordable or lost time incident or RQ events.	
		3	High impact or potential for "high" impact: For example, large scale environmental or safety impact, or potential for damage to reputation.	
		Suggested Corrective Action		
		BLUE	Document the event. Further resources for corrective action may not be warranted.	
		YELLOW	Investigate and implement corrective action in "reasonable "time.	
		RED	Investigate. Should receive top priority and urgency in implementing corrective action.	

4. CONCLUSION

Near miss incident management starts way before a near miss occurs. In a strategic sense, the near miss management pervades all layers of an organization and its interactions with external systems. Strong safety culture and trusting relationships are important components of an effective near miss management framework.

HAZOP Analysis and Debottleneck for Laboratory Operation in the Semiconductor Industry

Tianxing Cai
Dan F. Smith Department of Chemical Engineering
Lamar University, Beaumont, TX 77710, USA
tcai@lamar.edu

Qiang Xu
Dan F. Smith Department of Chemical Engineering
Lamar University, Beaumont, TX 77710, USA
qiang.xu@lamar.edu

Prepared for Presentation at
American Institute of Chemical Engineers
2013 Spring Meeting
9th Global Congress on Process Safety
San Antonio, Texas
April 28 – May 1, 2013

UNPUBLISHED

HAZOP Analysis and Debottleneck for Laboratory Operation in the Semiconductor Industry †

Tianxing Cai
Dan F. Smith Department of Chemical Engineering
Lamar University, Beaumont, TX 77710, USA
tcai@lamar.edu

Qiang Xu*
Dan F. Smith Department of Chemical Engineering
Lamar University, Beaumont, TX 77710, USA
qiang.xu@lamar.edu

Keywords: HAZOP Analysis, Semiconductor Industry, Laboratory Hazards, Operation Debottleneck

Abstract

Unlike the semiconductor manufacturing process with fully automatic equipment, the laboratory operation in the semiconductor industry will still involve multiple types of manual handling for chemical analysis, reliability test, material characterization and failure analysis. The test operators or lab engineers have different levels of exposure to a lot of chemicals and physical hazards in multiple phases. This paper will describe the straightforward HAZOP (Hazard and Operability Study) process to identify the hazards and assess the risks of laboratory operations. The analysis will be deployed in the ways of manpower, machine, materials, method and environment. The analysis output can be regarded as the aid to safer laboratory design, operator training and real-time fault diagnosis. This will help to debottleneck and achieve operation safety and occupational health in the semiconductor industry.

† For presentation at AICHE 2013 Spring Meeting ,9th Global Congress on Process Safety.

* All correspondence should be addressed to Prof. Qiang Xu (Phone: 409-880-7818; Fax: 409-880-2197; E-mail: Qiang.xu@lamar.edu).

1. Introduction

Semiconductor industry revenue in 2012 is expected to reach $323.2 billion, up a slight 3.3 percent from last year's revenue of $312.8 billion, according to an IHS iSuppli Global Manufacturing Market Tracker report from information and analysis provider IHS.
Based on the professional estimation, the growth in the semiconductor industry from 2013 to 2015 will average between a more encouraging 6.6 to 7.9 percent, as shown in the figure below, with total semiconductor revenue by 2015 rising to some $397.7 billion. In the coming several years, the strong semiconductor growth can be seen.

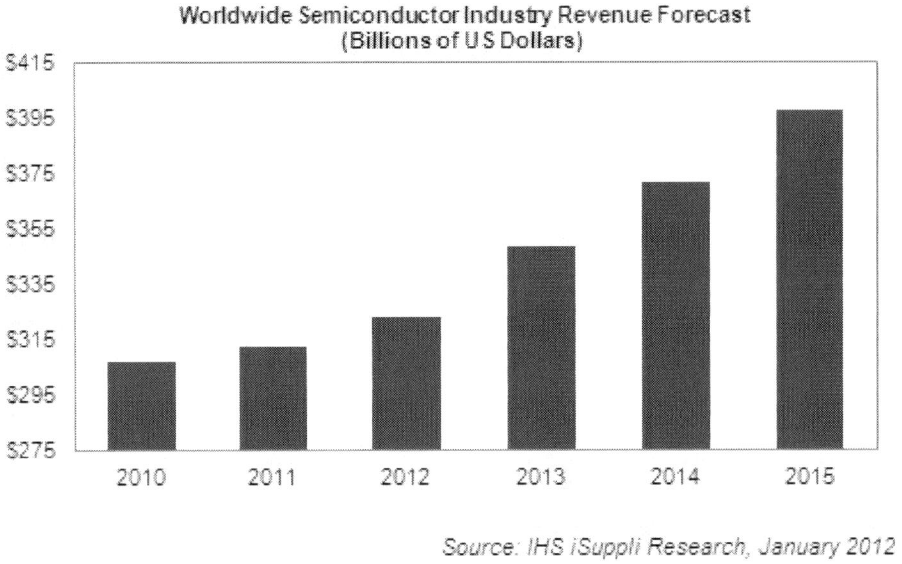

Figure 1: Worldwide Semiconductor Industry Revenue Forecast

Coupled with the fast development, the semiconductor employment has globally stretched. In USA, recently on Dec 11th, 2012, the Semiconductor Industry Association (SIA) announced that U.S. semiconductor industry jobs stretch across almost every region of the country and into the majority of states, according to an analysis of government data. California leads all states with 47,100 semiconductor jobs, followed by Texas (28,800), Oregon (23,400), Arizona (18,800), Massachusetts (10,100) and New York (7,600). According to BLS, the U.S. semiconductor industry's workforce grew by 3.7 percent over the previous year. In comparison, jobs throughout the broader U.S. economy increased by 1.2 percent over the same time period.

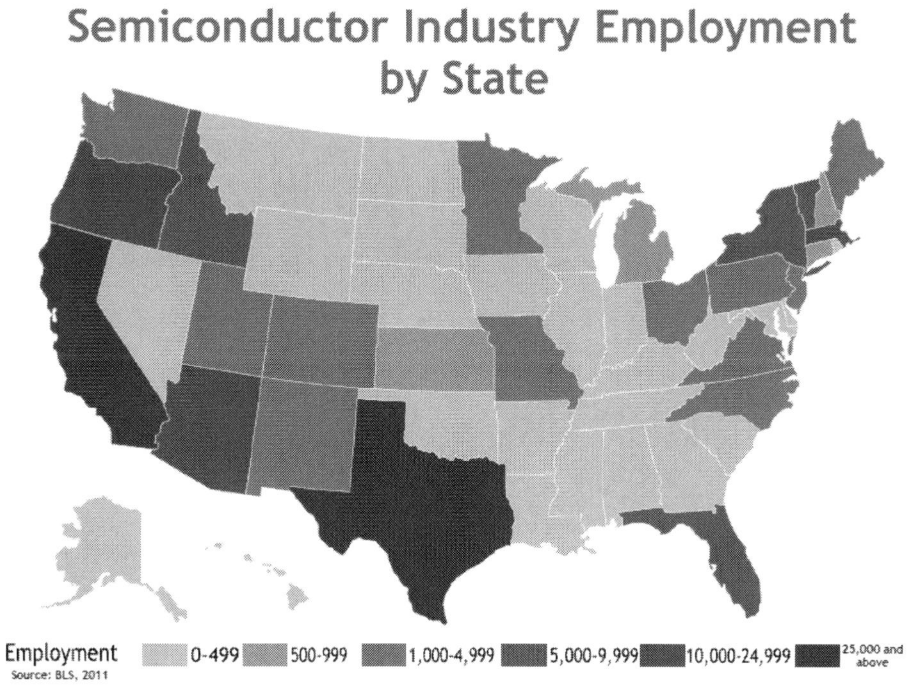

Figure 2: Semiconductor Industry Employment by State

Ever since the invention of transistor by the Bell Telephone Laboratories in 1948, the semiconductors industry has expanded greatly. The continuous fast development and innovations have expedited the increase of the semiconductor industry employment, which has further invoked an unavoidable concern of occupational safety to the semiconductor manufacturing operation.

Unlike the semiconductor manufacturing process with fully automatic equipment, the laboratory operation in the semiconductor industry will still involve multiple types of manual handling for chemical analysis, reliability test, material characterization and failure analysis. These operations will involve the handling of advanced and complex equipment or instruments such as Gas Chromatography (GC), Gas Chromatography-Mass Spectrometry (GC-MS), Fourier Transformed Infra Red Spectroscopy (FT-IR), Ultraviolet Visible Spectroscopy (UV/Visible), Liquid Chromatography (LC), Thermal and Elemental Analysis (TEA). Furthermore, due to rapid changes in this industry, manufacturing processes and their associated hazards may change completely every few years. These changes make hazard assessments more difficult to complete and require that they be conducted more often. Common hazards may include exposure to solvents, acid and caustic solutions, toxic metals, and radiation. Below Table 1 has listed the typical hazardous chemicals with their roles in the manufacturing process and potential health problems linked to the exposure. The test operators or lab engineers have different levels exposure to a lot of chemicals and physical hazards in multiple phases.

Table 1:Hazardous Chemicals in the Semiconductor Industry

Chemical name	Role in manufacturing process	Health problems linked to exposure
Acetone	Chemical-mechanical polishing of silicon wafers	Nose, throat, lung, and eye irritation, damage to the skin, confusion, unconsciousness, possible coma
Arsenic	Increases conductivity of semiconductor material	Nausea, delirium, vomiting, dyspepsia, diarrhea, decrease in erythrocyte and leukocyte production, abnormal heart rhythm, blood vessel damage, extensive tissue damage to nerves, stomach, intestine, and skin, known human carcinogen for lung cancer
Arsine	Chemical vapor deposition	Headache, malaise, weakness, vertigo, dyspnea, nausea, abdominal and back pain, jaundice, peripheral neuropathy, anemia
Benzene	Photoelectrochemical etching	Damage to bone marrow, anemia, excessive bleeding, immune system effects, increased chance of infection, reproductive effects, known human carcinogen for leukemia
Cadmium	Creates "holes" in silicon lattice to create effect of positive charge	Damage to lungs, renal dysfunction, immediate hepatic injury, bone defects, hypertension, reproductive toxicity, teratogenicity, known human carcinogen for lung and prostate cancer
Hydrochloric acid	Photoelectrochemical etching	Highly corrosive, severe eye and skin burns, conjunctivitis, dermatitis, respiratory irritation
Lead	Electroplated soldering	Damage to renal, reproductive, and immune systems, spontaneous abortion, premature birth, low birth weight, learning deficits in children, anemia, memory effects, dementia, decreased reaction time, decreased mental ability
Methyl chloroform	Washing	Headache, central nervous system depression, poor equilibrium, eye, nose, throat, and skin irritation, cardiac arrhythmia
Toluene	Chemical vapor deposition	Weakness, confusion, memory loss, nausea, permanent damage to brain, speech, vision, and hearing problems, loss of muscle control, poor balance, neurological problems and retardation of growth in children, suspected human carcinogen for lung and liver cancer
Trichloroethylene	Washing	Irritation of skin, eyes, and respiratory tract, dizziness, drowsiness, speech and hearing impairment, kidney disease, blood disorders, stroke, diabetes, suspected human carcinogen for renal cancer

In order to mitigate the potential risk during the chemical process operation, HAZOP study is always conducted. A hazard and operability study (HAZOP) is a structured and systematic examination of a planned or existing process or operation in order to identify and evaluate problems that may represent risks to personnel or equipment, or prevent efficient operation. The

HAZOP technique was initially developed to analyze chemical process systems, but has later been extended to other types of systems and also to complex operations and to software systems. A HAZOP is a qualitative technique based on guide-words and is carried out by a multi-disciplinary team (HAZOP team) during a set of meetings.
This paper will describe the straightforward HAZOP process to identify the hazards and assess the risks of laboratory operations. The analysis will be deployed in the ways of manpower, machine, materials, method and environment. The analysis output can be regarded as the aid to safer laboratory design, operator training and real-time fault diagnosis. This will help to debottleneck and achieve operation safety and occupational health in the semiconductor industry.

2. General Methodology

Several tools and methodologies are adopted to deploy the HAZOP Analysis and debottleneck for laboratory operation in the semiconductor industry. This has included Laboratory Operation Flow Diagram (LOFD), HAZOP Meeting, Parameters and Guide Words

2.1 Laboratory Operation Flow Diagram (LOFD)

The method applies to existing or planned laboratory operation for which design information is available. Similar with the commonly used process flow diagram, this commonly includes a laboratory operation flow diagram, which is examined in small sections, such as individual items of laboratory test instruments, glassware or chemicals between them. For each of these, a operation design intention is specified. For example, in a failure analysis laboratory, a standard solution of aqua regia (the combination of HNO_3 and HCl with the volume ratio of 1:3) may have been used. The intention to remove gold layer in the integrated circuit so that the further visual inspection can be conducted to check the failure mode. The HAZOP team then determines what are the possible significant deviations from each intention, feasible causes and likely consequences. It can then be decided whether existing, designed safeguards are sufficient, or whether additional actions are necessary to reduce risk to an acceptable level. Besides the operation safety, HAZOP can also give the detailed instruction on the opinion whether the potential deviation will affect the quality of experiment or test analysis result.

2.2 HAZOP Meeting

When HAZOP meetings were proceeded by the involved laboratory engineers, technicians and operators, they were generally scheduled for one to two hours per day. For a medium-sized laboratory where the total number of test items to be considered is 600 (items of equipment and pipes or other transfers between them), about 20 such meetings would be needed.

2.3 Parameters and Guide Words

The key feature is to select appropriate parameters which will apply to the design intention. The general words used in the laboratory operation are usually solution concentration, solution temperature, sampling volume, and instrument detection limit. In the above example, it can be seen that variations in these parameters could constitute deviations from the design intention of experiments. In order to identify deviations, a set of guide words is applied to each step of the experiment process. The current standard Guide Words are as follows:

Table 2: Guide Words for HAZOP Study of Laboratory Operation in the Semiconductor Industry

Guide Word	Meaning
NO OR NOT	Complete negation of the design intent
MORE	Quantitative increase
LESS	Quantitative decrease
AS WELL AS	Qualitative modification/increase
PART OF	Qualitative modification/decrease
REVERSE	Logical opposite of the design intent
OTHER THAN	Complete substitution
EARLY	Relative to the clock time
LATE	Relative to the clock time
BEFORE	Relating to order or sequence
AFTER	Relating to order or sequence

2.4 HAZOP Software

Various software programs are now available to assist in HAZOP meetings. Even though they are always used in the chemical industry, the principles will be quite applicable to laboratory operation in the semiconductor industry.

2.5 HAZOP Team

HAZOP is normally carried out by a team of people with roles as follows

Table 3: Team Members for HAZOP Study of Laboratory Operation in the Semiconductor Industry

Name	Position	Role
HAZOP Group Leader	Lab Manager	someone experienced in HAZOP and laboratory operation, to ensure that the method is followed carefully
Recorder	Lab Engineer 1	someone to ensure that problems are documented and recommendations passed on
Designer	Lab Engineer 2	someone to explain any design details or provide further information
User	Lab Operator	someone to consider it in use and question its operability, and the effect of deviations
Specialist	Lab Engineer 3	someone with relevant technical knowledge
Maintainer	Lab Technician	someone concerned with maintenance of the process.

3. HAZOP Management System

For the purpose of effective implementation of the HAZOP management policy and solid application of introduced general methodology, the laboratory HAZOP system should be set up. The system has coupled general management policy (operation specification and work instructions), environment control (temperature and humidity), operation safety (chemical storage, hazards assessment, chemical identification and MSDS), equipment management (PM, calibration, MSA), chemical control (data analysis and shelf life time control), training and certification (orientation, OJT, regular qualification). The whole system has covered multiple layers of analysis factors. The first layer will include manpower, machine, materials, method and environment.

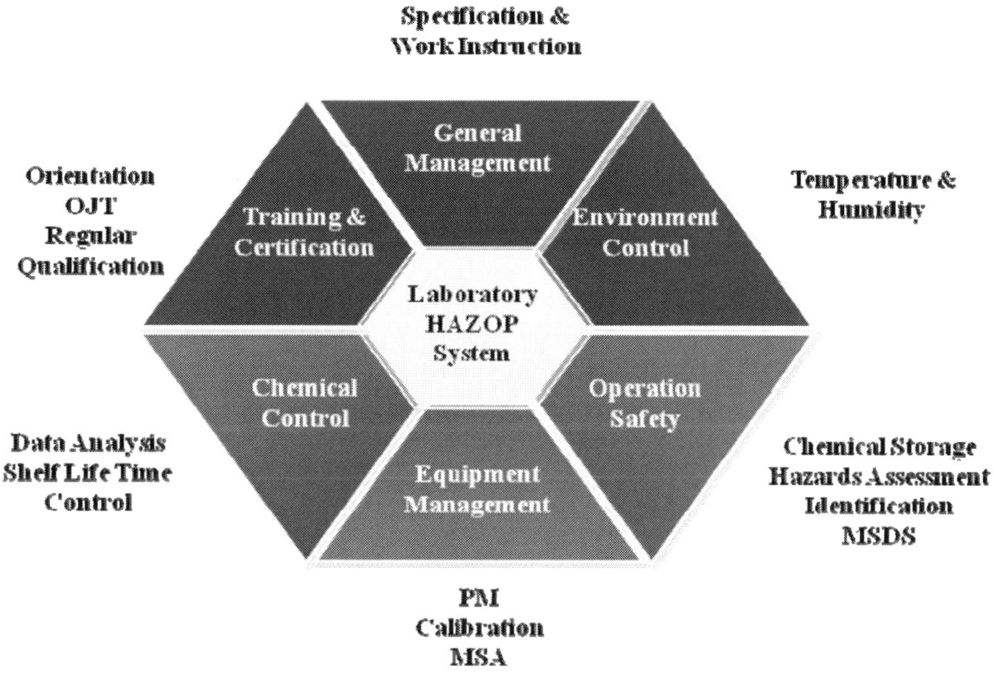

Figure 3: HAZOP Management System for Laboratory Operation in the Semiconductor Industry

These factors can be expanded with below fishbone diagrams so that the second layer of factors will be included into HAZOP matrix system: for example, sample collection and sample preparation will be considered in the material part; training, parallax, reproducibility, practice and ergonomics will be considered in the manpower part; samples, test methods, workmanship and standards will be considered in the method part; discriminations, repeatability, bias, linearity and calibration will be considered in the instrument part; vibration, lighting, temperature and humidity will be considered in the method part.

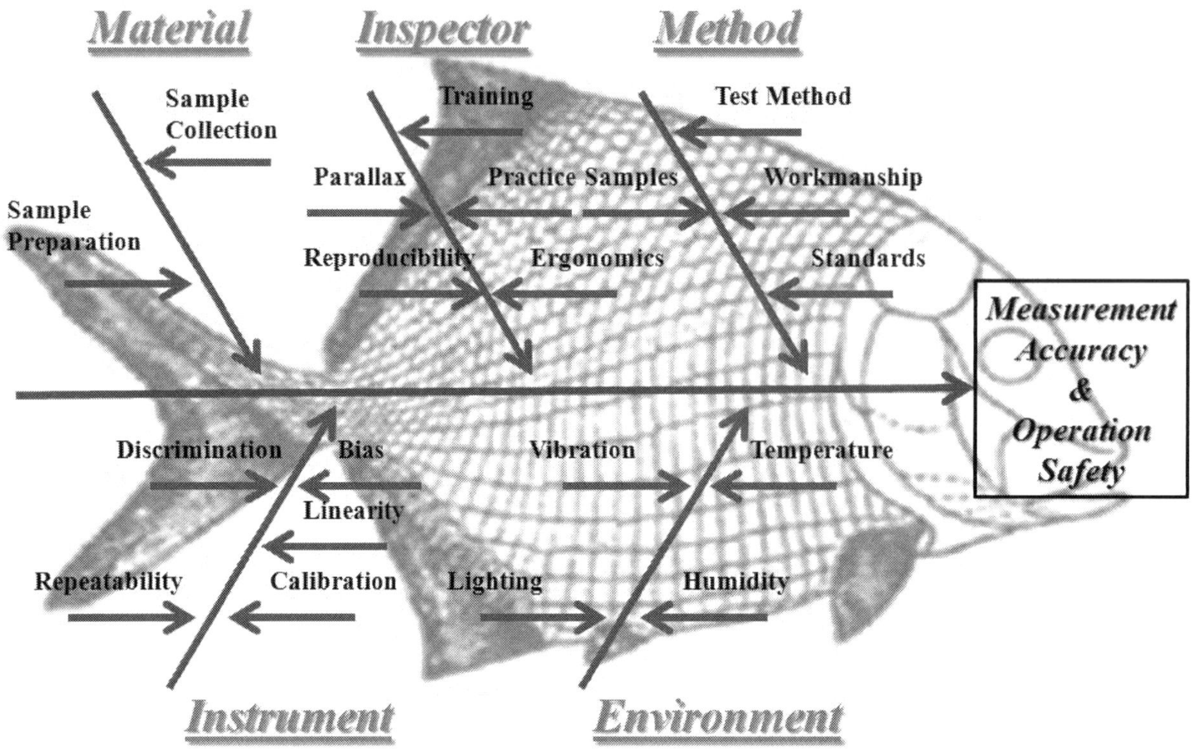

Figure 4: Fish Bone Diagram for HAZAP Study for Laboratory Operation in the Semiconductor Industry

Next, these practices can form a closed PDCA (Plan, Do, Check and Act) loop for continuous improvement of HAZOP activity. The close loop include four phases: plan, do, check and act. In the first phase of planning, the procedures and schedules to conduct the study should be identified and the input data should be collected. Then, the deviation should be identified by the application of key words and the causes and consequences should also be determined in the second phase. Next, the improvement actions need to be followed up. It is better to restudy the necessity of current HAZOP system. After the verification of the above items, the study report needs to be generated. Then the activity of HAZOP study should be evaluated to confirm whether the current scope, objective and responsibility need to be updated. Furthermore, the team should decide whether they need more team members to cover a larger scope.

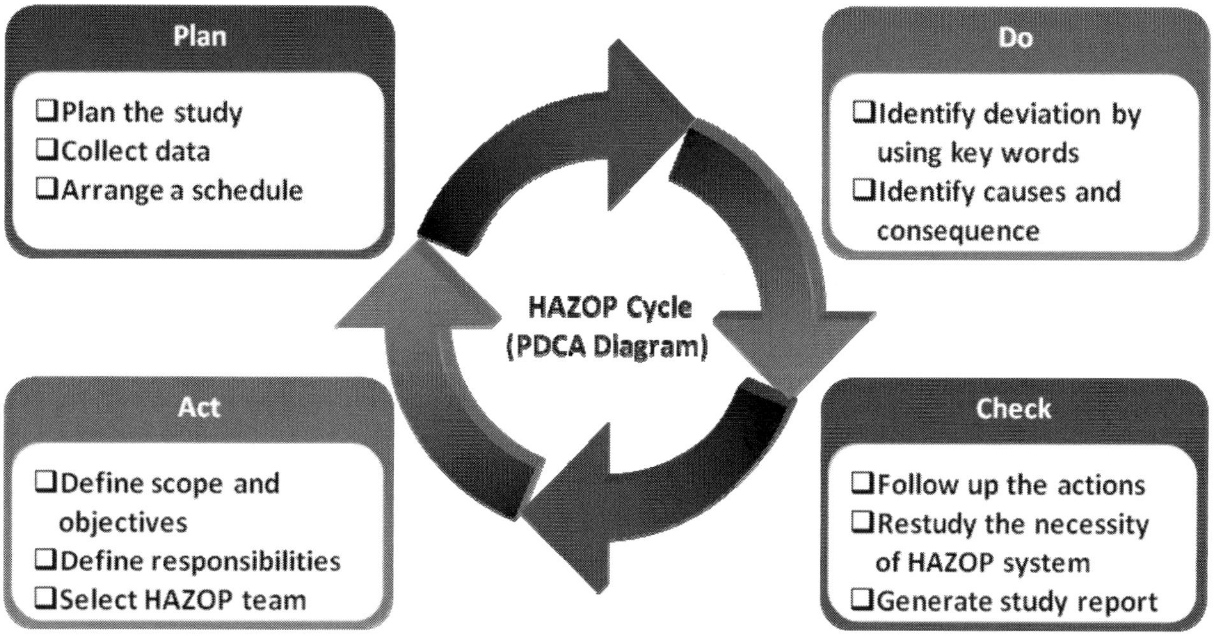

Figure 5: PDCA (Plan, Do, Check and Act) Diagram for HAZAOP Study for Laboratory Operation in the Semiconductor Industry

The analysis output can be regarded as the aid to safer laboratory design, operator training and real-time fault diagnosis. This will help to debottleneck and achieve operation safety and occupational health in the semiconductor industry.

4. Application Illustration

To demonstrate the efficacy of the developed systematic methodology, one case study has been conducted as the illustration of the methodology application in the real practice of plating bath monitoring of wafer bumping in the semiconductor industry. Wafer bumping is an advanced packaging technique that attaches solder spheres or gold bumps to the bond pads on a selected die. Gold bumping is generally used for Tape Automated Bond (TAB) assembly or Anisotropic Conductive Film (ACF) flip-chip assembly. Solder bumping is primarily used for reflow flip-chip assembly which unites the components face down, directly with the substrate or board through conductive bumps on the bond pads. These bumps deliver the electrical, mechanical and thermal interconnection which provides direct contact between the chip package and the device. Another process is called stud-bumping. Unlike gold and solder bumping, this process can be done at the die level. Wafer bumping also known as flip chip has traditionally been used for high-end niche applications. However in the last few years it has gone mainstream to include a wider range of consumer applications geared towards both the performance driven market and the form factor driven market.

Wafer bumping provides the electrical and mechanical connection between a die and substrate, which is one of the most critical elements of any flip chip package structure. Predominantly lead-tin and lead-free solders at the present time, these connections - or bumps - must exhibit superior adhesion to the die, minimal resistance, and result in high assembly yields. Solder bumps are formed by using either thin film metal deposition or ball loading techniques. Currently the

advanced semiconductor assembly and test company such as Amkor offers state-of-the-art capability in electroplated solder technologies including the electroplating capability to support fine pitch bumping and copper pillar structures.
Wafer bumping utilizes thin film deposition, photolithography and plating techniques to place solder or gold onto IC (Integrated Circuit) lead pads. The wafer solder bump process is to produce the solder bumping of the wafer pad. Then it is going to melt the bumping by heat in assembly. This technology can substantially reduce IC size and have the advantages such as high density, low sensitivity, low cost and effective heat dissipation. This bumping process is suitable for flip chip assembly for LCD, memories, microprocessors, and microwave RF ICs applications. The process includes sputter UBM (under Bump Metallurgy), Photolithography, Plating and Etching etc., This bumping is suitable to apply on TAB, Flip Chip. The detailed process flow chart has been illustrated in below flow chart.

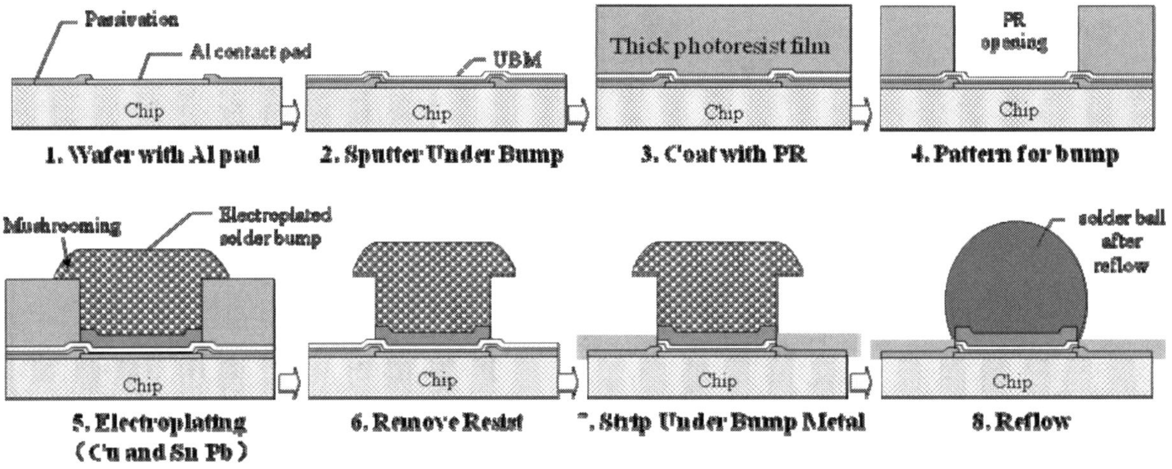

Figure 6: Process Flow Chart for Wafer Bumping Process

One of the key test item for wafer bumping operation is the electroplating bath monitoring and composition control of solder materials and additives. Plating solutions must be routinely analyzed in order to maintain the recommended bath formulation and to preempt the occurrence of problems related to improper levels of bath constituents. Contaminant levels in the solutions must also be monitored. Manufacturers of plating systems establish optimum specifications to ensure maximum solution efficiency and uniformity of deposits. The various factors that cause the concentrations of bath constituents to deviate from their optimum values are as follows: 1. drag-out; 2. solution evaporation; 3. chemical decomposition; and 4. unequal anode and cathode efficiencies. Regular analysis of plating solution becomes the responsibility of process engineering and quality control group.
The techniques employed for the quantitative analysis of plating solutions are classified as volumetric (titrimetric), gravimetric, and instrumental. Volumetric and gravimetric methods are also known as "wet" methods. The analyst must select the method that is best suited and most cost effective for a particular application. The wet methods outlined here are simple, accurate, and rapid enough for practically all plating process control. They require only the common analytical equipment found in the laboratory, and the instructions are sufficiently detailed for an average technician to follow without any difficulty. The determination of small amounts of

impurities and uncommon metals should be referred to a competent laboratory, as a high degree of skill and chemical knowledge are required for the determination of these constituents. Instrumental methods differ from wet methods in that they measure a physical property related to the composition of a substance, whereas wet methods rely on chemical reactions. The selection of an instrument for the analysis of plating solutions is a difficult task. Analysts must decide if the cost is justified and if the analytical instrument is capable of analyzing for the required substances with a high degree of accuracy and precision. Instruments coupled to computers can automatically sample, analyze, and record results. Mathematical errors are minimized and sample measurements are more reproducible than with wet methods. Instrumental methods are also extremely rapid when compared with wet methods. Analytical instruments frequently used in the analysis of plating solutions can be categorized as spectroscopic, photometric, chromatographic, and electro-analytical. Spectroscopic methods (flame photometry, emission spectrometry, X-ray fluorescence, mass spectrometry, and inductively coupled plasma) are based on the emission of light. Photometric methods (spectrophotometry, colorimetry, and atomic absorption) are based on the absorption of light. Chromatographic methods (ion chromatography) involve the separation of substances for subsequent identification. Electro-analytical methods (potentiometry, conductometry, polarography, amperometry, and electrogravimetry) involve an electric current in the course of the analysis. The instrumental methods, comprehensively reviewed below, are most applicable to plating environments.
Here, the manual titration procedures to determine copper concentration in the sample plating bath is used as illustration because the manual titration will have more potential safety concerns compared with the instrument analysis methods. Below is the procedure of the test.

PROCEDURE:

1. Preparation of EDTA (Ethylenediaminetetraacetic Acid) (0.01M). Dry EDTA for 2 hours at 80°C. GMW = 372.2., or check the reagent bottle. Dissolve needed grams in a 500 ml vol. flask. Store in a conditioned polyethylene container. EDTA solutions attack glass and deteriorate after 1 or 2 days.
2. Preparation of known copper solution. Place 0.2 g to 0.25 g of pure Cu. wire (or shot) (weighed on an analytical balance) in a 100 ml beaker. Add 15 ml conc. HNO3 and 1 ml conc. H2SO4. Place a watch glass over the beaker. Boil until dense white fumes of SO3 cease. Transfer carefully to a 200 ml vol. flask. Remember to rinse the watch glass. Dilute to the mark.
3. Preparation of unknown copper solution. Weigh 0.5 to 0.6 grams of unknown into a 200 ml vol. flask. Add 30 ml of 1:5 nitric acid. Heat until all particles are dissolved. Dilute to the mark.
4. Titration with EDTA - murexide indicator. Do this step for 4 samples of known and 4 samples of unknown.

KNOWN: Place 10 ml of Cu known solution (step 2) in 250 ml Erlenmeyer flask. Add pH buffer drop-wise until the solution turns completely blue and the precipitate first formed dissolves. Add about 100 ml. Add more buffer if solution becomes cloudy. Add 1 tablet of murexide indicator and titrate with EDTA until color changes from yellow (green) to deep violet. If the volume of EDTA is less than 10 ml, repeat using 25 ml liquid of Soln. 2.

UNKNOWN: Place 25 ml of Cu unknown solution (step 3) in 250 ml Erlenmeyer flask. Add pH buffer drop-wise until the solution turns completely blue and the precipitate first formed dissolves. Add 1 tablet of murexide indicator and titrate with EDTA until color changes from yellow (green) to deep violet. If the vol. of EDTA solution is less than 10 ml. use a larger liquid of Soln. 3.
Remarks: The copper-murexide complex is pure yellow, but the color at the starting point of the titration is greenish depending on the amount of copper present. The green is the mixture of yellow (indicator complex) and blue (copper-tetramine complex). As the titration proceeds the tetramine complex is transformed to the less intensely colored EDTA complex and the yellow of the indicator complex becomes permanent.
CALCULATION: Calculate the molarity of the EDTA. Calculate percent copper in the sample using the value of the Molarity of the EDTA found. Include the spread sheet for the calculation.

Based on the available procedures to determine the copper concentration in the sample solution of plating bath, HAZOP factors can be expanded so that the second layer of factors will be included into HAZOP matrix system: for example, sample collection and sample preparation will be considered in the material part; training, parallax, reproducibility, practice and ergonomics will be considered in the manpower part; samples, test methods, workmanship and standards will be considered in the method part; discriminations, repeatability, bias, linearity and calibration will be considered in the instrument part; lighting, temperature and humidity will be considered in the method part. These factors are listed as the column. Then, the HAZOP guide words will be listed as the row. The related information of the deviation stored in LOD(Laboratory Operation Database) and HPD(HAZOP Project Database) can be selected into the blank spaces from the drop-down list box in the above expanded table. Next, the data record of regular monitor results will be proceeded into data analysis to validate the impact of the above deviation . The data analysis result will couple with the expanded HAZOP table to quantitatively identify the potential deviation of operation safety and test accuracy.

5. Conclusion

The laboratory operation in the semiconductor industry will have more possibilities of manual handling for chemical analysis, reliability test, material characterization and failure analysis. Therefore higher attention should be paid to the laboratory operations. As illustrated in this paper, the methodology of HAZOP study procedure can be applied to identify the potential deviations in both experiment accuracy and operation hazards. Multiple layers of factor consideration by the application of 4M1E (Man, Machine, Material, Method and Environment) will help the brainstorm to give the aid to safer laboratory design, operator training and real-time fault diagnosis. This will help to debottleneck and achieve operation safety and occupational health in the laboratory operation semiconductor industry.

6. Reference

[1] Chepesiuk, R., "Where the Chips Fall: Environmental Health in the Semiconductor Industry." Environmental Health Perspectives, Volume 107, pp.452-457, 1999

[2] Swann, C. D., & Preston, M. L., "Twenty-five years of HAZOPs", Journal of Loss Prevention in the Process Industries, Volume 8, No 6, pp.349-353, 1995

[3] Amkor Technology Wafer Bumping Service Introduction, http://www.amkor.com /go/turnkey-services/wafer-bumping

(accessed Jan. 5, 2013)

[4] Nolan, D.P., "Application of HAZOP and What-If Safety Reviews to the Petroleum, Petrochemical and Chemical Industries". William Andrew Publishing/Noyes, 1994

[5] Lawley, H. G., "Operability studies and hazard analysis", Chemical Engineering Progress, Volume 70, No 4, pp.45, AIChE,1974

[6] Chemical Industries Association., "A Guide to Hazard and Operability Studies", 1977

[7] IHS iSuppli Research., "Semiconductor Industry Revenue to Ensure Slow Growth in 2012",http://www.isuppli.com/Semiconductor-Value-Chain/News/Pages/Semiconductor-Industry-Revenue-to-Endure-Slow-Growth-in-2012.aspx

(accessed Jan. 5, 2013)

[8] U.S. Bureau of Labor Statistics (BLS) Data Report,2011

[9] Semiconductor Industry Association, http://www.sia-online.org/

(accessed Jan. 5, 2013)

[10] Bolmen, Richard A., "Semiconductor Safety Handbook: Safety and Health in the Semiconductor Industry", Noyes Publications, 1998

Best Practice Management of Change & PSSR Process

Michael Bearrow
Optimized Systems and Solutions Inc.
8303 McHard Road, Suite 200, Houston, TX 77053
Michael.E.Bearrow@o-sys.com

Prepared for Presentation at
American Institute of Chemical Engineers
2013 Spring Meeting
9th Global Congress on Process Safety
San Antonio, Texas
April 28 – May 1, 2013

UNPUBLISHED

Best Practice Management of Change & PSSR Process

Michael Bearrow
Optimized Systems and Solutions Inc.
8303 McHard Road, Suite 200, Houston, TX 77053
Michael.E.Bearrow@o-sys.com

Keywords: MOC, process safety code, recommendation,

Management of Change (MOC) and Pre Start-up Safety Review (PSSR) are still the most challenging elements of OSHA's Process Safety Management (PSM) standard, the EPA's Risk Management Program (RMP) rule, and now, the US Department of Interior's Bureau of Safety and Environmental Enforcement's Safety and Environmental Management Systems (SEMS). Even though the PSM standard has been around since 1992 and the industry has been managing change for several decades, we still can get it wrong, sometimes with disastrous results. The mindful or diligent efforts of many working in concert are necessary to ensure that change is identified, analyzed and executed in a quality way. Just when we get it right on paper and get the workforce up-to-speed, we have employee turnover, neglect and sometimes regulatory change. New actors and a constantly changing script make it hard to manage change efficiently and effectively. This paper discusses how the chemical process industry has defined the MOC and PSSR best management practices and how they should be automated.

Management of Change (MOC) and Pre Start-up Safety Review (PSSR) are still the most challenging elements of OSHA's Process Safety Management (PSM) standard, the EPA's Risk Management Program (RMP) rule, and now, the US Department of Interior's Bureau of Safety and Environmental Enforcement's Safety and Environmental Management Systems (SEMS). Even though the PSM standard has been around since 1992 and the industry has been managing change for several decades, we still can get it wrong, sometimes with disastrous results. The mindful or diligent efforts of many working in concert are necessary to ensure that change is identified, analyzed and executed in a quality way. Just when we get it right on paper and get the workforce up-to-speed, we have employee turnover, neglect and sometimes regulatory change. New actors and a constantly changing script make it hard to manage change efficiently and effectively.

Many high-profile companies have addressed this challenge by creating and automating standard processes across their manufacturing or production facilities. MOC and PSSR are two of these major processes that have been successfully automated. Technology has been used to automate each step of the MOC workflow, enabling the collection of change ideas, formalizing the analysis of the ideas, and documenting the approval and execution of the change. Reports and dashboards can provide visibility of the health or current state of your MOC and PSSR processes. The use of software can enable the standardization of best practices. This paper discusses how the chemical process industry has defined the MOC and PSSR best management practices and how they should be automated.

What is a best practice?

"Best Practice" is an overused buzzword, like so many, and they are hard to define. Similar terms include Industry Standard, Best of Breed, Lesson Learned, Core competency, and Customer-centric, Downsizing, Drinking the Kool-Aid, Eating your own dogfood, Granular, Herding cats, Holistic, Low Hanging Fruit, Mindshare, Mission Critical, Management Visibility, Pain point, Return on Investment, Seamless, Touch point, Value-add and Visibility. Fortunately, there is a working definition of a best practice.

A **best practice** is a method, process or technique that consistently shows results superior to those achieved by other means. It is hard to prove that you have developed a standard process--like MOC--that enables your company, facility or organization to achieve repeatable superior performance, but it has been done.

Some other terms synonymous with best practices are good operating practice, good agricultural practices, good manufacturing practice, good laboratory practice, good clinical practice, good distribution practice, and **R**ecognized **A**nd **G**enerally **A**ccepted **G**ood **E**ngineering **P**ractices (REGAGEP). In the chemical process industry, the PSM standard relies on REGAGEP to ensure facilities adopt best practices. REGAGEP is specified in federal statutes from OSHA and EPA, and is also mentioned in the American Chemistry Council (ACC) Responsible Care® Process Safety Code.

How do you identify a Best Practice?

Everyone would like to be known as the best practice company. Even if you did not create the best practice, identifying them and using them can achieve the same results. If you can tick the box next to these five requirements, then you may have a best practice:

1. Gives a company or facility a tangible competitive advantage.
2. Takes advantage of technology.
3. Improves EHS performance, product quality, and also lowers cost.
4. Gives management more visibility, control and influence over outcomes.
5. Must be a working system; i.e. not a theoretical best practice

The benefits of having such a best practice might separate your company from your competition or keep you from costly business disruption, fines and/or law suits. Many of the best practices have common themes, including: ensuring consistency; standardization; simplifying processes; zero redundancies; leveraging technology and web-based software solutions; and single source for all data (one source of truth). It may be difficult for you to prove you have adopted/created a best practice. However, it is easy to identify a best practice when a confluence of large and small companies adopt the same technique or process and are successful.

In the context of this paper, we are focused on best practice MOC and PSSR automation. Now that we know what a best practice is, let's define change and discuss how to manage it.

What is Change & Why Do We Manage it?

"Change" means to transform or cause to be different. In the chemical process industry, change is omnipresent and required to stay competitive. William Edwards Deming, a famous American statistician, professor, author, lecturer and consultant, once said, "It is not necessary to change. Survival is not mandatory." In order to survive, a company must learn, change and continuously improve to survive. Successful change requires you to learn from your experience and the experience of your peers and make changes that will improve your competitiveness.

Change is a normal part of business, but uncontrolled or poorly managed change has resulted in some of the worst disasters in the chemical process, oil and gas and refining businesses. According to Ian Sutton, a noted PSM expert, *"The root cause of all accidents is uncontrolled change. Leaving aside sabotage and other malicious acts, all industrial facilities are designed and operated to be safe, clean and profitable - yet incidents continue to occur. In every case, the fundamental cause of the incident is that someone, somewhere lost control of the operation, i.e., they allowed operating conditions to deviate beyond their safe range. Hence, the proper management of change is the foundation of all safety and accident prevention programs; an effective Management of Change (MOC) creates an atmosphere of "no surprises".*

Whether it is Bhopal, Piper Alpha or Deepwater Horizon, time after time the difficulty of managing change confounds some of the most proven processes of the most successful companies in the world. As a consequence of these disasters and hundreds more, we have rules, standards, regulations and laws to help ensure that companies manage change better. Guidance documents on managing change include the following sources:

- *US EPA's Risk Management Programs for Chemical Accident Prevention (40 CFR PART 68) or EPA RMP*
- *US OSHA's Process Safety Management of Highly Hazardous Chemicals standard (29 CFR OSHA 1910.119) or PSM,*
- *US Department of Interior's Bureau or Safety and Environmental Enforcement (BSSE) - 30 CFR Part 250 Subpart S - Safety and Environmental Management Systems (SEMS),*
- *OHSAS 18001, 2007 revision - ANSI/AIHA Z10-2005.*
- *European Union Law Seveso II Directive (see Annex III of EU Council Directive 96/82/EC).*
- *API-691 - Risk Based Machinery Management (RBMM) (in development)*

What kinds of change are there?

All ideas for improvement require change. No one ever suggests that we should do exactly what we are doing today, tomorrow. This fact is lost on a segment of the chemical process industry who do not think of recommendations as changes. Recommendations should be treated as "in-kind" changes because they are. They do not have to follow a formal and rigorous MOC and PSSR process, but they are certainly recommendation for change.

All change can be separated into three categories: corrective actions; preventive actions; and continuous improvements. Corrective actions are meant to fix something that is broken, or at least to improve efficiency and functionality. These ideas for change originate from accident investigations, incidents and negative operating experiences. Preventive actions can also come from incidents, but are more commonly created from near misses, safety observations, audits, hazard reviews, inspections, emergency response critiques, and/or general surveillance or observations. These ideas for change help ensure that we address upstream issues or causes before a negative outcome (incident, fine, loss) results. The final kind of change is continuous improvement. These are generally ideas for increasing production, reducing manufacturing costs or increasing product quality.

No matter what kinds of changes you have, they all should be managed. No matter how good an idea is, it must be analyzed for impact (positive and negative) and executed expertly. Ideas can also be rejected or denied if the risk or cost of implementation exceeds the benefits or value of the change. If the change affects the design and/or operation of a facility, it is likely going to require formal change management or MOC. The diagram below illustrates three kinds of change in the idea universe.

Figure 1.0

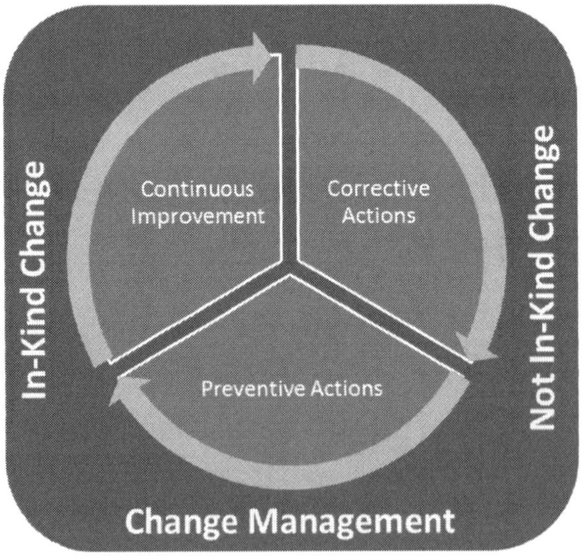

OSHA'S Two Categories of Change

The common way to refer to an idea that requires formal change management as defined by OSHA PSM or EPA RMP is an MOC or Management of Change (Not in Kind). MOCs are managed for changes to equipment, materials, technology, personnel, processes, procedures or anything that deviates from the original design, processes, operating procedures, or maintenance program.

In the chemical process industry, we refer to two categories of change. These are "Replacement In-Kind" and "Replacement Not In-Kind". These odd-sounding categories can have ambiguous meanings, but are explained by Ian Sutton, a noted process management expert, as follows:

"If an equipment item is to be replaced with one that is functionally identical, i.e., if the new item is built to the same specification as the old one, then the change is "in-kind". Otherwise it is "not-in-kind", and the MOC process has to be followed before the change can be implemented."

... The most challenging aspect of managing change is identifying that the proposed modification is in fact a change.

1. *Management of Change is needed if the change is Not-In-Kind.*
2. *A Not-In-Kind change is one where Management of Change is needed.*

In other words, the terms "Management of Change" and "Not-in-Kind" tend to be defined in terms of one another.

The second difficulty to do with the In-Kind/Not-In-Kind decision, noted above, is that all changes are, when analyzed deeply enough, not-in-kind."

What is the role of PSSR in the MOC process?

Pre-Startup Safety Review, or PSSR, is not actually part of the MOC process per se. It is intended to be a redundant check or review to ensure that an MOC has been implemented in a quality and safe manner. The PSM standard states:

The employer shall perform a pre-startup safety review for new facilities and for modified facilities when the modification is significant enough to require a change in the process safety information. Process safety information includes all information about the process. That covers about everything. The pre-startup safety review shall confirm that prior to the introduction of highly hazardous chemicals to a process:

- *Construction and equipment is in accordance with design specifications;*
- *Safety, operating, maintenance, and emergency procedures are in place and are adequate;*

- *For new facilities, a process hazard analysis has been performed and recommendations have been resolved or implemented before startup; and modified facilities meet the requirements contained in management of change, paragraph (l);*
- *Training of each employee involved in operating a process has been completed.*

When is Management of Change (MOC) Required?

OSHA requires the employer to establish and implement written procedures to manage changes (except for "replacements in kind") to process chemicals, technology, equipment, and procedures, as well as changes to facilities that affect a covered process. The MOC procedures shall assure that the following considerations are addressed prior to any change:

- The technical basis for the proposed change;
- Impact of change on safety and health;
- Modifications to operating procedures;
- Necessary time period for the change;
- Authorization requirements for the proposed change;
- Employees involved in operating a process and maintenance and contract employees whose job tasks will be affected by a change in the process shall be informed of, and trained in, the change prior to start-up of the process or affected part of the process;
- Updates to process safety information and operating procedures.

The industry has been practicing process safety for over 20 years. The definition of MOC and PSSR are well understood. Some companies have developed great procedures for addressing both of these elements in an integrated fashion using Internet-based web applications and powerful database engines. These software systems simplify the complex MOC process and enable standardization, ensuring a higher level rigor which is difficult to impose with paper-based execution. While the best practices for MOC and PSSR work practices continue to evolve, the following sections reveal what is currently accepted as best practice.

Best Practice MOC Part 1 – Collecting Change Ideas from the Idea Generators

As established above, continuous improvement is impossible without implementing change ideas that can come from anywhere. The problem is how can organizations empower their workforce to initiate ideas, capture and evaluate them then act on the ones that will improve the business? Can you imagine how successful your business would be if you could leverage all of the grey matter at your disposal to solve production problems?

In order to efficiently and effectively capture these ideas, we need to leverage technology. Using Internet–based software can make it easy for ideas to be collected at the source. The easier we make it for staff members to pass on their ideas, the more of these ideas we will receive with time to act on them. One best practice that has surfaced to address this need is to design an electronic suggestion or idea box. Web-based quick entry screens can be accessed using a browser on any company computer to easily suggest a change or improvement. Once these ideas are entered, they are automatically communicated to idea reviewers (supervisors) for vetting, categorization (in-kind or not in-kind) and prioritization. No matter what happens to these ideas, the reviewer must provide feedback to the idea initiators on what happened to their ideas. If an idea results in tangible savings or competitive advantage, a tangible reward will serve to prime the idea generator for the future.

You can visualize the best practice MOC process using a series of funnels (below). Changes begin as ideas and they can be entered by anyone in the company into a single quick entry screen and collected in the main funnel. The supervisor is automatically notified that a new idea has been entered for their unit or area. Each idea is meticulously reviewed and categorized as a recommendation or a MOC. Using electronic submission, this centralized collection funneling system is infinitely more efficient and effective than word-of-mouth, a paper system and/or multiple systems that collect and manage ideas. One common system for collecting, vetting, prioritizing and categorizing ideas is a best practice.

Figure 2.0

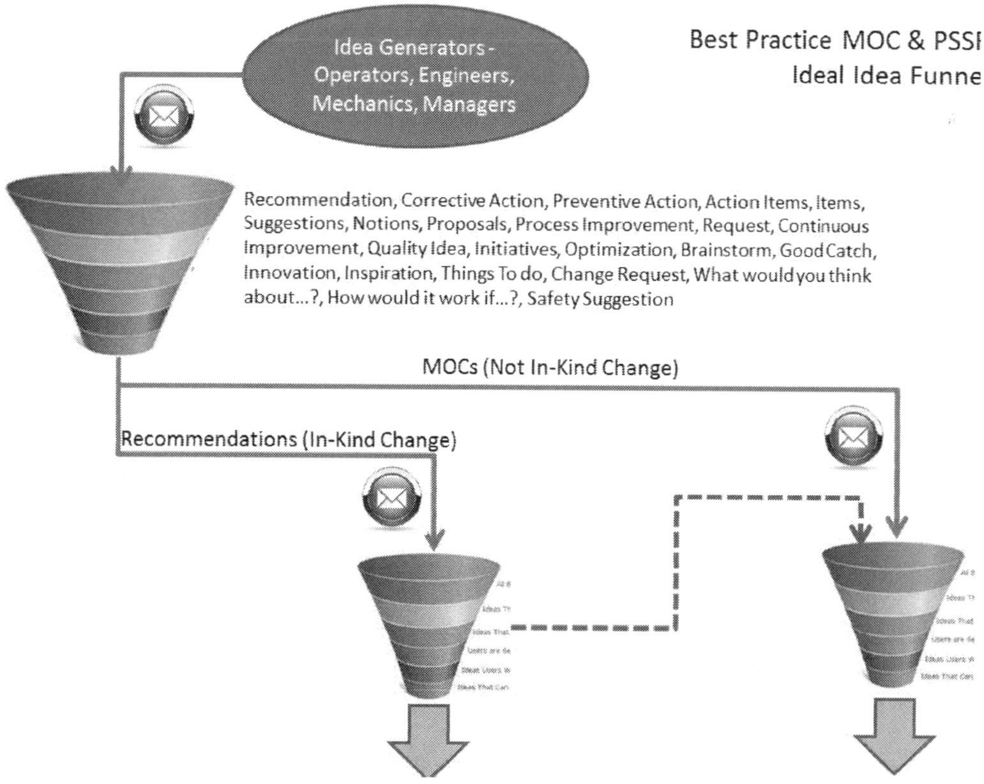

Best Practice MOC Part 2 – Managing the Recommendations (In-Kind Change)

Recommendations can come from audits, incidents, risk assessments, process hazard analysis, emergency response critiques, inspections, reviews, and surveillances. These recommendations are generally ideas for continuous improvement aimed at improving safety, environmental performance, quality or efficiency. Automating and standardizing the recommendation management process makes it repeatable, measurable and more readily understood. Utilizing software that illustrates the correct process and visually progresses users through the workflow is key to the acceptance and the eventual success of the MOC program.

Once the idea has been reviewed, assigned and categorized as a **recommendation**, or in-kind change, a simple workflow can be leveraged to ensure the idea is addressed in an efficient manner. A simple database can be used to document the steps in this process, but a best practice is to use software that illustrates the work process on screen as it is executed. The use of a visual workflow to navigate from step to step, showing progress with check marks or color-coding and enabling steps only when precursors (predecessor steps) are satisfied is a best practice. In addition, email notifications should be used to alert persons assigned to manage, approve and execute actions along the workflow. The more proactive the communication and visual the workflow is, the more intuitive it is to the end-user. The workflow presented below is used for hundreds of thousands of recommendations per year. The entire work process is automated from beginning to end.

Figure 3.0

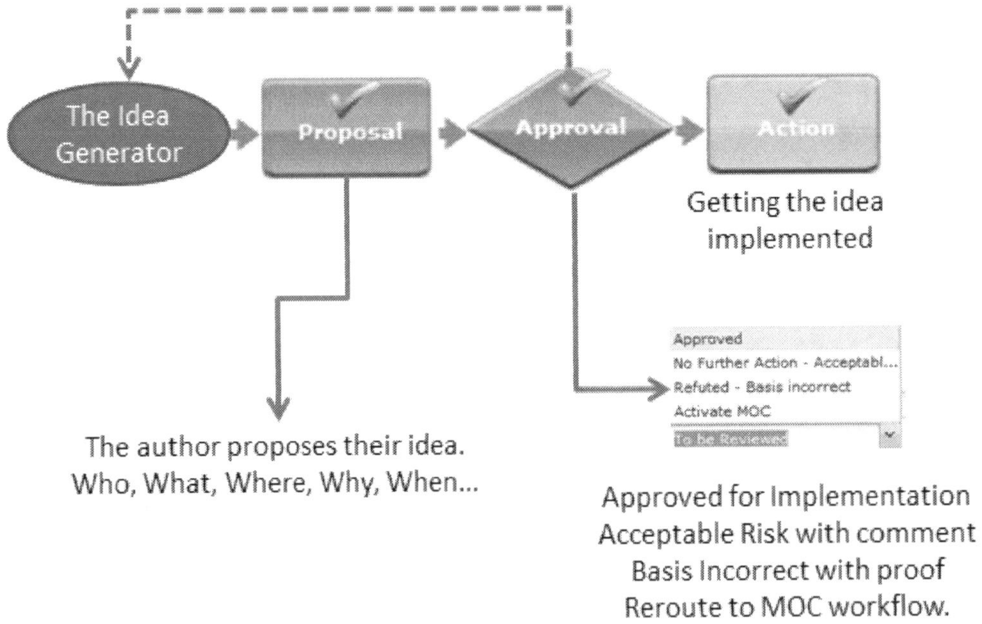

The workflow begins in the proposal step or in the "idea generator's" mind. The recommendation proposal must contain a title, a description (what needs to be done), a justification (why it is important) and a deadline (when it needs to be done) for getting this idea accomplished and a responsible person (RP). The reviewers of the proposal will ensure the accuracy and the resources are available to address the idea. The next step is to accept the idea into the system and automatically notify the RP. The acceptance is illustrated by a check or tick mark in the proposal icon.

After a review of the recommendation, the RP moves to the Approval step. It is in this step that they must decide whether to approve the idea, refute the idea, close the idea as an acceptable risk or decide it needs to be treated as an MOC. Approval of the idea will enable the creation of action items. Refuting a recommendation or addressing it as an acceptable risk should require proof or explanation. These comments must be documented for future reporting and analysis and automatically communicated to the idea generator.

Best Practice MOC Part 3 – Managing MOCs (Not In-Kind Change)

If the idea needs to be managed as a "Not in-Kind" change or MOC, a more complex workflow is required. Similar to the "In-Kind" change, it still requires a proposal, acceptance and assignment to a RP, but additionally will need a formal hazard review and planning. In almost all cases, a MOC also requires the pre-startup safety review (PSSR) step, a Startup Approval step and some post startup steps. Because of its complexity and the need to enforce a rigorous process, a "live" workflow showing what steps are necessary, which ones have been accomplished and which ones are left to do is essential. The workflow below is considered the best practice MOC and PSSR. There are no steps left out and no superfluous steps either.

Figure 4.0

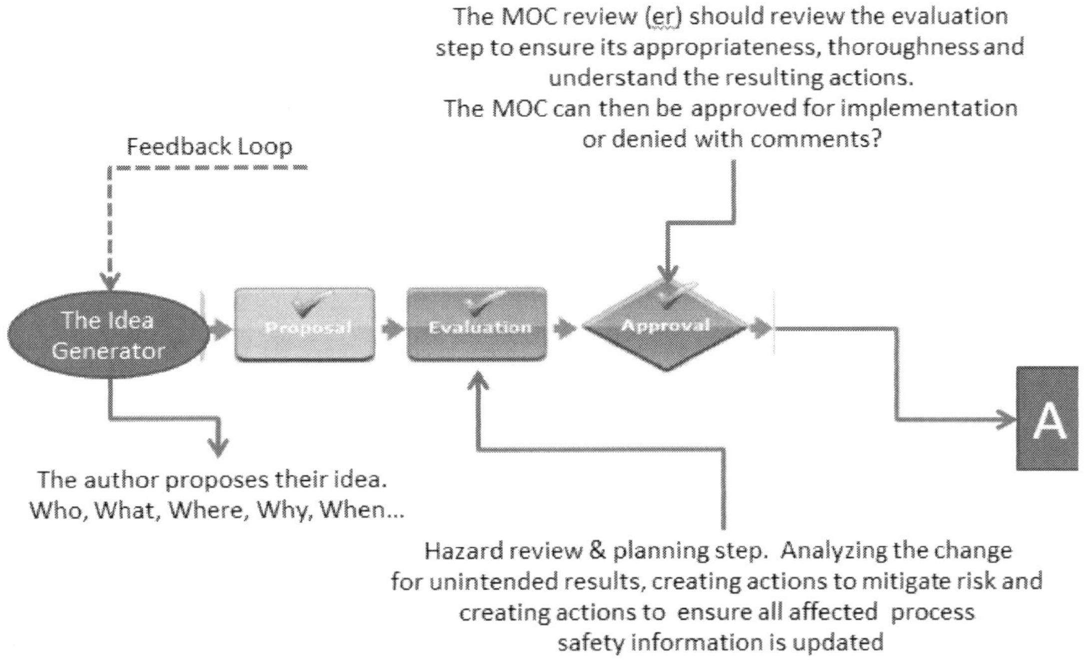

After the acceptance of the idea into the system, the RP receives a notification that they have been assigned a MOC to manage. The next step in the workflow is to "Evaluate" the MOC. This evaluation process is meant to ensure that the hazards of the change are reviewed, understood and mitigated, or at least a plan for mitigation is created. It is also meant to reveal the process safety information that will be affected and the training that will be necessary. Of vital importance is the creation of an action plan to make sure hazards are addressed and that appropriate training is done and all process safety information is completed. Just identifying the need falls very short of hitting the target. In fact, if you identify the need for action and then do not follow through, you have created the proverbial "smoking gun" and a red flag for auditors.

A smart checklist of questions should be presented to the RP in this step which are based on the category of the change. This eliminates the need to answer questions that have nothing to do with the change. Automation gives you the ability to tailor the evaluation process so that it is more effective and efficient to use. Asking evaluation questions unrelated to the change is very inefficient and can also make finding the right questions difficult. Even more automation can be leveraged to pre-assign reviewers to questions that only they should be answering and configuring automatic/predictable actions with associated questions. Here in this step we can make the comments and closure of all questions mandatory for moving to the next step. Until all questions are closed, we cannot move to the next step, the Approval step.

The Approval (to implement) step can be done by a person or groups of people. Their job is to look back at the MOC proposal, review the evaluation that was undertaken and judge whether it was done effectively by the right people. They should also take notice of the mitigating actions that resulted from that review. If more evaluation is required, the reviewer can loop back in the process and create more evaluation questions to satisfy. They should also ensure that actions have been tagged appropriately as pre- or post-startup. If the resulting actions or time have otherwise made this MOC a bad idea to execute, the MOC can be denied at this point with closure comments. Those comments are sent to the idea generator and should be documented in the system. If they are satisfied, then approval can be selected and we can begin implementing the MOC.

The MOC Action Step

Now the real work begins--updating procedures, marking up drawings, making sure all affected personnel are trained, writing work orders, getting an environmental permit, and all of the other actions required for completion to do a quality job. Once all actions (pre-startup) are completed, then you get a check mark over the action icon and we move on to the PSSR step.

Figure 5.0

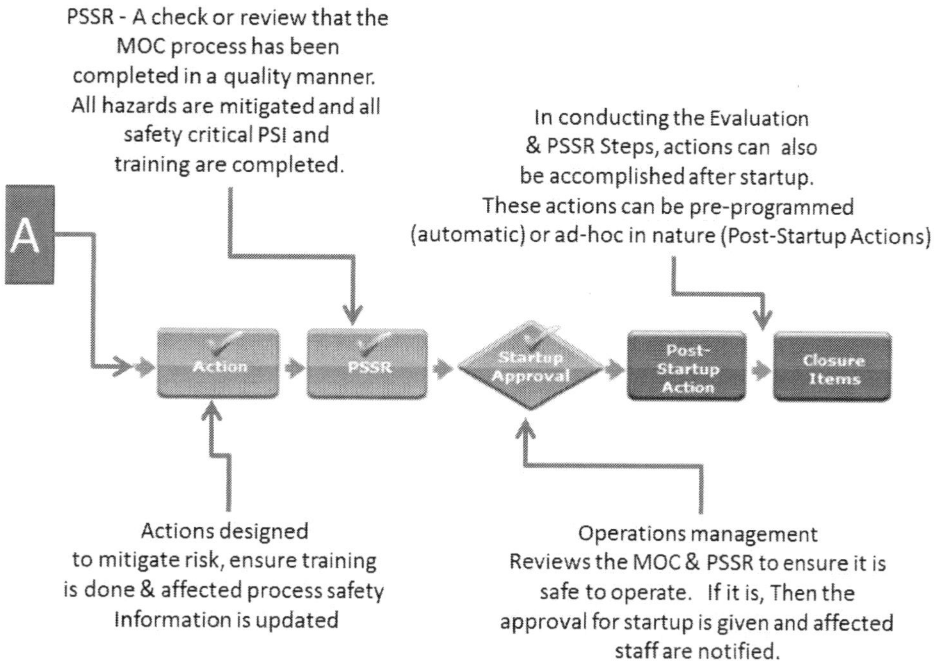

The PSSR Step

As previously stated, the PSSR is a redundant check to ensure the MOC has been properly executed. Like the Evaluation step, once we have chosen a category of change, automation can be used to ensure that the appropriate PSSR checklists/questions are presented to the RP. This increases the efficiency and effectiveness of this step as well. Once the PSSR is completed, we receive our check mark in that workflow step, and an automatic email is sent to the appropriate manager asking for permission to Startup (Startup Approval).

Startup approval is the formal acceptance of the change and the acceptance of the risk associated with it by operations or production. The person who completes this step should review all of the previous work steps in the workflow as well as physically review the work in the field.

The last two steps in the MOC workflow are closure items and post-startup actions. Although these are not referred to in any of the governing regulations, there will be non-safety related activities that are not required to be completed by startup and still need to be managed. This best practice will ensure these actions are accomplished and documented. Past Due email notifications are sent appropriately if the MOC becomes past due.

There are some distinct advantages to adopting the best practice MOC and PSRRs workflows discussed herein. As the best practice evolves, so must the users of the technology. Refresher training, upgrades and constant tuning and tweaking of the workflow are all part of continuous improvement of an MOC and PSSR program. This minimal investment is a small price to pay to avoid poorly managed changes or the opportunity cost of not collecting ideas from your idea generators.

How does Automation help?

Automation of the MOC and PSSR process helps in several ways. First, we can collect ideas at the source of the risk, whether it is the plant or on the platform. This is where the risk is and where continuous improvement is accomplished. Furthermore, it is also where most of the practical/achievable ideas for continuous improvement come from. The best practices must be useful and used where the money is made. Technology makes that possible.

Today's technology is perfectly suited for collecting, analyzing and sharing ideas. Accomplishing these actions without automation is nearly impossible. The days of paper-based MOCs are almost non-existent because automation makes these tasks more efficient and effective as well as less likely to be overlooked.

Reporting metrics and automatic alerts when MOC and PSSR actions are assigned, closed and past due ensure diligence and accountability. Technology can spread the news quickly to those who can create and execute a plan of action and do something with new ideas. Morse code, paper reports and snail mail do not get the job done like email communication and automatic reminders.

A central corrective and preventive action tracking software can help ensure that identified issues are dealt with in a safe and timely manner. Idea or recommendation management and action tracking can be accomplished with one database engine worldwide. That means one system to learn, maintain and continuously improve. It also means you have one global change management system to get things done, communicated and for generating reports. Managers can get the visibility they need to measure performance without bothering the people making the money.

An automated best practice MOC and PSSR system helps ensure that once you get it "right", you keep it "right". It also makes sure that change is appropriately identified, evaluated, approved and executed. It also makes poor change management easy to identify and correct through management surveillance.

Automation is not the goal; best practice is. Like reporting and management visibility and lots of other byproducts of this best practice, the goal is managing change efficiently and effectively.

Words of Caution

A cardinal sin of MOC & PSSR process automation is using a finance or maintenance technology that you already have to automate your best practice, because you already have it. This will sub-optimize a best practice faster than anything you can do, and it will likely defeat the purpose of the automation. Remember that 95% of your users are engineers, operators and mangers--not accountants, IT "geeks" or maintenance planners.

And finally, you can develop your own best practice MOC and PSSR, but to be successful as best practice software, you must have a full-time MOC best practice expert leading the development. Otherwise there will be several months of trial and error. Make no mistake; software development is difficult enough with unlimited resources, expert developers and patient end users. You cannot automate a business process you do not understand. Remember that managing change is a complicated management system for operations or production personnel. You cannot write a requirements document so thoroughly and clearly that a software development team can automate your MOC best practice workflow without constant oversight from the MOC experts.

Unifying Consequence Models for Natural Gas Pipelines in Petrobras Transpetro

Ana Paula Rodrigues Bahiense
PETROBRAS TRANSPORTE - TRANSPETRO
Av Presidente Vargas 328
Centro CEP 20091060
Natural Gas Division
Rio de Janeiro, Brazil
anapaulabahiense@petrobras.com.br

Maria Beatriz Winter Caputo
PETROBRAS TRANSPORTE - TRANSPETRO
Av Presidente Vargas 328
Centro CEP 20091060
Natural Gas Division
Rio de Janeiro, Brazil
Email: beatriz.caputo@petrobras.com.br

Prepared for Presentation at
American Institute of Chemical Engineers
2013 Spring Meeting
9th Global Congress on Process Safety
San Antonio, Texas
April 28 – May 1, 2013

UNPUBLISHED

Unifying Consequence Models for Natural Gas Pipelines in Petrobras Transpetro

Ana Paula Rodrigues Bahiense
PETROBRAS TRANSPORTE - TRANSPETRO
Av Presidente Vargas 328
Centro CEP 20091060
Natural Gas Division
Rio de Janeiro, Brazil
anapaulabahiense@petrobras.com.br

Maria Beatriz Winter Caputo
PETROBRAS TRANSPORTE - TRANSPETRO
Av Presidente Vargas 328
Centro CEP 20091060
Natural Gas Division
Rio de Janeiro, Brazil
Email: beatriz.caputo@petrobras.com.br

Keywords: pipeline, natural gas, consequence model

Abstract

Petrobras Transpetro has been contracting quantitative risk analysis for natural gas pipelines and their associated installations as a legal and internal requirement. These studies have been carried out by a series of consultants that estimate the magnitude of possible damage using different software available in the market. This fact led to a variety of results based on the premises adopted by consultants (most of them depending on the risk analyst) and differences regarding the peculiarities of mathematical models.

Petrobras Transpetro team, mainly field personnel, must use these data to carry out preventive actions inside vulnerable areas, and also to be aware of these ranges in case of a real emergency, thus providing evacuation and escaping actions, as necessary.

Based on this scenario, Petrobras Transpetro's Natural Gas Division had invested in purchasing a consequence-analysis software in order to model the effects of high-consequence loss events as an effort to unify consequence models for natural gas transportation pipelines at Petrobras Transpetro.

As per the above description, this paper aims to present results and benefits of the practical usage of unified consequence modeling information in risk management at Petrobras Transpetro's Natural Gas Division.

1. Introduction

Transpetro, a Petrobras wholly owned subsidiary, is the Brazilian leading logistics company dealing with transportation and storage of oil and byproducts, biofuels and natural gas. It operates 48 terminals (land & water), a fleet of 60 cargo ships, and a network of 15,000 kilometers of oil and gas pipelines interconnecting these facilities.) In addition, Petrobras Transpetro provides services to several distributing companies and to the petrochemical industry.
Although the contribution of natural gas to the country's energy matrix is still comparatively low, it is increasing in accelerated rate. On the other hand, the extent of the gas transmission pipeline network has doubled in the last 4 years. Nowadays, Petrobras Transpetro's Natural Gas Division is responsible for operating and maintaining 7.327km of onshore natural gas pipelines, 12 Compression Stations and 129 City Gates, all of them remotely operated by the National Operation Control Center (CNCO), located in Rio de Janeiro.
In Brazil, State & Federal Environmental Protection Agencies require every installation to carry out a Quantitative Risk Analysis including consequence and vulnerability analysis, frequency estimation and societal and individual risk calculation. These Agencies establish risk tolerability criteria and have, therefore, the responsibility of issuing operational licenses based on the risk results.
Beyond that, ANP (Agência Nacional do Petróleo, Gás Natural e Biocombustíveis), the Federal Agency responsible for regulating the oil & gas industry in Brazil, also requires risk assessments to be implemented for onshore pipelines, which may be either qualitative or quantitative, depending on the installation's complexity.
Based on governmental requirements and on internal procedures, Petrobras Transpetro has carried out Quantitative Risk Analysis (QRA) of natural gas pipelines, conducted by different Consultants, in different periods of time. The first QRA is often carried out when the Company requests an Environmental License, and then it's updated based on a regular basis. The review cycle for these studies depends on internal procedures and the License validity.
The results of consequence and vulnerability analysis, which are part of QRA studies, are the basis for pipelines' Emergency Plans. These results are used basically to assess whether there is population inside the vulnerable area and if it's necessary to evacuate people.
In case of onshore pipelines, that cross several communities and points of interest, such as schools, churches, hospitals and others, the information related to vulnerable areas is vital to adopt more adequate emergency response procedures.

2. Methodology

2.1 Software

Petrobras Transpetro uses PHAST - Process Hazard Analysis Software Tools, by DNV Risk Management Software (DNV RMS), to perform consequence and vulnerability analysis. PHAST contains discharge, dispersion and effects models to assess major hazards, foreseeing complex consequences from possible releases of hazardous materials.
The algorithm for evaluation of jet fire scenarios considers product characteristics (liquid, gas or two-phase), release rate and duration, height of release, orifice diameter and angle of the formed

jet (0° for horizontal jets and 90° for vertical jets). The software estimates flame dimensions and intensity.

For fireball scenarios, the model calculates its radius and duration. The critical parameters to be defined for a fireball model are: product mass involved in the fireball; strength of the thermal radiation source; and the fraction of the total heat that is released as thermal radiation.

The model simulates flash fires due to ignition of a vapor cloud, on an integrated basis, to the algorithm of atmospheric dispersion. The flammable cloud's size is used to model a transient front of flames that travels through the cloud. Thermal radiation in the receptor is calculated as a transient function due to variation of the flames' sizes inside the cloud.

This algorithm considers the speed, width and height of the flame, and concentration of the flammable vapors of the cloud during calculation time. The distance which corresponds to the Lower Flammable Limit (LFL) is assumed as the flame length. Result is calculated in terms of the distance reached by the flash fire.

2.2 Premises

2.2.1 Accidental Scenarios

Failure of a high-pressure natural gas pipeline can lead to various outcomes, some of which can pose a significant threat to people and property in the immediate vicinity of the failure location. For a given pipeline, the type of hazard, and the damage or injury potential associated with the hazard, will depend on the mode of line failure (i. e., leak vs. rupture), nature of the gas discharge (i.e., vertical vs. inclined jet, obstructed vs. unobstructed jet) and the time to ignition (i.e., immediate vs. delayed). The various possible outcomes are summarized in Figure 1. In mathematical simulations, methane was used as a representative for natural gas.

Figure 1: Possible outcomes from failure of a high-pressure natural gas pipeline

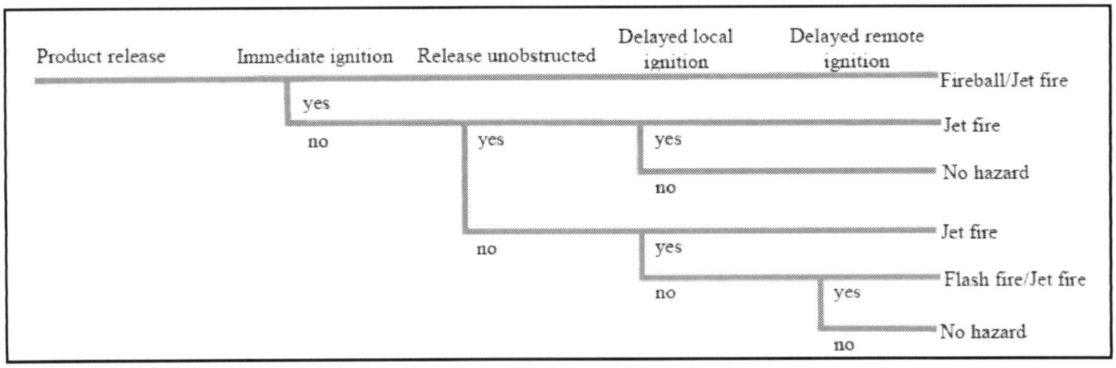

Although there is a great variety of outcomes that can result from failure of a high-pressure natural gas pipeline, the present study aims to analyze thermal effects due to jet fire.

2.2.2 Typical Pipelines

A set of 11 typical natural gas transportation pipelines was established to estimate physical effects due to natural gas possible scenarios. Typical pipelines considered in this study are listed on Table 1.

Table 1. Typical pipelines considered in this study

Pipeline	Pipeline Diameter	Operational Pressure
Pipe 1	8 in	70 bar
Pipe 2	10 in	100 bar
Pipe 3	12 in	100 bar
Pipe 4	14 in	100 bar
Pipe 5	18 in	100 bar
Pipe 6	20 in	120 bar
Pipe 7	22 in	74 bar
Pipe 8	26 in	100 bar
Pipe 9	28 in	100 bar
Pipe 10	38 in	100 bar

2.2.3 Meteorological data

The meteorological data considered in the mathematical simulations are presented in Table 2.

Table 2. Meteorological Data

Parameter	Day Condition	Night Condition
Ambient Temperature	25°C	20°C
Wind Speed	3 m/s	2 m/s
Atmospheric Stability classification*	C	E
Humidity	80%	80%
Soil Temperature	30°C	20°C

* Pasquill-Gifford Stability Categories

2.2.4 Physical Effects Considered

The Eisenberg Vulnerability Model, developed by the United States Coast Guard, was used for calculation of the exposed community vulnerability. This model uses probabilistic functions of the PROBIT type. It is a widely used approach to relate the phenomena intensity (like fire heat radiation) to their consequences.

Probit is a number related to the probability of damages, injuries or fatalities, expressed through the percentage of the population or physical installations affected.

It is relevant to emphasize that, conservatively, the vulnerable area corresponds to the area that could be affected in case of an accident scenario. The area predominantly affected will be influenced by local climatic and geographical conditions at the instant of the leak, such as interference by wind direction and speed, and may vary from person to person and equipment to equipment.

The following levels of interest were considered in this study:

1. Thermal radiation level of $4kW/m^2$ is frequently suggested in specialized literature as the pain limit after exposure of 30 seconds;
2. Thermal radiation level of $12.5kW/m^2$, which represents the limit to cause deaths - 1% fatality probability for an exposure time of 30 seconds;

The effects of thermal radiation arising from eventual jet fires were estimated for a height of one meter, understanding that this is a representative height for evaluating physical effects on people.

2.2.5 CETESB Requirements

CETESB is a State Environmental Protection Agency, from São Paulo State, publisher of the Quantitative Risk Analysis Manual: "Accidental risk from technological origin. Method for decision and reference terms", extensively used in Brazil, which establishes most of the premises adopted in this case study.

Although CETESB is a "local" Agency, it's considered as a benchmark all over the country. So, most of State Environmental Protection Agencies base their rules on CETESB requirements, including IBAMA, the National Environmental Protection Agency.

Premises based on CETESB Manual and used in this study are listed below:

- Simulations were carried out considering only pipeline buried stretches, calculating the effects of angular (45°) jet fires;
- Simulations also considered pipe rupture, considering the leak hole as 100% of pipe diameter.

3. Results

Results are presented in Table 3.

Table 3. Reaches (m) for Jet Fires Scenarios

Pipeline	Pipeline Diameter	Operational Pressure	Jet Fire Reaches (m)			
			Daily (kW/m^2)		Nightly (kW/m^2)	
			12,5	4	12,5	4
Pipe 1	8 in	70 bar	154	245	151	252
Pipe 2	10 in	100 bar	228	360	227	369
Pipe 3	12 in	100 bar	271	425	271	436
Pipe 4	14 in	100 bar	312	490	315	503
Pipe 5	18 in	100 bar	395	616	400	634
Pipe 6	20 in	120 bar	433	675	440	695
Pipe 7	22 in	74 bar	477	743	486	766
Pipe 8	26 in	100 bar	554	860	565	887
Pipe 9	28 in	100 bar	592	919	606	949
Pipe 10	38 in	100 bar	782	1209	803	1251

4. Conclusion

The results presented in item 3 show that the higher operational pressure and hole diameter, the higher would be the reaches of an accidental scenario of jet fire. The reaches shown in Table 3 could not be considered as an absolute value, instead it should be considered as a reference for the decision making process.

All models, including consequence models, present uncertainties. Input parameters that are modeled using the consequence analysis software may, and probably will, be different in an eventual emergency real case. Therefore, the results presented should be used as guidance to the emergency response team, so that adequate response is provided considering possible effects of

a jet fire on the pipe neighborhood, such as community evacuation, roads blocking, among others.
This study may be completed by simulating other possible outcomes, such as flash fires and fireballs, and by calculating thermal radiation levels.
As originally intended, the results present unified consequence modeling for jet fire scenarios involving natural gas pipelines, providing a quick reference to response preparedness and field personnel use in decision making process, in case of eventual emergencies.

5. References

[1] Manual de Orientação para a Elaboração de Estudos de Análise de Riscos, P4.261, CETESB, May 2003.

Additional references not cited:

[2] 8th Report of the European Gas pipeline Incident data Group - EGIG, December 2011.

[3] AMERICAN INSTITUTE OF CHEMICAL ENGINEERS (AICHE), 2000, "Guidelines for Chemical Process Quantitative Risk Analysis", New York, USA, 2nd Edition.

[4] "A Model for Sizing High Consequence Areas Associated with Natural Gas Pipelines", Gas Research Institute (GRI), 2000.

Automation Strategies: Avoiding Accidents During the Process of Draining Water From LPG Spheres

Carlos André Vaz Junior
Departamento de Engenharia Química, Escola de Química, Universidade Federal do Rio de Janeiro
Centro de Tecnologia, Bloco E, Sala E-209, Ilha do Fundão
cavazjunior@eq.ufrj.br

Maurício Bezerra de Souza Junior
Departamento de Engenharia Química, Escola de Química, Universidade Federal do Rio de Janeiro
Centro de Tecnologia, Bloco E, Sala E-209, Ilha do Fundão
mbsj@eq.ufrj.br

Diego da Silva Lione
Escola de Química, Universidade Federal do Rio de Janeiro
Centro de Tecnologia, Bloco E, Sala E-209, Ilha do Fundão
diegolione@ig.com.br

João Marcio Almeida Villaça
Escola de Química, Universidade Federal do Rio de Janeiro
Centro de Tecnologia, Bloco E, Sala E-209, Ilha do Fundão
joaovillaca@hotmail.com

Leonardo Salim Damian Aguiar
Escola de Química, Universidade Federal do Rio de Janeiro
Centro de Tecnologia, Bloco E, Sala E-209, Ilha do Fundão
leonardosdaguiar@gmail.com

Prepared for Presentation at
American Institute of Chemical Engineers
2013 Spring Meeting
9th Global Congress on Process Safety
San Antonio, Texas
April 28 – May 1, 2013

UNPUBLISHED

Automation Strategies: Avoiding Accidents During the Process of Draining Water From LPG Spheres

Carlos André Vaz Junior
Departamento de Engenharia Química, Escola de Química, Universidade Federal do Rio de Janeiro
Centro de Tecnologia, Bloco E, Sala E-209, Ilha do Fundão
cavazjunior@eq.ufrj.br

Maurício Bezerra de Souza Junior
Departamento de Engenharia Química, Escola de Química, UFRJ

Diego da Silva Lione
Escola de Química, UFRJ

João Marcio Almeida Villaça
Escola de Química, UFRJ

Leonardo Salim Damian Aguiar
Escola de Química, UFRJ

Keywords: LPG spheres, automation strategies, process of draining water

Abstract

The present paper has as its main objective to evaluate different strategies of automation, focusing on the reduction of risks on the water drainage procedures in spheres of liquefied petroleum gas (LPG). The condensation phenomenon of the fraction of the water found in gases during the storing process is largely observed in chemical plants. The condensed water will be accumulated in the bottom of the storing sphere and needs to be removed through a drainage procedure. Although being considered an operational routine process, such event was the main cause of at least two severe accidents: in 1966, in France, and in 1972, at the Refinaria Duque de Caxias (Duque de Caxias Refinery), in Rio de Janeiro, Brazil. One of the most critical decisions viewing to avoid that a new accident was to interrupt the drainage before the water could get completely drained. When the drainage continuous after all the water has been removed, the LPG begins to leak. The gas leaking phenomenon promotes the formation of a toxic and inflammable cloud, *par et passu* with a strong reduction of the temperature. In order to analyze the system, mathematical simulations of the drainage were carried out using the UniSim@ Design (Honeywell) simulator. Variations of the different properties (parameters) of the system were studied along the drainage. Understanding the dynamic of the process it is possible to develop automation systems which can preview and anticipate the best moment for the interruption of the water draining process. Traditional control architectures were suggested based on the PID control architecture. The "monitoring processes", that had been proposed by Seborg et al. [2], were tested as well. The monitoring system was considered more efficient, avoiding that natural fluctuations would anticipate and/or retard the action of the controller.

1. Introduction

The present study has as its objective to propose and to evaluate different controlling strategies for water draining systems in LPG (Liquefied Petroleum Gas) spheres. This analysis obtained its stimulus from the study of the accident that occurred in the Refinaria Duque de Caxias/Duque de Caxias Refinery, in Rio de Janeiro, in 1972. Such accident caused the explosion of a LPG sphere, resulting in several fatal victims and a large number of wounded people. The sphere's water draining system consisted, at the time, of a manual opening of the valve localized in the lower part of the reservoir. The employee in charge of such operation should be waiting for the estimated period of time so that practically all the water would have been removed. After this time, the employee should close the valve, interrupting the drainage and avoiding the gas leak. During the whole time of the drainage, the employee should stay in the area, following the process. In the accident of 1972, the employee left the place, not returning for the closing of the valve, causing the LPG leak which formed an inflammable cloud and the freezing of the equipment nearby. The low temperature generated by the liquefied gas leak promoted the freezing of the drainage valve, not allowing its closing. After the accident, the sphere's drainage system received several improvements, such as the inclusion of the blockage double system and safety instruments, even though the drainage procedures continued to be carried out essentially in a manual way.

Proposing to offer a more efficient and safe solutions for the drainage procedures of the water, this study used dynamic simulation. Through the simulator were made several drainage simulations, evaluating the behavior of several parameters of the process. Therefore, the automation strategy was based on the fact that the dynamics of the variables of the process can anticipate under a safe and exact way the ideal moment in which the end of the drainage should happen.

2. The Accident

The description made here is based on statements of the professionals that were working in the location at the time of the accident. The accident occurred in 1972, in the park of the LPG spheres of the Duque de Caxias Refinery, in Rio de Janeiro, Brazil. The storage of the LPG occurs under elevated pressure, allowing the maintenance of the gas in its liquefied state. Nevertheless, the elevated pressure also generates the formation of a condensed water layer in the bottom of the sphere. The valve and pipes localized in the lower part of the sphere are used for the water drainage. To remove the water is fundamental to ensure that the final product will meet the required specifications when in use. At the time of the accident, the space where the drainage pipes sat had a sole manual valve.

The conventional procedure to be carried out in 1972 by the employee on duty was to connect the hose to the drainage pipes. This hose was then positioned in a nearby ditch. The server would return to the place in order to open the valve. The drained water would be routed this way to the ditch. As the amount of water formed during the compression is known, as well as the flow of drained water, it is possible to estimate the necessary time for the water to be removed from the sphere.

According to instructions being followed at the time, it was expected that the employee on duty would stay in the location, following the process. Nevertheless, it is said that on the day of the accident the operator was absent at the exact moment of the end of the drainage, therefore not closing the valve as expected. Due to this fault, not only the water, but the gas as well, was sent to the draining pipes. Besides the formation of an inflammable cloud, the quick decompression of the liquefied gas promoted a strong reduction of the temperature of the line, causing the freezing of the valve. From this moment on, the operator became incapable to stop the leak. The emergency brigade of the company was called, but did not succeed in stopping the gas leak.

It was reported that a dense and large cloud of LPG was formed over the whole park of the spheres of the Refinery. The presence of an ignition source, never identified, detonated the cloud. This detonation, although causing light burning wounds in a few employees, was the main cause of the improvements that followed the accident. The detonation created a jet fire, thus forming a flame that touched the sphere, heating it.

Due to the seriousness of the accident and its duration, the fire was battled not only by the emergency brigade of the refinery, but also with the help of the employees of the companies of the neighborhood. Moreover, the first emergency teams were helped by the next shift of workers. This fact contributed to the increase of the number of victims of the accident.

After almost two hours of fire, the sphere that was leaking exploded in a phenomenon known as BLEVE. The explosion sent fragments over the other spheres, thus propagating the destruction. A sequence of explosions totally destructed the park of spheres of the refinery. Pieces of the reservoirs were found 800m away from their original place.

2.1 Implemented improvements

Similar to the accident reported, a few others occurred in other industrial plants in the same period, as the one in Feyzin, in France, in 1966 [1]. Although the task to remove the water formed in storage spheres of LPG continues to exist until today, several improvements were implemented. The ground of the spheres is now concreted including a slope, avoiding the accumulation of LPG under the sphere in case of a leak (pool fire). Other important changes were implemented in the drainage piping, which now present a larger number of valves, as shown in Figure 1. The opening-and-closing of the valve nearest to the sphere became remote, driven from the control room. The functioning from a distance has as its goal to allow that, even in case of fire, the valve will be controlled. The valve counts on a thermo-sensitive fuse, which promotes its closing in case the external temperature increases. Finally, the length of the drainage piping was also extended. The physical distance between the drainage point and the valves is set so that–even in a leaking scenario– the low temperatures would allow the freezing of the equipment. A larger number of valves in the system also contribute for a safer process. A project of "closed drainage" is being tested in Brazil, nowadays. This system should eliminate the gas release to the atmosphere, in case a leak happened. If a gas leak occurred during the drainage process, the system would send the gas to flare.

Figure 1: Drainage piping with a larger number of valves

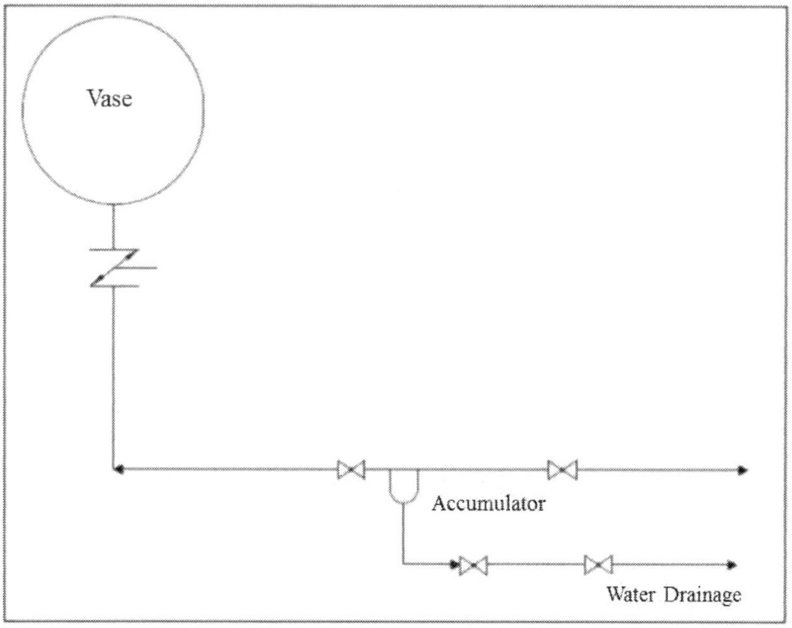

3. Simulator

Through the use of a dynamic simulator, it was possible to carry out several simulations of the drainage system. These results permitted the evaluation of the dynamic behavior of different variables of the process. From the behavior of the process parameters, different control strategies were proposed. The automation would allow the interruption of the drainage process in its ideal moment, thus avoiding the LPG leak.

In this paper we applied the software UniSim® Design. The simulation considers: the storage sphere, the valve directly connected to the bottom of the sphere, and the last blocking valve of the drainage line, as seen in Figure 2. For the simulation it was applied the Peng-Robinson's state equation. The LPG composition was presumed as the mixture formed by 70% of propane and 30% of butane. The amount of the condensed water in the bottom of the sphere, before the beginning of the drainage, was estimated in approximately 5% of the total volume. The diameter of the simulated sphere was 18.25 m. The diameter of the valve nearest to the sphere (VLV-100) was estimated to be 10 in, while the diameter of the most remote valve (VLV-101) was 1 in. The operating pressure of the sphere was presumed to be 8.5 kgf/cm^2 and its initial temperature, 29°C (environmental temperature). In order to simulate the drainage, the outlet flow (flow 5) pressure was adjusted for 1atm (atmospheric pressure). Finally, at the start of the drainage process the initial opening of both valves was presumed to be 100%.

Figure 2: System without PID control

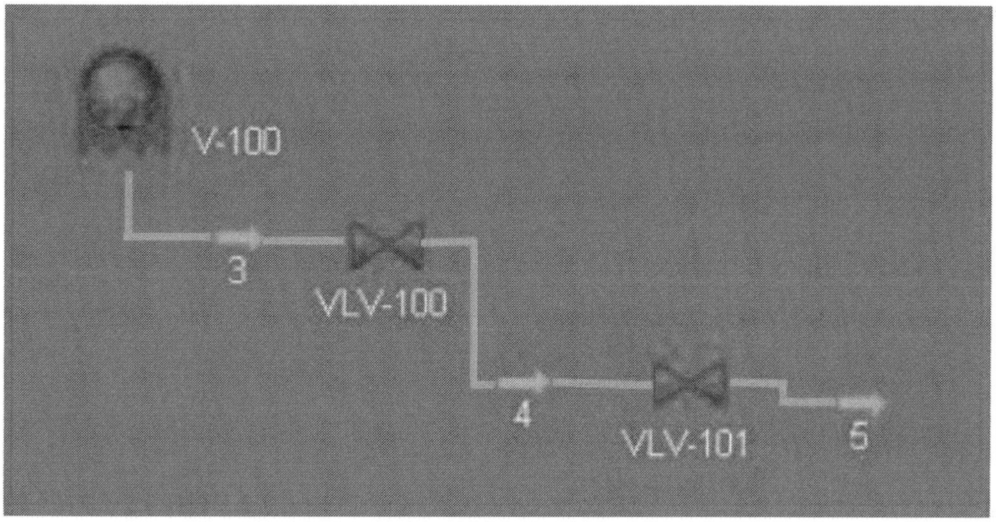

The system represented in Figure 2 does not have the PID controller to interrupt the process. This way, once both valves open, all the condensed water will be removed. After that, the gas release will occur in a situation similar to that observed in the 1972 accident.

After the simulations of the system represented in Figure 2, a PID controller was implemented (Figure 3). The control system sees, as its main goal, to allow that the drainage procedure be carried out safely and efficiently. To achieve this, it is necessary that the controller interrupts automatically the drainage before a leak of the LPG occurs.

Figure 3: System with PID control

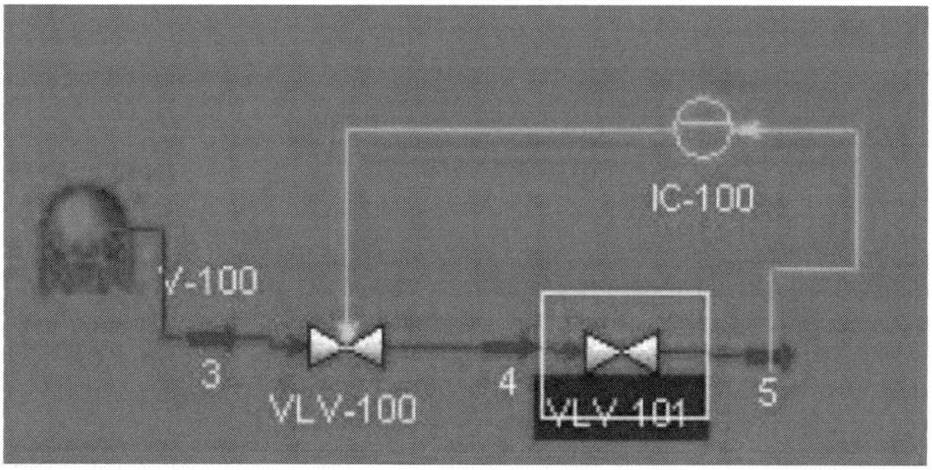

From the simulations study of the flow-diagram without PID controller (Figure 2), the controlled and manipulated variables were defined. The out-flow temperature (flow 5) was defined as being the controlled variable. Such choice was due to the fact that such process parameter presented high sensitivity to the beginning of the LPG leak, varying abruptly. The manipulated variable was defined as being the out-flow drainage of the tank, through the opening control of the valve (VLV-100).

4. Results

4.1 Process simulation without PID controller

First, the simulation of the system was done without the controller (Figure 2). In this simulation, the two valves were maintained opened during all the simulated time (1h40m). This time window was sufficient to simulate not only the water drainage, but the LPG leak as well.

The fractions of the phase water and LPG at flow 3 (closest to the sphere) are illustrated in Figure 4. One may notice that initially there is only a way-out for the water from the spherical tank. This occurs because the water is denser than the LPG, being accumulating so far at the bottom of the sphere. At instant t=1852 seconds, i.e., approximately 30 min. after the beginning of the simulation, the water fraction begins to decrease, initiating the way-out of the liquid LPG from the sphere. From this moment on, the water fraction of flow 3 decreases.

Figure 4: Fraction of the Phase x Time

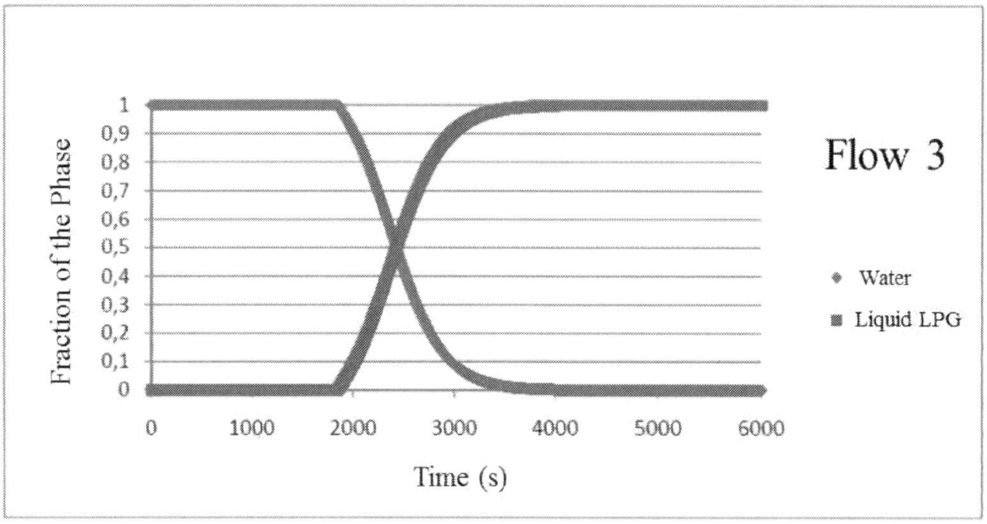

On the other hand, the fractions of flow 5 are represented in Figure 5. In a similar way to flow 3, one may notice that during the first moments of the simulation, the water fraction in the flow equals 1. From instant t=1855 seconds, the water fraction decreases, while the LPG fraction under a vapor state increases. The LPG presence under a vapor state occurs because this flow is being discharged on the environment, with an atmospheric pressure and an initial temperature of 29°C. Under such conditions, the LPG is found in a vapor phase. Therefore, the liquid LPG originated from the tank will vaporize after valve VLV-101.

As observed in the 1972 accident, the vaporization promotes a strong reduction of the line temperature, reaching -30°. C, as we can see in Figure 6. From instant t= 2170 seconds, the flow 5 temperature is already registering -23°C, and the LPG starts to leak in its liquid form. We shall highlight that, in a real scenario, this condition should be sufficient to promote the freezing of the nearest valve.

Figure 5: Fraction of the Phase x Time

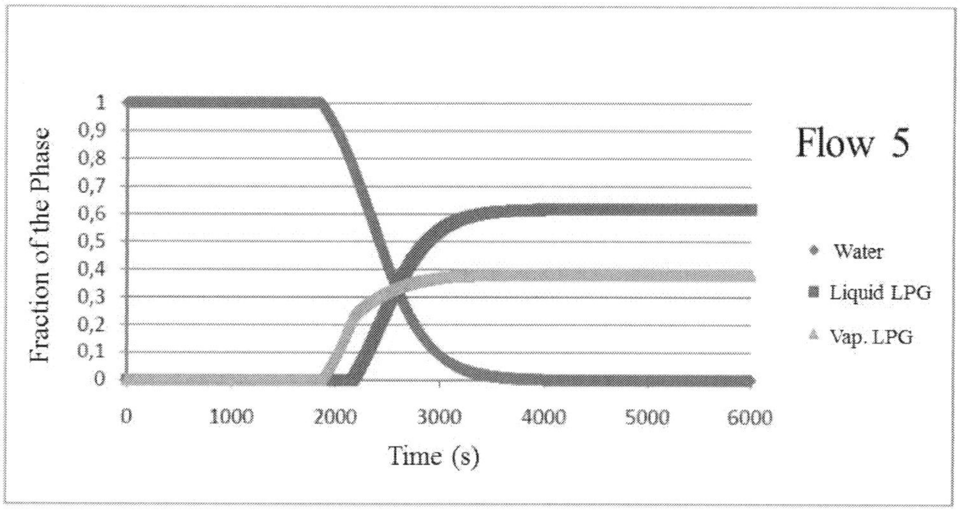

Figure 6: Temperature x Time

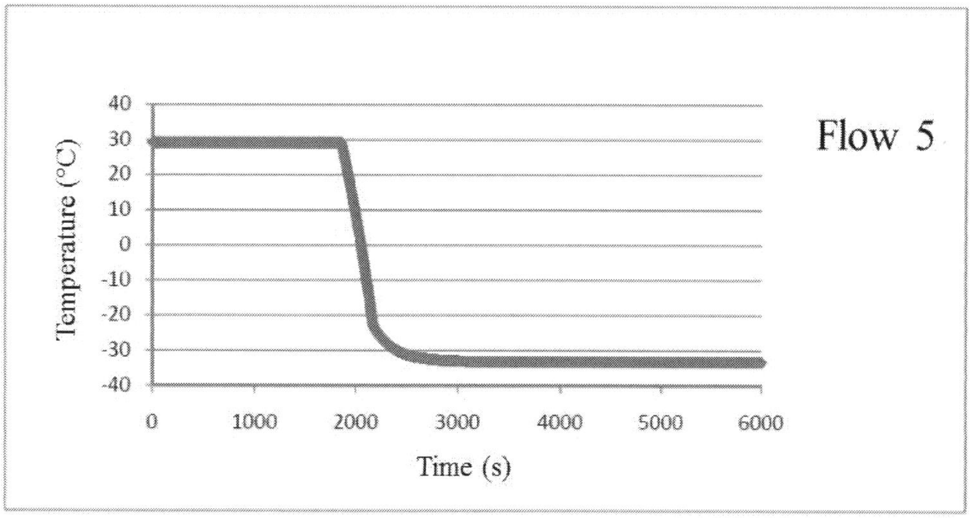

Through Figure 6 we can observe that the out-flow temperature (flow 5) is stable, approximately at the environmental temperature of 29° C, during the first 1855 seconds. In this period of time, only the water of the reservoir leaves it (Figure 5); from then on, the temperature decreases abruptly. At 1915 seconds, i.e., 1 minute after the beginning of the LPG leaking, the temperature was 20°.C, a fall rate of 10 degrees per minute. After 10 minutes, the temperature of the flow reached -30°C. This strong change of the temperature occurs due to the fast change of the phase suffered by the LPG.

From the simulation it was possible to observe that within the variables analyzed, the most sensitive to the beginning of the LPG leaking was the flow 5' temperature. This way, as mentioned, the control system was based on the temperature measures of flow 5.

4.2 Simulation of the process with different tunings of the PID controller

Intending to propose and test different automation strategies for the sphere's water drainage, the use of the PID methodology was initially investigated. Viewing to maintain the controller as simple as possible in a first approach, we opted for the use of a controller known as "P" (proportional). The tuning of the controller had, as its main goal, to maximize the withdrawing of the water from the spheres, thus avoiding the occurrence of a significant LPG leak. The chosen *modus operandi* for the controller was "direct": if the temperature starts to decrease the controller should be activated in order to reduce the flow by closing the valve.

Several simulations were made varying the gain (Kc) of the proportional controller. The set-point was adjusted for 19º.C, 10 degrees lower than the environmental temperature (29°C). In each simulation the following variables were evaluated: opening of the valve, the water fraction at flow 5 and the temperature in flow 5. The results are shown in Figure 7.

Figure 7: Water fraction in the flow 5 x Time

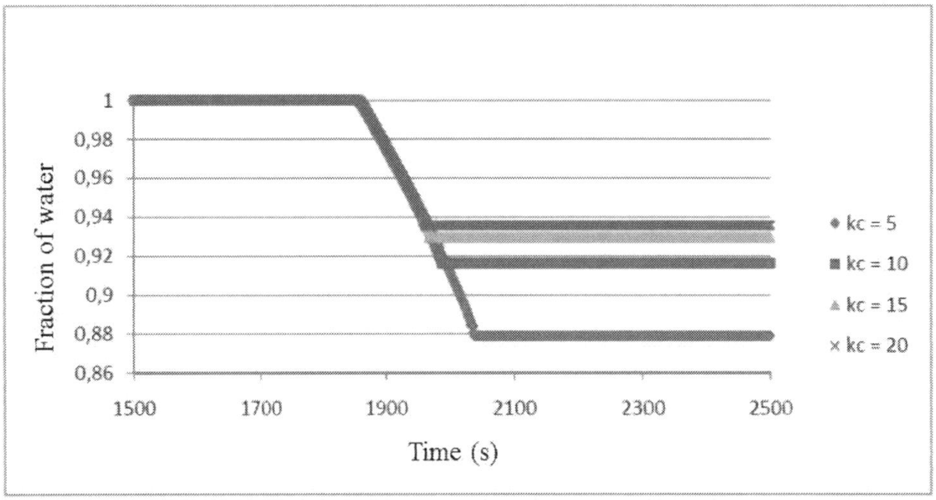

Figure 7 illustrates the behavior of the water fraction in flow 5 as time goes by, for different tunings of the control network. First of all, one may observe that only the water is being drained. From then on, a fall of this fraction occurs, indicating the beginning of the LPG leaking. As the water fraction decreases, the amount of LPG in the flow increases. As a consequence, the temperature decreases severely, according to Figure 8. It is possible to notice that the behavior of the temperature in the outflow follows the same profile of the water fraction in this flow.

Figure 8: Temperature of flow 5 x Time

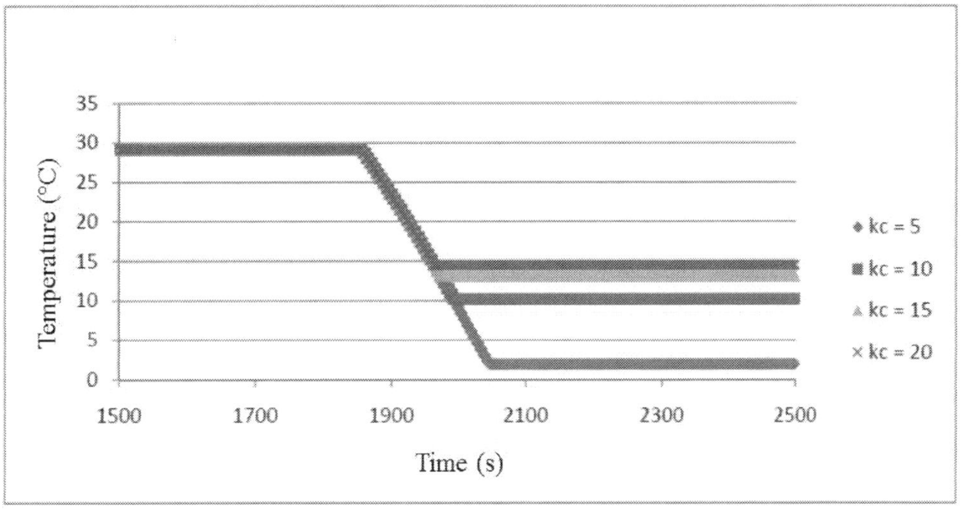

Figure 9 shows the percentage of the valve VLV-100 opening, as a function of time. Initially, while only the water is being removed from the sphere, the valve is kept in a totally open position (100%). As the simulation is carried out, such percentage decreases at the same time that the water content and the temperature decrease. The valve movement proceeds until the valve is totally closed (0% opening). The closing of the valve indicates that the controller is acting expecting the decrease of the leak, since temperature of flow 5 is coming down.

Figure 9: Opening of the valve x Time

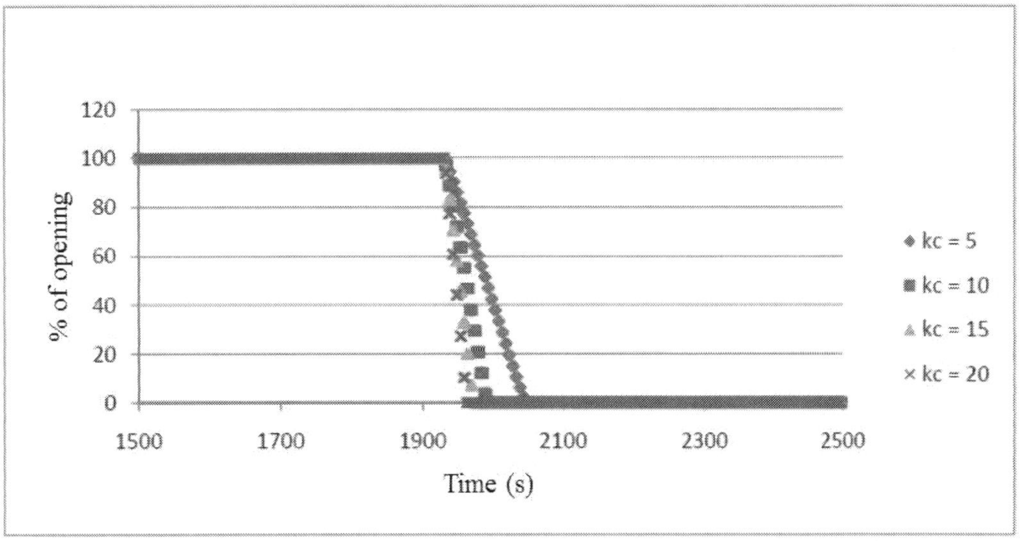

Comparing the behavior of the analyzed variables in relation to the values of the proportional controller gains (Kc), one may observe that the larger the gain, the quicker the valve closes and, therefore, the temperature and the water fraction are higher in flow 5 (Table 1). High gains minimize the amount of the LPG leaked, reducing the risks of frozen valves. Nevertheless, it is worth to highlight the fact that the closing time of the valve is still a function of the valve dynamics, which was not considered in this paper.

Therefore, for a given set-point, the increase of the gain of the controller promotes a quicker closing of the valve and drainage with a smaller LPG leak. It is interesting to notice that for a determined value of gain an increase on the value of the set-point would cause the same results.

Finally, we shall say that very elevated gains of the controller take the valve to act practically as a valve on-off. In this case, there are only two possible positions for the valve: totally closed or totally opened. Other types of control action (integral - I or derivative - D) were not tested, once the results based on the PID methodology took us to choose the on-off action.

Table 1. Dynamic behavior of the parameters of the process

Kc	Final Temperature - Flow 5 (°C)	Final Water Fraction - Flow 5
5	1.95	0.879
10	10.21	0.916
15	13.09	0.929
20	14.49	0.938
100	18.04	0.952
1000	18.74	0.955

4.3 Processes monitoring

In industrial processes, the maintenance of the operational conditions in general uses strategies of feedback or feed forward controller. A different approach was proposed by Seborg et al [2] and denominated "processes monitoring". This approach can be employed with the purpose of following the dynamical behavior of the variables. The monitoring, when associated to tools of supervision and control, guarantees that the variables stay within specific limits and rapidly detect abnormal conditions of operation, implementing corrective actions, etc. This way, the monitoring contributes to the reduction of the variability of the process. In given scenarios, this strategy may present advantages on the use of traditional approaches of control.

In the present study, the variable measured is the temperature of the out flow. The methodology of "process monitoring" is based on the realization of a series of measures of the temperature of the out flow carried out in the beginning of the draining. During the initial phase, the parameter stays practically stable, its average value being used as reference. In a real situation this initial value would correspond approximately to the environmental temperature. As the drainage proceeds and there is a space for a small fraction of LPG, the temperature decreases. The values measured are then compared in real time with the reference value and the differences are then added. When the sum overcomes a specific value, the control action will occur; in this case, the closing of the valve. The strategy of processes monitoring is known as "Cumulative Sum" [2].

It is important to highlight the fact that the temperature of flow 5 – before and at the beginning of the draining process – is subjected to natural fluctuations, such as alterations of the environmental temperature. This normal variation can reduce the efficiency of the traditional approach of control based in a fixed set-point.

In this study, through a simulation the average of the temperature of the out flow was measured during the first 10 minutes of drainage, thus obtaining the average value of 29.17ºC. Through previous simulations it was observed that the decrease of the temperature, caused by the beginning of the leak, was of approximately 0.5º.C This was the maximum value of the cumulative sum adopted to determine the closing of the valve.

Using the cumulative sum, two seconds after the beginning of the LPG leak, the cumulative term became larger than the value stipulated as maximum, this way determining the closing of the valve. The quick closing of the valve, immediately after the beginning of the LPG leak, prevents the freezing of the valves. This approach seems to be much more stable and efficient than the proportional controller.

So, it is important to highlight the fact that in real operations the existence of fluctuations noises (interference) and fluctuations of temperature, or the composition of the LPG yet, could well difficult these parameters. The adoption of higher values as the maximum allowed at cumulative sum could appear adequate.

5. Conclusion

The large accidents of the past reveal the catastrophic potential of the events involving large LPG storage spheres. The draining procedure of the water accumulated in the bottom of the spheres was the main objective of this paper. Focusing on the automation of the procedure, we tried to simulate such process, based on dimensions, parameters and typical industrial conditions. We searched for the identification of a variable that could be indicative of gas leaking, being able to be applied as reference to indicate the best time to interrupt the draining. The temperature of the out flow seemed to be the most adequate parameter.

Different tunings of the proportional controller were tested. It was noticed that the larger the gain value (Kc), the smaller was the amount the LPG liberated at the end of the drainage. This way, the results converged to an action of control of the type on-off. This approach seemed to be strongly influenced by noise (interference) on the variable measured and by the environmental temperature observed.

The use of the strategy of process monitoring was also tested, based on the cumulative sum method. The results obtained were higher than those found for the proportional controller. The drainage was immediately interrupted after the occurrence of a small leak of LPG through the out flow. A "cumulative sum" of only 0.5º C, was adopted as the best. A quick interruption of the drainage process would have reduced the risks of freezing in the line and the emission of LPG to the atmosphere. Therefore, it seems to be an interesting option to increase the safety of the procedure.

6. References

[1] Kletz, Trevor. *Learning from Accidents*. 3rd edition, Gulf Professional Publishing, Woburn, MA, 2001.

[2] Seborg, Dale; Edgar, Thomas; Mellichamp, Duncan, *Process Dynamics and Control*, New York, 2nd edition, Wiley, New York, 2003.

(44aa) Complex BMS Made Simple

Dave Sheppard, Emerson Process Management, Hazelwood, MO and Nikki Bishop, Industry Solutions Group, Emerson Process Management, Roswell, GA

Complex BMS systems with multiple fuels and multiple burners have long presented a struggle for engineers and operators alike. This presentation will cover the methods used to simplify the implementation of a complex BMS in DeltaV SIS. The approach to design and use of standardized sequence modules and graphics greatly increases operator awareness of the operating conditions of the BMS. Included in this presentation will be a description of the design philosophy, the configuration standards, and the approach to operator graphics.

A three step approach is implemented. Break the BMS sequences down in to incremental parts that are easy to understand. Develop standard modules that implement these parts individually. Use a graphical interface the represents these individual parts in an easy to understand way for the operator. This approach is highly scalable to meet various complex BMS implementations.

With this approach it is easy to take a complex BMS and break it down in to small and simple to understand pieces. This allows greater alignment and understanding between the safety engineer, configurator, and operator. Ease of use and troubleshooting for the operator is also greatly increased through a standard set of graphics that show actions needed and first outs for trips.

Through this approach it is possible to take the design of how a complex BMS needs to function and actually implement that design cleanly and simply. An accurate system to capture the first out for a trip or loss of a permissive and also indication of the current state and actions needed to progress forward is needed to increase operator knowledge of what is going on in the BMS at all times.

"Using ESD Valves as Safeguards, Myth or Reality?"

German Luna-Mejias, P. Eng.
ACM Facility Safety
Suite 700 940 6th Ave SW, Calgary, AB, Canada
gluna@acm.ca
guatire2@gmail.com

Prepared for Presentation at
American Institute of Chemical Engineers
2013 Spring Meeting
9th Global Congress on Process Safety
San Antonio, Texas
April 28 – May 1, 2013

UNPUBLISHED

"Using ESD Valves as Safeguards, Myth or Reality?"

German Luna-Mejias, P. Eng.
ACM Facility Safety
Suite 700 940 6th Ave SW, Calgary, AB, Canada
gluna@acm.ca
guatire2@gmail.com

Keywords: Process, Safety, Emergency, Shutdown, valve, bypass

Abstract

An Emergency ShutDown (ESD) valve is a safety device to be actuated during an emergency situation to prevent hazardous consequences. The ESD valves are to be designed not to create additional hazards when actuated. Sometimes, due to operational constraints ESD valves are bypassed which contradicts best practices and creates an incremental risk for the facility, people and environment. Lack of regulations sometimes drives ESD valves' users to overlook or ignore best practices with the associated incremental risk. This paper presents experiences during HAZOPs/field observations where installed ESD valves are claimed as safeguards but are not (can be bypassed). The intent of this paper is to call the attention of designers, operators and managers of the potential hazardous consequences they may face if ESD valves are not used as per recommended practices.

1. Introduction

Our police force has to be right every time to prevent worst case scenarios, the "bad guys" only have to be right once for a major impact to occur. Exactly the same happens with safety; safety devices have to work when required; only one failure is enough to cause a major impact.

The use, absence or abuse of safety devices –e.g. emergency isolation valves- have been the cause of several accidents in the industry:

- A.P. Green Refractories Company – Explosion at Brick Kiln during start-up due to a main gas isolation valve bypass line leaking with 2 injured. Georgia, USA 1985. [1]
- Oil & Gas, Sonat Exploration Company - Explosion and fire due to misalignment of manual valves when feeding a low pressure system with HP Gas with 4 fatalities. Louisiana, USA. March 1998.[2]

- Ammonia Plant: Fire and explosion with substantial damage to a reformer-furnace due to bypass of control valve -with shutdown function- inadvertently open. No injuries, India, May 1998.[3]
- Esso-Australia Gas Plant Explosion due to the improper use of bypass valve around safety valve as the safety valve was not operable with 2 fatalities and substantial public welfare impact. Victoria, Australia. Sept 1998.[4]

As per these four examples, problems can happen anywhere in the world. In the last 3 years, when performing Process Hazards Analyses (PHA), the PHA teams indicated no concerns about bypasses around ESD valves, because there are no regulations that prevent this. Common responses were: "this is not regulated, why worry about it" or even worse: "recommended practices are recommendations, they are not mandatory". These practices are not only recommended practices but common sense. Additionally, as AIChE members (including professional denominations), the Code of Ethics indicates that welfare of the public and environment is paramount.

2. Definitions

Emergency isolation valves, emergency block valves, Emergency Shutdown valves or Automated Shutdown valve are different names for the same type of function valve.

A paper presented to GCPS in 2012 [5] indicates that the Emergency Block Valves (EBV) are namely different as per different codes, regulations and companies.

2.1 Additional definitions:

Emergency Block valves are used to control a hazardous incident. These are valves for emergency isolation and are designed to stop the uncontrolled release of flammable or toxic materials [6]

The ASV (Automated Shutdown Valve) is most commonly use in the Pipeline Safety Act 2011 [7]. The concept of the ASV is to mitigate the release of flammable or toxic materials, not the prevention of it. Pipeline Safety concern is H&SE not Process Safety.

2.1 ESD valve:

A valve with the following characteristics:

- Actuated valve
- Actuate upon a detection of a dangerous event defined by designers/operators and confirmed during PHA study.
- Provide protection against possible harm to people, assets or environment.
- Not used for control

An Emergency Shutdown valve –for the scope of this paper- is a valve that will prevent a hazardous situation from occurring (i.e. closing to prevent the overpressure of the vessel downstream of the valve) or mitigating the impact after the situation occurred.

2.2 Bypass valve:

Definition of bypass:

- go around something
- channel carrying a fluid around a part and back to the main stream

In the oil and gas industry is sometimes common to find bypasses around control valves; this situation is commonly used when it is expected that the valve may need maintenance during the operation of the plant. The plant can continue to run for a short period of time with a manual controlled function using a globe valve. This is a common accepted practice.

In this paper, the word "bypass" is a key element of the study, and follows the second definition: "channel carrying a fluid around a part and back to the main stream"; more specific: "around the ESD valve".

2.3 Norms and Standards

Depending on the system under review, there are several norms, recommended practices, regulations, etc. associated with the operation and installation of ESD valves. Nevertheless, none of these documents indicates or dictates providing a bypass around an ESD valve. A list of these documents is provided in the appendix A.

3. Installation and Use of ESD valves

In Viviano's work [5], the use of the "5W" (What, Who, When, Where and Why) helps to identify the need for emergency isolation valves. In this paper, it is considered that all 5W are being taken in consideration and the evaluation is moving onto the next questions: Is it correctly installed? It is a valid layer of protection? Does it protect against a hazardous event? Does it mitigate the impact?

Is my safeguard device reliable? In the O&G business a spurious trip is also a problem, but does not affect public welfare. Once the ESD valve is included in the detail design, how do we ensure it is going to work in the different scenarios that led to the installation of the ESD valve?

The first phase of the analysis it to perform the risk evaluation or risk assessment.

3.1 Risk Analysis

During a PHA, analysis will be done to ensure the different ESD valves are valid safeguards to prevent the hazardous event from happening. In this phase of the study a qualitative analysis is performed and team members will agree, or disagree, on the protection being used for prevention. A good facilitator should challenge the team about the effectiveness of the safeguard (e.g. is there enough Process Safety Time, Is the actuation of the ESD valve creating another hazardous condition(s)?).

On the other hand, since the PHA is a qualitative analysis, each safeguard is considered to be of the same "weight" or of the "same importance". If required, an additional safeguards evaluation may be carried out: LOPA.

LOPA (Layers of Protection Analysis) is a semi-quantitative analysis to determine the PFD (probability of failure on-demand) of the safeguards to meet the tolerable frequency requirements of each company.

Based on LOPA an Emergency Shutdown Valve may require being a SIL rated device to achieve the tolerable frequency for a specified event. SIL rated means that the associated elements of the safety device has to be designed, constructed and maintain to a minimum level of "probability of failure when demanded to operate" [8]

Previously it was mentioned that ESD valves should not be used to perform a control function as the reliability or SIL rating of the ESD valve may be compromised. The PFD of a typical BPCS (Basic Process Control System) is 0.1 (1 failure every ten times it is demanded) and the ESD valve may require a higher reliability (e.g. SIL 2).

3.2 Use of ESD for Accident/Incident Prevention

When referring to Emergency Block Valves or Emergency Shutdown valves, two functions (or purposes) should be studied: 1. Use of ESD valves for preventing the hazardous situation from happening (preventive barrier). 2. Mitigating the consequence (acting after the hazardous event happens).

3.2.1 Accident/Incident Prevention

Valves in this category are mainly used in industries (e.g. Oil & Gas) to prevent the hazardous event from happening.

In several scenarios briefly described in the introductory section of this paper, ESD valves were installed to prevent a hazardous incident from occurring as per the original plant design concept.

3.2.2 Accident/Incident Mitigation

Valves in this category are mainly used to prevent public exposure to a hazardous event, after the initiating event occurs. These scenarios are largely described and analyzed in documents such as Pipeline Safety Act 2011 [7].

The concerns of the regulator are focused more on spurious trips that may create a public impact due to the effect of cutting the gas supply to communities. In this case, the regulator may recommend a manually activated Isolation Valve. For this scenario, a LOPA is recommended to ensure that the Isolation valve will have a low PFD (probability of failure on demand).

4. ESD Valve Configuration

Sometimes for "ease the operation" the bypassing of ESD valves are permitted as there are no regulations or norms against it but common sense. These "ease operation" scenarios may remove the protection intent of the ESD valve.

4.1 Creating additional hazards.

Installing an ESD valve to reduce the risk is a good idea unless this addition creates additional hazards. If the ESD valve closes, it will isolate two systems: one at high pressure and other at lower pressure. What happen if the high pressure system cannot withstand the pressure? This analysis should be part of the PHA analysis.

Sometimes, a risk analysis is improperly done just to complete a "check mark" in a paper that means the activity was performed. Therefore a proper evaluation is not accomplished and the operator may believe the facility is safe to operate and it is not.

Unfortunately in many projects the PHA is just a "check mark" to achieve the project approval and the PSM is not strong enough to ensure a proper PHA is accomplished. In an evaluation done by OSHA [9] it has been found that improper or inaccurate hazard assessment is one of the main deviations from the safety programs.

In many PHAs studies it has been found that as no regulations "mandate" the owner to avoid bypassing the ESD valves they are frequently bypassed as it makes the start-up much easier and faster.

If the ESD valve is designed to "Provide protection against possible harm to people, assets or environment" or "actuated upon a detection of a dangerous event" it should not be bypassed. A general rule of Safety indicates that: "the important thing to do is to be safe when nobody is watching"

4.2 Bypassing ESD valve for start-up.

After the PHA and LOPA team agrees to install an ESD valve with specific SIL rating and confirms that no added risk will be created if valve is actuated, what is next?

Even if no regulation "mandates it" common sense indicates that if a bypass is required around an ESD valve the bypass has to be equipped with similar protection or equivalent to prevent the inadvertent operation of bypass or minimize the possibility of leaking.

What are the problems associated with improper bypassing of ESD valves?

- Operator error may remove protection
- Valve leakage may remove protection
- Any of these scenarios will invalidate the PHA previously performed and "due diligence" may be difficult to justify if an accident happens.

Filling the lines with products before start-up is a common practice, occasionally done by opening the bypass around the ESD valve. The filling of lines is not a problem but the increased pressure on the downstream side of the ESD valve may be a problem. That is what happened in the Esso accident (Victoria, Australia) as a valve misalignment created an overpressure of the receiving vessel.

Use of CSC (car seal closed) valves in the bypass around ESD valves is a partial protection as the isolation valve may leak overtime without the operator knowing it. Pressure equalization is a bigger issue than flow.

In the incident that occurred in the Brick Kiln the "bypass valve" around the main isolation valve was "somehow inadvertently open"; it means that to start-up the facility the protection was removed.

5. Conclusions

Potential damages to public and environment can be serious, including fatalities, as it has been shown in several accidents around the world.

In the Pipeline Safety Act 2011 [7], the major concern is the amount of gas release and the impact on public after the incident happens. It does not direct the users to provide preventive safeguards.

As there are no specific regulations or codes, the bypassing of safety devices is a concern that is very likely to be overlooked.

6. Recommendations

6.1 General

- Accident prevention requires ACTIVE leadership by management of safety issues.
- Effective implementation of the right controls to manage, mitigate or eliminate hazards and reduce risk.
- Management systems and attitudes towards safety go hand-in-hand in creating robust defenses.

This paper is written with the intention that the reader perform "due diligence" to make sure the area under his/her responsibility is safe.

6.2 Safe Design

- As a general rule, ESD valves should be installed based on PHA/LOPA
- The actuation of ESD valves must not create additional hazards.
- The required reliability of an ESD valve (PFD) must be evaluated by a LOPA or similar analysis.
- Bypass around ESD valve should only be permitted if an additional ESD valve with the same functionality is installed in the bypass.

6.3 Operating practices

In the accidents reviewed at the beginning of this paper, in most of the cases the inadvertent opening or closing of a valve was a key element to the problem. It is highly recommended to implement a PSSR (Pre Start-up Safety Review) before any start-up.

7. Final Remarks

- Bypass of a safety device may save time for operator, but may cost lives!
- Follow PSM or equivalent because you want to be safe, not because a regulation requires it!
- A general rule of Safety indicates that: "the important thing to do is to be safe when nobody is watching"

8. References

[1] OSHA. Accident Report: Explosion in Brick Kiln. Accident # 14542070, report ID 0418100. Washington, DC. 1985

[2] CSB. Investigation Report: Catastrophic Vessel Overpressurization (4 deaths). Report # 1998-002-I-LA; Chemical Safety Board, Washington, DC. 2000

[3] R.M. Chopde & K.M. Patel ., "Ammonia Plants: Past accidents and lessons learnt", Krishak Bharati Cooperative LTD, India. Presented at International Fertilizer Industry Association (IFA) Technical Conference, Louisiana, 2000.

[4] A. Hopkins, PhD ., "Lessons from Esso's Gas Plant Explosion at Longford"; Australian National University, 2000.

[5] J. Viviano ., "Emergency Block Valves"; presented at VIII Global Congress on Process Safety, 2012.

[6] " API Recommended Practice API RP-553:"Refinery valves and accessories for Control and Safety Instrumented Systems", Chapter 7. American Petroleum Institute, 1998.

[7] "Pipeline Safety Act 2011" U.S. Congress, Law 112-90 – Jan 3, 2012.

[8] CCPS., "Layer of Protection Analysis: Simplified Process Risk Assessment", Center for Chemical Process Safety. New York, NY. 2001.

[9] Jordan Barab., "Learning from Industry Mistakes" presented at NPRA National Safety Conference, May 2010

APPENDIX A

Regulations, codes and recommended practices for the installation of emergency block valves.

- API RP-553 "Refinery valves and accessories for Control and Safety Instrumented Systems" American Petroleum Institute, 2007.
- NFPA 30 "Flammable and Combustible liquid Code", National Fire Protection Association, 2012
- NFPA 497 "Recommended Practice for the Classification of Flammable Liquids, Gases, or Vapors and of Hazardous (Classified) Locations for Electrical Installations in Chemical Process Areas" National Fire Protection Association, 2012
- API RP-500 "Electrical Area Classification" American Petroleum Institute, 2000
- NFPA-58 "Liquefied Petroleum Gas Code" National Fire Protection Association, 2011
- API RP-2510 "Design and Construction of LP-Gas Installations at Marine and Pipeline Terminals, Natural Gas Processing Plants, Refineries, Petrochemical Plants, and Tank Farms", 4th Edition, 1978. American Petroleum Institute; recommended practice.
- NFPA 85 "Boiler and Combustion System Hazards Code", National Fire Protection Association, 2011
- API RP-560 "Fired heaters for General Refinery Services" American Petroleum Institute, 2001
- API RP-556 "Instrumentation, Control and Protective Systems for Gas Fired Heaters", American Petroleum Institute, 2000.
- API RP-520 "Part I: "Sizing, selection, and installation of pressure-relieving devices in refineries". Part II – "Installation" American Petroleum Institute, 2000
- API RP-521 "Guide for pressure-relieving and de-pressuring systems", American Petroleum Institute, 1998
- Pipeline Safety Act 2011: U.S. Congress, Law 112-90 – Jan 3, 2012.

(44ac) Contours When Pipelines Meets Facilities

Karthick Deivasagayam, Loss Prevention, SNC Lavalin Inc, Calgary, AB, Canada

Pipelines involved in oil & gas sector carries risk of fires and explosion along with the actual pipelines. This risk of fire and explosion are vulnerable in the places it could affect and become even more vulnerable in places of proximity of facilities where the risk could turn out into initiating of a major catastrophic events.

This should be addressed in the way of doing things.

QRA done for the most of the projects are specific to their scope alone. Boundary limits of the QRA Study is the same as that as the contractual scope of work. Whereas the risk and hazard boundaries are not contractual but as long as it could affect.

This paper will discuss about few pipeline accidents their consequences and the probable effect if they occurred in a different location (close to the facilities) and from discussing about the potential of what a pipeline accident can do and the focus will shift to the way these risks are approached the way what we do.

This paper will also discuss the way the risks of pipeline projects are captured and also drawbacks and grey areas of the existing approach. This paper also recommends the proposed approach that could be followed for the future risk assessments.

(44ae) Ozone Flammability and Its Affect On the Lower Flammable Limit of Simple Hydrocarbons

Paul Osterberg, Thomas Johnson and Thomas Frawley, Fauske & Associates, Burr Ridge, IL

As a result of ozone's strong oxidizing capabilities and ease of generation on an as-needed basis, ozone applications in process operations are gaining momentum. Previous work has demonstrated that ozone is flammable at high concentrations with a lower flammable limit of 8.4 vol.% in oxygen. However, there is limited data regarding the effect of ozone on the lower flammable limit of other gases known to be flammable in air. This work investigated the lower flammable limit of methane, ethane, and propane in various concentrations of ozone at atmospheric pressure and temperature. The tests were conducted in a 2-L vessel using a fuse wire ignition source at 1 percent ozone intervals spanning from no ozone to the LFL of ozone in the vessel. It is anticipated that the addition of ozone will decrease the LFL of each flammable gas as the concentration of ozone increases.

9TH GLOBAL CONGRESS ON PROCESS SAFETY

Evaluate Plant-Wide Safety Of Your Interlock System

Principal Author: Mohammed Naved Khan
Ingenero Technologies India Pvt Ltd.
Mumbai, India
nkhan@ingenero.com

Presenter: Swapnil Pathak
Ingenero Technologies India Pvt Ltd.
Mumbai, India
spathak@ingenero.com

Co-author: Jim Brigman
Ingenero Technologies Inc.
Houston, Texas, USA
jbrigman@ingenero.com

Prepared for Presentation at
American Institute of Chemical Engineers
2013 Spring Meeting
9th Global Congress on Process Safety
San Antonio, Texas
April 28 – May 2, 2013

UNPUBLISHED

Evaluate Plant-Wide Safety Of Your Interlock System

Principal Author: Mohammed Naved Khan
Ingenero Technologies India Pvt Ltd.
Mumbai, India
nkhan@ingenero.com

Presenter: Swapnil Pathak
Ingenero Technologies India Pvt Ltd.
Mumbai, India
spathak@ingenero.com

Co-author: Jim Brigman
Ingenero Technologies Inc.
Houston, Texas, USA
jbrigman@ingenero.com

Keywords: IRM, interlock relationship, interlock-process interference, holistic, isolated information, PHA, mitigation

Abstract

Interlocks serve as important safety systems in industrial settings, where they protect equipment from damage and employees from toxic and harmful releases from that equipment arising out of unsafe conditions. The same safety systems can lead to damage of process systems, if not analyzed properly. In some of the cases we encountered, initiation of interlocks in equipment caused severe damage to other equipment (due to change in composition) or initiation of interlock caused damage to the same equipment it was intended to protect (due to incorrect sequencing of interlocks).

The information available with the interlock design documents (like cause and effect diagrams) is limited to the vicinity of that particular interlock. It does not give any information about the relationship of the interlock, under study, with other process equipment or even with other interlocks (except for initiation). The relationship between safety interlocks and process system arises out of the behavior of the process system (e.g. change in composition) to the actions taken by the interlock. This paper proposes how a holistic view for these behavioral responses can be obtained by developing the IRM (Interlock Relationship Matrix). IRM is a single process document (excluding instrumentation and logic details) that provides information on the effects of all interlocks on each element of process system. Developing IRM is crucial in evaluating the interlock-process interference (behavioral responses) and is beneficial in managerial decision making to assess the likely damage to equipment and provide mitigation thereof.

1. Introduction

Interlocks form the most critical and independent layer of safety in industrial settings. They are primarily designed for:

- Equipment - protect the equipment from damage due to unsafe conditions (eg. over speed) and the resulting economic loss arising out of replacement and downtime.
- Personnel - protect the personnel from exposure to harmful and toxic materials that may get released from equipment damage (e.g. Seal leaks from rotating equipment).
- Plant - maintain the integrity of the plant (e.g. flare mitigation methods employed to limit the flaring during global utility failures and avoid exposure of un-burnt hydrocarbons to the general public).

The design intention of the approaches employed for interlock systems are two-fold; first is the isolated view to protect the equipment at hand and second is the global view to maintain integrity of the plant. Although the approach seems to be looking fair, they (even the global interlock system) lack in considering the interlock-process systems relationships in designing interlock systems.

The very definition of an independent protection layer provided by interlock systems (whether a single or global interlock) has an inherent element of relationship between interlock actions and the behavior of process systems i.e. the actions of the interlock system are taken independent of the responses of process system. This relationship may or may not be relevant as far as the functions of that particular interlock is concerned, but in no case it should cause upset in other process units that could jeopardize their integrity and cause irreparable damage. In the following sections, we will look into some of the incidences that lead to irrevocable equipment damage as a direct result of the interlock-process interference and we will suggest an approach as to how a holistic view for these behavioral responses of the process systems to interlock actions can be obtained for effective mitigation. Also the major points of differences between global interlock systems and holistic view of individual interlock systems will be highlighted.

2. Interlock Design Information

The design of interlock systems starts with the development of interlock diagrams or narratives. They provide key information as to the analysis and identification of scenarios that can cause damage to particular equipment (initiating causes) and the actions that are required to be taken for the protection of that equipment. They are of the following three types:

2.1 Narratives

Narratives provide descriptive statements for the various elements of the interlock system. Narratives are a short description stating the primary function of the interlock, events that can lead to unsafe conditions (initiating causes) and finally actions required by the interlock to mitigate these events. They are written in a simple but precise manner without making use of detail identification tags.

2.2 Cause and Effect (C&E) Diagram

C&E diagrams are the next level of detail information about the interlock. C&E diagrams are presented in a tabular format, detailing the initiating causes and the actions taken in the form of individual tag numbers, usually initiating causes are activation of suitable switches due to unsafe conditions and corresponding actions taken (opening or closing a valve). Actions taken are segregated for each initiating cause. They also provide information about time delays. C&E diagrams provide detail accounting of each element of the interlock.

2.3 Logic and Ladder Diagram

The information provided in logic and ladder diagrams is specific to the instrumentation part of interlock system. They provide the circuit diagrams of the relay logic hardware used for the interlock system under consideration.

3. Limitations of Design Information

Interlock design information is limited to the vicinity of the particular interlock under consideration. The actions taken (or required) are bound by the isolated view of the equipment to be protected. They are limited in their ability to evaluate and analyze the effect of actions taken by the interlock systems on other process systems.

The structure of the C&E and ladder-logic diagrams is incompatible with identifying and analyzing such kind of relationships. Interlock narratives can incorporate this information in a piece-meal fashion (similar to the way information is presented in a narrative - one interlock at a time), however the utility of such disintegrated information is questionable due to the following facts –

- Language - The information presented in narratives is written in easy to understand language and hence lacks a structured and objective approach.
- Lack of specific information - They are prepared usually at a stage where system specific detail information is not available and lot of re-work might be required to make them consistent and compatible to receiving detail information.

4. Need for Holistic View

We will discuss two incidences where interlock activation in an equipment led to severe damage in the same or other equipment.

4.1 Incident one – Activation of Interlock in Acetylene Hydrogenation Reactor of Ethylene Unit

Ethylene cracker unit was operating with an acetylene hydrogenation reactor located between fourth and fifth stages of cracked gas compressor after Deethanizer and before the chilling train that produces tail gas (front end configuration). Tail gas, usually, is a methane-rich stream that is used as fuel in cracking furnaces. Fuel gas system for crackers also has backup from imported LPG to handle contingencies.

For this case, interlock was designed for shutting down, boxing up and depressurizing acetylene reactor to purge out reactive hydrocarbons from the catalyst bed and prevent runaway reactions that might cause a meltdown of the reactor. However, shutting down the reactor also meant stopping of tail gas flow to furnaces. Thus LPG from backup system was lined up, which changed the calorific value of fuel to the furnaces. The cracking furnace temperature controller was working without duty correction to account for change in composition and, therefore, was sluggish in taking appropriate corrective actions and closing the valves. This time gap led to increased heat input to the furnaces which resulted in a spike in the tube metal temperatures of the coils and subsequent coil damage. Thus to protect acetylene reactor, interlock actions lead to damage of cracking furnace coils due to inappropriate handling of fuel gas composition change.

In this case, the behavioral response of process system to the actions of interlock was change in fuel gas composition to the cracking furnaces. This information was not apparently or directly visible from interlock design information.

4.2 Incident two - Activation of Total Shutdown interlock of Cracking furnace of Ethylene Unit

Cracking furnaces in Ethylene units usually have two levels of interlock shutdown in case of upset, first is steam standby with isolation of hydrocarbon feed and operating the firebox in hot condition and second is the total shutdown where firebox is cooled.

Usually, first interlock is designed for faster restart after a partial shutdown by sustaining close to operating conditions inside the firebox and maintaining minimum inventory of BFW in the steam drum. The second interlock is designed to maintain integrity of the furnace and cool the firebox, which might be operating close to 1100 deg C, to a safe condition. Both interlocks have different functional requirements and are designed in a way so as to be independent of each other. However, from process point of view, initiation of first interlock also provides the necessary time lag for controlled rate heat release from the firebox, by maintaining minimum firing, before the cooling down of the firebox is initiated by total shutdown interlock.

For the case under consideration, in order to maintain minimum BFW inventory for first shutdown, the trigger was set at significantly low-level point. This resulted in activation of both interlocks with minimum time delay resulting in rapid cooling of firebox. The coke layer deposited inside the coils has very different coefficient of expansion than coil metallurgy leading to unequal contraction and subsequent mechanical failure of coils.

The incident gives the interlock interference or dependence that may be sometimes necessary due to process behavior. Design information of both interlocks, when observed in isolation, was sufficient to perform their designated tasks. However without a holistic view of looking at them, integrity of the equipment was jeopardized.

With the above-discussed instances, it is evident that sole reliance on interlock design information to provide the hindsight necessary to avoid damage to other equipment is insufficient. It requires a holistic view, encompassing all individual interlock systems, to study the interlock-process relationship for each element of the process system to better understand these relationships rather than piece-meal and isolated interlock design information.

One can argue that global interlock systems responsible for maintaining the integrity of the entire plant can take care of these relationships. We think that such a system is practically infeasible. The reasoning starts with the very definition of an interlock which states that it is the actuation of the initiating causes that triggers the interlock responses irrespective of the process. For global interlock systems to take care of interlock-process relationships, all the elements of the process system must be allowed to trigger them (along with the upsets in global utility networks).

Another approach could be to trigger specific interlock systems put in place to protect these relationships. The practicality of such an interlock system is bound to fail because of two reasons. First the investment required in putting up these extra interlock systems or triggers is huge as the permutations and combinations of initiating causes from all process elements is gigantic. Second it can lead to loss of earnings and operational prudence due to the higher cost of production for increased number of instances of unsteady plant operation due to spurious activations.

5. IRM

We propose to develop IRM as a tool to give the necessary holistic view for understanding the interlock-process relationship. IRM is an acronym for Interlock Relationship Matrix. It is the final outcome of the study for evaluation of interlock actions and is a single document that shows the effect of actions of each interlock systems on all elements of the process system.

It is prepared in a tabular format with interlock systems arranged in rows and process system elements arranged in columns. A typical IRM is shown in Table 1 for reference layout.

Table 1. Typical IRM Layout

Elements of Process System	Element 1	Element 2	Element 3	Element 4	...
Interlock Systems					
Interlock 1					
Interlock 2					
Interlock 3					
...					

The effect of each interlock action is documented in the corresponding column for process elements. For example, if the actions of Interlock 2 change the feed composition of Element 3, then "change in feed composition" is typed in cell corresponding to Interlock 2 and Element 3.

IRM should include only those effects of interlock actions that have the potential to cause significant damage to process elements. This will avoid ambiguity and redundancy of information. Populating IRM with all effects of interlock actions reduces the utility of the tool. Existing safeguards should also be considered for evaluating the potential damage to the equipment. If damage is possible after considering existing safeguards, only then IRM should be updated with the relevant effect. For earlier example, it can be analyzed whether change in feed composition can cause damage to the equipment if it is not corrected (by the absence of density or composition meter) for changed composition.

However, reference of existing safeguards, for effects that have potential for damage, can be made in IRM to ensure and document that the analysis of the interlock actions has been conducted.

6. Why IRM

There are generic approaches available for evaluating hazards for process system. They can be studied under two categories: Software Evaluation and PHA.

6.1 Software evaluation techniques

These techniques include Software Systems Hazards Analysis like Fault/Event Tree Analysis, Failure Mode and Effects Analysis, Software Failure Mode, Effects, and Criticality Analysis, etc. They are deemed to analyze and identify whether software system components' operation or failure (functioning or lack of functionality) could result in hazards for process systems. It begins when the components are designed sufficiently, and is updated as their design matures to take care of the potential hazards.

Functional difference between these analyses and IRM is that they are more concerned with the effects of software (interlock) systems on process elements in case the software goes rogue and malfunctions (lack of functionality). Moreover, the analysis of components' operation (functionality) on process systems is unable to understand interlock-process relationship as it is concerned only for segregating and avoiding intermixing interlock (safety actions) and process (control actions) systems.

Whereas, in the case of IRM, it analyzes the effects of interlock actions on process element considering that software (interlock) system has fulfilled its functionality and there is no sharing of functions from either side. It is not intended to find any design fault in software (interlock) systems but to evaluate and understand the probable responses of process system to software (interlock) system actions.

6.2 PHA

PHA techniques include operational and process hazards analysis like HAZOP, LOPA, HAZAN, etc. These techniques use guidewords in combination with process parameters to evaluate deviation scenarios from normal operation and verify the utility of existing safeguards. If the existing safeguards are not sufficient, new safeguards are suggested. Safeguards can include interlocks, process control, alarms, relief valves, operating procedures, etc.

Key difference between PHA and IRM is that PHA considers only the process parameters as the potential causes for deviation and subsequent consequences. To mitigate such consequences, PHA provides the scope for defining actions required by a safety system. However, it does not evaluate the actions of a safety system like an interlock as probable causes of deviation.

IRM fills this gap of analyses of actions of an interlock as potential causes of deviation for the process elements and gives insight to unearth latent harmful events of interlock actions.

7. Developing IRM

IRM can be developed in a structured way such that the aftereffects of interlock actions can be analyzed under the considerations of core parameters critical to the industrial setting. This approach will avoid ambiguity and subjectivity.

Simple IRM layout discussed in Table 1 can be modified to include critical parameters as shown in Table 2.

Table 2. Typical IRM Layout with critical parameters (cp)

Elements of Process System	Element 1	Element 2	Element 3	Element 4	...
Interlock Systems					
Interlock 1	*cp1:*	*cp1:*	*cp1:*	*cp1:*	...
	cp2:	*cp2:*	*cp2:*	*cp2:*	...
	cp3:	*cp3:*	*cp3:*	*cp3:*	...
	...	...	...	...	...
Interlock 2	*cp1:*	*cp1:*	*cp1:*	*cp1:*	...
	cp2:	*cp2:*	*cp2:*	*cp2:*	...
	cp3:	*cp3:*	*cp3:*	*cp3:*	...
	...	...	...	...	...
Interlock 3	*cp1:*	*cp1:*	*cp1:*	*cp1:*	...
	cp2:	*cp2:*	*cp2:*	*cp2:*	...
	cp3:	*cp3:*	*cp3:*	*cp3:*	...
	...	...	...	...	...
...	...	...	...	...	...

For the refining and petrochemical industry, critical parameters are pressure, temperature, flow and composition. Other parameters like chemical reaction, liquid level, etc. can be evaluated from these four parameters. Each action of an interlock system should be evaluated with respect to above four parameters for understanding the relationship with each process element.

We will prepare IRM for the two incidents discussed in Section 4 as case study (Table 3 and 4). These tables will form a subset of the overall IRM for the plant.

Table 3. IRM for interlocks and equipment involved in Incident 1

Elements of Process System	Cold box	Cracking Furnace
Interlock Systems		
Front End Acetylene Reactor shutdown interlock	*Pressure:* no change, cold box is boxed up	*Pressure:* no change, OSBL pressure controller is designed for backup
	Temperature: no effect, temperature will rise gradually due to ambient heating	*Temperature:* no effect
	Flow: no flow to cold box resulting in loss of tail gas and H2	*Flow:* flow of tail gas will be replaced by LPG

Elements of Process System	Cold box	Cracking Furnace
Interlock Systems		
	Composition: no effect on cold box streams composition	*Composition:* methane-rich tail gas will be replaced by propane-butane rich LPG increasingly the volumetric calorific value, no heat value correction available. Likely to increase firing duty due to sluggish temperature control of furnace coils.

Table 4. IRM for interlocks and equipment involved in Incident 2

Elements of Process System	Cracking Furnace
Interlock Systems	
First shutdown interlock of Cracking Furnace	*Pressure:* no change in firebox and furnace coil pressure
	Temperature: firebox temperature is maintained at lower level by minimum firing.
	Flow: less flue gas flow due to low firing
	Composition: NA
Second shutdown interlock of Cracking Furnace	*Pressure:* firebox pressure will rise to ambient due to unavailability of ID fan
	Temperature: firebox temperature will reduce rapidly to ambient due to fuel gas shutdown. Controlled rate heat release is required to avoid rapid cooling by maintaining minimum firing conditions (first shutdown interlock) for some duration or stopping flue gas instantly by closing the damper. However, neither the duration of firebox operation under first shutdown is sufficient, nor the damper will close to allow controlled rate heat release and avoid thermal shock. Likely damage to coils.
	Flow: no flue gas flow
	Composition: NA

8. Integration of IRM with PHA

A structured IRM described in Section 7 provides compatibility to integration with PHA techniques such as HAZOP, LOPA, etc. PHA technique like HAZOP analyzes each and every parametric deviation on process system that is applicable for an industrial setting. For refining and petrochemical industry, the parameters include pressure, temperature, level, flow, composition, chemical reaction, mixing, contaminants, special procedures, etc.

Development of IRM along with PHA study starts by identification of a subset of critical or primary parameters out of the PHA parametric set on which other parameters depend (like pressure, temperature, flow and composition). It then incorporates and analyzes actions of a safety system (interlock) as potential sources of deviation (in addition to the parametric deviations) for each process element (equipment or node) where the deviations of interlock actions are evaluated under the guidance of critical parameter subset.

9. Conclusion

Interlock design information, in the form of narratives, cause and effect diagrams or ladder/logic diagrams, is limited to the vicinity of the interlock system. It is unable to evaluate and analyze the effect of actions taken by the interlock systems on other process systems. This inability might result in irrevocable damage of equipment or potential personnel exposure to toxic materials as evident from the two case studies presented in Section 4. These damages can be avoided by understanding the interlock-process relationship to gauge probable process responses to interlock actions.

The interlock-process relationship is not visible explicitly in evaluation techniques like Software Systems Hazards Analysis (Fault/Event Tree Analysis, Failure Mode and Effects Analysis) and PHA (HAZOP, LOPA, HAZAN). These techniques intend to eliminate operational hazards caused by either malfunctioning of the software (interlock) system or deviations in process parameters. However, they do not take into consideration the probable causes that can occur in process elements after the actions of a software (interlock) system have successfully executed.

IRM (Interlock Relationship Matrix) can overcome the isolated nature of interlock design information by providing a holistic view of all interlock systems and process elements in the plant (in the form of a matrix) and evaluate critical interlock-process relationship under the guidance of critical parameters applicable for the industrial setting. It can also fill the gap in other evaluation techniques by analyzing actions of a safety system (interlock) as potential causes of deviation for the process elements. It provides efficient incorporation of new interlock systems or functions for existing process systems and vice-a-versa by giving a ready platform for analyzing the interlock actions on the process system. IRM is a managerial decision making instrument that provides valuable information to assess the likely damage to equipment and provide mitigation thereof.

Understanding interlock-process relationship is crucial to avoid jeopardizing mechanical integrity of the process system by the very interlock system installed to protect them and IRM makes it happen.

10. References

1. "Guidelines for Hazard Evaluation Procedures", 3rd edition, Center for Chemical Process Safety (CCPS), John Wiley & Sons, 2008.
2. "Lees' Loss Prevention in the Process Industries, Hazard Identification, Assessment and Control", Volume 1, 3rd edition, Sam Mannan, Elsevier Butterworth-Heinemann, 2005.

Additional references not cited:

1. "Instrumentation and Control systems", Princeton Plasma Physics Laboratory, ES&HD 5008 Section 2, Chapter 10, Revision 6, 2005.
2. "Code Walk-Through", 1984, Dunn, Robert, Software Defect Removal, McGraw-Hill, Inc.
3. "Event Trees and their Treatment on PC Computers", Limnious, N. and J.P. Jeannette, Reliability Engineering, Vol. 18, No. 3, 1987.
4. "A Guide to Hazard and Operability Studies", Chemical Industry Safety and Health Council of the Chemical Industries Association, Alembic House, London, UK.
5. "HAZOP and HAZAN", Klutz, T.V., Institution of Chemical Engineers, UK, 1986.
6. "Nuclear Surety Design Certification for Nuclear Weapon System Software and Firmware", Air Force Regulation 122-4, Department of the Air Force, 24 August 1987.
7. "Petri Net Theory and Modeling of Systems", Peterson, J.L., Prentice Hall, 1981.
8. "Procedures for Performing A Failure Mode and Effect Analysis", MIL-STD-1629A, Department of Defense, 24 Nov 1980.
9. "Generic Techniques in Reliability Assessment", Fussel, J., Noordhoff Publishing Co., Leyden, Holland, 1976.

Virtual Reality Applied to Training and Process Safety: A Semi-Industrial Boiler Case Study

Carlos André Vaz Junior
Departamento de Engenharia Química, Escola de Química, Universidade Federal do Rio de Janeiro
Centro de Tecnologia, Bloco E, Sala E-209, Ilha do Fundão
cavazjunior@eq.ufrj.br

Thiago Souza de Oliveira
Escola de Química, Universidade Federal do Rio de Janeiro
Centro de Tecnologia, Bloco E, Sala E-209, Ilha do Fundão
thiagowsousa@gmail.com

Maurício Bezerra de Souza Junior
Departamento de Engenharia Química, Escola de Química, Universidade Federal do Rio de Janeiro
Centro de Tecnologia, Bloco E, Sala E-209, Ilha do Fundão
mbsj@eq.ufrj.br

Prepared for Presentation at
American Institute of Chemical Engineers
2013 Spring Meeting
9th Global Congress on Process Safety
San Antonio, Texas
April 28 – May 1, 2013

UNPUBLISHED

Virtual Reality Applied to Training and Process Safety: A Semi-Industrial Boiler Case Study

Carlos André Vaz Junior
Departamento de Engenharia Química, Escola de Química, Universidade Federal do Rio de Janeiro
Centro de Tecnologia, Bloco E, Sala E-209, Ilha do Fundão
cavazjunior@eq.ufrj.br

Thiago Souza de Oliveira
Escola de Química, Universidade Federal do Rio de Janeiro

Maurício Bezerra de Souza Junior
Departamento de Engenharia Química, Escola de Química, Universidade Federal do Rio de Janeiro

Keywords: virtual reality, process simulation, training, 3D plant model

Abstract

The short-term time given to learn a particular task associated with the demands for higher levels of production to increase profits are potential reasons for the generation of accidents. Seeking to overcome this situation in training a large number of operators in a similar environment in which they will work, virtual applications were developed. A 3D reality, integrating phenomenological simulation developed in MATLAB, and Virtual Reality (VR) using Plant Design Management System© (PDMS), were applied in this work. Process simulators using virtual reality are applied in several areas, like flight simulators and firefighters' training. Additionally, the developed 3D plant model based tool allows different safety tasks, such as the optimum location of fire extinguishers, fire hoses and escape routes, among others. As a case-study, a semi-industrial boiler, located in Universidade Federal do Rio de Janeiro, Rio de Janeiro, Brazil, was chosen. The interaction between the user and 3D model was addressed by developing a process data interpreter, so that the model may been modified on the basis of actions taken by the operator and by the phenomenological simulation. As the developed tool improves the operator's understanding of the process, it may minimize rework, accidents and damage. Additionally, this methodology may be replicated in other industrial plants.

1. Introduction

Combine safety process, higher levels of production and environment cares is a permanent challenge to be overcome. Face this fact becomes much harder in the current days due to the increasing complexity of plant process, much ambitious quality range and the short-term time given to learn a particular task. In particular, the complexity of the chemical industry also affects in process safety.

Modern chemical processes use several electronic controls and layers of protection. However, even with all the complex components and systems, the human error still are the responsible by an expressive number of incidents in the industry [1]. It is understood as essential to the operator the full knowledge of the system in which he works. But with the growing demand for increased productivity, professionals have with this feature becomes increasingly rare.

For this scenario, industries are increasingly opting for the use of Virtual Reality (VR) training, where the VR can be defined as a computer-generated three-dimensional environment, updated in real time, allowing human interaction across multiple input-output devices [2]. Virtual reality is used in several studies that allow a greater understanding of safety and process control in many areas of engineering, such as analyzes of alarms allocation, analysis of electrical area classification, determination of routes of leakage, among others.

This work has as objective the development of a simulation tool of operational dynamics of industrial equipment, engaging in 3D graphical representation through an electronic model. To improve the study and the application, this tool was implemented to a plant containing a semi-industrial boiler, as a case study.

2. Material and Methods

In this work, the integration between process simulation and 3D graphical representation was done to provide the interaction with the user. Two specific tools were used: MATLAB and AVEVA Plant Design Management System (PDMS). The MATLAB was used to develop the phenomenological model of the boiler studied in the plant. The PDMS was used to develop the 3D model of the plant. PDMS also work as a bridge between data generated from process simulator (MATLAB) and the 3D model itself in PDMS. The system was used to train operators and safety projects.

2.1 virtual reality

The PDMS is a software engineering that has the capability of modeling three-dimensional CAD (Computer-Aided Design), allowing multiple concurrent users on a single model. The software can be used in several areas. He has a complete interactivity with the environment detailing dimensional, possessing style similar to that adopted by the software of the Windows environment.

The software has a number of benefits that among the most important, is the ability to extract a list of quantitative component containing its data of materials, weights, and other information. Importantly, such a list, depending on the size of the unit may have more than a thousand pages, which greatly increases the chance of inconsistencies, considered the possibility of creating this document manually. Thus, the software allows an important gain in terms of time, eliminating delays, the automated generation of various documents and reports, plans and two-dimensional views and isometric piping, and other documents. The user can check the information contained in the 3D model and compare them to equivalent 2D designs.

2.2 phenomenological simulation

The boiler operation was simulated in MATLAB software. To provide this, was used the model created by ÅSTRÖM & BELL [3], who presented a dynamic model for a boiler, based on basic principles of conservation of mass and energy. Justifies the adoption of this model here for its simplicity and because it has been validated with real data by the authors [3].

2.3 case study: boiler studied

The usability of the tool developed in this work has been tested and applied in a small boiler, model CV-500-VDM. This boiler is located in a pilot plant, which in turn is in the Utilities Central Laboratory of Chemical Engineering, Escola de Química, Universidade Federal do Rio de Janeiro, Rio de Janeiro, Brazil. In the Figure 1 will be shown a picture taken from the plant. All the related equipment was identified in it.

Figure 1: Studied pilot plant containing the boiler simulated

3. Results and discussion

3.1 process simulation

Using the MATLAB and the developed program was possible integrate "process simulation" and "virtual reality". To verify if the data is trustful was produced some graphics and analyze their behavior through steps given in some input variables. The Figure 2 represents the temperature behavior after some different steps given in heat flow, Q.

Figure 2: Temperature behavior after steps in heat, Q

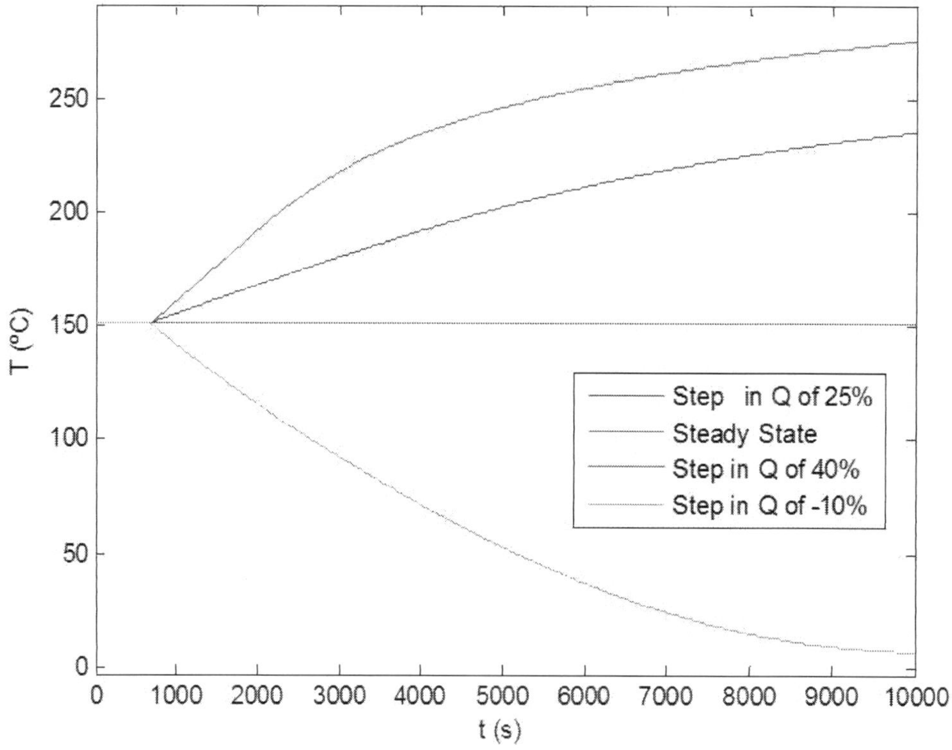

It is possible to see that if the heat is increased, the temperature consequently increase as well. By other side, if the heat is decreased, the temperature also decreases, following the dynamics of the boiler from energy and global mass balance. In the Figure 3, a similar result is obtained, where a step is given in Q and the pressure is observed after it.

Figure 3: Pressure behavior after steps in heat, Q

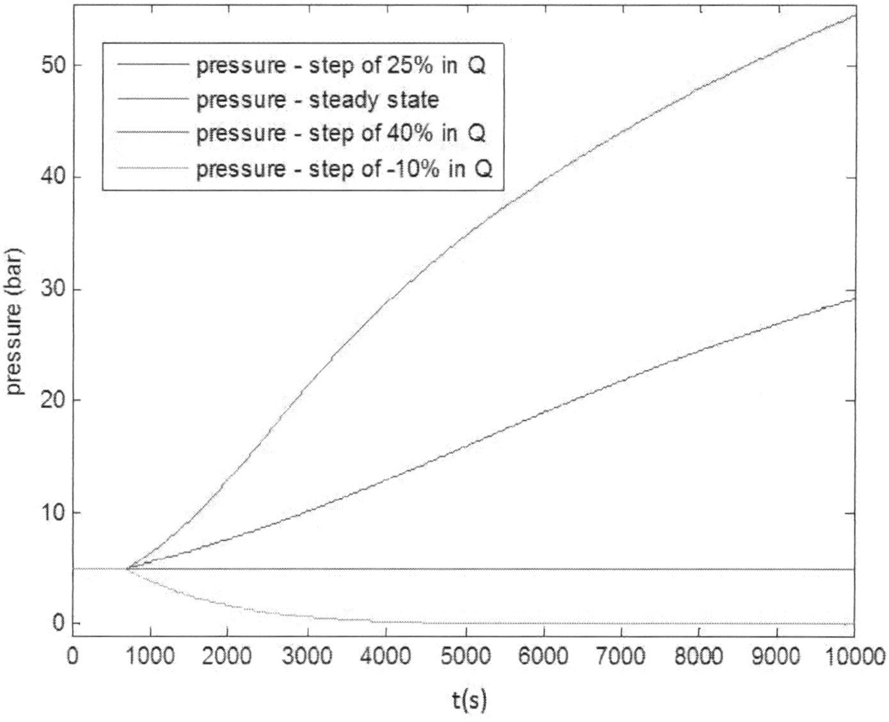

3.2 virtual reality tool

The developed tool has in summary three main functions as results. The first one is provide the integration between the simulations taken in MATLAB. The 3D model shows to the operator if the plant is operating in normal situation or if there is some equipment in alarm status. The model can shows data tables too. In the Figure 4, is possible to see some equipment analyzed by the integration system and showing if they are in normal operation (green), in maintenance (gray), in stand by (yellow) or in alarm (red). Beyond that, it is possible to see relevant data process, like pressure, temperature and level, directly in the model.

The second function is providing the training of the operator. In this work, the training was applied to "boiler starting". During the training the operator can see instructions and complementary texts. The 3D model change through the actions taken, interacting with the operator. The user has the free movement possibility during the training. In the Figure 5 is possible to see the virtual operator during the training. The water level and pump status are shown in the Figure 5 too.

Figure 4: Equipment analyzed by the integration tool

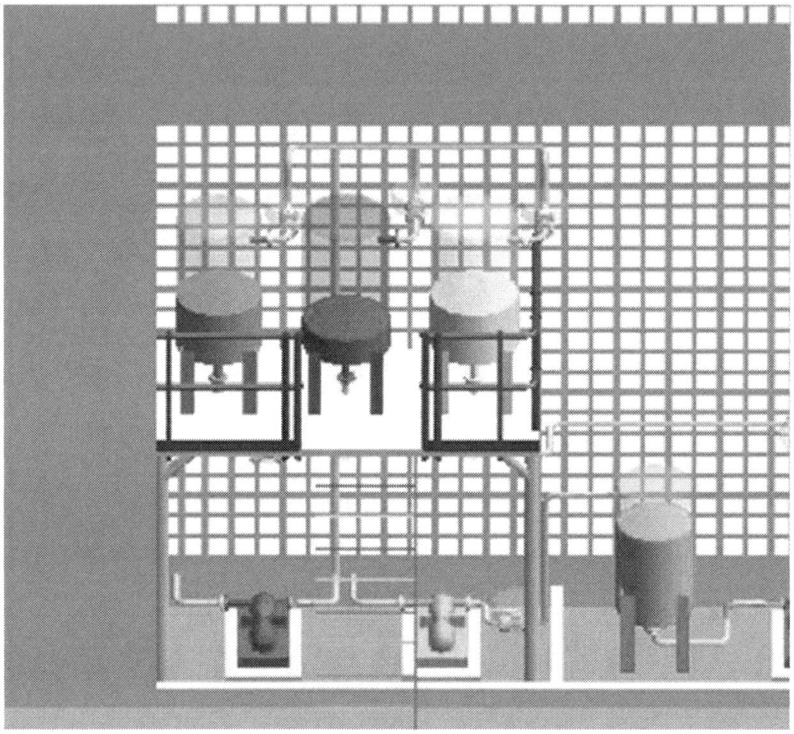

Figure 5: Model during the training

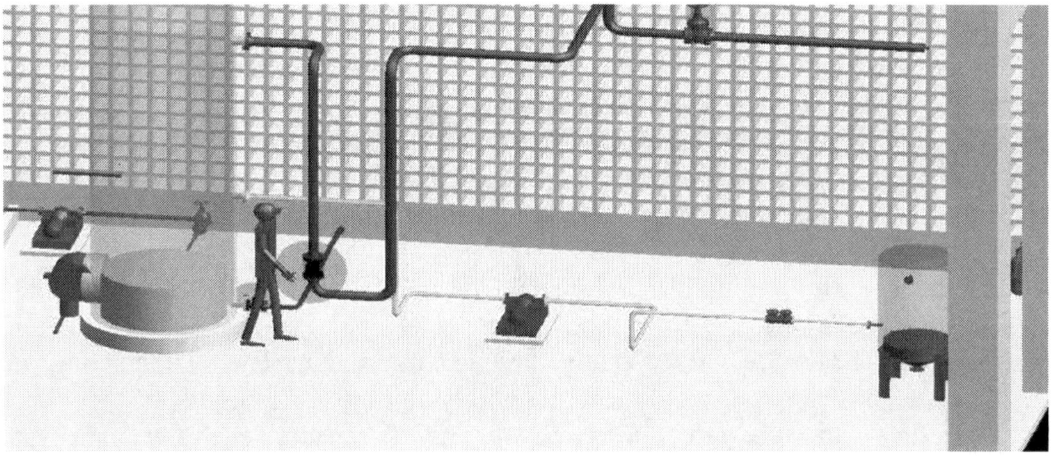

The third function is work as a "support tool", helping the safety analyzes in the plant. The safety analyzes can be applied during project of new process. The virtual reality can be used to design escape routes and emergency signaling. An example of the escape route chooses is showed in the Figure 6.

Figure 6: Virtual tool used to know the best exit to be taken in a case of emergency

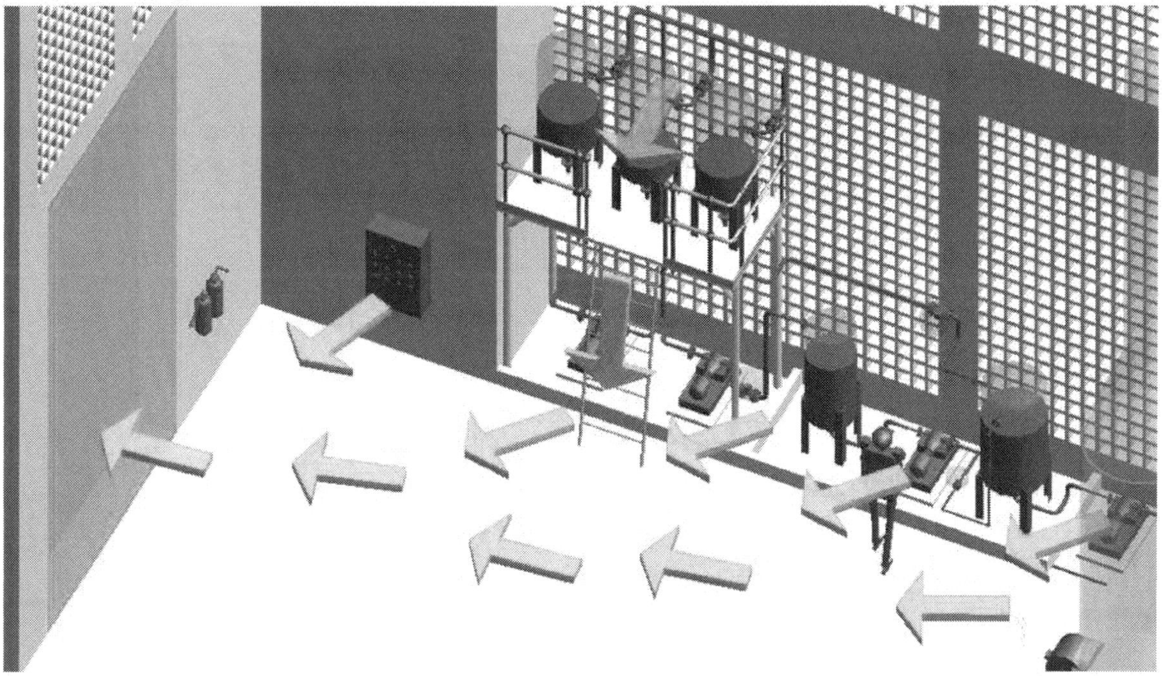

4. Conclusion

The integration of virtual reality and process simulation allow plant operators acquire knowledge about real process. This solution offers low cost and it's safer than use real equipment during operation training. In other words, performing training in the virtual reality tool developed, the user can get information about the plant without actually going to it.

About the development of a three-dimensional supervisory system prototype, it is concluded that it is valid and interesting for training purposes. With the training tool, it is possible to obtain many advantages, due to the operator interaction with the virtual three-dimensional model. So, it is concluded the development of a similar system for all industrial plants can facilitate the training of new operators, facilitating the determination of operations and/or maintenance actions in the plan. Through this, it is possible study the best way to access the site without the need to perform a complete analyzes of the activity at the site to be developed. Furthermore, the safety analyzes can be very useful for the operation team.

About study in general, it can be concluded the use of virtual reality, primarily for training purposes and in addition as a support tool to the operation unit, has much contributing to healthy plant operability. The implementation of virtual reality is becoming an increasingly viable option in view of the development of systems software and hardware capable of supporting the tools developed so that the cost of such equipment is not high as it would at the last decade of previous millennium.

5. References

[1] Lasala, K.P; Filgueiras, L.V.L; Gigley, H.; Ullman, R., *Designing Systems for Reliable Human Performance*, IEEE Tutorial Video, 1997.

[2] Pamplona Jr, A. R. S., *Hidrogeradora Virtual: Utilização de Técnicas de Realidade Virtual Desktop para o Estudo de uma Unidade Hidrelétrica de Energia*, Dissertação de Mestrado, Programa de Pós Graduação em Engenharia Elétrica, Centro Tecnológico, Universidade Federal do Pará, 2006.

[3] Åström, K. J.; Bell, R.D., *Drum Boiler Dynamics*. Automatica, v.36, n.3, p. 363-378, 2000.

Investigations on the influence of different inert gases on the laminar flame propagation in lycopodium/air mixtures

Hannes Kern
Montanuniversität Leoben
Franz-Josef Strasse 18, 8700 Leoben
hannes.kern@unileoben.ac.at

Bruno Wimmer, Harald Raupenstrauch
Montanuniversität Leoben
Franz-Josef Strasse 18, 8700 Leoben
bruno.wimmer@stud.unileoben.ac.at, harald.raupenstrauch@unileoben.ac.at

Prepared for Presentation at
American Institute of Chemical Engineers
2013 Spring Meeting
9th Global Congress on Process Safety
San Antonio, Texas
April 28 – May 1, 2013

UNPUBLISHED

Investigations on the influence of different inert gases on the laminar flame propagation in lycopodium/air mixtures

Hannes Kern
Montanuniversität Leoben
Franz-Josef Strasse 18, 8700 Leoben
hannes.kern@unileoben.ac.at

Bruno Wimmer, Harald Raupenstrauch
Montanuniversität Leoben
Franz-Josef Strasse 18, 8700 Leoben
bruno.wimmer@stud.unileoben.ac.at, harald.raupenstrauch@unileoben.ac.at

Keywords: dust explosion, flame velocity, limiting oxygen concentration

Abstract

The laminar flame velocity is a key parameter in the description of explosions. The determination of the flame velocity (laminar or turbulent) of dust/air flames is relatively complex due to the characteristics of dusts in general. In context of investigations on the limiting oxygen concentration of different powders, the influence of the oxygen concentration and different inert gases on the laminar flame propagation in lycopodium/air mixtures was investigated. In the course of this study, a new apparatus for the determination of the flame velocity was developed. The apparatus mainly consists of a vertical duct with a diameter of 140mm that is split in three main sections. The dust input system lies in the head section of the apparatus and is realized by a small conveyor screw. The bottom section of the apparatus houses the electrical spark ignition system and a gas inlet system. The gas inlet creates a steady flow of a chosen gas. The flame propagation is detected by IR photodiodes along the 2m long main section of the apparatus. The dust concentration is measured by light extinction. The determination of the flame velocity was carried out at dust concentrations of 100g/m³, 200g/m³ and 300g/m³ as well as around the lower explosion level of lycopodium. The oxygen concentration was varied in steps of 1Vol.-% from 21Vol.-% down to the limit where no ignition could be achieved. To investigate the influence of the inert gas on the flame propagation carbon dioxide and nitrogen were used as inert gases. The results of the recent studies are used for further studies on the limiting oxygen concentration of different dusts and the validation of a new CFD model for the simulation of dust explosions in large vessels.

1. Introduction

Dust explosion research mostly relies on sum parameters like the minimun ignition energy (MIE), the maximum explosion pressure or the limiting oxygen concentration (LOC) for the description of explosion processes and material characterization. For mathematical modeling these sum parameters show a lot of uncertainties due to the influence of experimental design. Therefore parameters like the laminar flame velocity are commonly used for the description of the combustion process itself. For combustible gases the determination of the flame velocity is relatively simple compared to combustible dusts. The biggest challenge concerning dusts is the creation of a homogeneous dust cloud with low turbulence.

In the course of this experimental study about the influence of the oxygen concentration on the flame velocity of lycopodium/air flames a new testing device was designed. With this apparatus it is possible to measure the flame speed within a 2m combustion tube, a new method for the calculation of the laminar flame speed from the results is yet to be developed. For current research we relied on present methods. Even if the laminar flame velocity can't yet be measured directly with this device, the influence of changing atmospheric conditions can be observed very well.

2. Experimental investigations

2.1 Method

Various apparatus for the measurement of the flame speed of dust/air mixtures are found in literature [1]. The testing apparatus used for current research consists of three major parts. As shown in Figure 1 the main sections of the apparatus are the top section, the main explosion tube and the bottom section. All elements are built out of a Plexiglas (PMMA) tube with a diameter of 140mm and a wall thickness of 5mm.

The head element houses the dust input system and is equipped with a flame arrestor (Figure 2). The dust input is carried out by a screw conveyor that can be regulated by the operating voltage. The input system allows the generation of dust concentrations from 50g/m³ to 500g/m³ (Lycopodium).

The bottom section contains a gas inlet system and the ignition source. The ignition source is an electric spark igniter (~200mJ), consisting of two 3mm steel electrodes and a high voltage generator. The gas inlet is designed as a circular tube with multiple nozzles to generate a homogeneous gas flow along the main section of the apparatus. The 2000mm long main section of the apparatus is equipped with 5 photodiodes to measure the flame propagation along the tube. The first diode is mounted 10cm above the ignition point, further diodes are placed every 40cm.

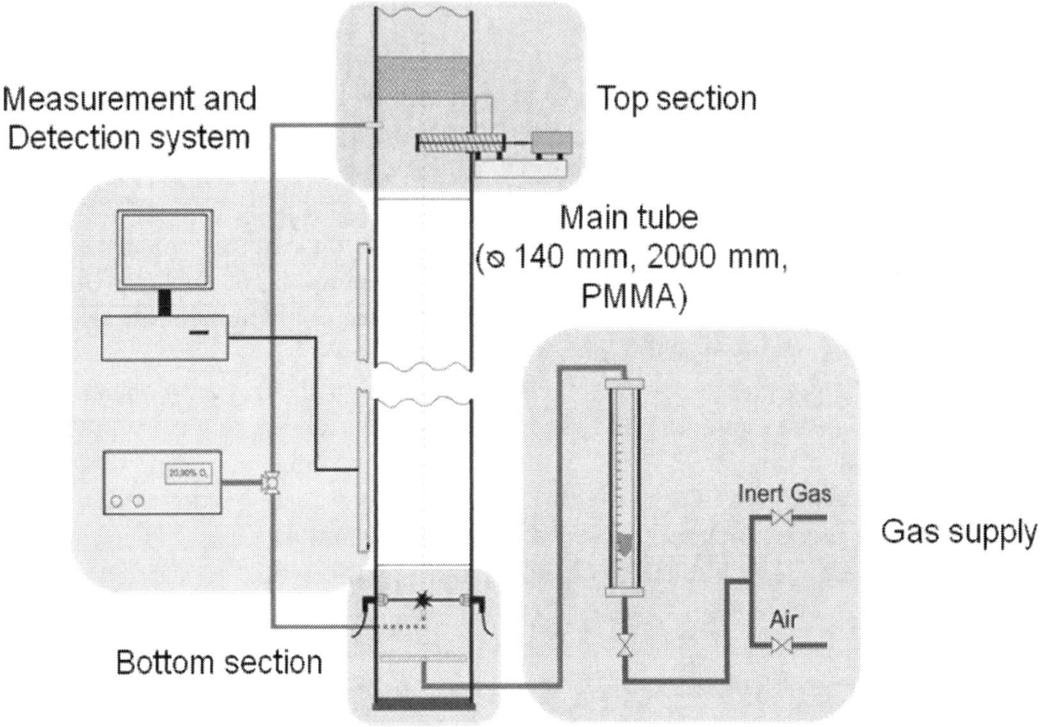

Figure 1: Experimental setup for the determination of the flame velocity

Two of the photodiodes are used to monitor the dust concentration in the tube before ignition by the extinction of IR light emitted by IR diodes. The emitter diodes are turned off shortly before the ignition by the electric spark.

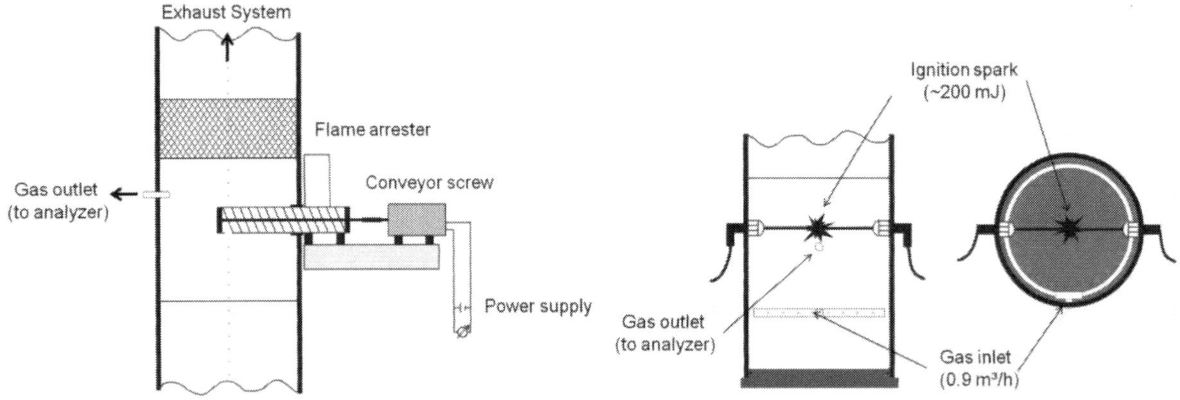

Figure 2: Top section (left) and bottom section (right) of the experimental setup

In the measuring mode a gas flow of 0,015m³/min is provided through the bottom to realize different atmospheric compositions. The dust concentration is observed throughout IR extinction. The oxygen concentration inside the tube is measured by an oxygen monitor at the head section of the apparatus. When the extinction reaches a constant level along the tube, the emitter diodes are switched off and the ignition source is activated.

2.2 *Analysis of flame propagation*

For dust explosion research in general, laminar flame velocity is a relevant parameter. Different methods for the determination of the laminar flame velocity are known, like using burners to create laminar dust flames or tube methods like described above. The data sets produced from all of these methods are more or less deviating from laminar conditions due to the experimental conditions. Figure 3 shows a sequence of pictures taken during the measurement of the flame velocity at a dust concentration of 300g/m³ and an oxygen content of about 15Vol.-% using nitrogen as inert gas. The picture shows a more or less turbulent combustion close to the ignition point and a relatively closed flame shape at around 800mm. At lower concentrations the flame tends to be more closed throughout the whole measuring distance. The flame therefore shows a more laminar character at lower concentrations. The turbulence effects may have occurred due to the falling particles themselves.

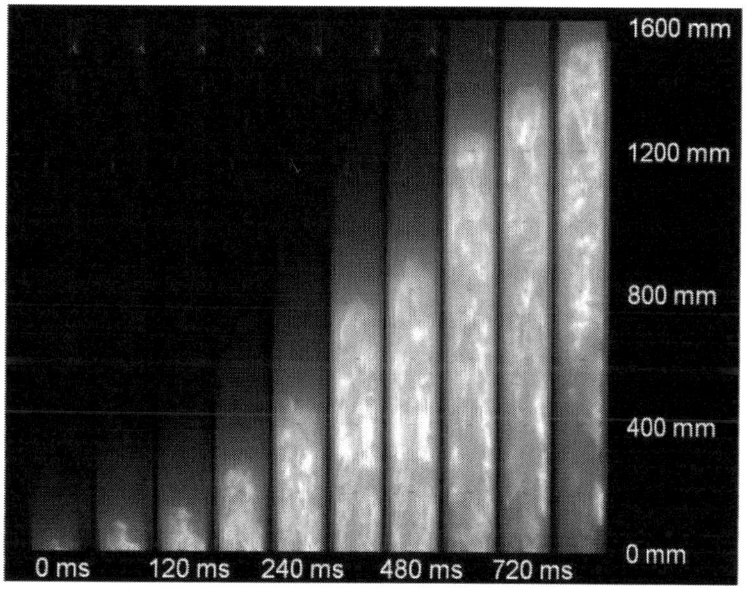

Figure 3: Flame propagation within the testing device

Andrews and Bradley [2] described a method for the calculation of the laminar flame velocity out of relations regarding shape and surface of the flame. Andrews and Bradley's approach (1) was taken into account for further investigations. A video analysis system was installed to determine the flame geometry.

$$S_L = \frac{A'}{A_{fl}} \cdot \left(S_{fl} - u\right) \qquad (1)$$

For a lycopodium concentration of 50g/m³, the flame surface (A_{fl}) and the area cross section (A') were interpolated and calculated by a Matlab code. The measured flame velocity (S_{fl}) was 0.62m/s. Calculations with Andrews and Brandly's relation resulted in a laminar flame velocity (S_L) of around 0.20m/s.

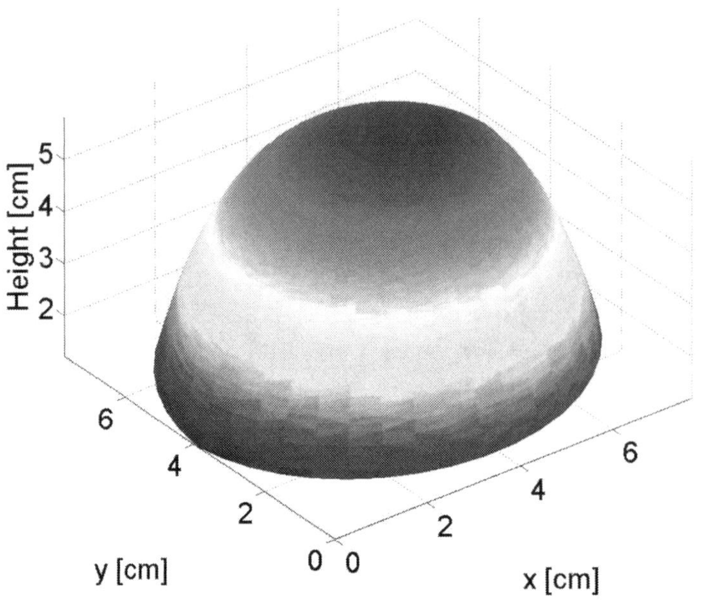

Figure 4: Flame shape at 50g/m³, surface area around 150cm²

Figure 4 shows an example for the flame shape of a 50g/m³ lycopodium/air flame. Compared to the results described by Eckhoff [1], the measured flame velocity is well in the range of velocities measured in other apparatus. For concentrations of around 50g/m³ the flame became very unstable in some measurements, so that higher concentrations were chosen for further research on the influence of the oxygen concentration. Due to the clear connection between actual and laminar burning velocity, only the actual flame speed was measured further on, to avoid deviations from the flame shape analysis.

2.3 Results

Measurements of the flame velocity (S_{fl}) have been carried out at three different dust concentrations. In order to keep the flame stable, concentrations of 100g/m³, 200g/m² and 300g/m³ were investigated. The oxygen concentration was lowered by nitrogen and carbon dioxide as inert gases. Measuring sequences started at ambient conditions (~21Vol.-% oxygen) and were lowered in steps of 1Vol.-% down to the point where no ignition could be reached.

Table 1: Flammability limits

	N_2	CO_2
Dust conc. [g/m³]	cO_2 [Vol.-%]	
100	12.0	15.0
200	12.0	15.0
300	13.0	15.5

As described in Table 1, no ignition could be achieved below 12Vol.-% during nitrogen inerting and 15Vol.-% when carbon dioxide was used. Bartknecht [3] described a relation between the oxygen concentration and the required ignition energy for lycopodium/air mixtures. Regarding the observed ignition limits during the experiments, the results show very good accordance with Bartknecht's values. For an oxygen content of around 12Vol.-% Bartknecht estimated an ignition energy between 150 and 250mJ which is well in the range of the ignition source used for the experiments. Figure 5 shows the development of the flame velocity for nitrogen inerting.

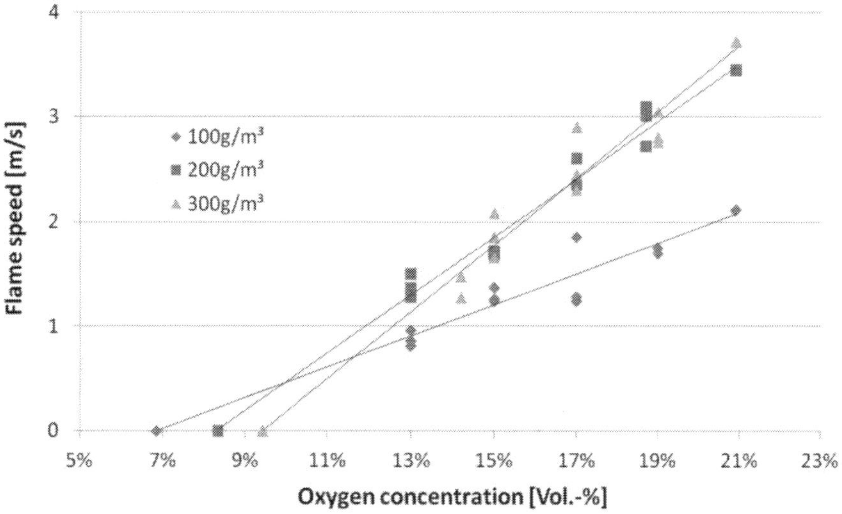

Figure 5: Development of flame speed during nitrogen inerting

At ambient conditions the flame velocity increases with the dust concentration. The increase of the flame velocity in the range from 100g/m³ to 200g/m³ is significantly bigger than from 200g/m³ to 300g/m³. This effect can be explained by stoichiometry [4].

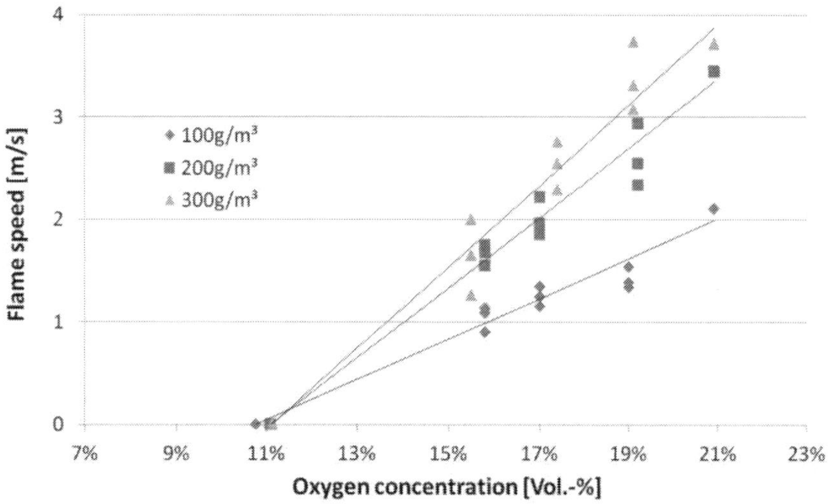

Figure 6: Development of flame speed during carbon dioxide inerting

At lower dust concentrations the combustible material is the limiting component for combustion while at higher dust concentrations oxygen becomes the limiting factor, which is why the flame velocity nearly stagnates. For all dust concentrations the flame velocity decreases with lower oxygen concentrations. The flame velocity drops faster at higher concentrations than at 100g/m³. That is also true for carbon dioxide inerting, as it can be seen in Figure 6. At Figure 5 the lines for 300g/m³ and 200g/m³ intersect at around 17Vol.-%. This is presumably due to problems with the gas measuring system and could not be observed at the carbon dioxide experiments. The values for zero flame speed are estimated.

As mentioned above values for zero flame speed were estimated to be able to compare possible results with the limiting oxygen concentrations of lycopodium. The extrapolated values are given in Table 2.

Table 2: Estimates limiting oxygen concentrations

	N_2	CO_2
Dust conc. [g/m³]	cO_2 [Vol.-%]	
100	6.8	10.8
200	8.4	11.1
300	9.4	11.1
average	8.2	11.0

Siwek [5] found a relation between the limiting oxygen concentrations (LOC) measured during nitrogen inerting and the LOC values for carbon dioxide inerting (2).

$$LOC_{CO2} = 1.36 \cdot LOC_{N2} \quad (2)$$

Due to the described problems with measurement attempts were made to recalculate the LOC for nitrogen inerting out of the measured values for carbon dioxide inerting. The estimation resulted in a value of 8Vol.-% for nitrogen inerting which is highly consistent with the 7.5Vol.-% published in BIA Report 12/97 [6] and the average value that was measured.

3. Conclusions

During the investigations on the flame speed of lycopodium/air mixtures a new apparatus for the determination of the flame velocity was developed. The results from the method regarding laminar flame speed showed good accordance with published values at ambient conditions. With the present experimental setup it was also possible to determine the influence of varying atmospheric conditions on the flame propagation in a vertical duct. Carbon dioxide and nitrogen were used as inert gases. Due to the relatively low ignition energy provided by the spark igniter, the flame propagation around the known limiting oxygen concentrations for lycopodium could not be reached. Extrapolations on the limiting oxygen concentrations based on the measured data have shown good accordance with values measured in the ISO 20L sphere. In addition to the the influence of inert gases on the flame propagation other deviations from atmospheric conditions will be investigated in future research. The results from the current experimental work are used for the validation of a new CFD model for the calculation of dust explosions in complex geometries.

4. References

[1] Eckhoff R. K. (3rd edition) (2003) Dust Explosions in the Process Industries. Gulf Professional Publishing

[2] G.E. Andrews, D. Bradley (1972), Determination of burning velocities: A critical review, Combustion and Flame, Vol. 18/1, Elsevier 1972

[3] Bartknecht W. (1993), Explosionsschutz – Grundlagen und Anwendung. Springer – Verlag

[4] Crowl D. A. (2003), Understanding Explosions, Center for Chemical Process Safety (CCPS) of the American Institute of Chemical Engineers, 1st edition, Wiley-AIChE, New York 2003

[5] Siwek R. (1989), Zuverlässige Bestimmung explosionsschutztechnischer Kenngrößen in der 20l-Laborapparatur, VDI Berichte 701, S215-262

[6] BIA – Report 12/97, Brenn- und Explosionskenngrößen von Stäuben, Hauptverband der gewerblichen Berufsgenossenschaften, November 1997

(44al) Fire Protection and Life Safety Issues in Semiconductor and Related Facilities

Jonathan Eisenberg, Rolf Jensen & Associates, Framingham, MA

Semiconductor and related process facilities contain numerous fire and explosion hazards that impact life safety and operations. This presentation discusses the major fire protection and life safety design aspects for such facilities, including specific requirements contained in the International Building & Fire Codes, as well as applicable NFPA codes and standards. Project examples will be used to illustrate how these aspects are addressed in facility and process design.

The presentation will include the following:

Location and design of Use Group H-5 occupancies – Major IBC Chapter 4, IFC Chapter 18, and NFPA 318 fire protection and life safety requirements, including vertical and horizontal location, occupancy separation, code approach for associated mechanical spaces including interstitial levels, electrical classification, ventilation, fire suppression (overhead and local), fire and gas/vapor detection, and emergency alarms.

Location and design of Use Group H-2, H-3 and H-4 storage areas – Building and fire code requirements for vertical and horizontal location, occupancy separation, electrical classification, ventilation, fire suppression, fire and gas/vapor detection, and spill control/secondary containment. Approaches and ideas on the number, location and function of flammable liquid, flammable and pyrophoric gas, and highly toxic gas storage areas, including piping and delivery philosophy.

Special suppression and detection – Fire protection system approaches for hazards found in a clean room/fab area, eg. wet bench chemistry involving flammable liquids, and processing that uses hydrogen and/or silane mixtures. Discussion of local suppression options, as well as specialty gas and flame detection. We will give our thoughts and experiences on how these systems interact with the rest of the building and the responding fire department.

Industrial Issues – Hydrogen detection; combustible dust hazards; NEC Article 500 electrical classification; explosion protection; and several examples from our industrial/chemical project experience

Licensing and permitting – Discussion of issues faced by owners of fab and clean room facilities that store and use hazardous materials, where the hazards are not clearly understood outside of the semi-conductor industry.

Lower Flammability Limits of Mixtures of Hydrogen and Methane at Atmospheric and Sub-atmospheric Pressures

Hai Le
Mary Kay O'Connor Process Safety Center, Artie McFerrin Department of Chemical Engineering, Texas A&M University
Room 200, Jack E. Brown Building, Texas A&M University, 3122 TAMU College Station, TX 77843-3122
haile@tamu.edu

Yi Liu, M. Sam Mannan
Mary Kay O'Connor Process Safety Center, Texas A&M University
Room 200, Jack E. Brown Building, Texas A&M University, 3122 TAMU
College Station, TX 77843-3122
mannan@tamu.edu, liuyi.chemical@mail.che.tamu.edu

Prepared for Presentation at
American Institute of Chemical Engineers
2013 Spring Meeting
9th Global Congress on Process Safety
San Antonio, Texas
April 28 – May 1, 2013

UNPUBLISHED

Lower Flammability Limits of Mixtures of Hydrogen and Methane at Atmospheric and Sub-atmospheric Pressures

Hai Le
Mary Kay O'Connor Process Safety Center, Artie McFerrin Department of Chemical Engineering, Texas A&M University
Room 200, Jack E. Brown Building, Texas A&M University, 3122 TAMU College Station, TX 77843-3122
haile@tamu.edu

Yi Liu, M. Sam Mannan
Mary Kay O'Connor Process Safety Center, Texas A&M University

Keywords: lower flammability limit, hydrogen, methane, pressure

Abstract

The knowledge of the LFLs of hydrogen mixtures with hydrocarbons is essential since hydrogen is present in various mixtures with hydrocarbons in many processes and applications. The lower flammability limits of various mixtures of hydrogen and methane in air were determined experimentally in a closed cylindrical stainless steel vessel (inner diameter 10.22 cm, length 100 cm) with upward flame propagation at room temperature (20 ^{o}C) and initial pressures of 1.0 atm, 0.5 atm, and 0.1 atm. It was found that the LFLs of mixtures of the same hydrogen fuel fractions did not change much when the initial pressure decreased from 1.0 atm to 0.1 atm. And pressure has more impact on the LFLs when there is more hydrogen in the fuel mixture. The application of Le Chatelier formula to predict the LFLs of the mixtures was also evaluated. Overall, Le Chatelier formula can predict the LFLs of mixtures of hydrogen and methane very well at all initial pressures.

1. Introduction

Lower flammability limit (LFL) is an important parameter used in the prevention of fire/explosion when handling combustible gas mixtures. By definition, LFL is the minimum concentration of gas in air capable of propagating flame upon ignition;[1] in other words, if the concentration of a gas is smaller than its LFL, it is no longer flammable. Therefore, LFL is widely used in the calculation of the level of inerting needed to prevent the formation of flammable mixtures.[2] In these calculations and other fire risk assessments, it is advised to select and use the proper value of LFL for the operating condition since LFL is affected by pressure, temperature, direction of flame propagation and other conditions.[1]

The literature has a vast number of LFLs of various gases determined with different apparatuses at both atmospheric and non-atmospheric conditions, e.g. high temperatures, pressures, etc. LFLs of pure gases are more readily available than LFLs of mixtures of gases. Whenever LFLs of mixtures of gases are not available, methods have been developed to predict them. One of the widely used methods is Le Chatelier formula. It is an empirical mixing rule to predict the LFL of a mixture of gases as below: [3]

$$LFL_{mix} = \frac{1}{\sum_i^n \frac{y_i}{LFL_i}}$$

where LFL_{mix} is the LFL of the mixture, y_i is the mole fraction and LFL_i is the LFL of the ith component. For mixtures at non-atmospheric condition (e.g. high pressure, temperature) the LFLs of the components at the same condition should be used to accurately estimate the LFLs of the mixtures.

The knowledge of the LFLs of hydrogen mixtures with hydrocarbons is important since hydrogen is present in various mixtures with hydrocarbons in many processes and applications. Examples of such mixtures are refinery gas streams which contain various percentages of hydrogen (up to 90%) and hydrocarbons (methane, ethane, C_3, C_{4+}.).[4] Hydrogen is also added to natural gas, consisting mostly methane, to reduce CO, CO_2 and NO_x emissions[5-7], improve flame stability,[6, 7] and extend the lean operating limit of spark ignited engines.[8] It is critical to understanding the LFLs of these hydrogen mixtures in order to handle them safely.

Reference to the literature shows that studies on the LFLs of mixtures of hydrogen and hydrocarbons at non-atmospheric conditions, particularly sub-atmospheric pressure conditions, are very limited. Therefore, to enhance our understanding about the LFLs of hydrogen mixtures, this study focuses on investigating the LFLs of various mixtures of hydrogen and methane at atmospheric and sub-atmospheric pressures. Specifically, the LFLs of mixtures of hydrogen and methane were determined experimentally at room temperature (20 ^{o}C) and initial pressures of 1.0 atm, 0.5 atm and 0.1 atm. All experiments were carried out in a closed cylindrical vessel with upward flame propagation. The application of Le Chatelier formula to predict the LFLs of mixtures of hydrogen and methane at atmospheric and sub-atmospheric pressures was also verified using the LFLs of pure hydrogen and methane generated by the same experimental method and conditions.

2. Experimental setup

The experimental apparatus and procedure is illustrated schematically in Figure 1 and briefly described below.

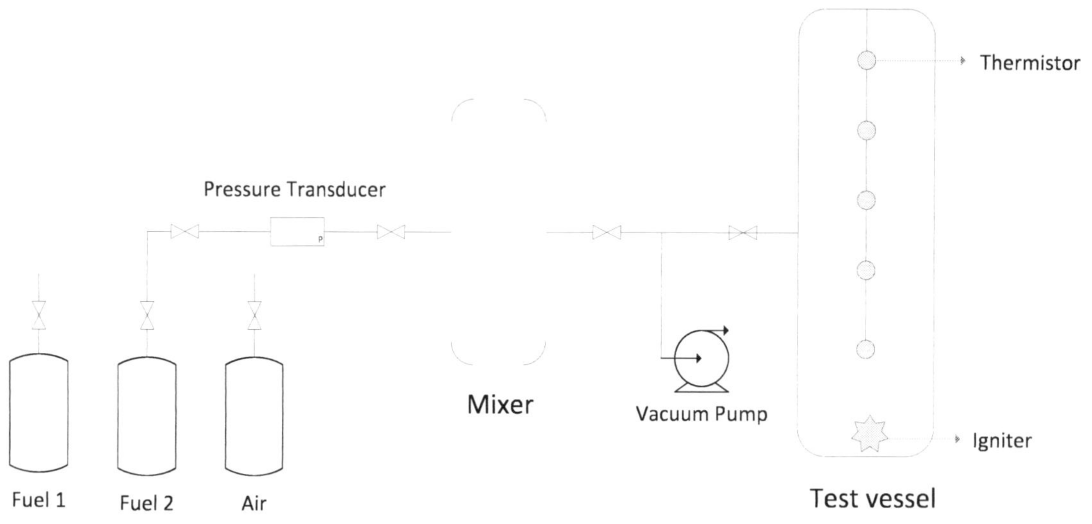

Figure 1: Experimental Setup

The LFLs of mixtures of hydrogen and methane were determined in a test vessel which was a closed stainless steel cylindrical vessel whose design was similar to that developed and used by the U.S. Bureau of Mines.[1] The vessel was 100 cm long with the inner diameter of 10.12 cm and outer diameter of 11.43cm. The igniter was installed at the bottom of the vessel and was capable of providing 10 J of energy consistently. The pressure inside the test vessel was measured by a pressure transducer placed at the top of the vessel. A set of 5 thermistors was positioned along the center axis of the test vessel to detect flame propagation. The distance between the top thermistor and the igniter was 75 cm. Further details about the configuration of the test vessel and its components can be found in our previous published work.[9]

High purity fuels and air from pressurized cylinders were loaded into the mixing vessel. The amount of the fuel and air were determined based on the partial pressure basis. The fuels and air were thoroughly mixed in a mixer which was a stainless steel vessel containing a Teflon block which slid along the vessel when it rotated lengthwise to enhance mixing. After the mixing, the test mixture was permitted to flow to the test vessel at the desired initial pressure measured by the pressure transducer. Then the mixture was ignited from the bottom, and the flame propagation was detected by the 5 thermistors inside the test vessel. Signals from the thermistors were obtained by a Keithley data acquisition card (Keithley® KPCI-3102, 225 signals per second at 0.05% accuracy) installed in a desktop computer, and the experimental process was controlled by a LabView® program. All experiments were carried out at room temperature (20 ^{o}C) and initial pressure of 1.0 atm, 0.5 atm and 0.1 atm. After each experiment, combustion products were purged from the test vessel which was then flushed with nitrogen and vacuumed to be ready for the next experiment.

A mixture is considered flammable if it can propagate flame to the top thermistor upon ignition;[9, 10] this behavior was defined as continuous flame propagation.[9, 10] For each mixture composition, 10 experiments were repeated and the probability of continuous flame propagation was recorded. Between two concentration values different by one experimental step size, one has the probability of continuous flame propagation equal or less than 50%, the other has the probability more than 50%. The former was chosen as the LFL, unless the former's probability is 0%, the later was chosen as the LFL. [9, 10]

3. Experimental Results and Discussion

3.1 Lower flammability limits (LFLs) of mixtures of hydrogen and methane at atmospheric and sub-atmospheric pressures

The experimentally obtained LFLs of various mixtures of hydrogen with methane at room temperature (20 °C) and initial pressure of 1.0 atm, 0.5 atm and 0.1 atm are shown in Figure 2. Overall, the LFLs of the mixtures of the same hydrogen fuel fractions did not change much when the initial pressure decreased from 1.0 atm to 0.1 atm. This is understandable since the LFLs of pure hydrogen and methane also changed very little with the initial pressure. When the fraction of hydrogen in the fuel mixture increased from 0% to 100%, the LFLs of the mixtures showed larger deviation from each other under the change of pressure from 1.0 atm to 0.1 atm. This suggests that pressure has more impact on mixtures of hydrogen and methane when there is more hydrogen in the mixture. However the impact is not significant since the degree of changes of the LFLs is small. It is worth to note that the LFLs of the mixtures was the lowest at 0.5 atm which means that there is an increased risk of fire/explosion when the initial pressure is reduced from 1.0 atm to 0.5 atm. Thus, more attention and precautions are recommended when handling mixtures of hydrogen and methane at pressures lower than 1.0 atm.

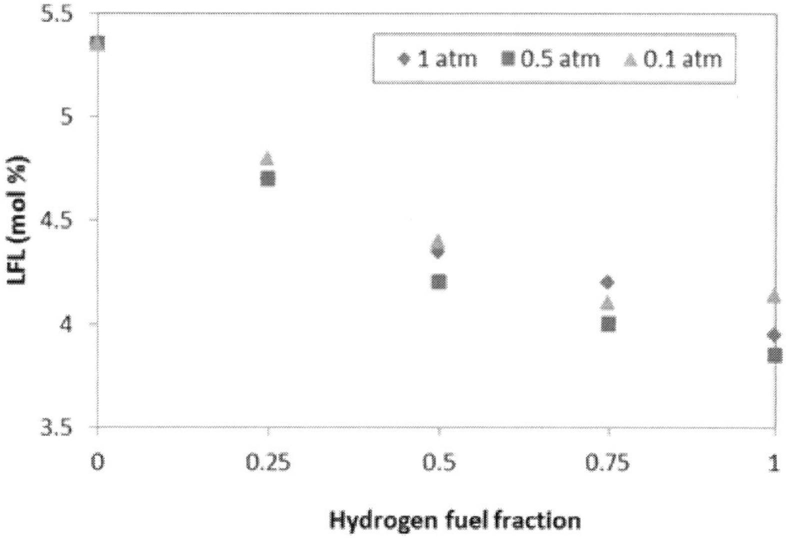

Figure 2: LFLs of mixtures of hydrogen and methane at atmospheric and sub-atmospheric pressures and room temperature (20 °C)

3.2 The application of Le Chatelier formula to the LFLs of mixtures of hydrogen and methane at atmospheric and sub-atmospheric pressures

The LFLs of mixtures of hydrogen and methane were calculated based on Le Chatelier formula using the LFLs of pure hydrogen and methane determined experimentally with the same experimental setup and conditions. Comparisons between the calculated LFLs and experimental LFLs are presented in Figure 3, Figure 4, and Figure 5 for 1.0 atm, 0.5 atm and 0.1 atm respectively. In general, Le Chatelier formula can predict the LFLs of the mixtures very well at all initial pressure conditions; in other words, the prediction of Le Chatelier formula for hydrogen and methane mixtures is not affected by changes in the pressure from 1.0 atm to 0.1 atm. The maximum deviation between experimental and calculated LFLs is 7.0 %. And the deviation is the largest when the fraction of hydrogen in the fuel mixture is 0.5.

The application of Le Chatelier formula to the upper flammability limits (UFLs) of the mixtures of hydrogen and methane will be evaluated in the future when we complete the determination of the UFLs of the mixtures experimentally.

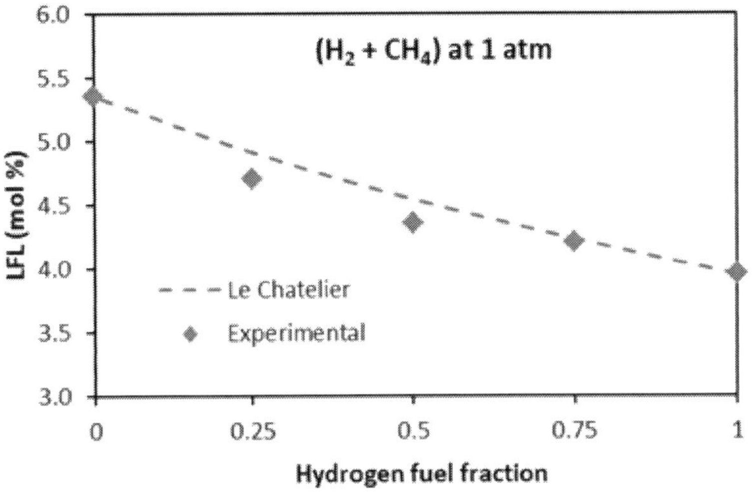

Figure 3: LFLs of mixtures of hydrogen and methane at room temperature and 1.0 atm (experimental and calculated by Le Chatelier formula)

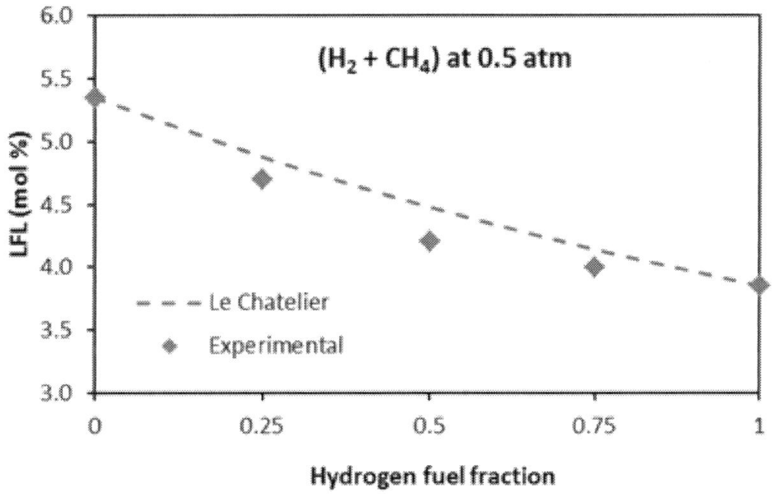

Figure 4: LFLs of mixtures of hydrogen and methane at room temperature and 0.5 atm (experimental and calculated by Le Chatelier formula)

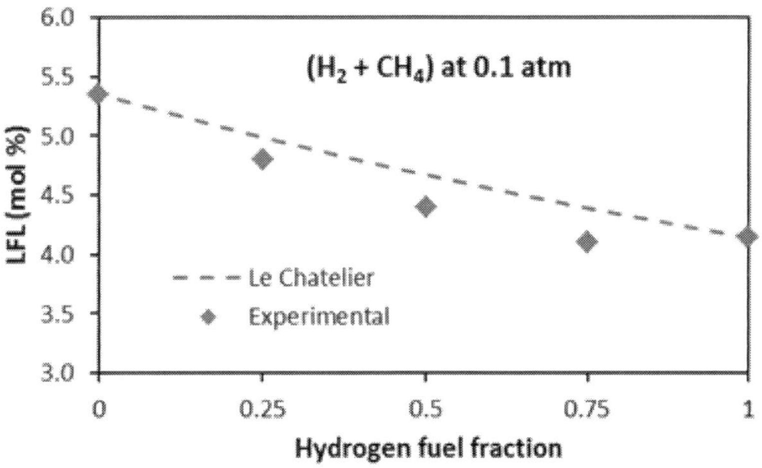

Figure 5: LFLs of mixtures of hydrogen and methane at room temperature and 0.1 atm (experimental and calculated by Le Chatelier formula)

4. Conclusion

The lower flammability limits (LFLs) of mixtures of hydrogen and methane in air were determined experimentally at room temperature and initial pressure of 1.0 atm, 0.5 atm and 0.1 atm. The application of Le Chatelier formula to predict the LFLs of the mixtures was also evaluated. It was found that:

- The LFLs of mixtures of the same hydrogen fuel fractions did not change much when the initial pressure decreased from 1.0 atm to 0.1 atm. Pressure has more impact on the LFLs of the mixtures when there is more hydrogen in the fuel mixture. The LFLs of the mixtures were the lowest at 0.5 atm which implies an increased risk of fire/explosion when the initial pressure decreases from 1.0 atm to 0.5 atm.

- Le Chatelier formula can predict the LFLs of mixtures of hydrogen and methane very well, and the prediction of Le Chatelier formula is not affected by changes in the pressure from 1.0 atm to 0.1 atm.

5. Acknowledgement

This research was sponsored by the Mary Kay O'Connor Process Safety Center, Artie McFerrin Department of Chemical Engineering, Texas A&M University.

6. References

[1] Coward, H.F. and G.W. Jones, *Limits of flammability of gases and vapors*, in *Bureau of Mines Bulletin,* 1952.

[2] Crowl, D.A. and J.F. Louvar, *Chemical Process Safety: Fundamentals with Applications, Second Edition*. 2001: Prentice Hall.

[3] Mashuga, C.V., *Determination of the combustion behavior for pure components and mixtures using a 20L sphere*, in *Chemical Engineering,* 1999: Michigan.

[4] Kaldis, S.P., G.C. Kapantaidakis, and G.P. Sakellaropoulos, *Simulation of multi-component gas separation in a hollow fiber membrane by orthogonal collocation — hydrogen recovery from refinery gases.* Journal of Membrane Science, 2000. **173**: p. 61-71.

[5] TerMaath, C.Y., et al., *Emissions reduction benefits from hydrogen addition to midsize gas turbine feedstocks.* Int J Hydrogen Energy, 2006. **31**(9): p. 1147-1158.

[6] Guo, H., et al., *The effect of hydrogen addition on flammability limit and NOx emission in ultra-lean counterflow CH4/air premixed flames.* Proc Combust Inst, 2005. **30**(1): p. 303-311.

[7] Hawkes, E.R. and J.H. Chen, *Direct numerical simulation of hydrogen-enriched lean premixed methane–air flames.* Combust Flame, 2004. **138**(3): p. 242-258.

[8] Bell, S.R. and M. Gupta, *Extension of the lean operating limit for natural gas fueling of a spark ignited engine using hydrogen blending.* Combust Sci Technol, 1997. **123**(1-6): p. 23-48.

[9] Le, H., S. Nayak, and M.S. Mannan, *Upper Flammability Limits of Hydrogen and Light Hydrocarbons In Air at Sub-atmospheric Pressures.* Ind. Eng. Chem. Res, 2012. **51**: p. 9396-9402.

[10] Wong, W.K., *Measurement of flammability in a closed cylindrical vessel with thermal criterion*, in *Artie McFerrin Department of Chemical Engineering, Texas A&M University,* 2006.

Informatics Rich Integrated Chemical Safety Risk Assessment (ICSRA) Platform

Muthukumarasamy Karthikeyan1, Renu Vyas2, HV Prasad3

1 Centre of Excellence in Scientific Computing and Digital Information Resource Centre, CSIR-National Chemical Laboratory, Pune, India, CSIR-National Chemical Laboratory, Pune, India
2 Chemical Engineering and Process Development, CSIR-National Chemical Laboratory, Pune, India, CSIR-National Chemical Laboratory, Pune, India
3 Chemical Engineering and Process Development, CSIR-Indian Institute of Chemical Technology, Hyderabad, Habsiguda, Hyderabad-500 607, India India

Abstract

Inherently Safer Practices (ISP) is a systematic innovative scientific methodology integrated with the process, design and operations for facilitating the improvement of process safety at all stages of process life cycle. ISP constitutes intensification (minimization), substitution, attenuation, limitation of effects and simplification as procedural options for implementation in process plants.

For this study we conducted literature data search focused on recent domestic and international incidents where the primary cause was related to chemical reactivity and runaway reactions; Several parameters were taken into consideration including chemical manufacturing (i.e., raw material storage, chemical processing, and product storage) and other industrial activities involving bulk chemicals, such as storage/distribution, waste processing, and petroleum refining.

For purposes of the incident search, only reactive incidents that caused serious consequences would be examined. Source of data: Four decades of data collection involve searching over 40 data sources, focusing on incidents where the primary cause was related to chemical reactivity. For the purposes of the investigation and analysis, an "incident" was defined as a sudden event involving an uncontrolled chemical reaction—with significant increases in temperature, pressure, and/or gas evolution etc.,

The proposed experimental and simulation platform will serve as a repository for plant-specific information, and facilitates the classification and ranking of Indian chemical industry - for the first time - through a Risk index.

Introduction

Computational and Simulation of Industrial chemical reactions to predict runaway reactions and conditions as safety measures. To our knowledge, no systematic studies on inherent safety are yet reported in India. The studies comprise important research areas e.g., process evaluation, estimation of reactivity, hazard identification and evaluation, assessment and prevention of thermal runaways, early fault diagnosis etc. The computational risk assessment platform intends to collect new chemical process safety and risk data, and plant-specific incident data from participating industries across the world for model building and simulation.

To integrate and include a dynamic interactive information platform, viz., chemical safety risk ranking and incident reporting system, by the application of High performance Computing (Cloud, Grid and Distributed systems) for simulation, modeling and response for use by Govt., Industry, Emergency Responders, Research and academic communities, and the public. Upgradation of Thermo chemical Laboratories to assess the proposed/existing industry from the context of implementing ISP, i.e., towards safer process chemistry including selection of solvents, safer process routes, safer process operating conditions for preventing Runaways, early detection of runaways, evaluation/selection of process and designs to handle lesser inventories etc.

Methodology

The methodology devised comprises recognition of the hazards and risks posed by the process, and efforts to reduce or control them to the lowest practical levels, keeping in view the principle objectives of the business. Here we studied data based on literature to identify and collect chemically significant industrially important data pertaining to physico-chemical, biological and toxicological profiles of chemicals (reactants, reagents, catalysts, solvents, products, side products) for their suitability and reusability in simulation and modelling. This includes development of suitable algorithms, chemical reaction component descriptors and methods to simulate experimental conditions and validation through experiments to improve the predictions of risk assessment and prevention of disaster in industrial conditions (Lab-scale, pilot plant and large scale manufacturing).

This process also includes mapping known literatures (Publications, patents and reports) to the existing knowledge for interactive risk assessment of chemicals and incidents in industrial environment. The computational approaches include classical methods, semi-empirical methods, and quantum methods to determine properties at the molecular level. We are developing a computational problem-solving platform and these tools, data and methods to build ICSRA a cloud-computing infrastructure for integrated chemical safety risk assessment of chemicals manufactured, imported and exported from world chemical sources. Proper documentation and accurate predictive models on chemicals of Indian origin would facilitate compliance with other global compliance on chemicals being exported especially in REACH, EPA, FDA and other global federal regulation context.

Development of QSPR (Quantitative Structure Property Relationships) models to relate macroscopic behaviour to molecular properties with detailed molecular computations to obtain thermo chemical data for reactive materials and to predict calorimetric data based on molecular properties. These open source based tools are integrated to this platform (ChemInfoCloud) includes identification and assessment of hazards due to chemical reactivity and risk assessment.

For this systematic study we considered major potential chemical group (CG) which are frequently encountered in industrial reagents like Organic peroxides, Organic Nitrites, Unsaturated Hydrocarbons, Epoxides, Hydrides and Hydrogen, Metal acetalides etc. Efforts to study and model major Chemical Reaction Types (CRT) include Decomposition, Polymerization, Pyrophoric, Peroxide Former, Water sensitive and Oxidation reactions. Molecular computing would help to predict heat of formation, electronic energy, core-core repulsion, dipole, no. of filled levels, ionization potential, molecular weight, topological, electronic and three dimensional descriptors for critical components of reactants, products, solvents.

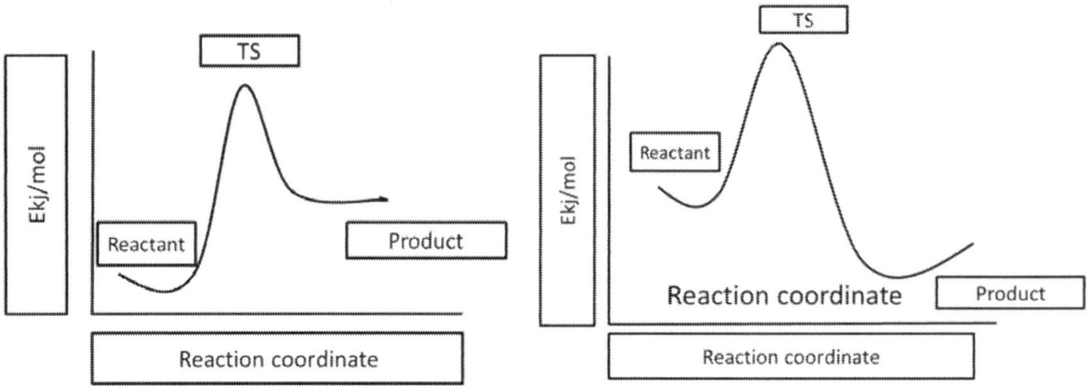

Figure-1: Computational Modeling generic types of chemical reactions along the reaction co-ordinates

Theoretical study of transition states of chemical reactions, computation of activation energies and experimental validation to predict that one reaction pathway is favoured over another reaction pathway. In order to predict the type of polymerization and runaway reactions enthalpy of reactions and kinetics was calculated using the semi-empirical quantum methods, molecular modelling and dynamics simulations of reactants under simulated diverse experimental conditions of Industrial Chemical Processes (ICP), including homogeneous and heterogeneous catalysis.

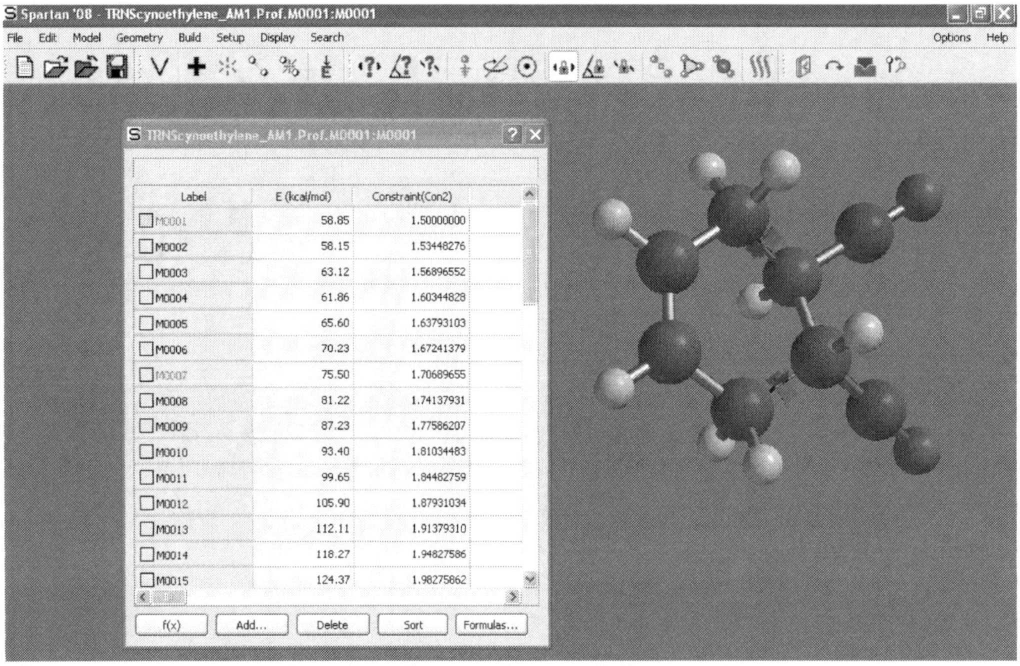

Figure 2: Spartan user interface for reaction modeling

Reactants **Transition State** **Product**

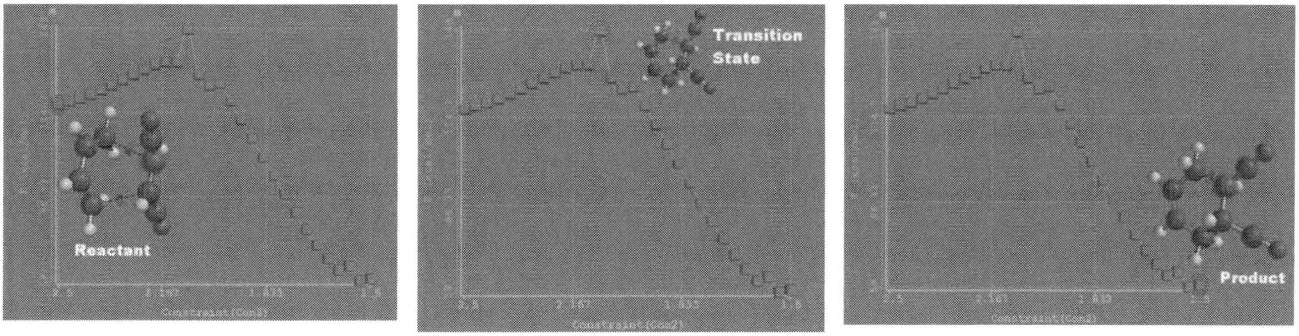

Figure: 3 Computational modelling of organic reactions (Exothermic)

Reactants	Transition State	Product
Heat of Formation: **489.651** kJ/mol Solvation Energy: SM5.4/A -9.308	Heat of Formation: **595.583** kJ/mol Solvation Energy : SM5.4/A -21.665	Heat of Formation: **246.229** kJ/mol

Table:1 Sample data of predicted Heat of formation for reactants, transition state and product

E (TS-Reactant)	E (Reactant-Product) ** heat generated!
E1=595.583-489.65 =105.933 kJ/mol =25.32 kcal/mol	E2=489.65-246.229 =243.421 kJ/mol =58.19 kcal/mol

Table:2 Computed energy difference

Digital mapping of incidental data concerning process safety and their availability to decision makers would help effective risk assessment and rapid response during unexpected events. During the presentation the detailed roadmap, current scenario and future direction of the chemical safety risk assessment platform will be discussed and the input from delegates for this meeting would be considered for integration into the tool if required.

References

1) Organic Reactions Series. John Wiley & Sons
2) http://downloads.wavefun.com/Spartan10Manual.pdf

IT Enablement For Effective Implementation And Monitoring of Process Safety Management

Ajay Kumar/ Pawan Mittal

Reliance Industries Limited

Jamnagar

Ajay.kumar@ril.com/ Pawan.mittal@ril.com

IT Enablement for various PSM elements

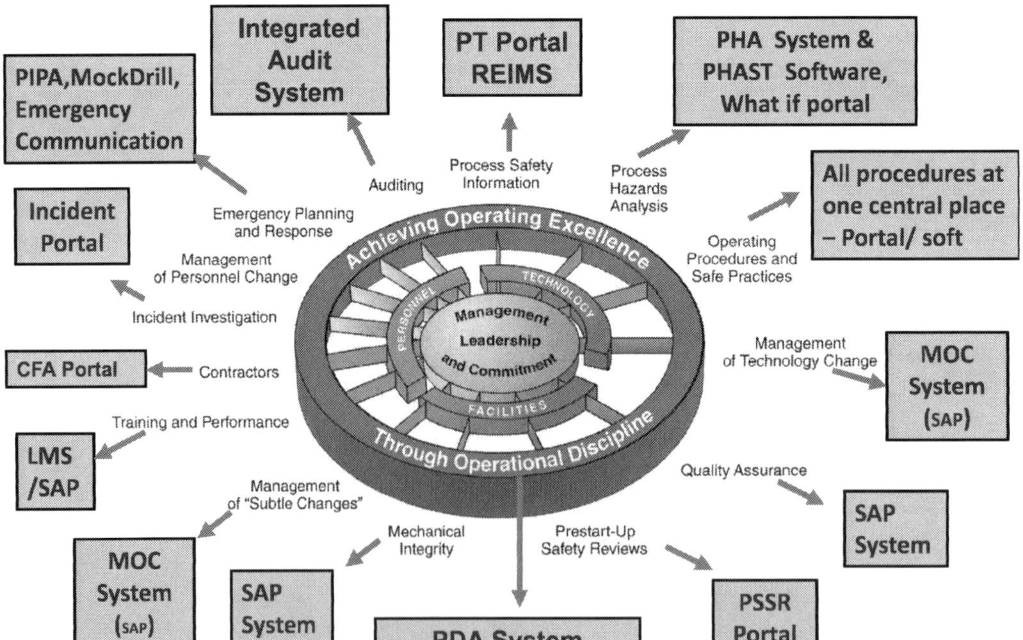

Governance and Control : Corrective Action Tracking System

Advantages of IT Enablement of PSM Elements

- **Process Technology (PT)**
  - ✓ Single repository for storage of all PT related documents
  - ✓ Easy access to PT documents to all concerned.
  - ✓ Easy updation of PT documents

- **Process Hazard Analysis (PHA)**
  - ✓ Automation of all 11 steps of PHA
  - ✓ Linkage of all relevant information required for PHA
  - ✓ Easy Tracking of PHA study progress
  - ✓ Consequence Modelling using PHAST software
  - ✓ Easy recordkeeping of PHA study reports & effective networking
  - ✓ Recommendation management in CAT system

Advantages of IT Enablement of PSM Elements

- **Management of Change Technology & Subtle (MOC – T & S)**
  - ✓ Single repository for recordkeeping of all changes
  - ✓ Controls provided for Hazard Analysis (Pre Screening PHA Checklist)
  - ✓ Tracking & recordkeeping of approval process
  - ✓ Tracking of PT document updating while MOC close out

- **Pre Start up Safety Review (PSSR)**
  - ✓ Automation of PSSR Checklist
  - ✓ Control provided for steps of PSSR e.g. Self Assessment , Level II & Level I
  - ✓ Pending checklist point tracking in CAT system
  - ✓ Permanent records of PSSR points & its close out status

Advantages of IT Enablement of PSM Elements

- **Learning Management System (LMS)**
  - Online nomination for the training program is available
  - Online training modules & validation sheets are available
  - Easy retrieval & tracking of employees' training records
  - Tracking of the pending training modules for planning training programs
- **Incident Reporting Investigation & Learning**
  - Reporting of incident in iRIL system.
  - Online team formation.
  - RCA tool interface
  - Uploading of investigation report in system.
  - Interface with CAT system for Recommendation till closure
  - LFI sharing to all concerned
  - Incident Investigation Quality
  - SAP based incident management system is under development.

Advantages of IT Enablement of PSM Elements

- **PIPA (Pre Incident Planning & Assessment)**
  - PIPA software used for management of emergency scenarios
  - Each plant selected credible scenarios & developed plan with the help of PIPA software
  - Rehearsal of PIPA scenarios & plans is ensured through emergency mock drills
- **Contractor Field Audits (CFA)**
  - Online scheduling & recordkeeping of field audits for various contractor agencies
  - Effective records of audit observations e.g. % Good Citizens observed during audit round, Score achieved by contractor agencies
  - Overall Tracking of Severity Index (Plant/Site wise)
  - Tracking of Contractor Field Audits e.g. Scope Vs. Completed

Advantages of IT Enablement of PSM Elements

- **Progressive Disciplinary Action System (PDA)**
  - ✓ Developed for recordkeeping of violation / negligence of safety rules & procedures
  - ✓ Easy retrieval of records
  - ✓ Uniformity in record maintenance of PDA at all RIL sites.
- **Corrective Action Tracking (CAT)**
  - ✓ Interface with iRIL, Audit Management, PSSR, Mock Drill
  - ✓ Tracking of Recommendation at one location
  - ✓ Mail Alert for Pending and Overdue Recommendation

Thanks

(44at) Sulphur Dust Explosions an Overview

Karthick Deivasagayam, Loss Prevention, SNC Lavalin Inc, Calgary, AB, Canada

This paper will focus on the sulphur dust, its explosion energy, available models and capabilities of modeling its explosion energy, the equipment susceptible for sulphur dust explosion, various prevention& mitigation strategies of sulphur dust explosion, and industrial applied technology to sulphur dust explosion mitigation. This paper will begin with increase of solid sulphur production from the refineries, its present forms of products and case studies.

The technologies and methodologies for sulphur dust explosion, frequency data, consequence modeling, prevention and mitigation of sulphur dust as applied in a present day sulphur plant.

Discussion for future developments will be discussed in this paper as well.

Simplified Uncertainty Analysis of LOPA Results

Raymond "Randy" Freeman
S&PP Consulting
12303 Lake Shore Ridge
Houston, TX 77041

rafree@yahoo.com

ABSTRACT

Layer of Protection Analysis (LOPA) is a semi-quantitative risk evaluation tool. The methodology is widely used in setting Safety Integrity Level (SIL) targets for Safety Instrumented Systems (SIS). The methodology is often used to evaluate the adequacy of existing protective systems against corporate risk targets. The LOPA method is based on the use of conservative estimates of the failure probability of protective systems.

Freeman previously presented a method for the quantification of uncertainty in a LOPA study based on the application of variance contribution analysis (VCA) techniques. The use of VCA to evaluate the uncertainty in a LOPA study requires the evaluation of the sensitivity of the LOPA equation to changes in each of the variables. Freeman used direct analytical computation of the partial derivatives of the LOPA equation for each of the uncertain variables. While mathematically correct, this use of a partial derivative analysis during an actual LOPA review is unlikely. Faster and easier methods are needed if LOPA uncertainty analysis is to be applied by practicing engineers.

This paper explores the use of approximations to compute the required uncertainty information without the use of the partial derivative computations. Numerical approximations (perturbation methods) are found to allow for the estimate of the partial derivatives with a maximum error of 16%. When evaluating simple safety systems designed per IEC 61511, the error is found to be no more than 10% for SIL-1 or SIL-2 capable systems. Compared to direct Monte Carlo simulation methods, the numerical approximation method is much more efficient. Monte Carlo simulation takes approximately 10,000 function evaluations to arrive at a result. For the base line problem studied, the numerical approximation method will take only 5 function evaluations.

BACKGROUND

Layer of Protection Analysis (LOPA) [Ref. 1] is a widely used method for the evaluation of the frequency of undesired events. LOPA is one of the methods proposed in the ANSI/ISA 84.00.01 standard [Ref. 2] for the determination of Safety Integrity Level (SIL) targets. The SIL target [Ref. 3] is used by process and instrument engineers for specification of equipment to reduce the frequency of undesired events such as overpressures, spills from tanks, releases to flare systems, fires, etc.

Freeman [Ref. 4, 5] developed a method for evaluating the uncertainty in a LOPA study using a mathematical method called Variance Contribution Analysis. The VCA method outlined in Freeman's paper require the evaluation of the partial derivative of the basic LOPA equation for each of the uncertain variables. There are several problems with the application of this method in an actual LOPA study, including:

1. Most engineers would not be comfortable evaluating a series of partial derivative functions while leading a LOPA study.
2. The LOPA team leader must develop the uncertainty parameters for each initiating event and each IPL during the analysis. A set of pre-calculated tabulations of the uncertainty parameters would be needed to speed the analysis.
3. The application of Monte Carlo simulation as used in Freeman's paper is not a topic familiar to most process engineers - nor is Monte Carlo simulation a simple methodology to implement for a real LOPA scenario.

The fundamental issue with the use of VCA in an actual LOPA study is the evaluation of the variance contribution from each uncertain parameter. This paper is directed at rectifying these deficiencies in the application of VCA to the evaluation of uncertainty in a LOPA study.

REVIEW OF LOPA METHODOLOGY

A LOPA [Ref. 1] review is a semi-quantitative risk evaluation of undesired events in a process. The LOPA begins with the definition of the set of undesired events (fires, explosions, toxic chemical releases, etc.). Next, potential safeguards (such as pressure relief valves, high/low level or pressure safety shutdown systems, emergency shutdown systems, etc) are reviewed to determine if there are independent protection layers (IPLs) that will prevent the undesired event from occurring. The frequency of the undesired event is then computed by combining the frequency of initiating event with the combined probability of failure on demand (PFD) of the IPLs as:

$$\text{Frequency of undesired event} = (\text{Initiating event freq}) \times \prod_{i=1}^{n} \text{PFD of the IPLs} \qquad \text{(Eqn. 1)}$$

Or in more compact form:

$$\text{Freq} = (\text{IE}) \times \prod_{i=1}^{n} P_i \qquad \text{(Eqn. 2)}$$

Where:

Freq = Frequency of the undesired event being studied by the LOPA method
IE = Initiating event frequency or the starting event frequency for the scenario under study

$P_i =$ Probability of Failure on Demand (PFD) of an Independent Protection Layer (IPL) counted as reducing the frequency of the undesired event.

$i =$ Subscript indicating the ith IPL

$n =$ Total number of IPLs counted in the LOPA study as satisfying the criteria for an IPL

Equation 1 is referred to as the basic LOPA equation. The basic LOPA equation of Eqn. 2 can be restated as:

$$Freq = (IE)\ P_1\ P_2\ P_3\ \ldots\ P_n \qquad \text{(Eqn. 3)}$$

All of the variables in Eqn. 3 are random variables with inherent uncertainty. The resulting calculated frequency of the LOPA scenario will therefore be uncertain.

APPLICATION OF VARIANCE CONTRIBUTION ANALYSIS TO THE BASIC LOPA EQUATION

Using variance contribution analysis, the mean and variance for the LOPA equation are found directly per the work of Freeman [Ref. 4, 5]. For the expression:

$$Y = F(x_i) \qquad \text{(Eqn. 4)}$$

Where:

$Y =$ dependent random variable computed from x_i

$F(x_i) =$ function that converts (maps) random variable x_i into Y

The mean of Y can be estimated using the following approximation:

$$E(Y) = F[E(x_i)] \qquad \text{(Eqn. 5)}$$

Where:

$E(Y) =$ expected value of random variable $Y =$ mean of Y

$E(x_i) =$ expected value of random variable $x_i =$ mean of x_i

The variance of Y can likewise be estimated as:

$$V(Y) = \sum_{i=1}^{n}\left[\frac{\partial Y}{\partial x_i}\right]^2 V(x_i) \qquad \text{(Eqn. 6)}$$

Where:

$V(Y)$ = variance of random variable Y
$V(x_i)$ = variance of random variable x_i

The application of the VCA method to the evaluation of the uncertainty in a LOPA calculation requires the evaluation of the partial derivative term in Eqn. 6. Previously, Freeman computed the partial derivative directly using the methods of calculus and partial differential equations. However, the direct analytical evaluation of each of the terms in a LOPA expression is time-consuming and prone to error.

The basic LOPA equation of Eqn. 3 is restated as:

$$Freq = (IE)\ P_1\ P_2\ P_3\ \ldots\ P_n \qquad \text{(Eqn. 3)}$$

Note that Freq is first order with respect to each random variable. Therefore, the partial derivative of the Frequency (Freq) with respect to a random variable will become a function of the form:

$$\frac{\partial Freq}{\partial P_1} = (IE)\ P_2\ P_3\ \ldots\ P_n \qquad \text{(Eqn. 7)}$$

The partial derivative may be evaluated numerically rather than analytically. The numerical approximation to the partial derivative may be found by simple evaluation of the Freq near the mean values of each random variable. For all cases to be reviewed in this paper, a 10 percent perturbation about the expected value of the random variable will be used to complete the numerical approximation calculations for the partial derivative term in a VCA study.

For the simple case where all of the IPLs are independent and share no common inspection or repair times, the numerical evaluation of the partial derivative of Freq with respect to the variable P_1 in Eqn. 3 may be estimated as:

$$\frac{\partial Freq}{\partial P_1} \approx \frac{\Delta Freq}{\Delta P_1} = \{[E(IE)\ 1.1E(P_1)\ E(P_2)\ldots E(P_n)] - [E(IE)\ E(P_1)\ E(P_2)\ldots E(P_n)]\}/0.1E(P_1)$$
(Eqn. 8)

Where:
$E(P_i)$ = expected value of random variable P_i = mean of P_i

The numerical estimate reduces to:

$$\frac{\partial Freq}{\partial P_1} \approx \frac{\Delta Freq}{\Delta P_1} = \left(\frac{1.1-1}{0.1}\right)\ [E(IE)\ E(P_2)\ E(P_3)\ \ldots\ E(P_n)] \qquad \text{(Eqn. 9)}$$

$$\frac{\partial Freq}{\partial P_1} \approx \frac{\Delta Freq}{\Delta P_1} = [E(IE)\,E(P_2)\,\ldots\,E(P_n)] \qquad \text{(Eqn. 10)}$$

Equation 10 is exactly the same as the analytical expression presented in Eqn 7. Thus, for simple LOPA expressions involving multiple independent protection layers, the result obtained by a numerical estimate will be the same as that obtained by the analytic expression obtained from the partial differentiation of the scenario basic LOPA equation. No error will be introduced by using the numerical approximation method to evaluate the partial derivative.

There are occasions in a LOPA review where the resulting basic LOPA equation may become non-linear. For example, if two or more IPLs are implemented in a safety instrumented system, the resulting LOPA equation becomes non-linear with respect to inspection frequency. For example a SIL-2 interlock design might involve two redundant instrumentation systems installed in parallel to reduce the frequency of an undesired event.

Appendix A to this paper evaluates the cases up to, and including, a SIL-4 capable design with 4 safety interlocks all tested on the same test period. The results of the analysis of Appendix A are summarized in Table 1. For most practical LOPA problems, a numerical analysis of the partial derivative term (Eqn. 6) will produce a result within 16% of that obtained using analytical methods.

Table 1.
Error Introduced in LOPA Uncertainty Calculations
by the Use of Numerical Estimates of the Partial Derivative Term

Case	Type of System	Error In LOPA Uncertainty* %
n = 1	SIL-1 Design or BPCS interlock	0
n = 2	SIL-2 Design or two SIL-1 interlocks	5
n = 3	SIL-3 Design or three SIL-1 interlocks	10
n = 4	SIL-4 Design or four SIL-1 interlocks	16

* Error = (Numerical Approximation Estimate – Analytical)/Analytical

The second problem with the application of the VCA method to LOPA uncertainty analysis is the need for pre-calculated tabulations of the mean and variance of the initiating event frequencies and IPL PFDs. Appendix B of this paper presents the method used as well as the results of calculating the needed uncertainty parameters using the data presented in the CCPS LOPA book. These tabulations may be used directly in a LOPA uncertainty study. The application of these simplifications is best illustrated using an example.

EXAMPLE PROBLEM

Previously, Freeman [Ref. 4, 5] presented an example problem to demonstrate the direct application of VCA to the uncertainty in the LOPA calculations. The same example is used for comparison in this work.

Consider the process system shown in Figure 1. The bottom sump level in separation tower (T-200) is controlled by a level controller (LIC-101) that regulates a control valve LCV-102. The tower is normally operating at a pressure of 500 psig and is designed for a maximum pressure of 750 psig. The liquid flow from T-200 goes to an intermediate vessel (Hold Drum D-205) prior to additional processing. D-205 is designed for a maximum pressure of 100 psig and normally operates at 75 psig.

Of concern is the failure of the level control system in T-200 to maintain the liquid seal at the bottom of the tower. If the liquid level in the T-200 sump is lost due to an instrumentation failure, the Hold Drum could be exposed to pressures as great as 500 psig. A pressure of 500 psig is a factor of 5 above the design pressure of the Hold Drum. Catastrophic failure of the Hold Drum is probable.

This problem is generically referred to as a high-low pressure interface problem or as a liquid blow-through case. Two protective systems are in place to prevent the Hold Drum from being over pressured.

1. The Separation Tower T-200 is provided with a SIL-1 low level interlock to prevent the loss of the liquid seal in the tower sump.
2. The Hold Drum D-205 is provided with a relief valve (PSV-206) sized to prevent overpressure if the tower liquid seal is lost.

LOPA EVALUATION OF THE EXAMPLE

For the example problem, the initiating event of concern is a basic process control system (BPCS) failure, resulting in loss of level control in the Separation Tower. From the CCPS LOPA book [Ref. 1], the initiating event frequency of a BPCS failure is 1 event in 10 years.

The low level protective interlock system was designed and installed in accordance with the ANSI/ISA 84.00.01 standard [Ref. 3] as a SIL 1 protective system. By definition, the probability of failure on demand (PFD) of a SIL-1 system is between 0.01 and 0.1. The LOPA book recommends using 0.1 as the PFD of a SIL-1 system.

The relief valve on the Hold Drum D-205 is designed for this hypothetical event. A value of 0.01 is suggested by the LOPA book as an estimate of the PFD for the relief valve of not opening on demand. However, for this analysis, a more conservative PFD value of 0.016 is used based on a review of the work by Bukowskia [Ref 6].

The LOPA formulation of the frequency of the Hold Drum being overpressured is:

$$\text{Freq} = \text{IE} \times \text{SIS failure} \times \text{PSV failure} \qquad \text{(Eqn 11)}$$

$$\text{Freq} = (0.1\ \text{event/yr}) \times (0.1) \times (0.016) = 1.6 \times 10^{-4}\ \text{events/yr} \qquad \text{(Eqn 12)}$$

In a LOPA, the frequency of 1.6×10^{-4} events per year would typically be compared with applicable risk acceptance criteria to determine if additional safeguards are needed. Many LOPA teams might recommend the addition of a critical high level alarm outside of the basic process control system, with adequate time for operator intervention, to reduce the risk of this liquid blow-through scenario.

Due to conflicts in turnaround schedules, the site management is interested in extending the test interval on the SIS system by 3 months, from 2 years to2.25 years. The management asks the LOPA team:

> "Will extending the SIS test interval by 3 months require the provision of additional IPLs to protect the system?"

At this point in the analysis, the engineer would not be able to explain to management the magnitude of the underlying uncertainty in the computed frequency of the Hold Drum overpressure. The next section of this paper attempts to answer this, and related, questions.

LOPA UNCERTAINTY

The basis for the evaluation of the LOPA uncertainty comes from two sources:

1. The LOPA model presented in Equation 11
2. The uncertainty in the parameters of the LOPA model

Rewriting Equation 11 in shorter terms, we obtain:

$$\text{Freq} = (\text{IE})\,(\text{SIS})\,(\text{PSV}) \qquad \text{(Eqn 13)}$$

Where:

Freq is the frequency of the Hold Drum D-205 being over pressured to the point of failure
IE is the initiating event frequency of the basic process control system failing in a dangerous state
SIS is the PFD of the SIL-1 low level interlock on the Separation Tower T-200
PSV is the PFD of the relief valve (PSV-206) opening and relieving the overpressure event

The variables in Equation 13 are all random variables. The variable SIS is actually the result of a simple calculation involving the equipment failure rate and test interval. For a negative exponentially distributed failure rate for a non-repairable device, the PFD is approximately:

$$\text{SIS} = \lambda\, T/2 \qquad \text{(Eqn 14)}$$

Where:

λ is the failure rate of the SIL-1 low level interlock on the Separation Tower T-200

T is the inspection or test interval for validation that the interlock is working and is functional

Equation 14 is an approximation of the negative exponential distribution. For $\lambda T/2 < 0.2$, the approximation is very good (error of 10%) and will be used throughout this paper.

Substituting the SIS failure rate and test interval into Equation 13 yields the basic LOPA equation for this scenario as:

$$\text{Freq} = (\text{IE})\,(\lambda)\,(T/2)\,(\text{PSV}) \qquad \text{(Eqn 15)}$$

The initiating event frequency in the basic LOPA analysis was taken directly from the LOPA book as 0.1 events per year. Discussions with site operations group indicates that the IE frequency is uncertain and might be as low as 0.05 events per year or as high as 0.5 events per year.

Likewise, the failure rate of the SIS interlock (λ) has significant uncertainty. The original plant design targeted this interlock as SIL-1 with a risk reduction factor of 50 based on a two-year testing interval, or a SIS failure rate of 0.02 failures per year. However, the failure rate may be as low as 0.01 events/year or as high as 0.1 events/year. From Appendix B (Table B-6) for a SIL-1 interlock designed with an RRF=50, the mean PFD is 4.3E-2 and the variance is 4.1E-4.

The test interval (T) may be targeted as once every 2 years. However, it might be as short as once every 1.5 years if the site takes advantage of an unexpected shutdown to test the system. It might be as long as 2.5 years, if the production run of the system is extended.

Appendix B has a discussion of relief valve PFDs. The Bukowskia [Ref 6] Process Safety Valve (PSV) PFD of 0.016 is higher than the CCPS LOPA book value and was used in the above LOPA calculation of the frequency of overpressure of the Hold Drum. For this uncertainty analysis, a mean PSV PFD of 0.016 with a standard deviation of 4.7×10^{-3} was used as the best estimate of the probability of failure of the relief valve to open on demand. The PSV variance would be $[4.7 \times 10^{-3}]^2$ or 2.2×10^{-5}.

These data may be used to establish the probability distribution for each of the uncertain variables in Eqn. 15. Assuming a triangular probability distribution (Ref. 7), the mean and variance may be calculated as:

$$\text{Mean} = (A + B + C)/3 \qquad \text{(Eqn 16)}$$

$$\text{Variance} = (A^2 + B^2 + C^2 - AB - AC - BC)/18 \qquad \text{(Eqn 17)}$$

Where:

A = Minimum Value

B = Mode

C = Maximum Value

The mean and variance for the initiating event (IE), the failure rate of the SIS (λ) and the inspection interval are calculated and presented in Table 2.

Table 2
Triangular Probability Distribution Parameters

Model Variable	Units	Minimum A	Mode B	Maximum C	Mean by Eqn 16	Variance by Eqn 17
Initiating Event (IE)	Events per Year	0.05	0.1	0.5	2.167E-01	1.014E-02
λ of SIS (From Table B-6 with RRF=50)	Failure per Year	0.01	0.02	0.1	4.3E-02	4.1E-04
Inspection Interval (T)	Years	1.5	2	2.5	2.000E+00	4.167E-02

The variance contribution for each uncertain (random) variable may now be computed using the numerical approximation for the partial derivative term presented above and derived in detail in Appendix A to this paper. Restating Eqn 15 as the basic LOPA equation for this scenario:

Freq = (IE) (λ) (T/2) (PSV) (Eqn 15)

Using a 10% perturbation about the means of each random variable, the partial derivative term may be estimated. For example the partial derivative portion of the contribution from the uncertainty in the initiating event may be estimated as:

$$\frac{\partial Freq}{\partial IE} \approx \frac{\Delta Freq}{\Delta IE} = \{[1.1\ E(IE)]\ E(\lambda)\ E(T/2)\ E(PSV) - E(IE)\ E(\lambda)\ E(T/2)\ E(PSV)\}/(0.1\ E(IE)) \quad \text{(Eqn 16)}$$

$$\frac{\partial Freq}{\partial IE} \approx \frac{\Delta Freq}{\Delta IE} = [1.1]\ E(\lambda)\ E(T/2)\ E(PSV) - E(\lambda)\ E(T/2)\ E(PSV)/(0.1) \quad \text{(Eqn 17)}$$

$$\frac{\partial Freq}{\partial IE} \approx \frac{\Delta Freq}{\Delta IE} = [(1.1 - 1)/0.1]\ E(\lambda)\ E(T/2)\ E(PSV) \quad \text{(Eqn 18)}$$

$$\frac{\partial Freq}{\partial IE} \approx \frac{\Delta Freq}{\Delta IE} = E(\lambda)\ E(T/2)\ E(PSV) \quad \text{(Eqn 19)}$$

Likewise the partial derivative contribution to the variance of the other random variables may be found as:

$$\frac{\partial Freq}{\partial \lambda} \approx \frac{\Delta Freq}{\Delta \lambda} = E(IE)\ E(T/2)\ E(PSV) \quad \text{(Eqn 20)}$$

$$\frac{\partial Freq}{\partial T} \approx \frac{\Delta Freq}{\Delta T} == E(IE)\ E(\lambda)\ (1/2)\ E(PSV) \qquad \text{(Eqn 21)}$$

$$\frac{\partial Freq}{\partial PSV} \approx \frac{\Delta Freq}{\Delta PSV} = E(IE)\ E(\lambda)\ E(T/2) \qquad \text{(Eqn 22)}$$

It should be noted that these results are identical to those obtained by taking the partial derivative of the basic LOPA equation for this scenario (Eqn 15) with respect to each of the random variables. With these estimates for the partial derivatives, the variance contribution from each of the uncertain random variables may now be computed (see Table 3). Note that the results in Table 3 are identical with those previously obtained by Freeman [Ref. 4, 5] using more formal mathematical methods. The same result could be obtained by simple numerical substitution. Equations 19 through 22 and those in Table 3 are presented to allow the reader to understand why the numerical results of a numerical approximation are the same as obtained using analytical methods.

Table 3.
Contribution of Each Model Variable
to Frequency of Drum Overpressure Variance

Model Variable	Units	Equation for Contribution	Variance Contribution	Percent of Total Variance
Initiating Event (IE)	Events per Year	$[E(\lambda)\ E(T/2)\ E(PSV)]^2\ V(IE)$	4.874E-09	41%
λ of SIS	Failure per Year	$[E(IE)\ E(T/2)\ E(PSV)]^2\ V(\lambda)$	4.874E-09	41%
Inspection Interval (T)	Years	$[E(IE)\ E(\lambda)\ (1/2)\ E(PSV)]^2\ V(T)$	2.351E-10	2%
Relief Valve (PSV)	none	$[E(IE)\ E(\lambda)\ E(T/2)]^2\ V(PSV)$	1.947E-09	16%
		Totals	**1.193E-08**	**100%**

Since the variance contribution and total variance are the same whether computed via an analytical method or by a numerical approximation method, the accuracy of the resulting will be the same as previously reported by Freeman [Ref 4, 5]. Table 4 presents a comparison of the computed results with those previously obtained by Freeman using both an analytical and a Monte Carlo method for estimating the frequency of overpressure of the Hold Drum in the example problem.

Table 4
Comparison of Results via Numerical Approximation,
Using Theoretical and Direct Monte Carlo Simulation of Hold Drum Overpressure

Method	Mean event/yr	Standard Deviation event/yr	Value of 70% Upper Limit event/yr	Value of 95% Upper Limit event/yr
Theoretical	1.50E-04	1.09E-04	2.07E-4	3.29E-04
Numerical Approximation	1.50E-04	1.09E-04	2.07E-4	3.29E-04
Monte Carlo	1.51E-04	1.19E-04	1.78E-4	3.90E-04
Basic LOPA	1.6E-04	Not Computed	Not Computed	Not Computed

With the results of the uncertainty analysis now available, we can answer the managements question in the example problem on the impact of extending the inspection and test interval by 3 months. As seen in Table 3, the variance contribution of the inspection interval is only 2% of the total. Increasing the inspection interval by 3 months will not appreciably impact the computed 95% upper limit on the frequency of the Hold Drum overpressure.

SIMPLIFIED METHOD FOR QUANTIFYING LOPA UNCERTAINTY

Based on the results obtained using the numerical approximations outlined in this paper, a simplification of the LOPA uncertainty method outlined by Freeman [Ref. 4, 5] is possible. With the exception of the method used to compute the partial derivatives for the variance contribution, the revised procedure will be the same as that which was previously presented. The change to the previously presented method is indicated in *italic font* below.

1. Complete the LOPA study using the methods outlined in the CCPS LOPA book [Ref. 1].

2. Write the basic LOPA equation for the scenario of interest in the form of the product of initiating event frequency and the PFD of the IPLs. An example of the basic LOPA equation is given in Equation 15.

3. Define the uncertainty in the terms of the expanded LOPA equation. The uncertainty can be given as the upper and lower range of the possible values (uniform probability distribution), as the upper, lower and recommended values (triangular distribution), or as a mean and standard deviation (normal distribution). See the example above for guidance in the evaluation of safety instrumented system interlocks.

4. Compute the expected value of each variable in the LOPA equation. The equations for the mean and variance for the uniform, triangular and normal probability distributions are presented for various probability distributions in Vose [Ref. 7]. *For commonly used LOPA initiating events and IPLs, the mean and variance are tabulated in Appendix B to this paper and may be used directly.*

5. Compute the expected value or mean of the LOPA equation using the mean value of each of the variables in the basic LOPA equation.

6. *Compute the sensitivity of the result from the LOPA equation by estimating the partial derivative of the basic LOPA equation with respect to each of the variables using a 10% perturbation about the mean or expected value of each random variable. See the problem worked above as an example of this step. The theory for this simplification is presented in Appendix A to this paper*

7. Compute the variance of the scenario frequency of the basic LOPA equation by use of the variance contribution of Equation 6. This entails multiplying the variance of each of the uncertain variables in the basic LOPA equation by the square of its sensitivity (obtained in step 6), as evaluated at the mean. Sum the resulting terms to obtain the overall variance of the result of the basic LOPA term.

8. Compute the 95% upper confidence limit on the computed frequency by use of the standard normal factor, Z, [Ref. 8] as:

   $$Z = [x_i - E(x)]/\sigma \qquad \text{(Eqn 23)}$$

   Where:

   σ = standard deviation of the frequency of the scenario of interest from the basic LOPA equation obtained from step 7. Note that the variance of a random variable is the square of the standard deviation of the random variable.

   $E(x)$ = the expected value of the frequency of the scenario of interest from the basic LOPA equation obtained from step 5

   For the 95% upper limit, Z = 1.64. Rearranging Eqn. 10 allows for the direct calculation of the corresponding value of the 95% upper confidence limit as:

   $$X_{95\%} = 1.64\,\sigma + E(x) \qquad \text{(Eqn 24)}$$

   Where

   $X_{95\%}$ = is the upper 95% limit on the computed frequency of the scenario of interest from the basic LOPA equation.

9. Compare the 95% upper confidence limit on the frequency of the scenario of concern with that obtained using the basic LOPA method (step 1). If the difference is greater than one order of magnitude, revise the LOPA study accordingly to incorporate better information obtained in the uncertainty analysis. Improve the integrity of the LOPA IPLs or identify additional IPLs to drive the risk to a tolerable level.

RESULTS AND CONCLUSIONS

A numerical approximation can be used to estimate the variance contribution of each uncertain variable in a basic LOPA. Use of a 10% perturbation about the mean or expected value of each random variable was found to be adequate to define the variance. The numerical approximation is exact for cases where all of the IPLs are independent. In the case where 4 IPLs share a common test or maintenance frequency, the resulting error in the uncertainty estimate will be less than 16%.

A simplified method has been developed to evaluate the uncertainty in the basic LOPA equation. Using available data, the simplified numerical approximation method estimates the mean and variance of the basic LOPA equation and then computes the uncertainty range in the resulting frequency of the scenario of concern. For the example studied with 4 random variables, a total of 5 evaluations of the basic LOPA equation were all that were required to characterize the uncertainty. The results are found to be in excellent agreement with those obtained using analytical methods. When compared to a Monte Carlo simulation of 10,000 function evaluations, the calculation speed improvement is approximately a factor of 2,000.

A set of tables for the mean and variance of initiating events (IE) and independent protection layers (IPLs) are presented in Appendix B of this paper. These can be used directly by the analyst to estimate the uncertainty in an IE or IPL.

The methodology presented in this paper is simple and direct. It may be programmed into any standard LOPA software package to allow the users the ability to estimate the uncertainty in their calculated result.

ACKNOWLEDGMENTS
This paper is an expanded version of a paper entitled ''Uncertainty in LOPA; Part 2 – Simplified Methods Applied to Safety Systems'' presented at the 2012 Mary Kay O'Connor Process Safety Symposium, October 23-25, 2012, College Station, TX. The author wishes to thank Kathy Kas, Angela Summers, Robert Johnson, Dan Crowl and other individuals who gave comments and suggestions for improving this article.

LITERATURE CITED

1. CCPS, *Layer of Protection Analysis – Simplified Process Risk Assessment*, American Institute of Chemical Engineers, New York, 2001
2. Instrumentation, Systems and Automation Society, Functional Safety: Safety Instrumented Systems for the Process Sector—Parts 1, 2, and 3, Standard ANSI/ISA 84.00.01, 2004. USA implementation of IEC 61511.
3. R.A. Freeman, "Using Layer of Protection Analysis to Define Safety Integrity Level Requirements, pp 185 – 194, *Process Safety Progress*, Vol. 26, No. 3, 2007
4. R.A. Freeman, "Uncertainty in LOPA Studies," pages 623-634, Proceedings of Mary Kay O'Connor Process Safety Center International Symposium held October 25-27, 2011 in College Station, Texas
5. R.A. Freeman, "Quantifying LOPA Uncertainty," pp 240 – 247, *Process Safety Progress*, Vol. 31, No. 3, 2012. Published in Wiley Online Library (wileyonlinelibrary.com), DOI 10.1002/prs.11493
6. Julia V. Bukowskia and William M. Goble, "Analysis of Pressure Relief Valve Proof Test Data," pp 24-29, *Process Safety Progress*, Vol 28, No.1, 2009
7. David Vose, *Risk Analysis – A Quantitative Guide, 3rd Edition*, John Wiley, 2008
8. Paul L. Meyer, *Introductory Probability and Statistical Applications*, John Wiley, New York, 1972

Appendix A – Numerical Approximation to Partial Derivatives

The basic LOPA equation of Eqn. 2 can be restated as:

$$\text{Freq} = (\text{IE})\ P_1\ P_2\ P_3\ \ldots\ P_n \qquad \text{(Eqn. A-1)}$$

Where:

Freq = Frequency of the undesired event being studied by the LOPA method

IE = Initiating event frequency or the starting event frequency for the scenario under study

P_i = Probability of Failure on Demand of the Independent Protection Layer (IPL) counted as reducing the frequency of the undesired event.

n = Total number of IPLs counted in the LOPA study as satisfying the criteria for an IPL

The probabilities of failure on demand of the IPLs may be restated in terms of a failure rate and a time between tests as:

$$\text{Freq} = (\text{IE})\ (\lambda_1)\ (T_1/2)\ (\lambda_2)\ (T_2/2)\ \ldots\ (\lambda_n)\ (T_n/2) \qquad \text{(Eqn A-2)}$$

Where:

λ_1 is the failure rate of the safety instrumented system, i

T is the inspection or test interval for validation that the interlock is working and is functional

i is a subscript indicating the ith IPL

If all the IPLs are tested or inspected at the same test interval

$$T_1 = T_2 = T_3 = T_n \qquad \text{(Eqn A-3)}$$

Forming a new parameter K as:

$$K = (\text{IE})\ (\lambda_1)\ (\lambda_2)\ \ldots\ (\lambda_n) \qquad \text{(Eqn A-4)}$$

The basic LOPA equation may now be re-written as:

$$\text{Freq} = K\ (T/2)^n \qquad \text{(Eqn A-5)}$$

When evaluated at the expected value of the random variables, the parameter K will not change and can be treated as a constant. How good is the numerical approximation for the partial derivative of Freq with respect to time as n increases from 1 to n?

The numerical approximation to a derivative requires the selection of a step size for the change in the variable of interest. In the limit as the step size goes to zero, the resulting answer will become the exact analytical expression for the derivative. However, from a practical point of view the selection of the numerical step size is constrained by two problems:

1. Too large and the function may change too much and the resulting numerical approximation will be very bad.
2. Too small and the numerical approximation will be limited by round-off error of the computer.

The step size of 0.1 T was chosen as the balance point between the two numerical problems. For the following cases (1 through 4), a step size of 0.1T was used as the basis for estimating the partial derivatives required for computations in the variance contribution analysis. Note that the step size for the evaluation of the derivative is arbitrary. A smaller step size could have been selected (say 0.01 T) and the resulting errors would decrease. However, the potential for numerical calculation errors would increase.

Case 1, n=1

For n = 1, restating the LOPA frequency becomes:

$$Freq = K\,(T/2) \qquad \text{(Eqn A-6)}$$

$$\frac{\partial Freq}{\partial T} = 0.5\,K \qquad \text{(Eqn A-7)}$$

The numerical approximation would become:

$$\frac{\Delta Freq}{\Delta T} = K\left(\frac{1.1^1 - 1}{0.1T}\right)(T/2)^1 \qquad \text{(Eqn A-8)}$$

$$\frac{\Delta Freq}{\Delta T} = K\left(\frac{0.1}{0.1}\right) 1/2 \qquad \text{(Eqn A-9)}$$

$$\frac{\Delta Freq}{\Delta T} = 0.5\,K \qquad \text{(Eqn A-10)}$$

This numerical approximation is exactly the same as the analytical result. For n =1, there will be no error in the computation of the partial derivative.

Case 2, n = 2

For n = 2, restating the LOPA frequency becomes:

$$Freq = K\,(T/2)^2 \qquad \text{(Eqn A-11)}$$

From basic calculus, the analytical expression for the partial derivative of equation A-5 when n=2 is:

$$\frac{\partial Freq}{\partial T} = 0.5\,K\,T \qquad \text{(Eqn A-12)}$$

The numerical approximation would become:

$$\frac{\Delta Freq}{\Delta T} = (\frac{1.1^2 - 1}{0.1T}) \text{ K } (T/2)^2 \qquad \text{(Eqn A-13)}$$

$$\frac{\Delta Freq}{\Delta T} = (\frac{0.21}{0.1}) (1/2)^2 \text{ K T} \qquad \text{(Eqn A-14)}$$

$$\frac{\Delta Freq}{\Delta T} = (\frac{2.1}{4}) \text{ K T} \qquad \text{(Eqn A-15)}$$

$$\frac{\Delta Freq}{\Delta T} = (0.525) \text{ K T} \qquad \text{(Eqn A-16)}$$

The error in the numerical approximation is:

Error = (Estimate - Actual)/Actual x 100 (Eqn A-17)

Error = (0.525 -0.5)/0.5 x 100 = 5% (Eqn A-18)

Thus, for a SIL-2 type of design, we would expect that no more than a 5% error in the LOPA uncertainty estimate would be introduced by the use of the numerical approximation for the partial derivatives.

Case 3, n =3

For n = 3, restating the LOPA frequency becomes:

Freq = K $(T/2)^3$ (Eqn A-19)

From basic calculus, the analytical expression for the partial derivative of equation A-5 when n=3 is:

$$\frac{\partial Freq}{\partial T} = \text{K } (3/8) \text{ T}^2 \qquad \text{(Eqn A-20)}$$

The numerical approximation would become:

$$\frac{\Delta Freq}{\Delta P_1} = (\frac{1.1^3 - 1}{0.1T}) \text{ K } (T/2)^3 \qquad \text{(Eqn A-21)}$$

$$\frac{\Delta Freq}{\Delta P_1} = (\frac{0.331}{0.1}) (1/2)^3 \text{ K T}^2 \qquad \text{(Eqn A-22)}$$

$$\frac{\Delta Freq}{\Delta P_1} = (\frac{3.31}{8}) \text{ K T}^2 \qquad \text{(Eqn A-23)}$$

$$\frac{\Delta Freq}{\Delta P_1} = (0.3875)\ K\ T^2 \quad \text{(Eqn A-24)}$$

The error in the numerical approximation is:

Error = (Estimate - Actual)/Actual x 100 (Eqn A-25)

Error = (3.31/8-3/8)/(3/8) x 100 (Eqn A-26)

Error = (3.31-3)/(3) x 100 = 10% (Eqn A-27)

Thus, for a SIL-3 type of design, we would expect no more than a 10% error in the LOPA uncertainty estimate to be introduced by the use of the numerical approximation for the partial derivatives.

Case 4, n =4

For n = 4, restating the LOPA frequency becomes:

$$Freq = K\ (T/2)^4 \quad \text{(Eqn A-28)}$$

From basic calculus, the analytical expression for the partial derivative of equation A-5 when n=4 is:

$$\frac{\partial Freq}{\partial T} = (1/4)\ K\ T^3 \quad \text{(Eqn A-29)}$$

The numerical approximation would become:

$$\frac{\Delta Freq}{\Delta P_1} = \left(\frac{1.1^4 - 1}{0.1T}\right) K\ (T/2)^4 \quad \text{(Eqn A-30)}$$

$$\frac{\Delta Freq}{\Delta P_1} = \left(\frac{0.464}{0.1}\right) (1/2)^4\ K\ T^3 \quad \text{(Eqn A-31)}$$

$$\frac{\Delta Freq}{\Delta P_1} = \left(\frac{4.64}{16}\right) K\ T^3 \quad \text{(Eqn A-32)}$$

$$\frac{\Delta Freq}{\Delta P_1} = (0.290)\ K\ T^3 \quad \text{(Eqn A-33)}$$

The error in the numerical approximation is:

Error = (Estimate - Actual)/Actual x 100 (Eqn A-34)

$$\text{Error} = (0.29 - 0.25)/(0.25) \times 100 = 16\% \qquad \text{(Eqn A-35)}$$

Thus, for a SIL-4 type of design, we would expect no more than a 16% error in the LOPA uncertainty estimate to be introduced by the use of the numerical approximation for the partial derivatives.

Table 1 presents a summary of the results of these calculations. In summary, the use of a numerical approximation for estimating uncertainty in a LOPA analysis will introduce no greater than a 16% error in the results. A 16% error in the LOPA uncertainty is well within the accuracy of the Variance Contribution Analysis method for estimating the LOPA uncertainty. This result eliminates the need for time-consuming evaluation of partial derivatives during a LOPA uncertainty analysis. Instead, a simple numerical analysis using a 10% perturbation about the mean of the inspection frequency is adequate to evaluate the partial derivatives needed in a LOPA uncertainty study. The results of this appendix are general and will apply to any basic LOPA equation formulated up to a 4th degree in one or more random variables.

Appendix B – Derivation of Tables for LOPA Uncertainty Analysis

The use of Variance Contribution Analysis to estimate the uncertainty in an estimate of the frequency of an accident requires a source for estimating the statistical parameters of the initiating events (IEs) and independent protection layers (IPLs). This appendix presents a methodology for estimating the required statistical parameters using available published data.

Initiating Events

The initiating event frequencies commonly used in a LOPA study are presented in Table 5.1 CCPS LOPA book [Ref. 1]. Included in Table 5.1 are ranges for the frequencies of each initiating event as determined from a literature search during development of the book. The uncertainty in the frequency data is expressed as a range (minimum to maximum), with a particular value assigned for use during a LOPA review. This type of uncertainty can be described by the triangular distribution [Ref. 7]. The triangular distribution has three parameters:

A = Minimum value
B = Mode
C = Maximum value

For a triangular distribution, the mean and variance are:

$$\text{Mean} = (A + B + C)/3 \qquad \text{(Eqn. B-1)}$$
$$\text{Variance} = A^2 + B^2 + C^2 - AB - BC - AC \qquad \text{(Eqn. B-2)}$$

For the data in Table 5.1, the minimum value becomes A, the maximum value becomes C, and the value used in LOPA becomes the mode, B. The mean and variance of the initiating event frequencies of the CCPS LOPA book Table 5.1 are calculated using Equations B-1 and B-2 and are presented in Table B-1 below. Note that the triangular distribution parameters (A, C and B) are defined in Table B-1 as the Min A, Max C, and Used in LOPA B respectively.

Table B-1. Mean and Variance of LOPA Initiating Events

Initiating Event	Units	Min A	Max C	Used in LOPA B	Mean per unit	Variance per unit^2
Pressure Vessel residual failure	per year	1E-07	1E-05	1E-06	3.7E-06	5.0E-12
Piping residual failure - 100 meter - full breach	per year	1E-06	1E-05	1E-05	7.0E-06	4.5E-12
Piping Leak - 10% section - 100 meter	per year	1E-04	1E-03	1E-03	7.0E-04	4.5E-08
Atmospheric Tank Failure	per year	1E-05	1E-03	1E-03	6.7E-04	5.4E-08
Gasket Packing Blowout	per year	1E-04	1E-02	1E-02	6.7E-03	5.4E-06
Turbine / diesel engine overspeed with casing breach	per year	1E-04	1E-03	1E-04	4.0E-04	4.5E-08
External impact by backhoe, vehicle, etc.	per year	1E-04	1E-02	1E-02	6.7E-03	5.4E-06
Crane Drop	per lift	1E-04	1E-03	1E-04	4.0E-04	4.5E-08
Lightning strike	per year	1E-04	1E-03	1E-03	7.0E-04	4.5E-08
Safety valve operates spuriously	per year	1E-04	1E-02	1E-02	6.7E-03	5.4E-06
Cooling water failure	per year	1.E+00	1E-02	1E-01	3.7E-01	5.0E-02
Pump seal failure	per year	1E-02	1E-01	1E-01	7.0E-02	4.5E-04
Unloading / loading hose failure	per year	1E-02	1E-01	1E-01	7.0E-02	4.5E-04
BPCS failure	per year	1E-02	1E-01	1E-01	7.0E-02	4.5E-04
Regulator Failure	per year	1E-01	1E+00	1E-01	4.0E-01	4.5E-02

The CCPS LOPA book Tables 6.3, 6.4, and 6.5 present the probability of failure on demand of passive, active mechanical and human operator IPLs. Using the same of data as for the initiating events allows for the uncertainty of IPLs to be calculated using Equations B-1 and B-2. The results of these calculations are summarized in Tables B-2, B-3, and B-4 below.

Table B-2. Mean and Variance of LOPA Passive IPLs

IPL Description	Min A	Max C	Used in LOPA B	Mean per unit	Variance per unit^2
Dike	1E-03	1E-02	1E-02	7.0E-03	4.5E-06
Underground Drainage System	1E-03	1E-02	1E-02	7.0E-03	4.5E-06
Open Vent (no valve)	1E-03	1E-02	1E-02	7.0E-03	4.5E-06
Fireproofing	1E-03	1E-02	1E-02	7.0E-03	4.5E-06
Blast Wall / Bunker	1E-03	1E-02	1E-03	4.0E-03	4.5E-06
Flame /Detonation Arrestors	1E-03	1E-01	1E-02	3.7E-02	5.0E-04

Table B-3. Mean and Variance of LOPA Active Mechanical IPLs

IPL Description	Min A	Max C	Used in LOPA B	Mean	Variance
Relief Valve*	1E-05	1E-01	1E-02	3.7E-02	5.1E-04
Rupture Disk	1E-05	1E-01	1E-02	3.7E-02	5.1E-04
BPCS	1E-02	1E-01	1E-01	7.0E-02	4.5E-04

Note the asterisk (*) in Table B-3 for relief valves The issue of relief valves not opening when needed is s topic that has been discussed extensively by Bukowskia [Ref. 6]. Bukowskia found that there is a significant probability that a relief device is not working when installed. A PFD value of 0.016 was suggested as an estimate for the PSV not opening on demand due to initial installation problems. Table 1 in the paper by Bukowski presents an estimate for the uncertainty in the PFD of the relief valve in terms of the standard deviation. For the data sets evaluated, the mean PFD of the PSV may be taken as 0.016 with a standard deviation of 4.7×10^{-3}. The corresponding PSV variance would be $[4.7 \times 10^{-3}]^2$ or 2.2×10^{-5}. The data of Bukowski indicate a PFD of about half that of the mean based on the range of the CCPS book Table 6.4 with a smaller variance. The LOPA value of 0.01 is lower than the value obtained using Bukowski's paper. The analyst completing a LOPA uncertainty study should carefully evaluate which data set to use for representing the uncertainty in the relief valve PFD.

Table B-4. Mean and Variance of LOPA Human Operator IPLs

IPL Description	Min A	Max C	Used in LOPA B	Mean	Variance
Human action with 10 minutes response time	1E-01	1E+00	1E-01	4.0E-01	4.5E-02
Human action with 40 minutes response time	1E-02	1E-01	1E-01	7.0E-02	4.5E-04

Safety Instrumented Systems

Safety instrumented systems are often provided in process design to control undesirable events such as overpressure, runaway reactions, or loss of level interface control. Note for the purposes of this paper the use of the phrase "safety interlock" means:

> "A Safety Instrumented Function (SIF) implemented in a Safety Instrumented System (SIS) to achieve a target Safety Integrity Level (SIL) designed and installed in accordance with the requirements of ANSI/ISA S84 – 2004 [Ref. 2]"

The S84 standard defines a range of acceptable performance operation of the interlocks in terms of a safety integrity level (SIL). The various SIL levels are defined by the probability of failure on demand (PFD) of the interlock, where the RRF is related to the PFD as:

$$RRF = 1/PFD \qquad \text{(Eqn. B-3)}$$

For example, an interlock designed for SIL-1 must have a PFD between 0.1 and 0.01, or by Equation B-3, an RRF between 10 and 100. Table B-5 presents the defined range of PFDs and RRFs for various SIL classes.

Table B-5. Probability of Interlock Failure on Demand of Various SIL Classes

IPL Description	Min PFD	Max PFD	Used in LOPA	Min RRF	Max RRF
SIL-1	1E-02	1E-01	1E-01	10	100
SIL-2	1E-03	1E-02	1E-02	100	1,000
SIL-3	1E-04	1E-03	1E-03	1,000	10,000
SIL-4	1E-05	1E-04	1E-04	10,000	100,000

There are two strategies available for the interlock design engineer to use:

1. Complete the design using rules of thumb, evaluate the RRF, and accept the design if the RRF falls in the desired range (as specified by the standard).
2. Complete the design to a desired risk reduction factor (RRF)

These two different design strategies will result in different probability distributions for the resulting uncertainty in the performance of the interlock. In the first case, any design that falls in the desired RRF range will be acceptable. The resulting uncertainty will be uniformly distributed from the minimum to maximum acceptable RRF. The uniform probability distribution may be characterized by two parameters, A and C:

A = Minimum value
C = Maximum value

The mean and variance of the uniform distribution are:

Mean = (A + C)/2 (Eqn. B-4)

Variance = $((C-A)^2)/12$ (Eqn. B-5)

In the second case, the design is targeted to a particular value of the RRF. This RRF becomes the mode (parameter B) of the triangular probability distribution of outcomes. Equations 1 and 2 may be used to compute the mean and variance of the resulting distributions.

Since a range of RRFs are available for each selected interlock SIL, a set of three possible targets was selected to represent the SIL range. These are:

Target RRF = 25 percent of allowed range (Eqn. B-6)

Target RRF = 50 percent of allowed range (Eqn. B-7)

Target RRF = 75 percent of allowed range (Eqn. B-8)

The uncertainty in the interlock performance was calculated for both design strategies for SIL-1, SIL-2, SIL-3 and SIL-4 at each of the three targeted SIL levels (Eqns. B-6, 7, and 8). The results of these calculations are presented in Tables B-6, B-7, B-8, and B-8.

For design strategy 1, Table B-6 presents the uncertainty parameters for each interlock SIL class when no target RRF is assigned prior to design. For example, such a case could occur if the design engineer were told to simply provide a SIL 2 interlock with no specification of the corresponding RRF.

Table B-6. Uncertainty Parameters for Interlock PFD for Various SIL Classes
(No target RRF set prior to design)

IPL Description	Mean PFD	Variance of PFD
SIL-1	5.5E-02	6.8E-04
SIL-2	5.5E-03	6.8E-06
SIL-3	5.5E-04	6.8E-08
SIL-4	5.5E-05	6.8E-10

For design strategy 2, the interlock RRF is selected prior to the start of design. This selected RRF then determines the design target PFD, which becomes the mode (parameter C, Equations B-1 and 2) for the triangular distribution. Tables B-7, B-8 and B-9 present the results for the computation of the mean and variance of the PFD of interlocks designed for various RRFs.

Table B-7. Uncertainty Parameters for Interlock PFD for Various SIL Classes with Target RRF at 25% of Allowed Range

IPL Description	Min PFD A	Max PFD C	Used in LOPA	Design RRF	Mode B	Mean PFD	Variance of PFD
SIL-1	1E-02	1E-01	1E-01	25	4.0E-02	5.0E-02	3.5E-04
SIL-2	1E-03	1E-02	1E-02	250	4.0E-03	5.0E-03	3.5E-06
SIL-3	1E-04	1E-03	1E-03	2,500	4.0E-04	5.0E-04	3.5E-08
SIL-4	1E-05	1E-04	1E-04	25,000	4.0E-05	5.0E-05	3.5E-10

Table B-8. Uncertainty Parameters for Interlock PFD for Various SIL Classes with Target RRF at 50% of Allowed Range

IPL Description	Min PFD A	Max PFD C	Used in LOPA	Design RRF	Mode B	Mean PFD	Variance of PFD
SIL-1	1E-02	1E-01	1E-01	50	2.0E-02	4.3E-02	4.1E-04
SIL-2	1E-03	1E-02	1E-02	500	2.0E-03	4.3E-03	4.1E-06
SIL-3	1E-04	1E-03	1E-03	5,000	2.0E-04	4.3E-04	4.1E-08
SIL-4	1E-05	1E-04	1E-04	50,000	2.0E-05	4.3E-05	4.1E-10

Table B-9. Uncertainty Parameters for Interlock PFD for Various SIL Classes with Target RRF at 75% of Allowed Range

IPL Description	Min PFD A	Max PFD C	Used in LOPA	Design RRF	Mode B	Mean PFD	Variance of PFD
SIL-1	1E-02	1E-01	1E-01	75	1.3E-02	4.1E-02	4.3E-04
SIL-2	1E-03	1E-02	1E-02	750	1.3E-03	4.1E-03	4.3E-06
SIL-3	1E-04	1E-03	1E-03	7,500	1.3E-04	4.1E-04	4.3E-08
SIL-4	1E-05	1E-04	1E-04	75,000	1.3E-05	4.1E-05	4.3E-10

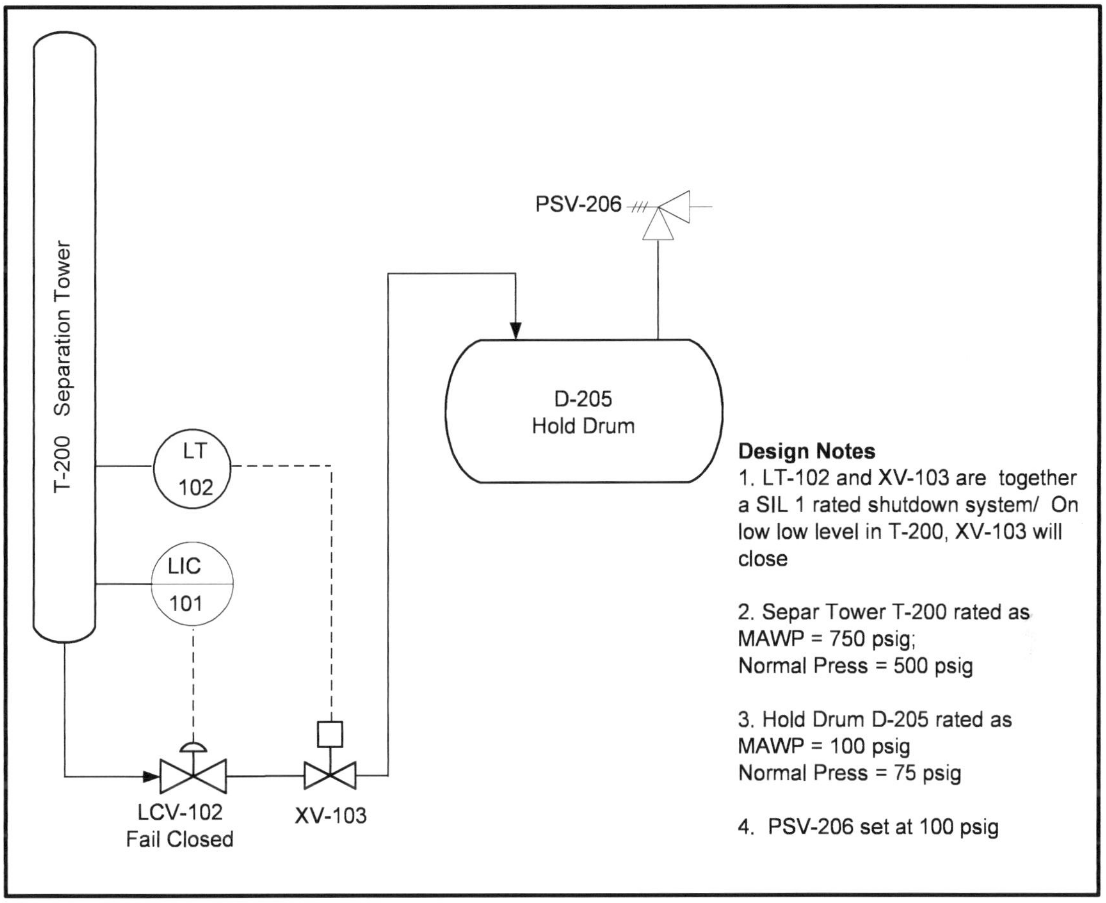

Figure 1. Piping and Instrument Diagram for Example

The Impact of Ozone on the Lower Flammable Limit of Hydrogen in Vessels Containing Savannah River Site High Level Waste

Paul Osterberg
Fauske & Associates, LLC
16W070 83rd Street
Burr Ridge, IL 60527
Osterberg@fauske.com

Carol Sherburne
Savannah River Remediation, LLC
Savannah River Site, Aiken, SC
Carol.Sherburne@srs.gov

Tom Johnson, Thomas Frawely
Fauske & Associates, LLC
16W070 83rd Street
Burr Ridge, IL 60527
Johnson@fauske.com, Frawely@fauske.com

Prepared for Presentation at
American Institute of Chemical Engineers
2013 Spring Meeting
9th Global Congress on Process Safety
San Antonio, Texas
April 28 – May 1, 2013

UNPUBLISHED

The Impact of Ozone on the Lower Flammable Limit of Hydrogen in Vessels Containing Savannah River Site High Level Waste

Paul Osterberg
Fauske & Associates, LLC
16W070 83rd Street
Burr Ridge, IL 60527
Osterberg@fauske.com

Carol Sherburne
Savannah River Remediation, LLC
Savannah River Site, Aiken, SC
Carol.Sherburne@srs.gov

Tom Johnson
Fauske & Associates, LLC
16W070 83rd Street
Burr Ridge, IL 60527
Johnson@fauske.com

Thomas Frawely
Fauske & Associates, LLC
16W070 83rd Street
Burr Ridge, IL 60527
Frawely@fauske.com

Keywords: hydrogen, ozone, flammability, lower flammability limit

Abstract

The Savannah River Site, in conjunction with AREVA Federal services, has designed a process to treat dissolved radioactive waste solids with ozone. It is known that in this radioactive waste process, radionuclides radiolytically break down water into gaseous hydrogen and oxygen, which presents a well defined flammability hazard. Flammability limits have been established for both ozone and hydrogen separately; however, there is little information on mixtures of hydrogen and ozone. Therefore, testing was designed to provide critical flammability information necessary to support safety related considerations for the development of ozone treatment and potential scale-up to the commercial level. Since information was lacking on flammability issues at low levels of hydrogen and ozone, a testing program was developed to focus on filling this portion of the information gap. A 2-L vessel was used to conduct flammability tests at atmospheric pressure and temperature using a fuse wire ignition source at 1 percent ozone intervals spanning from no ozone to the Lower Flammable Limit (LFL) of ozone in the vessel, determined as 8.4%(v/v) ozone. An ozone generator and ozone detector were used to generate and measure the ozone concentration within the vessel *in situ*, since ozone decomposes rapidly on standing. The lower flammability limit of hydrogen in an ozone-oxygen mixture was found to decrease from the LFL of hydrogen in air, determined as 4.2 % (v/v) in this vessel. From the results of this testing, Savannah River was able to develop safety procedures and operating parameters to effectively minimize the formation of a flammable atmosphere.

1. Introduction

Currently, in support of the Federal Facilities Agreement Act, tanks at the Savannah River Site (SRS) used to collect high level nuclear waste are being emptied, cleaned and grouted; and the waste permanently dispositioned in a series of specialized treatment processes. The majority of this waste removal is accomplished by a Bulk Waste Removal (BWR) campaign which empties the tank of easily accessible liquid supernatant and other liquid, followed by salt dissolution/removal and, finally, heel removal. At the end of the BWR campaign, fewer than 20,000 liters of sludge (less than 2 inches depth) remains in the tank. Currently, this final residual material is cleaned from the tank during a process called Bulk Oxalic Acid Cleaning (BOAC), whereby heated, 6 weight percent oxalic acid is added to the tank to a sufficient level to enable mixing. While BOAC has been effective in sludge dissolution and removal, the metal oxalates formed during oxalic acid dissolution cause downstream complications during evaporation (e.g. foaming) in addition to requiring excessive sludge washing (thereby greatly increasing the overall volume of waste to be dispositioned) to meet feed qualifications of the Defense Waste Processing Facility. This prompted research, testing and acceptance of a treatment process that utilized the proven effectiveness of oxalic acid for sludge dissolution and mobilization, yet removed the offending metal oxalates from the solution without adding liquid volume or contaminants. The Enhanced Chemical Cleaning (ECC) process was thus developed to replace BOAC stage of waste removal.

The Enhanced Chemical Cleaning (ECC) process uses ozone to promote the oxidation of metal oxalates produced during the dissolution of sludge in the Savannah River Site (SRS) waste tanks. The process is held at 70°C to improve decomposition kinetics and 10 psig to improve ozone solubility. A simplified depiction of the ECC process is shown in Figure 1, below.

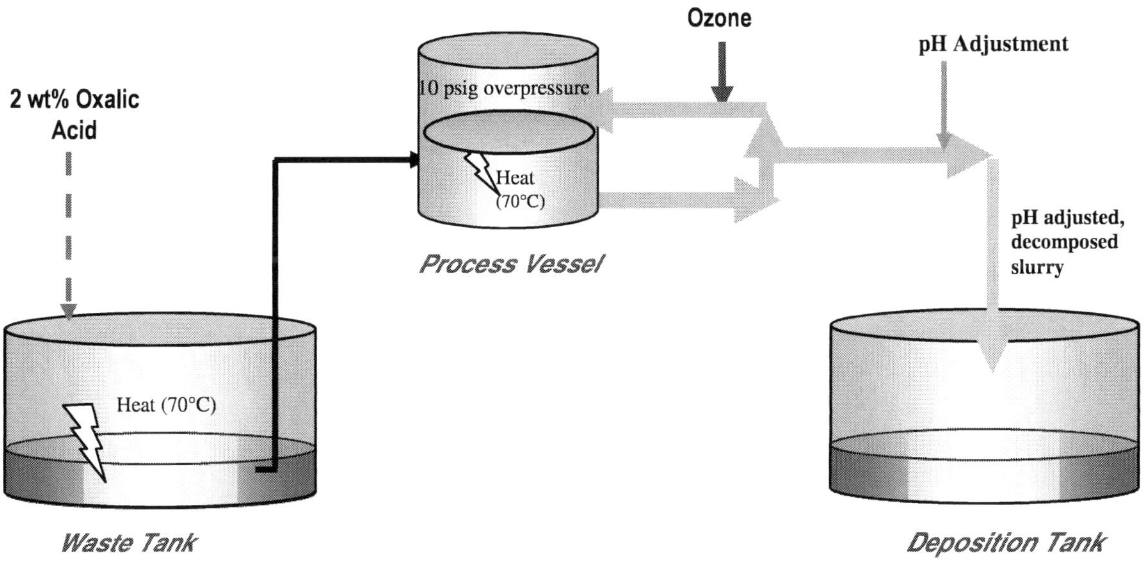

Figure 1: Schematic of the Enhanced Chemical Cleaning Process

In this process, the ozone reacts with the metal oxalates to form metal oxide and hydroxide precipitants, and the CO_2, O_2, H_2O and any unreacted O_3 gases are discharged into the vapor space. The major chemistry employed in the ECC process vessel is the reaction of ozone with the metal oxalates to form metal oxides, metal hydroxides, CO_2 and H_2O by the reactions shown below:

$$O_3 + H_2O \rightarrow O_2 + 2OH\bullet$$
$$Fe_2(C_2O_4)_3 + 6\ OH\bullet \rightarrow 6CO_2 + Fe_2O_3 + 3H_2O$$
$$H_2C_2O_4 + O_3 \rightarrow 2CO_2 + H_2O + O_2$$

While only the Iron oxalate reaction is listed above, formation of the powerfully reactive hydroxyl radical initiates many oxidizing reactions, effectively putting most waste constituents into their highest oxidation state.

In addition to the non-radioactive metals in the waste, however, the SRS radioactive waste also contains a variety of radionuclides, hence, hydrogen gas is also present in the vapor space of the ECC system. In order to prevent a buildup of hydrogen, and, hence, a flammability/explosion concern, waste tanks are purged and the hydrogen concentration monitored to ensure that the vapor space of each waste tank and all related processing equipment remains below the Lower Flammability Limit (LFL) of hydrogen. Furthermore, a comprehensive flammability control program is utilized to ensure that the LFL of the vapor space of a waste tank will not be attained for at least 7 days without forced ventilation. Because the use of ozone/oxygen in the processing equipment may also introduce ozone/oxygen into the vapor space of the waste tanks, the impact of this gas stream on the LFL of hydrogen must be understood for all possible operating scenarios of both normal and off-normal situations, with particular emphasis at the elevated temperatures and pressures of the typical process operating conditions. Oxygen is a known accelerant in combustion reactions, but while there are data associated with the behavior of hydrogen/oxygen environments, relevant studies addressing the effect of ozone on the flammability limit of hydrogen proved scarce. Further, discussions with industry experts verified the absence of data in this area and indicated that laboratory testing, specific to defined operating parameters was needed to comprehensively address the issue. Testing was thus designed and commissioned to provide the data necessary to support safety related considerations for the ozone treatment process.

2. Experimental

2.1 Flammability Apparatus

In order to ensure consistency between this testing and other documented flammability testing, the test methodology utilizes the guidelines established in ASTM E-918-83 (2005), "Standard Practices for Determining Limits of Flammability of Chemicals at Elevated Temperature and Pressure" [1]. Using this standard as the testing basis, a test setup was developed and is given in Figure 2.

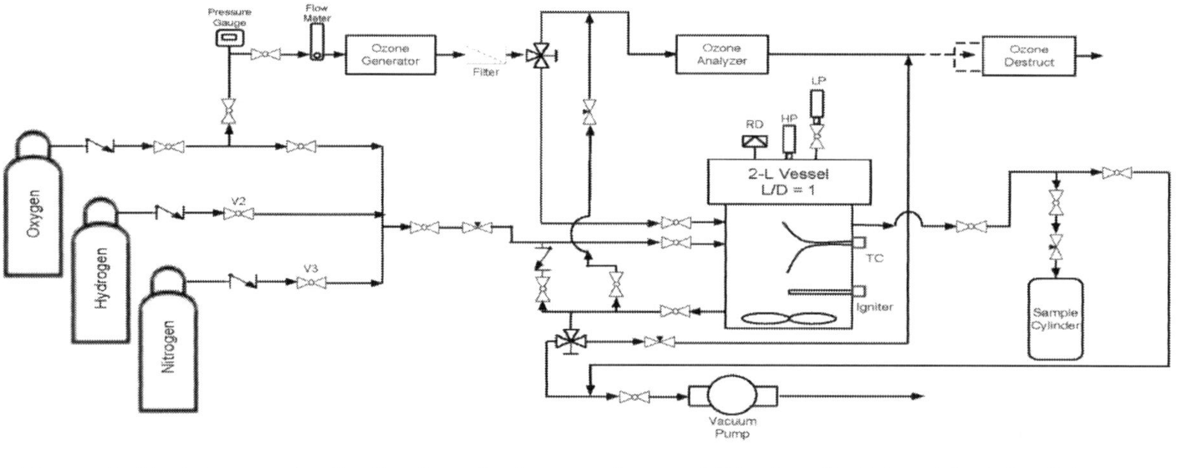

Figure 2: Test Apparatus for Testing the Lower Flammability Limit of Specified Gas Mixtures

The testing was conducted in a 2-L vessel rated to 3,000 psi at 500°C with an L/D ratio of one. There are two pressure transducers mounted on the vessel: a high accuracy (0.05% full scale) low range pressure transducer (0-25 psia) used for preparing the gas mixtures by partial pressure and a high accuracy (0.05% full scale) high range pressure transducer (0-200 psia) used to monitor the pressure after the mixture is ignited. High purity research grade gases were used for the preparation of the gas mixtures: nitrogen (99.9999%), oxygen (99.999%), and hydrogen (99.9999%). Two thermocouples were placed in the vessel to measure the temperature at the center and at the upper area of the vessel. A stir bar was placed at the bottom of the vessel to provide mixing of the gases. The gas mixture was ignited using an exploding fuse wire positioned a third of the way from the bottom. The energy provided by the fuse wire was approximately 10 joules. An ignition energy of 10 J was chosen because it produces very little wire spatter and a cylindrical shock wave is formed upon the vaporization of the wire because it is above the 5 J/cm threshold [2].

Ozone is generated on-site using a high ozone concentration generator. The ozone concentration of the ozone/oxygen stream is analyzed with an in-line UV absorption unit designed to enable accurate measurements of ozone from 0.01 wt. % ozone in oxygen to 20 weight percent.

In order to minimize undesired ozone decomposition, the materials used for this testing were carefully chosen. The test vessel, tubing, and fittings are all made of 316 stainless steel and passivated per ASTM A967 [3]. The packing material and/or seals for the valves were made of PTFE or fluorocarbon FKM.

2.2 *Experimental Method*

Prior to testing, the thermocouples, pressure transducers, ozone analyzer, and data acquisition instrumentation are checked to ensure they are functioning properly. The 2-L vessel is opened and wiped down with acetone. A 10mm 40 gauge tinned copper wire is wrapped around the igniter electrodes. The vessel is sealed and checked for leaks through pressurizing the vessel. This is followed by a vacuum and pressure purging procedure with a diluent gas (N_2 or O_2) depending on the test conditions. The gas manifold and piping are vacuumed and pressure purged with the respective gas prior to addition into the vessel. The purging helps minimize trace contamination of other gases as well as to remove residual moisture.

By assuming that these gases behave ideally, the gas concentration is determined based on partial pressure addition. Prior to gas filling, the vessel is evacuated down to approximately 0.2 psia. The magnetic stirrer is turned on to provide mixing during the gas filling step. Each gas is added slowly to the desired concentration based on the partial pressure calculation. During addition, there is a slight temperature rise due to the compression of the gases; therefore, after each gas addition, the mixture is allowed to reach equilibrium. The gas manifold and piping are purged with the subsequent gas prior to being added to the mixture.

For the ozone tests, oxygen is fed into a corona discharge ozone generator. A potentiometer on the unit controls the power to the elements. Thus, by adjusting the oxygen flowrate through the generator and tuning the potentiometer, the precise ozone/oxygen ratio specified in the test matrix can be produced. The ozone concentration is measured with an in-line ozone analyzer that is placed on the outlet line of the vessel.

Early scoping tests performed to optimize the test setup indicated that if ozone was added to the test vessel while the vessel was under a vacuum, the ozone decomposed very rapidly, making it difficult to establish the desired test conditions. By feeding the ozone/oxygen mixture through the process vessel at atmospheric pressure, then monitoring the ozone concentration at the discharge, an accurate determination of the ozone concentration in the test vessel, just prior to the addition of hydrogen, may be made. The test vessel can then be evacuated just enough to add the prescribed volume of hydrogen for the test, minimizing loss of ozone.

Once the mixture has been prepared, the magnetic stirrer is turned off. The mixture is allowed to relax prior to ignition so that a quiescent mixture is tested rather than a turbulent mixture. The low pressure transducer valve is closed and the data collection is started. The mixture is manually ignited using the fuse wire providing a nominal 10 joules. The data acquisition records the pressure and temperature time traces in the vessel. A mixture is considered to be flammable if the pressure rise after the ignition event is greater than or equal to 7% in accordance with ASTM E918. The procedure is repeated until there are a minimum of two demonstrations conducted at the lowest fuel concentration to be considered flammable and three at the highest fuel concentration considered non-flammable.

3. Results and Discussion

3.1 Hydrogen Lower Flammable Limit

The lower flammable limit for initially quiescent hydrogen-air mixtures was determined in order to validate the test equipment setup and the procedure. The LFL for hydrogen in air was determined as 4.2 volume percent based on the ASTM 7% pressure rise criteria. A comparison of this result to data obtained by other researchers is summarized in Table 1.

Table 1: Hydrogen LFL in Air, this testing compared to others

Source	Hydrogen Concentration (vol. %)	Comments
LFL using 5% pressure rise criteria		
SAFEKINEX [4]	4.1	2.8-L vessel
	3.9	6-L vessel
	4.1	14-L vessel
	4.7	40-L vessel
	5.8	1250-L vessel
	4.1	2000-L vessel
LFL using 7% pressure rise criteria		
Present Work	4.2 ± 0.1	2-L vessel (L/D = 1)
Crowl and Jo [5]	5.7	20-L spherical vessel
Cashdollar et al. [6]	5.0 ± 0.5	8-L chamber
	6.0 ± 0.5	20-L chamber
	6.5 ± 0.5	120-L spherical vessel
LFL using a visual criterion		
Coward and Jones [7]	4.0	Flammability Tube

The experimentally measured LFL is in close agreement with the accepted lower flammable limit of 4 vol. % which was obtained through visual criterion [7]. Many other researchers have shown that it is difficult to obtain this value using the 7% pressure rise criteria. This has been explained through hydrogen's high diffusivity as compared to oxygen and nitrogen [6].

The majority of tests using the pressure rise criteria are performed with the ignition source in the center of the vessel. Central ignition is used for the determination of the explosion severity of flammable mixtures because it gives the maximum rate of pressure rise and the maximum overpressure. However, central ignition may not be useful when determining the limits of flammability. It has been shown that ignition source location has an impact on the flammability limits of methane [8]. As the location of the ignition source is lowered, a larger percentage of the combustible mixture participates in the reaction with minimal heat loss to the vessel wall. This results in a higher explosion overpressure, thereby widening the flammable region. The present work uses an ignition source placed approximately one-third from the bottom of the vessel.

3.2 Ozone Lower Flammable Limit

Initially, tests were conducted to determine the extent of ozone decomposition over time inside the combustion vessel at ambient conditions. This was accomplished by preparing an ozone-oxygen mixture and holding the mixture in the combustion vessel for approximately 30 minutes while stirring. After the specified hold time the ozone concentration was measured. It was found that there were negligible amounts of ozone decomposition that occurred in the vessel.

The LFL of ozone in oxygen was determined as 8.4 volume percent. Previous studies on the flammability of ozone in oxygen are presented in Table 2.

Table 2: Ozone LFL in Oxygen, this testing compared to others

Source	Ozone Concentration (vol. %)	Comments
Present Work	8.40 ± 0.17	2-L vessel (L/D = 1)
Mizutani et al. [9]	10.5	1-L cylindrical vessel, Electric spark
Koike et al. [10]	10.0	1-L cylindrical vessel, 1 J electric spark
Cook et al. [11]	9.5 ± 0.5	10 cm i.d. glass bulb, 0.47 J electric spark

The present work reveals that the explosion limit of ozone in oxygen is consistent with previously reported values; however, it is lower than other literature data which can be attributed to the differences in the experimental method and procedure. The lower explosion limit obtained in this work is most likely a result of a combination of factors: the type of ignition source, location of the ignition source, and vessel size and geometry. In these tests, the ignition source energy was setup to give approximately 10J which is much greater than that used in previous work. Flammability limits are a function of ignition energy, thus, by increasing the strength of the ignition source the flammable limits tend to widen [12]. There also tends to be some variation in the flammable limits which is influenced by the type of ignition source: a fuse wire as compared to an electric spark [13]. This may be an effect of the energy density provided by each type of ignition source and the amount of energy actually being transferred to the mixture. As discussed previously, the ignition source location may be playing a part in the lower value obtained.

The differences in the size and dimensions of the explosion vessels could have affected the lower flammable limit [13]. As the vessel size is increased, the heat losses to the wall become negligible. Through minimizing heat losses to the vessel wall, more heat is transferred to the combustion reaction, therefore, promoting flame propagation. This will result in a widening of the flammable region and combustion can occur at lower temperatures.

3.3 Lower Flammable Limit of Hydrogen in Ozone/Oxygen Mixtures

The experimental results for the lower flammable limits of hydrogen in the ozone/oxygen are shown in Figure 3. The solid circles represent a flammable mixture and the open circles represent a non-flammable mixture.

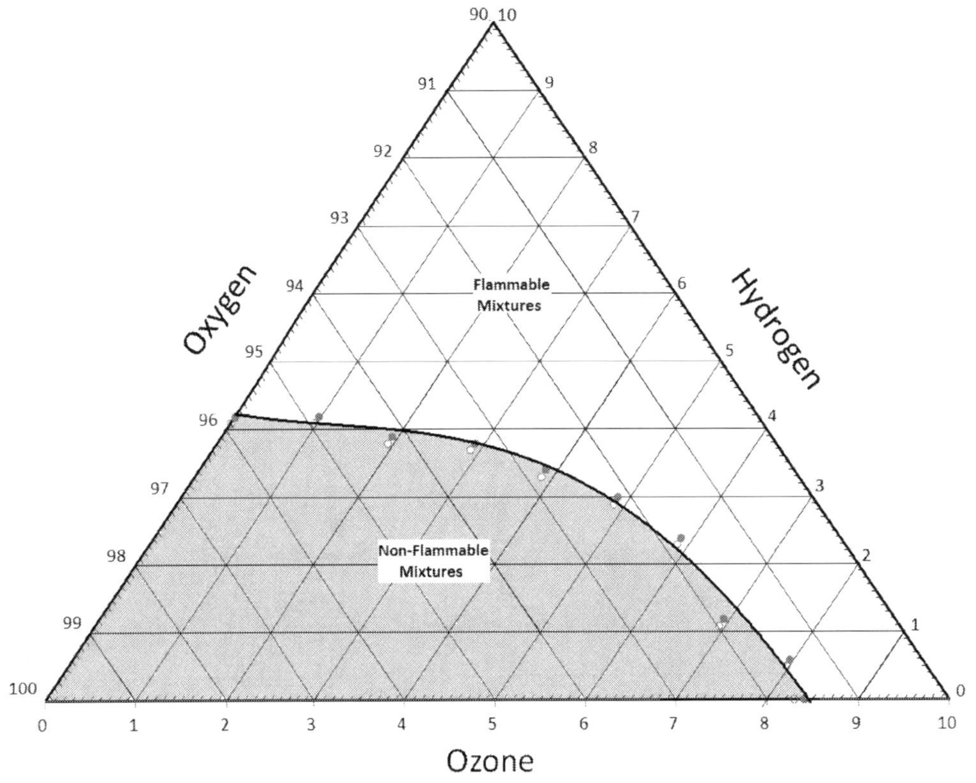

Figure 3: Flammability envelope for hydrogen/ozone/oxygen mixtures (all values are in volume percent).

Typically, it is of interest to estimate the flammability limits of multi-component mixtures. In the absence of experimental data, estimation methods can be useful if the flammable limits of pure components are known. Le Chatelier [14] developed an empirical equation for predicting the flammable limits of fuel mixtures,

$$LFL_{MIX} = \left(\sum_{i=1}^{n} \frac{y_i}{LFL_i} \right)^{-1} \qquad \text{Eq. [1]}$$

where y_i is the mole fraction of the ith component on a combustible basis and LFL_i is the lower flammable limit of the ith component in volume percent.

A comparison of the LFL predictions to the experimentally determined LFL of hydrogen in various ozone-oxygen mixtures is shown in Figure 4. It was found that the LFL of hydrogen decreases as the ozone concentration in the mixture increases. This result is consistent with the values approximated through Le Chatelier's rule; however, the experimental results are higher than the predicted calculations. Previous studies have shown that Le Chatelier's rule provides close approximations for many simple hydrocarbons; however, it is less accurate for mixtures with hydrogen and generally not applicable for substances that undergo decomposition [15]. However, from a safety stand point, Le Chatelier's rule provides a conservative estimate of the LFL.

The experimental data was further compared to flammability predictions from the calculated adiabatic flame temperature (CAFT) method. The flame temperature criteria used to assess whether a mixture is flammable or not was established by first calculating the adiabatic flame temperature at the determined LFL of hydrogen in air and ozone in oxygen. Three flame temperatures were used – the flame temperature of 8.4% ozone in oxygen, the flame temperature of 4.2% hydrogen in air and an average of these two flame temperatures. The LFL of hydrogen in various concentrations of ozone/oxygen was then calculated from these flame temperatures.

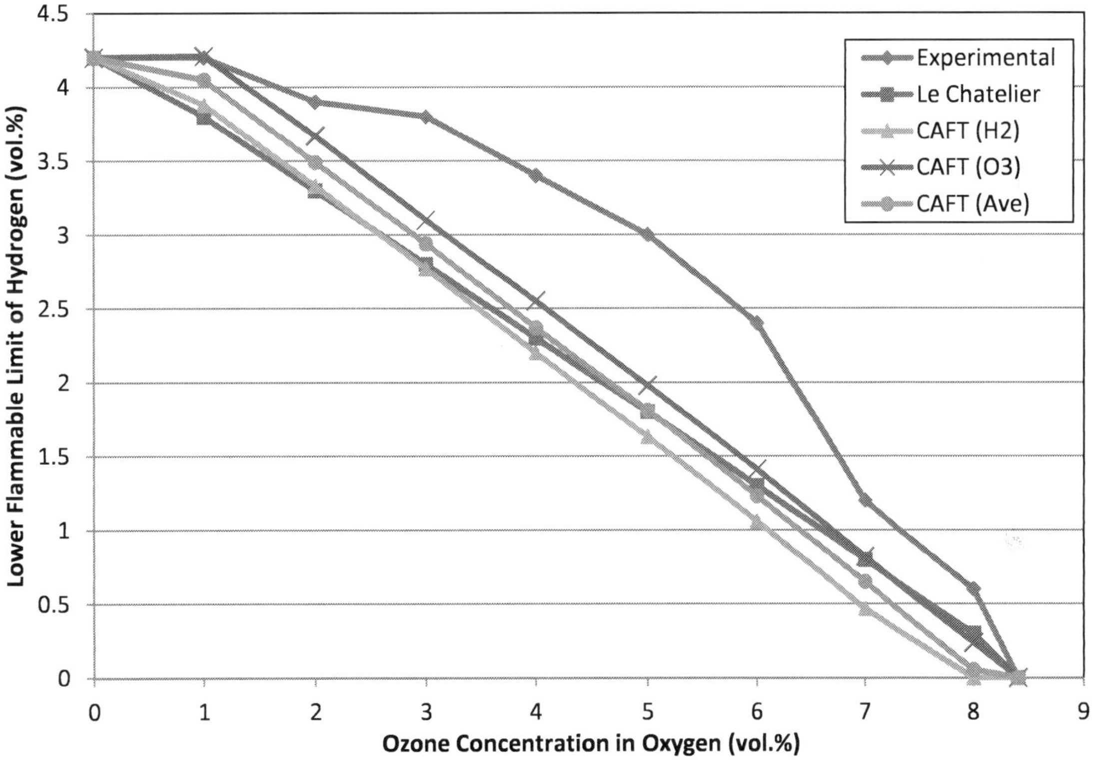

Figure 4: Comparison of lower flammability limit of hydrogen in ozone/oxygen to calculated adiabatic flame temperature predictions.

The calculated adiabatic flame temperature (CAFT) predictions are very similar to the predictions given by Le Chatelier's empirical equation. The deviations of the experimental data from the estimations in the flammability limits of a hydrogen/ozone/oxygen mixture lie within the unpredictability of ozone. This unpredictability comes from the instability of ozone, which tends to readily decompose to form oxygen. Furthermore, there are many different reactions that could be taking part in these experiments: H_2-O_2, O_3 decomposition, O_3-O_2, and H_2-O_3. The data seems to suggest that as the ozone concentration is increased, the reaction mechanism that takes precedence changes.

The deviation between the predicted and the experimental values may be two-fold. First, the reaction mechanism taking preference during the combustion reaction changes as the ozone concentration is increased. Secondly, the concentration of ozone participating in these reactions may be changing as well. Tests conducted on ozone decomposition in oxygen have shown that only 5.3% of the ozone decomposes for a gas mixture containing 6.7 mole % ozone and as high as 90% decomposition when the ozone concentration was 11.2% in the mixture [11]. In any case, additional work is needed to fully understand the nature of these experiments.

3.4 Temperature and Pressure Effect on the Lower Flammable Limit of Hydrogen-Ozone-Oxygen Mixtures

It has been reported previously that the ozone decomposition rate increases at the temperature of its environment increases [16]. Therefore, stability tests were performed again to determine the extent of ozone decomposition at the elevated test conditions. This time it was found that the ozone decomposition rate could not be considered negligible. With initial mixtures containing 5 vol.% ozone in oxygen at approximately 75°C, the ozone decomposition rate was approximately 0.1 vol.% ozone every 10 minutes.

The results from the tests performed at elevated temperatures and pressures are shown in Figure 5. It was found that as the temperature is increased the LFL decreases and the opposite was observed when the pressure was increased.

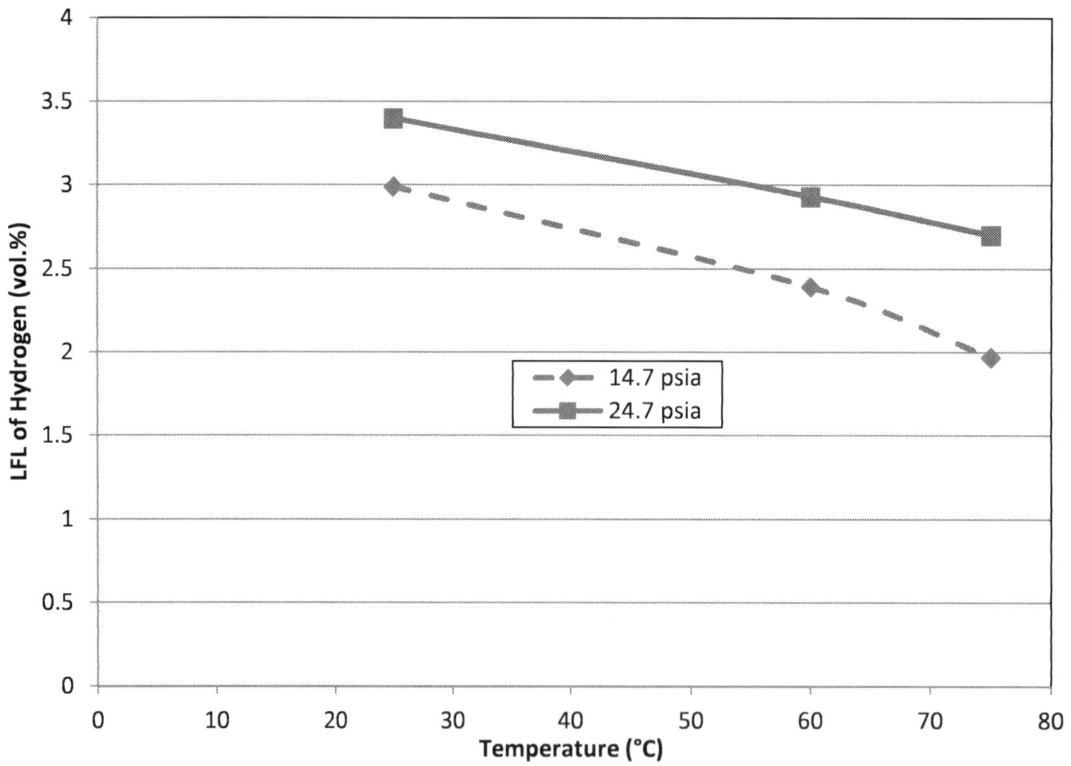

Figure 5: Effect of temperature and pressure on the flammability limits of hydrogen-ozone-oxygen mixtures. Tests were performed with 5 vol.% ozone in oxygen.

The increase in the LFL as a result of the increase in pressure is generally not characteristically observed for flammable gases; however, this is a characteristic of hydrogen flammability [17]. This result leads to the conclusion that the reaction mechanism responsible for the increased LFL as the pressure is increased for hydrogen is still the dominate mechanism when ozone is added. Furthermore, it is noticed that as the temperature is increased the decrease in the LFL is fairly linear throughout the range of temperatures tested. The LFL results from these tests show that the LFL is more susceptible to elevated temperatures when compared to results of hydrogen-air or hydrogen-oxygen. One explanation for this is that at the increase temperatures more of the ozone participates in the reactions that occur in the vessel. Previous research on ozone decomposition has found that only a part of the ozone decomposes when the ozone concentration is below its LFL. At the higher temperatures, the reaction rates increase as well as the instability in the ozone molecule.

3.5 Effect of Relative Humidity on the Lower Flammable Limit of Hydrogen-Ozone-Oxygen Mixtures at Elevated Temperatures and Pressure.

Tests conducted with the addition of water vapor have shown a slight increase in the LFL as shown in Figure 6. Since there is an additional component (water vapor) added to the mixture, the results should be compared on a total combustibles (hydrogen plus ozone) basis rather than just on the concentration of hydrogen in the mixture. This is because as more water vapor is added to the mixture, the concentration of ozone decreases; therefore, even if more hydrogen was needed to be added to the mixture, the total amount of combustibles may have remained constant to produce a flammable mixture.

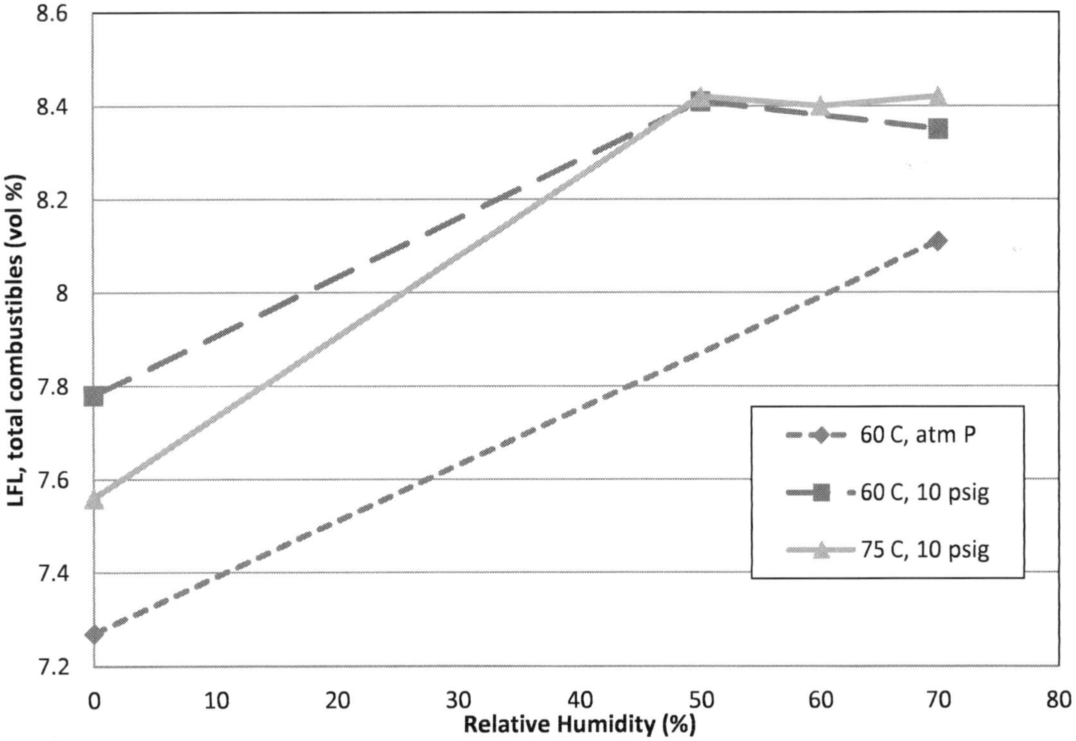

Figure 6: Effect of relative humidity on the LFL of hydrogen-ozone-oxygen mixtures. Tests were performed with 5 vol.% ozone in oxygen.

The slight increase in the LFL could be a result of either the addition of the water vapor or decrease in ozone concentration. The addition of water vapor to the mixture helps extract heat from the reactions occurring in the vessel (combustion or decomposition), thereby, quenching the flame propagation by lowering the reaction rate. Furthermore, the observed increase in the LFL could be a result of a decrease in the ozone concentration. It is known that both hydrogen and water act as a catalyst for ozone decomposition; therefore, the actual concentration inside the vessel prior to igniting the mixture is lower than the reported concentrations.

4. Impact on Operation

This testing was designed to provide data under the conditions considered most optimal to produce a deflagration. The geometry and materials of construction of the testing vessel; the location of the fuse wire; the magnitude of the supplied energy; the careful minimization of diluents and other contaminants; and meticulous procedural detail to maintain integrity of the ozone to the maximum extent practical, result in data that reflect not the expected process conditions, but those that enhance the possibility of flame propagation. For this reason, there is believed to be considerable conservatism in the indicated results. Per the vendor, the maximum possible ozone concentration producible by the planned ECC Ozone generator is 8 volume percent (typical maximum expected concentration is 6.8 vol%), less than the 8.4 minimum volume % concentration shown to be flammable in a pure O_2 environment at the optimally conservative conditions established in this testing. As both water vapor and nitrogen, together constituting approximately 6.5 % of the actual feed stream, are powerful diluents, it is believed not credible that deflagration can occur at this maximum feed concentration condition. In addition, once the ozone stream contacts the waste stream, the many simultaneous oxidizing reactions will rapidly decompose available ozone to well below flammable levels. Further, because the radiolytically generated hydrogen quantity is negligible compared to the supplied ozone/oxygen stream (0.0004 moles per minute H_2 vs 76 moles per minute ozone/oxygen), even the total H_2/O_3 mixture, without crediting decomposition reactions, does not approach flammable concentrations. Finally, even at the "end" of the ECC batch cycle, when most of the metal oxalates have been decomposed, testing has indicated that the ozone concentration in the vapor space of the ECC process vessel reaches a concentration of no more than 3 vol%, remaining well below concentrations of concern.

The major issue for the ECC operation established by this testing is the impact of the data when applied to off normal conditions. While it is possible to discontinue ozone addition to the reaction vessel at any time, the radiolytic hydrogen generation rate continues, varying slightly as ambient pressures and temperatures change. Relative to the data generated and analyzed in this testing, the ECC hazards analysis team will re-evaluate off normal conditions (e.g. those during which process exhaust ventilation is lost) such that issues involving mixtures of hydrogen and ozone in the vapor space can be appropriately controlled.

5. Conclusion

The LFL of hydrogen in air was determined and is in good agreement with the literature data. Ozone-oxygen mixtures were found to be flammable at concentrations above 8.3 vol.% based on the ASTM E918 7% pressure rise criteria for flame propagation. This result is lower than previously reported values which can be explained through the variations in the test setup and procedure. Tests performed with hydrogen in various concentrations of ozone in oxygen have shown that the LFL of hydrogen decreases as the concentration of ozone in the mixture increases.

The estimations of the LFL for various mixtures of hydrogen and ozone in oxygen was performed using Le Chatelier's rule as well as the CAFT method. It was found that both methods underestimate the LFL for mixtures of hydrogen and ozone.

The effects of temperature, pressure, and relative humidity were found for hydrogen-ozone mixtures. The trends observed through these tests were consistent with the general trends in reported results for hydrogen. Overall, the difficulty with some of these tests ultimately lies with the unpredictability of ozone.

6. References

[1] ASTM. (2005). ASTM E 918-83 Standard Practice for Determining Limits of Flammability of Chemicals at Elevated Temperature and Pressure. Philadelphia, PA: American Society for Testing and Materials.

[2] Mashuga, C. V. (1999). *Determination of the Combustion Behavior for Pure Components and Mixtures using a 20-Liter Sphere.* Phd Thesis. Michigan Technological University. Ann Arbor: UMI Company.

[3] ASTM. (2005). ASTM A967-05, Standard Specification for Chemical Passivation Treatments for Stainless Steel Parts. Philadelphia, Pa: American Society for Testing and Materials.

[4] Project SAFEKINEX, Contract No. EVG1-CT-2002-00072, Deliverable No.8, *Report on experimentally determined explosion limits, explosion pressures and rates of explosion pressure rise – Part 1: methane, hydrogen, and propene,* March 2006.

[5] Crowl, D. A., & Jo, Y.-d. (2009). A Method for Determining the Flammable Limits of Gases in a Spherical Vessel. *Process Safety Progress , 28* (3), 227-236.

[6] Cashdollar, K. L., Zlochower, I. A., Green, G. M., Thomas, R. A., & Hertzberg, M. (2000). Flammability of methane, propane, and hydrogen gases. *Journal of Loss Prevention in the Process Industries , 13*, 327-340.

[7] Coward, H. F., & Jones, G. W. (1952). *Limits of Flammability of Gases and Vapors.* BuMines, 503.

[8] Van den Schoor, F., Norman, F., & Verplaetsen, F. (2006). Influence of the ignition source location on the determination of the explosion pressure at elevated initial pressure. *Journal of Loss Prevention in the Process Industries* , 459-462.

[9] Mizutani, T., Matsui, H., Sanui, H., & Yonekura, M. (2001). Decomposing detonation and deflagration properties of ozone/oxygen mixtures. *Journal of Loss Prevention in the Process Industries*, *14*, 559-565.

[10] Koike, K., Nifuku, M., Izumi, K., Nakamura, S., Fujiwara, S., & Horiguchi, S. (2005). Explosion properties of highly concentrated ozone gas. *Journal of Loss Prevention in the Process Industries, 18*, 465-468.

[11] Cook, G. A., Spadinger, E., Kiffer, A. D., & Klumpp, C. V. (1956). Explosion Limits of Ozone-Oxygen Mixtures. *Industrial and Engineering Chemistry*, *48* (4), 736-741.

[12] Bartknecht, W. (1981). *Explosions, Course Prevention Protection.* New York: Springer-Verlag.

[13] Project SAFEKINEX, Contract No. EVG1-CT-2002-00072, Deliverable No. 2, *Report on the experimental factors influencing explosion indices determination*, September 2003.

[14] Le Chatelier, H. (1891). Note on the Determination of Firedamp by Flammability Limits. *Ann. Mines*, *19*, 388-395.

[15] Babrauskas, V. (2003). *Ignition Handbook.* Issaquah: Fire Science Publishers.

[16] Koike, K., Inoue, G., Suzuki, M., Saida, J., & Komatsu, I. (1998). Decomposition Characteristics of Concentrated Ozone. *Journal of Chemical Engineering of Japan*, *31* (2), 195-200.

[17] Schroeder, V., & Holtappels, K. (2005). Explosion Characteristics of Hydrogen-Air and Hydrogen-Oxygen Mixtures at Elevated Pressures. *International Conference on Hydrogen Safety.* Pisa.

Process Safety Considerations for New DOE Depleted Uranium Hexafluoride Conversion Plants

Peter M. Burban, Ph.D.
Department of Energy
Portsmouth Paducah Project Office
Peter.Burban@lex.doe.gov

Prepared for Presentation at
American Institute of Chemical Engineers
2013 Spring Meeting
9th Global Congress on Process Safety
San Antonio, Texas
April 28 – May 1, 2013

UNPUBLISHED

Process Safety Considerations for New DOE Depleted Uranium Hexafluoride Conversion Plants

Peter M. Burban, Ph.D.
Department of Energy
Portsmouth Paducah Project Office
Peter.Burban@lex.doe.gov

Keywords: depleted uranium hexafluoride, dry conversion, fluidized bed, uranium oxide, hydrofluoric acid, layers of protection, independent safety system, defense in depth, basic process control system, HF detectors, passive safety systems, Department of Energy

Abstract

The Department of Energy (DOE) OH & KY depleted uranium hexafluoride (DUF6) facilities are first of a kind operating plants designed to convert DUF6 into a safer form for reuse and/or disposal. Safety played a lead role in process design with HAZOP results guiding implementation of layers of protection to reduce risks associated with several materials including UF_6, H_2 and HF. A stand-alone independent safety control system functions to provide independent monitoring and control capability for those processes that have significant safety considerations.

1. Introduction

The Department of Energy Piketon, OH and Paducah, KY DUF6 facilities, Figure 1, are new fluid bed operating plants designed to convert approximately 700,000 metric tons of depleted uranium hexafluoride into a safer form for reuse and/or disposal. Depleted uranium hexafluoride has been generated in the United States since the mid-1940's as a byproduct of enriching uranium for both civilian and military applications at three gaseous diffusion plants located near Oak Ridge, Tennessee; Paducah, Kentucky; and Piketon, Ohio. This paper will highlight important process safety design considerations implemented that are critical to ongoing operations. Key aspects of the process hazards analysis and some lessons learned will be shared regarding chemical hazards associated with HF.

Figure 1. Portmouth (top) and Paducah DUF6 facilities

2. DUF6 Conversion Process and Safety Considerations

2.1 Process Description

The source DUF6 material, stored at both plants in approximately 63,000 large steel cylinders, is fed direct from the storage cylinders. A DUF6 feed cylinder, heated in a specially designed autoclave, vaporizes DUF6 which is fed with steam and hydrogen, blended with nitrogen, to a high temperature fluidized bed reactor for conversion to uranium oxide powder and hydrofluoric acid vapor. Some elements of the chemical process are described by Urza [1] and a general schematic of the process is given on the internet [2]. The uranium oxide powder transfer system loads powder into modified feed cylinders. The HF recovery system condenses the water and HF vapors in the process offgas, producing aqueous HF acid which is temporarily stored in tanks and later loaded for shipment to HF vendor facilities. The environmental green aspects of this approach include reuse of the feed DUF6 cylinders, regeneration of KOH used in scrubber operation and recovery of aqueous HF for commercial sale.

2.2 Overall Process Safety

A major safety advantage of the conversion process is low pressure reactor operation. The 'dry' conversion process uses superheated steam and nitrogen diluent to ensure safe reactor hydrogen concentrations. Administrative control of the feed cylinders ensures only depleted UF_6 is processed, eliminating any nuclear criticality event.

Hazard evaluations were performed to develop accident scenarios with the potential for exposure to HF, the greatest chemical hazard. Unmitigated consequences and risks were estimated for the credible events and compared to DOE risk guidelines. This paper focuses how safety significant engineered features and administrative controls are integrated in layers of protection to prevent or mitigate accidents associated with the major HF risks [3].

An Integrated Control System (ICS) controls and monitors the facility automated systems and associated components during normal operations. The system also mitigates accidents involving the conversion process and any of the systems or activities that have an interface with the process. The ICS consists of BPCS (Basic Process Control System) and ISS (Independent Safety System) subsystems. The purpose of the BPCS is to provide monitoring and control for normal operations from the Control Room and/or local operator workstations. The purpose of the ISS is to provide independent monitoring and control capability for those processes that have significant safety considerations. The ISS does not rely on the ISS human machine interface or BPCS to perform interlock actions.

3. Safety Features

3.1 Feed System Safety Features

A major component of the feed system is the autoclave, which serves to vaporize DUF6. The autoclave also serves as a pressure containment boundary (at several multiples of the supply pressure) to mitigate the consequence of an accidental release of DUF6 from a breached feed cylinder/piping in the autoclave. Since UF_6 reacts with atmospheric moisture to generate HF vapor, both HF and UF_6 releases have similar hazards classifications.

In the event of an over pressurized autoclave leading to activation of the pressure relief, the vent gases are directed to a controlled plenum. Automatic interlocks shut off DUF6 flow and/or electrical power to the cylinder heating system for an autoclave high pressure condition, as well as the following abnormal operating conditions: cylinder over-temperature, DUF6 header high-pressure, and an open autoclave door. The ISS monitors UF_6 and HF detectors located in the autoclave room which, upon alarm, terminate DUF6 flow to the downstream piping via an autoclave isolation valve. Additional administrative controls lead to shutdown of the reactor system. These layers of protection provide 'defense in depth' to reduce worker exposure from an abnormal UF_6 feed cylinder/system release.

3.2 Reaction System Safety Features

Each reactor serves as a primary containment boundary for reactants and products including DUF6 and HF. The ISS monitors UF_6 and HF detectors located in the reactor system room and adjoining HF condenser room which, upon alarm, terminate DUF6 flow to the downstream piping via the autoclave isolation valve. The BPCS in-line process oxygen monitors alarm on detection of oxygen in the process off gas. Additional administrative controls lead to shutdown of the reactor system.

Upon significant UF6 or HF leak detection, specific administrative controls instruct control room operations to terminate hydrogen flow via an automatic hydrogen feed isolation valve (a redundant manual isolation valve is also closed). Additional administrative controls lead to shutdown of the reactor system. These layers of protection provide 'defense in depth' to reduce worker exposure resulting from breach of the reactor system.

3.3 HF Recovery System Safety Features

Off gases from each reactor pass through individual partial condensers which produce an aqueous HF stream collected in individual primary receivers located in the HF recovery room. Off gases continue flowing to first a water and then to a KOH scrubber. All scrubber outlet streams merge and flow to a final backup KOH scrubber prior to venting out the facility stack. Multiple scrubbers reduce the likelihood of stack HF emissions. A secondary overflow HF receiver for each primary mitigates primary receiver level control problems. The HF recovery system units are skid mounted and reside in a common secondary containment dike mitigating the effect of HF leaks.

The ISS monitors several HF detectors located in the HF recovery room which, upon alarm, terminate DUF6 flow to the specific reactor via the autoclave isolation valve. The ISS also monitors HF detectors located in the HF recovery area for HF leaks in the HF transfer line which, upon alarm, terminates flow from the HF receiver tanks. Upon significant HF leak detection, specific administrative controls instruct control room operations to terminate DUF6 and hydrogen flow via an automatic hydrogen feed isolation valve (a redundant manual isolation valve is also closed). Additional administrative controls lead to shutdown of the reactor system. These layers of protection provide 'defense in depth' to reduce worker exposure resulting from breaches in the HF recovery system.

An important series of lessons learned include proper maintenance of HF process lines to eliminate chronic leaks. In addition, care was taken for stress relief of all piping flange connections, and modifications were made to address piping areas experiencing vibration. The use of HF vapor highly sensitive paint was helpful to identify the onset of leaks. Additional isolation valves were installed to improve the safety and ease of corrective maintenance.

3.4 HF Storage Tank Farm Safety Features

The HF storage area includes several passive, active and administrative safety controls. Each HF storage tank is situated in a secondary liquid dike that can hold the entire capacity of the 10,000 gallon tank in case of a leak. The tanker truck and rail loadout areas have a gravity drain equipped with a pumping system to transfer spilled HF to either a containment dike or a storage tank. This minimizes the spill area to reduce the potential on and off site consequences of an HF leak.

The ISS also monitors HF detectors located in the HF storage area which, upon alarm, terminates flow via the HF storage tank automated isolation valve. In addition, the ISS terminates HF flow from the HF receiver tanks. Specific administrative controls instruct an operator to be in required attendance for all transfers. This operator observes each HF storage tank to mobile tanker transfer and can terminate the transfer via an emergency button which stops HF transfer flow even before HF detection by the ISS. Upon alarm, following administrative control, the operators in the control room isolate the HF recovery system and then shutdown the reactor system.

These layers of protection provide 'defense in depth' to reduce worker exposure resulting from leak accidents at the HF storage tank system.

3.5 Additional Safety Features

Several fire suppression systems minimize the likelihood of a fire causing a cylinder breach and release of HF and/or DUF6. Strategically placed vehicle barriers add "defense in depth" to reduce consequences associated with collision with storage tanks and staged feed cylinders. Safety design elements were also incorporated for natural event accident initiators so that process buildings and equipment meet design criteria for forces such as seismic and high winds. Safety management requires the use of appropriate personal protection equipment for all work activity with potential HF exposure. These layers of protection provide 'defense in depth' to reduce worker exposure resulting from HF leak accidents in various systems described above.

4. Discussion

The process safety approach at the DUF6 facilities blends active, passive and specific administrative controls. Considering the other chemical and radiation hazards at the facility, the ability to maintain a safe operation depends on reliable machines and human performance of required operator actions. A rigorous preventative maintenance program is administratively controlled to maintain the reliability of all active safety systems at the facility in line with DOE safety requirements. The Integrated Safety Management Program at the DUF6 facility includes systematic approach to training for Operations. Training includes preparing each individual with the toolbox needed to address all specific administration controls, responses to alarms and recovery from abnormal operating conditions. Monitoring of human performance is an essential element for maintaining safe operations and continuous improvement.

Major facility start up challenges included accurate procedure development and updates, initial training and continuous training as the Operations knowledge base grew. Unique lessons were learned for operations in the four season climates of Ohio and Kentucky, necessitating adaptation of procedures, and emphasis on minor equipment changes to minimize the need for corrective maintenance. Both sites have had a good start to their multiple decade operations campaigns built on the cornerstone of a defense in depth process safety approach.

5. References

[1] Urza, Inaky J., "Conversion of uranium hexafluoride to uranium dioxide" US Patent 4,830,841

[2] "The DUF6 Process," www.bwconversionservices.com, accessed on Feb. 11, 2013

[3] DOE, Final Environmental Impact Statement for Construction and Operation of a Depleted Uranium Hexafluoride Conversion Facility at the Portsmouth, Ohio, Site, DOE/EIS-0360, prepared by Argonne National Laboratory for U.S. Department of Energy- Oak Ridge Operations, Office of Environmental Management, Oak Ridge, Tenn. June 2004. energy.gov/nepa/downloads/eis-0360-final-environmental-impact-statement

(44b) Minimum Ignition Energy Study of Flowing Heat Transfer Fluid Aerosols: Experimental and Theorectical Approach

Szu-ying Huang[1], Xinrui Li[1] and M. Sam Mannan[2], (1)Mary Kay O'Connor Process Safety Center, Artie McFerrin Department of Chemical Engineering, Texas A&M Univerisity, College Station, TX, (2)Mary Kay O'Connor Process Safety Center, Artie McFerrin Department of Chemical Engineering, Texas A&M University, College Station, TX

Liquid-gas aerosol has been a significant concern in process industries for its complexity and unpredictability of formation, ignition probability, combustion mechanisms, and flame propagation with impact on onshore/offshore facilities. Heat transfer fluid, specifically, has been proved to have the possibility to form aerosol from pressurized leaking, spraying, and potential liquid impingment from surfaces. Having more combustible matter and higher energy density, the high-flash point aerosol may induce more hazardous fires and explosions even operated or stored under the safety limits. Among the indicators describing flammability, minimum ignition energy (MIE) is one of the most widely used and studied ones. It is capable to reflect the internal chemical properties of the material, as well as to make connection to analyze the possible facility ignition sources. This paper demonstrates MIE experimental result of commercial heat transfer fluid aerosols, and develops COMSOL-Multiphysics models as theoretical approach on MIE prediction of similar chemical species. The influence of aerosol droplet sizes, initial spray velocity, fuel-air equivalence ratio, and fluid temperature are analyzed. The characteristic ignition temperature, which is the "flash point" of aerosol system, is also developed to compare with single-phase material. Results of the study may be used for industry liquid handling, working fluid selection, mitigation design, and ignition elimination to prevent fire and explosion of aerosols.

Domino Effect Model Applied to the Hertfordshire Oil Storage Terminal (Buncefield Oil Depot)

Manuscript ID: 295031

Esteban Bernechea
Centre for Technological Risk Studies (CERTEC). Department of Chemical Engineering.
ETSEIB. Universitat Politècnica de Catalunya.
Av. Diagonal, 647. 08028-Barcelona (Spain).
esteban.bernechea@estudiant.upc.edu

Xavier Seguí
Centre for Technological Risk Studies (CERTEC). Department of Chemical Engineering.
ETSEIB. Universitat Politècnica de Catalunya.
Av. Diagonal, 647. 08028-Barcelona (Spain).
xavier.segui@upc.edu

Josep Arnaldos
Centre for Technological Risk Studies (CERTEC). Department of Chemical Engineering.
ETSEIB. Universitat Politècnica de Catalunya.
Av. Diagonal, 647. 08028-Barcelona (Spain).
josep.arnaldos@upc.edu

1. Introduction

On December 11, 2005, a series of explosions occurred at the Buncefield Oil Storage Terminal, located in Hemel Hempstead, Hertfordshire, England. The explosions resulted in large fires which spread throughout the installation, causing more than 40 injuries, the destruction of most of the installation and the emission of large clouds of black smoke to the atmosphere; the accident also caused significant damage to residential and commercial properties in the vicinity of the terminal, the evacuation of population centers near the terminal, and an overall economic loss estimated at close to £ 1 billion [1].

The Buncefield disaster is a perfect example of domino effect in the oil industry; the initial explosion, caused due to the overflow of one of the tanks, caused fires that engulfed more than 20 storage tanks, and resulted in the destruction of a significant part of the terminal. The domino effect is defined as a cascade of events in which the consequences of previous accidents are increased both spatially and temporally by following ones, thus leading to a major accident [2]. This definition describes perfectly the situation that occurred in the Buncefield terminal.

One of the characteristics that made this accident the object of many researches was that the overpressure wave ensuing from the first explosion was much higher than would be expected from a Vapor Cloud Explosion (VCE). However, the aim of this manuscript is not to model the exact explosion, or to determine the reasons that caused the overpressure resulting from it to be higher than expected; much works have been dedicated to this endeavor, producing satisfactory conclusions to the phenomenon. The objective of this work is rather to use the Buncefield Terminal layout to test a domino effect model proposed by the authors [3]; to achieve this, possible accident sequences after an explosion will be simulated, to ascertain if they are similar to the chain of accidents that took place during the real event. Also, Quantitative Risk Assessments (QRAs) will be performed using the layout: one in a traditional fashion [4], without accounting for domino effect, and another taking domino effect into account through the application of the proposed model. This is done to analyze the change in the iso-risk curves obtained for both QRAs.

First, a more detailed description of the accident is presented, followed by an explanation of the proposed domino effect model. After this, the way in which the model has been applied to the layout of the installation is described and the results obtained are presented and discussed. Finally, the conclusions reached by the end of the study are shown.

2. Description of the accident

A brief abstract of the incident as described in Volume 1 of The Final Report of the Major Incident Investigation Board [1] is presented in this section.

On Saturday 10, December 2005, a delivery of unleaded petrol started to arrive at Tank 912 in bund A. At about 05:30 am on 11 December, the safety systems that prevented the overfilling of the tank failed to operate. Petrol cascaded down the side of the tank and was collected in Bund A. As the released continued, a vapor cloud mixture of fuel and vapor was formed and started dispersing westwards. Up to 300 tonnes of petrol were released, and about 10% turned to vapor and mixed with air, eventually reaching dangerous concentrations. At 6:01 am on Sunday 11, December 2005, the first, in a chain of explosions, occurred. The explosions caused a huge fire which engulfed more than 20 tanks and burned for 5 days, destroying great part of the terminal.

3. Domino effect model

The model used in this work [3] was developed to assess the impact that the domino effect could have on the frequencies of the accidents used to perform QRAs on storage installations; however, it can also be used to produce stand-alone accident sequences.

The model is an algorithm that has been programmed in MATLAB; it follows the steps used to perform a QRA in a cyclical manner to produce accident sequences which, in the end, will produce frequencies of accidents in all the tanks that are part of the installation.

The first step to obtain the final frequencies of accidents is to input some initial data that describes the installation; these are: the number of tanks in the installation, the substances stored and their properties, the mass contained in each tank, the sizes of the containment dikes, the types of tanks (atmospheric, pressurized), the initial frequencies of accidents per tank and the distances between tanks.

After the characteristics of the installation have been defined, the initial Loss of Containment events (LOCs) that will occur in each of the units, along with their frequencies have to be specified. Using these LOCs, and knowing the substance stored, event trees can be used to propose sets of possible accidents, together with their probability of occurrence; when these probabilities are multiplied by the frequency of the LOC, the initial frequency of accident is obtained. The initial LOCs and the event tree used in this work are those proposed in "The Purple Book" [4]. The LOCs for atmospheric storage tanks and their frequencies are presented in Table 1.

Table 1: Initial LOCs in atmospheric storage units

Code	Definition	Frequency (y^{-1})
G.1	Instantaneous release of the complete inventory.	5×10^{-6}
G.2	Continuous release of the complete inventory in 10 min at a constant rate of release.	5×10^{-6}
G.3	Continuous release from a hole with an effective diameter of 10 mm.	1×10^{-4}

The steps followed up to this point are the same as for a regular QRA. To produce the accident sequences, one of the tanks will be selected to suffer the initial LOC, and the effects and consequences of the accidents in this unit will be calculated for all other tanks. Depending on the consequences, a level of damage, corresponding to a new LOC will be assigned to each of the affected units. This was done using the threshold values presented in Table 2.

Table 2: Threshold values applied in the model

Accident	Effect	Threshold Values		Consequence
		Pressurized Tanks	Atmospheric Tanks	
BLEVE	Thermal radiation (kW/m2)	≥ 100	≥ 100 [5]	G.3 type release
		< 100	< 100 [5]	No consequence
Explosion	Overpressure wave (kPa)	≥ 100	≥ 25 [6]	G.1 type release
		53 < ΔP < 100 [7]	20 < ΔP < 25 [7]	G.2 type release
		30 ≤ ΔP < 53 [7]	14 ≤ ΔP < 20 [7]	G.3 type release
		ΔP < 30 [5]	ΔP < 14 [8]	No consequence
Jet fire	Jet length (m) Heat Radiation (kW/m2)	dj ≥ d [5]	dj ≥ d [5]	G.1 type release
		dj < d & I ≥ 40 [5]	dj < d & I ≥ 10 [5]	G.3 type release
		dj < d & I < 40 [5]	dj < d & I < 10 [5]	No consequence
Flash fire	-	-	-	No consequence
Pool fire	Thermal radiation (kW/m2) Pool duration (min)	dp ≥ d [5]	dp ≥ d [5]	G.1 type release
		60 ≤ I & tp ≥ 10 & dj < d [9]	15 ≤ I & tp ≥ 10 & dj < d [9]	G.2 type release
		I ≥ 8 & dj < d	I ≥ 8 & dj < d	G.3 type release
		I < 8 & dj < d	I < 8 & dj < d	No consequence
dj = jet length; dp = pool diameter; d = distance between tanks; tj = duration of jet fire; tp = duration of pool fire				

Once new LOCs resulting from an initial accident have been defined in the affected units, the effects and consequences of subsequent events on surrounding remaining units will be estimated; this will lead to other LOCs in other tanks, producing more and more branches of possible events. For each of the levels of propagation, the frequencies of accidents will be calculated, using the frequency of the previous accident as the initiator and the same event tree for the probabilities of possible accidents. Once one of the accidents in a branch reaches a frequency lower than 1×10^{-9}, the sequence will stop, as it is deemed that this frequency is too low to be taken into account in the QRA.

Figure 1. Domino effect model algorithm

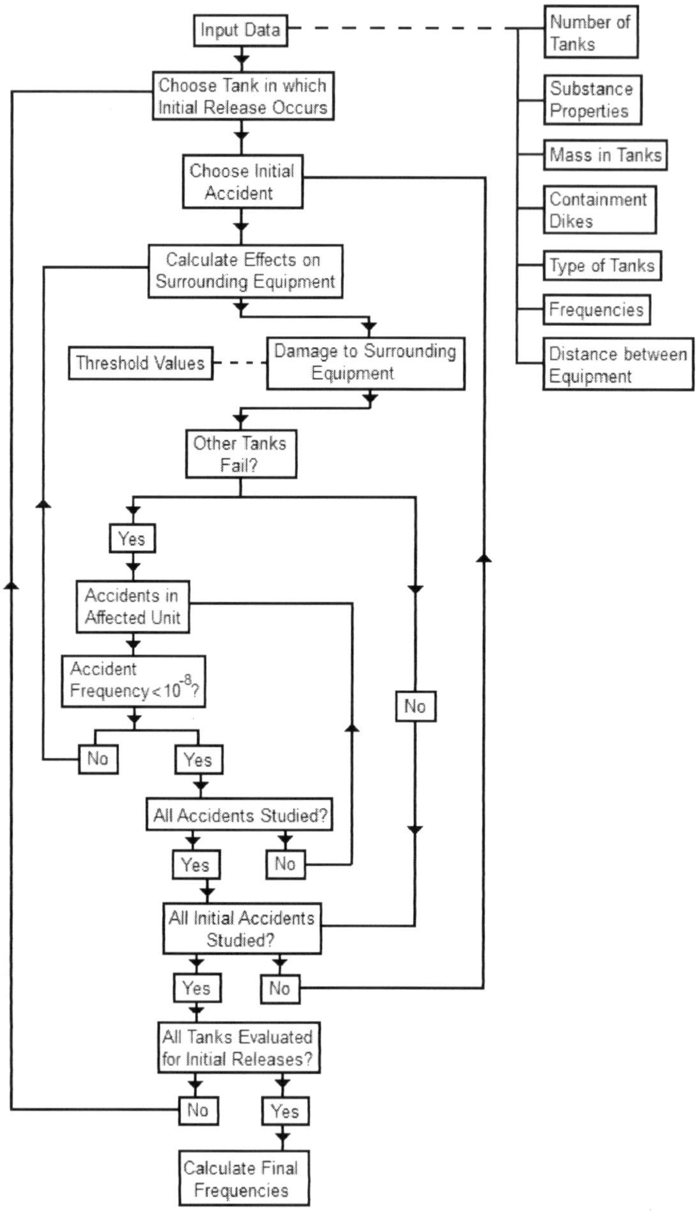

The process has to be repeated in loops, until all sequences have been explored, and all the tanks have been selected as initiators (suffering the initial LOC). In each step, the frequencies with which the different accidents occur in the different tanks are stored. Finally, these frequencies are added, by accident and by tanks, (for example, adding all the times that "accident 1" possibly occurs in "Tank 1"), whether due to an initial LOC event or due to the effect of an accident in another tank; these frequencies are referred to as final, and are the ones used in the QRA. A diagram of the algorithm that is followed in the methodology is presented in Figure 1.

The results obtained using the model can be compared against those obtained from a traditional QRA in order to assess the impact of domino effect on the risk associated to the installation. The model can also be used to produce a singular accident sequence, resulting from a specific accident in one of the units.

Figure 2. State of the terminal and proposed layout

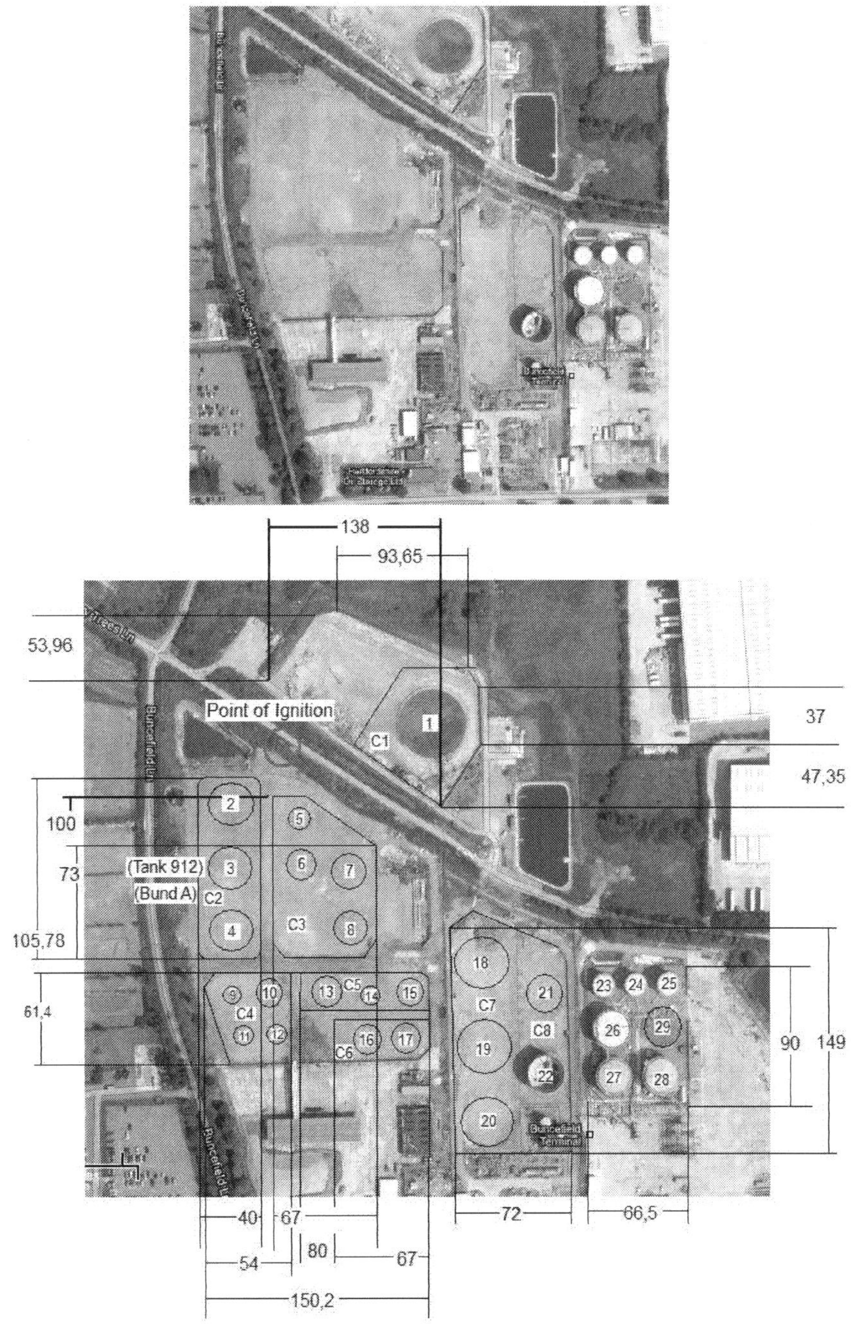

4. Applying the model to the Buncefield terminal

In this section the domino effect model will be applied to the terminal. First, to produce an accident sequence in a situation similar to the real accident and later, to produce iso-risk curves in two different QRAs.

4.1 Modeling the installation

To model the installation, the "Google Maps" tool was used to view the current state of the terminal, and obtain measures similar to those of the tanks that were destroyed and of the tanks that are still standing, as well as the distances between the tanks and the dimensions of the containment bunds. Figure 2 shows, on top, the latest image found of the state of the installation and on the bottom, the distances and dimensions used to model the installation. After estimating the dimensions of the installation, the tanks and bunds were renumbered for the application of the model; for example, Tank 912 becomes Tank 3, and Bund A is called C2.

4.2 Approaches to Accident Modeling

The vapor cloud that resulted in the explosion in the Buncefield Terminal is thought to have ignited at the emergency pump house located close to the containment bund A (in which Tank 912 was located), as shown in Figure 2. Since the domino effect model works by going through a fixed set of possible accidents in each of the tanks, the approach followed to produce the accident sequences was to define a fictional tank "Number 30" in the area in which the ignition of the flammable cloud is thought to have occurred; this tank is modeled containing a mass close to 150,000 kg of pentane.

This approach was followed to produce the accident sequence, but not to develop the QRAs, which were completed using the layout proposed in Figure 2, with all the tanks filled at full capacity.

4.3 Accident Sequence

Figure 3 shows the most probable incidents in the sequence produced when the domino effect method is applied to the case of an explosion derived from a G.2 type release in Tank 30 (the fictional tank created to model the initial explosion). The tanks are represented inside circles, while the possible accidents are shown inside diamonds (E stands for explosion, P for pool fire). The lines generated from the accident diamonds represent the level of damage caused by an event; these lines later divide, reaching different tanks, which means that the tank suffers the associated type of release after the previous incident. All the lines shown in the figure represent possible accidents identified by the methodology. The continuous lines represent the most probable accidents, which derive in the destruction or heavy damage of the affected unit, while the dotted lines represent accidents that do not cause catastrophic damages, or that are less probable; however, they are presented to show other possible scenarios produced when applying the algorithm. Once a tank is reached by a continuous line, it means that it has suffered a level of damage that will result in its destruction.

When an explosion (from a G.2 LOC) occurs in tank 30, the overpressure wave affects tanks 2, 3, 5, 6 and 7 in a catastrophic way; tank 1 suffers damages that lead to a G.2 release and tanks 4 and 8 are affected in a minor way. This is the first sequence of accidents, and it is understood that the failures and later releases occur almost simultaneously. After the mentioned tanks fail due to the explosion, other accidents will generate in each of them, specifically, explosions, flash fires or pool fires; since these last type of accident will be the most probable one (according to the event trees used to produce the sequences), it will be shown for each of the affected tanks.

Figure 3. Accident sequence developed using the domino effect model

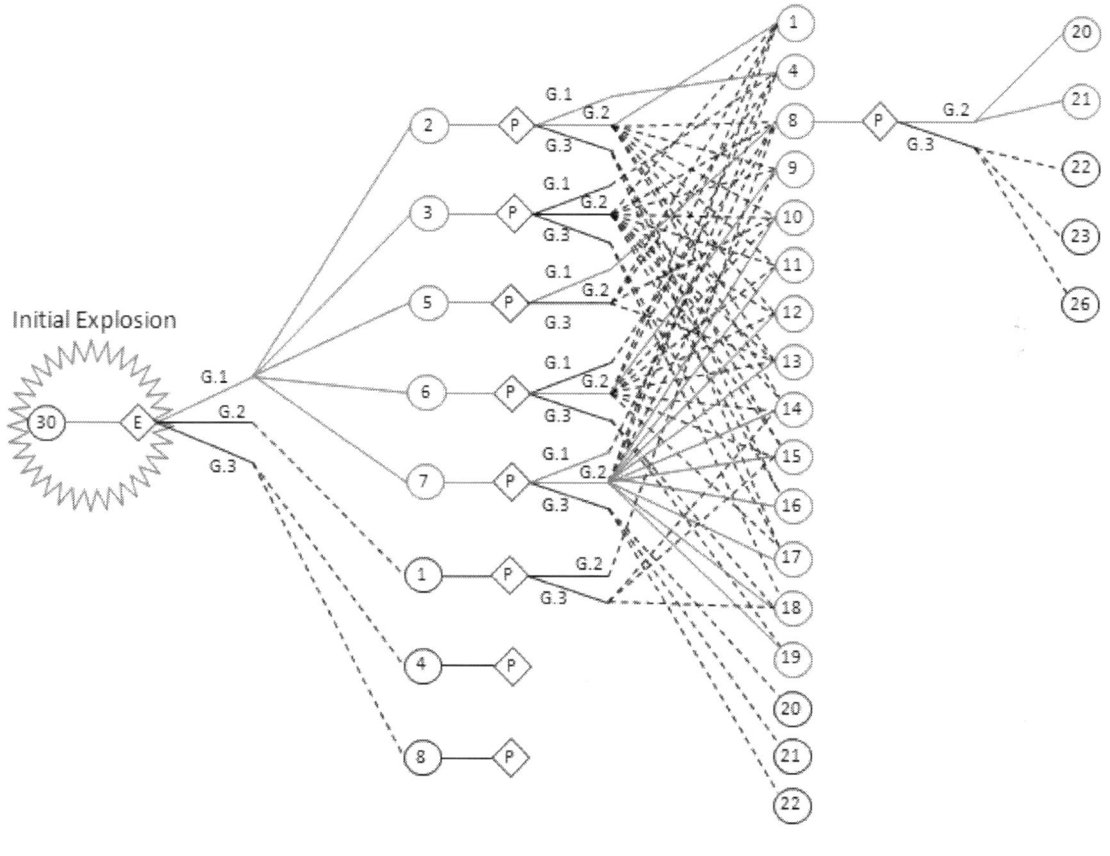

In the case that a pool fire occurs in Tank 2, Tank 4 will fail catastrophically due to the thermal radiation load it receives, while Tank 1 will also be affected in a major way, and also fail. There are other affected tanks, but these possible paths are not explored in detail here. It has to be noted that by the time Tank 2 fails, Tank 3 is also failing. A catastrophic failure in Tank 3 would also result in a pool fire that would further the chain of accidents, but none of these occurrences is explored. Pool fires in tanks 5, 6 and 7 would, according to some paths of the sequence obtained, result in the failure, and grievous damage, of tanks 8 through 19. If the sequence is furthered following a pool fire in Tank 8, tanks 20 and 21 would be damaged beyond repair, while tanks 22, 23 and 26 would suffer minor damages. It has to be noted that in the sequence explored, tanks 24, 25, 27, 28 and 29 do not suffer damages.

The sequence of accidents explored, which is the one that presented higher frequencies in each of its events, results in the destruction of 21 out of the 29 tanks that are part of the installation; this is in accordance to the level of damage in the real Buncefield terminal after the situations resulting from the explosion were controlled. It has to be said that this is one of many possible sequences produced by the algorithm, several of which are different from the real occurrence; however, branches of the possible sequences could be trimmed, in order to choose occurrences that are deemed to be more probable.

4.4 Application to QRA

Two QRAs were performed using the layout presented in Figure 2, the first using the normal procedure and the regular frequencies obtained from the original LOCs and the event trees; the second, using the frequencies of accident obtained from applying the domino effect model. The iso-risk curves obtained for the studies are presented in Figures 4 and 5.

Figure 4. Regular iso-risk curves

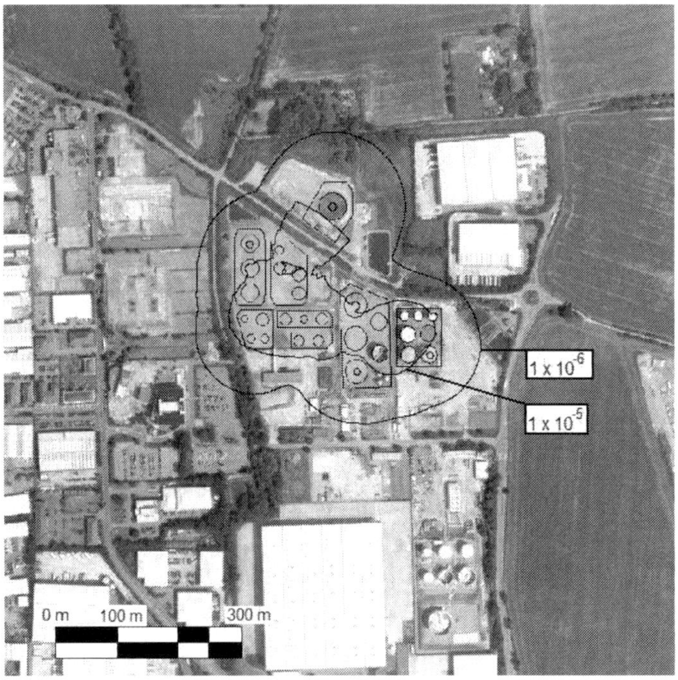

It is clear when observing the figures presented above that the individual risk associated to the installation increases when the domino effect methodology is applied. When regular frequencies are used, only 10^{-5} and 10^{-6} curves appear, while a 10^{-4} curve appears when using the modified frequencies; it encompasses the complete layout of the installation, and in some points, extends some 80 m from the limits of the terminal. The 10^{-4} increases it size so much because many lesser accidents with low frequencies appear in the accident sequences generated by the algorithm. The 10^{-5} curve also increases its area significantly, growing to the point of covering the complete evaluated installation, and reaching close to a 100 m from the limit of the terminal on its furthest point. The 10^{-6} curve is the one that presents a less appreciable change; to the north of the installation the curve suffers almost no alteration, to the south, it increases its length some 30 m and to the east and west, it increases some 20 m in each direction.

Figure 5. Iso-risk curves accounting for domino effect

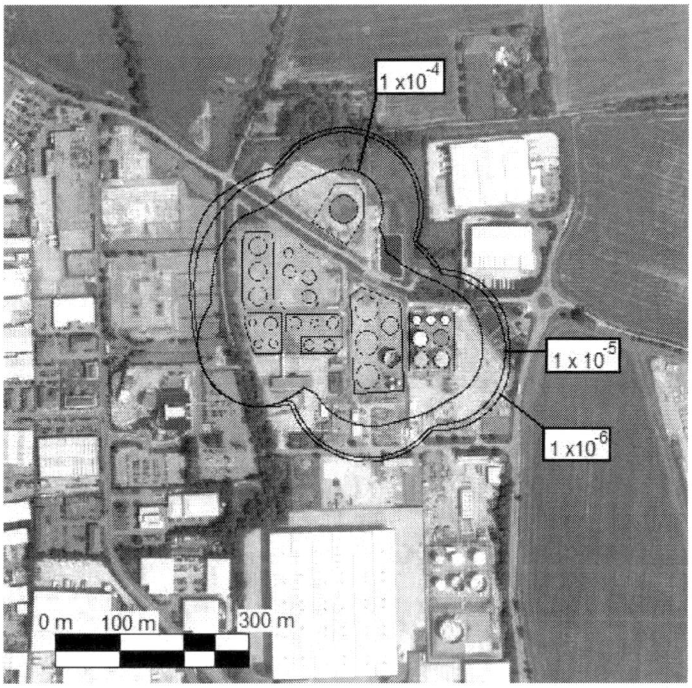

It is interesting to note that when performing a QRA of a storage installation composed of atmospheric tanks, the accidents with highest frequencies of occurrence are the pool fires, followed by flash fires and explosions. This means that the iso-risk curves related to the type of installation studied will not extend over great distances, as occurs when evaluating installations that store pressurized flammable liquids or toxic gases. Therefore, although the increase in the 10^{-6} curve, which, in many European countries is the most representative at the moment of evaluating external affectation, seems marginal, it is important, due to the type of installation that has been evaluated.

5. Conclusions

The domino effect model applied to a layout similar to that of the old Buncefield terminal produced many accident sequences; one of them is thought to be similar to the real accident that occurred in the facility in 2005. This means that the model is indeed suitable to be applied on real layouts, and that it can be used to help assess the risk related to a projected or existing installation, taking into account the possibility of domino effect occurrence.

The model can be applied with a minimum cost of time, and the frequencies obtained can be introduced in commercial software like "Riskcurves" without additional cost; QRAs could be performed taking domino effect into account, without the need to spend more time and money in the development of the study.

An increase in the risk associated to the installation was observed when the domino effect frequencies were used. The 10^{-6} curve does not suffer a significant increase in size due to the nature of the installation and its equipment.

Domino effect should be taken into account when performing QRA of industrial installations, as it can have an appreciable effect on the risk associated to the plant. This could lead to an impact on the Land Use Planning of the zone in which the facility is located.

References

[1] The Buncefield Incident 11 December 2005. The Final Report of the Major Incident Investigation Board. Volume 1, 2008. Buncefield Major Incident Investigation Board.

[2] Delvosalle, C, "Domino Effects Phenomena: Definition, Overview and Classification", First European Seminar on Domino Effects, Leuven, 1996.

[3] Bernechea, E.J., Vílchez, J.A., Arnaldos, J., 2012. A model for estimating the impact of the domino effect on accident frequencies in quantitative risk assessments of storage facilities. Process Safety and Environmental Protection, http://dx.doi.org/10.1016/j.psep.2012.09.004

[4] CPR 18E, 2005. Guidelines for Quantitative Risk Assessment (Purple Book). Directorate General for Social Affairs and Employment, The Hague, Netherlands.

[5] G. Antonioni, G. Spadoni, V. Cozzani, Application of domino effect quantitative risk assessment to an extended industrial area, Journal of Loss Prevention in the Process Industries, 22, 614-624, 2009.

[6] F. I. Khan, S. A. Abassi, Estimation of probabilities and likely consequences of a chain of accidents (domino effect) in Manali Industrial Complex, Journal of Cleaner Production, 9, 493-508, 2001.

[7] V. Cozzani, E. Salzano, The quantitative assessment of domino effects caused by overpressure: Part I. Probit models, Journal of Hazardous Materials, 107, 67-80, 2004.

[8] V. Cozzani, E. Salzano, Threshold Values for domino effects caused by blast wave interaction with process equipment, Journal of Loss Prevention in the Process Industries, 17, 437-447, 2004.

[9] G. Landucci, G. Gubinelli, G. Antonioni, V. Cozzani, The assessment of the damage probability of storage tanks in domino events triggered by fire, Accident Analysis & Prevention, 41, 1206-1215, 2009.

Proven Approach to Investigating Near Misses

Greg Smith, Senior Engineer
Process Improvement Institute, Inc. (PII)
7502 Chad Road
Harrison, TN 37341
Phone: (423) 413-0038
Fax: (865) 671-4944
e-mail: gsmith@piii.com

Gerald D. Burch, Senior Engineer
Process Improvement Institute, Inc. (PII)
e-mail: gburch@piii.com

William G. Bridges, President
Process Improvement Institute, Inc. (PII)
e-mail: wbridges@piii.com

Prepared for Presentation at
9th Global Congress on Process Safety
San Antonio, TX
April 28-May 1, 2013

UNPUBLISHED

9TH GLOBAL CONGRESS ON PROCESS SAFETY

Proven Approach to Investigating Near Misses

Steven (Greg) Smith, Senior Engineer
Process Improvement Institute, Inc. (PII)
7502 Chad Road
Harrison, TN 37341
Phone: (423) 413-0038
Fax: (865) 671-4944
e-mail: gsmith@piii.com

Gerald D. Burch, Senior Engineer
Process Improvement Institute, Inc. (PII)
e-mail: gburch@piii.com

William G. Bridges, President
Process Improvement Institute, Inc. (PII)
e-mail: wbridges@piii.com

Keywords: investigation, near miss, accident, incident, root cause, RCA

Abstract

Getting near misses reported is a major hurdle for most companies, one reason being is that the investigation system is not ready for the large influx of reports. Workers and management respond by being reluctant to potentially overload the investigation system (and overload themselves). This paper explains approaches to manage efficiency and effectiveness in near miss investigations that have been successful for handling near misses reporting within both large and small companies and to ensure high value from the investment of reporting and analysis. This paper builds on the updated paper presented in 2012 at GCPS on "Gains from Getting Near Misses Reported."

1. Introduction

Much attention has been given to the detection and correction of early warning signals as a means of improving safety performance.

Process safety events often have entirely different causal factors (human error, equipment malfunction, natural causes) and root causes (management system weaknesses) from those found in more traditional personal injuries.[1,2] Process safety incidents are more technical and complex in nature, usually involving multiple failures of protective barriers. The human actions and errors which cause them are often several steps removed from the incident itself. The human factors involved are not as obvious to investigators; the causes are more systemic than individual in their nature.

Even though personal injury rates have declined, significant injuries and financial losses have continued in a series of high profile, catastrophic (multiple injuries) process safety events. Because of inconsistencies in reporting standards and definitions, it is difficult to tell if these events are actually increasing in frequency or if their visibility has simply increased due to media attention and heightened public concerns. Either way, improved understanding and enhanced tools are required to improve our process safety performance.

In the process industries one such tool is near miss reporting, the early warning system equivalent to unsafe behaviors for process safety. Companies which have successfully developed the ability to recognize, report, investigate, and learn from near misses have shown dramatic reductions in loss events, losses, and injuries due to process incidents. In some cases, loss reductions of more than 90% have been achieved.[2]

2. Definitions

The following definitions are used throughout this article:

> *An incident is either an accident or a near miss.*
>
> *An accident is a sequence of unplanned events and conditions that result in harm to people, environment, process, product or image.*
>
> *A near miss is an unplanned sequence of events that could have caused harm if conditions were different or is allowed to progress, but did not in this instance.*
>
> *A causal factor is a human error (typically an error by the at-risk employee performing a task/job in the process) or a component fault/failure. Note that these human errors and component failures are probably caused by other humans making mistakes, and all errors are controlled by management systems. An incident typically has multiple causal factors. Natural phenomena can also be a causal factor.*
>
> *A root cause is a management system weakness that results in a causal factor. A casual factor typically has multiple root causes.*

The definitions above are the same as used in *Guidelines for Investigating Chemical Process Incidents, Second Edition* (CCPS, 2003).

3. Importance of reporting and learning from near misses

Investigating near misses is critical to preventing accidents, because near misses share the causes and root causes of accidents; they are one or two barriers away from the loss/accident. We are very likely preventing many apparently unrelated accidents when we prevent the ones that are obviously related. Figure 1 illustrates a hypothetical relationship between causal factors and root causes of accidents and those for near misses. From our experience, this relationship appears to be valid; see the case study section of this paper to see some of the proofs of this concept.

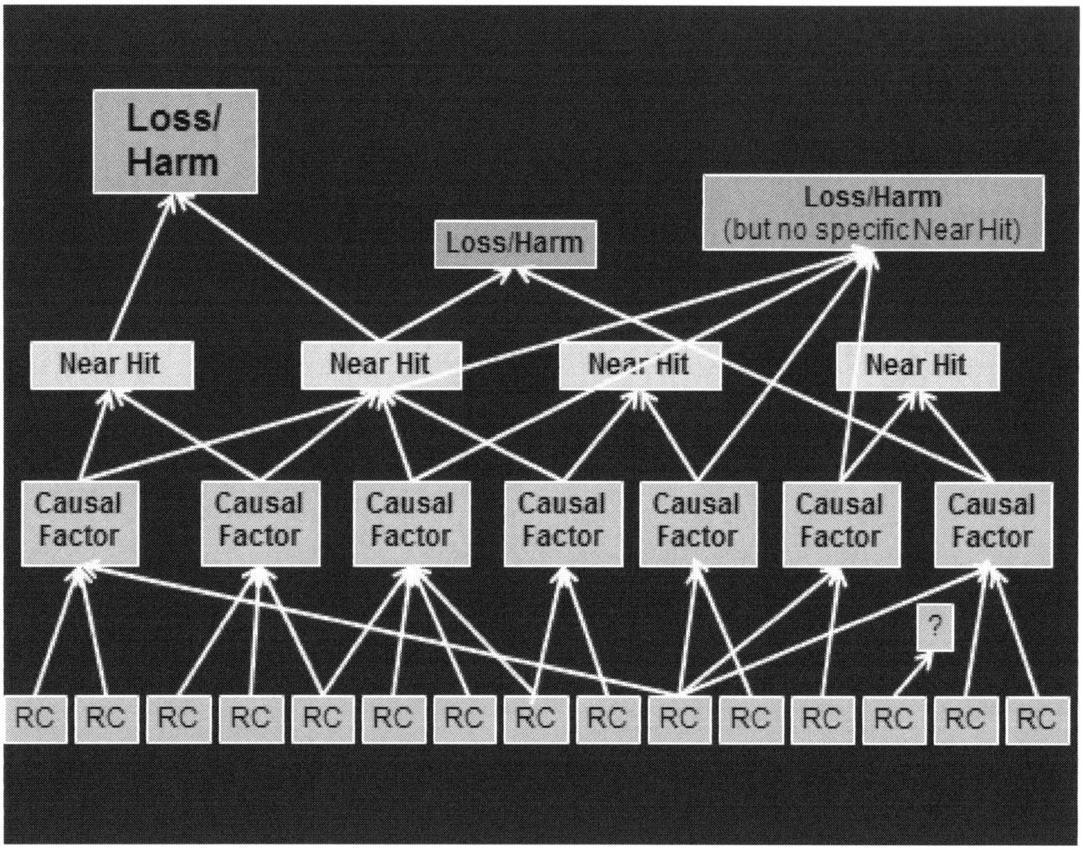

Figure 1 Interrelationship between the causes of Accidents (Losses) and the causes of Near Misses

As a brief explanation of Figure 1, root causes (management system weaknesses) make it more likely for a causal factor to occur. Combinations of causal factors (or in rare cases, perhaps a single causal factor) result in near misses. Additional causal factors result in losses (accidents) in increasing severity. A root cause can increase the likelihood of seemingly unrelated causal factors.

Formal and informal organizational surveys[3] detail the difficulties of implementing an effective near miss reporting system. In this paper, a number of barriers are outlined and potential solutions are proposed. The identified barriers are:

1. *Fear of disciplinary action*

2. *Fear of teasing by peers (embarrassment)*
3. *Lack of understanding of what constitutes a near miss versus a non-incident*
4. ***Lack of management commitment (no training provided on investigation techniques and procedures) and lack of follow-through once a near miss is reported (time is not allocated to investigate near misses, or corrective actions not implemented)***
5. ***An apparently high level of effort is required to report and to investigate near misses compared to low return on this investment***
6. ***There is No Way to investigate the thousands of near misses per month or year***
7. *Disincentives for reporting near misses (e.g., reporting near misses hurts the department's safety performance)*
8. ***Not knowing which accident investigation system to use (or confusing reporting system)***
9. *Company discourages near-miss reporting due to fear of legal liability if these are misused by outsiders*

4. Primary Barriers to Effective Investigation of Near Misses

Four of the barriers (4, 5, 6, and 8) tie directly to the organization's ability to effectively and efficiently investigate and resolve near miss incidents.

Of the four barriers related to getting near misses reported, the two that are most related to effective near miss investigations are **Barriers 6 and 8.**

Overcoming these barriers is described below by looking at the experience of organizations as they implement these concepts.

*4.1 **Barrier 6. There is No Way to investigate all of the thousands of near misses per month or year!***

Normally, when the discussion of having a huge number of near misses reported, such as four per worker per month, the reaction will be shock and then a statement such as Barrier 6. This barrier is closely related to Barrier 5*: An apparently high level of effort is required to investigate near misses compared to the small gain perceived.* If a site has 500 staff as operators and maintenance craftsmen, then likely 25,000 near misses could be reported. At first glance, it can appear impossible to cope with, let alone investigate that number of near misses (incidents).

Maybe the most important solution to Barrier 5 and 6 is to first decide which near misses and losses/accidents need to be investigated. The best solution is normally stated as:

> ***Let front line foremen or supervisors decide if a near miss or accident needs to be investigated to root causes; the decision is made on the apparent Learning Value of the incident. How do you decide it a high or low learning value? The foremen or supervisors will easily decide based on their hands-on experience.*** *For instance, an additive pump failure may directly result in a*

minor and temporary impact to quality or production rates but does not appear to have the potential to cause a large loss. So the foreman of the shift decides to designate this as a low learning value incident and therefore not do an immediate RCA. But, then other similar pumps fail over a six month period. A query shows that after 12 months, there have been 3 failures of the same additive pump and 8 failures of 5 other similar pumps in different services. Based on the number of failures and cumulative losses, the company performs an RCA of the 11 failures, resulting in findings which when applied to all similar pumps and services, greatly reduces the production/quality impacts. The RCA of the grouping of incidents proved very cost effective.

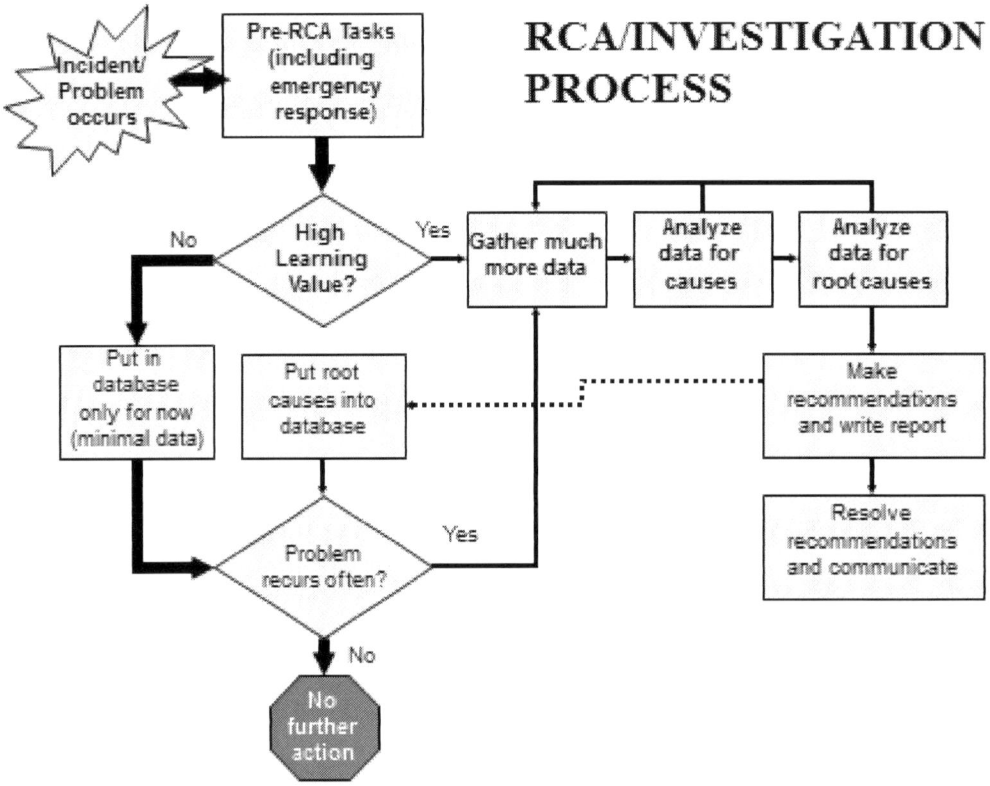

Figure 2: RCA/Investigation Flow Process

Figure 2 illustrates the process flow for an investigation system than can handle a large volume of near misses and losses/accidents. For this process to work:

- Be prepared for investigations by having enough staff trained in root cause analysis methods (or to help in the analysis, such as being able to interview peers).
- When the near miss, etc., is first noticed or reported by staff, let the frontline supervisor or foreman decide if it has high learning value.
- For high learning value incidents, investigate now. For low learning value incidents, put in the database now, along with the little data you have and any obvious causes. Do not investigate yet!

- Query the database every one to six months and perform Pareto or similar analysis to help decide which recurring events need to be analyzed in more detail.
- Take the root causes from investigations/RCAs and put in the database as well.
- Query the database every one to six month

4.2 Barrier 8. Not knowing which accident investigation system to use

One consideration that is not related to any of the barriers mentioned above, except marginally to Barrier 3 (lack of understanding of what a near miss is), is the scope of the investigation program. Some companies have one investigation system for occupational safety incidents, another one for process safety incidents, another for environmental releases, another for reliability issues, and yet another for quality and customer services issues. We have found that the same investigation approach and investigator training works well for incidents in any facet of a business. We believe there is merit in combining the systems and, in particular, in combining the incident databases. Combining the incident systems will require more work on defining near misses and in determining success in reporting near misses.

Consider having ONE incident reporting system with ONE approach for teaching employees the definition of a near miss and with ONE approach for doing incident investigations (including one approach for root cause analysis).

5. Effective and Efficient Incident Investigation and Root Cause Analysis

The incident investigation and root cause analysis (II/RCA) process should enhance organizational learning, reduce accidental losses, invest wisely in protective and mitigative systems, better protect people and equipment from harm, and continuously improve EHSS performance. An efficient and effective process ensures that scarce organizational resources are used appropriately in activities that matter.

The overall II/RCA process consists of the following steps:

- Data collection
- Causal factor analysis
- Root cause analysis
- Develop recommendations
- Reporting and follow through

5.1 Data collection

Every investigation begins by collecting, organizing, and preserving all relevant data as quickly as possible after the incident has been stabilized. Data comes in the form of personnel interviews, physical evidence, and documentation. All data is fragile and susceptible to loss or damage if it is not properly collected and preserved in a timely manner. All front line supervisors should be trained in the II/RCA process with particular emphasis on initiating the investigation, interviewing people, and collecting the physical evidence. Provide front line supervisors with an investigation tool kit containing a camera, barrier tape, collection bags and bottles, and appropriate forms and checklists.

5.2 Causal factor analysis

Once collected, data must be organized and analyzed to determine what happened – the chain of events (causal factors) that led to an undesirable occurrence. The most effective and proven methods for analyzing data are Event and Causal Factor Charting (ECFC) supplemented, if needed, with a Simplified Fault Tree Analysis for complex, chronic, or acute events. These are only effective when the rules are followed for their use – but with training and practice most people can be taught to identify causal factors using these basic tools for most of the incidents they will encounter.

Event and Causal Factor Charting: When an investigation team begins a root cause analysis (investigation), the first step is to prepare an Event and Causal Factors Chart (ECFC).

- Arranges event building blocks (sticky notes) to graphically depict cause-effect relationships between known events
- Allows for testing of "necessity" and "sufficiency"
- Best method when timing of events is important

Events make up the backbone of the ECFC. Event statements describe specific occurrences (e.g., "4-12 shift operator filled Tank 123," or "Control room operator acknowledged level alarm for Tank 123"). Conditions usually provide descriptive information (e.g., "Pressure was 1000 psig" or "The operator did not have hands-on training in this task") as opposed to stating action (e.g., "Operator placed Valve ABC into open position"). Primary events form the basic sequence in the diagram. Secondary events are actions that impact the primary events, but which are not directly involved in the situation. Figure 3[3] provides the steps for developing a casual factor chart.

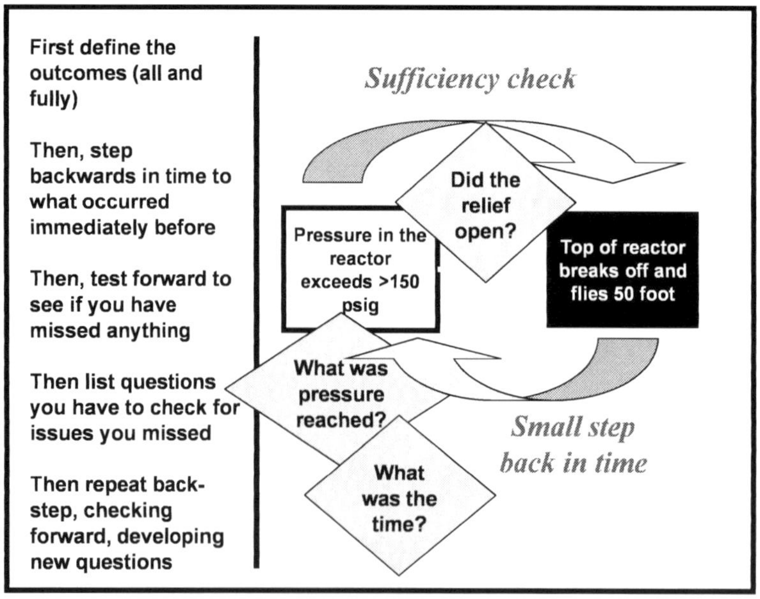

Figure 3: Steps for Casual Factor Charting

If applied correctly, ECFC will aid the investigators in generating a great many new questions. These will lead to more data gathering. Above are a couple of examples of how we show these questions on an ECFC. The answers to the questions sometimes become new events or conditions on the ECFC, as illustrated in Figure 4[3]:

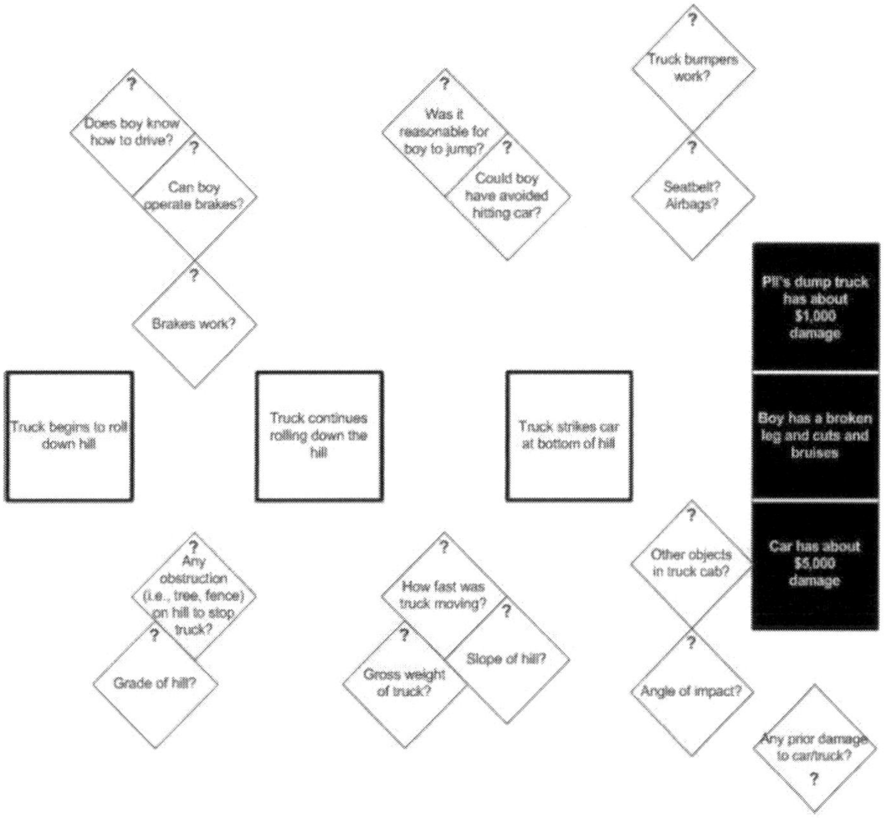

Figure 4: Example of a Partial Casual Factor Chart

The investigators will go back and forth between doing the ECFC, generating new questions, getting new data to answer the questions, and taking more steps backwards in time and sequence. This process continues until all of the data is exhausted and/or the investigators fully understand the sequence of the events and all of the mistakes and failures that led to the final ending condition or loss.

After the ECFC has been completed, the investigators are in a good position to identify the negative factors that influenced the course of events. These components are labeled <u>causal factors</u>. Causal factors, which may be in the form of events or conditions, are those items that are considered to be major contributors to the incident and as defined earlier, are either component failures or human errors or severe environmental problems.

The logic rules for constructing an ECFC are critical to success. These govern how small of step backwards the investigators take backwards in time and how many questions they generate as they test to ensure they understand each pair of building blocks.

5.3 Root cause analysis

When all the causal factors have been identified, the investigation team should begin asking "Why?" each causal factor occurred. Using a combination of the "5 Whys" and a Root Cause Chart™ (derived from the management oversight and risk tree process developed for US DOE's Savannah River Laboratory) shown in Figure 5, step down each of the root cause paths by:

- Select a causal factor from the causal factor chart or fault tree
- Work through the map for each causal factor
- Ask Why to help gathering the right data to get to the lowest level (root causes)
- Step down paths, identifying:
  - causal factor type (equipment, personnel, other)
  - root cause category
  - near root cause
  - root cause
- Record results on forms, at each step (see next page)

The Root Cause Chart™ structures the reasoning process for identifying root causes, identifies detailed root causes for each causal factor, facilitates consistency across all investigations, supports trending of "root causes" and "categories", and gives more thorough coverage of root causes by reminding investigator of all possible root causes.

Figure 5: Root Cause Chart™

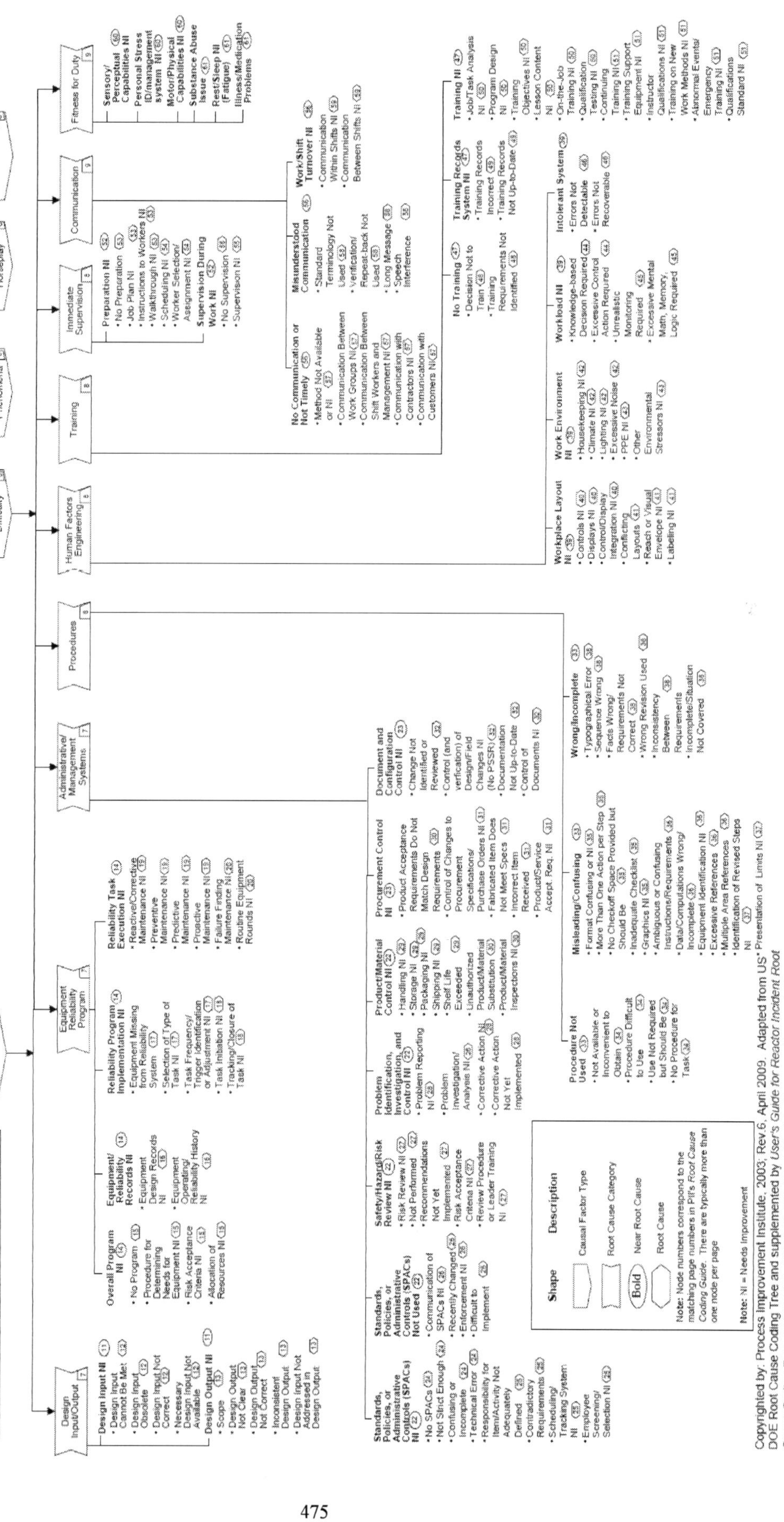

5.4 *Develop recommendations*

The investigation team should develop recommendations to remedy each root cause of each causal factor. Recommendations should describe the general objective to be accomplished and the consequences if not completed. Recommendations can also include possible solutions considered by the team and an example of how this could be accomplished.

5.5 *Reporting and Follow Through*

5.5.1 Simplify the reporting of the investigation/RCA results to the bare minimum needed

Part of the reason for the belief that it is impossible to investigate large numbers of near misses stems from the large reports currently required by the company for investigating incidents. Some companies insist on producing what they call "professional" reports of accidents, and these grow to 50 or 100 pages (half of the pages are attachments). Why produce such a large report? What is the use of that large of a report? What makes "size" of a report equivalent to "professionalism" of a report? Think about every aspect of the report and make sure it is needed. Normally, all that is needed is a:

- Cover sheet that includes the date, time, location, one or two sentence description of the near miss or incident, and a title that summarizes the incident at a glance. The cover sheet should also list the team members.
- Forms that have the causal factors filled in and the root causes filled in, with perhaps one or two sentences that explain the root cause. These forms can also contain the recommendations necessary to correct the root causes.
- That's It!

So, most near miss investigation results will be two pages or so and most loss/accident reports will be four pages or so.

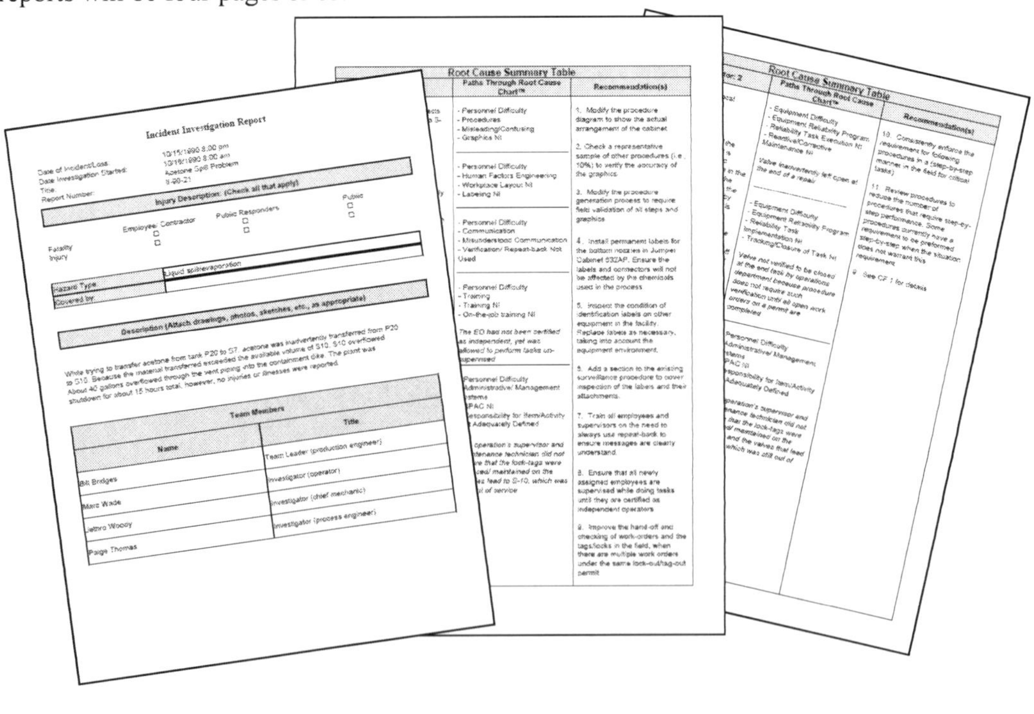

The complexity of results reporting has grown from the legacy of only investigating losses/accidents. When an organization gets a large ratio of near misses report and therefore a large number of investigations going on, the reports must shrink. This is a good thing. However, if you still have a major accident (which you won't have if you get a large number of near misses reported and investigated), then add more documentation to meet the needs related to litigation, regulatory interface, etc.

5.5.2 Deciding which incidents should be investigated

All near miss incidents should be reported and cataloged with basic information and apparent causes, but not all near miss incidents will warrant a full investigation. Only those incidents with either high potential impact or high learning value should be investigated using the II/RCA process. This is an excellent opportunity to engage staff and give them control over the process by granting front line supervisors the authority to decide which incidents have high learning value. For example, establish a shift target of investigating 15-20% of the near miss incidents reported and then let them decide which ones they want to investigate to meet their goal.

6. CASE STUDY: One Company's Experience with II/RCA

A large pharmaceutical company (more than 50,000 employees) initiated a drive to report and analyze large numbers of near misses. Before this initiative, they had about 3 near misses reported per loss event. Their loss categories included quality impacts, reliability issues, safety mishaps, and environmental issues. Also before the initiative, they had five incident reporting systems, investigation protocols, and databases. One of the first steps the company took towards the stated goal was to consolidate the five reporting and investigation systems into one system. They tried to use one of the existing database systems for this consolidation, but decided that all were deficient for this purpose. So, they modified the reliability incident database to make it suitable for all types of incidents. This took several months of effort (not full time though).

In parallel, they adopted a new investigation approach, which is the one described in this paper. This involved changing the database structure to provide pull-down menus for each branch on the Root Cause Chart. The chart was not modified. Next, a training course for lead investigators was customized to meet their new needs and to use their examples in the exercises and case studies. Then, this course was presented seven times in the USA and UK, thereby training an initial group of 200 investigators but more importantly training 7 instructors within the company. The eventual goal was to train 15% of the operating staff how to perform root cause analysis (RCA) of near misses and losses. The new database and consolidation of reporting and analysis systems (developed in parallel with the development of the customized training materials), was rolled out at the same time as the initial training courses. The reporting system was incorporated into their Lotus Notes e-mail system. The reporting forms were very simple and any worker could report a near miss or incident. Initial reports were routed automatically by the e-mail system to the worker's supervisor and viewable by any qualified investigator. A lead investigator on the same shift rotation and from the site (but not from the same group) was assigned to each incident; investigations typically began within one hour and on average were completed within the next hour.

These initial efforts were extremely successful and within about one year, the company achieved a ratio of more than 100 misses being reported per loss event. All near misses and loss events were captured in the database of incidents and about 20% were high value and immediately investigated. Typical reports were 2-3 pages. After the first year, the sites performed queries of the databases and based on Pareto analysis, performed RCA on the highest number of repeated "similar" incidents. In subsequent years, the queries were performed every three months.

This new system was much more productive in reducing losses than the previous collection of five systems. Overall, production and quality losses were reduced about 95% with little or no increase in time spent in RCAs (because the number and magnitude of losses decreased dramatically and so there was much less time spent on fire-fighting and major investigations).

7. Conclusions

Near miss reporting and investigation has been very effective in reducing losses due to process safety incidents. However, implementation has proven difficult for many companies due to several implementation barriers. The proven methods described will improve both the efficiency and the effectiveness of the II/RCA process. Consistently applying these methods will enable near miss reporting, providing early detection and correction of process safety warning signals, thereby reducing losses and improving process safety performance.

8. References

1. Busch, Carsten, *Heinrich's Common Cause Hypothesis: A tool for creating safety*, July 30, 2012, http://www.predictivesolutions.com/safetycary/heinrichs-common-cause-hypothesis-a-tool-for-creating-safety/
2. Wright, Linda and van der Schaaf, Tjerk, *Accident versus near miss causation: a critical review of the literature, an empirical test in the UK railway domain, and their implications for other sectors*, Journal of Hazardous Materials 111 (2004) 105–110.
3. Bridges, William, *Gains from Getting Near Misses Reported,* AIChE 8th Global Congress on Process Safety, 2012.
4. *Incident Investigation and Root Cause Analysis Course Materials*, Process Improvement Institute, Inc., 2003-2013.

9TH GLOBAL CONGRESS ON PROCESS SAFETY

Key Strategies for Implementing a Corporate Fatigue Risk Management System

David E. Cummings
E.I. DuPont de Nemours & Co.,Inc.
1007 Market Street, Wilmington DE 19898
david.e.cummings-1@dupont.com

Prepared for Presentation at
American Institute of Chemical Engineers
2013 Spring Meeting
9th Global Congress on Process Safety
San Antonio, Texas
April 28 – May 1, 2013

UNPUBLISHED

Key Strategies for Implementing a Corporate Fatigue Risk Management System

David E. Cummings
E.I. DuPont de Nemours & Co.,Inc.
1007 Market Street, Wilmington DE 19898
david.e.cummings-1@dupont.com

Keywords: fatigue, fatigue management, ANSI/API 755, human factors

Abstract

The First Edition of ANSI/API Recommended Practice 755 "Fatigue Risk Management Systems for Personnel in the Refining and Petrochemical Industries" was published in April 2010 as a result of the BP Texas City catastrophe of March 2005. In the last two years, many companies in the petroleum and chemical processing sectors have been working to implement the basic tenants of this new practice. An effective fatigue management program involves a multi-disciplined team approach to engage various functions of an organization, including PSM/EHS, HR, Medical, and Operations. This paper will explore the process, associated elements, and key issues integral to the design, development, and implementation of a formal fatigue management system across DuPont as a global manufacturing organization with a diversity of high hazard processes. This includes internal policies, training and development, standardized procedures and tools, and internal collaboration strategies.

1. Introduction

ANSI/API Recommended Practice 755 [1] was issued in 2011 based on a recommendation issued to the American Petroleum Institute by the U.S. Chemical Safety Board as an outcome of the BP Texas City 2005 incident investigation report. Prior to that time, there was little industry guidance or standards on the subject of fatigue management (FM). There is and has been substantial literature published and technical resources developed on the general issues and concerns with fatigue management in terms of circadian rhythms and the effects on human performance variables, and many industries such as aviation and commercial transportation have adopted specific standards with accompanying regulatory requirements.

E.I. du Pont de Nemours & Co. is a global science-based products and services company with over 200 manufacturing units across 70 countries involving nine business segments. It is engaged in technology solutions and innovation across a range of disciplines, including agriculture and industrial biotechnology, chemistry, biology, materials science and manufacturing, offering a wide range of products and services for markets including agriculture and food, building and construction, electronics and communications, general industrial, and transportation. Manufacturing technologies and site operations involves a myriad of process hazards, including flammable liquids and gases, highly toxic materials, reactives, combustible dusts, corrosives, asphyxiants, and thermal/mechanical energies. DuPont's philosophy and current practice is to apply process safety management systems and requirements to all manufacturing operations and locations using one global set of internal elements and technical requirements. Based on the company's current risk profile, process safety systems that help maintain a high degree of operational discipline and reduce the potential for human error, including those that could be related to employee fatigue, are key elements of an integrated risk approach.

DuPont's process safety management model and system has included a long standing element and requirements for "Training and Performance" where the performance portion of the element requires a specific focus on physical capabilities (fitness for duty) as well as diminished capacity (mental and psychological capabilities for process safety sensitive jobs, including the effects of fatigue) for operating and maintenance personnel working at all manufacturing locations. DuPont updated the "Training and Performance" element to specifically include an internal fatigue management policy as an integral part of the global process safety management program in January 2011. Initial implementation of the new technical requirements has been progressing since that time. The program embraces the following key elements of fatigue management as described in ANSI/API 755:

- Maximum hours of service
- Training for employees and line management
- Staff workload balance
- Work environment evaluations
- Medical Evaluations (personal risk factors)

Implementing a fatigue management system was integrated into corporate process safety management systems; however, active involvement of other key functions within the organization is important in this area. Within DuPont, the process safety competency serves as the corporate "owner" on FM, while other key functions such as Human Relations (HR), Integrated Health Services, and Occupational Health (EHS) all have played a meaningful and contributing role in developing and implementing portions of the FM process in the initial design stages, during implementation, and in an ongoing role as part of the "run and maintain" phase of this work. This collaboration has contributed greatly to the overall results.

2. Foundational Activities for FM Implementation

Building an internal fatigue management system based on the elements in the API standard requires a number of key strategic decisions and initiative planning prior to formal roll-out, as well as ongoing support, training, and tools during the active implementation phase. The following provides examples of many questions that were posed as part of DuPont's initial design strategy for fatigue management:

- Should the fatigue management program be a global or U.S. based only focus?
- What is the current state at all sites (does anyone already have an effective FM system)?
- What will be the major impacts on Operations at each site (resources, costs, tracking tools)?
- Who are the key stakeholders who should be involved in the design of the system?
- Which sites will be required to implement a local fatigue management system?
- How will fatigue management requirements be "codified" in corporate policies?
- What elements are consider mandatory versus suggested practices/guidance?
- What site roles and positions will be included in FM (and not included)?
- What is the required internal timing for implementation and how will it be tracked?
- When and how will the PSM auditing process incorporate FM elements during site reviews?
- How will interpretations and guidance on FM be captured and distributed?
- Will the program be required for contractors, and if so, which ones?
- Are there leveraged solutions and tools that can be developed to assist sites for effectiveness and efficiency?
- Who has primary responsibility to ensure compliance with Table 1 (employees or line managers)?

2.1 Site Surveys to Define Current State FM Practices

The first step that was taken to develop a fatigue management program across DuPont was to conduct a detailed survey across all manufacturing locations to accurately determine the present state relative to the main elements of ANSI/API 755. This included the following self-assessment parameters:

- Existence of site level fatigue management written programs and policies
- Existence of local policy limits on overtime and maximum consecutive work hours
- Past and present fatigue management training
- Existence of tools or tracking systems to measure work hours and overtime by work group and employees
- Local site systems and capabilities to integrate fatigue management into existing medical programs
- Forecasted requirements and needs with local labor organizations and contracts for fatigue management

Once the site survey was completed, the data was collated and utilized as inputs for the development of corporate policies, timing and implementation plans, management training, and development of internal FM training and guidance documents. The survey was collected and analyzed in a simple excel® tool similar to that depicted in Figure 1.

Site Fatigue Management Survey – Current State

What is your site's current level of implementation of Fatigue Management requirements? List all numbers that apply. 1. Leadership is aware. 2. The site is tracking hours. 3. Employees have received general awareness training. 4. The site is actively managing hours for Shutdowns/TurnArounds, Special Circumstances, or Normal Operation. 5. A Fatigue Management policy has been written. 6. An Exception Management policy has been developed.	If you listed No. 2 above, how is the site tracking hours? Is the tracking historical or real time? What system or tools are being used?			If your answered No. 4 above, when and how is the site managing hours?		Have you considered which employees would be included in a Fatigue Management program? If yes, how many at your site? Does this number include employees in LHO's? If yes, how many of the total are in LHO's?	How are your shifts set up, ie. 8 hour, 10 hour, 12 hour, rotating, 5 days, 7 days, etc.?	If your site has changed any policies, procedures, tracking, people's work times, etc. to move toward compliance, can you describe what you have done?	Site Changes
Level of Implementation	Tracking Hours	Tracking Hrs.?	How Tracked ?	Managing Hours	Mgt. Hrs.?	Which Employ.	Shifts	Changes	Changed Schedule/ Pay?

Figure 1: Sample Site Survey Baseline Questions and Spreadsheet

2.2 Corporate PSM Standard Policies and Requirements

DuPont drafted and included a new fatigue management section in the global process safety management standard as part of the" Training and Development" element as depicted in Figure 2. Specific new content in the corporate standard was intended to incorporate the minimum mandatory requirements involving fatigue management elements described in ANSI/API 755, and not a be a re-statement of the guideline's full technical content. It was also defined with the standard scope that new internal fatigue management requirements would be applied to vital PSM roles within DuPont sites as well as to specific types of contractors, but only at higher hazard site operations (DuPont classifies all global manufacturing locations as higher hazard processes or lower hazard operations). The standard changes went through a broad internal peer review and was published in January 2011.

Figure 2: DuPont's Process Safety Management Model

2.3 Education and technical guidance to drive consistency

Internal webinars were developed and delivered which were targeted to several different key audiences, including site PSM leaders, operations leadership, site HR leaders, and contract administrators in order to develop foundational knowledge of the new FM program elements across the organization. The initial ANSI/API 755 standard as published is a performance based standard with specific prescriptive FM requirements such as Table 1 maximum hours of service limits, however, it became apparent early in the FM implementation process that numerous technical and interpretive questions existed both on the standard, as well as within DuPont's local site implementation efforts. In order to attain as much consistency as possible, a series of Frequently Asked Questions (FAQ's) were collected and formally documented with specific guidance. Once vetted and resolved, these FAQ's were distributed to sites and posted on the internal PSM website for retention and routine access. Additional questions and needs for interpretations have continued to surface from sites and functions and been added to the FAQ "evergreen" document even after almost two years of implementation progress. There is currently no effective platform within industry to collect and address these types of interpretations in a consistent manner.

As another step to implement consistency across all locations, DuPont developed and distributed a standard site based written fatigue management program or template which can be applied by each site as a base document and modified to add in local responsibilities, procedures, and line management practices.

There are also unique situations that may develop for plant operations that do not fit neatly into the existing definition and guidance in ANSI/API 755 involving "outages" as it applied to extended consecutive work shifts in Table 1. The existing definition for outages includes the following:

> *"Planned or unplanned interruption in the normal operations of a unit or plant, including mobilizing or de-mobilizing. Outages include but are not limited to such things as turnarounds, unit shutdowns, operational responses, etc."*

This definition does not cover all situations where an extended work set and shift schedules may be applicable and necessary to minimize fatigue, including commissioning new plants or substantive MOC projects where construction, PSSR's, start-ups, and lining out the process may be necessary involving operations, technical and R&D resources, response to natural disasters (e.g., hurricanes, earthquakes, etc.), mothballing or dismantlement activities, new product trials in pilot plants, etc. Each company should define when and how these types of situations should be managed in terms of unique conditions and the application of extended work hours and shift schedules with potential fatigue factors.

DuPont adopted an internal policy at the site level to develop a formal exception process for situations where the Hours of Service Limits (ANSI/API Table 1) may not be met due to specific circumstances. An exception process is discussed in the standard in Section 4.8.5 with general criteria, however; the specific procedures and details should be developed by each company. For DuPont, the exception approval process has been designed to be executed at the local level for a case-by-case exception by site or unit management. For longer term, permanent, or "standing" exceptions, off-site leadership approvals are required. In both cases, exceptions will document the situation, the rationale, and any relevant alternative controls or methods to be taken to mitigate the potential effects of fatigue. This includes use of a standardized form and line management approvals for each exception situation.

3. Shift Schedules and Maximum Hours of Service

API 755 includes specific requirements for maximum hours of service provided in Table 1 of the standard. The hours of service element in FM has received the most internal discussion, review and implementation work. A detailed analysis of existing shift rotation schedules at all U.S. site operations was conducted by DuPont HR to ascertain if any inherent conflicts existed in the current schedule structures that were not be fully compatible with Table 1. This included a specific emphasis on the number of consecutive shifts and the total hours of recuperative time for both day and night shift rotations. This review revealed that DuPont had over one hundred (100) different shift rotation schedules in use at all U.S sites for 8, 9, 10, and 12-hour operations and maintenance shifts. In a very limited number of cases, an inherent conflict in the site shift structure existed which did not fully meet the maximum "hours of service" in Table 1 (mostly involving the transition from day to night 12 hour shift schedules). Figure 3 describes the shift schedule analysis results at all US sites, and figure 4 provides an example of the detailed shift schedules which were generated and reviewed at a typical multi-unit site.

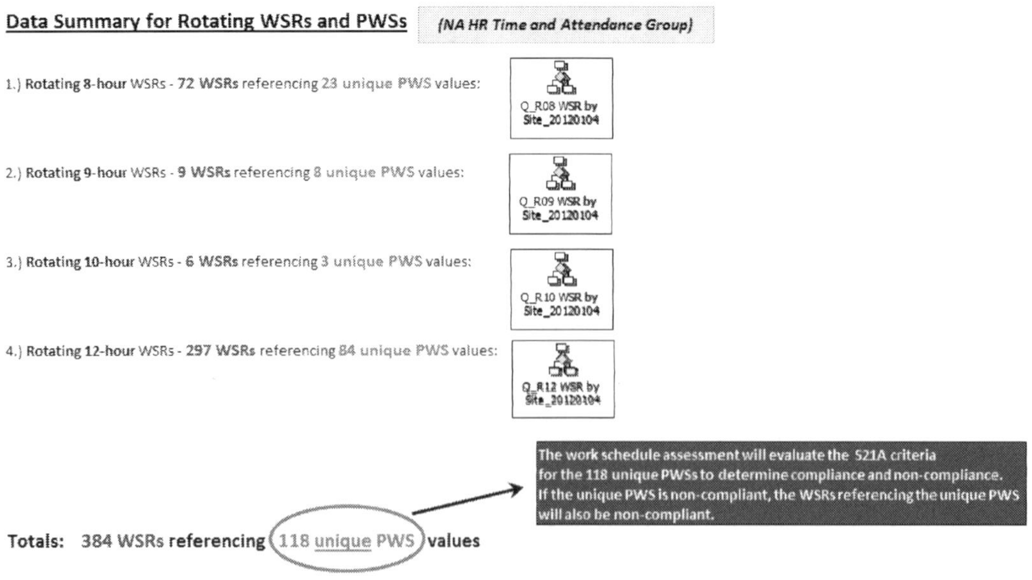

Figure 3: Data Summary for Plant Rotating Work Schedules

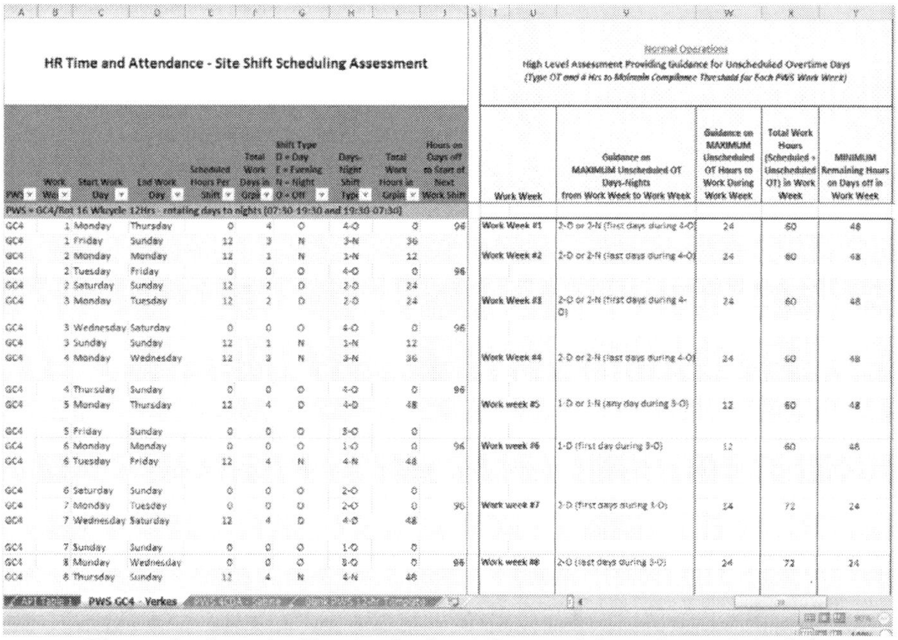

HR Time and Attendance - Site Shift Scheduling Assessment

Normal Operations
High Level Assessment Providing Guidance for Unscheduled Overtime Days
(Type OT and # Hrs to Maintain Compliance Threshold for Each PWS Work Week)

PWS	Work We	Start Work Day	End Work Day	Scheduled Hours Per Shift	Total Work Days in Grp	Shift Type D = Day E = Evening N = Night O = Off	Days-Night Shift Type	Total Work Hours in Grp	Hours on Days off to Start of Next Work Shift	Work Week	Guidance on MAXIMUM Unscheduled OT Days-Nights from Work Week to Work Week	Guidance on MAXIMUM Unscheduled OT Hours to Work During Work Week	Total Work Hours (Scheduled + Unscheduled OT) in Work Week	MINIMUM Remaining Hours on Days off in Work Week
PWS = GC4/Rot 16 Wkcycle 12Hrs - rotating days to nights (07:30-19:30 and 19:30-07:30)														
GC4	1	Monday	Thursday	0	4	O	4-O	0	96	Work Week #1	2-D or 2-N (first days during 4-O)	24	60	48
GC4	1	Friday	Sunday	12	3	N	3-N	36						
GC4	2	Monday	Monday	12	1	N	1-N	12		Work Week #2	2-D or 2-N (last days during 4-O)	24	60	48
GC4	2	Tuesday	Friday	0	0	O	4-O	0	96					
GC4	2	Saturday	Sunday	12	2	D	2-D	24						
GC4	3	Monday	Tuesday	12	2	D	2-D	24		Work Week #3	2-D or 2-N (first days during 4-O)	24	60	48
GC4	3	Wednesday	Saturday	0	0	O	4-O	0	96					
GC4	3	Sunday	Sunday	12	1	N	1-N	12						
GC4	4	Monday	Wednesday	12	3	N	3-N	36		Work Week #4	2-D or 2-N (last days during 4-O)	24	60	48
GC4	4	Thursday	Sunday	0	0	O	4-O	0	96					
GC4	5	Monday	Thursday	12	4	D	4-D	48		Work week #5	1-D or 1-N (any day during 3-O)	12	60	48
GC4	5	Friday	Sunday	0	0	O	3-O	0						
GC4	6	Monday	Monday	0	0	O	1-O	0	96	Work week #6	1-D (first day during 3-O)	12	60	48
GC4	6	Tuesday	Friday	12	4	N	4-N	48						
GC4	6	Saturday	Sunday	0	0	O	2-O	0						
GC4	7	Monday	Tuesday	0	0	O	2-O	0	96	Work week #7	2-D (first days during 3-O)	24	72	24
GC4	7	Wednesday	Saturday	12	4	D	4-D	48						
GC4	7	Sunday	Sunday	0	0	O	1-O	0						
GC4	8	Monday	Wednesday	0	0	O	3-O	0	96	Work week #8	2-D (last days during 3-O)	24	72	24
GC4	8	Thursday	Sunday	12	4	N	4-N	48						

PWS GC4 - Verkes

Figure 4: Sample Site Analysis – Time and Attendance Schedules

A significant part of DuPont's initial evaluation of existing shift schedules and site data survey questions focused on the types of tracking and software (scheduling) tools that were either already in use or could require development to track work hours, including scheduled and unscheduled overtime call-ins. In most cases, site systems tracking overtime rates for work groups but not specific employees other than annual salary statistics. Site level tracking systems which were cataloged that could be modified and used to monitor maximum hours of service and off site recuperation times included the use of an HR work hours and pay status, web based tools, security gate electronic logs, and access data bases. None of these existing systems as structured were deemed an acceptable and reliable means to proactively schedule, manage, and track individual employee work hours. A six sigma black belt resource and project was initiated in order to review each of the work hour and overtime tracking system attributes, and develop a standardized solution that could be used at all sites to help line management ensure that the maximum hours of work in ANSI/API 755 Table 1 maximum hours of work. As part of the six sigma project, the main attributes of an overtime tracking system were captured as illustrated in figure 5.

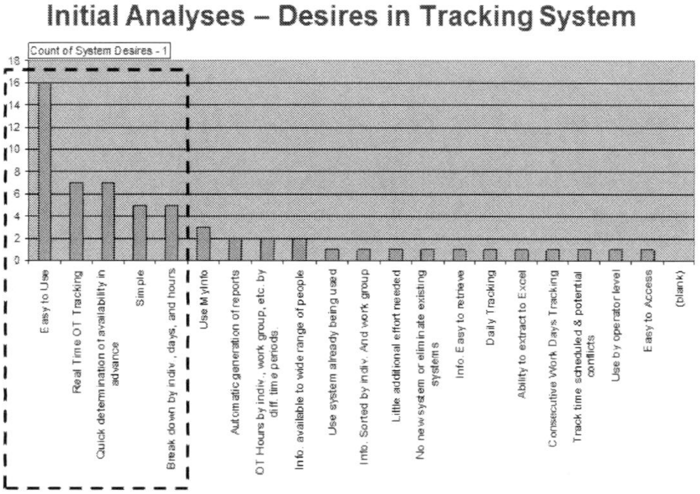

Figure 5 : Site Survey Data – Desirable Attributes for Electronic Scheduling and Tracking

4. Training

DuPont developed a standardized and leveraged training package which could be applied at each site in order to provide initial and refresher training on fatigue management. Two packages were developed ; one to target employees who were subject of FM (operators and mechanics), and one for line managers who supervise these employees. The guidance and criteria in ANSI/API 755 was used to create the technical content of each package. Initial training was designed to be one hour of information and cover the following topics:

- Basics of sleep and fatigue including causes, symptoms, and prevention strategies
- Understanding risk of on the job fatigue and impacts on safety
- Overview of the site fatigue management system including maximum hours of service criteria

DuPont occupational health resources have also previously partnered with external groups such as the Porcupine Health Unit and the Sudbury & District Health Unit in Canada to develop additional sources of fatigue management training such as the Shiftwork like Clockwork (SWLCW) training modules illustrated in Figures 6 and 7 . These materials provide significant details over nineteen (19) modules that can be used for both initial training of new employees and refresher training for existing employees. DuPont has adopted an internal FM refresher training frequency of three years.

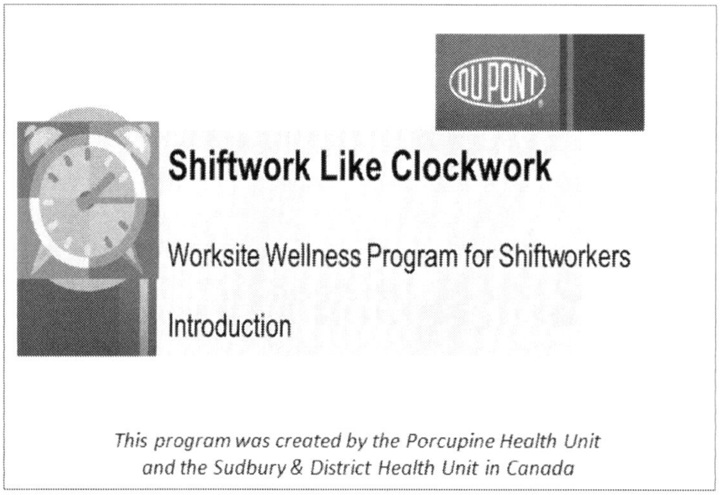

Figure 6: "Shiftwork Like Clockwork" Training Program

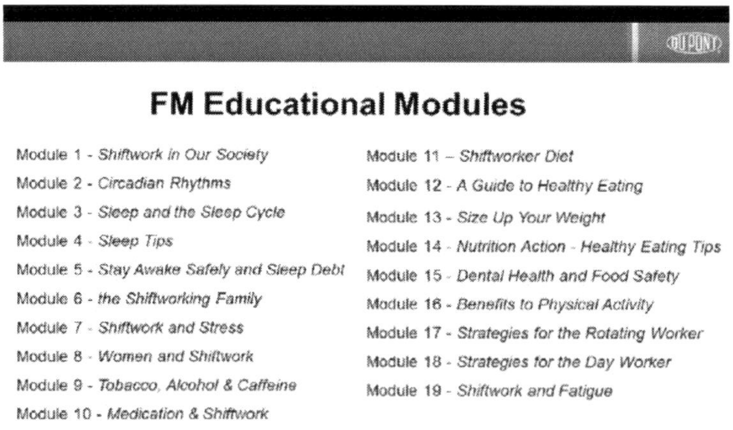

Figure 7: "Shiftwork Like Clockwork" Training Modules

5. Staff Workload Balance

Another key tenant of the ANSI/API 755 standard includes a focus to help ensure a staff workload balance is achieved and maintained within each operating area at each site. This should include an analysis of work activities on each shift, and a focus on considerations and impacts related to employee turnover, long term disability, extended time off due to personal reasons (vacation, leave of absence), staffing vacancies, and other sources of long term absenteeism. DuPont has adopted a policy that each site will conduct and document an annual review of staff workload balance within each higher hazard operating unit to periodically evaluate the status and performance in this area. A standardized set of evaluation criteria for the annual review was developed and distributed to all sites as a leveraged tool to help ensure consistency across all locations.

6. Work Environment

DuPont has utilized a series of PHA related tools and checklists to conduct baseline, cyclical and MOC related evaluations for over the past twenty years. A human factors checklist has been one of these historical tools. DuPont's strategy to address FM work environmental factors involved adding content to these existing tools in order to help ensure that fatigue management workplace factors were considered and evaluated during all cyclic and project related PHA's. Integrating this review into existing PHA processes has proven to be a more sustainable and effective practice rather than creating a "stand alone" fatigue management work environment assessment process.

7. Medical Evaluations (Personal Risk Factors)

DuPont Integrated Health Services, in coordination with the process safety competency, developed a fatigue management medical screening approach and screening form which could be used as a voluntary tool during periodic health exams for shift employees who may be subject to fatigue risks. The health services group also provided targeted FM training to site medical professionals (nurses, doctors, and NP's) in order to consistently apply FM medical aspects, including both the proactive (voluntary) approaches as well as "for cause" evaluations where potential fatigue has been observed by line management in the work place.

DuPont Medical also worked with the corporate HR, labor relations, and legal functions to develop and communicate consistent medical guidance and case management methods where employee fatigue concerns have been identified. This is an area where emerging issues, specific interpretations, and case law should be considered by each entity such as the following (examples only):

1) Where additional FM medical testing is prescribed based on an employee's physical status and personal risk factors (obesity, concerns about sleep apnea) which could relate to fatigue in the workplace, should the costs for further medical evaluations be borne by the employee as part of personal health care or as part of the occupational medical program for a given company?
2) How should an employee be managed if they refuse to participate in additional medical screening or specific diagnostic tests that focus on fatigue management?
3) Are certain medical conditions that can cause fatigue in the work place subject to the 1990 Americans with Disabilities Act (ADA) in the United States and if so, which ones, and what may be the expectations for reasonable accommodations?

8. Contractors

DuPont has also adopted internal process safety requirements for a fatigue management program applied to selective contractors who work in higher hazard process areas which typically is limited to resident maintenance service providers, test and inspection consultants, and specialized services resources who may be on-site only during major turnarounds.

Contractors are required to develop their own internal programs with the basic elements of fatigue management as described in ANSI/API 755. DuPont will typically audit their systems and performance at the site level on a periodic basis to verify the program is adequately implemented. Many major contractors have embraced an enhanced focus to manage fatigue risks. In one example, a resident maintenance contractor developed a modified approach for their employee work scheduling during an annual shutdown with a DuPont site in Protection Technologies in order to improve fatigue management practices. The shutdown typically lasts 15 days involving 12 hour shifts and the contractor reduced the shift duration to 11 hours and inserted a day off for each employee during the process. Feedback from the employees was very positive, and this modifications were managed so that there were no additional costs or delays in the shutdown schedule. This unique approach serves as a good example where the owner/operator (DuPont) worked in collaboration with contactor partners to achieve a common objective which also supported the business and operations.

9. Incident Investigation Processes

DuPont maintains a global internal incident investigation reporting, classification and investigation policy for both EHS and process safety incidents which addresses communications, investigation teams, root cause failure analysis (RCFA), recommendations tracking, and overall documentation. The existing policy does not specifically include or consider fatigue management components as they may relate to contributing factors and human error as part of the overall RCFA process. At some point in the future, the policy will be updated to include a focused consideration of fatigue during all formal investigations, particularly when human error has been identified. This may include a standardized review of employee work shift schedules and overtime for at least the previous two weeks prior to the incident date.

10. Conclusion

Based on the "lessons learned" from the BP Texas City incident in 2005 and the "new" ANSI/API 755 Recommended Practice, internal company initiatives to integrate fatigue management into existing risk management systems is a prudent and necessary step to help minimize the potential for human error and the associated episodic incidents in high hazard process operations. The process safety function should be the logical "process owner", however, the initial systems design and implementation activities involving fatigue management necessitate a collaborative effort between other key company and site level resources and functions, including operations, human relations, health services and occupational health, in order to develop an integrated system and achieve maximum risk management benefits. Since the ANSI/API RP 755 standard includes primarily performance based content, implementation also may involve a substantial number of internal corporate or local policy decisions, and potential issues which require focused discussions and interpretations to promote consistency with internal applications at each site.

The initial FM implementation process may span two to three years in order to plan, educate, develop, implement and monitor site level organizations and accompanying results. Organizations that include fatigue management requirements in existing PSM policies should recognize that implementation progress may vary at each site depending on size, complexity, shift schedules, the existence of labor unions with collective bargaining agreements, and other factors. These potential variations on the rate of progress require a simplified process to track implementation across all applicable operating sites (see figure for an example of tracking based on the six main program elements). Integration of fatigue management evaluation criteria into existing PSM audit protocols and systems is typically the last step in a long term FM implementation and performance measurement process.

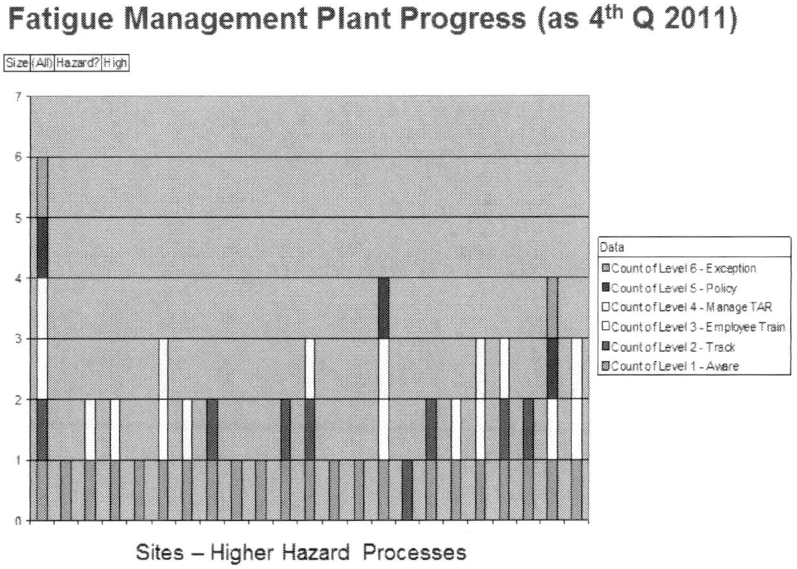

Figure 8: Example of Visual Management Board – Site Progress

11. References

(1) American National Standards Institute (ANSI)/American Petroleum Institute (API) Fatigue Risk Management Systems for Personnel in the Refining and Petrochemical Industries - RP 755 (First Edition, April 2010)

Consequences calculation for boilover and storage tank fires

Jaime E. Cadena
Chemical Engineering Department, Universidad de los Andes
Cra 1 Este No 19A – 40, Bogotá, Colombia
je.cadena21@uniandes.edu.co

Alejandra Cano[1], Juan Carlos Díaz[2], Alexis Vignes[3], Felipe Muñoz Giraldo[4]
[1,2,4]Chemical Engineering Department, Universidad de los Andes
Cra 1 Este No 19A – 40, Bogotá, Colombia
[3]INERIS
Technologique ALATA, 60550, Verneuil-en-Halatte, France
[1]na.cano744@uniandes.edu.co, [2]jc.diaz111@uniandes.edu.co, [3]Alexis.VIGNES@ineris.fr, [4]fmunoz@uniandes.edu.co

Prepared for Presentation at
American Institute of Chemical Engineers
2013 Spring Meeting
9th Global Congress on Process Safety
San Antonio, Texas
April 28 – May 1, 2013

UNPUBLISHED

Consequences calculation for boilover and storage tank fires

Jaime E. Cadena
Chemical Engineering Department, Universidad de los Andes
Cra 1 Este No 19A – 40, Bogotá, Colombia
je.cadena21@uniandes.edu.co

Alejandra Cano[1], Juan Carlos Díaz[1], Alexis Vignes[2], Felipe Muñoz Giraldo[1]
[1]Chemical Engineering Department, Universidad de los Andes
Cra 1 Este No 19A – 40, Bogotá, Colombia
[2]INERIS
Technologique ALATA, 60550, Verneuil-en-Halatte, France

Keywords: Storage tank fires, boilover modeling, process safety, emergency response

Abstract

During the past decade, high loss incidents have been recorded due to fires in hydrocarbon storage facilities worldwide. An example is the storage tanks fire at a refinery in Amuay, Venezuela (August 25 2012, under investigation), which led to the loss of 650000 barrels of naphtha, around 41 fatalities and 80 injured. These incidents have high magnitude consequences, therefore an a-priori consequences analysis of these events become a key element for the decision-making processes to prevent, control and mitigate fire risks. The risk evaluation and mitigation for fire scenarios usually follows a prescriptive approach, therefore this work presents a tool for obtaining quick and conservative calculation of the thermal risks due to fire scenarios of hydrocarbons storage tanks facilities. These intensities are translated into consequences, emergency response requirements and key information for Contingency Plans development. To complement the fire scenarios, the tool also includes the calculation of effected distances for boilover phenomena. The results obtained by the fire tool were within an acceptable margin of error but it still conservative in comparison with real data. For the boilover tool different models were implemented among them the only one conservative for all the validation cases was the one proposed by Casal et. al. [1] for the feedback heat and the one proposed in the book Risk analysis in industrial facilities [2] for the boilover time.

1. Introduction

Storage tanks are a key element of any Oil & Gas operation since they constitute intermediate steps in the upstream operation e.g. in a refinery or a pipeline operation, and also in the downstream operations in which they are the source of final products to be delivered to the clients and customers. Nevertheless, even when hydrocarbons are stored without physico-chemical changes, risks are still present such as fire and boilover risks [3]. A fire in a storage tank is usually assumed as a pool fire in the surface of its contents, while a boilover is the ejection of the stored HazMat due to the fire and posterior boiling of a liquid with lower boiling point and higher density than the HazMat [2]. Storage tank fires and boilover scenarios represent low probability but high consequences events, making it necessary to apply available knowledge and strategies to analyze, assess, control and mitigate the related risks in order to avoid high capital and human loss as illustrated in Table 1 [4], which highlight that regardless of location or date, such events continue to occur. This makes it a key topic of study to avoid major hazard events and irreparable loss, especially for a region with a high industrial development such as Latin America [5, 6], in which the deployment of new Oil & Gas operations not only present emerging risks but also modify the existing ones [7].

Table 1. Storage tank fires and/or boilovers and related loss

Date	Location	Event	Related loss
13/12/1982	Bogotá, Colombia	Esso oil storage facility fire [8]	• 8000 people evacuated • Loss of 3 storage tanks
19/12/1982	Caracas, Venezuela	Explosion and subsequent oil tank fire [9]	• 153 fatalities • 7 missing • Severe damage to the facility
24/02/1986	Thessaloniki, Greece	Fire at an oil storage tank facility [10]	• 10 out of 12 tanks destroyed • US$330 M
11/12/2005	Buncefield, UK	UVCE and fire at hydrocarbon storage facility [11, 12]	• 43 injured • Over 2000 people evacuated • Low and moderate damage to surrounding buildings • Total destruction of most of the complex • Possible Detonation to Deflagration Transition (DDT) [13]
26/10/2009	Bayamon, PR, USA	Caribbean Petroleum Company fire (currently under investigation) [14]	• 1 injured • Costs around US$6.4 million • Around 5.6 million of product loss • Around 600 people evacuated • Damage to buildings in a radius over 1 mile (1.6 km)
25/08/2012	Amuay, Venezuela	PDVSA Amuay refinery explosion and fire (currently under invetsigation) [15]	• 48 fatalities (latest information) • Over 80 injured • Stop of the refinery with 640000 bpd capacity • Damage to at least two contiguous storage tanks, generating fire • 18 national guard and 15 civilian fatalities • Damage to around 200 residential buildings and 11 businesses • 33 families displaced • Damage to a military base

Although the subject has been addressed through different approaches [16, 17], in the field of process safety it is fundamental to understand the phenomena underlying the events of interest. For tank fires and boilover events this has been carried out for several decades, and a wide variety of empirical, semi-empirical and CFD models have become available to model such events [1, 18-23]. This work aims to develop a methodology to apply a set of the available models to calculate the consequences of the events of interest through its implementation in a user-friendly computational tool that allows obtaining quick and conservative results for either a preliminary risk analysis stage in which critical scenarios are identified or during actual emergencies in which knowledge of the possible outcomes is needed to establish mitigation strategies.

2. Methodology

In order to achieve the objectives stated above, a methodology (Figure 1) was proposed for the development and validation of the tool. First step is defining the scenario to evaluate, the tool use variables that influence tank fire and a boilover like some specific hydrocarbon properties such as viscosity, initial and final boiling point temperatures and combustion heat as input variables, next the tool implements models for the calculation of heat released by the fire the feedback heat and boilover and duration of fire times. Based on the last calculations the tool use models to determine the safe distances for both phenomena (boilover and fire) and in the case of the fire the tool also returns as output variables some mitigation strategies like refrigeration rings.

Figure 1. Methodology

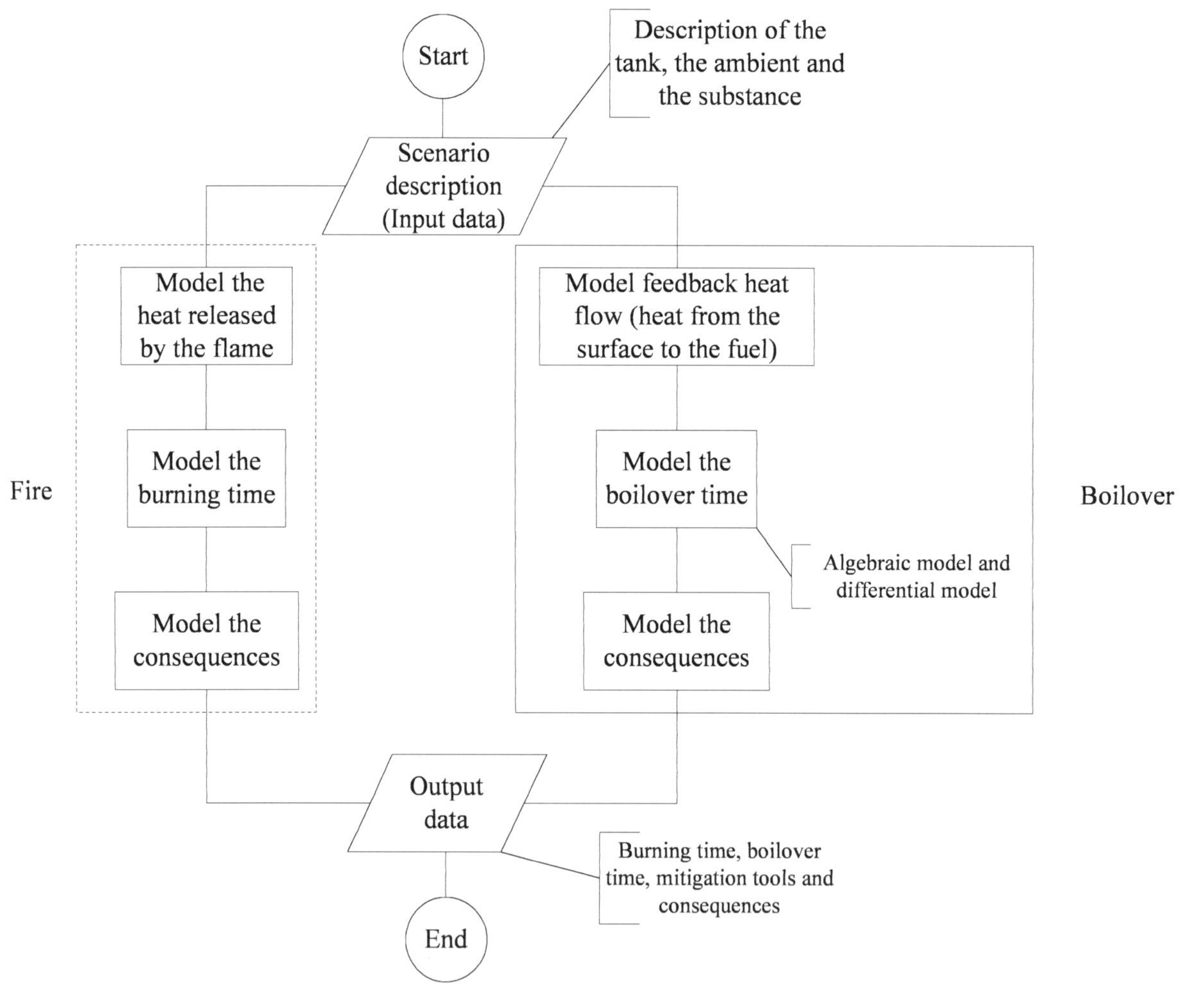

3. Tool implementation

The tank fire tool was developed in excel (visual basic) using macros for both, the heat released by the fire and the burning time. For the boilover there are multiple approximations, for the feedback heat there is an average value reported [2] and models for calculating it using a fraction of the total heat released by the fire, the difference among the last ones is the calculation of the total heat released that can be done using models with combustion heat and experimental corrections depending on the fuel [24] or heat balances between the tank and the surroundings [1], for the boilover time there is an algebraic approximation derived from a heat balance [2] and differential approximations; varying among them the initial and boundary conditions [1] . Some of them were implemented in excel and the ones involving differential approximations were implemented in matlab as this software allows multiples ways for solving differential equations in a short and easy way. Both Excel and Matlab have a user-friendly interface. The models selected for the two tools are presented on Table 2.

Table 2. Models used for the tool implementation.

		Model	Assumptions	Ref
Fire	Time calculation	$t_b=4V/\pi D^2 v$ $v=m''/\rho$	-	[24]
Fire	Heat calculation	$Q = m''\Delta H_{c,eff}(1-e^{-\kappa\beta D})A_{dike}$ $D=(4\ A_{dike}/\pi)^{1/2}$	- Experimental correction depending on the fuel	[24]
Boilover	Time calculation	1) $t_{boilover}= \rho\ c_p\ h_{HC}\ (T_{hw} - T_a)/(Q_f - m''(H_v + c_p\ (T_{0av} - T_a)))$ $T_{0av}=(\ T_{0min}\ T_{0max})^{1/2}$		[2]
Boilover	Time calculation	2) $\delta^2T/\ \delta y^2 = (1/\alpha)\delta T/\delta t$	-Unidimentional	[21]
Boilover	Heat calculation	1) $Q_f\approx 60$ kW	-	[2]
Boilover	Heat calculation	2) $Q_f= (4\chi/\pi)\ \rho_\infty\ c_{p\infty}\ (T_\infty\ g\ (T_f - T_\infty)\ D)^{1/2}$ $\chi = H_v/H_c$	- A heat conduction mechanism is assumed, no conduction in the fuel layer -The radiation is fully absorbed at the surface	[21]

The values for T_f in the heat calculation for the boilover, change depending on the assumptions used for the estimation, some authors use the adiabatic flame temperature as an approximation and others recommend approximated values like 1100 K [1] or values ranging from 1600 to 1900 K depending on the substance.

To solve the differential equation for time of boilover the following boundary conditions were established.

- Initial condition. [1]

$$\mathbf{T(y,0) = T_\infty} \quad \text{[Eq. 1]}$$

- Boundary condition 1.[1]

$$\mathbf{T(y(t),t) = T_s} \quad [\text{Eq. 2}]$$

Where **y(t)**is defined by the following equation.

$$\mathbf{\delta y(t)/\delta t = m'' / \rho_f} \quad [\text{Eq. 3}]$$

- Boundary condition 2.

$$\mathbf{-k_f * \delta T / \delta y |_{fuel} = -k_w * \delta T / \delta y |_{water}} \quad [\text{Eq. 4}]$$

To achieve boundary condition 2, the modeling of the water layer is needed and it can be solved by the same equation for the fuel reported on Table 2. The first boundary condition for the water layer is stated on Equation 4 and the second boundary condition stated in Equation 5. This boundary condition implies assuming adiabatic conditions in the limit of the water layer with the bottom of the tank.

$$\mathbf{\delta T / \delta t |_{y=0} = 0} \quad [\text{Eq. 5}]$$

3.1 Input and output variables

For the fire tank tool is needed from the tank the diameter, the height, the type of roof and the capacity; the tool have all the characteristics for the fuel, it only need to be specified with the volume present in the tank. In addition if the studied facility has more storage tanks near to the one in fire (storage farm tanks) the distances from the one in fire to the others must be known. The output variables are the duration of the fire, the area involved on the fire, the zone classification (hot, warm, cold) and the mitigation and prevention actions [24] like if the refrigeration ring should be activated.
For the boilover the input values are the diameter of the tank the ambient conditions, specifying the fuel involved in the fire, the initial height of the fuel in the tank before the ignition the temperature of the heat wave when the boilover occurs and for one of the models the flame temperature. The output data consists of the boilover time and the distances at which lethally and irreversible consequences occur.

3.2 Consequence estimation

The consequences for both tank fire and boilover were calculated using Excel and both have dependence from the heat irradiated from the fire in the case of the tank fire and for the resulting fireball in the case of the boilover.
The consequence of the fire depends on the heat radiation determining the hot, warm and cold zone. In the hot zone the equipment will fail and the heat release will be lethal for human beings. The warm zone can be accessed with protection equipment, the firemen attack the fire there. Finally the cold zone is the one where the equipment is displayed, there is no need of protection equipment in this zone. The hot zone is determined by values where the heat is over 10 kW/m^2, the warm zone is the one with heat values between 2,5 kW/m^2 and 10 kW/m^2 and the cold zone is the one with values lower than 2,5 kW/m^2 [24]. The consequences of a boilover event are fireballs and in the ejection of ignited fuel around the tank. According to the radiation heat

release of the fire there can be determined lethal and irreversible (serious burns) effect distances corresponding to radiation heats of 1000 kW $m^{-2}$and 600 kW m^{-2} each one, they are estimated using the conservative assumption that the entire fuel in the tank participates to the fireball. The models used for the safe distances and zones are reported on Table 3 [2].

Table 3. Models used for consequence calculations.

	Consequence	Model	Ref
Fire	Irradiated heat	$d_{hotzone}=(QXr/4\pi\, q''_{hotzone})^{1/2}$ $d_{hotzone}=(QXr/4\pi\, q''_{warmzone})^{1/2}$	[24]
Boilover	Fireball	$d_{lethality}=k_{lethality}W^{0.45}$ $d_{irreversible}=k_{irreversible}W^{0.45}$	[2]

4. Results and analysis

To validate the tool it was compared with selected cases for each event (fire and boilover), the case selection was made using two criteria, the first one is searching for cases that are not affected by other events after the fire starts and the second one is searching for cases with complete information about the outputs and inputs for the tool mentioned above. The cases and the comparison with the tool results are described in this section with the comparison among the boilover time and the burning time.

4.1 Tank fire tool

For the tank fire tool the case selected was an accident that occurred at ConocoPhillips Company's Glenpool South tank farm, Glenpool, Oklahoma on April 7 2003. The event involved the tank 11 of the facility during the transport of Diesel from another tank [25] . The results obtained by the tool are reported on Table 4 and Table 5.

Table 4. Results related to the tank and the fire variables

Burning Tank Name:	11
Type of Roof:	CONE
Product:	Diesel
Nominal Volume (m3):	8420,00
Contents (m3):	1338,67
Height (m):	14,63
Diameter (m):	33,22
Area of Fire (m2):	1526,84
Time of Fire (hr):	8,39
Time of Fire (day):	0,35
Refrigeration Range (m):	71,89

Table 5. Results related to the consequences of the fire

	From (m)	To (m)
Hot Zone	0	64,298
Warm Zone	64,298	128,596
Cold Zone	128,596	N/A

The variable to compare among the case and the tool was the burning time because that is the variable reported in the few cases that also have all the input parameters. The percentage of error among the real case and the results reported by the tool is lower than 5%. In the case of the burning time the percentage of error reported may be due to the model, because the variables involved in its calculation (volume of the fuel and dimensions of the tank) don't require any assumptions and are directly specified by the case to evaluate. The possible source of error for this validation case may be directly related to it, as the tank was being filled at the moment of the fire the initial volume of fuel is not an accurate value.

4.2 Boilover tool

For the boilover tool different models were used, a comparison among them is also included in this section.

Case 1:

For the boilover, a large scale experiment was used for the study of boilover in large tanks using crude oil [26]. The results obtained by the tool with different models are presented on Table 6 and the distance effects are reported on Table 7. They are independent from the model used for heat and time of boilover so there is just one value for $d_{lethality}$ and one for $d_{irreversible}$ reported for the case.

Table 6. Case 1 results for the boilover time with the models proposed on Table 2.

	Tflame (K)		tbo (s)		treal (s)	%error	
Heat model 1	NA	NA	4067,67		4728	13,96637056	
Heat model 2	1100	1900	1127,07357	692,031935	4728	76,1617264	85,3631147
Time model 2	NA	NA	60604		4728	NA	

Table 7. Safe distances for the boilover consequences for case 1

dletal (m)	14,7314678
dirrev (m)	20,0281754

Case 2:

A fire on tank 3 containing crude at the Czechowice-Dziedzice Refinery on June 26, 1971 at 19:50 occurred due to a lightning hit [27]. For this case the results obtained by the tool are reported on Table 8.

Table 8. Case 2 results for the boilover time with the models proposed on Table 2.

	Tllama (K)		tbo (hr)		treal (hr)	%error	
Heat model 1	NA	NA	29,77614629		5,3	461,8140809	
Heat model 2	1100	1900	2,4275687	1,64560344	5,3	54,196817	68,9508784
Time model 2	NA	NA	1068,095		5,3	NA	

Table 7. Safe distances for the boilover consequences for case 2

dletal (m)	350,936098
dirrev (m)	477,115369

For case 1 the most accurate value reported by the tool was the one calculated by using an approximated value of 60 kWm^{-2} for the feedback heat, while the most accurate values in case 2 were the ones calculated by using the heat model 2, for this case the value using the heat model 1 is not conservative. This can be explained by the fact that the heat proposed by literature is very small compared with the one calculated by the second heat model, it worked well for a tank with an small amount of fuel but with great amounts of fuel that heat is not enough and will take more time to increase the temperatures in fuel layer and to take the water to its boiling point, making the boilover time bigger. So the second model may be better for modeling large tanks with big amounts of fuel, keeping the time in the same order of magnitude and conservative compared to the real time.

In the case of the differential model mentioned before for both cases the boilover time calculated with the tool is four orders of magnitude or more above the real time, this can be explained by the assumption around the model, first the model is one dimensional assuming that all the heat is transferred from the top to the bottom of the tank when it exist also a radial heat transference, this can also be seen by comparing the results from case 1 with the results from case 2 where the difference among the real time and the calculated time is proportional to the amount of fuel and the tank dimensions. Also the thermal diffusivities along with the rest of the fluid properties were assumed constant during the fire, despite the changes of temperatures in the fuel and water layers. One of the assumptions of the model implies assuming only a heat conduction mechanism, no convection inside the layer occurs, this last phenomena implies the change of density in different heights of the fuel layer and this difference in densities implies that the fuel in the top of the tank will eventually move to the bottom and the fuel in the bottom will eventually move to the top making the heat transference quicker and the boilover time smaller.

4.3 Comparison among boilover time and burning time

The case 2 described for the validation of the boilover tool have the complete information to use also the tank fire tool and can be used to compare the time of burning if the boilover doesn't occur with the time at which boilover occurs. Giving a little notion of what fraction of the total burning time takes the boilover to occur.

The results obtained by tool for the fire are presented on Table 8 and the consequences on Table 9.

Table 8. Results related to the tank and the fire variables for case 2

Burning tank name:	1
Type of roof:	cone
Product:	Crude Oil

Nominal volume (m3):	80000,00
Content (m3):	10142,89
Height (m):	14,7
Diameter (m):	33,00
Area of fire (m2):	1523,99
Time of fire (hr):	93,33
Time of fire (day):	3,89
Refrigeration rate (m):	57,11

Table 9. Results related to the consequences of the fire for case 2

	From (m)	To(m)
Hot zone	0,000	51,083
Warm zone	51,083	102,167
Cold zone	102,167	N/A

The boilover results are reported in the above section, the comparison was made using the real value reported in literature for the case and the only conservative value reported by the tool that was calculated using the heat model 2. The percentages reported on Table 10 are the fraction that takes the boilover to occur with respect to the total time of fire.

Table 10. Percentage boilover time/ burning time

	%
Real Value	5,67877424
Heat model 2	2,59294975

The boilover time is a little fraction of the total burning time, showing how important is to estimate the first one, modeling only the fire will give more time for mitigation actions but if a boilover occurs this time is reduced on a 95 % and being conservative on a 97%, the consequences are also greater for the boilover where the safe distances are superior than 477,11 m while the cold zone for the fire is above 102,167 m, about a fifth times less the boilover distance. Making the boilover a more dangerous event than the fire itself and a matter to be worried about.

4. Conclusions

The fire tool's validation results are within an acceptable error margin but still conservative, which implies that the used model describes the event and can be used for real calculations and being also simple to manipulate for the user because there is no need of complicated interfaces or calculations. In the case of the boilover the recommended model based on the results reported on this work is the one reported by Casal et. al. [1] for the feedback heat and the algebraic approximation for time because with them there were obtained for both cases used for validation a conservative result. This implies that some of the proposed models can actually be implemented on tools that can be used in real life cases helping to generate and design mitigation strategies.

5. Perspectives

For future work the differential solution for time will be studied calculating parameters like the thermal diffusivities depending on the temperature profiles and solving a two-dimensional model. Also validating the tool against more cases including cases that involve alternative fuels.

Nomenclature

Q: Pool fire heat release rate (kW)
m": Mass burning rate of fuel per unit surface area (kg $m^{-2}sec^{-1}$)
$\Delta H_{c,}$: Effective heat of combustion of fuel (kJ/kg)
$A_f = A_{dike}$:Surface area of pool fire (area involved in vaporization)
$\kappa\beta$: Empirical constant (m^{-1})
D: Diameter of the pool fire (m^2)
t_b: Burning duration of pool fire (sec)
V: Volume for liquid (m^3)
D: Pool diameter (m)
v: Regression rate (m sec^{-1})
ρ:Liquid fuel density (kg m^{-3})
c_p: Specific heat of the fuel at T_a (kJ $kg^{-1}K^{-1}$)
h_{HC} : Height of the fuel in the tank after ignition (m)
ΔH_v : Vaporization rate atT_{0max} (kJ kg^{-1})
T_a : Ambient temperature (K)
T_{hw} : Heat wave temperature when the boilover occurs (K)
T_{0av} : Average boiling temperature of the fuel (K)
Q_f: Thermal flow entering the fuel from its surface (kW m^{-2})
T_{0min}:
T_{0max}:
α: Crude thermal diffusivity
ρ_∞ : Air density (kg m^{-3})
$c_{p\infty}$: Specific heat of the air (kJ $kg^{-1}K^{-1}$)
T_∞: Ambient temperature (K)
T_f: Average flame temperature (K)
Xr:
$q''_{hotzone}$: Heat per unit area that humans and equipment can't tolerate (10 kW/m^2)
$q''_{warmzone}$: Heat per unit area that can be tolerated with protection equipment (2,5 kW/m^2)
W: Mass of the fuel at the beginning of the fire.
$k_{lethality}$: Lethality constant depending on the fuel.
$k_{irreversible}$: Irreversible consequences constant, depends on the fuel.

N. References

[1] J. Hristov, E. Planas-Cuchi, J. Arnaldos, and J. Casal, "Accidental burning of a fuel layer on a waterbed: a scale analysis of the models predicting the pre-boilover time and tests to published data," *International Journal of Thermal Sciences,* vol. 43, pp. 221-239, 2004.
[2] J. Casal, *Análisis del riesgo en instalaciones industriales*: Edicions UPC, 2001.
[3] K. Trevor, "Chapter 5 - Storage Tanks," in *What Went Wrong? (Fifth Edition)*, ed Boston: Butterworth-Heinemann, 2009, pp. 97-121.
[4] B. Zheng and G.-h. Chen, "Storage tank fire accidents," *Process Safety Progress,* vol. 30, pp. 291-293, 2011.

[5] E. Corportation, "2012 The Outlook for Energy: A view to 2040," 2012.
[6] CIA. (2011, 11/01/2012). *The World Factbook :: Oil Production*. Available: https://www.cia.gov/library/publications/the-world-factbook/fields/2173.html
[7] G. Gómez, S. Rodríguez, J. Cadena, and F. Muñoz, "Kletz's legacy for developing countries: Simple systems based on inherently safer design," *Journal of Loss Prevention in the Process Industries,* vol. 25, pp. 843-847, 2012.
[8] AP, "8000 flee Colombia gas fire," in *The Telegraph*, ed, 1982, p. 1.
[9] S. Mannan, F. P. Lees, and Knovel (Firm). (2005). *Lee's loss prevention in the process industries hazard identification, assessment, and control (3rd ed.)*. Available: http://www.knovel.com/knovel2/Toc.jsp?BookID=1470
[10] J. I. Chang and C.-C. Lin, "A study of storage tank accidents," *Journal of Loss Prevention in the Process Industries,* vol. 19, pp. 51-59, 2006.
[11] D. Bradley, G. A. Chamberlain, and D. D. Drysdale, "Large vapour cloud explosions, with particular reference to that at Buncefield," *Philosophical Transactions of the Royal Society A: Mathematical, Physical and Engineering Sciences,* vol. 370, pp. 544-566, February 13, 2012 2012.
[12] B. Investigation Board, "The Buncefield Incident," Health and Safety Commision, HSC2005.
[13] B. J. Lowesmith, G. Hankinson, and D. M. Johnson, "Vapour cloud explosions in a long congested region involving methane/hydrogen mixtures," *Process Safety and Environmental Protection,* vol. 89, pp. 234-247, 2011.
[14] U. S. C. S. a. H. I. B. CSB, "CSB Conducting Full Investigation of Massive Tank Fire at Caribbean Petroleum Refining; Investigative Team Plans to Thoroughly Examine Facility Safety Practices," in *News Room* vol. 2012, ed. Washington DC, USA, 2009.
[15] K. Vyas and E. Minaya. (2012) Explosion at Amuay refinery in Venezuela kills 41; fire still burning. *Hydrocarbon Processing*. Available: http://www.hydrocarbonprocessing.com/Article/3080380/Environment-and-Safety/Explosion-at-Amuay-refinery-in-Venezuela-kills-41-fire-still-burning.html
[16] C. Crippa, L. Fiorentini, V. Rossini, R. Stefanelli, S. Tafaro, and M. Marchi, "Fire risk management system for safe operation of large atmospheric storage tanks," *Journal of Loss Prevention in the Process Industries,* vol. 22, pp. 574-581, 2009.
[17] C. D. Argyropoulos, M. N. Christolis, Z. Nivolianitou, and N. C. Markatos, "A hazards assessment methodology for large liquid hydrocarbon fuel tanks," *Journal of Loss Prevention in the Process Industries,* vol. 25, pp. 329-335, 2012.
[18] C.-I. Lin and H.-F. Wang, "Research on Safety and Security Distance of Flammable Liquid Storage Tank," *Procedia Engineering,* vol. 11, pp. 51-60, 2011.
[19] M. Arai, K. Saito, and R. A. Altenkirch, "A Study of Boilover in Liquid Pool Fires Supported on Water Part I: Effects of a Water Sublayer on Pool Fires," *Combustion Science and Technology,* vol. 71, pp. 25-40, 1990/05/01 1990.
[20] T. Inamura, K. Saito, and K. A. Tagavi, "A Study of Boilover in Liquid Pool Fires Supported on Water. Part II: Effects of In-depth Radiation Absorption," *Combustion Science and Technology,* vol. 86, pp. 105-119, 1992/11/01 1992.
[21] I. M. Shaluf and S. A. Abdullah, "Floating roof storage tank boilover," *Journal of Loss Prevention in the Process Industries,* vol. 24, pp. 1-7, 2011.
[22] M. Muñoz, J. Arnaldos, J. Casal, and E. Planas, "Analysis of the geometric and radiative characteristics of hydrocarbon pool fires," *Combustion and Flame,* vol. 139, pp. 263-277, 2004.
[23] M. Muñoz, E. Planas, F. Ferrero, and J. Casal, "Predicting the emissive power of hydrocarbon pool fires," *Journal of Hazardous Materials,* vol. 144, pp. 725-729, 2007.
[24] NFPA, *SFPE handbook of fire protection*, 3rth ed., 2002.

[25] "Storage Tank Explosion and Fire in Glenpool, Oklahoma," National Transportation Safety Board, Washington DC2003.
[26] H. Koseki, Y. Natsume, Y. Iwata, T. Takahashi, and T. Hirano, "Large-scale boilover experiments using crude oil," *Fire Safety Journal,* vol. 41, pp. 529-535, 2006.
[27] H. Persson and A. Lönnermark, "Tank Fires Review of fire incidents 1951–2003," SP Swedish National Testing and Research Institute.

A practical view of risk reduction management of hardware and software for safety critical applications

Luis Duran
Product Marketing Manager Safety Systems
ABB
3700 West Sam Houston Parkway South suite 600
luis.m.duran@us.abb.com

Mike Krywonos
Safety Execution Center Manager
ABB
3700 West Sam Houston Parkway South suite 600
mike.krywonos@us.abb.com

Prepared for Presentation at
American Institute of Chemical Engineers
2013 Spring Meeting
9th Global Congress on Process Safety
San Antonio, Texas
April 28 – May 1, 2013

UNPUBLISHED

A practical view of risk reduction management of hardware and software for safety critical applications

Luis Duran
ABB
3700 West Sam Houston Parkway South suite 600
luis.m.duran@us.abb.com

Mike Krywonos
ABB
3700 West Sam Houston Parkway South suite 600
mike.krywonos@us.abb.com

Keywords: Safety, Risk Reduction, Safety Systems, Independent Protection Layers, Common Cause Failures, Functional Safety Management System, IEC61508, IEC61511, TÜV, third party certification, Human Factors

Abstract

Best in Class companies (both National and Global) continue to invest in safety systems; this in most cases is because safety is at the core of their production process and among their key performance metrics, closely linked to their productivity.

Many unfortunate industrial accidents such as the recent explosion of the Deep Water Horizon with the largest marine oil spill in history are evidence of the negative consequences they have on people, environment, production assets and corporations.

This paper discusses the accepted definition of safety and risk reduction in the industry and will discuss the best practices applied by engineers in the design of products for mission critical functions and the implementation of those products in projects.

Both in product design and implementation, this session will dedicate special attention to methodologies or work processes used to reduce potential common cause failures that might impair the operation of the safety system and ultimately expose the facility to a higher risk; and the human factor considerations recommended to ensure proper operation upon a critical condition or an emergency in the industrial process.

1. Introduction

Best in Class companies worldwide, defined as such for their high percentage of Overall Equipment Effectiveness (OEE) and low injury frequency rate; continue to show strong initiatives intended to expand the safety of their operation. From executive sponsored programs at the corporate level to the definition of proactive risk management strategies, these companies are investing in safety systems and processes; simply because in most cases, safety is identified among the core values of those companies, at the center of their production process and among their key performance metrics, closely linked to their productivity.

There are numerous publications and records of Industrial Accidents, a large number of them with terrible consequences, including not only loss of human life or environmental impact but also affecting the production assets in the site. As an example, in researching for this paper, the authors found record of at least 20 "notable" oil and gas offshore blowouts between 1980 and 2010.

As an example for this paper, the authors compared two serious incidents in the Oil & Gas sector; Piper Alpha operated by Occidental Petroleum and Deep Water Horizon leased to BP. Although both incidents are 22 years apart from each other, occurred in different geographies and have many differences in causes and consequences. They illustrate the risk and the potentially catastrophic dimension they can take in this industry and the impact on people, environment and corporations, which might cease to exist after these incidents.

2. Defining Safety and Risk

Safety is defined in the industry as a reduction of existing risk to a tolerable or manageable level while risk is a combination of the probability of a harmful incident and magnitude of the harm. Traditional design practices allocate the risk reduction across different and independent protection layers, the rationale behind it is simple "any system that can fail will fail" so the engineering best practice is to distribute the risk reduction tasks across multiple independent functions or systems one of these systems is a Safety Instrumented System.

2.1 Safety Instrumented Systems one of many Independent Protection Layers to Reduce Risk

A Safety Instrumented Systems (SIS) is a mission critical system designed following international design practices such as IEC61508 [1] Functional Safety Standard to reduce risk to the people in and around the production environment, the environment, the production asset and the business. In many cases the Safety Instrumented Systems are the last resource to prevent disaster.

SIS performance is measured by Safety Integrity Level (SIL) SIL 1 low, SIL 3 high o Risk Reduction Factor.

As mentioned earlier, the Safety Instrumented System is one of many functionally independent systems each intended to perform a task, as shown on Table 1 typically referred to as Independent Protection layers [2]

Table 1: Expected Functionality of Independent Protection Layers

Layer of Protection	Expected Functionality
BPCS:	Keeping the process under control
Alarms:	Alerting the operator of abnormal conditions and providing guidelines for appropriate operator response
SIS:	Automatically taking the process to a safe condition in case the abnormal condition goes out of control and the operators can't take any corrective action in time
Other layers:	Intended to mitigate the consequence of the hazard

The concept of Independent protection layers (IPL) can also be illustrated in Figure 1

3. Common Cause Failures

The idea of functional independent systems is an attempt to avoid Common Cause Failures (CCF).

Common Cause Failures [3] are failures that might affect more than one of these protection layers at the same time. As engineers, the tendency will be to focus on elements such as Heat, Humidity, Shock, and radio interference among others. However this paper also considers with special interest the human elements related to the design of safety critical systems which can contribute to additional failures.

As with other systems, some SIS problems are related to the Commercial Off the Shelf (COTS) products designed for a specific function (i.e. hardware limitations or poor documentation) while other problems are related to the use of those COTS product such as misapplication, user application programming or poor maintenance practices. Both areas can be addressed by implementing appropriate design best practices to reduce risk.

The industry has conceived these "best practices" as a series of steps that must be performed before putting one of these systems in operation (design, installation and commissioning). This paper will describe the two areas:

- Product Design and Implementation of the Commercial Off-the-Shelf (COTS) Product
- Application Design

Figure 1: Independent Protection Layers

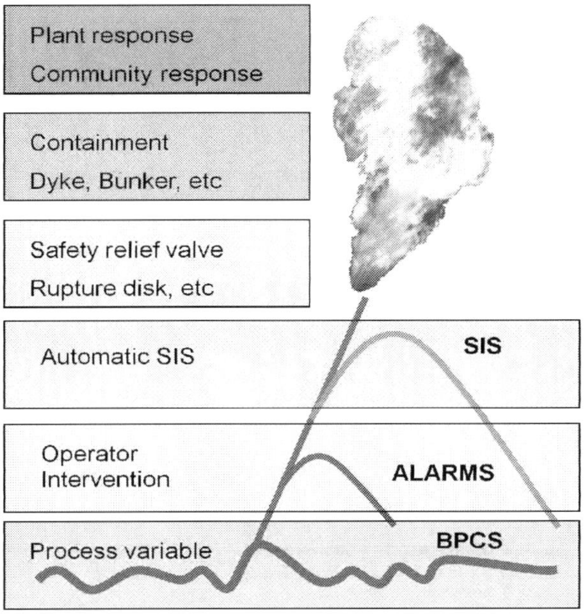

3.1 Reducing Common Cause Failures by Design

Traditionally or historically, Systems designed for SIS functions have relied on physical separation and redundancy [4] to reduce common cause failures. The authors identified technology changes across multiple generations of SIS over a period of 30 years. However over the same time frame the industry applied lessons learned from incidents to develop best practices in design and implementation of systems, the international industry standards used more often are IEC61508 and IEC61511 [5] (Figure 2)

3.2. Hardware Fault Tolerance (1st Generation SIS)

1st Generation systems rely on Hardware Fault Tolerance or redundancy to achieve reliability and availability as required for these applications, this practice is rooted in the late 1970s and early 1980s technology. As the majority of these systems were designed prior to the release of the Functional Safety standards, they do not follow the best practices found with regards to design, documentation and testing found in those standards.

Figure 2: Evolution of National Prescriptive Standard to International Performance Base Standards

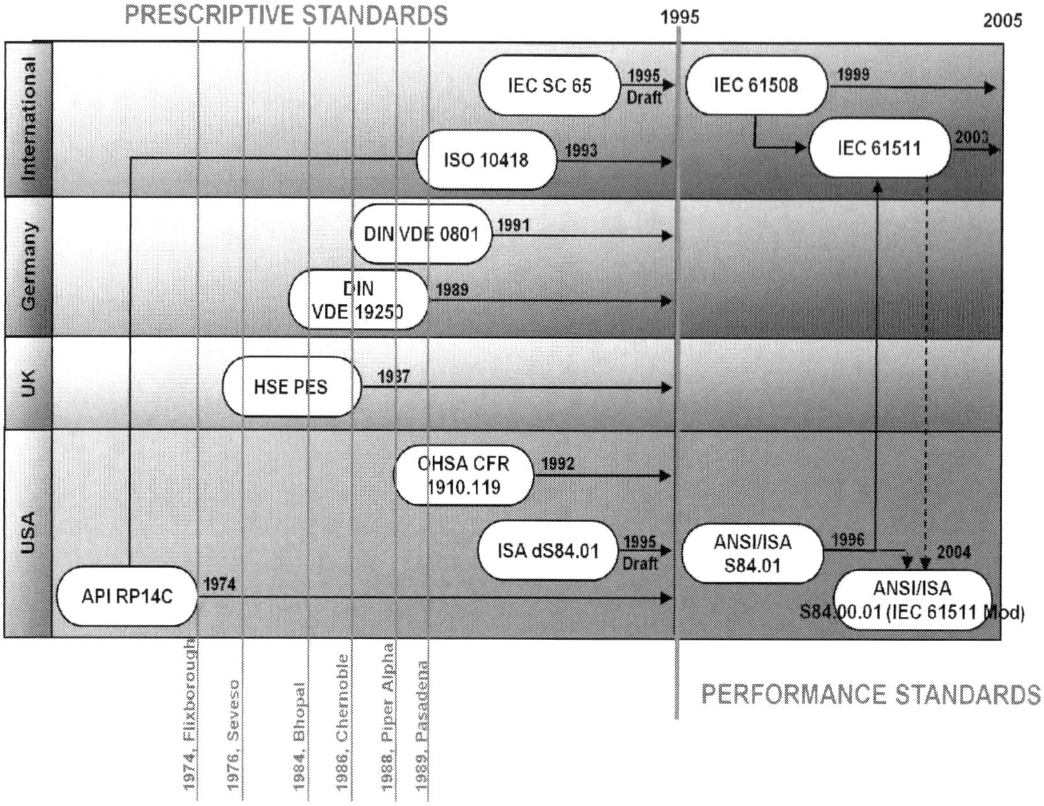

3.3. 2nd Generation Systems

As the standards became available (mid 1990s), vendors started to apply the design best practices and to pursue certification to those standards. This generation of systems was produced having third party assessment and certification but continued to rely on Hardware Fault Tolerance to satisfy the performance requirements. The authors found that a large portion of the systems applied as SIS in the market today, although designed to satisfy Functional safety standards haven't change dramatically in their use of hardware fault tolerance and basic software and hardware diagnostics.

3.4. Diverse Architecture and Implementation a 3rd Generation Systems

A more recent and different type of system, uses diverse redundancy, diverse implementation and active diagnostics [6] and can deliver not only reliability and availability required for the application while minimizing common cause failures found by the use of the same technology but also introduce additional protection to systematic failures normally related to human factors

3.4.1 Hardware Design – Diverse Hardware

This newer generation of systems use diverse processing hardware (multiple technologies) such as diverse operating systems and diverse hardware and diverse redundancy, both enhanced with diverse implementation (different implementation teams).

3.4.2 Application Execution in a SIL Compliant Environment

The use of diverse operating systems (use of different technologies) including the use of third party certified COTS and different execution path with different compiler rules extends the risk reduction found in the hardware design to the firmware/software environment in the system, including the use of limited variability software functions and provision of systematic capabilities as specified in the Functional Safety standards for critical applications.

3.4.3. Systematic Capabilities and Human Error

The concept of systematic capabilities and its associated performance measurement was introduced in the latest revision of IEC61508 standard (didn't exist as such in 1st Generation systems and was emerging in 2nd generation systems).

3.5. Safety and Network Security

1st and 2nd generation Safety Systems were not designed to reside on a networked plant, for the most part these systems allowed some communication interface or gateway but were not intended for integration, therefore in general are not equipped to counter security threats as those the industry has experienced in recent years.

A brief analysis of the definitions will show a clear differences Safety issues are random in nature and statistic analysis is possible security intentional in nature and statistical analysis will not measure performance.

The authors concur with industry standards and experts in the field. Network security may affect system performance and the safety of the installation. A system can't be safe without it being secure.

3.6. Design Best Practices and V-Model

As indicated earlier, industry standards in Functional Safety introduced the concept of Functional Safety Management System (FSMS) which includes a series of steps in the design, documentation and testing of the system, and should include Network Security as well.

Figure 3: Product Development V-Model

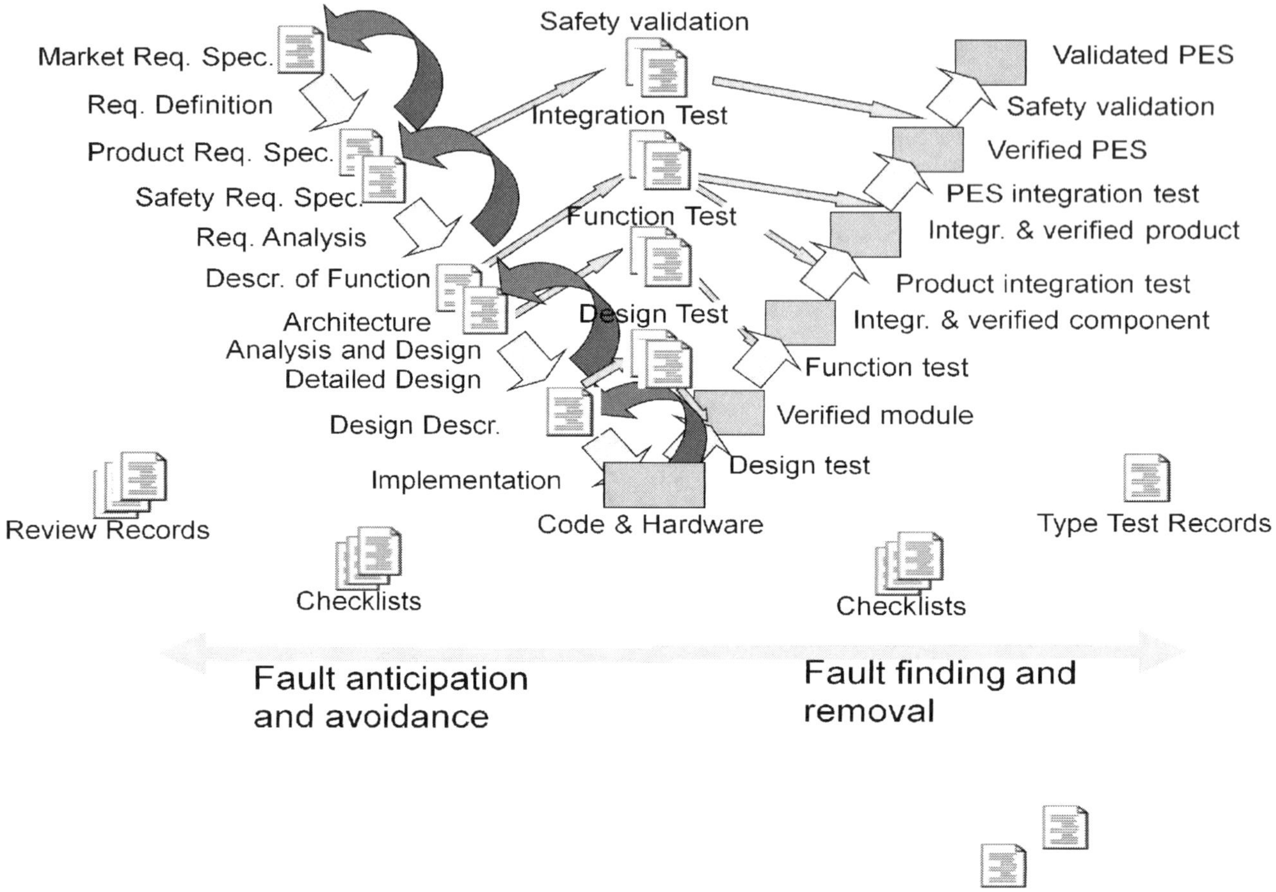

Today product development or design organizations responsible for releasing COTS products (hardware and software) intended for safety applications use design best practices as the V-Model (Figure 3) and are evaluated by a third party organization (i.e. TUV) and their FSMS is approved prior to the assessment of their products.

4. Engineering best practices to reduce risk in application design

As indicated previously, best practices in design apply both in the design of a Commercial Off the Shelf (COTS) Product and the Application Design or the application of a COTS tailored to reduce a particular risk in an industrial installation.

Best Practices in Application Design will apply a model similar to the V-Model used in the development of a product (Figure 3) and reinforced by a structured Quality Management system appropriate for Functional Safety applications.

Application Design starts with Hardware Design, which starts with a design that follows the COTS Product's Safety Manual and considers the particularities of the application as described on the Application Safety Requirements Specification to design a system that is less susceptible to product failures by considering aspects such as

- Power feeds
- Power Supplies
- CPUs
- I/O Modules Hardware
- Application Software
- Networks

4.1 Hardware

It's clear that hardware is not perfect, it's subject to fails, and the application design must consider fail recovery modes or additional risk reduction methodologies or options including diverse methods of performing shutdowns as:

- Hardwired wired pushbutton,
- Hardwired logic systems (with appropriate SIL rating)
- Solid State relays
- Etc.

4.2 Software

One critical aspect that has gained visibility in recent years is the software design of the application. Efforts in this area include the development of a well documented Software Functional Design, appropriate selection of software tools and libraries that meets the Safety Integrity Level (SIL) required for the application, Code Review practices including Independent Validation and Verification depending on the SIL, competence of the programmer and other elements described later in section 5.1. The features available in the COTS product to address the systematic capabilities requirements are a valuable asset for the programmer and contribute to reducing potential errors that will later affect the safety of the installation.

4.3 Functional Testing and Periodic Proof Testing

Another critical aspect is Functional Testing and Periodic Proof Testing, particularly because Not ALL System Faults are Self-Revealing and Covert Faults that may inhibit SIS action on Demand can only be detected by testing the entire system, this requires not only a Full Functional Test prior to commissioning, typically conducted as an integration test in the vendor or System Integrator facility and later in the final instrumentation room in the plant; but Periodical Functional Tests using a documented procedure to Detect Covert Faults and covering the entire SIS.

Functional Testing should Record and Analyze activation of SIS functions, and spurious activation of an Emergency Shutdown Valve due to a Process Shut Down, but this does not test the Entire Function of the same valve during an ESD action.

5. Automation can't check for human intervention

Once the design is complete, all of these systems will have different degrees of interaction with users, and different type of users with different levels of competence depending to their role (i.e. operation, maintenance personnel and engineers)

The most sophisticated automation can't prevent human error, industrial plants are designed with the highest accuracy in mind, with several separate safety loops often checking the integrity of process systems. This apparatus however can't check for human intervention and a small human error could cause an enormous catastrophe. 70% of reported incidents in the oil and gas industry worldwide are attributable to human error and account for in excess of 90% of the financial loss to the industry.

Results from research conducted by the Health and Safety Executive and published in the book Out of control: Why control systems go wrong and how to prevent failure [7] show the impact of the human element in industrial incidents throughout project lifecycle. This is shown on Figure 4.

International Functional Safety Standards (IEC 61508 and IEC61511) and previously ISA 84 introduced the concept of the Safety Lifecycle, describing the phases that should take place from concept to design to implementation and operation of a Safety Instrumented System. The Safety Lifecycle is a step in the direction to reduce the impact of human factors by establishing the proper design best practices, documentation reviews and validation and verification steps in the execution of a safety project. Additionally the standards introduce two important elements: Competence of Personnel mentioned briefly in Section 4.2 and Functional Safety Management System (FSMS).

Recent changes to International Functional Safety Standard IEC61508 have turned the requirements on FSMS and Competence of Personnel into normative clauses of mandatory compliance instead of a recommendation as was presented in the previous version of the same standard.

5.1 Functional Safety Management System (FSMS) and Competence Requirements

A Functional Safety Management System (FSMS) consists of proven and validated procedures, methodologies, templates and report outlines, covering those essential elements of hazard and risk assessment; SIL determination; SIS design and SIL capable hardware, SIS installation, SIS commissioning and validation and SIS operations and maintenance. A FSMS is used in the design of a COTS product and during the Application Design. It's typically an extension of the existing Quality Assurance/Quality Management process and lives from the conception to decommission of the SIS and beyond to become part of the organization best engineering practices and work processes.

Figure 4: Root Cause of Control System Failures

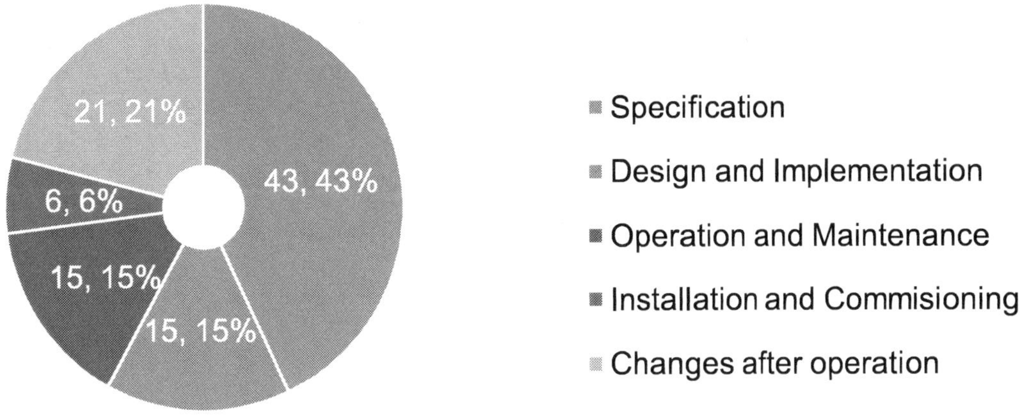

As indicated, the safety standards call for competence of personnel. This requirement shouldn't be interpreted as a need for safety experts in every aspect of the design, commissioning or operation and maintenance of a SIS, but instead as the need for an evaluation of the appropriate expertise and experience required to perform a particular job function in any phase of the safety lifecycle. In practical term those involved with activities related to a safety system should have demonstrated "competence" either by formal training, certification or accreditation or on-the-job relevant experience.

The reader might conclude that the standards today require competent and independent review groups because system designers don't trust product designers or other application designers and therefore we would rather be safe than sorry.

6. Summary

In summary

- Best in Class companies link safety to their success and invest in programs and systems to reduce their risk
- Engineers have attempted to reduce risk by minimizing the potential for common cause failures

- Common cause failures can occur in products (hardware or software) or the implementation of the application
- The industry has conceived best practices (i.e. FSMS) to minimize the impact of human error
- Human factors can't be ignored in the design, particularly in the application design
- Enforcement of these best practices via formal FSMS has proven to be a way to reduce the risk introduced by engineers… it's vital!

References

[1] IEC 61508 Functional safety of electrical/electronic/programmable electronic safety-related systems International Standard 2010

[2]Integrating Control and Safety: Where to draw the line.
Robin McCrea-Steele, TÜV FSExpert, Invensys-Premier Consulting Services

[3] Common Cause and Common Sense, Designing Failure Out of Your SIS
Angela E. Summers, Ph.D. and Glenn Raney, Oct 2000

[4] Integrated but separate, Advances in integrated and safety control,
Roger Prew, ABB , June 2009

[5] IEC 61511 Functional safety - Safety instrumented systems for the process industry sector
International Standard 2003

[6] Providing Independent Layers of Protection with Integrated Safety Systems
Luis M. Duran, ABB and Ron Johnson Dow Chemical, Oct 2009

[7] Out of control: Why control systems go wrong and how to prevent failure

HSE Books ISBN 0-7176-2192-8

(44bj) Establishing a 3 Credit Course in PSM for Undergraduate Engineers

Deborah L Grubbe, Executive Mgmt, Operations and Safety Solutions, LLC, Chadds Ford, PA

Considering academic direction and industrial needs, the Purdue Chemical Engineering faculty and an industrial alumni team jointly sponsored, developed, executed and audited a 3 credit course in Process Safety Management for undergraduates in Engineering. This "team taught" course is now a required in the Purdue ChE curriculum, and has received enthusiastically high marks from the students. Originally conceived in 2010, the course went from concept to required in less than 24 months.

One of the alumni team members, with over 30 years in the process industries, will describe each step of the process: creating the support, drafting the syllabus, designing the course, executing the lectures, planning the ancillary support activities, and auditing the results of the 15 week, 45 lecture series. The final exam was the simulation of an emergency situation, which was modeled after a real life process safety incident. Actual student feedback will be shared. An eBook text has been developed for this course, and the speaker will conduct a short demonstration of this product as part of the overall presentation.

Investigations on the self heating potential of garage waste as basis for the hazard evaluation of bulk storage facilities

Hannes Kern
Montanuniversität Leoben
Franz-Josef Strasse 18, 8700 Leoben
hannes.kern@unileoben.ac.at

Christoph Spijker, Harald Raupenstrauch
Montanuniversität Leoben
Franz-Josef Strasse 18, 8700 Leoben
christoph.spijker@unileoben.ac.at, harald.raupenstrauch@unileoben.ac.at

Prepared for Presentation at
American Institute of Chemical Engineers
2013 Spring Meeting
9th Global Congress on Process Safety
San Antonio, Texas
April 28 – May 1, 2013

UNPUBLISHED

Investigations on the self heating potential of garage waste as basis for the hazard evaluation of bulk storage facilities

Hannes Kern
Montanuniversität Leoben
Franz-Josef Strasse 18, 8700 Leoben
hannes.kern@unileoben.ac.at

Christoph Spijker, Harald Raupenstrauch
Montanuniversität Leoben
Franz-Josef Strasse 18, 8700 Leoben
christoph.spijker@unileoben.ac.at, harald.raupenstrauch@unileoben.ac.at

Keywords: self heating, garage waste, differential reactor

Abstract

The self heating of bulk material is a major point to be considered in hazard analysis especially for waste management and waste treatment companies. The self heating potential for bulk material mainly depends on the composition of the material itself and on systemic influences like the geometry of the storage area, weather conditions or the storage temperature. Besides microbiological activity, chemical oxidation at low temperatures is the main reason for the self heating and auto ignition of bulk material. Data from the measurement of the chemical and biological oxidation kinetics can be used as basis for the simulation of storage areas by computational fluid dynamics. The measurement of the low temperature oxidation kinetics is carried out in a differential reactor, which is characterized by a small amount of sample material, isothermal conditions and very low turnover rates. Due to the fact that chemical oxidations reactions at low temperatures show very low turnover rates themselves, the experimental setup is laid out as a cycle process. The oxidation kinetics are measured by the analysis of the carbon monoxide (CO) and carbon dioxide (CO2) production rate at different temperatures. Usually a temperature range from 40°C up to 120°C is investigated. Waste materials from workshops contain significant amounts of grease and oil as well as residuals of metal dusts or chips dispersed on the surface of paper or cloth material used for cleaning purposes. The large surface combined with the fine dispersion of grease and oil accelerates the chemical oxidation significantly. To determine the influence of the oil and grease content on the oxidation kinetics, measurements of paper and cloth samples with different contents of oil were carried out in a differential cycle reactor. For the investigations biological cutting oil and oil with a petrochemical origin was investigated. Furthermore the influence of metal dusts on the oxidation characteristics of the different materials was investigated. The results of the investigations were compared with the oxidation kinetics of garage waste material with a mean diameter of 0- 40mm.

1. Introduction

The self heating of bulk material is a major point to be considered in hazard analysis especially for waste management and waste treatment companies. Besides safety issues the auto ignition of bulk storage areas is also a serious economical threat and occurs quite frequently. In industrial branches where material with an inhomogeneous composition (like waste treatment facilities) are processed the problem of self heating is even more present. Especially at residual derived fuel storage areas fires are also a serious ecological threat, due to composition of the materials themselves. The measurement of the low temperature oxidation kinetics combined with CFD [1] modeling used for the prediction of the self heating potential of bulk storage areas is an important element of hazard analysis and loss prevention. In the present study the self heating potential of garage waste was investigated. Garage waste shows a high tendency of self heating and therefore auto ignition is a major hazard. In the course of this study different influencing parameters on the self heating potential of garage waste were investigated, to be able to evaluate the risk of auto ignition. Self heating in general is a complex process that mainly depends on the parameters presented below [2]:

Material specific parameters

- composition
- chemical kinetics
- particle size distribution
- bed porosity
- heat conductivity

Apparatus specific boundary conditions

- geometry of the storage area
- size surroundings

Operating conditions

- temperature
- gas velocity

The chemical oxidation at low temperatures is besides microbiological activity the main reason for the self heating and auto ignition of bulk material. Basicly two main oxidation reactions occur at low temperatures:

$$a_{Fuel(s)} + b_{O_2(g)} \rightarrow c_{CO(g)} + d_{\mathrm{Rest}(s)} \tag{1}$$

$$e_{Fuel(s)} + f_{O_2(g)} \rightarrow g_{CO_2(g)} + h_{\mathrm{Rest}(s)} \tag{2}$$

The reaction of the solid fuel with oxygen gives carbon monoxide (1) and carbon dioxide (2). In a simple kinetic approach it is assumed that the solid fuel can be seen as bulk with a constant concentration compared to the oxygen concentration. After some simplifications the relations (3) and (4) can be derived from the Arrhenius relation for chemical reaction kinetics:

$$r_{CO} = k_0 \cdot \exp\left(\frac{-E_A}{R \cdot T}\right) \cdot y_{O_2}^n \quad (3)$$

$$r_{CO} = k_0 \cdot \exp\left(\frac{-E_A}{R \cdot T}\right) \cdot y_{O_2}^n \quad (4)$$

For further research it is assumed that the postulated reactions undergo only chemical limitation, because chemical limitation exceeds transport limitation by several orders of magnitude. The determination of the kinetic parameters is carried out by the observation of the CO and CO_2 production rates.

2. Experimental investigations

2.1 Sample Material

Waste materials from workshops contain significant amounts of grease and oil as well as residuals of metal dusts or chips dispersed on the surface of paper or cloth material. The large surface combined with the fine dispersion of grease and oil accelerates the chemical oxidation significantly [3]. To determine the influence of the oil and grease content on the oxidation kinetics, paper and cotton cloth samples with different contents of oil were chosen as model substances for further investigations. The calorific values as well as the chemical composition of the investigated materials can be seen in Table1. The oil content of the garage waste sample material was determined by Soxhlet extraction with cyclohexane as solvent. The sample had an oil content of around 15% and a particle size of up to 30mm.

Table 1: Composition of the sample material

	Carbon [%]	Hydrogen [%]	Nitrogen [%]	Calorific Value [kJ/kg]
Paper	40.41	6.70	0.61	16277
Cotton cloth	41.81	6.80	0.35	16484
Oil (arecal)	78.7	12.1	0.01	35911
Oil (WD 40)	85.7	13.3	0.04	42258

The product “arecal” is a cutting oil based on rape oil while “WD40” is a cutting oil derived from fossil hydrocarbons. Pre tests in the differential reactor described in 2.2 showed that arecal is far more reactive than WD40. That may be explained by the high content of unsaturated carbon bonds in rape oil, which are more reactive in terms of oxidation processes. Further investigations on the oxidation behavior have been carried out only with the more reactive arecal oil.

2.2 Method

The measurement of the low temperature oxidation kinetics is performed in a differential reactor (Figure 1), which is characterized by a small amount of sample material, nearly isothermal conditions and very low turnover rates. Due to the fact that chemical oxidations reactions at low temperatures show very low turnover rates themselves, the experimental setup is laid out as a cycle process. The oxidation kinetics is measured by the analysis of the carbon monoxide (CO) and carbon dioxide (CO_2) production rate at different temperatures. Usually a temperature range from 40°C up to 120°C is investigated.

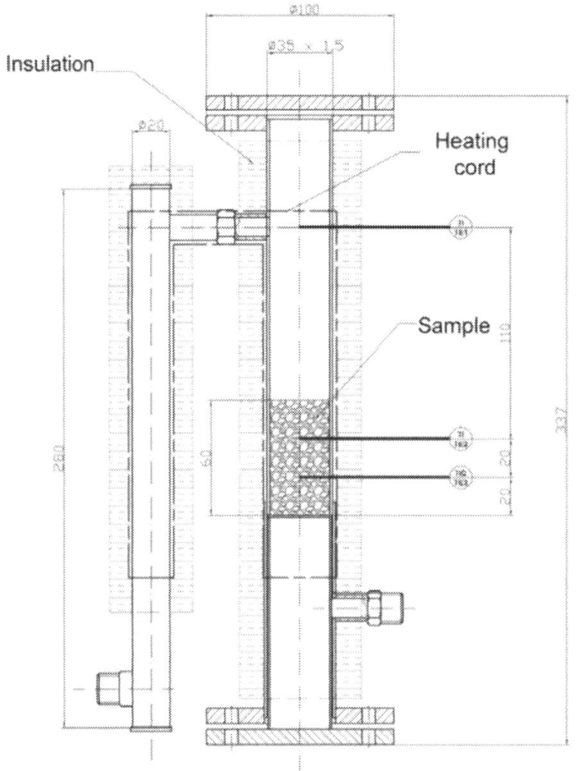

Figure 1: Principal layout of the differential reactor [2]

Figure 2 shows a typical of temperature development and gas production (e.g. carbon dioxide). At the beginning of the measurements the sample material inside the reactor is heated up above the highest planned measuring temperature to desorb carbon dioxide and carbon monoxide that may be adsorbed at the surface of the sample material. After the desorption process the sample material is cooled to the first measuring temperature and constantly purged with synthetic air (80%N_2, 20%O_2). When the sample reaches a constant temperature value the measurement process is started and the measuring cycle is closed. The areas of constant CO_2 increase in the concentration graph of Figure 2 represent these measurement cycles at different temperatures. At higher temperatures an increase of carbon dioxide production can be observed, which is directly linked to oxidation kinetics. The calculation of the kinetic parameters is carried out by the Arrhenius relation after the measuring procedure. A measuring cycle normally takes about 30min for one temperature.

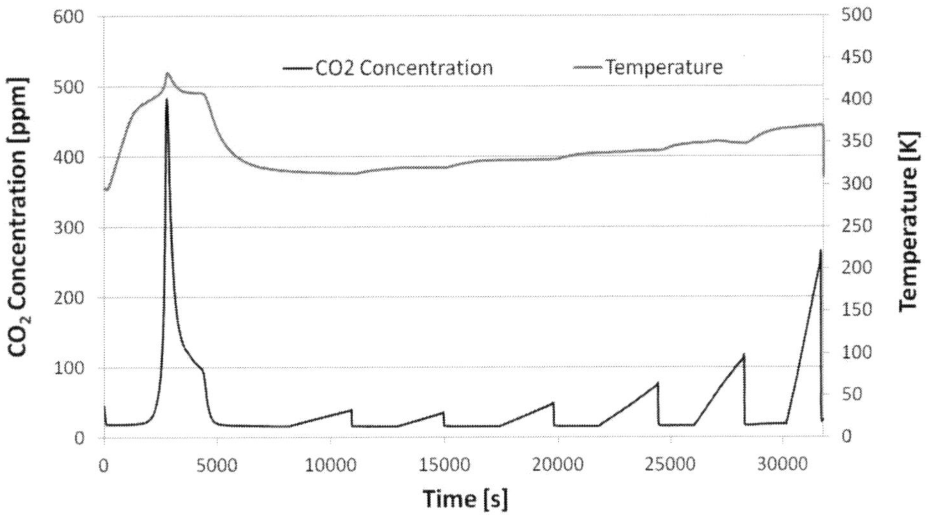

Figure 2: Principal layout of the differential reactor

2.3. *Paper*

Figure 3 below shows the CO production rates of paper with different oil contents. The "0% Oil content" line represents paper material not contaminated with arecal oil. The CO production rate increases with the oil content of the sample. The progress of the CO production rate with an oil content of 6.9% is widely comparable to the results of garage waste (~15% oil content).

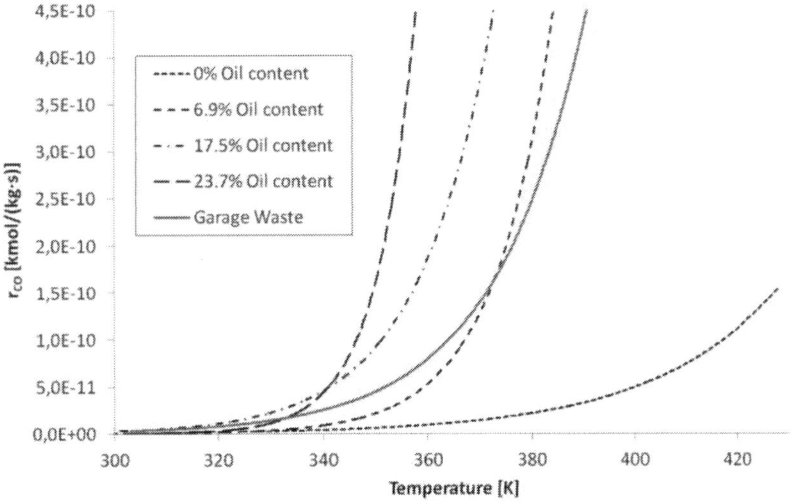

Figure 3: CO production rates of paper with different oil contents

For CO_2 production, as shown in Figure 4, the results were a bit different. The CO_2 production rate of garage waste was higher than all other rates measured. Interestingly the CO_2 production rate for the paper with an oil content of 17.5% was higher than for 23.7%. That may be explained by kinetic effects and is maybe linked to the significant amounts of CO that were produced by this paper sample.

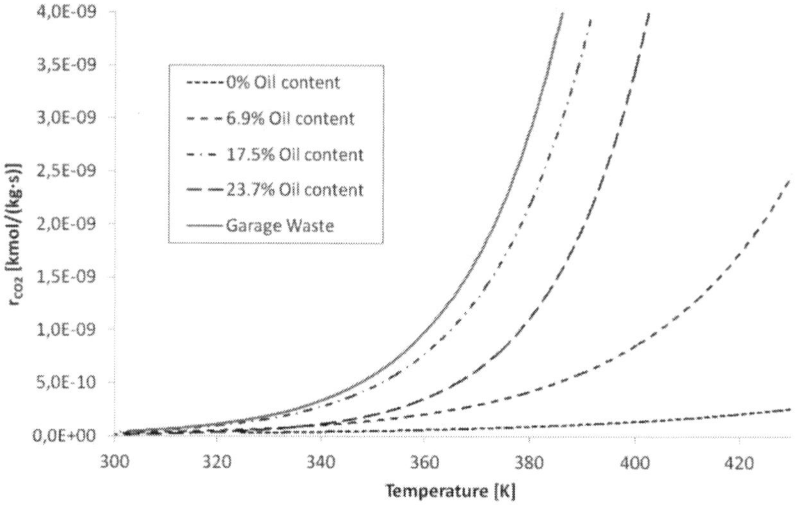

Figure 4: CO_2 production rates of paper with different oil contents

The kinetic data resulting from the calculations based on the Arrhenius relation are presented in Table 2.

Table 2: Kinetic data of the paper sample material

	CO		CO_2	
	E_a [kJ/mol]	K_0 Kmol/kg*s	E_a [kJ/mol]	K_0 [Kmol/kg*s]
Paper 0.0% Oil cont.	4.89E+04	3.25E-09	2.52E+04	9.74E-12
Paper 6.9% Oil cont.	8.70E+04	7.45E-03	3.45E+04	6.63E-10
Paper 17.5% Oil cont.	6.69E+04	3.17E-05	4.94E+04	3.84E-07
Paper 23.7% Oil cont.	1.02E+05	1.91E-00	5.48E+04	1.04E-06
Garage waste	5.59E+04	3.35E-07	5.22E+04	1.14E-06

2.4 Cotton cloth

During the preparation of the cotton sample material oil contents that were similar to the paper samples have been tried to be reached. Although there are smaller deviations in the oil content, the results are comparable.

Figure 5 shows the results for the investigation of the cotton cloth material. The results do not show good accordance with the garage waste material. But again it can be observed, that the sample materials with a high oil contents also show higher reactivity than the sample with lower contents.

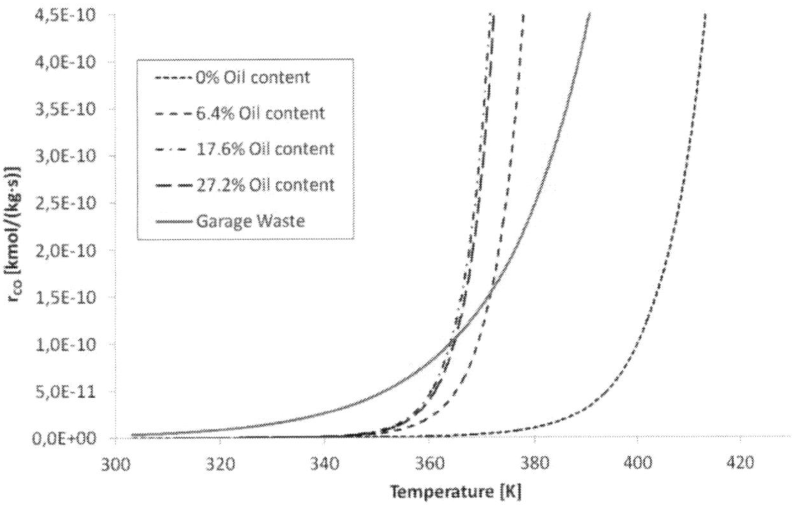

Figure 5: CO production rates of cotton cloth with different oil contents

The production rates from Figure 6 do not show a big deviation regarding the oil content, but the samples contaminated with oil clearly show a higher reactivity than the cotton material itself. In general all CO_2 production rates are comparable with the original garage waste sample material, which again indicates a strong link between oil content and reactivity.

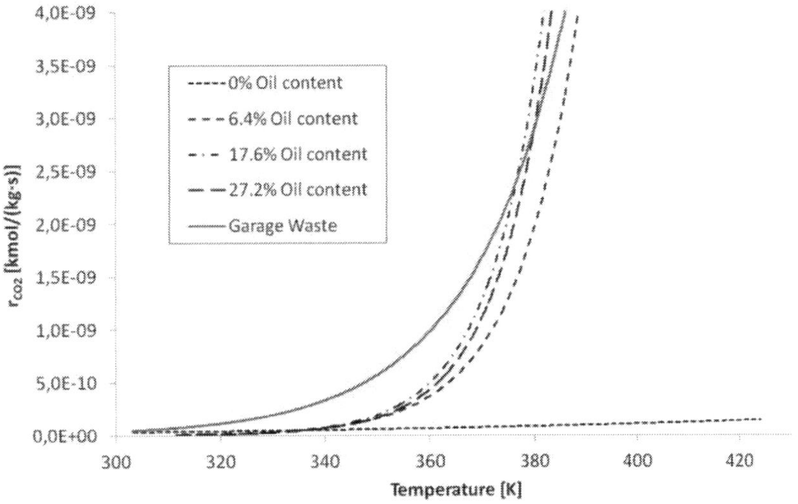

Figure 6: CO_2 production rates of cotton cloth with different oil contents

The kinetic data of the cotton cloth samples is presented in Table 3.

Table 3: Kinetic data of the cotton cloth sample material

	CO		CO_2	
	E_a [kJ/mol]	K_o Kmol/kg*s	E_a [kJ/mol]	K_o [Kmol/kg*s]
Cotton 0.0% Oil cont.	3.68E+04	1.66E-10	1.34E+04	1.99E-13
Cotton 6.4% Oil cont.	1.90E+05	2.90E+12	9.01E+04	1.52E-01
Cotton 17.6% Oil cont.	1.27E+05	3.14E+04	9.96E+04	4.82E+00
Cotton 27.2% Oil cont.	1.02E+05	1.80E+00	5.48E+04	9.78E-07
Garage waste	5.59E+04	3.35E-07	5.22E+04	1.14E-06

2.5 Influence of iron dusts

Due to normal workshop operations garage wastes always contain significant amounts of metal residues. To investigate the influence of metal residues on the oxidation kinetics small amounts of iron dust were added to a cotton cloth sample with an oil content of around 11%.

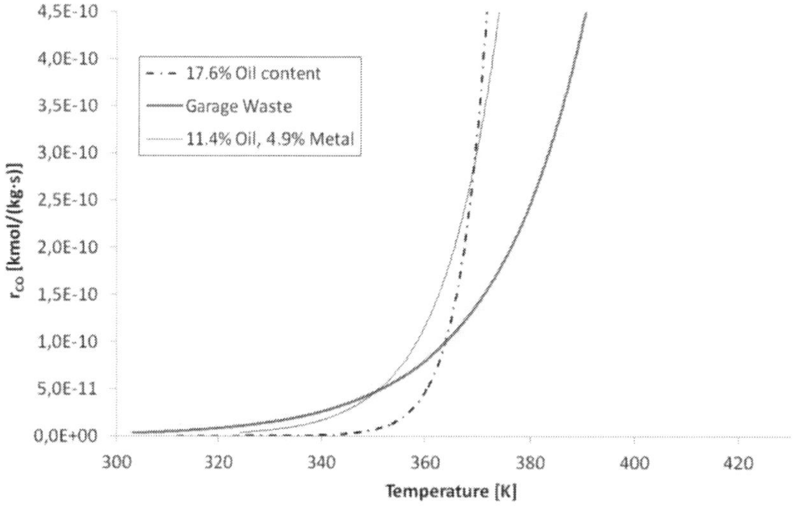

Figure 7: CO production rates of cotton cloth contaminated with oil and iron dust

Figure 7 shows the production rates for the cotton/oil/iron dust sample and also the results for pure cotton cloth with 17.6% oil content. The results show that the iron dust has an influence on the reactivity of the material. Both materials show comparable CO production rates even though the sample containing iron dust contains less oil. The same effect can be observed for the CO_2 production rate in Figure 8. The CO_2 production rate in general shows good accordance to the results of the garage waste sample which indicates the representativeness of the chosen sample materials.

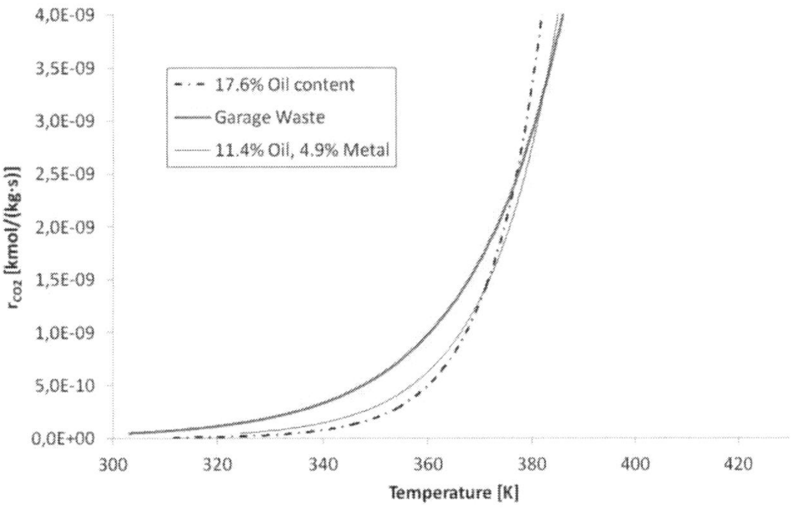

Figure 8: CO_2 production rates of cotton cloth contaminated with oil and iron dust

3. Conclusions

The results of the investigations of garage waste and different reference materials show a clear connection between the oil content of garage waste and its reactivity. Garage waste in general should be considered as highly reactive. Field investigations show, that a significant temperature rise can be observed within days from the beginning of the storage period. High contents of oil and metal residues lead to a significant fire risk. To determine the permissible storage conditions for garage waste CFD models could be applied. The results from measured kinetics can be used as the basis for CFD modeling in combination with the layout of the storage area and the further surrounding conditions [4].

4. References

[1] Wallner, S. (2004). Ein Modell zur Untersuchung des Selbsterwärmungsverhaltens von Schüttgütern. Dissertation, Technische Universität Graz

[2] Walkner, R. (2005) Untersuchung des Selbsterwärmungsverhaltens eines Lagers für aufbereiteten Substitut Brennstoff (ASB). Diplomarbeit, Technische Universität Graz

[3] Wallner, S., Walkner, R., Raupenstrauch, H. (2004). Untersuchung des Selbsterwärmungsverhaltens gelagerter Kunststoffmaterialien mittels numerischer Simulation. Endbericht für das FFF- Projekt „Entwicklung neuer Brennstoffqualitäten aus Abfall". Technische Universität Graz

[4] Raupenstrauch, H. (1997) Gasdurchströmte, chemisch reagierende Schüttschichten. Habilitationsschrift, Technische Universität Graz

9TH GLOBAL CONGRESS ON PROCESS SAFETY

Estimate Vibration Risk for Relief and Process Piping

Georges A. Melhem, Ph.D.
ioMosaic Corporation
93 Stiles Road | Salem, NH 03079
melhem@iomosaic.com

Prepared for Presentation at
American Institute of Chemical Engineers
2013 Spring Meeting
9th Global Congress on Process Safety
San Antonio, Texas
April 28 – May 1, 2013

UNPUBLISHED

Estimate Vibration Risk for Relief and Process Piping

Georges A. Melhem, Ph.D.
ioMosaic Corporation
93 Stiles Road | Salem, NH 03079
melhem@iomosaic.com

Keywords: Acoustic induced vibration (AIV), flow induced vibration (FIV), vibration risk, resonance, singing relief valve.

Abstract

Current API, AIChE/CCPS, and AIChE/DIERS pressure relief and flare systems guidelines and standards do not formally address vibration risk. They do not offer specific guidance on velocity limitations other than backpressure calculations and they do not offer guidance on acoustic induced or flow induced piping vibration fatigue failure. This paper provides a summary of experience based methods for the estimation of vibration risk in relief and process piping. In addition, this paper extends the applicability of the experience based methods to two-phase flow.

1. Introduction

Fatigue failure of relief and/or process piping caused by vibration can develop due to the conversion of flow mechanical energy to noise. Factors that have led to an increasing incidence of noise vibration related fatigue failures in piping systems include but are not limited to (a) increasing flow rates as a result of debottlenecking which contributes to higher flow velocities with a correspondingly greater level of turbulent energy, (b) frequent use of thin-walled piping which results in higher stress concentrations, particularly at small bore and branch connections, (c) design of process piping systems on the basis of a static analysis with little attention paid to vibration induced fatigue, (e) and lack of emphasis of the issue of vibration in piping design codes. Piping vibration is often considered on an ad-hoc or reactive basis. According to the UK Health and Safety Executive (HSE), 21 % of all piping failures offshore are caused by fatigue/vibration. Typical systems at risk include large compressor recycle systems and high capacity pressure relief depressuring systems. For relief and flare piping, flow induced turbulence and high frequency acoustic excitations are key concerns.

2. Flow Induced Vibration (FIV)

Fluid flow in pipes generates turbulent energy (pressure fluctuations). Dominant sources of turbulence are associated with flow discontinuities in the piping systems (e.g., partially closed valves, short radius, mitered bends, tees or expanders). The level of turbulence intensity is a function of pipe size, fluid density, viscosity, velocity, and structural support. High noise levels are generated by high velocity fluid impingement on the pipe wall, turbulent mixing, and if the flow is choked, shock waves discontinuities downstream of flow restriction. This can lead to high frequency excitation / vibration risk.

3. Acoustic Induced Vibration (AIV)

High frequency acoustic energy is often generated by a pressure reducing device such as a relief valve, control valve, or orifice plate. Acoustic induced piping failure is of a particular concern for safety related systems (e.g. relief and blowdown/depressuring). The severity of high frequency acoustic excitation is primarily a function of the pressure upstream and downstream of the pressure reducing device, pipe diameter, and the fluid volumetric flow. Acoustically induced piping failures are known to occur at non-asymmetric discontinuities in the downstream piping such as small bore and branch connections and welded supports.

4. Additional Causes of Vibration

Additional causes of piping vibration include mechanical excitation, pulsation, vortex shedding, surge or momentum changes associated with valves, cavitation, and flashing. Mechanical excitations are often associated with pipes connected with reciprocating compressors, pumps, or rotating machinery. Such connection machines cause vibration of the pipe and its support structure. Thermowells are intrusive fittings and are subject to static and dynamic fluid forces. Vortex shedding is the dominant concern as it is capable of forcing the thermowell into flow-induced resonance and fatigue failure. This is particularly important at high fluid velocities.

5. Relief and Depressuring Systems

Depressuring systems are often subjected to acoustic energy (rapidly fluctuating pressure forces) generated by flow turbulence which is accentuated by flow restricting devices within the flow path. The magnitude of pressure fluctuations depends on the mass flow rate, speed of sound, and density. For choked flow, intense noise due to large shock discontinuities and pressure fluctuations is generated. The generated noise is non-periodic due to the randomness of the pressure fluctuations. Choked flow typically leads to a wide frequency spectrum with peak values than can exceed 1000 Hz. Vibrations associated with small fittings and branch connections are of special concern because they introduce discontinuities and stress concentration points.

In many situations resonance can onset which can lead to magnification of static piping loads up to a factor of 50 times. The presence of discontinuities such as tees and welded pipe supports can further increase these loads.

6. Noise Generation

A pressure reducing device or relief device controls flow by converting internal energy into kinetic energy. Some energy is converted to heat through friction (viscous forces) by intense turbulence and shock formation. Some of the energy is also transferred to the pipe wall as vibration, and a portion of this is radiated as noise. The primary noise generating mechanism is the confined jet of fluid formed between the upstream and downstream locations. Flow noise can be modeled as noise of a confined jet. As a result, the noise-generation mechanisms are turbulent mixing, turbulence boundary interaction, shock, shock/turbulence interaction and flow separation.

Since the noise is generated downstream of a flow restriction, most of the acoustic energy is radiated to the downstream piping, which becomes the transmitting medium. As the noise travels downstream along the inside of the pipe in the fluid it radiates through the pipe wall along its entire length.

7. How to Calculate Sound Power Level

We can calculate the sound power level in a fundamental way by assigning an efficiency factor to provide the fraction of flow mechanical energy that is converted to noise:

$$W = \eta \frac{1}{2} \dot{m} u^2 \quad (1)$$

$$L_W = 10 \log_{10} \left[\frac{W}{10^{-12}} \right] = 10 \log_{10} W + 120 = 10 \log_{10} \left[\eta \frac{1}{2} \dot{m} u^2 \right] + 120 \quad (2)$$

where W is the flow mechanical power or energy, L_W is the sound power level in dB, and η is defined as the acoustical efficiency factor. If the flow is choked, then W becomes:

$$W = \eta \frac{1}{2} \dot{m} u_{sonic}^2 \quad (3)$$

If a value of η can be estimated for single and multi-phase flow, then the sound power level can be easily calculated not only for pressure reducing devices but also for pipe flow. Attenuation due to friction and temperature changes can be calculated from pipe flow equations in a more detailed manner. Computer codes such as SuperChems can then calculate the sound power level at every axial location for flow piping for single and multi-phase flow. For incompressible flow, the value of flow velocity for an ideal nozzle can be calculated from the mechanical energy balance:

$$u = \sqrt{\frac{2}{\rho_l} \Delta P} \quad (4)$$

Substituting the above equation for u in Equation 1 yields the following Equation for W for liquid flow through an ideal nozzle:

$$W = \eta \dot{m} \frac{\Delta P}{\rho_l} \quad (5)$$

For liquid flow, a typical value of η is approximately 10^{-8}.

All gas flow acoustic efficiencies have been reported to approach 1 % of the total flow mechanical energy for rockets[i]. This is illustrated in Figure 1. When the acoustic efficiency is plotted against the flow mechanical power Figure 1 is obtained. The measured curve indicates that the acoustic efficiency falls off as the flow mechanical energy gets larger while the calculated curve[ii] indicates increasing acoustic efficiencies.

Figure 1. Acoustic efficiency trends

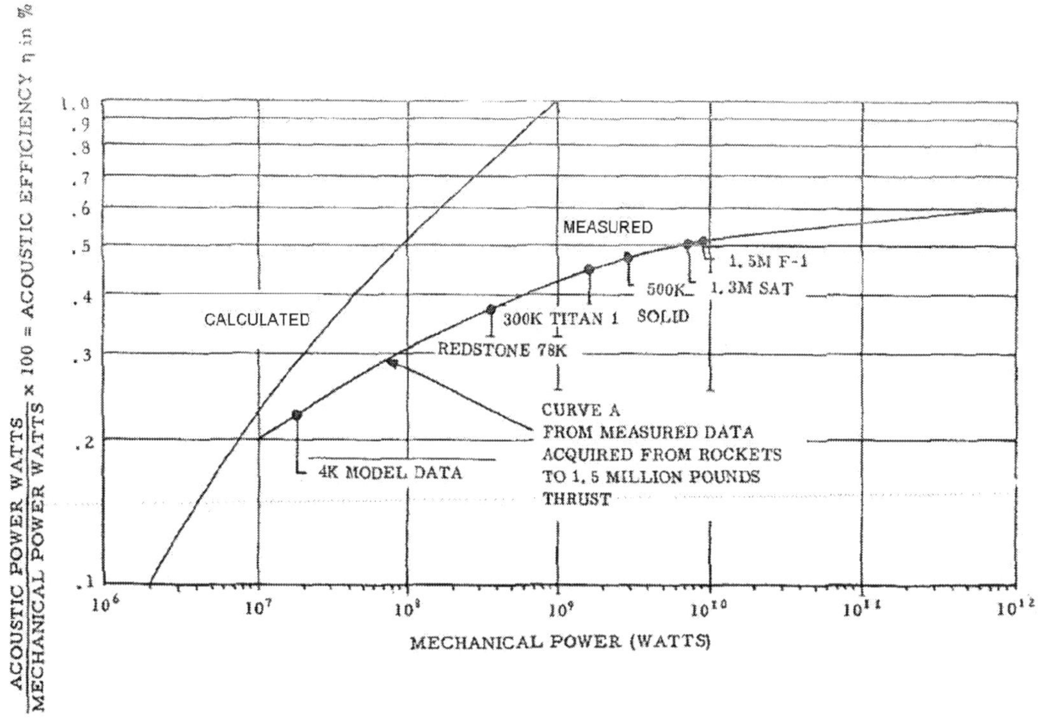

Two-phase flow sound power level can also be calculated using Equation 1. Recent work by Singh et al.[iii, iv] demonstrated that more attenuation of noise is exhibited by two-phase flow. Therefore, using the gas acoustic efficiency values will over predict sound power and noise levels for two-phase flow.

8. Existing Methods and Guidance

Current API, AIChE/CCPS, and AIChE/DIERS pressure relief and flare systems guidelines and standards do not formally address vibration risk. They do not offer specific guidance on velocity limitations other than backpressure calculations and they do not offer guidance on acoustic induced or turbulence induced piping vibration fatigue failure. The Marine Technology Directorate Limited (MTD) has published in 1999 "Guidelines for the Avoidance of Vibration Induced Fatigue in Process Pipework"[v]. A second edition of these guidelines were published in 2008 by the Energy Institute[vi]. The methods outlined in these guides have been incorporated in the SuperChems Expert ioVIPER modules. The MTD/Energy Institute guidelines provide qualitative and quantitative methods for the assessment of piping vibration failure risk and depending on the calculated risk level they provide generic guidance for the mitigation of vibration risk.

Many operating companies have established their own internal guidance for evaluating and minimizing piping vibration risk. Although these criteria vary from company to company, they all in general include a limit of flow velocity in some form.

A common criteria used in relief systems is to limit the value of flow Mach number to a value ranging from 0.3 to 0.9:

$$0.3 \leq M = \frac{u}{u_{sonic}} \leq 0.9 \tag{6}$$

Another common criteria often encountered is to limit the dynamic pressure component or kinetic energy of a flow stream to a value of 100,000 Pascals for gas flow:

$$\rho u^2 \leq 1 \times 10^5 \ \ \mathrm{kg/m/s^2} \tag{7}$$

and 50,000 for two-phase flow:

$$\rho u^2 \leq \frac{1}{2} \times 10^5 \ \ \mathrm{kg/m/s^2} \tag{8}$$

The 2006 fifth edition of the NORSOK Process Design Standard P-001 limits the flow velocities for all flare lines to $\rho u^2 \leq 200{,}000 \ \ \mathrm{kg/m/s^2}$ for single and multiphase flow. The piping for flare headers and sub-headers are designed for a maximum Mach number of 0.6 and lines downstream of pressure relief valves to the first sub-header are designed for a maximum Mach number of 0.7. For process lines the maximum design velocity for gas pipes is limited to 60 $\mathrm{m/s}$ or $u \leq 175 \frac{1}{\rho^{0.43}}$ whichever is less. The maximum velocity for two-phase lines is limited to $u \leq 183 \frac{1}{\rho^{0.5}}$ or:

$$\rho u^2 \leq 183^2 \leq 33{,}489 \ \ \mathrm{kg/m/s^2} \tag{9}$$

9. The Method of Carucci and Mueller

Carucci and Mueller have published guidance for the estimation of sound power levels for control valves and pressure reducing devices[vii, viii].

$$L_W = 10 \times \log_{10} \left[\left(\frac{P_1 - P_2}{P_1} \right)^{3.6} \times \dot{m}^2 \times \left(\frac{T_1}{M_w} \right)^{1.2} \right] + 126.1 \tag{10}$$

$$=10\times\log_{10}\left[4\times\left(\frac{P_1-P_2}{P_1}\right)^{3.6}\times\dot{m}^2\times\left(\frac{T_1}{M_w}\right)^{1.2}\right]+120 \tag{11}$$

$$=36\log_{10}\left(\frac{P_1-P_2}{P_1}\right)+20\log_{10}(\dot{m})+12\log_{10}\left(\frac{T_1}{M_w}\right)+126.1 \tag{12}$$

where P_1 is the upstream pressure or source pressure, P_2 is the downstream pressure, T_1 is the source temperature, $\dot{m}$ is the mass flow rate, and M_w is the gas molecular weight.

Attenuation of noise due to friction and heat conduction losses is estimated from the following equation:

$$L_{W,At}=0.06\frac{L}{D_i} \tag{13}$$

where L is the pipe length and D_i is the internal pipe diameter. For an L/D of 50 the attenuation loss is 3 dB. Abrupt changes in flow area (expansion) in the piping can also lead to attenuation. The decrease in sound power level can be estimated from the following equation:

$$L_{W,Ex}=2\left(\frac{D_2}{D_1}-1\right) \tag{14}$$

where $D_2 > D_1$. A 3 dB reduction is typically applied to the flow leaving a tee in each direction or entering into a header or a large drum or vessel. The Carucci and Mueller Equation 12 is used by the Energy Institute Guidelines for the assessment of failure likelihood for high frequency acoustic excitation [vi]. Note that the Carucci and Mueller equation cannot be easily applied to complex piping systems, multi-phase systems, and relief piping with multiple chokes. Some companies also add 6 dB to the sound power level estimate when sonic flow exists at a branch connection to account for amplified dynamic strain response.

9.1 Analysis of the Carucci and Mueller Data Set

Careful analysis of the Carucci and Mueller data set (see Table 1) indicates that the source pressures were high enough to produce choked (sonic) flow through a flow limiting orifice or valve upstream of the failure point further downstream, often at a branch or line connection. This important point was missed by Eisinger who used the downstream (non-choked) flow velocity to establish his Mach number based failure criterion. We were also able to reproduce Eisinger's estimates of the original Carucci and Mueller data set based on his published paper.

Carucci and Mueller provided enough data in their original paper to allow the calculation of the upstream flow limiting flow area and choke point conditions (this is possible because they reported the actual flow rates) as well as the conditions downstream of the flow limiting device in the discharge piping. Note that the sound power level estimate upstream of the choke points should be based on the pressure difference (or pressure ratio) of the source pressure and the choke point. The sound pressure level downstream of the choke point (the primary cause of acoustic induced fatigue failure in downstream piping of the choked point) should be based on the difference (or pressure ratio) of the choke point and the shock discontinuity pressure downstream of the choke point. The original equation proposed by Carucci and Mueller used the pressure difference (or ratio) between the source pressure and the exit pressure downstream of the choke point. This yields an overestimate of the sound pressure. This overestimate is somewhat tempered by the fact that the pressure difference ratio to the source pressure is raised to the 3.6 power.

The following information was provided by Carucci and Mueller regarding the five failures and two high vibrations cases observed in Table 1.

A. Failure occurred during startup. Sonic velocity is achieved at the 6 inch branch connection of a 24 inch pipe run downstream of the recycle valve. The sound power level estimate considers the combined acoustic energy generated by the letdown valve and the sonic condition at the branch connection. The upstream pressure at the branch connection was estimated to be 98.8 psia.

B1/B2. This system consisted of a control valve letting down pressure to a safety valve / flare header system. The failure occurred after five to ten hours of operation as a crack at a 10 inch branch connection to 28 inch header. Sonic velocity was achieved at both the control valve and the 10 inch branch connection to the 28 inch flare header. As such, two pertinent data points are established:

1. Acoustic energy within the 10 inch line as generated by valve (not the failure point), and
2. Combined acoustic energy within the 28 in header as generated by valve and sonic condition at the 10 in branch connection (the failure point).

C. High vibrations, no failures.

D. Failure during startup. Cracks were observed in a 24 inch line downstream of the compressor recycle valve after twelve hours of operation. The cracks were near the branch connections of a 6 inch line and a 3/4 inch drain valve and at an I-beam support welded to the pipe at the elbow immediately downstream of the control valve.

E. High vibration, no failures.

F. Failure at severely undercut weld on a 300 mm (12 in.) line made to fasten conduit support clip. Points of high stress concentration were later eliminated.

G. This system consisted of six, parallel, high pressure steam letdown valves, each with a downstream, three pass contra-flow attemporator and an in-line silencer. Failure of the 18 inch attemporator shells occurred after four hundred hours of operation.

H. This system included four parallel desuperheaters located downstream of control valves. Several cracks developed in this system after two to three months of operation. Two to three of the four letdown valves were discharging steam through a 10 inch connection to a 20 inch header which swaged up to a 30 inch diameter. Longitudinal cracking occurred at the bottom of the 20 inch header at a transverse guide 0.5 meters downstream of the fourth 10 inch branch connection. In addition, a 1 inch bypass line for a block valve downstream of the third letdown valve cracked and the 20 inch header cracked around a pressure tap downstream of the transverse guide.

Table 1 compares the acoustic efficiencies implied in the Carucci and Mueller method (at the upstream choke point) vs. acoustic efficiencies established using IEC methods described later on in this paper. The acoustic efficiencies implied by the Carucci and Mueller method at the upstream choke point ranged from 0.1 to 5 percent while the IEC acoustic efficiency ranged from 0.67 to 1 percent. The Carucci and Mueller method is based on acoustic energy theories encompassing both jet and choked flow noise as well as test data from work performed by Exxon research and engineering. As discussed in later sections, the experience based failure criteria originally developed by Carucci and Mueller can only be used with sound power levels calculated by Equation 12.

The implied acoustic efficiency by Equation 12 is proportional to the mass flow rate and inversely proportional to u^2:

$$\eta = 8\left(1-\frac{P_2}{P_1}\right)^{3.6}\left(\frac{T_1}{M_w}\right)^{1.2}\left(\frac{\dot{m}}{u^2}\right) \quad (15)$$

If we consider the case of an ideal gas undergoing isentropic choked (sonic) flow through a restriction orifice, we can establish the following expressions for variables of interest at the choke point:

$$T_2 = 2\frac{T_1}{\gamma+1} \quad (16)$$

$$P_2 = P_1\left(\frac{2}{\gamma+1}\right)^{\frac{\gamma}{\gamma-1}} \quad (17)$$

$$\rho_2 = \frac{P_2 M_w}{R_g T_2} \quad (18)$$

$$u_2 = \sqrt{\frac{\gamma R_g T_2}{M_w}} \quad (19)$$

$$\dot{m} = \rho_2 u_2 A_o \quad (20)$$

Substituting the values of mass flow rate $\dot{m}$, choked flow velocity u_2, and choke pressure P_2 in Equation 15, we obtain the following simplified expression for the Carucci and Mueller acoustic efficiency after some algebraic manipulations:

Table 1. Vibration risk data set used by Carucci and Mueller and reproduced by Melhem

CASE	SERVICE	T_1, K	P_1, bara	T_2, K	P_2, bara	D_2, m	u_2, m/s	Exit Mach Number, M_2	Mass Flow Rate, kg/s	Heat Capacity Ratio, C_p/C_v	Restriction Orifice / Valve Flow Speed, m/s	Carucci and Mueller Acoustic Efficiency, %	IEC Acoustic Efficiency, %	PWL_cm, dB	PWL_mel, dB	PWL_cm - PWL_mel, dB	PWL_mel (IEC), dB
A	Propylene Compressor Recycle	319.26	20.55	238.91	1.59	0.610	32.50	0.137	29.74	1.13	224.30	0.65	0.85	168.5	165.0	3.5	161.5
B1	Natural Gas Pressure	318.71	42.75	226.00	6.27	0.254	125.60	0.399	48.26	1.22	322.99	2.54	0.91	170.9	168.1	2.8	163.6
B2	Natural Gas Letdown to Flare	318.71	5.17	270.38	2.07	0.711	58.20	0.159	48.26	1.22	357.00	0.41	0.69	171.4	168.9	2.5	163.3
C	Natural Gas Letdown to Flare	308.15	47.57	195.68	4.14	0.406	64.40	0.218	47.63	1.23	311.29	4.17	0.95	171.8	169.8	2.0	163.4
D	Nitrogen Compressor Recycle	310.93	6.96	280.31	3.72	0.610	93.80	0.257	119.70	1.20	324.76	1.05	0.79	168.3	168.2	0.1	167.0
E	Nitrogen Compressor Recycle	310.93	16.82	267.59	7.03	0.457	60.10	0.154	84.80	1.21	328.35	0.79	0.66	168.8	165.6	3.3	164.8
F	Natural Gas Compressor Recycle	330.37	51.09	220.53	13.86	0.305	28.80	0.093	35.28	1.45	272.52	0.69	0.82	164.9	159.5	5.4	160.3
G	Desuperheater Steam	802.59	99.97	480.32	10.34	0.457	49.80	0.093	36.89	1.29	619.65	2.91	0.94	175.5	173.1	2.4	168.2
H	Desuperheater Steam	658.15	43.09	476.11	10.34	0.508	49.60	0.093	45.99	1.29	586.00	0.86	0.83	163.9	168.3	-4.4	168.9
1	Process Gas Compressor Recycle	308.15	13.17	204.98	1.03	0.406	122.80	0.461	27.09	1.19	256.81	2.22	0.95	165.9	164.2	1.7	160.6
2	Process Gas Compressor Recycle	311.48	38.75	265.88	10.00	0.305	14.20	0.048	13.86	1.13	258.54	0.32	0.83	156.6	151.8	4.8	155.9
3	Propylene Compressor Recycle	316.48	20.55	262.14	3.24	0.406	15.70	0.055	12.60	1.11	209.52	0.81	0.91	156.0	153.5	2.5	154.0
4	Process Gas Compressor Recycle	300.37	2.48	257.91	1.03	0.508	145.60	0.450	34.78	1.21	325.00	0.26	0.77	161.3	156.7	4.6	161.5
5	Process Gas Compressor Recycle	310.93	15.86	215.12	2.21	0.406	68.70	0.226	25.33	1.23	357.01	1.21	0.91	165.2	162.9	2.3	161.7
6	Process Gas Compressor Recycle	300.93	39.44	250.67	15.51	0.254	26.10	0.075	20.66	1.24	330.10	0.20	0.78	158.3	153.6	4.7	159.4
7	Propylene Compressor Recycle	310.93	16.65	233.88	1.38	0.356	131.30	0.575	38.05	1.13	216.35	3.92	0.95	166.6	165.4	1.2	159.3
8	Propylene Compressor Recycle	310.93	16.69	234.72	1.38	0.508	59.20	0.259	35.03	1.13	216.35	3.61	0.95	165.9	164.7	1.2	158.9
9	Waste Steam Dump	403.71	2.76	370.82	1.03	0.406	245.10	0.566	18.90	1.10	436.60	0.15	0.77	160.5	154.4	6.1	161.4
10	Process Gas Compressor Recycle	302.59	10.69	207.90	1.03	0.406	95.80	0.357	20.92	1.19	295.15	1.49	0.94	163.2	161.3	1.9	159.3
11	Process Gas Compressor Recycle	308.15	54.81	263.43	27.51	0.305	7.20	0.016	9.15	1.29	441.77	0.10	0.79	150.5	149.7	0.8	158.5
12	Natural Gas Compressor Recycle	308.15	3.79	282.54	2.48	0.914	28.40	0.075	39.69	1.26	315.95	0.19	0.73	155.4	155.7	-0.3	161.6
13	Natural Gas Compressor Recycle	308.15	6.48	279.42	4.00	0.914	17.30	0.046	39.44	1.25	334.77	0.24	0.75	157.2	157.2	0.0	162.2
14	Natural Gas Compressor Recycle	308.15	11.45	282.82	6.76	0.762	14.40	0.039	38.98	1.19	346.05	0.28	0.76	158.1	158.1	0.0	162.5
15	Natural Gas Compressor Recycle	308.15	21.79	266.72	11.65	0.660	10.10	0.027	36.04	1.30	359.11	0.33	0.78	159.4	158.8	0.6	162.6
16	Natural Gas Compressor Recycle	308.15	40.75	272.56	21.99	0.610	6.30	0.017	35.91	1.25	344.77	0.34	0.78	159.2	158.6	0.6	162.2
17	Natural Gas Compressor Recycle	308.15	80.12	271.44	40.89	0.508	4.80	0.013	35.41	1.23	324.89	0.40	0.78	160.1	158.8	1.3	161.6
18	Natural Gas Compressor Recycle	308.15	172.99	240.53	79.98	0.406	3.80	0.010	35.41	1.47	326.41	0.71	0.82	161.5	161.3	0.2	161.9
19	Natural Gas Compressor Recycle	308.15	3.86	280.10	2.34	0.914	27.60	0.074	37.17	1.24	340.74	0.24	0.75	157.3	157.1	0.2	162.1
20	Natural Gas Compressor Recycle	308.15	6.21	281.91	3.93	0.914	16.50	0.044	36.79	1.24	325.37	0.20	0.74	155.9	155.9	0.0	161.6
21	Natural Gas Compressor Recycle	308.15	10.62	274.09	6.27	0.762	14.30	0.038	36.04	1.29	345.37	0.26	0.76	157.4	157.4	0.0	162.1
22	Natural Gas Compressor Recycle	308.15	19.72	266.35	10.62	0.660	10.10	0.027	33.01	1.31	359.23	0.30	0.78	158.4	158.0	0.4	162.2
23	Natural Gas Compressor Recycle	308.15	36.82	263.14	19.79	0.610	6.40	0.017	32.89	1.34	346.54	0.31	0.78	158.4	157.9	0.5	161.9
24	Natural Gas Compressor Recycle	308.15	64.47	270.22	36.96	0.508	4.90	0.013	32.82	1.31	319.94	0.32	0.77	157.3	157.2	0.1	161.1
25	Natural Gas Compressor Recycle	308.15	139.48	320.30	64.54	0.406	4.20	0.012	32.76	0.95	311.67	0.56	0.80	160.7	159.5	1.2	161.1
26	Propane Compressor Recycle	308.15	12.07	234.24	0.97	0.457	143.50	0.644	50.40	1.12	217.76	4.82	0.96	168.9	167.6	1.3	160.6
27	Steam Desuperheater	658.15	43.09	522.78	10.34	0.762	22.00	0.041	45.99	1.19	564.55	0.51	0.83	161.2	168.3	-7.1	167.9

$$\eta=(4.7950\gamma-2.5882)\left(\frac{M_w}{T_1}\right)^{0.3}P_1A_o \tag{21}$$

where A_o is the effective restriction orifice flow area in m^2, P_1 is the upstream pressure in bara, T_1 is the upstream temperature in Kelvin, and η is the calculated Carucci and Mueller acoustic efficiency in percent. This expression shows that the Carucci and Mueller acoustic efficiency is directly proportional to upstream pressure, and the restriction orifice flow area (flow rate). It can be shown that the Carucci and Mueller acoustic efficiency produces unrealistic values for high pressure systems, large flow area, and/or mixtures with high molecular weights.

If we consider the case of methane discharging through a restriction with an effective flow area of 0.1 m^2, upstream temperature of 373.15 K, and an upstream pressure of 100 bara, the calculated acoustic efficiency is:

$$\eta=(4.7950\times1.282-2.5882)\left(\frac{16}{373.15}\right)^{0.3}\times100\times0.1 \tag{22}$$

$$=3.559\times0.388\times100\times0.1=13.8\% \tag{23}$$

The same value can also be calculated by calculating $\dot{m}$, u_2, P_2, and η directly from the above ideal gas flow equations and Equation 15:

$$T_2=2\frac{T_1}{\gamma+1}=2\frac{373.15}{1.282+1}=327.05\text{ K} \tag{24}$$

$$P_2=P_1\left(\frac{2}{\gamma+1}\right)^{\frac{\gamma}{\gamma-1}}=100\left(\frac{2}{1.282+1}\right)^{\frac{1.282}{1.282-1}}=54.9\text{ bara} \tag{25}$$

$$\rho_2=\frac{P_2M_w}{R_gT_2}=\frac{54.9\times100000\times16}{8314\times327.03}=32.306\text{ kg}/\text{m}^3 \tag{26}$$

$$u_2=\sqrt{\frac{\gamma R_gT_2}{M_w}}=\sqrt{\frac{1.282\times8314\times327.03}{16}}=466.76\text{ m}/\text{s} \tag{27}$$

$$\dot{m}=\rho_2u_2A_o=32.306\times466.76\times0.1=1507.63\text{ kg}/\text{s} \tag{28}$$

$$\eta=8\left(1-\frac{P_2}{P_1}\right)^{3.6}\left(\frac{T_1}{M_w}\right)^{1.2}\left(\frac{\dot{m}}{u^2}\right) \tag{29}$$

$$=8\times(1-0.549)^{3.6}\times\left(\frac{373.15}{16}\right)^{1.2}\times\left(\frac{1507.63}{466.76^2}\right) \tag{30}$$

$$=8\times0.0568\times43.78\times69.2\times10^{-4} \tag{31}$$

$$=0.1376\text{ or }13.76\% \tag{32}$$

A similar equation to 21 can be derived for the acoustic efficiency downstream of the choke point:

$$\eta = (68.63 - 4.836\gamma)\left[1 - (1.0397 + 0.6096\gamma)\frac{P_b}{P_1}\right]^{3.6}\left(\frac{M_w}{T_1}\right)^{0.3} P_1 A_o \tag{33}$$

where P_b is the superimposed backpressure downstream of the choke point. Applying Equation 33 to the same example above we calculate an acoustic efficiency of 227 percent at a backpressure of 1 bara:

$$\begin{aligned}\eta &= (68.63 - 4.836\times1.282)\left[1 - (1.0397 + 0.6096\times1.282)\frac{1}{100}\right]^{3.6}\left(\frac{16}{375}\right)^{0.3} \times100\times0.1 \\ &= 62.430\times0.9359\times0.388\times100\times0.1 \\ &= 226.70\%\end{aligned} \tag{34}$$

It is evident from the actual data reported by Carucci and Mueller and from the above theoretical proof that the acoustic efficiency used by Carucci and Mueller in their equation can produce unrealistic values, well in excess of 1 percent.

10. Failure Criteria

Several methods are now available for screening and analyzing the potential failure risk of piping caused by vibration. These methods include:

- Experience Based - D/t method or D Method
- Experience Based - Mach number method (Not recommended)
- MTD methods
- Detailed structural dynamics methods

The experience based methods center around correlating likelihood of failure based on actual experience and/or test data. The most widely cited reference is that of Carucci and Mueller (see vii and viii) for steel pipe.

Figure 2 illustrates the failure criteria developed originally by Carucci and Mueller and later modified by Eisinger. This failure criteria establishes a design sound power level vs. the ratio of pipe diameter to thickness. This criteria is fundamentally better than the other criteria based on pipe diameter only since thicker wall pipes are stronger than thinner wall pipes.

The original work by Carucci and Mueller suggests a limit provided by the following equation (Figure 2):

$$L_{W,limit} = 184.6 - 0.215\frac{2D_i}{D_o - D_i} = 184.6 - 0.215\frac{D_i}{\delta} \tag{35}$$

where δ is the pipe thickness.

The lower allowable limit developed by Eisinger also shown in Figure 2 is given by the following equation:

$$L_{W,limit} = 173.6 - 0.125\frac{2D_i}{D_o - D_i} = 173.6 - 0.125\frac{D_i}{\delta} \tag{36}$$

The NORSOK standard published guidance in 2006 using the same equation for $L_{W,limit}$ as proposed by Eisinger above.

Note that the above limits are based on sound power levels calculated using the source pressure and the downstream exit pressure. We have recalculated the sound power levels based on the correct pressure ratios downstream of the choke point. This data is shown in Figures 3, 4, and 5.

Another design limit criterion that was originally proposed by Carucci and Mueller is based on pipe diameter only for steel pipe. This criterion applies to pipe diameters ranging from 10 inch to 36 inch (approximately between 200 mm to 800 mm) and with wall thicknesses ranging from 0.219 in to 0.439 in (approximately 5.5 mm to 10 mm). The design limit is shown in Figure 3 for the corrected data and can be approximated by:

$$L_{W,limit} = 192.8 - 9.8\ln D \tag{37}$$

where D is the nominal pipe diameter in inches and L_W is the sound power limit in dB.

Recent theoretical and experimental work by Chiyoda Engineering[ix] indicates that AIV risk will increase for smaller pipe diameters at the same diameter to thickness ratio. As a result, the limiting criteria established above should be decreased by 6 dB for an 8 inch pipe, or by 20 $\log_{10}$ (16/D).

The design limit correlations developed in this paper can be used with computer programs such as SuperChems Expert which calculates the sound power level at every axial locations for single and multiphase flow to decide if the steel piping inside diameter to thickness ratio exceed the allowable limits as shown by Figure 6.

11. Other Methods for Calculating η

The simplest method for calculating η is to assume the same efficiency as an expanded jet (see Figure 7) and apply it equally to single and multi-phase flow. The efficiency shown in Figure 7 is the same as what is currently used by API for the estimation of flare noise.

The flow regime types are defined by varying jet shapes in the area upstream and downstream of the vena contracta. These jets change their shape when certain differential pressure ratios are exceeded. In regimes II to IV, higher Mach numbers arise downstream of the vena contracta. Yet, M at the vena contracta itself remains unchanged at 1. The IEC method produces equivalent values of efficiency to the data shown in Figure 7 (as $F_L \rightarrow 1.0$) as shown in Figure 9. Note that Figure 9 uses $x = 1 - P_2 / P_1$ as the X axis while Figure 7 uses $P_1 / P_2 = \frac{1}{1-x}$.

Figure 2. Piping failure limit as originally proposed by Carucci and Mueller and by Eisinger for steel pipe

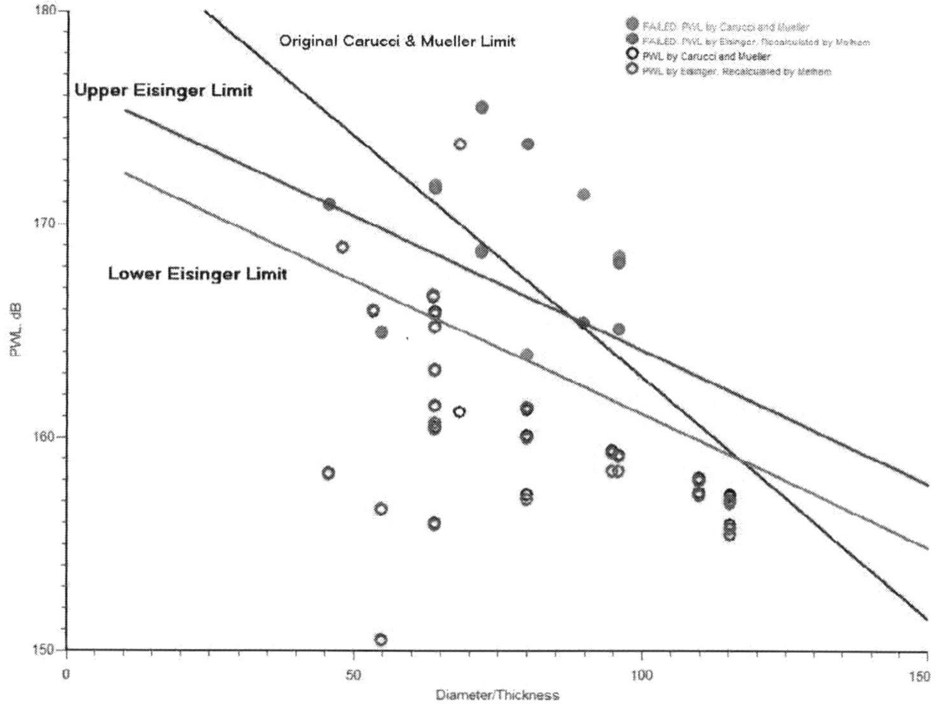

The efficiency is related to the flow pressure ratio and the flow regime as well. IEC calculates the efficiency based on five different flow regimes (see Figure 8) depending on the value of the downstream pressure, P_2:

Regime I The flow is sub-sonic. The sound generation has the character of a dipole jet. The highest Mach number is reached at the vena contracta, not exceeding Mach 1 at the maximum. Downstream of the vena contracta, the jet expands, leading to partial pressure recovery (F_L factor).

Regime II Sonic and supersonic flows exist together, which means that strongly turbulent flow and shock cell structure dominate. Pressure recovery drops until the top limit of regime II is reached.

Regime III The rise in pressure is non-isentropic. The flow is supersonic and shear turbulence predominates.

Regime IV The shock cells disappear and a Mach disk forms. The dominant mechanism is the interaction between shock cells and turbulence.

Regime V The acoustical efficiency is constant.

Figure 3. Piping failure correlation developed by Melhem using Carucci Mueller acoustic efficiency and corrected sound power level estimates for steel pipe. PWL vs. D

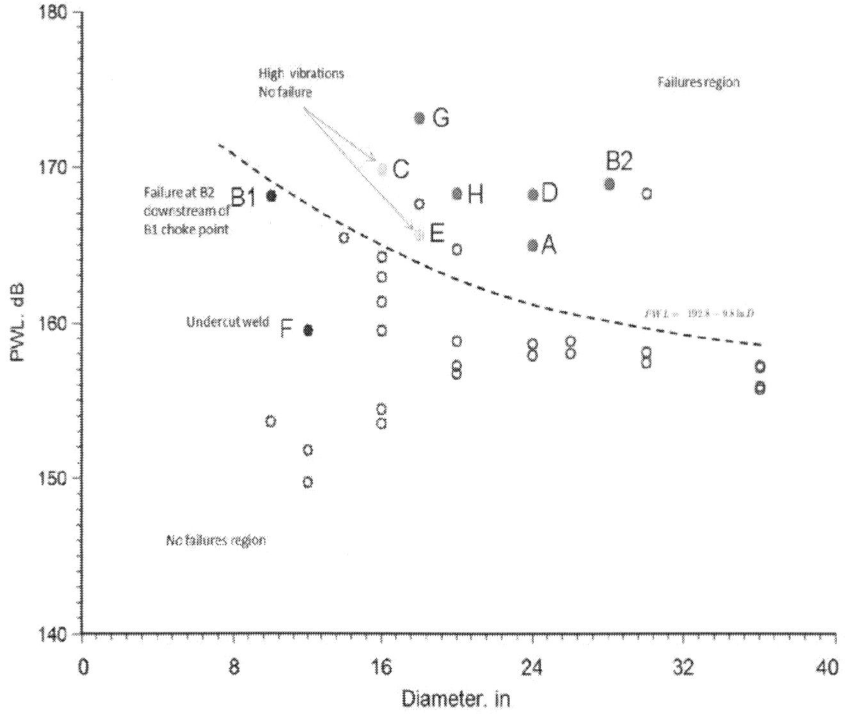

Figure 4. Piping failure correlation developed by Melhem using Carucci Mueller acoustic efficiency and corrected sound power level estimates for steel pipe. PWL vs. D/t

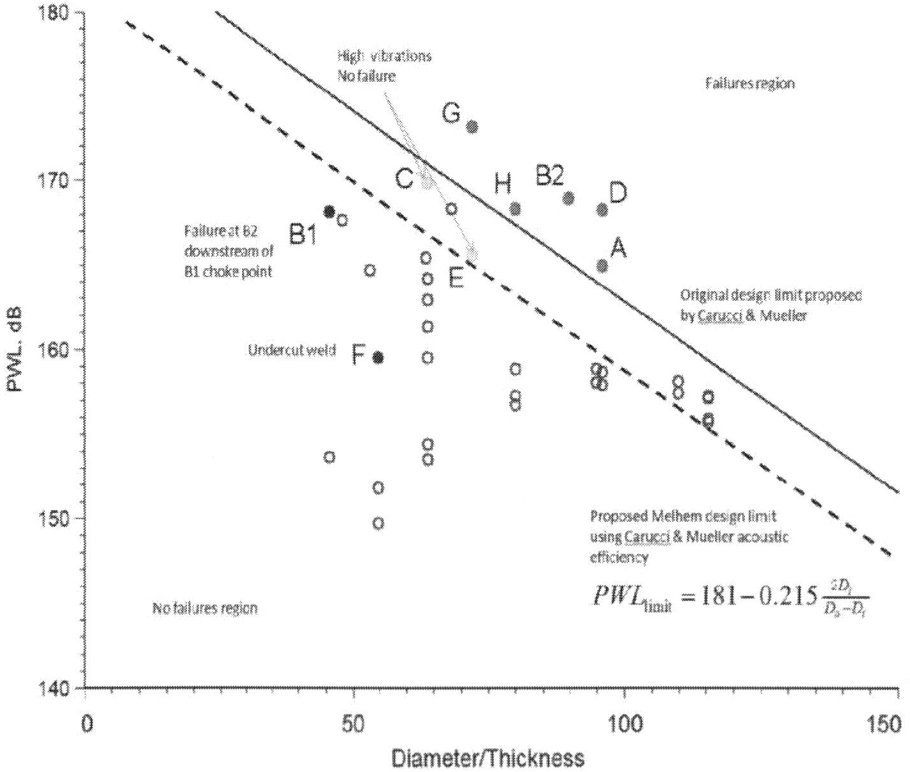

12. Vibration Frequencies

As a fluid moves through piping components there is typically a separation of the fluid from the constraining wall as the fluid changes flow direction. As a result a vortex is formed and then swept into the main stream. This vortex shedding occurs at fairly well defined dimensionless frequencies. The strength of the vortex varies but does not need to be very strong to cause damage especially if the shedding frequency couples with the natural frequency of the piping system. The shedding peak frequency for a vortex for subsonic and sonic flows (Regimes I and II up to a Mach number of 1.4) is given by:

$$f_p = \frac{N_{Str} u}{D} = 0.2 \frac{u}{D} \tag{38}$$

Figure 5. Piping failure correlation developed by Melhem using IEC acoustic efficiency and corrected sound power level estimates for steel pipe. PWL vs. D/t

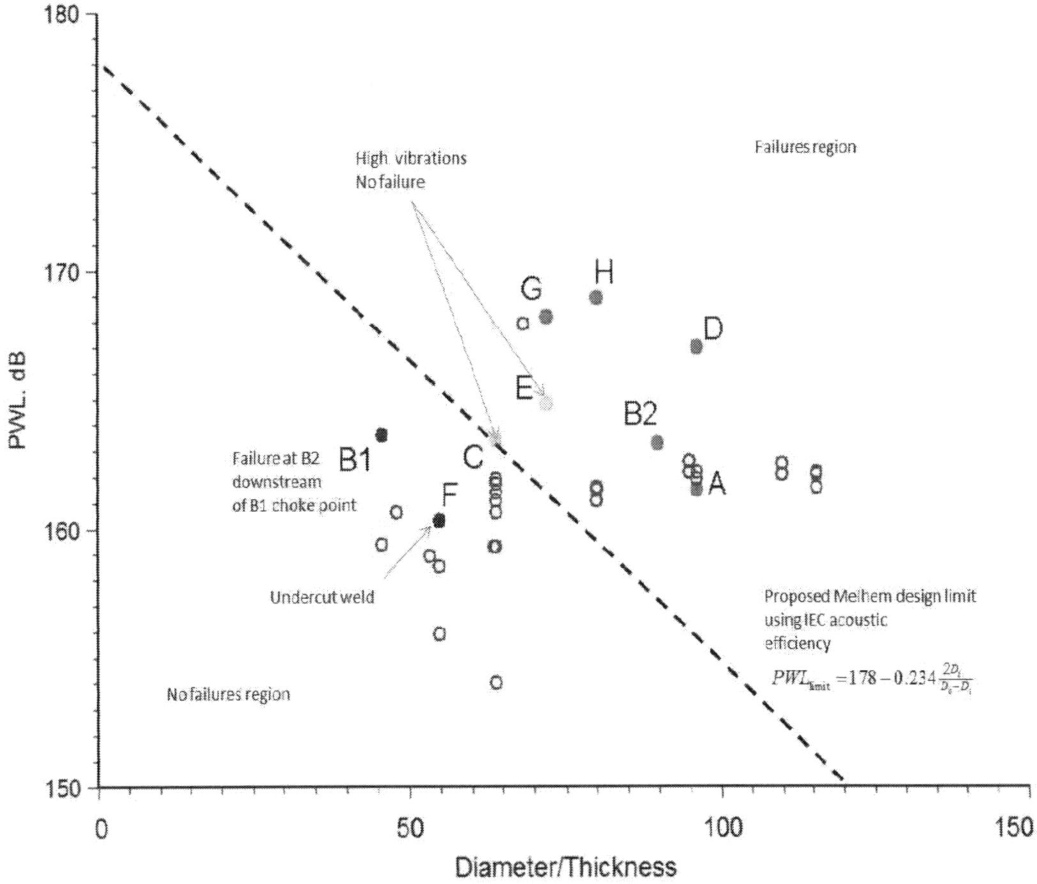

where u is the flow velocity in m/s, D is a characteristic flow dimension (perpendicular to flow), f_p is the peak frequency in Hz and N_{Str} is the Strouhal Number which varies depending upon the geometry causing the separation of the boundary layer. For a circular cylinder its value is 0.2 over a wide range of Reynolds numbers. It is usually between 0.1 and 0.3. Vortex shedding frequencies are generally more than 30 Hz [upper limit for most piping system natural frequencies].

For flows with Mach numbers larger than 1.4 ($M > 1.4$, Regimes III to V), the peak frequency f_p is given by:

$$f_p = \frac{0.4 u_{sonic}}{1.25 D \sqrt{M^2 - 1}} \tag{39}$$

For pipe flow, D is equal to the inside flow diameter. For a control valve, D is given by:

$$D = 0.0046 \sqrt{\frac{C_v F_l}{N_o}} \tag{40}$$

Figure 6. Sample piping sound power level calculated by SuperChems Expert v6.4mp vs. experience based allowable limit for steel pipe

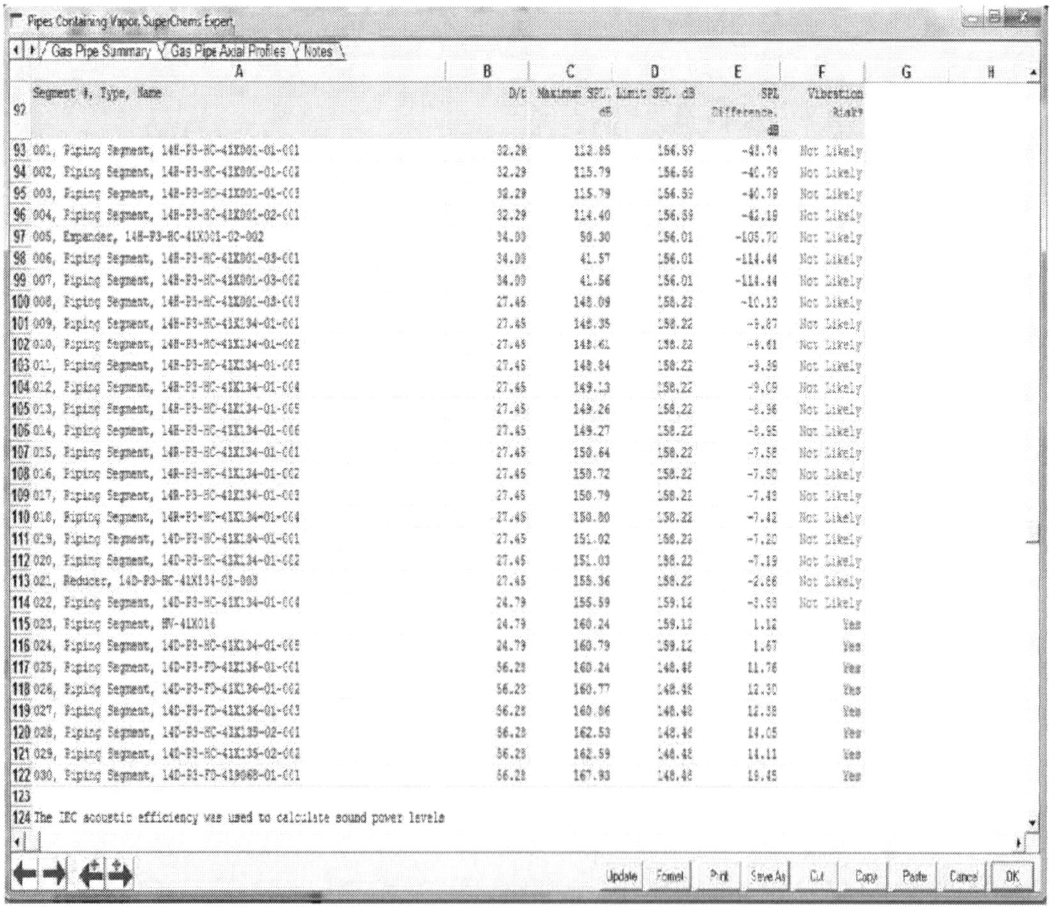

	A	B	C	D	E	F
92	Segment #, Type, Name	D/t	Maximum SPL, dB	Limit SPL, dB	SPL Difference, dB	Vibration Risk?
93	001, Piping Segment, 14H-P3-HC-41X001-01-001	32.28	112.85	156.59	-43.74	Not Likely
94	002, Piping Segment, 14H-P3-HC-41X001-01-002	32.29	115.79	156.58	-40.79	Not Likely
95	003, Piping Segment, 14H-P3-HC-41X001-01-003	32.29	115.79	156.58	-40.79	Not Likely
96	004, Piping Segment, 14H-P3-HC-41X001-02-001	32.29	114.40	156.59	-42.19	Not Likely
97	005, Expander, 14H-P3-HC-41X001-02-002	34.09	50.30	156.01	-105.70	Not Likely
98	006, Piping Segment, 14H-P3-HC-41X001-03-001	34.09	41.57	156.01	-114.44	Not Likely
99	007, Piping Segment, 14H-P3-HC-41X001-03-002	34.09	41.56	156.01	-114.44	Not Likely
100	008, Piping Segment, 14H-P3-HC-41X001-03-003	27.45	148.09	158.22	-10.13	Not Likely
101	009, Piping Segment, 14H-P3-HC-41X134-01-001	27.45	148.35	158.22	-9.87	Not Likely
102	010, Piping Segment, 14H-P3-HC-41X134-01-002	27.45	148.61	158.22	-9.61	Not Likely
103	011, Piping Segment, 14H-P3-HC-41X134-01-003	27.45	148.84	158.22	-9.39	Not Likely
104	012, Piping Segment, 14H-P3-HC-41X134-01-004	27.45	149.13	158.22	-9.09	Not Likely
105	013, Piping Segment, 14H-P3-HC-41X134-01-005	27.45	149.26	158.22	-8.96	Not Likely
106	014, Piping Segment, 14H-P3-HC-41X134-01-006	27.45	149.27	158.22	-8.95	Not Likely
107	015, Piping Segment, 14H-P3-HC-41X134-01-001	27.45	150.64	158.22	-7.58	Not Likely
108	016, Piping Segment, 14H-P3-HC-41X134-01-002	27.45	150.72	158.22	-7.50	Not Likely
109	017, Piping Segment, 14H-P3-HC-41X134-01-003	27.45	150.79	158.22	-7.43	Not Likely
110	018, Piping Segment, 14H-P3-HC-41X134-01-004	27.45	150.80	158.22	-7.42	Not Likely
111	019, Piping Segment, 14D-P3-HC-41X134-01-001	27.45	151.02	158.22	-7.20	Not Likely
112	020, Piping Segment, 14D-P3-HC-41X134-01-002	27.45	151.03	158.22	-7.19	Not Likely
113	021, Reducer, 14D-P3-HC-41X134-01-003	27.45	155.36	158.22	-2.86	Not Likely
114	022, Piping Segment, 14D-P3-HC-41X134-01-004	24.79	155.59	159.12	-3.53	Not Likely
115	023, Piping Segment, HV-41X014	24.79	160.24	159.12	1.12	Yes
116	024, Piping Segment, 14D-P3-HC-41X134-01-005	24.79	160.79	159.12	1.67	Yes
117	025, Piping Segment, 14D-P3-FD-41X136-01-001	56.28	160.24	148.48	11.76	Yes
118	026, Piping Segment, 14D-P3-FD-41X136-01-002	56.28	160.77	148.48	12.30	Yes
119	027, Piping Segment, 14D-P3-FD-41X136-01-003	56.28	160.86	148.48	12.38	Yes
120	028, Piping Segment, 14D-P3-HC-41X135-02-001	56.28	162.53	148.48	14.05	Yes
121	029, Piping Segment, 14D-P3-HC-41X135-02-002	56.28	162.59	148.48	14.11	Yes
122	030, Piping Segment, 14D-P3-FD-419068-01-001	56.28	167.93	148.48	19.45	Yes
123						
124	The IEC acoustic efficiency was used to calculate sound power levels					

where D is in meters, N_o is the number of separate flow passages, C_v is the valve flow coefficient, $\frac{C_v}{N_o}$ is the channel flow coefficient, and F_l is the valve recovery factor. The pressure recovery factor is defined as:

$$F_l = \sqrt{\frac{P_1 - P_2}{P_1 - P_{vc}}} \tag{41}$$

where $P_1 - P_2$ is the pressure differential across the valve and P_{vc} is the pressure at the vena contracta (note that P_2 is larger than P_{vc} since the pressure at the exit of the valve recovers):

$$P_{vc} = P_1 - \frac{P_1 - P_2}{F_l^2} \tag{42}$$

For liquid flow, P_{vc} can reach the vapor pressure of the liquid and cause choked flow.

$$P_{vc} = F_f P_v \approx \left[0.96 - 0.25\sqrt{\frac{P_v}{P_c}}\right] P_v \tag{43}$$

Figure 7. Acoustic efficiency of shock noise generated by choked jets, η vs. jet pressure ratio P_1 / P_2x

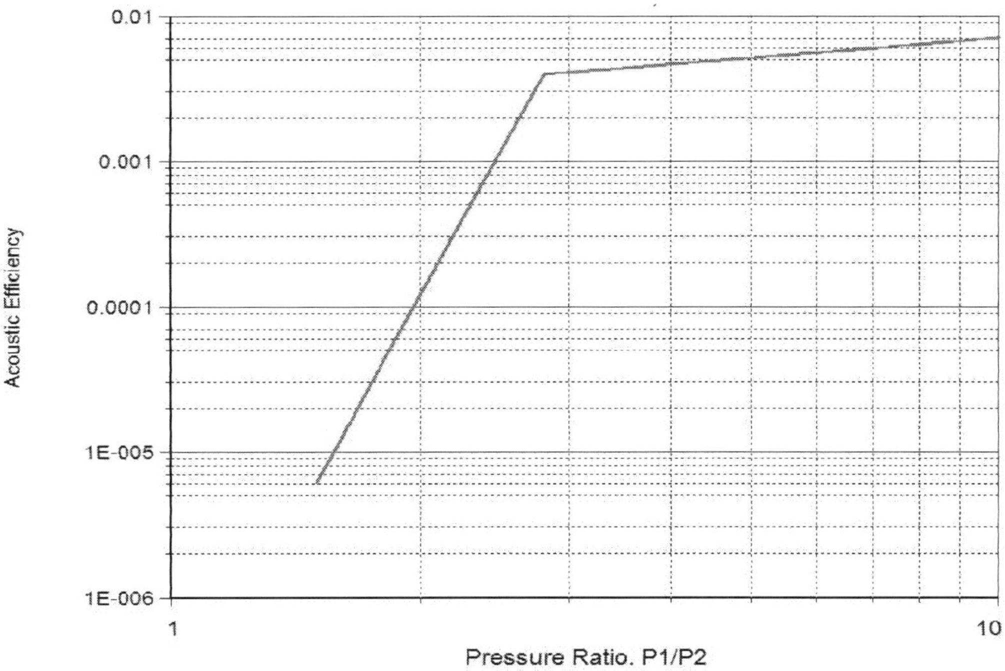

where F_f is the liquid critical pressure factor, P_v is the liquid vapor pressure, and P_c is the liquid critical pressure.

The lowest possible value of P_{vc} in a valve flowing liquid would be vacuum or 0. For this particular limiting case, Equation 41 can be solved for the pressure ratio P_1 / P_2:

$$P_d = \frac{P_1}{P_2} = \frac{1}{1-F_l^2} \tag{44}$$

where P_d is the damaging pressure ratio. If the valve is operated at a pressure ratio exceeding P_d, the flow will be choked, noisy, and subject to excessive vibration. Table 2 shows typical valve values for F_l and P_d.

Within every flowing pipe there will also be a standing wave moving axially back and forth in the pipe. The frequency of this wave depends on the effective acoustic length of the pipe and the sonic velocity of the fluid in the pipe. The effective acoustic length of the pipe is the distance between obstructions or acoustic barriers. Examples of obstructions would be valves, pumps, and orifices. An acoustic barrier would be an opening into a larger pipe, a reservoir, the end of a pipe run such as a 'T' intersection where the branch of interest requires a right angle turn. Piping components such as expanders or reducers could be an obstruction. An assessment of vibration risk should consider both the frequencies with and without the presence of expanders as obstructions.

The frequency of the standing wave can be calculated as shown below and then compared to the natural frequencies of valve components and the piping system to determine if there is a potential for vibration and/or resonance.

Closed End Pipe $f = \frac{i*u_{ac}}{4L}$

Open End Pipe $f = \frac{i*u_{ac}}{2L}$

where $i = 1, 2, ...$ is the wave number, L is the length between acoustic barriers, and u_{ac} is a characteristic acoustic speed throughout the pipe contents defined as follows:

$$u_{ac} = \frac{u_{sonic}}{\sqrt{1+\frac{dK}{\delta E}}} \tag{45}$$

where u_{sonic} is the speed of sound in the fluid, K is the bulk modulus of elasticity in the fluid, d is the pipe diameter, δ is the pipe thickness and E is the pipe material of construction modulus of elasticity.

The above equation is derived from a more general form that depends on the elastic properties of the pipe:

$$u_{ac} = \frac{u_{sonic}}{\sqrt{1 + \frac{K}{E}\psi}} \tag{46}$$

where ψ is a function of the elastic properties of the pipe. Typical expressions of ψ are shown in Table 4. Typical material properties for ψ are shown in Table 3. Note that materials properties change with temperature as shown in Figure 10.

The speed of sound will change as a function of pressure, temperature, and composition. The presence of dissolved gas nuclei (such as air or other gases) in compressed liquids can significantly reduce the sonic velocity following a pressure drop which leads to the formation of gas bubbles. Up to 40 % reduction in sonic velocity has been observed[xi].

To control the vibration caused by a standing wave it is necessary to change the magnitude and/or the frequency of the standing wave or to change the natural frequency of the pipe or components being excited by the wave. The best approach is to address the magnitude of the standing wave. The magnitude is related to the fluid turbulent energy that is enforcing the wave. The most dominant source of this turbulence is the kinetic energy generated by the fluid jet exiting the valve trim. Thus a valve change with a trim that reduces this jet energy will eliminate this wave influence. Trying to change the frequencies is usually not beneficial. There is such a wide range of frequencies present in the turbulent flow that excitation can continue to establish a strong wave at the new frequency and continue the piping vibration.

Figure 8. Flow regimes considered for the estimation of acoustical efficiency

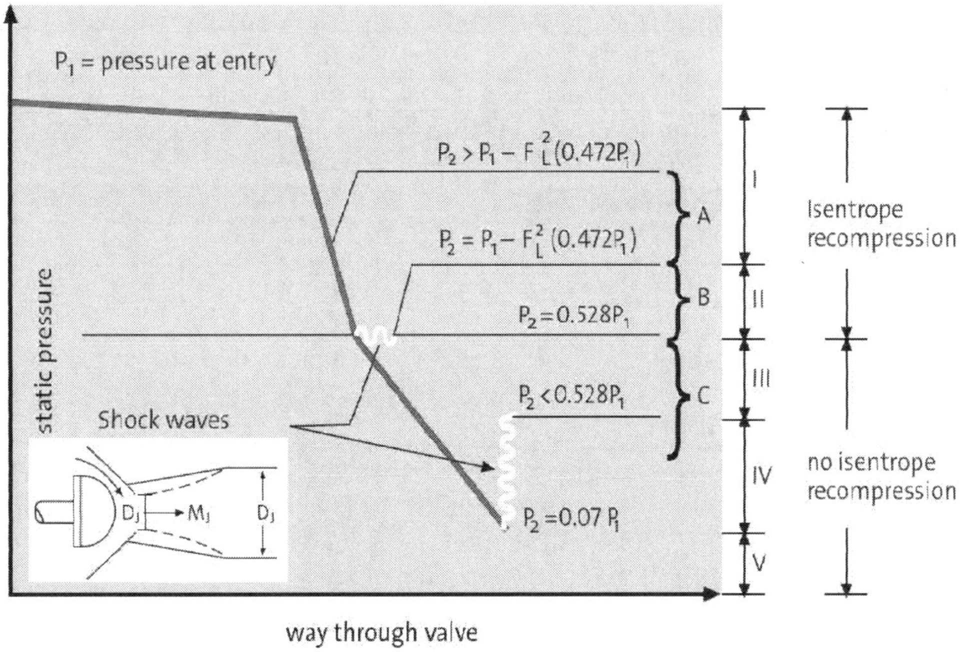

Figure 9. IEC flow acoustical efficiency as a function of F_L and differential pressure ratio

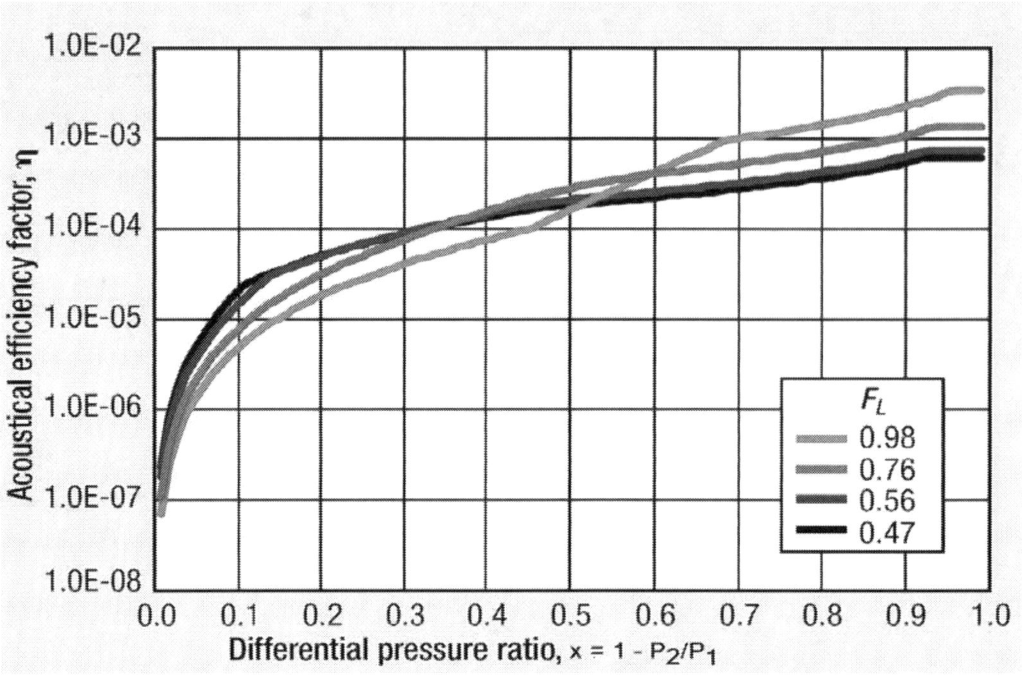

13. The Singing Safety Relief Valve Problem

As a result of increasing steam flow rates, several boiling water reactor (BWR) nuclear power plants have recently experienced the excitation of acoustic standing waves in closed side branches, e.g., safety relief valves (SRVs), due to vortex shedding generated by steam flow in the main stream lines. Flow past a valve entrance cavity excites a standing wave, resulting in noise and vibration[xii]. A similar tone is produced when air is blown across the mouth of a glass bottle.

Table 2. Typical valve values for F_l and P_d

Body	Trim	Flow Direction	F_l	$P_d = \frac{P_1}{P_2}$
Single Seat Globe	Cage	Open	0.90	5.3
	Cage	Closed	0.80	2.8
	V Plug	Open	0.90	5.3
	V Plug	Closed	0.90	5.3
	Contoured	Open	0.90	5.3
	Contoured	Closed	0.80	2.8
Double Seat Globe	V Plug		0.90	5.3
	Contoured		0.85	3.6
	Standard Bore		0.55	1.4
	Characterized		0.57	1.5
Angle	Cage	Open	0.85	3.6
	Cage	Closed	0.80	2.8

Ball	0.8 dia. Orifice		0.55	1.4
Butterfly	60°	Open	0.68	1.8
Butterfly	90°	Open	0.55	1.4

The amplitude of the acoustic pressure waves can be several times higher than the dynamic pressure present in the system (see Figure 12). The acoustic waves propagate in the steam lines, eventually reaching sensitive components such as steam dryers and turbine stop valves. In addition, the acoustic waves generated in the side branches may generate vibration problems locally and may lead to complications such as valve-seat wear. Therefore, the structural components are subjected to high-cycle fatigue loads, which over time may severely impact those components' functionality and safety.

Resonance occurs when the vortex shedding frequency coincides with the acoustic frequency of the standpipe or the valve components. The natural frequency of the standpipe/valve combination for a closed end pipe is given by the following equation:

$$f_a = \left(\frac{2n-1}{4}\right)\frac{u_{ac}}{L+L_e} = \left(\frac{2n-1}{4}\right)\frac{u_{ac}}{L+0.425d} \tag{47}$$

where n is the mode number (1 for 1st mode, 2 for 3rd mode, etc.), u_{ac} is the acoustic speed through the pipe contents as defined earlier, and L_e is an end correction corresponding to Rayleigh's upper limit.

The frequency of pressure oscillations (sound) created by vortex shedding, the energy source for the standing waves, is given by the following equation:

$$f_s = N_{str}\frac{u}{d+r} \approx 0.33(n-0.25)\frac{u}{d+r} \tag{48}$$

Typically peak oscillations occur at a Strouhal Number around 0.4 as shown in Figure 12. Note that the root mean square pressure amplitude shown in Figure 12 is the ratio of pressure oscillations divided by dynamic pressure ($\frac{1}{2}\rho u^2$). RMS begins increasing at a specific onset Strouhal Number and flow velocity depending on acoustic speed, pipe diameter, and pipe length, reaches a peak value and then decreases.

There are many similar installations of pressure relief valves in the process industries where the valves are mounted on large process lines such as overhead lines for distillation columns. In order to avoid fatigue failure from resonance caused by the coupling of normal flow vortex shedding frequency and the acoustic frequency of the standpipe ($f_s = f_a$), the normal flow velocity in the main line has to be limited to less than this critical value:

$$u_{max} < \frac{f_s}{N_{st}}(d+r) < \frac{1}{4}\left(\frac{u_{ac}}{L+L_e}\right)\left(\frac{d+r}{N_{str}}\right) \tag{49}$$

$$< \frac{1}{4}\left(\frac{u_{ac}}{L+0.425d}\right)\left(\frac{d+r}{0.6}\right)$$

As shown in Figure 12 the pressure fluctuations start to increase at a Strouhal number of 0.6 and then decrease after they reach a peak value around a Strouhal number of 0.4. The same approach can be applied to the flow through the inlet and/or discharge line of a pressure relief valve:

$$u_{max} < \frac{1}{2}\frac{u_{ac}}{L}\frac{D}{0.6} \quad (50)$$

where L is the acoustic length of the inlet or discharge line. Resonance can also be checked by comparing an open pipe/contents frequency with the natural frequency of the pressure relief valve, f_n:

$$f_n = \frac{1}{\tau_n} = \frac{\omega_n}{2\pi} = \frac{1}{2\pi}\sqrt{\frac{K_s}{m_D}} \quad (51)$$

$$\tau_n = \frac{2\pi}{\omega_n} = 2\pi\sqrt{\frac{m_D}{K_s}} \quad (52)$$

where τ_n is the undamped natural period in s, and f_n is the undamped natural frequency in Hz where one Hz equals 1 cycle/second, K_s is the spring constant in $\mathrm{N/m}$, m_D is the mass of the valve disc and moving parts in kg, and ω_n is the undamped circular natural frequency in $\mathrm{radians/s}$.

14. Conclusions

A general method for the estimation of vibration risk for piping systems for single and multiphase flow is developed and implemented. An experience based vibration risk failure criteria is outlined using the work of Carucci and Mueller for gas flow. Because the Carucci and Mueller sound power level equation can produce unrealistic values of acoustic efficiency, well in excess of 1 percent for high pressure systems and/or systems with large mechanical flow energy, the revised experience based failure criteria developed in this paper (see Figure 5) based on the IEC acoustic efficiency is recommended for multiphase flow.

Figure 10. Temperature effects on material of construction

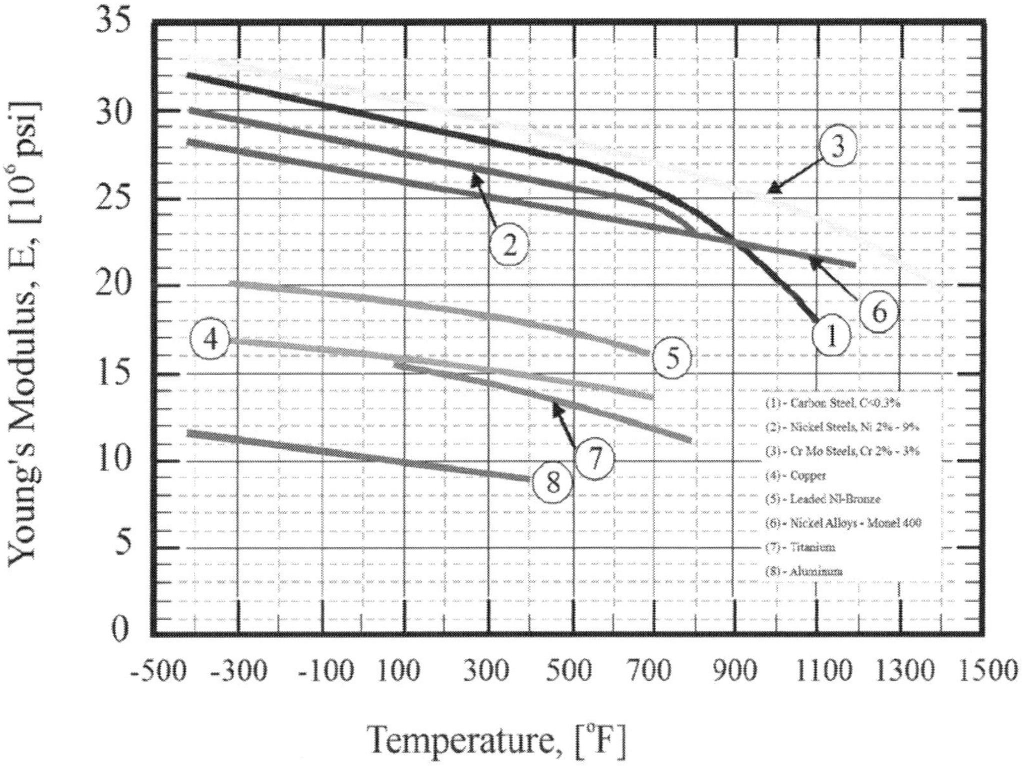

Figure 11. The singing safety relief valve[xiii]

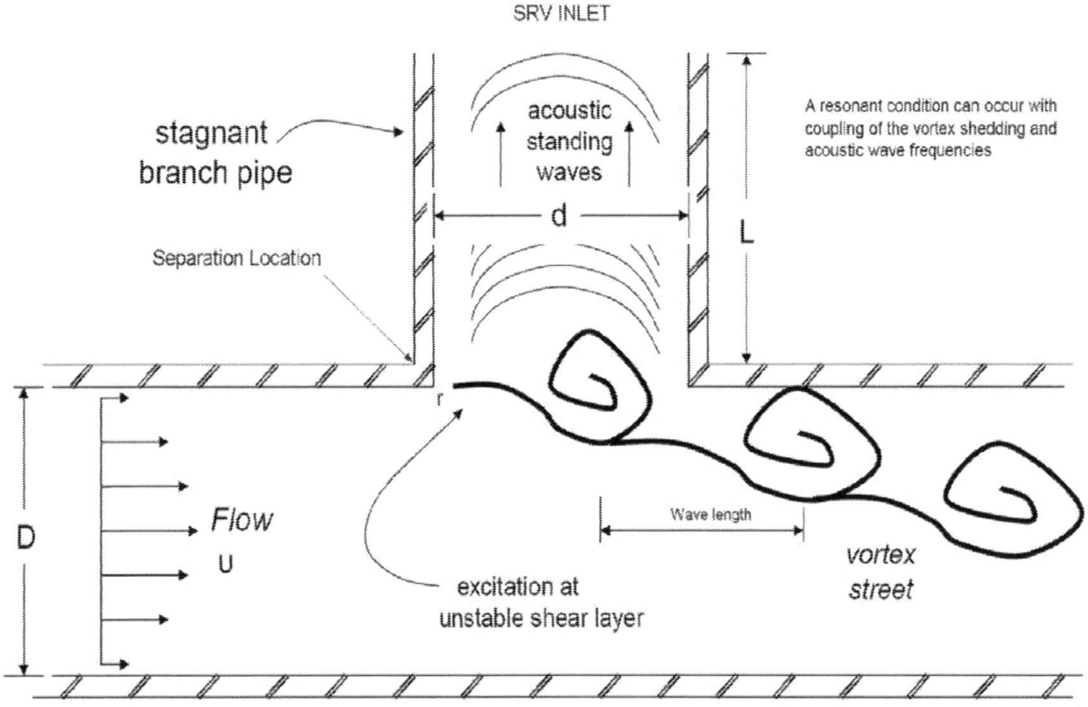

Figure 12. Comparison of calculated and measured pressure fluctuations as a function of Strouhal number[xiv]

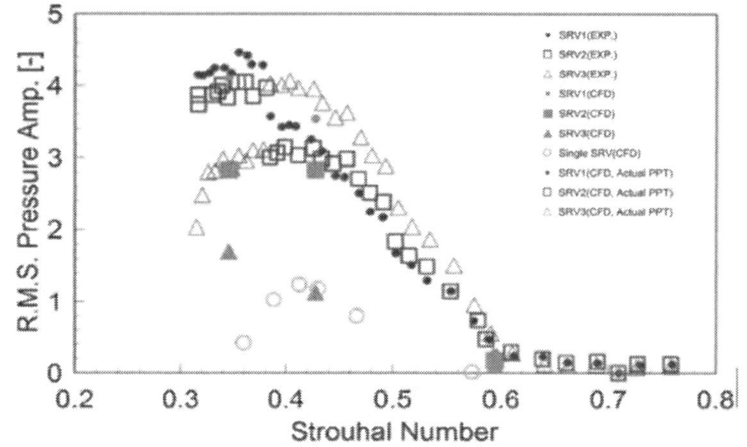

E is typically referred to as Young's modulus of elasticity

G is typically referred to as modulus of torsion, $G = \frac{1}{2}\frac{E}{1+\nu}$

Table 3. Typical data used in the estimation of sonic velocity in pipelines

Material	E (GPa)	Poisson's Ratio ν	$K = \frac{1}{\kappa}$ (GPa)	ρ in kg / m3
Aluminum	69	0.33		
Brass	78-110	0.36		
Carbon steel	202	0.303		
Cast iron	90-160	0.25		
Concrete	20-30	0.15		
Copper	117	0.36		
Ductile iron	172	0.30		
Fibre cement	24	0.17		
High carbon steel	210	0.295		
Inconel	214	0.29		
Mild steel	200-212	0.27		
Nickel steel	213	0.31		
Plastic / Perspex	6.0	0.33		
Plastic / Polyethylene	0.8	0.46		
Plastic / PVC rigid	2.4-2.75			
Stainless steel 18-8	201	0.30		
Water - fresh			2.19	999 at 20 C
Water - sea			2.27	1025 at 15 C

Table 4. Typical expressions for ψ

Pipe condition	ψ
Rigid	0
Anchored against longitudinal movement through its length	$\frac{d}{\delta}\left(1-\nu^2\right)$
Anchored against longitudinal movement at the upper end	$\frac{d}{\delta}\left(1.25-\nu\right)$
Frequent expansion joints present	$\frac{d}{\delta}$

References

[i] W.A. Skipwith. Acoustics technology. Survey SP-5093, NASA National Aeronautics and Space Administration, 1970.

[ii] C.M. Harris. *Handbook of Noise Control.* cGraw-Hill Book Co., Inc., 1957.

[iii] N.R. Miller, G.M. Singh, E. Rodarte and P.S. Hrnjak. Modification of standard aeroacoustic valve noise model to account for friction and two-phase flow. In *ACRC TR-162*, pages 1-11. Mechanical and Industrial Engineering Department, University of Illinois, 2000.

[iv] N.R. Miller, G.M, Singh, E. Rodarte and P.S. Hrnjak. Prediction of noise generated by expansion devices throttling refrigerant. In *ACRC TR-163*, pages 1-13. Mechanical and Industrial Engineering Department, University of Illinois, 2000.

[v] MTD. *Guidelines for the Avoidance of Vibration Induced Fatigue in Process Pipework.* Marine Technology Directorate Limited (MTD), 1999.

[vi] Energy Institute. *Guidelines for the Avoidance of Vibration Induced Fatigue in Process Pipework.* The Energy Institute, London, England, 2008.

[vii] V.A. Carucci and R.T. Mueller. Acoustically induced piping vibration in high capacity pressure reducing systems. In *82-/WAPVP-8*, pages 1-13. American Society of Mechanical Engineers, ASME, 1982.

[viii] F.L. Eisinger. Designing piping systems against acoustically-induced structural fatigue. In *PVP-VOL 328, Flow-Induced Vibration*, pages 397-404. American Society of Mechanical Engineers, ASME, 1982.

[ix] Hisao Izuchi, *Investigation of Pipe Size Effect Against Vibration Risk*, Presentation to ioMosaic Corporation, March 2013, Houston.

[x] I. Heitner. How to estimate plant noises. *Hydrocarbon Processing*, 47, 67-74, 1968.

[xi] V.L. Streeter and E.B. Wylie. Water hammer and surge control. *Annual Reviews in Fluid Mechanics*, pages 57-74, 1974.

[xii] T.M. Mulcahy, S.A. Hambric and V.N. Shah. Flow-induced vibration effects on nuclear power plant components due to main stream line valve singing. In *Ninth NRC/ASME Symposium on Valves, Pumps, and Inservice Testing, NUREG/CP-0152,* volume 6, pages 3B:49-3B:69. NCR/ASME, 2006.

[xiii] P. Sekerak. Potential adverse flow effects at nuclear power plants. In *16th International Conference on Nuclear Engineering*. ICONE16-48900, 2008.

[xiv] K. Okuyama, F. Inada, Y. Ogawa, R. Morita, S. Takahashi and K. Yoshikawa. Evaluation of acoustic and flow induced vibration of the BWR main stream lines and dryer. *Journal of Nuclear Science and Technology,* 48 (5): 759-776, 2011.

(44bt) Performance Metrics Development Management and Reporting Reaching Sustainability by Design

Robert C. Smith, ABB, Inc., Houston, TX

Overview

The saying is, "You can't manage, what you can't measure". This presentation will discuss various performance management measurement frame-works; How to use organizational goals and objectives to develop appropriate performance measures; Design effective performance measures to implement your organization's goals and strategies; Link performance measures to a variety of management processes; Develop a performance management framework that drives results. Presentation goals

Create a clear understanding of setting and managing key performance metrics surrounding highly hazardous environments.

- Create a practical base knowledge of leading and lagging metrics and how to practically apply them into your environment.
- Learn about the most common errors people make when selecting metrics and you'll gain new ideas around how metrics can be visualized better.
- This session will give you an in-depth look at the often confusing and misunderstood world of performance metrics. You'll learn the truth about metrics, and how an effective performance measurement system can help you make better decisions.
- Recognize management's role in leading and implementing performance metrics.
- Have a foundation to build metrics and scorecards by using the tips, techniques, and tools acquired in the presentation.

Who should attend?

This presentation is intended for any executive, manager or employee with responsibility for implementing or maintaining score carding performance management systems, and is interested in learning more about how to establish and measure metrics. What you'll learn:

- Detailed knowledge of leading and lagging metrics
- Measuring program success
- Key points continuous improvement and sustainability of measures

- Understand the purpose of process management
- How to analyze an existing metric system
- How to create well designed metrics that support strategy
- How to steer away from some of the classic pitfalls of metric development

(44bu) Alarm Management - the Human Factors Challenge

Robert C. Smith, ABB, Inc., Houston, TX

Refinery alarm systems are supposed to support the operator, yet in several notable accidents they have proven to be ineffective. Part of the problem is that these systems are not being designed or managed with human needs in mind. Do alarm systems help the operator or are they a hindrance? This paper is a short review of how human factors contributed to several incidents including:

- July 1994 explosion at Texaco, Milford Haven, UK
- September 1998 explosion at Esso, Longford, Australia
- March 2005 explosion at BP, Texas City, US.

Human operators are needed to ensure plants and processes continue to operate safely. The control room and control system interfaces must therefore be built with human needs in mind. Includes a review of how implementation of ISA 18.2 and/or EEMUA 191 can improve alarm management practice.

Dust Flame Propagation in Industrial Scale Piping
Part 2: CFD Study of a Conveying Vessel-Pipeline System

Jim Vingerhoets
Fike Europe
Toekomstlaan 52, 2200 Herentals, Belgium
jim.vingerhoets@fike.com

Thomas M. Farrell
Fike Corporation, Industrial Protection Group
704 SW 10th Street, Blue Springs, MO 64015, USA
tom.farrell@fike.com

Jef Snoeys
Fike Europe
Toekomstlaan 52, 2200 Herentals, Belgium
jef.snoeys@fike.com

Prepared for Presentation at
American Institute of Chemical Engineers
2013 Spring Meeting
9th Global Congress on Process Safety
San Antonio, Texas
April 28 – May 1, 2013

UNPUBLISHED

Dust Flame Propagation in Industrial Scale Piping Part 2: CFD Study of a Conveying Vessel-Pipeline System

Jim Vingerhoets
Fike Europe
Toekomstlaan 52, 2200 Herentals, Belgium
jim.vingerhoets@fike.com

Thomas M. Farrell
Fike Corporation Industrial Protection Group
704 SW 10th Street, Blue Springs, MO 64015, USA
tom.farrell@fike.com

Jef Snoeys
Fike Europe
Toekomstlaan 52, 2200 Herentals, Belgium
jef.snoeys@fike.com

Keywords: Dust explosion, flame propagation, vented vessel, pipelines, CFD

Abstract

The use of pipes to connect vessels and to transport hazardous materials, such as flammable gases or dusts, is a common practice in the process industry. The potential of fires and explosions has been recognized for decades and is addressed in standards such as NFPA 68 [1] and NFPA 69 [2]. Unlike for gases, the experimental evidence on dust flame propagation is scarce, and this study is particularly aimed at expanding the knowledge in this field. The purposes of this paper are to present the results of a parametric computational fluid dynamics (CFD) study validated by experimental evidence and to further develop a detailed understanding of combustion propagation in industrial scale piping.

The results of an ongoing research program which involves a piping system connected to a 5 m3 vented vessel are presented in the accompanying paper [3]. CFD was used to broaden the field of application beyond the tested configurations, and to study the influence of vessel volume and pressure, fuel characteristics and flow conditions.

Upstream and downstream explosion propagation behavior is studied as a function of fuel reactivity and type, vessel fuel concentration, pipeline fuel concentration, flow velocity, vessel volume, vent area, vent location, ignition location, pipe diameter and length.

1. Introduction

The dust explosion hazard continues to represent a constant threat to process industries that manufacture, use and/or handle powders and dust of combustible materials. In the process industry, the majority of process equipment is connected via pipes or ducts, and accidents have shown that the propagation of the explosion through the connection pipe work worsens the consequences of the explosion. Understanding the mechanisms of explosion flame propagation is a prerequisite for designing effective explosion protection solutions.

The problem of dust explosions in large-scale connected equipment has been much less investigated than isolated equipment, mainly due to cost and complexity to conduct experiments. The information in the public domain on dust explosion propagation is scarce; refer to the bibliography section at the end of the paper for relevant contributions. An alternative to large scale testing is the use of mathematical modeling and simulating the problem using CFD - computational fluid dynamics. CFD uses numerical methods and algorithms to simulate the combustion phenomena defined by boundary conditions. Although the use of CFD in gas and dust explosion applications is relatively new, this technique is irrefutably gaining interest.

The present research investigates, by experiment and CFD, flame propagation from a vented vessel through connected pipes. The aim is to examine the mechanisms that govern the velocity and acceleration of flame through pipelines, thus providing valuable information for the design of explosion protection solutions and systems. Once sufficient correlation between experimental and numerical results was demonstrated, the scope was extended with a numerical sensitivity study to quantify the influence of several application parameters on the flame propagation behavior. These parameters include:

- fuel type and reactivity,
- fuel concentration in vessel and pipelines,
- flow velocity and direction,
- vessel volume, pipe diameter and length,
- vent area, location,
- ignition location.

2. Simulation code

The numerical explosion simulations have been performed using DESC - Dust Explosion Simulation Code. DESC resulted from a research project supported by the European Commission under the Fifth Framework Program and was reviewed by Skjold in 2007 [4]. The main purpose of the project was to develop a simulation tool based on computational fluid dynamics (CFD) that could predict the potential consequences of industrial dust explosions in complex geometries. DESC is presently based on the existing Flame Acceleration Simulator (FLACS) code for gas explosions. The dust particles are modeled as gaseous components that follow the flow of the other gaseous components in the system. The combustion model used in FLACS and adopted by DESC is the so-called β flame model developed by Arntzen [5]. The turbulent burning velocity used by the β model is a function of the root mean square of the turbulent velocity fluctuations and the Karlovitz stretch factor. More specifically, the turbulent burning velocity is formulated as:

$$S_T = 15.1 S_L^{0.784} u'^{0.412}_{rms} l_I^{0.196}$$

with S_T the turbulent burning velocity, S_L the laminar burning velocity, u'_{rms} the root mean square of the turbulent velocity fluctuation, also called the turbulence intensity and l_I the turbulence integral length scale. The inputs for the combustion model are laminar burning velocities extracted from the measured K_{st} and P_{max} values in standardized 20 L sphere tests. The code has been validated against large scale tests reported in the public domain. It was found that DESC is sensitive to the numerical grid resolution, the presence of bends and interconnected vessels in general. The grid resolution in the current research has been chosen based on recommendations from this validation, and bends were excluded from the study.

3. Experimental setup

A test rig was built at the test site of Fike Corporation consisting of an inlet pipeline, a vessel, and an outlet pipeline, arranged as shown in the following image. The inlet and outlet pipelines were oriented 90° to each other, with the outlet pipeline aligned with the major axis of the vessel. All explosion venting was constrained to a single panel on the vessel, on the wall opposite the outlet pipeline interconnection. The system was equipped with a fan, enabling adjustable pipe flow velocities. By using separate constant mass flow rate dust injectors on the pipelines in addition to traditional dust injectors on the vessel, the vessel dust concentration and pipeline dust concentration could be varied independently of each other.

Figure 1: Photo of experimental setup

In order to cover several industry branches, different fuel types were used to conduct the research. The properties of the fuels used in this program are listed in Table 1.

Table 1: Closed vessel explosion characteristics of the dust fuels used

Dust type	Dust concentration (g/m³)	20 L sphere	
		K_{st} (bar.m/s)	P_{max} (barg)
Wood flour	250	50	5.7
	500	108	7.0
	750	109	7.2
Phenolic dust	250	203	7.3
	500	217	6.9
	750	200	6.6
Corn Starch	250	93	6.6
	500	199	8.7
	750	205	8.6

Many tests were conducted to determine explosion pressures and flame arrival times as a function of venting area, vessel fuel concentration, pipeline fuel concentration and conveyance velocity. More information can be found in the accompanying paper [3].

4. Simulation setup

Figure 2 shows the basic setup of the explosion simulations. It represents as much as possible the experiment setup. The vessel itself consists of two identical cubes of 2.55 m^3 each, attached to each other without a partitioning wall in between. The cube assembly is equipped with dust jet injectors, 4 vent panels, a 10 kJ igniter and flanges to connect two pipelines. Each of the 4 vent panels can be closed off (to vary the total vent area) or opened at a controlled burst pressure. In some simulation scenarios, the location of the vent panels was moved to another wall of the vessel.

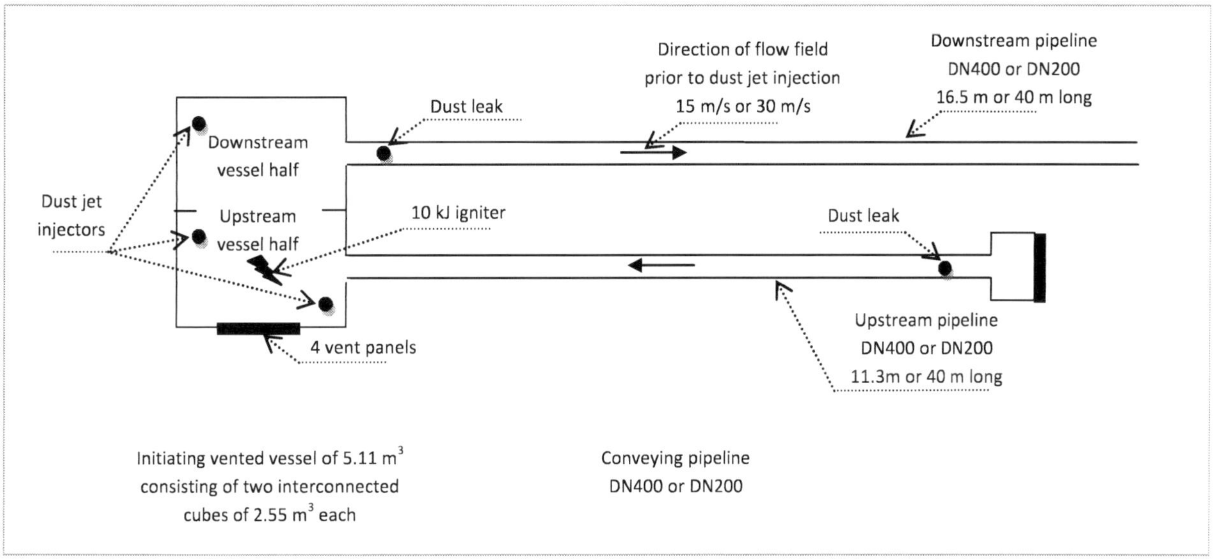

Figure 2: Sketch of basic setup of the explosion simulations.

Figure 3: Three dimensional CAD drawing picture of the simulation setup

The default location of the igniter is in the centre of the upstream vessel half, but in several simulation scenarios the igniter was moved to the centre of the downstream vessel half or to the interconnection with one of the pipelines. The delay time between start of dust jet injection and ignition was identical to the experimental program. The pipelines were either 400 mm or 200 mm in diameter and 11.3 m, 16.5 m or 40 m long. In order to restrict the simulation domain and associated computer processing time, the downstream pipeline was rotated 90° with respect to the experimental setup. Flow of 15 m/s or 30 m/s was established in the pipelines, depending on the simulation scenario. Dust leaks were programmed to fill the pipelines with a dust/air mixture of known concentration. Other simulations were done in a 62 m^3 vessel with proportionally scaled vent panels.

All simulation scenarios have been numerically solved on cubical grid cells with optimized size conforming to DESC guidelines.

5. Comparison between experiments and numerical simulations

Figure 4 to Figure 6 display comparative plots of vessel pressure and flame position as a function of time for three example scenarios including each of the three dusts of the program. Flames propagating towards the blower are represented by negative pipeline distances.

Table 2: Comparison between experiments and simulations giving the main characteristic values of the explosion propagation

Comparison of main characteristic values		Example 1		Example 2		Example 3	
		Exp	Sim	Exp	Sim	Exp	Sim
Vessel peak pressure		3.2 psig	2.0 psig	4.2 psig	6.7 psig	8.6 psig	2.3 psig
Time to reach 1,5 psig in vessel		64 ms	161 ms	46 ms	56 ms	54 ms	110 ms
Average flame speed in first 10 meters of pipe	Downstream	90 m/s	36 m/s	90 m/s	116 m/s	154 m/s	116 m/s
	Upstream	79 m/s	55 m/s	116 m/s	143 m/s	221 m/s	112 m/s
Time to reach 2,7 meters in the pipe	Downstream	110 ms	382 ms	114 ms	117 ms	88 ms	214 ms
	Upstream	127 ms	233 ms	106 ms	95 ms	85 ms	135 ms

The simulations under-predict the explosion peak pressure and pressure rise time in the vessel for two out of the three scenarios. The modeled explosion peak pressure for example 2 is higher (6.7 psig) than its experimental equivalent (4.2 psig), with similar pressure rise time (1.5 psig @ 56 ms and 1.5 psig @ 46 ms). Similar observations are made for the flame front pipeline propagation: the modeled explosions for example 1 and 3 propagate at lower speed (36 m/s downstream, 55 m/s upstream and 116 m/s downstream, 112 m/s upstream respectively), whereas the modeled dust explosion of example 2 propagates faster. Finally, the modeled explosions of example 1 and 3 enter the pipes later whereas the modeled dust explosion pipe entrance time of example 2 is similar to the experiment.

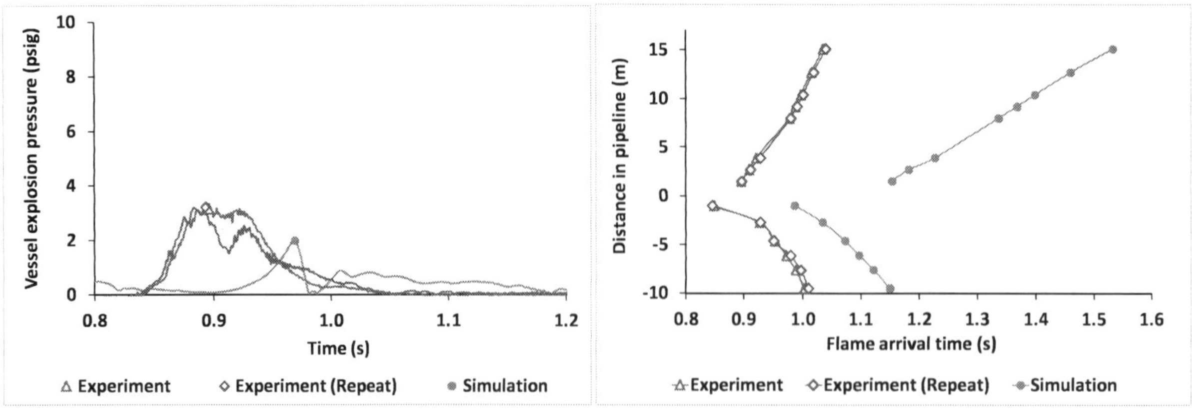

Figure 4: Comparison experiment and simulation: example 1

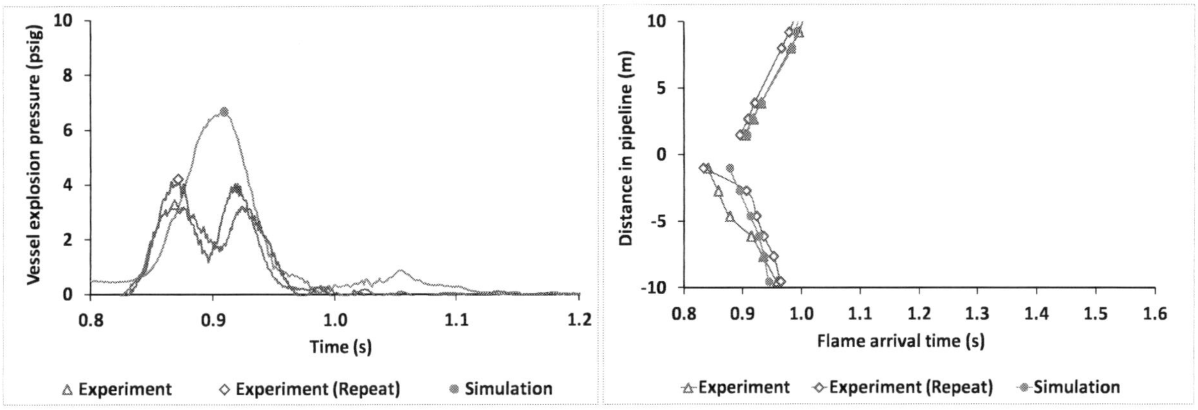

Figure 5: Comparison experimental and simulation: example 2

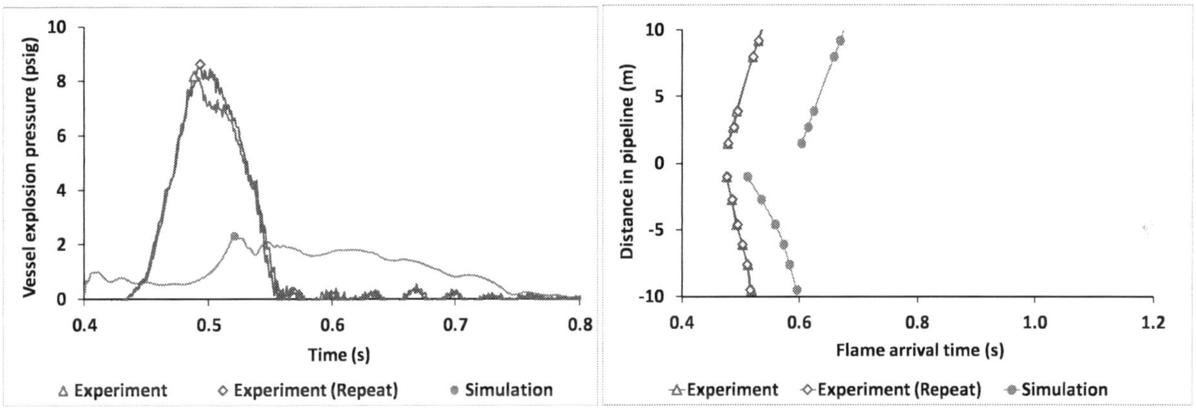

Figure 6: Comparison experimental and simulation: example 3

A likely explanation for the divergence between experimental and simulated results is that the modeled dust jet injectors generate different turbulence conditions than the injectors used in the experiments. This is a plausible explanation since the dust particles are modeled as gaseous phase fuel. Figure 7 shows the turbulence intensity in the centre of the upstream vessel half during simulated dust injection for different vessel dust concentrations and for a pipe flow of 15 m/s. The simulated dust jet injectors create an average turbulence intensity level between 2.0 and 3.4 m/s in the first 200 ms after ignition. 200 ms is the typical duration of an explosion in a 5 m^3 vessel. Although slightly higher than the reported turbulence in 20 L and 1 m^3 tests, the numbers are quite reasonable. Unfortunately, the turbulence could not be measured during the experiments.

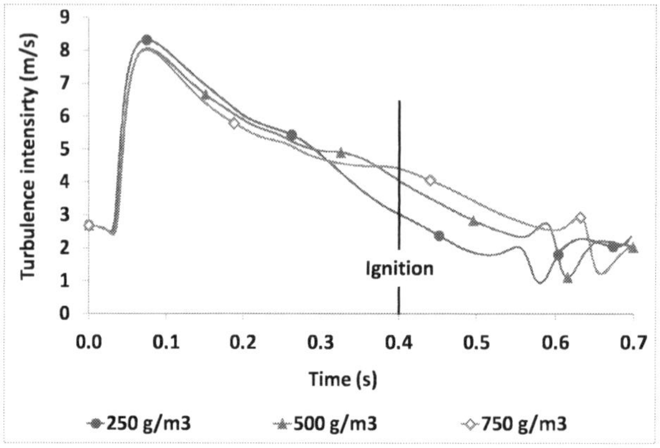

Figure 7: Turbulence intensity in the centre of the upstream vessel half during simulated dust injection for different vessel dust concentrations; initial pipe flow = 15 m/s

Another possible reason is that the 20 L flame data used by DESC do not scale up well to larger volumes. Several researchers postulated scaling problems related to turbulence and the powerful igniter used in the closed vessel 20 L and 1 m^3 standardized tests. This problem was investigated by additional explosion experiments in a closed 1m^3 vessel with varying ignition delay [3]: the K_{st} of two of the three dusts were more or less independent of volume, although strongly affected by ignition delay. The other dust showed an opposite behavior and is relatively insensitive to decreased ignition delay.

Figure 8 shows the effect of increasing the laminar burning velocities which was used as a surrogate for boosting the turbulence. The agreement becomes much better which indicates that the prediction potential of DESC is improved, if both the fundamental combustion parameters (the laminar burning velocity derived from K_{st} and P_{max}) and the initial process turbulence are estimated correctly.

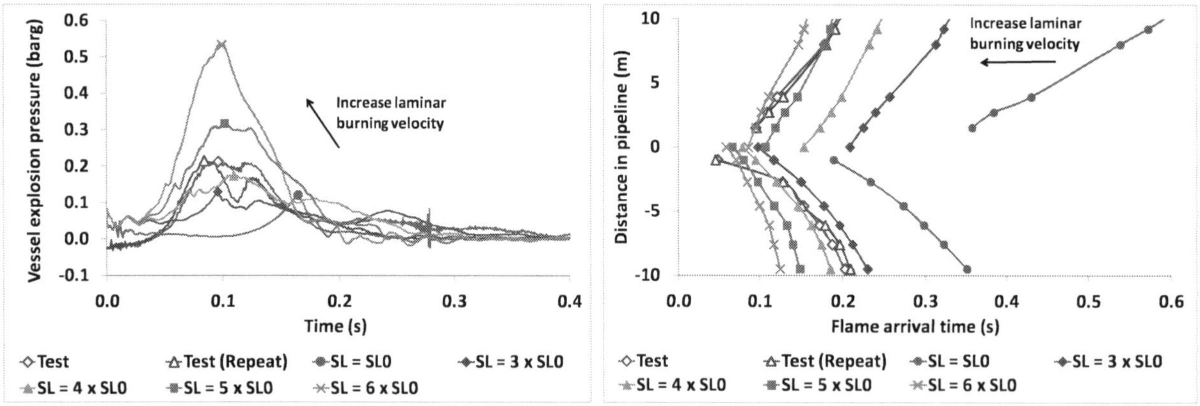

Figure 8: Sensitivity of vessel explosion pressure and flame propagation to the modeled laminar burning velocity and comparison with experiment (example 1)

In order to deal with the uncertainties in turbulence and combustion parameters for different vessel volumes and injector assemblies, simulations were performed in the reference test volume with all vent panels and connections to the pipelines closed off, and the K_{st} and P_{max} values were recorded. The results obtained by DESC should be referenced to K_{st} and P_{max} as obtained by DESC for the closed vessel. By following this approach, together with additional specific considerations as pointed out further, it is believed that simulations can represent industry reality with similar level of confidence as experiments.

6. Mechanisms of flame propagation

A thorough analysis of 94 explosion simulations for different fuel types, vessel and pipeline fuel concentrations, vent areas, flow velocities, significantly contributes to our understanding of the fundamental mechanisms which govern the propagation of dust flames through pipelines which are connected to a vented vessel. For all these scenarios the vessel and pipeline explosion pressures have been recorded as well as the flame arrival times at several distances along the pipelines. Besides pressure and flame, other physical variables, such as flow velocity and dust concentration have also been recorded. In addition to dust fuels listed in Table 1, also other dust fuels were used in the simulation study, so covering a wider range of K_{st} values, from 28 bar.m/s up to 281 bar.m/s.

In all scenarios the flame propagated with high speed up until the end of both pipelines, including the ejection of a large fireball, even against strong initial flows of 30 m/s. The highest recorded flame speed was 434 m/s. This was at the end of the downstream pipeline for the dust and concentration with highest reported K_{st} value and with the smallest venting area programmed. The lowest recorded flame speed was 34 m/s, for the dust and concentration with lowest K_{st} value and the largest vent area programmed.

It was confirmed that flame enters both pipes and propagates through a dust cloud that is pushed from the vessel into the pipes, as shown in Figure 9 below. In this scenario, the pipelines were not filled with dust fuel prior to ignition.

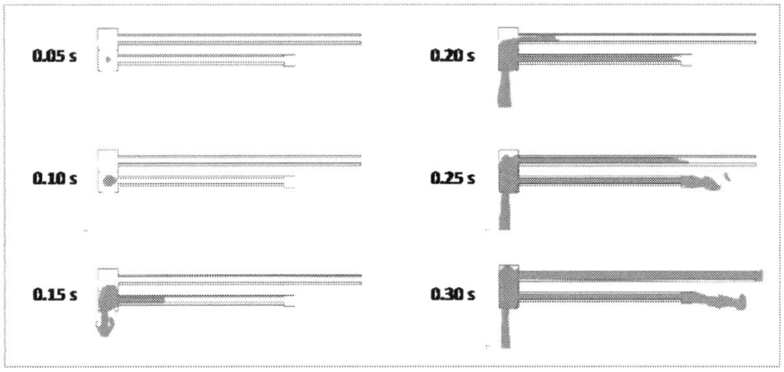

Figure 9: Flame propagation shown at subsequent time steps during an example explosion. No flow through pipe, no dust in pipe. 2 vent panels open. Vessel dust concentration = 500 g/m^3.

Figure 9 also illustrates that flame accelerates as it enters the pipe: It takes 0.1 seconds for the flame to reach the pipe and in only another 0.1 seconds it manages to travel the full pipeline distance.

In the vessel the freely expanding flame front propagates at a moderate speed of about 5 to 10 m/s. When the explosion flame enters the connected pipes, it will accelerate because the flame front will travel along with the combustion gases flowing at high speed out of the vessel:

$$S_{\mathrm{f,\,pipe}} = S_{\mathrm{g}} + S_{\mathrm{T}}.\,E$$

The pipe gas flow velocity induced by the explosion pressure in the vessel (S_g) is often a lot larger than the speed of turbulent burning and associated local expansion ($S_T.E$) and can reach up to 300 m/s. For low to moderate explosive reactive materials the vessel pressure induced gas flow will dominate the pipe flame speed.

Figure 10 shows the correlation between gas flow velocity in the pipelines and flame front velocity, for three different fuels and for three different dust concentrations. The more reactive the fuel and the larger the gas flow velocity, the more the flame velocity exceeds the gas velocity. This is because the turbulent burning velocity increases with fuel reactivity and with increased flow turbulence. The flow turbulence is higher at higher gas velocities.

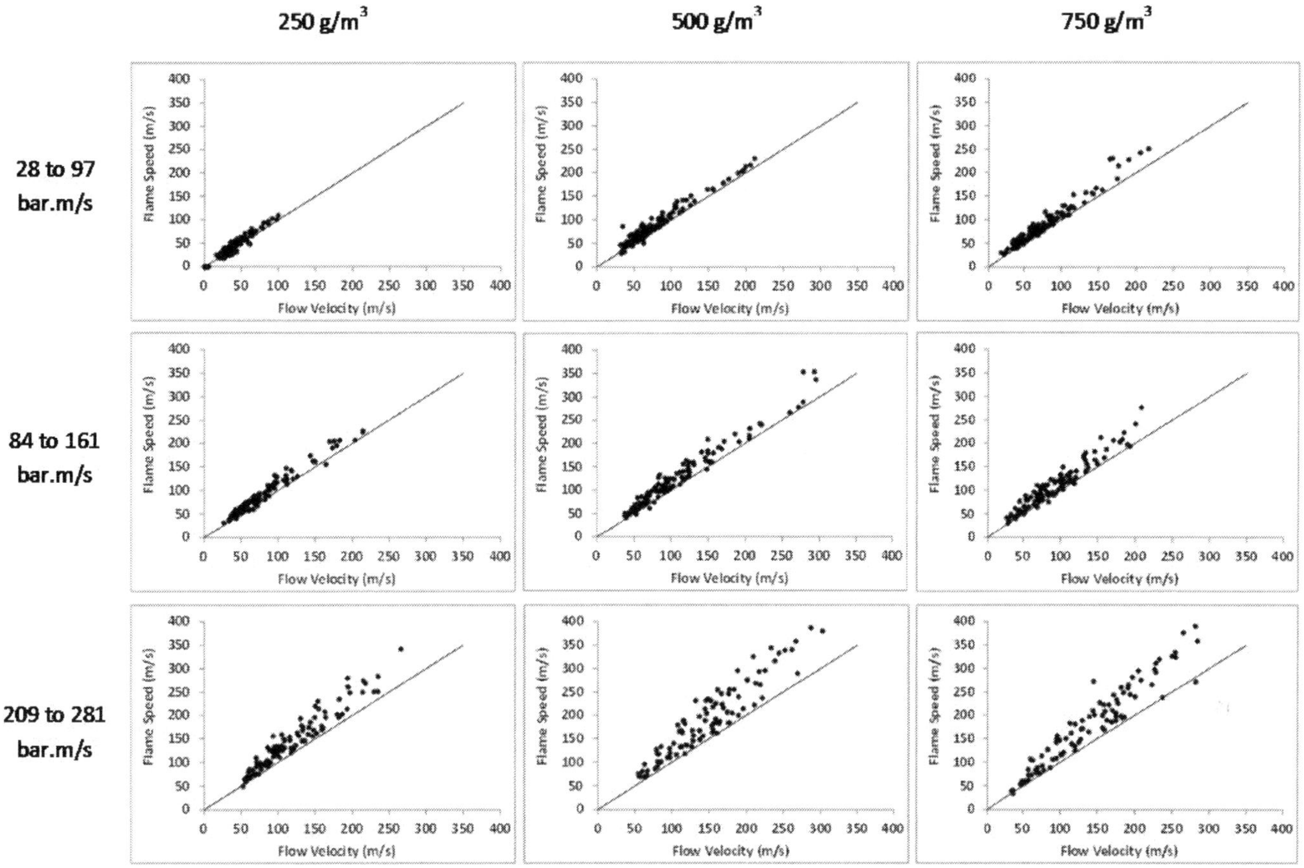

Figure 10: Correlation between flame speed and explosion induced gas velocity in the pipelines for different fuel types and dust concentrations

7. Sensitivity study

To enable extrapolation of the findings to real industry applications, 172 additional scenarios have been simulated in which the flow direction, vessel volume, vent area, vent location, ignition location, pipe diameter and length were varied.

7.1. Effect of vent area

In all cases, the flame speed decreases with larger vent areas. Figure 11 shows example simulation results for two dust types. This finding was consistently confirmed by experiments [3] and is attributed to lower vessel reduced pressures associated with larger venting areas. The decrease in flame speed is less pronounced in the upstream pipe, because the ignition was located closer to the upstream pipe, so that the flame could enter the upstream pipe before the vent panel opened.

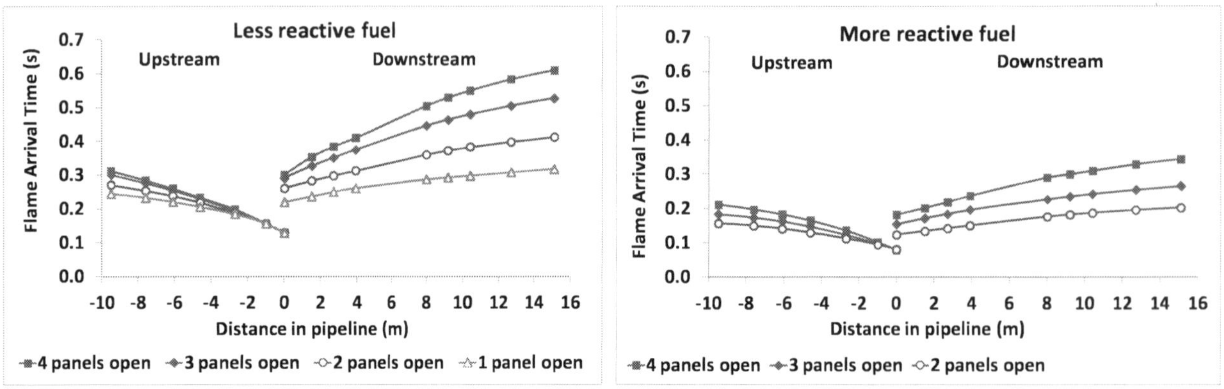

Figure 11: Effect of vent area on flame propagation. Flow = 15 m/s; Vessel and pipeline dust concentration = 750 g/m^3; Ignition location = centre of upstream vessel half.

7.2. Effect of pipeline dust concentration

Many different types of dust transport exist in the industry, including dust aspiration and dust extraction and lean to dense phase transport. Some dusts will stick to the pipe wall and build layers that may become airborne by the explosion propagating through the pipeline, resulting in combustible dust-air cloud generation in transport systems that typically operate at or below the minimum explosible concentration. It is therefore interesting to study the effect of pipeline dust concentration on the explosion propagation. Figure 12 shows examples for two dust types with the pipeline dust concentration changed from 500 g/m^3 to 50 g/m^3. The vessel dust concentration was kept constant at 500 g/m^3. Although the fuels are more reactive at the higher dust concentration, flames are slower at high pipeline dust concentrations. This effect is attributed to the change of the inertia of the fluids in the pipelines. Also this finding was confirmed by the experiments [3].

Figure 13 demonstrates that the combustion in the first meters of the pipeline is fueled by dust ejected from the exploding vessel and not so much by the pipeline dust.

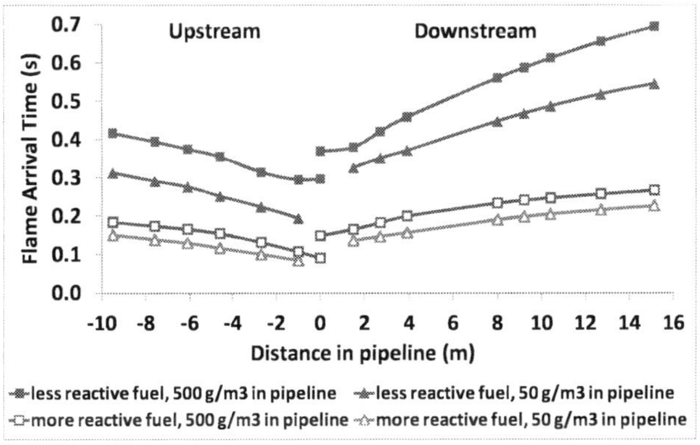

Figure 12: Effect of pipeline dust concentration on flame propagation for two dust fuels; Vessel dust concentration = 500 g/m^3.

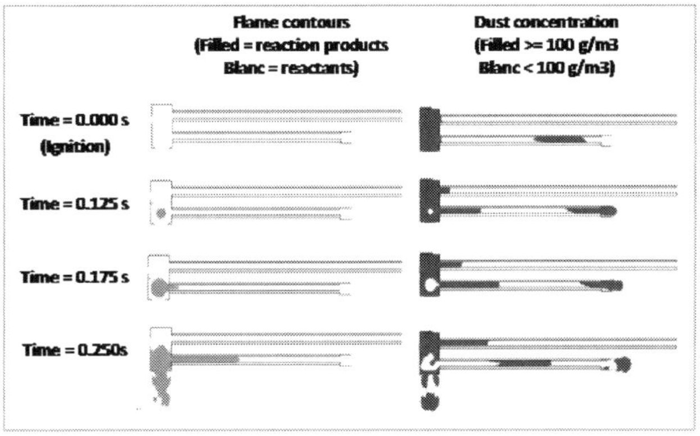

Figure 13: Illustration of how dust is ejected out of the initiating vessel before the flame front arrives for an example explosion; Vessel dust concentration = 500 g/m3; Pipeline dust concentration = 50 g/m^3.

7.3. Effect of ignition location

Scenarios testing the effect of ignition location were run with the igniters located in the three positions indicated in Figure 14 below, including ignition near the vent (1), at the interconnection with the upstream pipeline (2), and in the center of the downstream vessel half, far from the vent (3)

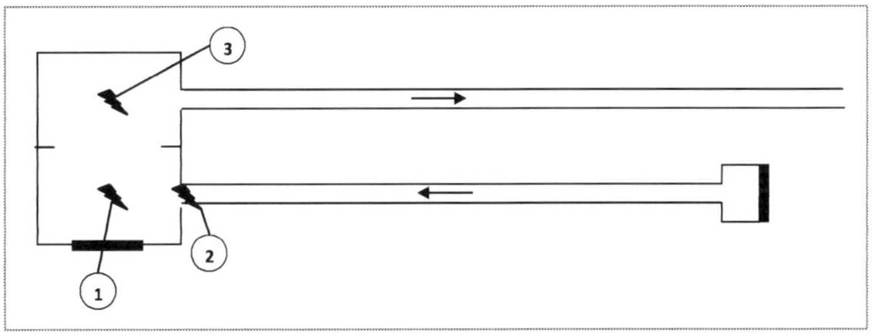

Figure 14: Ignition locations

The effect of the ignition location is twofold: the closer the ignition point is located towards the pipe entry, the sooner the explosion flame will enter the pipeline; and the further the ignition point is located from the pressure venting device, the higher will the explosion pressure in the vessel be, with faster flames as a result. These phenomena are illustrated in Figure 15, where for examples with two dust types, the flame arrival times are compared for three explosion scenarios with varying ignition location. It is clearly shown that flames enter the pipeline earlier for pipeline ignition (2), and that ignition far away from the vent panel (3) generates significantly faster flames

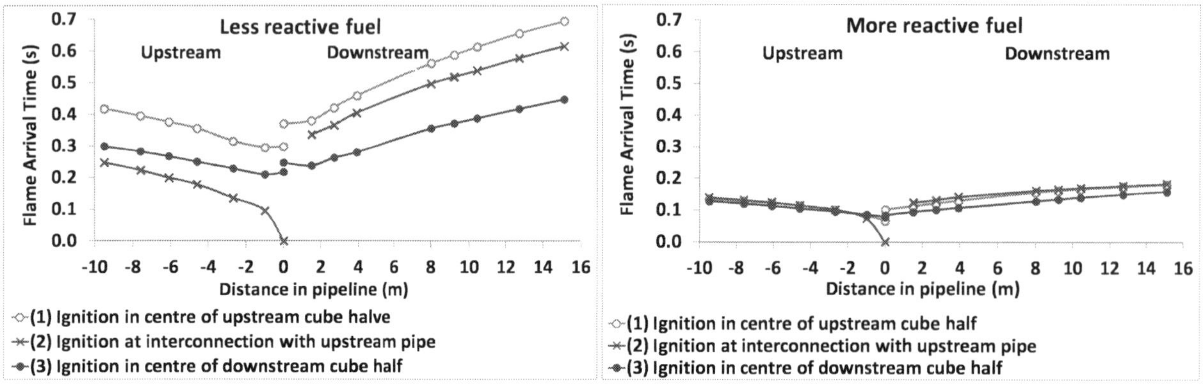

Figure 15: Effect of ignition location on flame propagation. Examples for two dust fuels.

7.4. Effect of vent location

Ideally, the explosion vent panel should be positioned as close as practically possible to the expected ignition location. This will provide an early escape of the combustion gases and will limit the resulting explosion overpressure in the vessel and consequently the flame speed in the pipeline. This is illustrated in Figure 16 and Figure 17: the further the vent panel is moved away from the ignition position (– with vent location alternative 3 further than alternative 2, and alternative 2 further than alternative 1), the faster the explosion flame propagates in both pipelines. The driving factor is the increased explosion intensity in the initiating vessel.

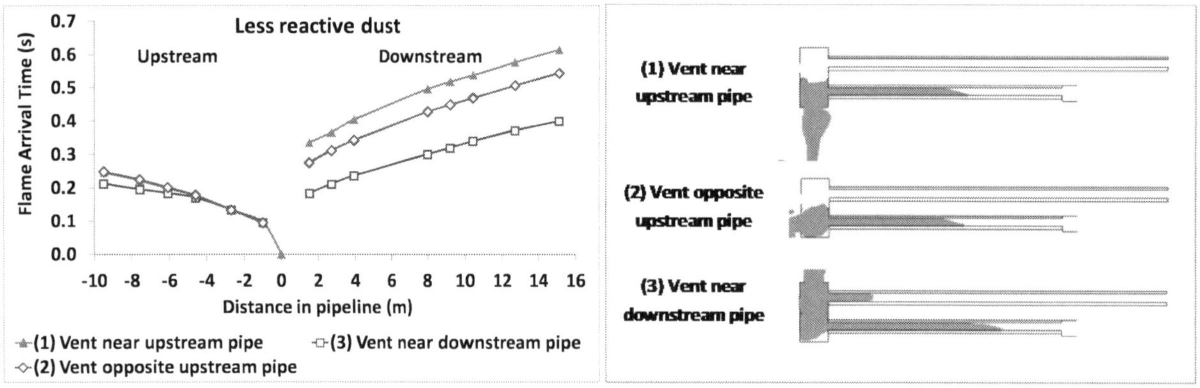

Figure 16: Effect of vent location on flame propagation for an example explosion scenario. Ignition location = interconnection of vessel with upstream pipeline. Left: flame arrival times. Right: flame contours at 0.2 s after ignition.

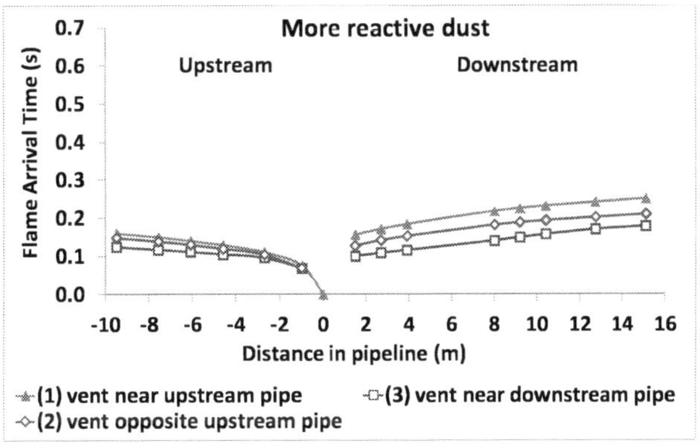

Figure 17: Effect of vent location on flame propagation for another example explosion scenario. Ignition location = interconnection of vessel with upstream pipeline.

7.5. Effect of pipe length

Longer pipes will increase the flow resistance and gas column inertia, resulting in a less aggressively accelerating flame along the pipeline. The effect is most pronounced for cases with low vessel explosion pressure, where flow resistance and gas column inertia play a relatively stronger role, and for cases with low K_{st} value, where the flame velocity is stronger linked with the gas flow velocity. Figure 18 shows the simulated flame arrival times for three example explosions, where the length of the piping was varied from the 11.3/16.5 m of the basic configuration to 40 meters.

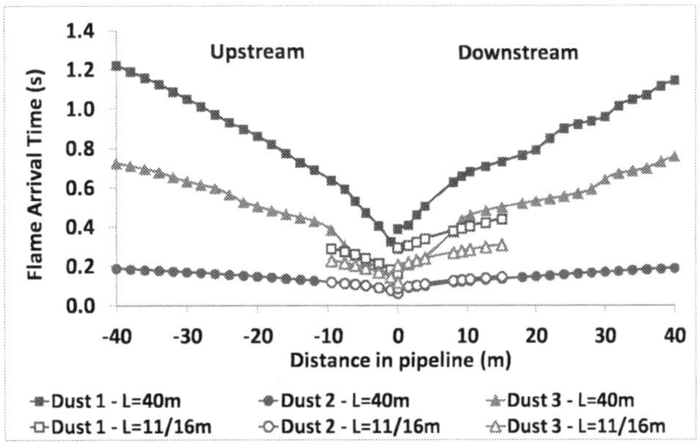

Figure 18: Effect of pipe length on flame propagation for three example explosions.

7.6. Effect of pipe diameter

As with the pipe length, changing the pipe diameter affects the flow resistance. But, in addition to this, the pipe diameter also changes the Reynolds number and thus the gas turbulence in the pipeline. Flames will accelerate faster in smaller pipes as a result of higher turbulence. Note that heat losses of the hot burning gases to the cold pipe wall material have not been modeled. Figure 19 shows the effect on flame propagation when the pipe diameter is decreased from 0.4 m to 0.2

m. For the example with lowest fuel reactivity and generating relatively low vessel explosion pressures, the flame propagates slower in the smaller diameter pipes than in the larger diameter pipes. For the example with highest fuel reactivity and generating relatively high vessel explosion pressures, the flame propagates faster in the smaller diameter pipes than in the larger diameter pipes. For the example with medium fuel reactivity, it is seen that in the initial phase of the explosion, the flame propagates slower in the smaller diameter pipe geometry, but later on, when the explosion pressure develops and the fluid turbulence enhances, the flame will propagate faster.

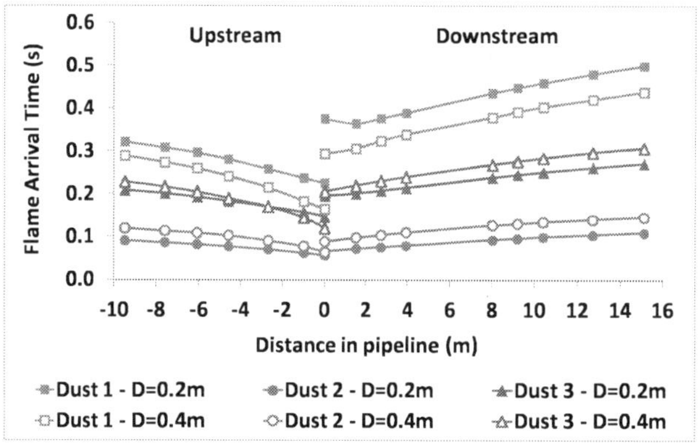

Figure 19: Effect of pipe diameter on flame propagation for three example explosions.

7.7. Effect of vessel volume

When the volume of the initiating vessel is increased, delayed pipeline flame propagation can be expected, because the explosion pressure rise in the vessel will be slower. This is clearly illustrated in Figure 20, where for three example explosions the flame propagation is compared for the cases when initiated in a 5 m^3 vessel and in a 62 m^3 vessel, all other parameters either kept identical or scaled with the volume. The ignition was located in the centre of the 5 m^3 upstream vessel half, which is 0.68 m away from the interconnection with the upstream pipeline, which is then of course out of the centre for the 62 m^3 vessel half. To make the comparison explosion pressure independent, the vent area on the 62 m^3 vessel was increased by factor of 4, which approaches the factor $\left(\frac{62}{5}\right)^{2/3} = 5.36$ which would be required theoretically. Initially the pressure rise is slower which leads to a delayed flame entry into the pipes, but as soon as the vessel pressures in the 62 m^3 exceed the ones recorded in the 5 m3, the flame acceleration rate for the 62 m^3 vessel cases becomes higher.

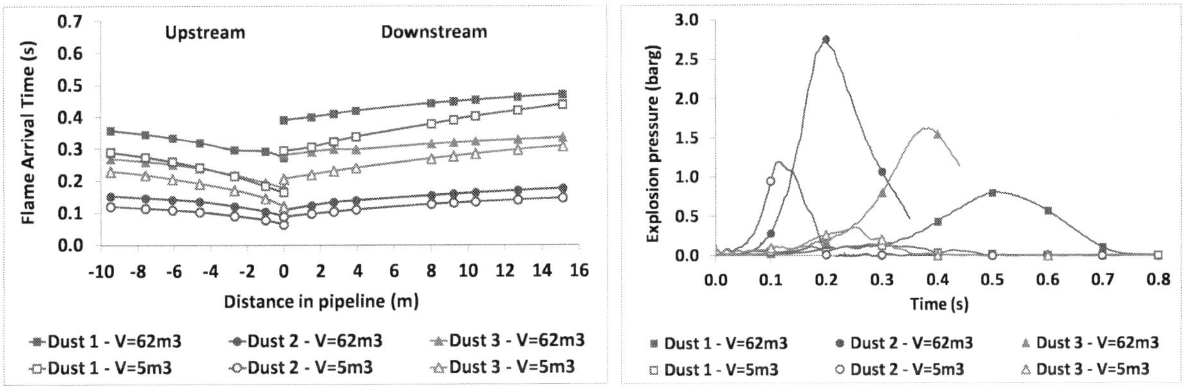

Figure 20: Effect of vessel volume on flame propagation for three example explosions. Left: flame arrival times in pipelines. Right: vessel explosion pressures.

8. Summary

DESC, a CFD code with an integrated combustion model, has been used to study the behavior of propagation of dust explosions through pipelines connected to vented vessels. A 5 m^3 vented vessel in which explosions were initiated, was connected to two DN400 pipelines, respectively 11.3 m and 16.5 meter long. The system was equipped with a fan to establish a flow field prior to dust injection and explosion ignition. Ignition always occurred in the vessel; neither the present study nor the accompanying empirical program [3] considered ignition in the pipeline. Input for the combustion model are laminar burning velocities extracted from K_{st} and P_{max} values determined by standardized 20 L tests.

Comparison between experimental and simulation results revealed that for two out of the three dust fuels, the experimental dust explosions were more intense than the simulated equivalents. A method was devised to improve the confidence level of DESC simulations. By following this approach, and by understanding the underlying physics, it is believed that simulations can represent industry reality with similar level of confidence as experiments.

A total of 94 explosion simulations have been performed in the basic configuration, where the fuel type, vessel and pipeline fuel concentration, vent area and flow velocity were varied. The effects of the parameter variations on flame arrival times in the pipelines were fully in line with the experimental results [3]. In all cases the flame propagated along the full length of the pipes, both upstream and downstream. A high degree of correlation was revealed between explosion induced expansion flow and flame speed.

172 additional simulations were performed in which flow direction, vessel volume, vent location, ignition location, pipe diameter and length were varied. This extended scope enabled to confirm the pronounced effect of the vessel reduced explosion pressure and pipeline geometry.

9. Conclusion

Understanding the mechanisms of explosion flame propagation is paramount for designing effective explosion protection solutions for interconnected process equipment that handle flammable substances. The present research has demonstrated that dust explosions do propagate through pipelines, even against strong flows and that many different application parameters besides K_{st} and P_{max} considerably influence the speed of dust flame propagation. When these parameters are not taken into account, the protection system may fail to protect against the explosion hazard. It has been irrefutably demonstrated that controlling the explosion pressure in the initiating vessel is very effective in controlling the speed of flame propagation. The explosion protection design should therefore always consider the explosion protection problem in its totality: a combination of vessel explosion pressure protection and pipeline explosion propagation protection.

10. References

[1] NFPA 68, Standard on Explosion Protection by Deflagration Venting, 2007

[2] NFPA 69, Standard on Explosion Prevention Systems, 2008

[3] T.M. Farrell, J. Vingerhoets, J. Snoeys, J.E. Going, "Dust Flame Propagation in Industrial Scale Piping - Part 1: Experimental Study in a Conveying Vessel-Pipeline System", Submitted for 9th Global Congress on Process Safety, San Antonio, April 28 - May 1, 2013

[4] T. Skjold, "Review of the DESC project", *Journal of Loss Prevention in the Process Industries*, Volume 20, p291-302, 2007

[5] Arntzen, B.J., "Modeling of turbulence and combustion for simulation of gas explosions in complex geometries", Dr. Ing Thesis, NTNU, Trondheim, Norway, 1998

Additional references not cited:

[6] A.E. Dahoe, J.F. Zevenbergen, S.M. Lemkowitz, B. Scarlett, "Dust explosions in spherical vessels: the role of flame thickness in the validity of the cube-root law", *Journal of Loss Prevention in the Process Industries*, Volume 9, p33-44, 1996

[7] A.E. Dahoe, R.S. Cant. and B. Scarlett, "On the decay of turbulence in the 20-litre explosion sphere", *Turbulence and combustion*, Volume 67, p159-184, 2001

[8] A.E. Dahoe, K. van der Nat, M. Braithwaite and B. Scarlett, "On the sensitivity of the maximum explosion pressure of a dust deflagration to turbulence", *KONA*, Volume 19, p178-195, 2001

[9] G.A. Lunn, "Dust explosions in totally enclosed interconnected vessel systems", *Journal of Loss Prevention in the Process Industries*, Volume 9, p45-58, 1996

[10] P. Holbrow, S. Andrews, G.A. Lunn, "Dust explosions in interconnected vented vessels", *Journal of Loss Prevention in the Process Industries*, Volume 9, p91-103, 1996

[11] M. Roser, A. Vogl, S. Radandt, W. Malalasekara, R. Parkin, "Investigations of flame front propagation between interconnected process vessels. Development of a new flame front propagation time prediction model", *Journal of Loss Prevention in the Process Industries*, Volume 12, p421-436, 1999

[12] A. Vogl, "Explosionsübertragung aus Behältern in Rohrleitungen", *VDI Berichte*, Volume 1272, p215-236, 1996

[13] J. P. Pineau J.P., G. Ronchail, "Propagation of dust explosions in ducts", Symposium on the control and prevention of dust explosions, Bâle, 1982

[14] A. Vogl, S. Radandt, "Explosionsübertragung durch dünne Rohrleitungen", FSA Report F-05-9903, 2005

[15] U. Bielert, M.Sichel, "Numerische Simulation von Staubexplosionen in pneumatischen Saug-Flug-Förderanlagen", *VDI Berichte*, Volume 1601, p449-472, 2001

[16] W. Jian, L. Xinguang, Z. Shengjun, S. Randandt, W. Fuli, R. Chunli, "Flame propagation through potato starch/air mixtures in pipe of interconnected vessels", *Measurement and Control of Granular Materials*, Volume 8, p376-379, 2009

(44c) The Use of CFD to Evaluate the Interactions Between Multiple-Leak Sources and to Assess Effectiveness of Integral Modeling Techniques in a Multiple-Leak Scenario

Joshua Richardson, Ha Nguyen and M. Sam Mannan, Mary Kay O'Connor Process Safety Center - Texas A&M University, College Station, TX

In many practical applications within the chemical industry, it is of vital importance to know the concentration profile that may occur from the dispersion of a material. For example, effect footprints for the release of a toxic material are needed to properly assess risk to both a facility and its surrounding communities and are useful for emergency response planning. Likewise, knowing the concentration gradient of a flammable gas can affect, among many other things, the layout of a facility to prevent or lessen the consequence of explosion and fire scenarios or for the placement of detection systems.

There are two main tools for the modeling of dispersion scenarios. Computational fluid dynamics (CFD) is a useful tool for accurately assessing the dispersion of materials, whether toxic, flammable, pollutant, or otherwise. It has found extensive use in academic as well as industrial settings when scenarios and geometries are well-defined because it takes into account the surroundings in the release environment. However, it is relatively slow and computationally expensive to use.

Also used extensively in the chemical industry are integral consequence modeling tools (such as Phast) which are generally more user-friendly and have the advantage of delivering results in a more time-efficient manner, but cannot take into account complex geometry considerations. A particularly interesting problem to consider in practice is that of multiple simultaneous release sources. This could be caused by a common failure of two units causing concurrent release, a domino effect where an event occurs that causes another event leading to multiple releases, or even an intentional event such as a terrorist attack. While CFD has the capability to resolve the possible interactions between the multiple release sources, integral modeling does not.

A simple current approach to the problem of multiple-source scenarios is to model them independently and superimpose the results upon one another. This may be a good approximation in some cases, but it does not take into account the complex interactions that may take place between the sources. The practical question then becomes whether it is acceptable in certain cases to model a certain set of simultaneous releases as an independent set of sources or whether it could be possible to model them as a single combined source.

This study aims to utilize CFD modeling of a set of varying scenarios to classify multiple-release dispersion scenarios into one of three different categories:

- Releases that are independent and may be modeled as such
- Releases that may be combined and modeled as one source
- Releases that must be considered together due to inherent interactions between them

Based upon criteria such as material, distance between sources, dispersion rate and duration, weather conditions, and geometry considerations, a general methodology for classifying multiple-release scenarios into these three different categories will be proposed so that these releases can ultimately be modeled in the appropriate manner.

(44ca) Understanding, Managing, and Communicating High Consequence Risk

Charles A. Soczek, DuPont Sustainable Solutions, Wilmington, DE

This presentation discusses the relationship between the tools used to identify, understand, mitigate and communicate high consequence risks. The understanding and reporting of high consequence risks, along with providing management, senior leadership, and the Board of Directors with the information necessary to ensure successful risk management, are fundamental outputs to a risk management process. Tools such as the Risk Matrix, Bow Tie Analysis, and Risk Register will be discussed along with key issues such as risk mitigation and risk information flow from the operating level to senior leadership to the Board of Directors. Fundamental to this discussion is the concept that all hazardous events are managed to an acceptable risk through engineering and administrative controls, and a robust Process Safety Management system.

Frequencies of Accidents as Functions of Design Variables

Manuscript ID: 295029

Esteban Bernechea
Centre for Technological Risk Studies (CERTEC). Department of Chemical Engineering.
ETSEIB. Universitat Politècnica de Catalunya.
Av. Diagonal, 647. 08028-Barcelona (Spain).
esteban.bernechea@estudiant.upc.edu

Cristian Massaguer
Department of Chemical Engineering. ETSEIB. Universitat Politècnica de Catalunya.
Av. Diagonal, 647. 08028-Barcelona (Spain).
cristian.massaguer@gmail.com

Josep Arnaldos
Centre for Technological Risk Studies (CERTEC). Department of Chemical Engineering.
ETSEIB. Universitat Politècnica de Catalunya.
Av. Diagonal, 647. 08028-Barcelona (Spain).
josep.arnaldos@upc.edu

1. Introduction

The mainstream definition of Risk is that it is a measure of the damage caused on humans, the environment or material property, in terms of the probability of occurrence of an accident and the magnitude of the damage it causes. This can be expressed mathematically as follows:

$$Risk = Frequency \cdot Consequences \quad \text{[Eq. 1]}$$

Risk is then, a function of the frequency of occurrence and consequences of a possible accident.

Often, a Quantitative Risk Assessment (QRA) of a specific equipment or installation has to be performed; in Europe the Seveso Directive makes it compulsory for installations that work with a certain amount of dangerous materials. In order to perform a QRA it is necessary to define initial Loss of Containment Events (LOCs) in the studied equipment, to later estimate the frequencies of the accidents that may result from these releases, and the consequences of said events. These values can finally be combined in order to obtain a measure of the risk that the installation poses over individuals or communities.

There is a wide variety of techniques that can be applied to estimate the frequency of an accident, like historical analysis, failure trees or event trees, subjective estimation, Bayesian analysis or Monte Carlo simulations. In Catalonia and Spain it is common to use the values of frequency provided by "The Purple Book" (CPR18E) [1] for the possible LOCs that can occur in different equipment. Later, event trees which depend on the probabilities of occurrence of certain events (like the immediate or delayed ignition of a flammable release) are used to obtain the final frequencies of the set of accidents that will be studied in the QRA.

However, the base frequencies presented in The Purple Book do not explicitly include the possibility that the LOCs will occur due to a domino effect; this phenomenon is defined as a cascade of events in which the consequences of previous accidents are increased both spatially and temporally by following ones, thus leading to a major accident [2]. The complexity of domino effect makes it difficult to model, and more so to include in the calculations performed in a QRA. This leads to the problem of how to take into account a phenomenon that can have a significant impact on the risk associated to the installation, but that is virtually impossible to include (without significantly increasing the number of scenarios to be considered). In some cases in which there are various similar equipment, the common criteria is to select a representative scenario, and multiply its frequency by the number of units and a factor of 2. In this way, domino effect may be included in the calculations, but not in a rigorous way, which could lead to the under or over-estimation of risk.

In this work, a methodology proposed by the authors to produce accident sequences in a domino effect has been exploited to obtain equations that describe frequencies of accidents in a specific installation. The studied plant stores pressurized liquefied propane in horizontal cylinders. The equations will depend on the number of units and the mass that is stored in the installation; they will also depend on the layout, which in turn depends on the size of the tanks, and therefore on the mass and number of equipment. Distances between tanks were defined by using the NFPA58 standard [3].

By finding equations that can be used to estimate the frequency with which a certain accident will occur in the installation, the time required to include domino effect into a QRA could be decreased importantly; only the frequency obtained through the equations, applied to a representative scenario would be necessary to take the phenomenon into account.

The main objective of this work is to use a mathematical model to find equations that relate the frequency of accidents in an installation to some of its design variables. This could be useful not only at the moment of performing QRA, but also at the moment of designing the installation; also, it can help understanding how the frequency of the diverse accidents is affected by the design of the plant.

In the next sections, the model that allows developing the accident sequences is briefly explained, after this, the design of the experiment, which is how the mentioned model was exploited, is presented; later, the results obtained are shown, and finally, some conclusions are reached.

2. Model for the development of accident sequences

The model used to develop the domino effect sequences [4] is based on the use of event trees in a sequential way; it evaluates the damage suffered by one tank in the case that an accident occurs in another, and establishes the possible accidents that may occur afterwards. The LOCs and event trees used to develop the sequences are those presented in the Purple Book [1], and are shown in Table 1 and Figures 1 and 2.

Table 1: LOCs used to define initial accidents

Code	Definition
G.1	Instantaneous release of the complete inventory.
G.2	Continuous release of the complete inventory in 10 min at a constant rate of release.
G.3	Continuous release from a hole with an effective diameter of 10 mm.

The first step to produce the accident sequences is to select one of the units, choose an initial LOC and define the possible accidents that may occur. After this, one of the accidents must be selected, and its effects must be calculated on all surrounding units and compared against the threshold values presented in Table 2.

Depending on the effects of the previous accident each of the affected tanks will suffer a specific LOC, which will in turn, generate new possible accidents; the sequence will continue to unfold until all the tanks have suffered catastrophic events, or the frequencies of the accidents become lower than 10^{-9} y^{-1}. This procedure has to be repeated for each of the possible accidents and for every one of the initial LOCs, starting the sequence in each of the tanks. Finally, the frequencies of all the occurrences of a specific accident in all the sequences, (for example, a flash fire derived from an instantaneous release), has to be summed, in order to know the number of occurrences of the incident.

The frequencies of the accidents down the sequence are calculated using the event trees and probabilities of accidents; each time an accident occurs caused by a previous incident, its initial frequency will be that of the event that caused its failure, multiplied by its probability of occurrence.

Figure 1. Event Tree for the Instantaneous Release of a Pressurized Flammable Gas

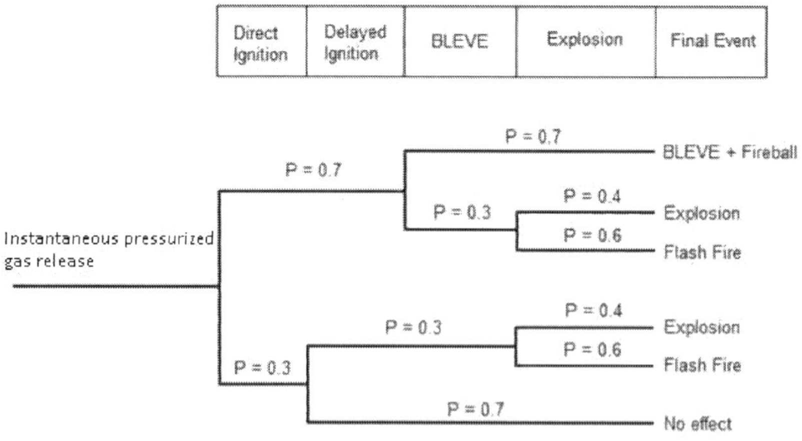

Figure 2. Event Tree for the Continuous Release of a Pressurized Flammable Gas

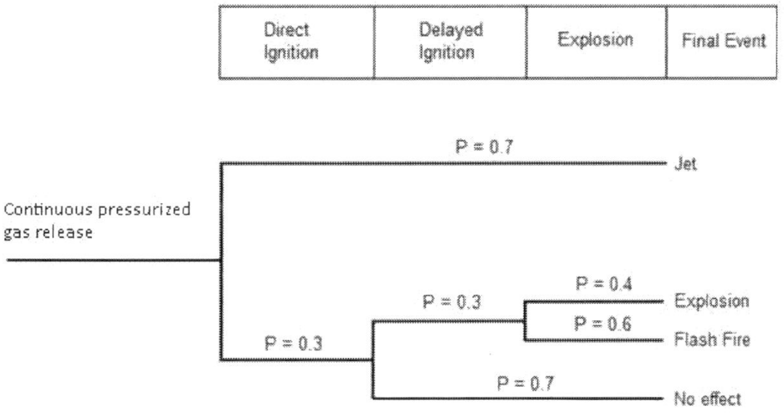

3. Experimental design

The equations that describe the frequencies of the different accidents will be developed for storage terminals of pressurized liquefied propane that use horizontal cylinders. The equations will be functions of the number of tanks used and the total mass of propane stored in the terminal. The number of tanks will be varied from 1 to 4, while 5 values of stored mass will be used: 5,000 kg, 10,000 kg, 20,000 kg, 50,000 kg and 100,000 kg; these possibilities for number of equipment and stored mass will be used to develop different layouts (for example 3 tanks which store 50,000 kg of propane) in which the total mass stored will be equally divided amongst all the tanks (all of which will have the same dimensions). The propane will be said to be stored at the conditions presented in Table 3.

Table 2: Threshold values for tank damage and associated LOCs

Accident	Effect	Threshold Values		Consequence
		Pressurized Tanks	Atmospheric Tanks	
BLEVE	Thermal radiation (kW/m2)	≥ 100	≥ 100 [5]	G.3 type release
		< 100	< 100 [5]	No consequence
Explosion	Overpressure wave (kPa)	≥ 100	≥ 25 [6]	G.1 type release
		$53 < \Delta P < 100$ [7]	$20 < \Delta P < 25$ [7]	G.2 type release
		$30 \leq \Delta P < 53$ [7]	$14 \leq \Delta P < 20$ [7]	G.3 type release
		$\Delta P < 30$ [5]	$\Delta P < 14$ [8]	No consequence
Jet fire	Jet length (m) Heat Radiation (kW/m2)	$dj \geq d$ [5]	$dj \geq d$ [5]	G.1 type release
		$dj < d$ & $I \geq 40$ [5]	$dj < d$ & $I \geq 10$ [5]	G.3 type release
		$dj < d$ & $I < 40$ [5]	$dj < d$ & $I < 10$ [5]	No consequence
Flash fire	-	-	-	No consequence
Pool fire	Thermal radiation (kW/m2) Pool duration (min)	$dp \geq d$ [5]	$dp \geq d$ [5]	G.1 type release
		$60 \leq I$ & $tp \geq 10$ & $dj < d$ [9]	$15 \leq I$ & $tp \geq 10$ & $dj < d$ [9]	G.2 type release
		$I \geq 8$ & $dj < d$	$I \geq 8$ & $dj < d$	G.3 type release
		$I < 8$ & $dj < d$	$I < 8$ & $dj < d$	No consequence
dj = jet length; dp = pool diameter; d = distance between tanks; tj = duration of jet fire; tp = duration of pool fire				

As the frequency of the accidents will depend, not only on the number of tanks and the mass stored, but also on the distances between the units, this parameter has also been taken into account; this was done by using the distances proposed in the NFPA58 [3], which depend on the dimensions of the tank.

Figure 3 shows an example of the used layouts. In it, L is the length of the cylinder, while ϕ stands for the diameter; B represents the sum of the diameters of the tanks and the distances between them (d). The complete width of the facility will come from multiplying B by a factor of 1.1, and its length, from multiplying L by an equal factor.

Table 3: Atmospheric and Storage Conditions Used for the Application of the Model

Parameter	Value
Ambient Temperature (K)	298.15
Relative Humidity (%)	70
Atmospheric Stability Class	D
Wind velocity (m/s)	5
Ground Roughness Coefficient (cm)	10
Tank Pressure (bar)	9.51
Tank Temperature (K)	298.15

Since the substance stored in the facility is a liquefied pressurized flammable gas, the LOCs presented in Table 1, as defined in The Purple Book are used. The base frequencies of the different accidents for each tank are obtained by using the event trees presented in Figures 1 and 2 for the instantaneous and continuous releases.

Figure 3. Layout example

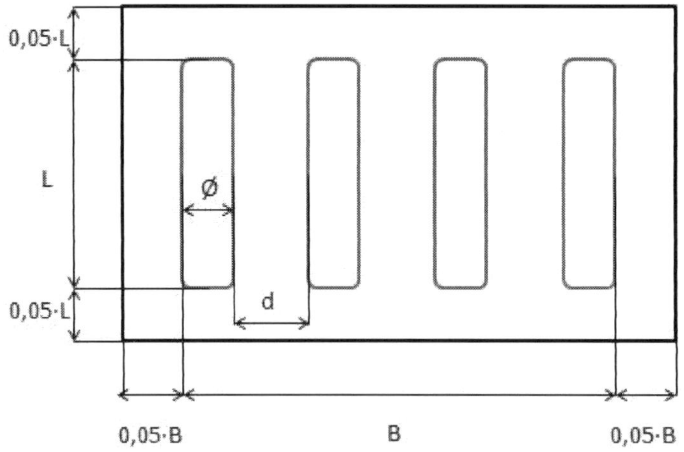

4. Results

It was necessary to use the model a total of 180 times to develop accident sequences for every possible initial accident, combining the number of tanks and values of stored mass. However, since all the proposed layouts are symmetrical, it was only necessary to perform 90 simulations.

Figure 4. G.1 accident sequences for two tanks storing 50,000 kg each

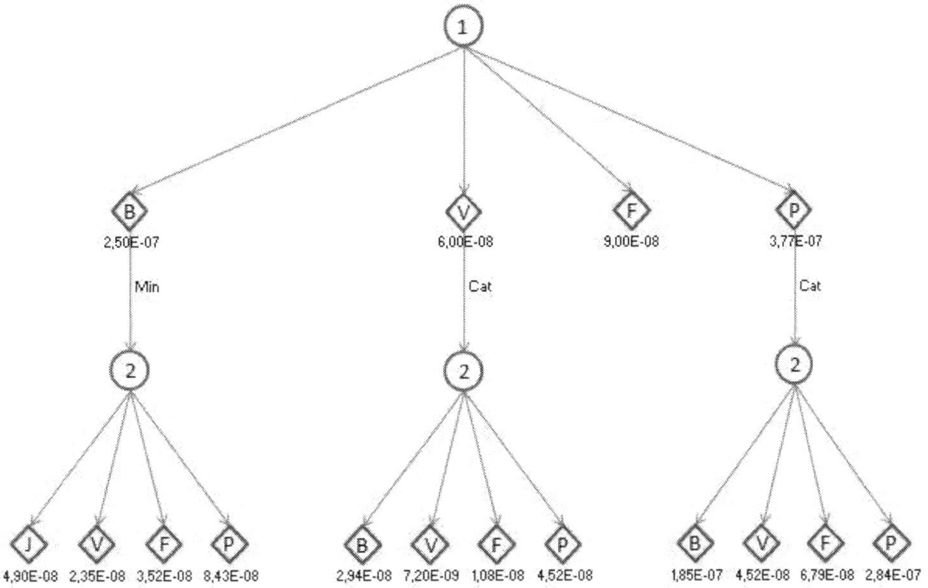

Figure 4 presents the possible accident sequences obtained by applying the model for the case of an instantaneous release from one of two tanks, each storing 50,000 kg of propane. In it, it can be seen that four possible accidents can occur after the G.1 release, a BLEVE (B), a VCE (V), a flash fire (F) or a pool fire (P); the frequencies with which these accidents occur is presented, as well as the level of damage they cause on the second tank, and the possible accidents that may occur in it, due to the initial ones.

The final values of frequency obtained for the case of BLEVE, depending on the mass and number of tanks are presented in Table 4. They serve as examples of the complete results obtained when the model is applied.

After the final frequencies for all possible combinations of design variables have been obtained, they can be plotted in 3D graphs, (in which one of the axis represents the number of tanks, another, the stored mass, and the last one, the frequency of occurrence). The plots obtained for all the studied accidents are presented in Figures 5 through 7.

Table 4: Final frequencies for the BLEVE case

Mass (kg)	Number of Tanks			
	1	2	3	4
5,000	$1.75x10^{-7}$	$1.90x10^{-6}$	$5.77x10^{-6}$	$1.04x10^{-5}$
10,000	$1.75x10^{-7}$	$1.90x10^{-6}$	$4.12x10^{-6}$	$6.89x10^{-6}$
20,000	$2.45x10^{-7}$	$3.44x10^{-6}$	$5.13x10^{-6}$	$6.87x10^{-6}$
50,000	$2.45x10^{-7}$	$3.44x10^{-6}$	$9.95x10^{-6}$	$2.21x10^{-5}$
100,000	$2.45x10^{-7}$	$3.44x10^{-6}$	$1.18x10^{-5}$	$2.10x10^{-5}$

Figure 5. Frequency vs. Number of tanks and Stored Mass for G.1 accidents

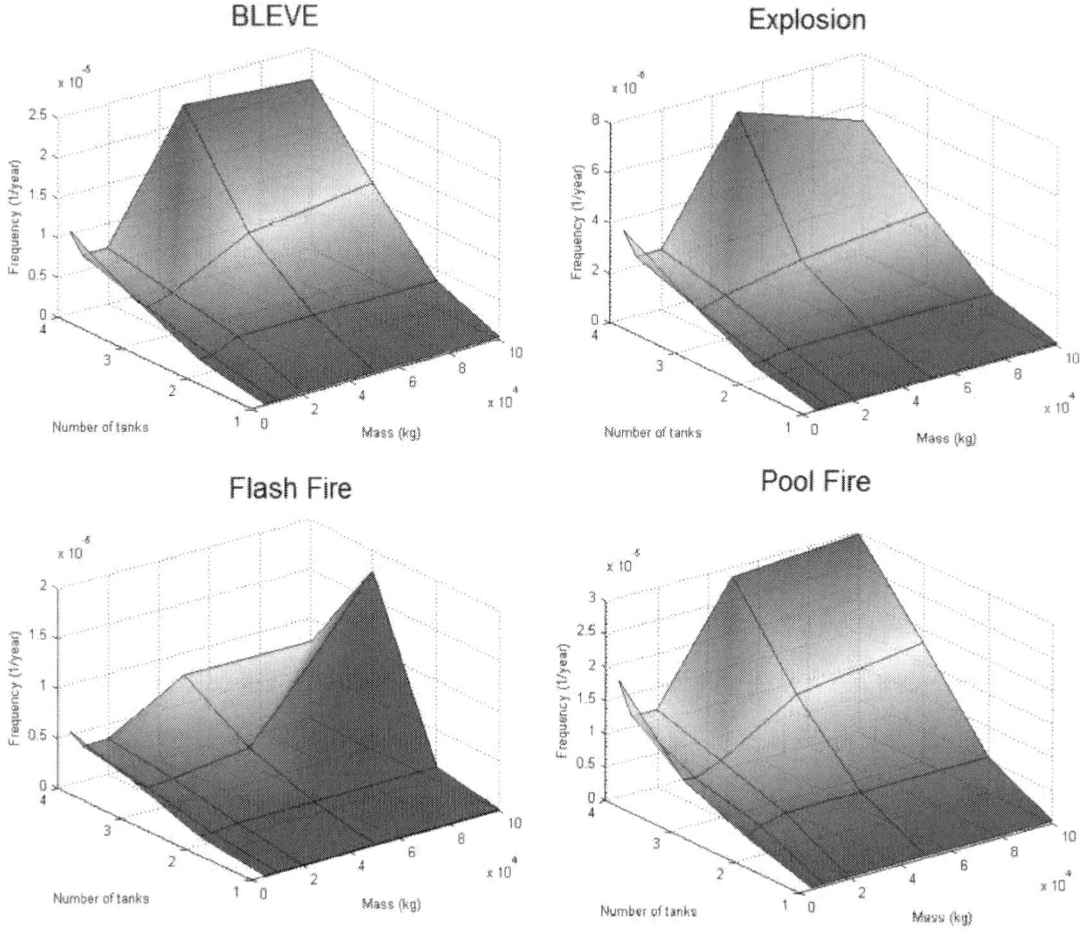

After the plots were obtained, they were fitted using different types of equations, like power, exponential or logarithmic among others. Finally, a Gaussian equation of the form presented in Eq. 2 proved to be the best fitting.

$$f = A \cdot exp\left(\frac{(n-B)^2}{C} + \frac{(m-D)^2}{E}\right) \quad \text{[Eq. 2]}$$

Where n is the number of tanks, m the stored mass in kg and A, B, C, D and E are dimensionless coefficients; these coefficients for all the studied accidents, as well as the values of R^2 obtained for the fitting are presented in Table 5, and the plots obtained when using the Gaussian equation with the presented coefficients are shown in Figures 8 through 10.

Figure 6. Frequency vs. Number of tanks and Stored Mass for G.2 accidents

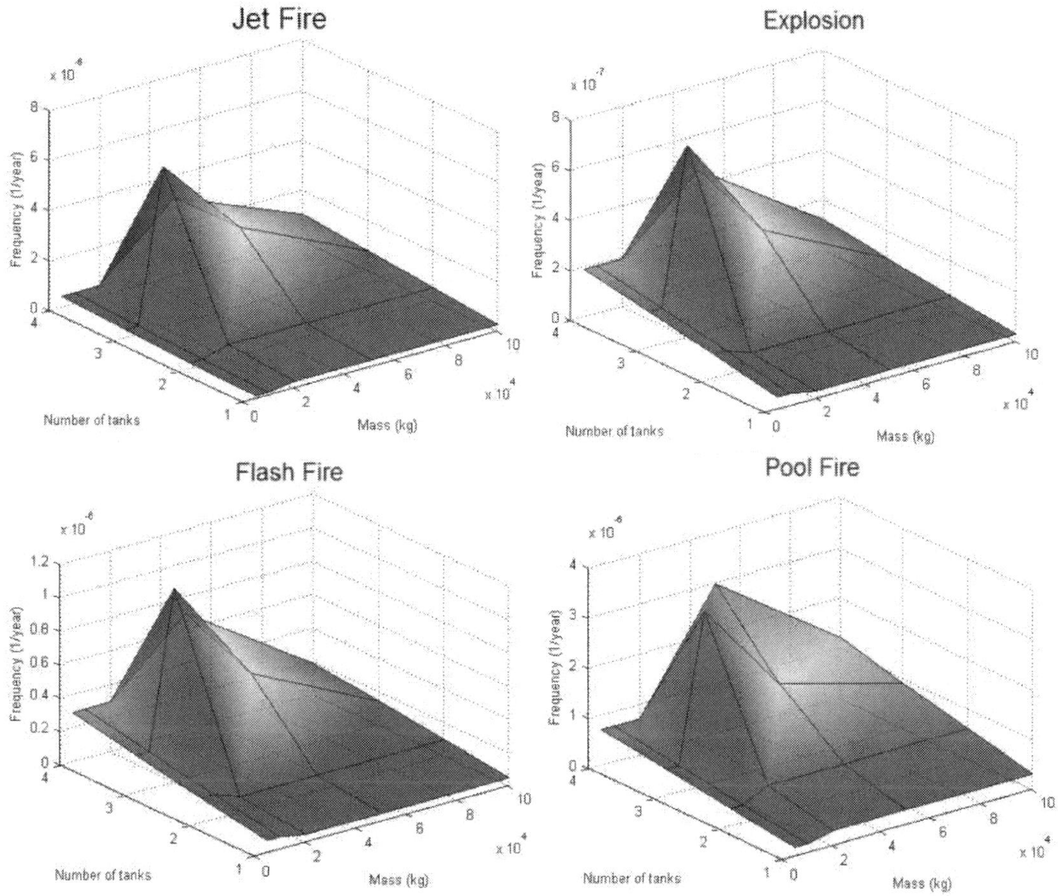

Table 5: Coefficients for the Gaussian equation for all accidents

Accident	Coefficients					R2
	A	B	C	D	E	
BLEVE	2.8x10-5	4.79	-4.15	7.7x104	-4x109	**0.94**
G.1 Explosion	9.9x10-6	5.37	-5.10	7.0x104	-4x109	**0.91**
G.1 Flash Fire	2.8x10-2	3.33	-0.8	1.0x102	-1.2x1011	**0.85**
G.1 Pool Fire	3.4x10-5	4.57	-3.96	8.3x104	-6.3x109	**0.95**
G.2 Jet Fire	2.2x10-5	3.14	-0.56	3.4x104	-1.5x108	**0.84**
G.2 Explosion	1.1x10-6	3.28	-0.82	3.3x104	-3.9x108	**0.77**
G.2 Flash Fire	1.6x10-6	3.27	-0.82	3.3x104	-3.9x108	**0.77**
G.2 Pool Fire	2.7x10-6	3.53	-2.17	5.5x104	-2.3x109	**0.57**
G.3 Jet Fire	1.5x10-4	6.13	-0.15	2.2x106	-1.8x1012	**0.98**
G.3 Explosion	7.8x10-5	6.04	-0.15	2.3x106	-1.9x1012	**0.99**
G.3 Flash Fire	9.2x10-5	6.10	-0.15	2.1x106	-1.7x1012	**0.98**
G.3 Pool Fire	2.2x10-4	6.10	-0.15	2.1x106	-1.7x1012	**0.99**

Figure 7. Frequency vs. Number of tanks and Stored Mass for G.3 accidents

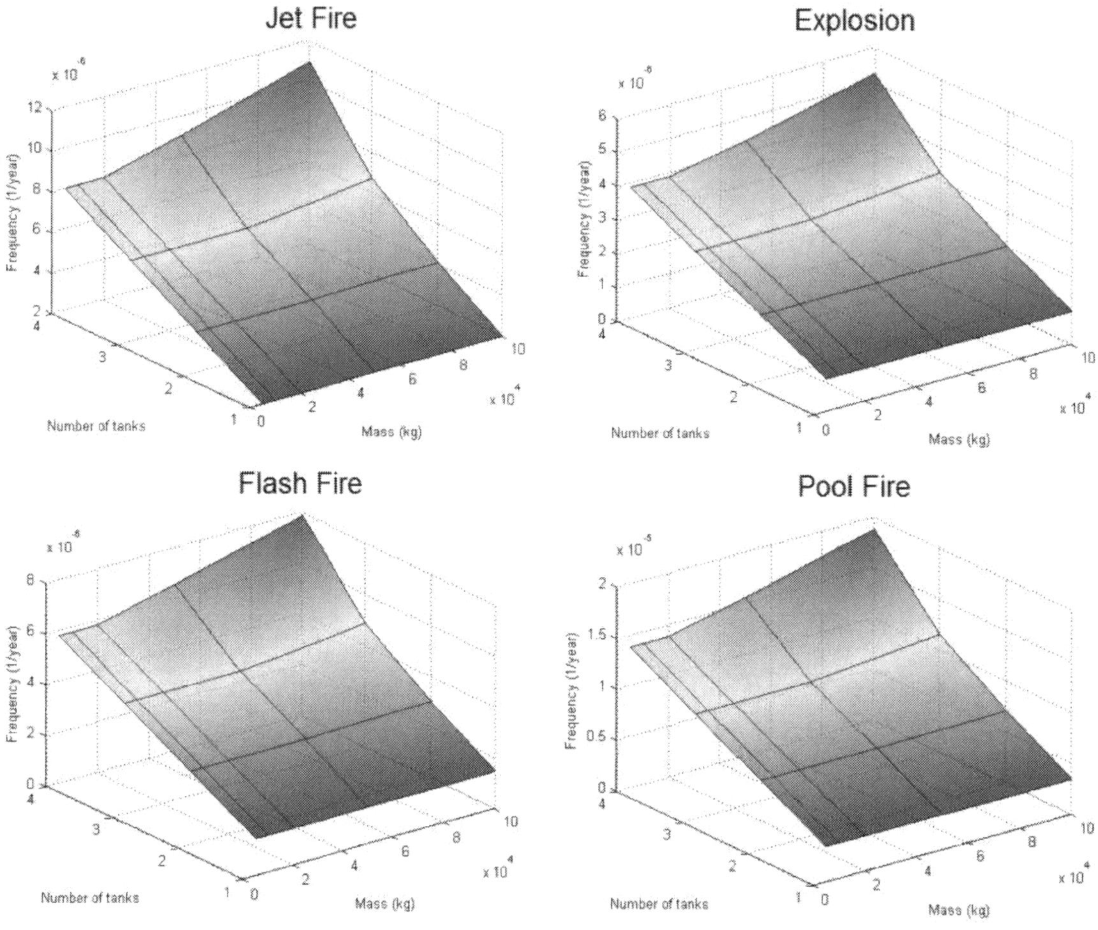

5. Conclusions

Equations that depend on the number of tanks and mass stored in the installation have been found, which, in many of the cases, present good tendencies in replicating the data obtained in the experiment.

The equations obtained could be used to take domino effect into account when performing a QRA on a storage installation of characteristics similar to those presented in the experiment, without increasing significantly the time required to complete the study.

Also, equations of this kind could be used during the design stages of a storage installation to help decide the number of equipment and quantities of substance stored; they could be used to know how risk behaves depending on different designs, and which design is inherently safer.

Figure 8. Frequency vs. Number of tanks and Stored Mass for G.1 accidents using the Gaussian Equation

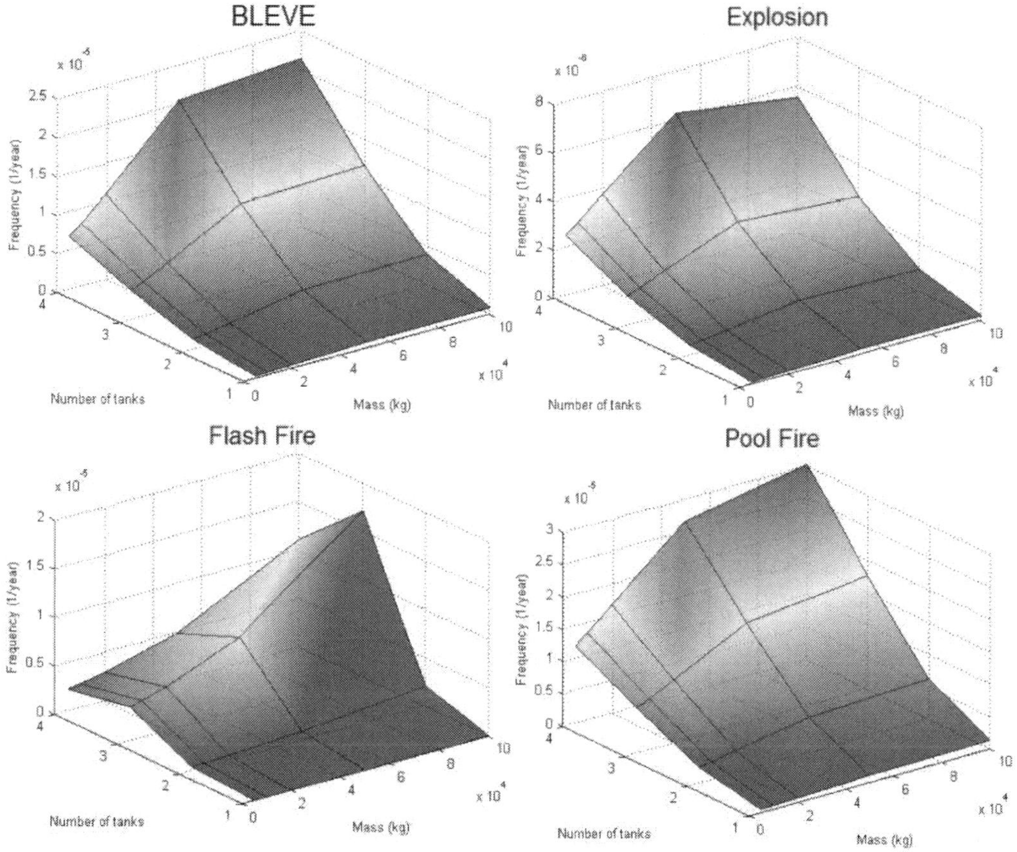

Figure 9. Frequency vs. Number of tanks and Stored Mass for G.2 accidents using the Gaussian Equation

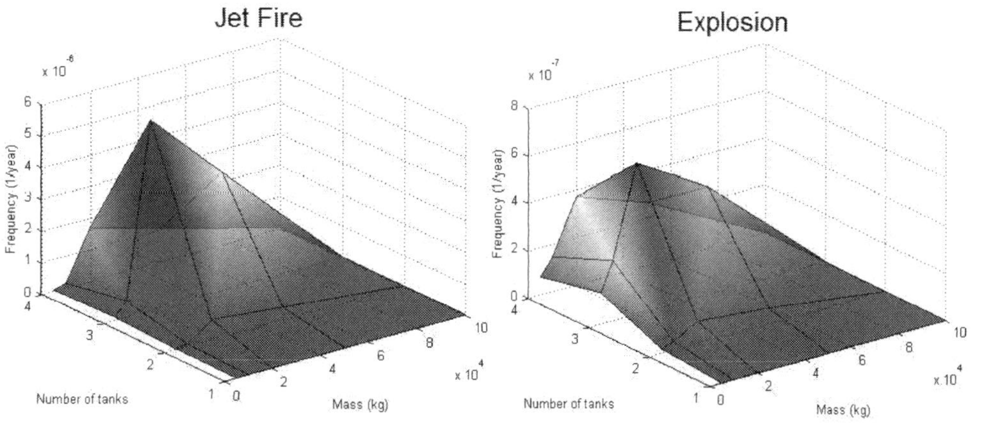

Figure 9. Frequency vs. Number of tanks and Stored Mass for G.2 accidents using the Gaussian Equation (Cont.)

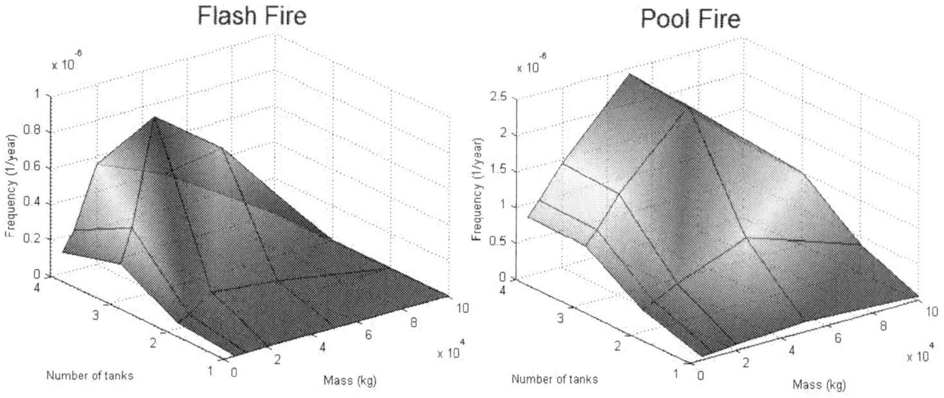

Figure 10. Frequency vs. Number of tanks and Stored Mass for G.3 accidents using the Gaussian Equation

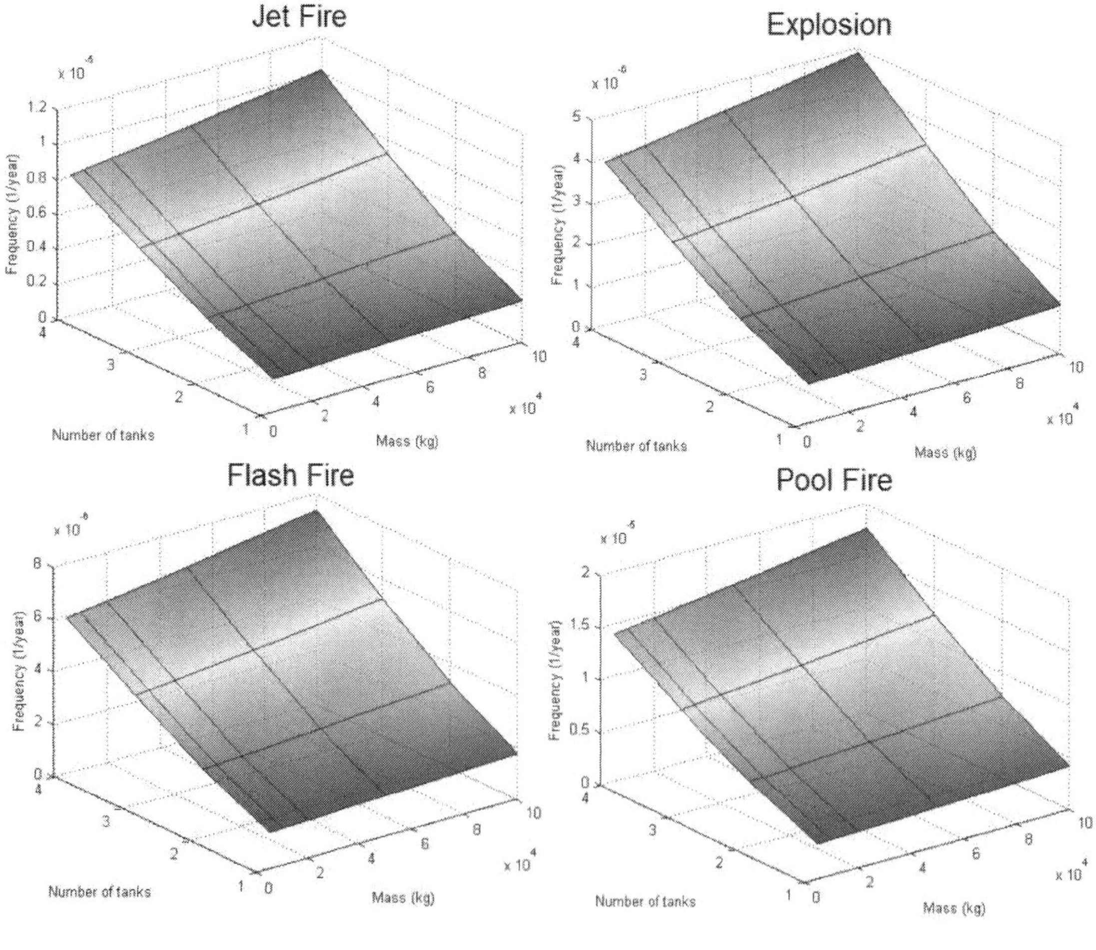

References

[1] CPR 18E, 2005. Guidelines for Quantitative Risk Assessment (Purple Book). Directorate General for Social Affairs and Employment, The Hague, Netherlands.

[2] Delvosalle, C, "Domino Effects Phenomena: Definition, Overview and Classification", First European Seminar on Domino Effects, Leuven, 1996.

[3] "Liquified Petroleum Gas Code". National Fire Protection Association (EEUU), 2001.

[4] Bernechea, E.J., Vílchez, J.A., Arnaldos, J., 2012. A model for estimating the impact of the domino effect on accident frequencies in quantitative risk assessments of storage facilities. Process Safety and Environmental Protection, http://dx.doi.org/10.1016/j.psep.2012.09.004

[5] G. Antonioni, G. Spadoni, V. Cozzani, Application of domino effect quantitative risk assessment to an extended industrial area, Journal of Loss Prevention in the Process Industries, 22, 614-624, 2009.

[6] F. I. Khan, S. A. Abassi, Estimation of probabilities and likely consequences of a chain of accidents (domino effect) in Manali Industrial Complex, Journal of Cleaner Production, 9, 493-508, 2001.

[7] V. Cozzani, E. Salzano, The quantitative assessment of domino effects caused by overpressure: Part I. Probit models, Journal of Hazardous Materials, 107, 67-80, 2004.

[8] V. Cozzani, E. Salzano, Threshold Values for domino effects caused by blast wave interaction with process equipment, Journal of Loss Prevention in the Process Industries, 17, 437-447, 2004.

[9] G. Landucci, G. Gubinelli, G. Antonioni, V. Cozzani, The assessment of the damage probability of storage tanks in domino events triggered by fire, Accident Analysis & Prevention, 41, 1206-1215, 2009.

(44ce) Routes Optimization for Hazardous Materials Transportation in Bogotá, Colombia

Jorge Victoria, Universidad de los Andes, Bogota, Colombia, Nubia Velasco, Universidad de los Andes, Eliecer Gutierrez, Industrial Engineer, Universidad de los Andes, Bogota, Colombia and **Felipe Munoz**, Universidad de los Andes, Bogota D. C., Colombia

The present work presents an optimization approach to design the set of routes for the transport of hazardous materials (HazMats) at Bogotá. HazMats transportation importance lies in the social, environmental and economic impact that might happen when an accident, during the transportation occurs. This paper is focused in developing a decision support system supported on optimization tools with the purpose of find the routes, -from gas distributor to gas stations-, that minimize: the total travel distance of heterogeneous tank trucks, the social risk and the infrastructure risk around the routes created.

The problem is modeled as a vehicle routing problem and it is solved using the principle of Route first, Cluster second. In a first stage the algorithm generates a big tour where all clients are visited, ignoring the vehicle capacities. Secondly, thanks to a Split procedure the tour is segmented considering the capacities. Then routes are improved with a MultiObjective Metaheuristic called MultiObjective Variable Neighborhood Search (MO-VNS). The algorithm is tested on two set of instances, the first one randomly generated and the second from Bogotá Geographic Information System (GIS). Several approaches to generated solution are proposed.

(44cf) Mejores Prácticas En Los Procesos De Gestión Del Cambio Para Seguridad De Procesos y La Evaluación De Seguridad Previa Al Inicio (PSSR)

J. Luis Aguirre, KMS, Optimized Systems and Solutions, Houston, TX

La Gestión del Cambio continúa siendo el elemento más retador/complicado de la Gestión de Seguridad de Procesos (PSM), del Programa de Gestión de Riesgos (RMP) y de los requerimientos regulatorios de los Sistemas de Gestión de Medio Ambiente y Seguridad (SEMS). A pesar de que el estándar PSM ha existido desde 1992 y que la industria ha gestionando cambios desde hace mucho tiempo, aún es común no obtener el resultado deseado. Los esfuerzos conscientes de muchas personas trabajando en conjunto son necesarios para garantizar que el cambio sea identificado, analizado y ejecutado con calidad. Justo cuando logramos capturarlo en papel y mantener informado al equipo de trabajo involucrado, se presentan cambios, negligencias o las regulaciones cambian. Nuevos involucrado y responsabilidades dificultan la gestión del cambio como sabemos que debe hacerse.

Sin embargo, la tecnología ayuda en todas las etapas del flujo de la Gestión del Cambio (MOC) al permitir la recopilación de ideas, formalizando el análisis de las mismas, su aprobación y ejecución del cambio. También puede ayudar a proporcionar visibilidad del estado actual y análisis de tendencias, así como la identificación de peligros a los que están expuestos los diferentes equipos, sitios, y bienes. Correctamente aplicada, la tecnología podría ayudar a predecir el futuro. Además, la tecnología facilita y ayuda a implementar sus estándares resultando en Mejores Prácticas de Gestión del Cambio, Evaluación de Seguridad Previa al Inicio (PSSR) y el seguimiento al cumplimiento de acciones, notificaciones de alerta y el cierre.

(44cg) Lesha – Multi-Layer Progressive Stress & Impact Assessment On Health & Behavior

Salvador Ávila Filho, PEI TECLIM, UFBA, Salvador, Brazil

The chemical process industry seeks, constantly, to achieve the demands social and environmental (mandatory and voluntary) within the international economic chain. The Integrated Management Systems meet issues in the areas of quality, environment, safety, health, social responsibility, now including risk management (ISO31000). This excess of valuation in integrated management system just prioritizes management functions in place of the functions related to core activity. By integrating management activities into the routine of work, the task will be performed in a more committed form avoiding the domain of management system in place of production skills.

Mandatory Standards for work in the areas of health, hygiene and safety are concerned in dealing the most damage, searching through punishment, correct managerial failure, or correct the ignorance. The preventive tools are not fully developed leading to leaders to be worried about maximum allowable values in the areas of safety, health, hygiene, and also the environment.

The Voluntary activities are not developed due to difficulties initiating deployment of management systems that were born in a TOP-DOWN philosophy. The human and operational reliability was borned in a shape BOTTOM-UP avoiding the occupation of operator, most part of time, doing management systems reports. Operational risks are analyzed including the role of man, the role of group in the planning and execution of tasks. The techniques involve subjectivities as effect of cultures fight in the body of work(social - mass culture, regional culture, organizational, and security). The analysis of dominant cultures involves identifying the various vectors impacting cultural influences in the workplace. Psychological stress causes physical and behavioral changes (affecting the decision) and in the body, both, can cause the loss of quality of life.

Various techniques were developed and presented in GCPS to assess the hazard and the risk of social conflicts and that can cause human error. This error initiates or increases the process of failure in equipment or processes through the intervention performed on the steps of the task. Thus, social conflict in new business (APRSH), cognitive traps and human error in critical task (Social HAZOP), causality analysis of cognition / decision, task and process losses (COGNOSP), analysis of organizational and cultural factors as root cause for accidents, decision analysis operation in progressive stress (LODA), and analysis of cognitive failure of operator in the control of equipment and components (FMEAH).

At the present moment we present a technique that try to study the effect of progressive psychological and physical stress on the body and quality of life, here dubbed as LESHA (Stress Analysis & Progressive Effect on Human Health and Behavior). This study highlights the importance of analyzing the life cycle of faults that impact on health by studying the trend of biological indicators resulting from psychological and physical stress. We also emphasize the search for signs of failure before occupational diseases being installed, showing the most suitable installation of barriers against failure, bringing the barrier closer to the root cause.

The LESHA technique therefore intends to monitor the level of physical and psychological stress of environmental factors, and stress from events related to task, which can turn a pathogen resident in symptom (Holnagel and Reason) and hence in occupational disease.

The events that initiate failures, and the stressed environments can cause stress progressive state on individuals, activating, or causing the trigger of failure through the stages: pathogen resident; uncontrolled organic process; symptom; disease; replication of the disease; and fatality or permanent labor shortage. Each level of stress is analyzed for signs and / or indicators of failure that impact over worker health permitting taking preventive or corrective action (biological safety barriers).

The following levels of analysis are proposed: Level 1 - Identification of Human Type, and the respective pathogen resident or signs that may cause the future disease in the body and in the behavior. Level 2 - initiators of failure in the task, or, bad social relationship that can cause the lack/ fault of organic metabolism, generating growth signals, variables still toward the disease. Signs of lost productivity interlinked with organic symptoms ... Level 3 - Shares of the social environment (society) or technical (technical knowledge), and secondary causal events and their impact on the body by increasing the metabolism of organic lack/ fault of symptom onset with fixed initial stage of the disease. Signs of lost productivity interlinked with organic symptoms...

Level 4 - Emergence of social ills (irritation, and communication difficulties), emergence of occupational diseases (slips related to the task and can cause accidents), and biological diseases like ulcer, high blood pressure and other diseases related to workplace using indirect biological ways (early onset of diabetes or other). Level 5 - Replication of the disease by adopting bad habits in the workplace where colleagues, "introject" behavior and even biological processes of "friends." Level 6 - fatality or permanent labor shortage causing low morale in the team, and developing or activating resident pathogen not active or nonexistent.

Understand the signs of this biological process is challenge of safety, hygiene and occupational medicine in conjunction with studies of human reliability. Different disciplines come together to prevent the disease from the body (regardless of the root cause) had been evident. The diseases are originate from labor, directly, or resulting from the influence of the job, and, from society influence. One should avoid firing the trigger of failure and the development of the pathogen resident.

The result of this analysis allows build the programs to avoid diseases. Prevention and correction programs depend on the intensity of the interference signals and the stress level reached in this process. It is proposed for each level and intensity of the biological and behavior signals following through risk mapping:
* Intensity Level X frequency.

9TH GLOBAL CONGRESS ON PROCESS SAFETY

Building Design for Resilience Beyond Design Basis Blast Loads

Michael A. Polcyn
Baker Engineering and Risk Consultants, Inc.
3330 Oakwell Court, Suite 100
San Antonio, TX 78218
MPolcyn@BakerRisk.com

Raymond H. Bennett, Ph.D., P.E.
Baker Engineering and Risk Consultants, Inc.
3330 Oakwell Court, Suite 100
San Antonio, TX 78218
RBennett@BakerRisk.com

Prepared for Presentation at
American Institute of Chemical Engineers
2013 Spring Meeting
9th Global Congress on Process Safety
San Antonio, Texas
April 28 – May 1, 2013

UNPUBLISHED

Building Design for Resilience Beyond Design Basis Blast Loads

Michael A. Polcyn
Baker Engineering and Risk Consultants, Inc.
3330 Oakwell Court, Suite 100
San Antonio, TX 78218
MPolcyn@BakerRisk.com

Raymond H. Bennett, Ph.D., P.E.
Baker Engineering and Risk Consultants, Inc.

Keywords: blast resistant design, consequence-based design, risk-based design

Abstract

One of the guiding principles of API RP-752 is that buildings intended for occupancy are to be designed, constructed, installed, modified, and maintained to protect occupants against explosion, fire, and toxic material releases. Owners/operators may use either a consequence-based approach or a risk-based approach to evaluate the siting of both new and existing buildings for blast loads. While a consequence-based approach should be based on a maximum credible event (MCE) for each building and type of hazard considered, a risk-based approach may consider numerous scenarios, accounting for both the consequences and the frequencies of the scenarios. However, when designing either new buildings or upgrades to existing buildings, designers are typically provided only a single set of blast loads and a single response criteria (for example, medium response per ASCE "Design of Blast-Resistant Buildings in Petrochemical Facilities"). A lower response level implies lower occupant vulnerability, while higher response levels imply higher occupant vulnerability. A design for higher magnitude loads to low response criteria would result in very conservative design. On the other hand, a design for lower magnitude loads to higher response limits would result in a building without capacity to protect occupants in the event of a more severe explosion.

A better approach would be to consider multiple levels of blast loads when developing a design. Lower magnitude loads (usually from higher probability events) should require designs to meet the lower response limits, while higher magnitude loads (usually from lower probability events) should require designs to meet the higher response limits. This paper demonstrates this approach by evaluating typically designed building systems for a range of blast loads commonly encountered at refining and chemical plants. Recommendations are provided for the design of new buildings or upgrades to existing buildings. By including the concept of resilience in the design process, the designer may provide protection from future yet-to-be-determined hazards or scenarios.

1 Introduction

The structural design of buildings for conventional loads, including wind, snow, and seismic loads, is a well-defined process that is followed by engineers on a routine basis. Loads are specified by various codes or design manuals (e.g., International Building Code [1] or ASCE 7-05 [2]). Similarly, the design approach for determining the capacity of various types of structures is defined in corresponding design manuals. For example, the American Concrete Institute publishes "Building Code Requirements for Structural Concrete" [3], and the American Institute of Steel Construction publishes the "Manual for Steel Construction" [4].

In contrast, the design of buildings subjected to blast loads is not as formalized. Design case blast loads are not specified in any guidance documents and are left for the owner to define appropriate loads based on the processes and materials in place at their facilities. There are several guidance documents that describe approaches for designing and analyzing structures subjected to blast loads, which include dynamic analysis methodology, material properties, and structural response limits. Response or deformation limits, such as those defined by the American Society of Civil Engineers [6], are given ranges of response, where Low Response refers to a component with no more than slight visible permanent damage, Medium Response corresponds to a component with some permanent deflection, but is generally repairable, and High Response indicates that the component has not failed, but has undergone significant permanent deflections and is likely unrepairable following the design case loads. It is up to the owner to select the response level, and thus the level of protection to building occupants, for their building design.

This paper provides an overview of the design process, highlighting the differences between a consequence-based design approach and a risk-based design approach. The discussion continues by presenting features to the design that will enhance the structure's ability to be usable following an explosion event and to provide some degree of protection to occupants for events exceeding the selected design load cases. It should be noted that only explosion hazards are addressed in this paper; building and upgrade designs should also consider hazards from fire and toxics.

2 Dynamic Analysis Method

When designing buildings or building upgrades for blast loads, structural components are usually assessed using nonlinear dynamic Single-Degree-of-Freedom (SDOF) analytical methodology. The SDOF methodology summarized in Biggs [6] is a widely accepted approach for estimating damage to structural components subjected to time-varying blast loading. The SDOF approach is a government and industry accepted methodology that has been described in many guidelines such as UFC 3-340-01 [7], UFC 3-340-02 [8], and ASCE [5].

The dynamic response of structural components from prescribed blast loads is determined by modeling the components as an SDOF system, such as the equivalent spring-mass system shown in Figure 1. Structural components such as walls, windows, beams, doors, and panels will deform and vibrate when loaded with a shock wave. The SDOF model for each component is constructed using the component's physical structural properties so that the model will exhibit the same displacement history as the point of maximum deflection in the component. The displacement history of the SDOF model is obtained with numerical integration techniques using a time-stepping computer program to solve the equation of motion of the equivalent system at discrete time steps.

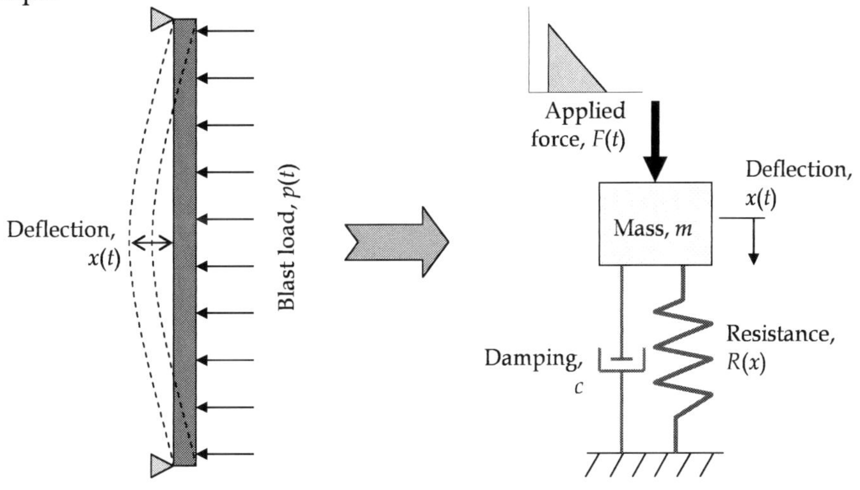

Figure 1. Equivalent Spring-Mass SDOF System

The calculated peak deflection (see Figure 2 for example) is used to obtain the support rotation and ductility ratio, which represent the deformation limit criteria most commonly used in blast design. As shown in Figure 2, the support rotation is the angle between the original un-deflected shape of a component and a straight-line segment between the point of maximum deflection and the support. The ductility ratio expresses the maximum deflection in terms of the minimum deflection of the component when it reaches its ultimate capacity. Therefore, ductility ratios that are greater than unity (1.0) indicate that permanent deformations have been sustained.

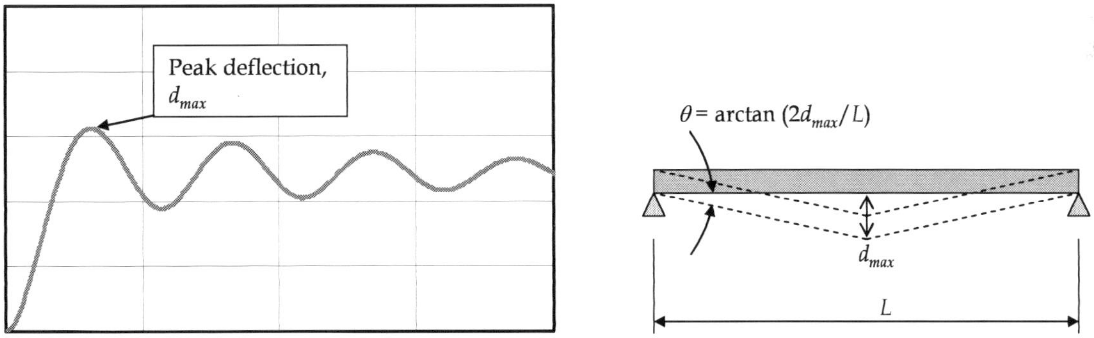

Figure 2. Calculation of Support Rotation from Peak Deflection

The design process for blast loads is an iterative process, where the pertinent design blast loads and response criteria are defined, "trial" structural members are selected based on the building geometry, and the structural components are analyzed using the dynamic analysis methods discussed above. If the response of the component does not meet the required response criteria, a heavier section is evaluated or the building geometry is modified until the analysis shows that the response criteria are met. Connections are typically designed to resist the maximum end reactions predicted from the dynamic analysis.

Pressure-impulse (P-i) diagrams are commonly used to assess building or component response to multiple sets of blast loads. P-i diagrams are plots of constant damage levels for various combinations of blast overpressure and impulse. Each individual curve represents the suite of all pressure and impulse pairs that will result in the same structural damage to a given component. The combination of the three individual curves forms a complete component P-i diagram to define Low, Medium, and High Damage Levels. An example of a P-i diagram is provided in Figure 3. As shown, the region above the upper curve would imply component failure.

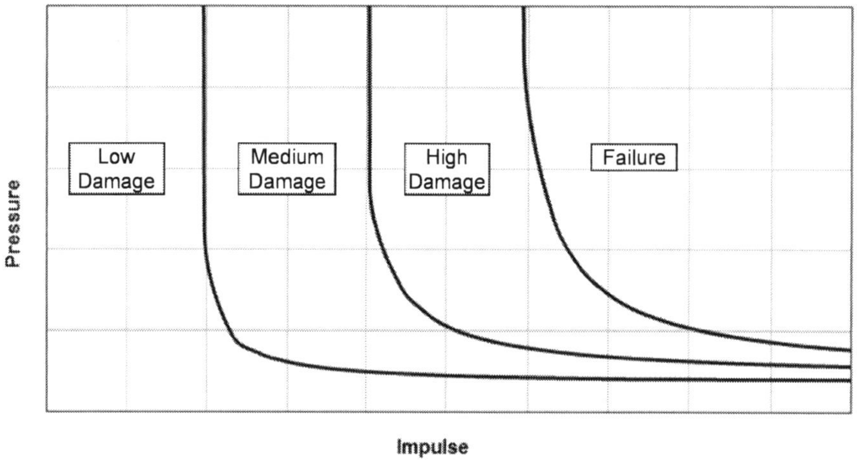

Figure 3. Example of a Pressure-Impulse Diagram

The use of P-i diagrams allows for quick consideration of a variety of blast loading situations. If the blast load of interest falls below or to the left of a particular damage curve, then the given component is predicted to receive damage less than or equal to the level represented by that P-i curve. In addition, component curves can be combined to develop a set of building P-i damage curves, or BDL curves.

3 Consequence-Based vs. Risk-Based Design

Traditionally, design of buildings or upgrade to buildings has been performed using a consequence-based approach where the frequency of events is not considered. According to API RP-752, a consequence-based approach is to be based on maximum credible events (MCEs) for each building. For example, a precast concrete wall for a control room could be designed to meet the ASCE Medium Response criteria when subjected to the applied blast loads from the MCE determined for that building. The resulting P-i curves for the wall would be similar to those shown in Figure 4.

The figure also demonstrates the concept of resiliency in the design. Four pressure-impulse pairs are shown, corresponding to two Design Load Cases and two Potential Overload Cases. The potential overload is 1 psi greater than the design pressure for both cases, while the impulse is held constant. In both Design Cases, the component just meets the Medium Response criteria. However, the curves show that for Design Load 1, the component has the capacity to resist the overload, albeit with larger deformations, but without failing (the component undergoes a high level of response). For low probability events, such behavior might be acceptable. However, the implied resiliency of the component cannot be determined solely by setting the response limits. For example, in Design Case 2, the corresponding overload will result in the failure of the component, even though the Design Case load is farther below the boundary between medium and high response. Thus the true resiliency of a design has to be evaluated in light of the potential for an overload.

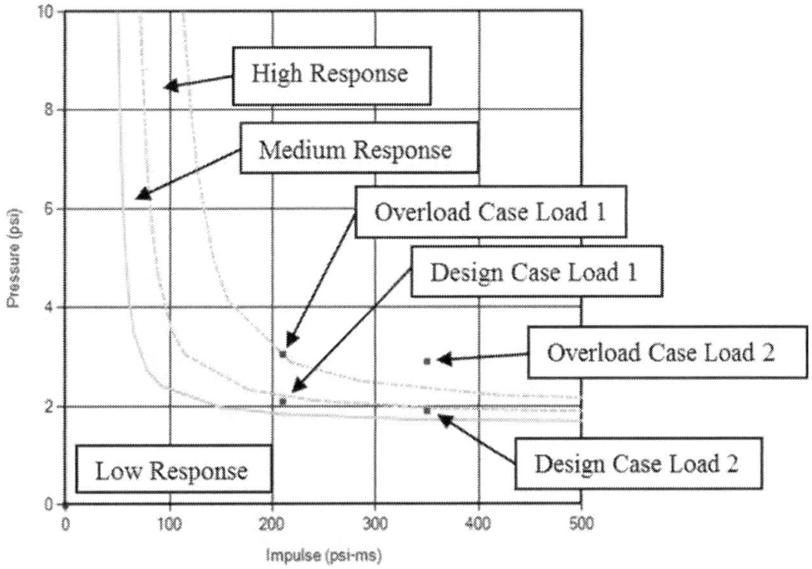

Figure 4. Component P-i Curves

In more recent years, companies have conducted quantitative risk assessments (QRAs) in support of facility siting study efforts. Such studies include estimates of the expected frequency and consequences of potential accidents, typically resulting in an estimate of risk, either as an aggregate risk, an individual risk, or a relationship of the expected frequency to the consequence (e.g., an F/N curve). This approach, which considers a large number of possible explosion scenarios, can be extended to a risk-based design approach for buildings or building upgrades.

When using a risk-based approach, it is necessary to consider the overall building damage rather than the response of the individual components independently. By doing so, the predicted building damage for each scenario can be related to the expected occupant vulnerability [9], which is used to perform the risk calculations. For example, the Building Damage Level (BDL) curves in Figure 5 show that the building is expected to undergo moderate damage, or BDL 2a, when the components are designed to meet the ASCE Medium Response criteria. Note that the inflection points in the BDL curves are caused by a change in the governing building component. For example, the upper left portion of the BDL 3-4 curve may be set by roof purlin response and the lower left portion may be set by a roof girder response. If the design case loads are defined by the MCE for all of the postulated explosion scenarios, the BDL plot in Figure 6 can be generated showing the loads from "all" postulated scenarios with the BDL curves. As shown, none of the scenarios will result in greater than moderate damage (BDL of 2a), and the resulting risk will be very low. It can be concluded that such a design has significant resiliency to withstand loads greater than those postulated.

Alternatively, the building could be designed for higher response limits under the design load cases, possibly allowing partial building failure under the most severe load cases. The BDL curves in Figure 7 demonstrate a case where the components were designed to meet the ASCE High Response limits for the MCE and ASCE Medium Response Limits for the majority of the load cases. As shown, the expected damage for the MCE approaches higher BDLs, but building collapse (BDL of 4) is not expected. In this case, the expected risk to building occupants would still be relatively low, although somewhat higher than the risk demonstrated in Figure 6. However, there is less resiliency in the design should there be an overload case. Note particularly that if the scenario shown in Figure 5 were 0.5 psi greater, the BDL for the building portrayed in Figure 6 would still be 2A, while the BDL for the design represented in Figure 7 would be increased to BDL of 3.

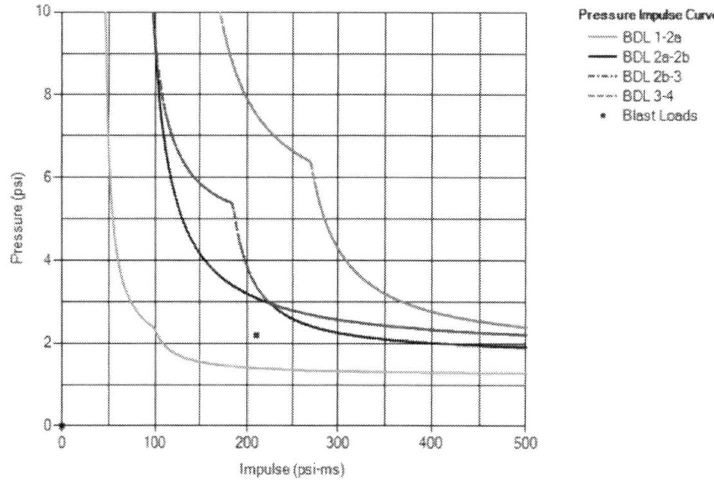

Figure 5. Building Damage Level (BDL) Curves

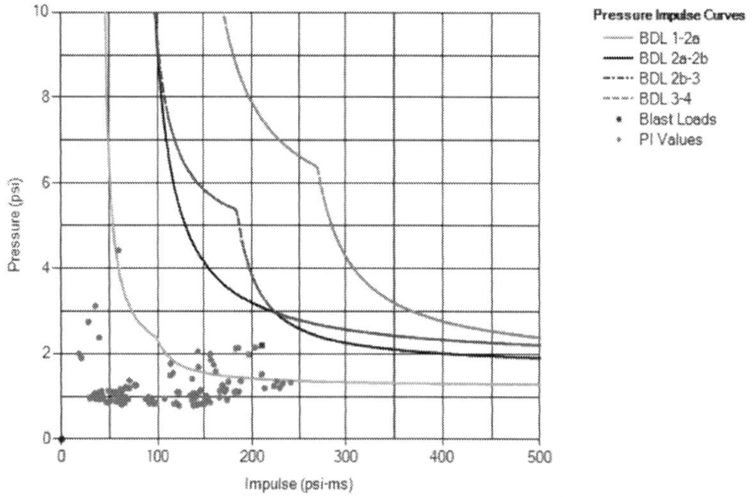

Figure 6. BDL Curves with Loads from "All" Postulated Scenarios

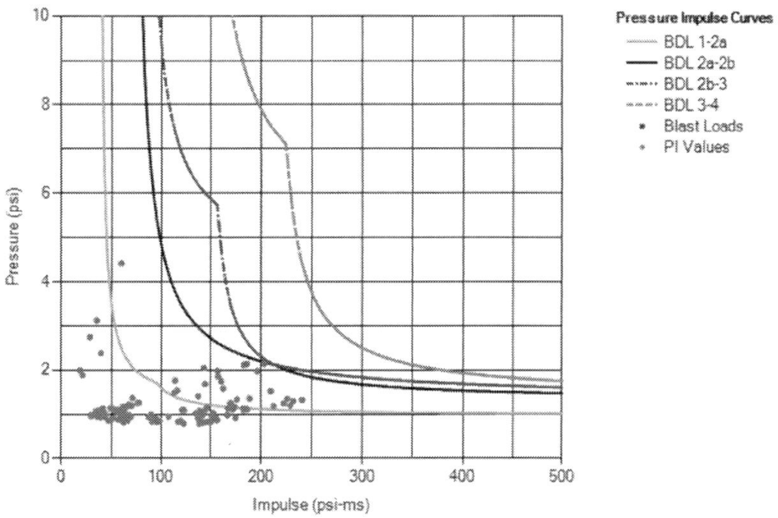

Figure 7. BDL Curves for "Weaker" Building

4 Resiliency in Design of Occupied Buildings

The examples discussed in the previous section demonstrate three approaches to providing a resilient structural design:

1. Structural systems that can develop a high degree of ductility are greatly preferred over structural systems that undergo a more brittle response. Most materials commonly used for blast resistant structures (steel and concrete) can undergo significant plastic deformations, typically well beyond those associated with published response limits, before failing. This plastic deformation allows the building to absorb the energy applied by the blast load. In addition, structural components designed to undergo large plastic deformations will have lower support reactions than those designed to behave elastically or with small plastic deformations, thus resulting in lower demands for connections and the building lateral load-carrying systems.

2. When selecting response limits, the likelihood of the selected design loads should be considered. Typically, high probability scenarios result in lower magnitude loads, while more severe load cases are the result of lower probability scenarios. This would imply that higher response limits could reasonably be used when designing for low probability, high magnitude load cases. Similarly, when designing for higher probability / lower magnitude loads, or loads with considerable uncertainty associated with their prediction, lower response limits should be considered.

3. Since one of the objectives of a blast resistant building is to protect building occupants, the layout of the building and the location of occupants relative to the sources of the potential explosions should be considered. By placing unoccupied areas of the building on sides facing the explosion source, localized failure of non-loadbearing components may be acceptable.

5 Conclusion

The design of blast resistant buildings does not follow an approach that is as prescriptive as the approach used for conventional design. Blast loads must be defined based on the conditions at the facility. Appropriate response limits must be selected for the components comprising the building. In addition, process conditions and building use may change to meet future requirements. Due to the uncertainty that is inherent in this design process, it is important to provide a building design that protects building occupants with a degree of resiliency. This can be developed by selecting the building structural system, building layout, and required response limits in a manner that is consistent with the postulated loads and the expected potential for these loads. In addition, the potential for more severe loads should also be considered in the design process.

6 References

[1] 2012 International Building Code, published by the International Code Council, Inc., May 2011.

[2] ASCE/SEI 7-05, "Minimum Design Loads for Buildings and Other Structures," published by the American Society of Civil Engineers, 2006.

[3] ACI 318-11, "Building Code Requirements for Structural Concrete and Commentary," reported by ACI Committee 318, published by the American Concrete Institute, 2011.

[4] "Steel Construction Manual," 14th Edition, Second Printing, published by the American Institute of Steel Construction.

[5] "Design of Blast Resistant Buildings in Petrochemical Facilities, Second Edition," Task Committee on Blast Resistant Design, American Society of Civil Engineers, 2010.

[6] Biggs, J., Introduction to Structural Dynamics, McGraw-Hill, Inc. New York, NY, 1964.

[7] UFC 3-340-01, "Design and Analysis of Hardened Structures to Conventional Weapon Effects," June 2002. (Formerly Technical Manual for Army TM 5-855-1, Air Force AFPAM 32-1147(I), Navy NAVFAC P-1080, and Defense Special Weapons Agency DAHSCWEMAN-97, August 1998.)

[8] "Structures to Resist the Effects of Accidental Explosions," UFC 3-340-02, 5 December 2008.

[9] Oswald, Charles J. and Baker, Quentin A., "Vulnerability Model for Occupants of Blast Damaged Buildings," presented at the 34th Annual Loss Prevention Symposium, March 6-8, 2000, Session 3.

(44g) Reliability Analysis of Man-Machine Response to an Emergency Situation Based On a Detailed-Action Plan

Najmeh Vaez, North America Risk, Environmental Resources Management (ERM), Houston, TX; Environmental Science and Safety Department, University of Houston Clear Lake, Houston, TX, Farshad Nourai, Aftab Imen Parto Consulting Engineers Co., Tehran, Iran, Yaneira Saud, North America Risk, ERM Southwest, Houston, TX and Suresh Yelisetty, Americas Risk Practice, Environmental Resource Management (ERM), Houston, TX

The primary emergency response actions in oil and gas industries are the functions of the process safety control system and the actions completed by the control room operators. The performance of the safety control system and the control room operating crew has a significant role in proper emergency response. To overcome each critical scenario, an emergency response operating procedure is provided with specific instructions, that in this research paper it is named a detailed action plan (DAP). This term is recommended for this purpose by the Center for Chemical Process Safety (CCPS). The reason of using a DAP is to reduce the probability of error in the emergency response of control room operators by providing scenario-specific written instructions. However, most of the DAP's are developed during design phase when there is no opportunity to test them through emergency drills. Therefore, a framework is proposed to analyze the reliability of the DAP's. This framework should be able to determine reliability of every single human and equipment action as well as the combined human-machine reliability, modeled by means of a Reliability Block Diagram to calculate the reliability of integrated automatic-operator emergency actions. The framework focuses on incorporating the operational and cognitive operator's errors in the process/equipment reliability analysis. For this purpose, a human reliability analysis method named SPAR-H was utilized. The proposed framework was applied for analyzing DAPs of an emergency well control operation in order to check and improve the applicability of this framework. The presented framework facilitates the DAP assessment process by performing task analysis modeling the detailed tasks, and completing a human-machine reliability analysis. Besides, the application of the presented framework, its findings and recommended actions enables a considerable improvement in the reliability of DAPs and therefore proves its efficiency and effectiveness.

(44i) Hazards of Gasoline Contamination – Case Study of a Tanker Truck Explosion Resulting From Switch Loading

Scott G. Davis, GexCon US, Bethesda, MD and Kees Van Wingerden, GexCon AS, Bergen, Norway

When a tanker truck is drained of a Class I liquid (e.g., gasoline) and refilled with a similar liquid, there is no flammable mixture on the surface of the rising fluid because Class I fluids create mixtures at their surface that are too fuel rich to burn. Similarly, if a tanker truck is drained of Class II liquid (e.g., diesel) and refilled with a Class II liquid, there is no flammable mixture created because Class II liquids do not produce ignitable mixtures at ordinary temperatures. An extremely hazardous condition occurs, however, if the liquids are ***switch loaded***, which is when a Class II or III liquid is loaded into a tank vehicle that previously contained a Class I liquid. Under these conditions, the atmosphere above the fluid surface can become ignitable. An explosion can occur if a spark occurs from the fluid surface to another object, igniting the flammable mixture in the headspace of the tank.

This paper describes a recent incident involving an explosion during the filling of a tank of a transport truck with diesel fuel. This particular tank was previously filled with gasoline and was not completely emptied of the gasoline prior to the switch loading. After only filling one-quarter of the programmed load, a spark ignited the flammable mixture within the headspace. The resulting explosion and overpressure caused significant damage to the truck and neighboring tanks. The study will address issues and lessons learned regarding spark promotion with the tank, reduction of static generation by the fluid being filled into the tank, and elimination of the flammable mixture within the tank prior to switch loading.

(44j) Is Your MOC Process Doing All That It Should Be?

William Fink and Carolina Herrera, Sage Environmental Consulting, L.P., Austin, TX

OSHA 1910.119 Process Safety Management of Highly Hazardous Chemicals was the driver for refining and chemical industry to implement Management of Change (MOC) programs some 20+ years ago. However, companies have now recognized the benefits of implementing and maintaining an effective MOC program beyond regulatory compliance. Effective MOC programs enhance a company's ability to respond to changing "market" or processing needs with up to date process safety information. Management of Change has even been incorporated into staffing levels and personnel expertise by some Companies.

Management of Change programs typically focus on changes that are "not in kind" changes. Significant effort is placed on proper evaluation of these changes and the potential impacts on the operation. This presentation emphasizes both the typical and non-typical areas of MOC that have been found to be overlooked. MOC programs rigorously evaluate the potential to introduce new hazards of the process such as fire, explosion, toxicity, and environmental impacts. Is that far enough to ensure that you are meeting all of your performance needs?

Focus on environmental impacts such as water contamination, air pollutants, and hazardous waste generation may not be going far enough. Gradual increases in highly hazardous material inventories may cross the lines of "threshold quantities" and introduce new OSHA or EPA requirements. Does your MOC program look at this potential? In addition, management of "in kind" changes may also not be keeping pace with the changes required by the EPA Leak Detection and Repair (LDAR) requirements or the Benzene Waste Operations NESHAP (BWON) requirements. These requirements as applied to MOC are further explained in this presentation. Several "gaps" in MOC programs will also be examined and suggestions for improvements provided.

(44l) Beyond DSC, Improved Thermal Hazards Screening Methodology

Peter J. Ralbovsky IV, Netzsch Instruments North America, LLC, Burlington, MA

Thermal Hazards Screening is a primary part of any chemical process safety program. Differential Scanning Calorimetry (DSC) is used extensively in studying the thermal properties of materials and is often used to screen for thermal hazards. This method is quick and inherently safe as it can be done with a small amount of material (typically milligrams). These two benefits cannot be understated as there may be hundreds of screening test which a lab may have to do in a given year and sometimes little or nothing is known about its thermal behavior, its toxicity, or the toxicity of the decomposition products. DSC also has some short comings as it is unable to measure pressure, which can be a significant hazard, even when the heat release is relatively low. Also it can be difficult to handle multiphase samples or samples which react together at room temperature.

The Multiple Mode Calorimeter (MMC) is an excellent tool to be used in conjunction with the DSC to provide a more comprehensive look at thermal hazards. The MMC uses samples sizes up to about 3 grams so it is possible to do injection and stirring. In the MMC temperature and pressure are recorded as a function of time. Sample energies measurement of both endotherms and exotherms is possible. Heat capacity can be calculated as a function of temperature. This method offers a quick and inexpensive way to perform isothermal tests which can be used to help design larger scale tests done in 2L+ reaction calorimeters or pilot plant operations.

The Multiple Mode Calorimeter consists of an electronic base and a detachable calorimeter assembly. There are multiple types of calorimeters available which can be operated by the electronic base. This allows the user to select the right calorimeter for the type of experiment that is needed. The calorimeters can also be tethered to the electronic base with an electronic cable. This allow for the calorimeter to be placed remotely from the electronics. The calorimeter can then be easily placed in the hood and/or behind explosion barriers or even placed inside a freezer for sub-ambient operations. This instrument has been described in detail elsewhere.

There are a number of calorimeters available; the best two for Thermal Hazards Screening is shown in Figure 1 & 2. The first calorimeter (Figure 3) is capable of operating as a low phi adiabatic calorimeter but more importantly, for thermal hazards screening, can run in constant power or constant temperature rate mode. The MMC external heater module does not run in adiabatic mode but can be operated in scanning mode. Both MMC modules incorporate the VariPhi technology and can run true isothermal calorimetry tests. These MMC modules can measure endotherms and exotherms as well as calculating heat capacity during those periods where no transitions or reactions are occurring. The MMC measures sample pressure as a function of time/temperature which is critical in screening for thermal hazards.

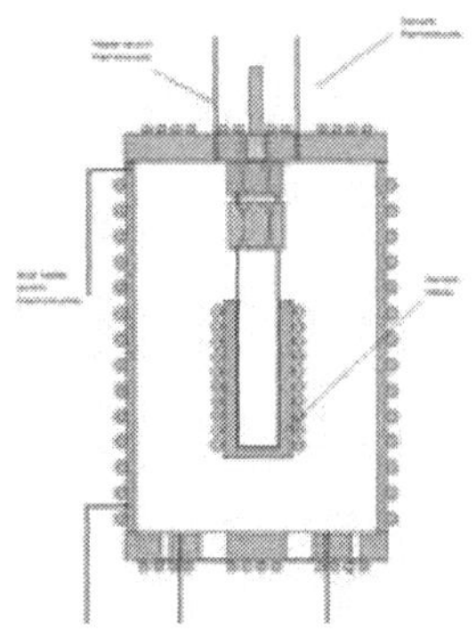

Figure 1

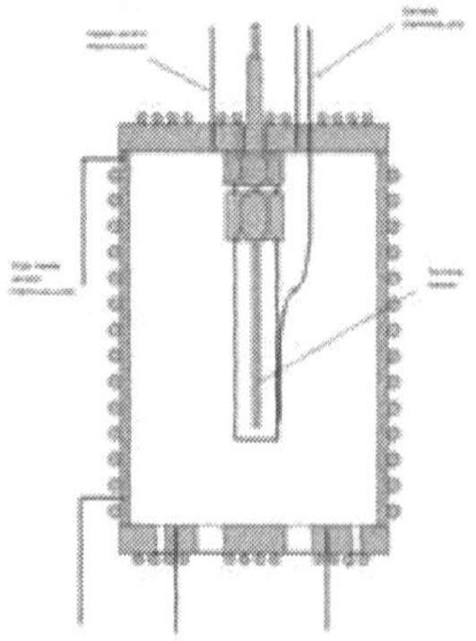

Figure 2

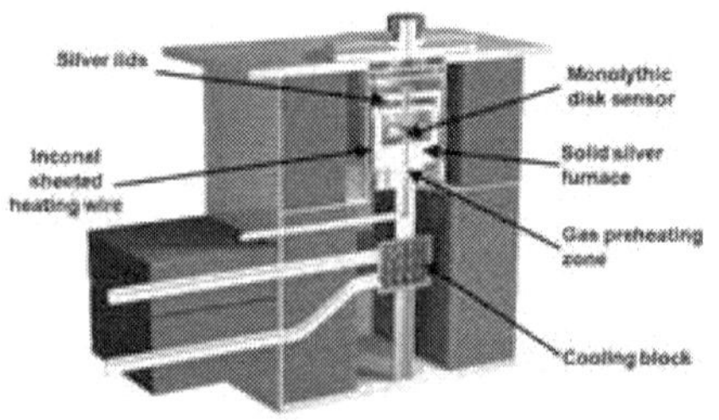

Figure 3

MMC VariPhi Calorimeter MMC External Heater Netzsch DSC 200 F3 *Maia®*

DSCs, like shown in Figures 3, have been used routinely as an initial step in the thermal hazards screening process. They have some unique advantages which make them suitable for this task. Scanning tests can be done very rapidly and with an auto-sampler they can be used to screen many samples efficiently. They sample size is small which makes them intrinsically safe in case of explosion or toxic gas production. These systems are small, inexpensive and have low operating costs. There are numerous DSCs available on the marketplace and they are used in many different applications. Two features are important from the thermal screening application. The first is an auto-sampler as discussed earlier. The second is the material of construction used in the DSC. It is likely that a DSC system used for thermal hazard screening will see explosions and the release of corrosive materials. DSC systems that can run hundreds of polymer melt tests without fail may have a limited life in the thermal hazards screening field. This is often unlooked. Figure 5 shows the cut-away of a robust DSC system.

Their continues to be a need to be able to safely and effectively screen thermal hazards for organizations involved in the manufacture, processing, storage or transportation of chemicals. Numerous screening tools are available in the marketplace, perhaps the most widely used is the Differential Scanning Calorimeter (DSC). This method has many advantages but the largest disadvantage is that the sample pressure is not measured. Thermal hazards are normally associated with pressure increases and these pressure increases are what ultimately provide the hazard to equipment and personnel. A storage drum does not melt due to a thermal runaway it ruptures due to the pressure increase because of the thermal runaway. Screening tools should be sensitive enough to see small exotherms that in near adiabatic conditions can fuel a runaway, they should be easy to use, provide quick results, and be as flexible as possible to handle different scenarios.

[1] Chippett, A New, Low Φ Factor, Scanning, Adiabatic Calorimeter, 18th Annual CCPS International Conference, 2003

[2]Chippett, US Patent Number 7,021,820 B2, April 4, 2006

Using Explicit Finite Element Analysis to Simulate the Effects of External Chemical Explosions on Single and Double-Walled Storage Tanks

Phillip E. Prueter
The Equity Engineering Group Inc., Shaker Heights, Ohio
peprueter@equityeng.com

David J. Dewees, P.E.
The Equity Engineering Group Inc., Shaker Heights, Ohio
djdewees@equityeng.com

Prepared for Presentation at
American Institute of Chemical Engineers
2013 Spring Meeting
9th Global Congress on Process Safety
San Antonio, Texas
April 28 – May 1, 2013

UNPUBLISHED

Using Explicit Finite Element Analysis to Simulate the Effects of External Chemical Explosions on Single and Double-Walled Storage Tanks

Phillip E. Prueter
David J. Dewees, P.E.
The Equity Engineering Group Inc., Shaker Heights, Ohio
peprueter@equityeng.com

Keywords: Explicit Finite Element Analysis, Blast Loading, CONWEP, Accidental Explosions, Storage Tanks

Abstract

Accurately simulating the overpressure or shock wave associated with a given far-field chemical explosion is extremely valuable in assessing the structural response and possible failure modes of critical process equipment such as storage tanks, piping, or pressure vessels. Furthermore, assessing the potential damage from an overpressure wave can provide valuable information about protecting structures from external explosions and improving designs that advocate blast damage mitigation. Such analyses become all the more important should the contents of the tank or vessel exposed to such a potentially catastrophic event pose a risk to humans or the surrounding environment. This paper discusses the underlying theory and examines the practical application of multiple finite element based explicit computational techniques for simulating the load acting on a structure due to an external blast. Explicit three-dimensional blast analysis of single and double-walled storage tanks that carry an extremely high consequence of failure is performed. The structural response of the tanks due to postulated accidental explosions is investigated and likely failure modes are discussed.

The two explicit computational blast loading methods discussed in this paper are the incident wave loading model and the Conventional Weapons Effects Blast Loading Model or CONWEP. The CONWEP model is appropriate for detonations of conventional explosions, and in this case automates many crucial features of analysis, such as developing the overpressure time history (positive and negative phases), defining overpressure spatial decay, and accounting for spatially varying reflection effects. Large vapor cloud explosions are known to behave very differently from conventional explosives, and for these cases, the more manual, but also general, incident wave loading approach permits the user to define their own time-varying overpressure amplitude. Additionally, this method also allows for coupled structural-acoustic analysis (should reflection effects need to be considered in detail). While it is not necessary to model the fluid medium using acoustic elements when using the incident wave loading model, it does become essential if reflection effects dominate the response of the given problem. If the fluid medium is modeled (air in this case), the acoustic elements can be coupled to the structural elements and the loading surface becomes the boundary of the acoustic mesh closest to the blast source. While this approach can be accomplished, it is extremely resource and time intensive, and is generally not necessary (which will be discussed later in this paper). Three-dimensional examples of the above blast loading approaches are presented in this paper; incident wave, and CONWEP. Comparisons are made between these methods and the corresponding computational results.

Furthermore, commentary on the structural response and likely failure modes of the storage tanks is provided and a discussion regarding design features that supplement blast damage mitigation and process safety is rendered. The advanced computational methods discussed herein permit realistic evaluation of complex, three-dimensional process equipment subjected to accidental explosions. Simulating and understanding the possible failure modes due to blast loading of critical process equipment in the petrochemical and related industries is very beneficial in not only improving process safety, but also providing plant operators and safety personnel with crucial information regarding the possibility of equipment failure should a catastrophic explosion occur.

1. Introduction

An explosion in air creates a blast wave as a result of the atmosphere surrounding the explosion being pushed back. In general, the pressure of the compressed air at the blast wavefront decays as it moves away from the blast source. As discussed in references [1-6], a typical blast wave, as observed at a location removed from the center of a chemical explosion, reaches a peak value of overpressure and begins to decay exponentially (as shown in Figure 1), eventually decreasing below atmospheric pressure. This peak overpressure decreases as the distance from the explosion source increases. The amount of time it takes the wavefront to reach a given point is known as the arrival time. The overpressure profile can be divided into the positive pressure portion (positive phase) and negative pressure portion (negative phase). Furthermore, the amount of time it takes the peak overpressure to decay below atmospheric pressure is known as the positive phase duration and is closely related to the damage capabilities of a given blast wave. Similarly, the amount of time it takes for the negative pressure to return to atmospheric pressure is the negative phase duration [1]. The area under the curve during the positive phase duration is the positive blast wave impulse.

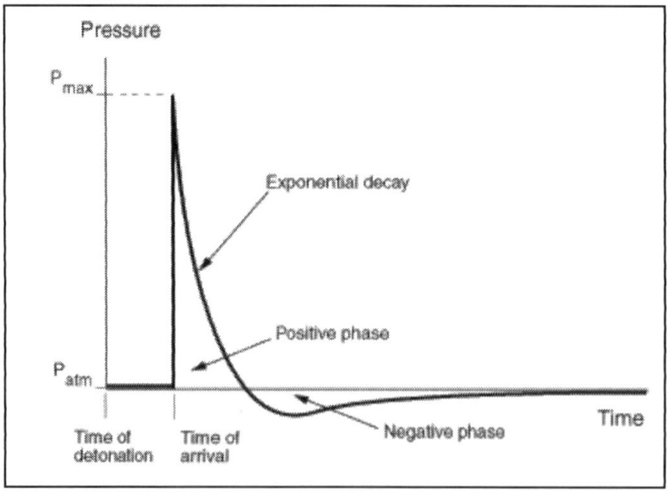

Figure 1: Typical pressure-time amplitude for a chemical explosion [7].

In some cases, the positive phase impulse is calculated and an equivalent triangular impulse with the same peak overpressure is used to approximate the realistic overpressure amplitude shown in Figure 1. In many cases, this approximation is appropriate because the positive phase portion of the incident pressure wave is typically the most damaging. Generally, structural response is more sensitive to capturing the correct incident wave reflection. Accurately simulating a realistic blast wavefront on a structure for a given explosive scenario is challenging because the peak overpressure, positive phase duration, amount of explosives, and distance from the explosion source all affect the overpressure amplitude. Furthermore, accounting for reflection effects on a loading surface adds further complication to the blast loading model.

Two different explicit finite element techniques used to simulate blast loading due to external chemical explosions in air will be discussed herein; incident wave loading and the Conventional Weapons loading model (CONWEP) [8]. CONWEP is well suited for simulating the overpressure wave associated with the detonation of conventional explosives while incident wave loading permits reasonable analysis of vapor cloud explosions. The underlying methodology of each approach will be discussed as well as the advantages and applications of each. Furthermore, both computational methods will be compared for the explicit dynamic simulations of realistic explosive overpressure waves (from a postulated concentrated explosion source and a vapor cloud) impacting single and double-walled storage tanks that carry an extremely high consequence of failure.

2. Blast Curve Generation

In order to develop the blast parameters required to generate realistic overpressure curves for conventional explosives, the equivalent TNT method is most commonly used. TNT equivalency of an explosive is simply the ratio of the mass of TNT to the mass of explosive such that both yield equal pressure and impulse [9]. In order to compare shock wave parameters to available published data, a scaled distance from an explosion source to the loaded structure may be used. As discussed in [10], the most common form of blast scaling is the Hopkinson-Cranz or cube-root scaling method following

$$Z = R/W^{(1/3)}$$

where Z is the scaled distance, R is distance from the explosion source to the loading point, and W is the weight of TNT. This formulation implies that self-similar blast waves are produced at identical scaled distances when two different sized explosive charges (of similar geometry and chemical composition) are detonated in the same atmosphere. Using published empirical data, this scaled distance from the explosion source to a loading surface can be used to determine specific blast parameters. The CONWEP blast loading model discussed herein is based on a TNT explosion (assumed surface blast). Thus, when using CONWEP, it is necessary to convert the mass of explosives in question to an equivalent amount of TNT.

As outlined in [11], a major limitation associated with the equivalent TNT method is that the blast yield is a function of the amount of explosives and not the combustion mode; making this method not preferred when analyzing vapor cloud explosions. As discussed in References [12-

16], vapor cloud explosions differ from conventional explosive detonations in that flame accelerations, which can be exacerbated by confinement due to intervening structures, define the overpressure associated with a given explosion more so than simply the amount of fuel ignited. In order to quantify blast curves associated with vapor cloud explosions several widely accepted methods have been developed; TNO [15] blast curves and Baker-Strehlow [17] blast curves among others. In this study, triangular impulses of varying peak overpressures are used for simplicity when analyzing the structural response of a double-walled storage tank subjected to a postulated vapor cloud explosion (discussed below). This also permits investigation into the sensitivity of structural response to varying blast wave parameters.

3. CONWEP and Incident Wave Loading

A significant challenge associated with computational blast modeling is accounting for reflection effects. The peak blast pressure that a structure is subjected to is greater than the peak overpressure and is known as the reflected pressure. This pressure is a function of the peak overpressure and the angle of incidence (the angle between the loading direction and the surface normal of a point on the loading surface). Neglecting incident wave reflection can yield non-conservative results, particularly for high peak overpressures, while assuming the maximum reflection effect at all points on a (cylindrical) structure can potentially be overly conservative.

The CONWEP blast loading model is based on a collection of conventional weapons effects calculations from the equations and curves of TM 5-855-1 [8] and is implemented into the commercial finite element program, Abaqus. The main advantage of using CONWEP over the incident wave loading approach is that realistic overpressure amplitudes (including positive and negative phase) and other blast wave parameters are inherently calculated based on a defined amount of TNT at a given distance from the explosion source. Furthermore, the CONWEP blast loading model does not require modeling the fluid medium (air) to account for reflection effects.

Incident wave loading is an explicit finite element technique in which an incident wave overpressure at a point can be defined as a function of space and time. This pressure amplitude is defined at a standoff point, located between the detonation point and the loaded structure. When simulating blast loading, realistic overpressure amplitudes need to be defined by the user. An incident wave can be defined as planar or spherical. A schematic showing the incident wave loading model in Abaqus is given in Figure 2 below. For a more rigorous overview of the underlying theory behind both the CONWEP blast loading model and the incident wave loading as well as a benchmark of the overpressure reflection calculations of CONWEP, refer to Reference [18].

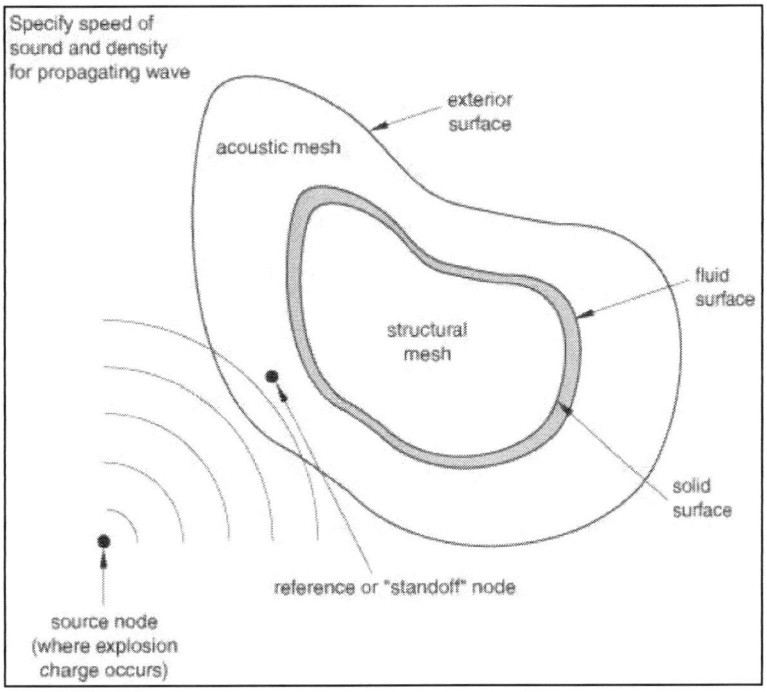

Figure 2: Incident wave loading model [7].

4. Explicit Analysis of a Storage Tank Subjected to a Vapor Cloud Explosion

As discussed in [19, 20] the consequences of storage tank failure in chemical or petrochemical facilities can be dire. Furthermore, it is estimated that 85 percent of storage tank accidents are due to fire and explosion [21]. Analyzing the structural response of storage tanks subjected to potential blast loading from conventional detonations or vapor cloud explosions is of great interest to the process safety community. The explicit finite element analyses discussed below provide real-world examples of the usefulness that computational blast analysis provides in the petrochemical and related industries.

The first computational blast analysis examined herein relates to a double-walled storage tank subjected to a postulated external vapor cloud explosion. The rough size of the outer tank is a 140 foot diameter with a shell height of approximately 125 feet. A complete, elastic-plastic, three dimensional finite element model was constructed using linear shell elements. Figure 3 shows the half-symmetry model with detailed geometric features. The baseline finite element model includes all major structural features, including the suspended deck, bracket supports, and the lower support straps. Quadratic beam elements are used to model the suspended deck hanger rods and the strap supports, and both the inner and outer tanks are modeled for completeness. Contact is defined between the suspended deck and the shell wall.

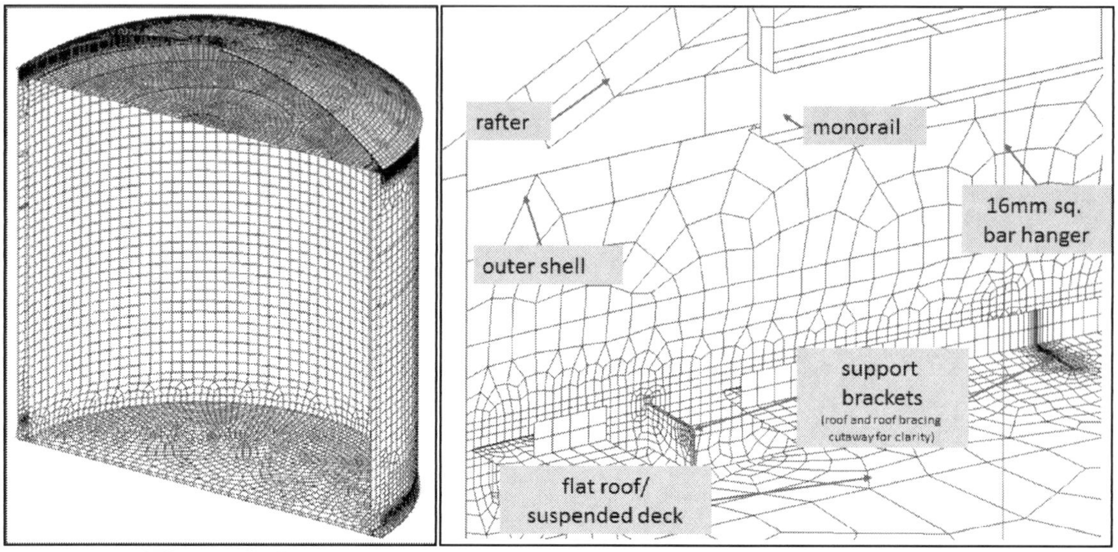

Figure 3: Vapor cloud explosion analysis finite element model.

A simplified, lumped-mass model is also analyzed. In this model, the mass associated with the suspended deck is lumped into the domed roof and the mass of the inner tank shell is lumped into the outer tank shell. The mesh is coarsened in the simplified model to aid in computational efficiency. Incident wave loading is used to analyze a range of peak overpressures (assumed to be peak reflected pressure because acoustic elements are not used) applied as planar triangular impulses. The effects of gravity (self-weight) are included.

In this case, because of the range of potential vapor cloud explosive scenarios, iterations are performed to determine the maximum permissible overpressures and loading durations at the tank to ensure containment and safe operability. That is, the focus is on establishing a sort of blast rating for the structure independent of a specific explosion scenario. This provides an essential characteristic of the structure that can be referred to for any future scenarios or concerns that might arise, and is similar in some sense to the maximum allowed working pressure (MAWP) of a pressure vessel. Peak reflected pressures of 5, 8, and 11 psi are applied for an impulse duration of 50 milliseconds to the lumped model. Figure 4 below compares maximum deflections and stresses in the lumped model from the domed roof (opposite impact side) for peak applied overpressures of 5, 8 and 11 psi. This figure demonstrates the relatively constant nature of the largest displacements. In the 5 psi case, maximum deflections are generally less than 12 inches with relatively little accumulation in plastic strain.

These results indicate that a peak reflected pressure of 5 psi is near the limit to achieve tank containment following blast loading. Furthermore, for the 8 and 11 psi overpressure cases, results indicate that plastic collapse of the domed roof and tank shell would be likely and containment of the vessel contents would be jeopardized. Figure 4 also shows an example transient deformation corresponding to the full model results for 5 psi. In general, there is reasonable agreement between the simplified, lumped parameter model and the complete model with the suspended deck across cases. However, the suspended deck is a critical feature of the tank, and is also an indicator of safe operability. As Figure 4 illustrates, the compete model indicates that the suspended deck is one of the weakest points of the tank internals; results from the complete model series of analyses show that a maximum peak reflected pressure of 2 psi is advisable to ensure safe tank operability (again, impulse duration assumed to be 50 milliseconds). This corresponds to a maximum side-on overpressure of 1 psi (uniform/constant reflection coefficient of 2).

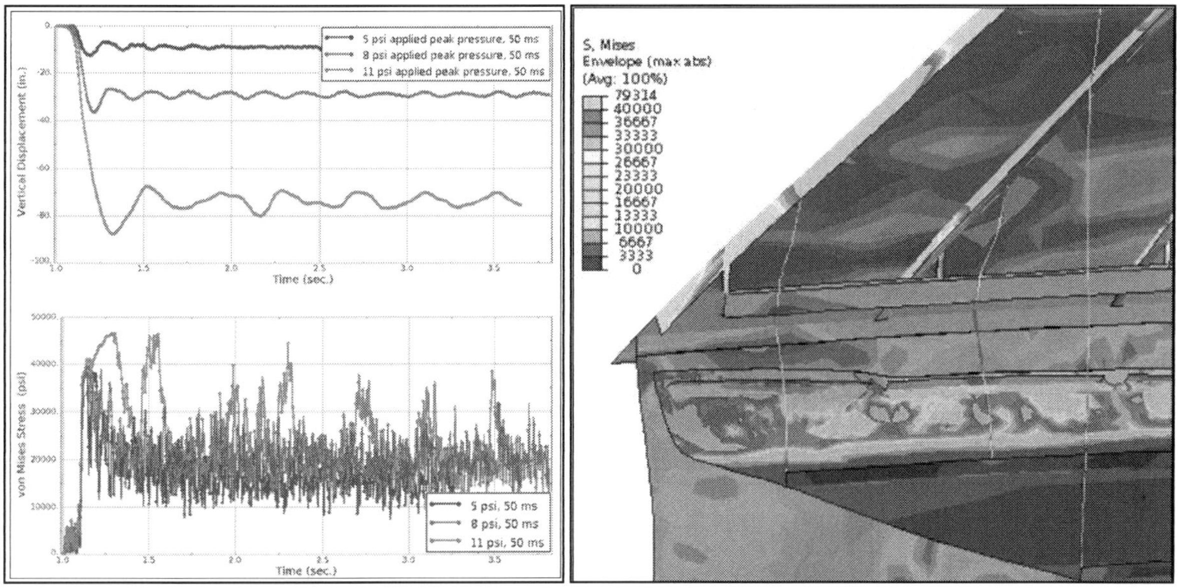

Figure 4: Deflection/stresses in domed roof opposite of impact side for varying peak overpressures and suspended deck deformation.

Additionally, in order to compare the effect of varying blast wave impulse durations, a simulation at 11 psi peak applied overpressure with an impulse duration of 10 milliseconds is carried out. Figure 5 compares maximum domed roof deflections and stresses for the 50 and 10 millisecond loading durations. As expected the duration of the impulse has a significant effect on the overall structural damage. It is noted that much lower peak overpressures acting at longer durations could also be damaging.

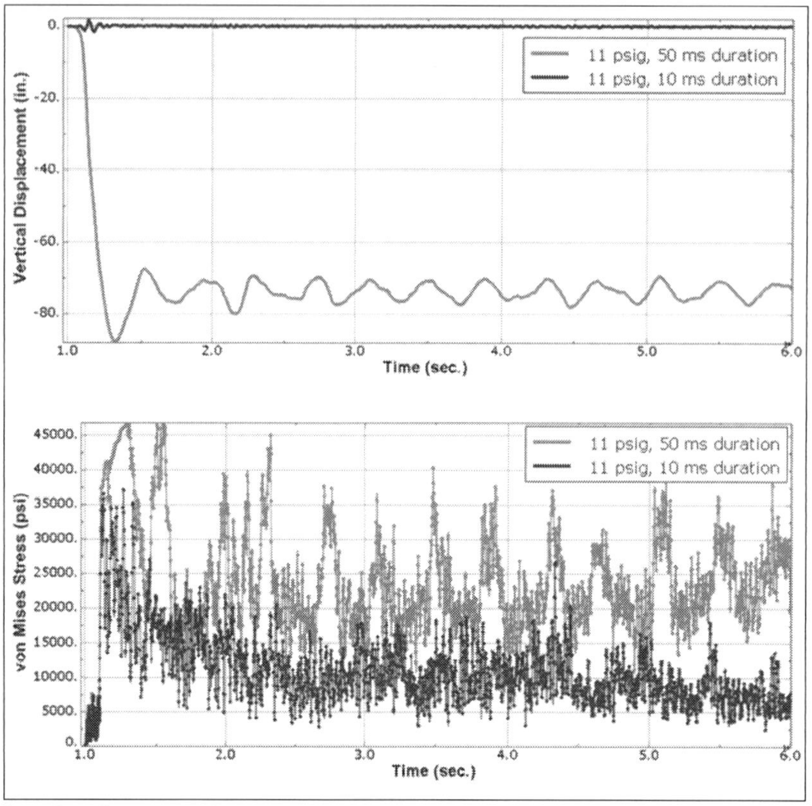

Figure 5: Deflection/stresses in domed roof opposite of impact side for varying impulse duration.

5. Explicit Analysis of a Storage Tank Subjected to Conventional Explosion

A different double-walled storage tank is analyzed; this time the postulated blast loading is due to accidental detonation of a known amount of concentrated, combustible fuel. The approximate outer tank diameter is 100 feet, and the outer tank height is roughly 75 feet. Again a complete, elastic-plastic, three dimensional finite element model is constructed using linear shell elements. Figure 6 shows a cutaway view of the model with contours of shell element thickness. The inner tank has four stiffening rings that protrude into the annular space between the inner and outer tank. The outer tank roof support structure consisting of channels and angles is also modeled in detail and included in the finite element model. Contact is modeled in between the tank structures and the effects of self-weight are included.

An initial analysis is performed on just the outer tank to investigate the plastic strains and wall deflections due to the applied blast wave. Both the CONWEP and incident wave loading models are compared (the equivalent TNT method is used to determine the overpressure profile in accordance with CONWEP). Due to the size of the finite element model and the potentially high computational cost, the fluid medium is not explicitly modeled for the incident wave simulation. An overpressure amplitude that accounts for reflection effects is used (peak reflected pressure applied in the incident wave model). Furthermore, in order to mimic the behavior of the CONWEP simulation as much as possible, the overpressure profile defined for the incident wave simulation as a function of time does, in fact, consist of positive and negative phases and decays exponentially once it reaches its peak value. The side-on reflection coefficient for the overpressure magnitude in this case is 2.

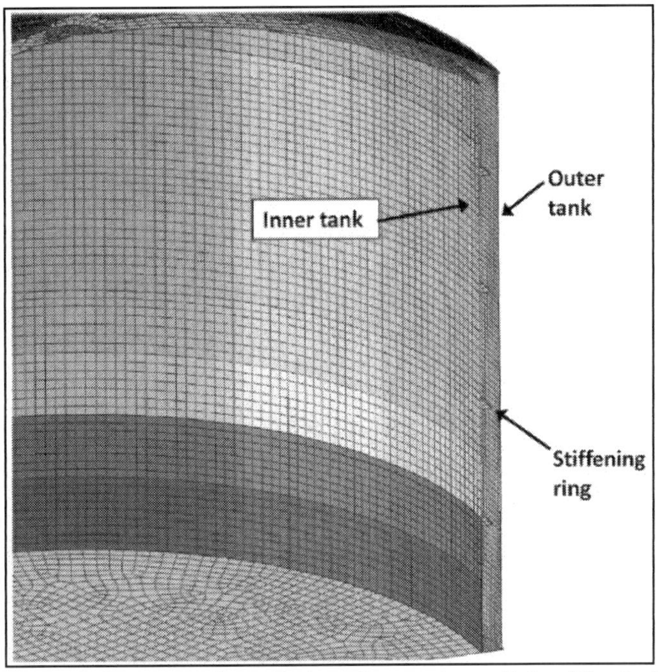

Figure 6: Cutaway view of the complete finite element model with shell thickness contours shown.

Even for the relatively small peak reflected overpressure of roughly 1.9 psi, the outer tank wall deflections prove to be significant, to the point of local collapse. A static post-buckling analysis (using the Riks solver), with surface pressure applied using an analytical field definition to match surface pressure in the dynamic simulations confirms that the outer tank wall buckles before the peak reflected pressure is reached. Riks post-buckling analyses are discussed in more detail below.

For comparison, Figure 7 shows contours of outer tank plastic deformations (identical scale) due to the overpressure wave for both methods. As expected, the incident wave approach produces higher deflections than CONWEP because the applied peak reflected overpressure does not decrease as angle of incidence increases. However, the general agreement between methods is actually quite good. This suggests that the simplification of a single maximum reflection coefficient in this case for the incident wave approach is not excessively conservative, and that reasonable results were obtained without modeling the entire acoustic domain (in which case reflection effects would be rigorously accounted for).

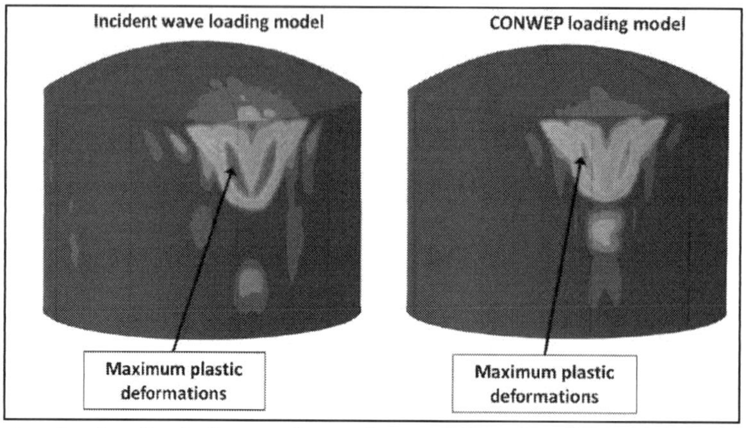

Figure 7: Outer tank plastic deformations due to blast loading using the CONWEP and incident wave loading models (contour scales identical).

For the blast analysis of the complete finite element model that includes both the inner and outer tanks with contact defined between shells, the outer wall deflections are much less significant because the outer tank impacts the inner tank stiffeners. Essentially, the inner tank stiffening rings absorb the impact from the outer tank wall and plastically deform. Furthermore, the deflections in the inner tank wall are minimal. This is demonstrated in Figure 8, which shows the maximum deflection during blast loading from the CONWEP simulation of the inner tank and the outer tank. The contour scale for the inner tank corresponds to 10 percent of the outer tank contour scale. Additionally, the maximum inner tank wall deflection is less than 10 percent of the maximum outer tank wall deflection, and the maximum outer tank deflections occur in the regions in between the inner tank stiffening rings. The outer tank wall never directly impacts the inner tank wall. The minimal inner tank deflections suggest that the sloshing effects of the liquid contained by the inner tank should also be minimal.

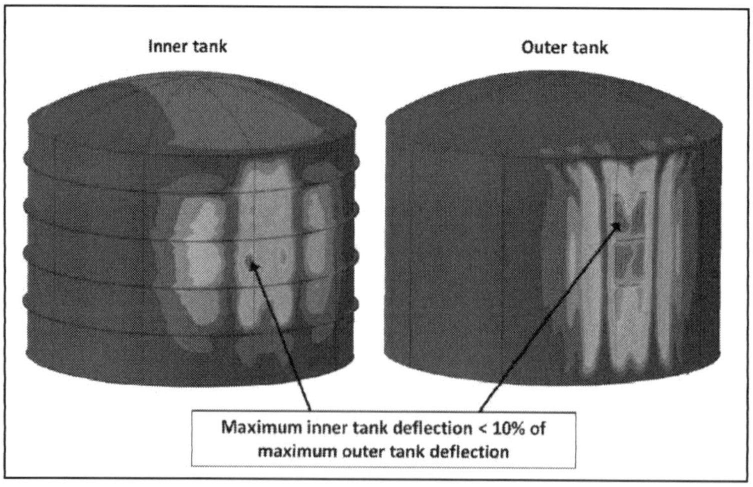

Figure 8: Maximum deflection of the inner and outer tanks due to blast loading from the CONWEP model (inner tank contour scale corresponds to 10 percent of outer tank scale).

These results indicate that from a blast damage mitigation perspective, the four inner tank stiffening rings prove to be very beneficial in reducing the severity of the impact between the inner and outer tanks. While both explicit computational techniques used to analyze this complete, three dimensional model give reasonably similar results, from a convenience standpoint, CONWEP proves to be advantageous for conventional detonations because the appropriate overpressure and spatial decay are automatically calculated for the given equivalent amount of TNT. Additionally, due to more rigorous accounting for reflection effects, some conservatism is removed in the CONWEP case, making it the natural choice for well-defined, conventional explosive detonation scenarios.

6. Explicit Analysis of a Cluster of Single-Walled Storage Tanks

The postulated blast source in this case is the same source as discussed in the previous example (conventional detonation). However, this cluster of single-walled tanks is located much closer to the blast source, meaning peak overpressures are much higher than the above example. All three tanks are roughly 30 feet in diameter and 35 feet high. The tanks also have self-supporting roofs and are anchored to their respective foundations via hold down clips and hold down bolts. As is the case for the aforementioned storage tanks in this study, these tanks carry an extremely high consequence of failure due to their process contents.

Three dimensional, elastic-plastic finite element models of each tank are constructed using linear shell elements and explicitly analyzed using CONWEP. Peak reflected overpressures at each tank as a function of time are provided in Figure 9 below. Again, the effects of gravity are included. A node at the center of each hold-down clip is fixed in all degrees of freedom and then coupled to the surrounding bolt hole to simulate the restraint of the bolts as shown in Figure 10. A rigid surface representing the tank foundation is constructed and included in the model, and contact between this rigid surface and the tank bottom is defined. This ensures that the tank will not unrealistically sag due to gravity alone (essentially penetrate the foundation due to self-weight and induce non-realistic stresses in the tank floor/walls) while allowing the tank bottom to physically lift off its foundation when the blast load is applied. In order to account for the weight of the process fluid, an equivalent pressure is applied to the tank bottom that corresponds to the weight of fluid up to the maximum fill height. This pressure is applied throughout the dynamic simulation (sloshing effects of the fluid are not considered; this assumption is discussed below).

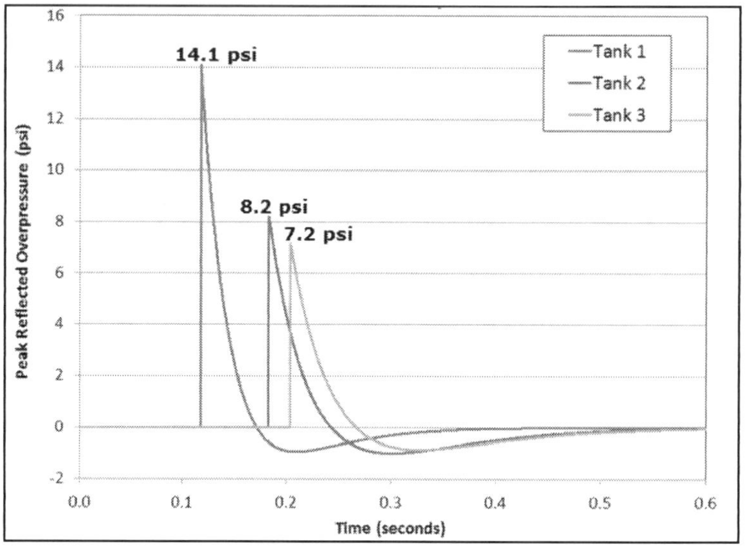

Figure 9: Peak reflected overpressures at each tank.

After the onset of the blast wave, the wall and the roof of each tank significantly deform. In order to better visualize the tank behavior following the onset of the blast wave impact, Figure 11 shows contours of Tank 1 deflection as time elapses (time scale given is seconds after detonation). These large deflections are more indicative of plastic collapse like behavior than a dynamic or vibratory failure. Furthermore, the stresses observed in the tank shortly after blast wave impact are significant.

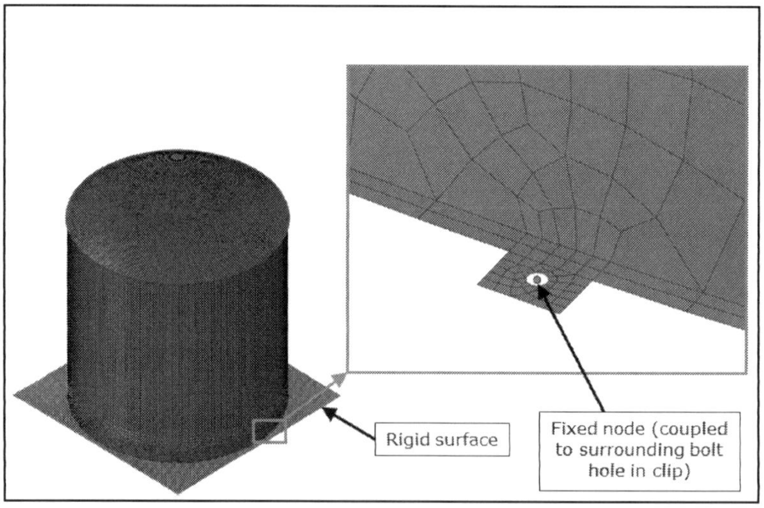

Figure 10: Finite element model of Tank 1.

Additionally, the stresses in the tank at the hold-down clips and reaction forces at the assumed bolt locations predict likely failure of either the bolts themselves or the hold-down clips. Figure 12 shows equivalent plastic strains in Tank 1 after the impact of the blast wave (contour scale capped at 5%). The hold-down clips and portions of the tank wall near the hold-down clips exceed 100% equivalent plastic strain. Results for Tanks 2 and 3 are similar.

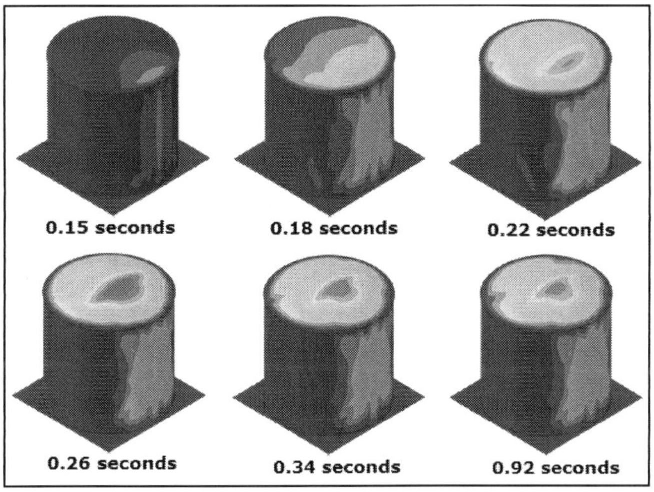

Figure 11: Tank 1 deflections during blast loading.

In order to further investigate the possibility of plastic collapse or buckling, static Riks post-buckling analyses are performed on each tank (a static pressure distribution is applied to mimic peak blast loading pressures). A Riks analysis is useful for simulating non-linear instability problems, and includes the effects of large deformations and elastic-plastic material behavior. It can also be used to look at post-buckling or collapse shapes. The applied load in a Riks analysis is actually a variable and part of the solution for each increment, and is always relative to the current deformed configuration of the structure. If no converged solution is possible for an incremental increase in load, then the equilibrium solutions can be searched for with a decreased load. This is the mechanism that allows post-buckled shapes to be predicted and visualized. In extreme cases, no equilibrium solution can be reached, even with decreased load, and only load reversal is predicted to lead to stability. This is actually the case for the tanks analyzed here, where once collapse has initiated, displacements are predicted to be essentially unbounded for any amount of positive load and thus, surface pressure becomes negative before peak overpressure values are achieved.

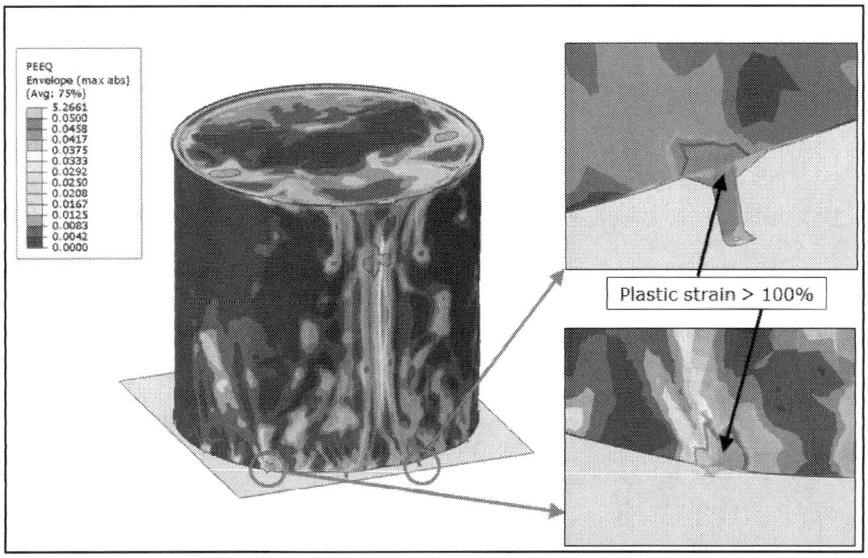

Figure 12: Equivalent plastic strain in Tank 1 due to blast loading (scale capped at 5%).

Based on the results discussed above, the three storage tanks subjected to the assumed postulated blast load are likely to fail via several possible failure modes. The first likely failure mode is plastic collapse of the tank walls and/or roofs. All three tanks see large deflections (the most severe occurring in Tank 1). Riks post-buckling analyses confirm that the magnitude of deflections seen in the tank walls would likely result in unstable collapse behavior before the peak blast overpressure is achieved. A second likely failure mode is extreme damage in the tanks at the hold-down bolts and/or regions of the tanks that are connected to the bolts. In all three tanks, equivalent plastic strains exceed 100% in these regions. Failure of multiple hold-down bolts or hold-down clips could cause a given tank to slide off its foundation and/or overturn. Furthermore, sloshing effects of the fluid inside the tanks are neglected in the simulations discussed herein, but would increase the likelihood of failure.

In short, the close proximity of all three tanks to the postulated blast source causes the corresponding overpressures acting on each tank to be relatively severe. Design changes to the tanks to supplement blast damage mitigation would likely have to be drastic in order to achieve a significant reduction in the probability of failure due to the postulated explosion considered in this study. Finally, in order to achieve practical simulation times, all of the explicit analyses discussed herein are conducted across multiple CPUs (often 16 or more) simultaneously. Utilizing parallel computing permits evaluation of complex three dimensional structures, such as the tanks discussed above, in reasonable amounts of time.

7. Conclusions

This study compares two different explicit computational methods for modeling blast loading on structures; incident wave loading and the CONWEP charge property. The underlying methodology for each method is discussed. CONWEP is well suited for conventional detonations and uses the equivalent TNT method, and it inherently calculates the appropriate blast wave parameters for a given explosive event. Additionally, CONWEP achieves realistic overpressure amplitudes, spatial decay, and reflection effects without having to model the fluid medium, as is the case for incident wave loading. Incident wave loading provides the user with the flexibility to define a time-varying overpressure amplitude. This is useful when performing sensitivity studies to blast loading parameters or when attempting to characterize the blast wave associated with a vapor cloud explosion that does not behave like a conventional detonation. The development and implementation of the CONWEP and incident wave blast loading models into commercial finite element programs makes accurately analyzing the response of complex three dimensional structures subjected to blast loading feasible.

Additionally, explicit, three dimensional finite element analyses of both single and double-walled storage tanks that carry an extremely high consequence of failure are discussed herein. Incident wave loading is used iteratively to determine the maximum permissible overpressure for a double-walled storage tank subject to a postulated vapor cloud explosion. The CONWEP blast loading model is used to analyze single and double-walled storage tanks subjected to a postulated chemical explosion (conventional detonation). The explicit computational methods, such as the ones examined herein, can provide valuable insight into the blast resistance of existing structures and can guide design decisions regarding how to mitigate blast damage for new structures in the petrochemical and related industries. Gaining a better understanding of the possible failure modes due to blast loading of critical process equipment is very beneficial in not only improving process safety, but also in enhancing safety protocol for catastrophic accidental explosions in industrial environments.

8. References

1. DOD, "Unified Facilities Criteria (UFC) 3-340-02: Structures to Resist the Effects of Accidental Explosions," *Department of Defense, United States of America*, 2008
2. Smith, P., and Hetherington, J., "Blast and Ballistic Loading of Structures, 1st ed," *Butterworth-Heinemann*, Oxford, 1994.
3. Kinney, G., and Graham, K., "Explosive Shocks in Air, 2nd ed," *Springer-Verlag*, Berlin, 1985.
4. Dusenberry, D., "Handbook for Blast-Resistant Design of Buildings, 1st ed," *John Wiley and Sons, Inc.*, Hoboken, New Jersey, 2010.
5. ASCE, "Design of Blast-Resistant Buildings in Petrochemical Facilities, 1st ed," Task Committee on Blast-Resistant Design, *American Society of Civil Engineers.*, Reston, Virginia, 1997.
6. Bulson, P., "Explosive Loading of Engineering Structures, 1st ed," *Taylor and Francis Group*, London, 1997.
7. Abaqus 6.10-1, DSS Simulia, "Analysis Users' Manual," 2010.
8. Hyde, D., "Users' Guide for Microcomputer Programs CONWEP and FUNPRO – Applications of TM 5-885-1," *U.S. Army Engineer Waterways Experimental Station*, Vicksburg, VA, 1988.
9. Huntington-Thresher, W., and Cullis, I., "TNT Blast Scaling for Small Charges," *19th International Symposium of Ballistics*, 2001.
10. Baker, W., "Design Considerations for Toxic Chemical and Explosives Facilities – Chapter 1, Blast Pressure Effects: An Overview," *American Chemical Society*, 1987.
11. Tang, M.J., and Baker, Q.A., "A New Set of Blast Curves from Vapor Cloud Explosion," *Process Safety Progress*. Vol. 18, No. 3, 1999.
12. American Institute of Chemical Engineers (AiChE), "Guidelines for Evaluating the Characteristics of Vapor Cloud Explosions, Flash Fires, and BLEVEs," *Center for Chemical Process Safety of the American Institute of Chemical Engineers*., New York, 1994.
13. Lees, F.P., "Loss Prevention in the Process Industries," *Butterworth-Heinemann*, Oxford, 1996.
14. Mercx, W.P.M., van den Berg, A.C. and van Leeuwen, D., TNO report PML 1998-C53, "Application of correlations to quantify the source strength of vapor cloud explosions in realistic situations, Final report for the project: GAMES," *Netherlands Organization for Applied Scientific Research (TNO)*, 1998.
15. Van den Bosch, C.J.H. and R.A.P.M. Weterings, eds., "Methods for the calculation of physical effects: due to releases of hazardous materials (liquids and gases) ("TNO Yellow Book"), CPR 14E, Parts 1 and 2, 3rd Edition," Director-General for Social Affairs and Employment, *Committee for the Prevention of Disasters*. The Hague, 1997.

16. Woodward, J. L., "Estimating the Flammable Mass of a Vapor Cloud," *Center for Chemical Process Safety of the American Institute of Chemical Engineers*, New York, 1998.
17. Strehlow, R.A., Luckritz, R.T., Adamczyk, A.A., and Shimpi, S.A. "The Blast Wave Generated by Spherical Flames," *Combustion and Flame*, Vol. 35, 297-310, 1979.
18. Prueter, P.E., Dewees, D.J., and Brown, R.G. "Examining Computational Techniques to Simulate the Effects of External Chemical Explosions Using Explicit Finite Element Analysis," *In ASME Pressure Vessels and Piping Conference*. PVP2012-78566, 2012.
19. Maremonti, M., Russo, G., Salzano, E., and Tufano, V., "Post-Accident Analysis of Vapor Cloud Explosions in Fuel Storage Areas," *Trans IChemE*, 77(B), 1999.
20. Buncefield Investigation Board, "The Buncefield Incident 11 December 2005: The Final Report of the Major Incident Investigation Board," Volume 1, 2008.
21. Chang, J., and Lin, C., "A Study of Storage Tank Accidents," *Journal of Loss Prevention in the Process Industries*, 19(1) 51-59, 2006.

(44o) The Hidden Risks of Deposits, by-Products, and Residues On Exhaust and Waste Streams

Carlos A. Barrera and Abid Kemal, Thermal Sciences, Exponent, Menlo Park, CA

The safety risks associated with reactive compounds are usually captured and addressed during the conceptual design and pre-construction process safety reviews as, and for well-established chemistries these risks are well understood for the main processes and reactions. Lest time, if any, is devoted to the analysis of the reactive nature of the compounds present waste streams or exhaust lines. These lines are often overlooked during HAZOPs because of their secondary nature, or because they go to a scrubber or flare, which erroneously is considered a "catch-all" device that will "take care of everything." Waste streams and exhaust lines from certain physicochemical processes contain highly reactive compounds in the form of residues or by-products that in the right conditions can accumulate and form dangerous deposits. Some examples of such deposits are pyrophoric, shock-sensitive, and corrosive compounds. The purpose of this paper is to create awareness in the process safety community about the importance of a complete understanding of the chemistry of the reactive species present in these streams and the possible physical mechanisms than can lead to the formation of dangerous deposits, as well as the importance of including these streams in the process hazard analysis. Examples drawn from multiple processes will be presented.

(44q) Explosion Characteristics of Carbon Nano-Fibers

Jiaqi Zhang, Artie McFerrin Department of Chemical Engineering, Texas A&M Univerisity, Mary Kay O'Connor Process Safety Center, College Station, TX, Yi Liu, Mary Kay O'Connor Process Safety Center, Department of Chemical Engineering,Texas A&M University, College Station, TX and Dr. M Sam Mannan, Artie McFerrin Department of Chemical Engineering, Mary Kay O'Connor Process Safety Center, College Station, TX

Dust explosion is a hazard with serious consequences. While the main principle of dust explosion is still being exploring, the gap of dust explosion between particles of micro-size and those of nano-size also needs tremendous efforts. The explosion characteristics of CNFs are studied in Mary Kay O'Connor Process Safety Center. These experiments are conducted in a 36-L explosion vessel, which has been calibrated with a standard 20-L explosion vessel. Among the experiments, different kinds of CNFs of different grades and different conditions (milled and unmilled) are used. Compared with collected published data of graphic particles of micro-scale, the experimental data shows that the carbon nano-fibers create a higher Kst than the micro-scale ones. The results also show that explosion of the XT-PS grade (Pyrolytically Stripped) is much stronger than that of the XT-LHT grade (Low Temperature Heat-treated) and XT-HHT grade (High Temperature Heat-treated). Also, the explosion of milled CNFs is stronger than the unmilled ones. Further study to seek the causes is undergoing.

(44r) High-Fidelity Combined Dynamic Modeling of Depressurizing Vessels and Flare Networks: Accurately Assess the Effects of Process Changes On Mechanical Integrity and System Capacity

James Marriott, VP PSE Oil & Gas, Process Systems Enterprise Ltd, London, United Kingdom

The use of high-fidelity dynamic modeling for combined depressurization and flare network analysis can result in identification of additional capacity in flare networks without compromising process safety, presenting potentially significant opportunities for reducing or saving capital expenditure (CAPEX). The same analysis can simultaneously provide accurate assessment of temperatures in vessels and pipework, to identify areas where close approach to or violation of minimum metal temperatures may occur, thus compromising vessel and pipework mechanical integrity.

The flare networks for major plants represent a non-negligible part of the overall capital investment. Current industrial practice for their design is primarily based on steady-state analysis, as described in standards such as API 521 and supported by widely-used software tools. However, it is a widely recognized fact that the application of steady-steady considerations to what is fundamentally a dynamic system inevitably requires the use of conservative assumptions which often result in significant oversizing of flare headers and other components of the network. Another major contributor to capital cost is the use of special materials for the parts of the system that may be exposed to low-temperature fluids – typically the tail pipes attached to high-pressure process vessels – and which are therefore at risk of embrittlement. The key to being able to limit the additional capital expenditure is the accurate estimation of the length of piping that is subject to "abnormal" temperatures.

In addition to dynamic flare network analysis, the detailed dynamic modeling and simulation of the rapid depressurization ("blowdown") of high-pressure vessels is a key factor in determining both the load imposed on the pressure relief system (e.g. flare network) and also the temperatures of the vessel and downstream pipework walls. Current methods typically use 'lumped' equilibrium approaches that do not adequately predict these quantities; indeed, in contrast with the current over-conservative practice in flare network design, they may provide non-conservative solutions. By contrast, the high-fidelity approach incorporates a multi-phase non-equilibrium representation of the vessel fluid contents, as well as a 3-dimensional model of the metal walls that takes into account the transfer of heat between regions of the wall in contact with different phases. This allows a much more accurate estimation of the wall temperatures, and the direct computation of thermal stresses.

In the past it has been difficult to assess these important design criteria effectively because of the challenges in modeling such complex systems. This paper describes how the combination of high-fidelity depressurization capabilities and an advanced model-based system for flare system network design, coupled with evolving best practice in using these tools, represents a transformation for process safety management. The new techniques allow accurate assessment of the effects of process changes on mechanical integrity, thus enabling informed safety and CAPEX decision support.

An Engineering Method to Mitigate the Impact of Regulatory Focus on Relief System Installations by Prioritizing Risk

Dustin Smith
Smith & Burgess
5700 NW Central Dr. Ste 301 Houston, TX 77092
dustin.smith@smithburgess.com

Prepared for Presentation at
American Institute of Chemical Engineers
2013 Spring Meeting
9th Global Congress on Process Safety
San Antonio, Texas
April 28 – May 1, 2013

UNPUBLISHED

An Engineering Method to Mitigate the Impact of Regulatory Focus on Relief System Installations by Prioritizing Risk

Dustin Smith
Smith & Burgess
5700 NW Central Dr. Ste 301 Houston, TX 77092
dustin.smith@smithburgess.com

Keywords: Pressure Relief Valve Installation, API 3% Rule, Chatter, Relief Device Stability

Abstract

Since around 2005, regulators in the United States have put greater emphasis on relief device installations meeting the 3% rule. Spending large amounts of money to "fix" relief device installations that pose no safety risk decreases the plant's overall safety. To date, more incidents have occurred during facility construction / modifications than due to relief device chatter. This paper presents a method to assist engineers in determining if relief devices are susceptible to chatter. The methodology in this paper provides an engineering study to determine if existing installations are safe, which are allowed in the relevant engineering standards (API STD 520), and shows the research data by which it was validated. The model is used as a screening method that places the relief devices into two categories: (1) those installations that may chatter and (2) those installations that need no further review. The goal of any experimental comparison is that the model will error on the side of predicting chatter, but will be reliable enough to screen valves.

In addition to presenting the model, this paper will compare instances of known chatter to research conducted by API and work done by the Electric Power Research Institute in the 1980s. Thus far, based on research and acquired information, the method predicted all instances of chatter known to the authors. By providing a screening methodology that is supported by experimental data, plants can focus their spending on fixing real safety issues by identifying which relief installations are not expected to chatter. The paper will close by giving a brief explanation of the on-going research in relief valve stability.

1. Introduction

In November 2011, *Hydrocarbon Processing* published a paper that documented a method to determine if relief devices were susceptible to chatter. Other methods are being developed to determine the chances of chatter for a specific installation; however, the model discussed in the published paper is the only screening method that places the relief devices into two categories: (1) those installations that may chatter and (2) those installations that need no further review. The goal of any experimental comparison is that it will error on the side of predicting chatter, but will be reliable enough to screen valves. Since the publication of that article, the Oil & Gas industry has continued to struggle with the issue of relief device stability, so much so that API delayed issuance of *API STD 520 Part II Sizing, Selection, and Installation of Pressure-Relieving Devices in Refineries-Part II, Installation.* This paper compares instances of known chatter to research conducted by API, and uses this comparison to evaluate the model. Thus far, based on research and all acquired information, the method predicted all instances of chatter known to the authors.

This paper summarizes the analysis methods and validation for vapor service valves. The November 2011, *Hydrocarbon Processing* paper documents analysis methods for two-phase and liquid services. The focus on vapor / gas service valves is based on the validation data for these services analyzed herein.

2. Summary of the Stability Screening Method

The following is a summary of items used to confirm if chattering is likely to occur. Refer to the published paper for an extensive discussion of the methodology:

1. Confirm that the inlet line length is within the criteria to ensure that the pressure wave decay is not high enough to cause valve instability. The following equation (from Frommann & Friedel, Source 4) is used for this criteria:

$$L_i < 45{,}390 \frac{d_i^2}{w_{\%O}} \left(\frac{P_S - P_{RC}}{P_S} \right) (P_S - P_B) t_o$$

Equation 1

2. Confirm that the system's frictional and acoustic losses are less than the difference between the operating pressure and the valve's closing pressure. The formula for estimating the acoustical losses (from Singh, Source 5) is presented below:

$$\Delta P_{Acoustic} = \frac{L w_{PSV}}{12.6 d_i^2 t_O} + \frac{1}{10.5\rho} \left(\frac{w_{PSV} L}{c d_i t_O} \right)^2$$

Equation 2

3. Ensure that stability will not be compromised due to vortex shedding in the inlet line (this phenomenon occurs during normal operation and is generally not associated with relief conditions).

$$L_i < \frac{d_i c}{2.4U}$$ **Equation 3**

4. Ensure that the relief rate or system size is at sufficient capacity to confirm that chattering will not occur from being oversized. For vapor service relief devices, this is assumed to not occur if the open-close cycle time for the system is greater than 1 second. This cycle is broken into two components: the time it takes for the system pressure to increase from the relief devices' blowdown/closing pressure to the set pressure and the time it takes for the system to depressure through the relief valve from the set pressure to the blowdown/closing pressure. This is represented by the following equation.

$$1 > t_{cycle} = t_{P_{RC} \rightarrow P_S} + t_{P_S \rightarrow P_{RC}}$$ **Equation 4**

5. Ensure that the relief device is installed per the manufacturers' and/or general installation guidelines. These guidelines are shown in the catalogs, ASME B&PVC Section VIII, API STD 520, etc.

6. Review the valve's operational history to ensure that there is no known history of chatter.

For a complete discussion on the methodology of this analysis, the reader is referred to the work published by Smith Et. al, as the discussion is not further expanded in this paper.[1] Note that equation 17 does not follow the verbiage, and the *less than* sign should be replaced with a *greater than* sign.

3. Acceptance Criteria

This model is the only screening method to place the relief devices into two categories: (1) those installations that may chatter and (2) those installations that need no further review. The goal of any experimental comparison is that it will error on the side of predicting chatter, but will be reliable enough to screen valves. The author considers this model to be reliable as an engineering screening tool if the following criteria are met:

- *High Correlation to Experimental Data* – If the method predicts that valves will chatter and valves won't chatter most of the time, then it is a viable screening tool given the low cost of performing the analysis compared to making piping modifications.

- *No False Negatives* – The screening tool will be considered valid so long as it does not predict any installations as safe (no chatter) when in fact chatter was experimentally found.
- *Limited False Positives* – Given that the method is a conservative screening tool, some level of false positives are to be expected. This includes instances when the model would indicate the possibility of chatter when experimental results show valve stability.
- *Screens Valves as Acceptable* – The model will only be of value if the results of the model result in existing installations not requiring further modification. Based on the results presented in the referenced paper, 50% of the installations with inlet pressure losses greater than 3% were deemed acceptable as is.

The model, therefore, will be deemed conservative and reliable if the previous four criteria are met.

4. Comparisons / Validations

To confirm the validity of the procedure, the procedure has been reviewed against all cases of chatter known to the authors. The cases are divided into three different categories:

1. ***Known Installations*** – These installations are those that are known by the authors to have failed with a loss of containment and those installations that have sufficient information to perform the analysis. The information for some of the installations is not available in the public domain, as the confidentiality of these installations has been preserved.
2. ***ASME Studies*** – In the 80's, Zahorsky performed a set of research for the nuclear industry in which they experimentally set the blowdown as the minimum percentage needed for stable operation[2]. In order to meet the acceptance criteria, therefore, this method should determine that these installations are unstable. The results were re-checked with an increased blowdown of 2% above what the author listed to see what would happen.
3. ***API PERF Studies*** – In 2011 the results of the API PERF study were presented at the API 520 Committee meeting. In this work, the API subgroup tested 18 different valves with three different inlet line lengths to determine when chatter would occur. The results of the 54 trials were introduced in their presentation (some combinations were not tested).

5. Results

The following are the results of the comparisons:

Table 1 lists the results of the model comparison with the valves that have had a loss of containment where inlet line chatter was the cause or a contributor to the incident. The model predicted the potential for chatter for each of these installations.

Table 1: Summary of known installations that have chattered

Installation	Service	Chatter Predicted	Data Source
Refinery, North America	Liquid	Yes	Internal company incident investigation
Gas Plant, Middle East	Vapor	Yes	Internal company incident investigation
Refinery, North America	Liquid	Yes	Internal company incident investigation
Refinery, North America	Liquid	Yes	API Published Document [3]

Table 2 lists the blowdown at which the relief devices do not chatter as experimentally determined by Zahorsky. The table also lists the blowdown at which the model would predict stability. The difference between the two values can be taken as an estimate to which the model is conservative. Since the blowdown was experimentally set as the minimum percentage, the model is accurate, as it predicts false positives for these devices.

Table 2: Comparison of minimum predicted blowdown to the results experimentally derived by Zahorsky

Run Case	Experimentally Determined Blowdown	Predicted Acceptable Blowdown	(Δ Blowdown) Predicted – Expermintal)
1	3.9%	4%	0.1%
2	3.9%	5.6%	1.7%
3	5.6%	9.7%	4.1%
4	8.4%	16.7%	8.3%
5	8.3%	12.6%	4.3%
6	4.3	5.3	1.0%

Table 3 lists the comparison of the model results to the experimental PERF results. The model had agreement ~72% of the time if the cases that were not tested were excluded. If one assumes that the results for the untested cases would have chattered, the agreement is ~76%. Of the cases that the model did not agree with the experimental results, the model always predicted chatter on stable valves; thus, the model is accurately screening and is slightly conservative.

Table 3: Comparison of the model to the API PERF Study Results

Model Correlation	**PERF Results**	**Model Prediction**	**No. Of Cases**
Agreement	Chatter	Chatter	9
Agreement	Stable	Stable	25
False Negative	Chatter	Stable	0
False Positive	Stable	Chatter	15
Agreement[1]	Not Tested	Chatter	7

Note 1: There are a number of cases that were not tested, but were assumed to chatter as the reason for not being tested was not included.

6. Conclusion

The model results always predicted when valves may chatter (need modifications). There are no instances of valves that have chattered where the model predicted stability.

- *High Correlation to Experimental Data* – There is a 70% to 75% agreement with the API PERF Studies and all industry installations with known chatter where identified with the model. This model predicted if a valve would chatter accurately 75% of the time (see the next item on false negatives).
- *No False Negatives* – No cases of chatter were found when the model predicted stable operation.
- *Limited False Positives* – 25% of the PERF study valves that were stable indicated that chatter was possible. Also, the work by Zahorsky indicated that the model predicts stable operation with an inlet blowdown generally 2% to 4% greater than needed.

The model, therefore, is a reliable method to screen relief devices based on the comparison to 60+ experimental and industrial data points.

7. Notation

The equations presented in this paper are dimensional and the following listing explains the variables and unit set.

c = speed of sound (ft/s)
d = diameter (in)
L = length (ft)
P = pressure (psig)
t = time (s)
U = process fluid velocity as it passes the PSV nozzle (ft/s)
w = mass flow rate (lb/s)

<u>Greek Letters</u>
ρ = fluid density (lb/ft^3)

Subscripts
B = backpressure on relief device
i = inlet
o = opening
PSV = process safety valve
RC = valve reclosing pressure
S = relief device set pressure
%O = flow rate at the valves percent open

8. References

[1] Smith, Dustin P.E., John Burgess P.E., and Craig Powers PhD. "Relief Device Inlet Piping: Beyond the 3% Rule." *Hydrocarbon Processing.* 2011.

[2] Zahorsky, J.R. "Testing and Analysis of Safety/Relief Valve Performance." The ASME Winter Annual Meeting. November 15-19, 1982. Atlanta, GA

[3] Myers, DF, Conoco. "Denver Refinery Fire." *API.* 1986.

[4] Frommann, O. and Friedel, L. "Analysis of safety relief valve chatter induced by pressure waves in gas flow." *Journal of Loss Prevention in the Process Industries.* 1998.

[5] Singh, A. "On the Stability of a Coupled Safety Valve-Piping System." The ASME Winter Annual Meeting. November 15-19, 1982. Atlanta, GA.

Making Relief Load Estimates Match Reality

Nicholas N. Cristea
Smith & Burgess, LLC
5700 Northwest Central Dr. Ste 301 Houston, Texas 77092
Nicholas.Cristea@smithburgess.com

Dustin Smith
Smith & Burgess, LLC
5700 Northwest Central Dr. Ste 301 Houston, Texas 77092
Dustin.Smith@smithburgess.com

Prepared for Presentation at
American Institute of Chemical Engineers
2013 Spring Meeting
9th Global Congress on Process Safety
San Antonio, Texas
April 28 – May 1, 2013

UNPUBLISHED

Making Relief Load Estimates Match Reality

Nicholas N. Cristea
Smith & Burgess, LLC
5700 Northwest Central Dr. Ste 301
Houston, Texas 77092
Nicholas.Cristea@smithburgess.com

Dustin J. Smith, P.E.
Smith & Burgess, LLC

Keywords:
Relief load estimates, relief system analysis, process simulations, reduced duty, inlet control valve failure, column systems

Abstract

A pressure relief system designer must make some assumptions when looking at a relief systems analysis. Some assumptions may result in costly modifications of the system, so knowing when to apply certain assumptions is paramount with the ability to accurately size a relief valve. When looking at systems, industry standards should be consulted before taking certain credits based on the given scenario. For example, the designer may be able to reduce the relief load by looking at a reduced reboiler duty and a temperature pinch during an overpressure event. When evaluating an inlet control valve failure, some credits may be taken for the minimum turndown for the outflow. However, when looking at how a downstream system may respond to the overpressure event, credit for flow continuing may not be acceptable and flow may be assumed to be lost. The application of assumptions is common practice in relief systems analysis; however, an assumption that does not match the true operational limits may lead to problems with the relief system that do not actually exist. These invalid assumptions could cost millions of dollars to fix the problem.

1. Introduction

When doing a Relief Systems Documentation Analysis, the ability to accurately model a "worst case" relief scenario is paramount to ensuring the system is protected from the overpressure event. The engineer is undoubtedly limited to the availability of equipment data and accuracy of the process simulation developed to predict upset conditions. However, if some conservative yet unrealistic assumptions are used, the relief device can be labeled as providing inadequate protection in the event of the initiating event. [1] Physically modifying the system to fix the problem, as well as the loss of profit from having to shut down the unit to perform the update, can cost the facility large sums of money. However, performing a detailed analysis on the system may better predict the "worst case" scenario, avoiding unnecessary modifications. The ability to make relief load estimates match what may happen out in the field can still be accomplished, but knowing the difference between a realistic conservative assumption and an unrealistic conservative assumption is paramount. The pressure relief system designer should consult industry standards (e.g. API 521or NFPA) to see how to estimate relief loads for given overpressure scenarios and what assumptions for mitigating items/actions are acceptable. This paper will look at two cases where relief load estimates were reduced and one where by matching the physical constraints of the system to reality resulted in a larger relief load estimate.

2. When Realistic Assumptions Lower Relief Estimates

Oftentimes relief estimates are based on normal operating flow rates, heat exchange, and/or flow paths continuing during the upsets. With the amount of optimization and debottlenecking, these assumptions may not be realistic. Equipment capacities that are right on the design limit may not be able to produce the same flow rate or heat transfer under upset conditions. Flow paths that are available under normal operation may be isolated or otherwise not be available during upsets. The following are three examples that show these effects in a typical plant situation. The first two are examples of assumptions that tended to increase the relief rate while the last example illustrates an optimistic assumption that under-predicted the relief requirements.

2.1 Reduced Reboiler Duty

When looking at a column system, the primary relief loads tend to consist of a combination of continued heat input with the loss of cooling. The relief load estimates can be complex due to the very nature of the system. The ability to do a detailed analysis of the column may be able to better predict upset conditions. The authors have seen many cases where a detailed column analysis prevented the relief device modifications from being made based on inaccurate sizing basis. The ability to accurately model the relief scenario is paramount because an error could lead to events such as an unnecessary facility modification or a loss of containment from an undersized relief device. When a pressure relief system designer looks at the loss of overhead cooling on a column system, the scenario is a transient event. The case can start with the loss of overhead condensing with the reflux continuing for a short duration. After some time, the loss of condensing and continued overhead liquid product output results in the loss of level and reflux flow. To estimate this relief condition, the column system is modeled as two steady state cases to bracket the relief load. Under either of these conditions, the bottoms temperature may increase due to the increase in system pressure, but this phenomenon may be offset by the reduction in separation; therefore, a reduced duty calculation should be performed for both calculation sets to ensure that the reboiler is capable of transferring the same heat into the column system as during normal operating conditions. When looking at the loss of overhead cooling with the reflux continuing, the reboiler duty may not be able to be reduced due to increased light component traffic in the tower (as compared to the top tower reflux failure) when compared to the high temperatures in the reboiler and lower duties expected during the top-tower reflux failure scenario due to the loss of the light ends liquid traffic in the column. Taking credit for the reduced duty is still a reasonable assumption as an upper conservative limit when the assumption is based on the maximum duty possible for the installed reboiler and is limited by equipment capacity. The design equation for heat exchangers is given below [2]:

$$Q = UA\Delta T_{LM}$$

Reduced duty calculations, limited to process/process or process/utility exchangers (e.g. shell and tube exchangers), are not typically performed on fired heater reboilers. The exchanger data sheet, if up-to-date, is the best resource for determining the heat transfer characteristics of an exchanger. Most process simulators can simulate the duty of shell and tube exchangers by specifying the clean UA and, if applicable, the utility stream (e.g. 50# steam). The duty for the exchanger may change as reflux is lost to the system, which the upset may result in hotter bottoms temperature to the reboiler. The reduced duty may reduce the relief load estimate, as there is less vapor generation as compared to using the process duty under normal conditions. Figure 1 shows a screenshot from a process simulator that estimates the rate for a column system with continued feed and heat input, but a loss of cooling and reflux.

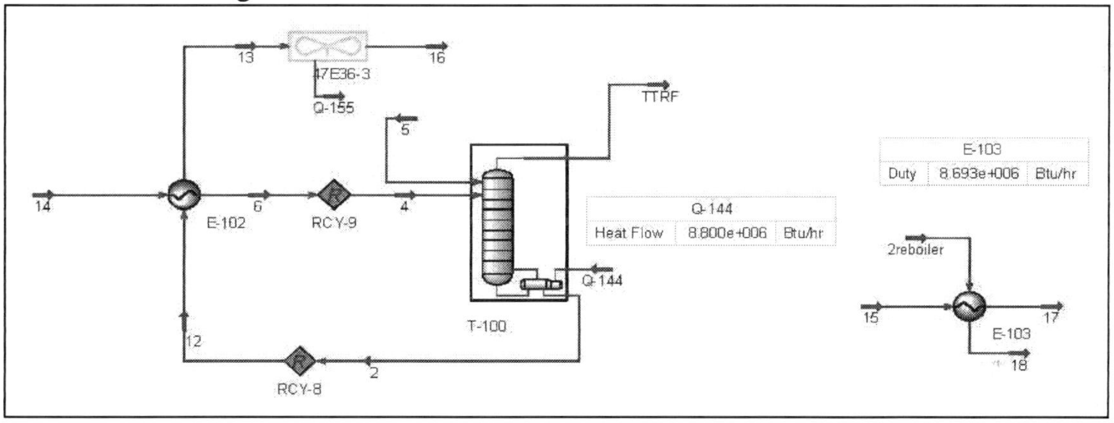

Figure 1: Example of a simulation for a column with loss of cooling and top-tower reflux failure with a reduced duty calculation performed.

2.2 *Using the correct composition and physical properties*

Generally the capacity of the reboiler (e.g. the reduced or LMTD pinched duty) will vary greatly if the calculation is based on the normal bottoms product as compared to the overhead reflux/liquid product. In the case of system upsets, the actual liquid inventory in the bottom of the tower may be lighter than normal. The pressure relief system designer is cautioned to use a composition in the reboiler to estimate the reduced duty that is consistent with the overpressure scenario being reviewed.

2.3 *Inlet Control Valve Failure Credits*

Another common overpressure scenario is the failure of automatic controls. Per API 521, this scenario is the complete failing open of a control valve (irrespective of its fail safe position).[3] If this upset condition results in overpressure, a relief device may be required to protect downstream equipment. When looking at an inlet control valve failure, the first common assumption pressure relief system designers make is to estimate the full capacity of the control valve and then verify if the relief valve can pass that fluid at the appropriate overpressure. The designers verify if the capacity of the relief device is greater than the control valve. If this estimation results in insufficient relief valve capacity, then some credits may need to be taken to make the relief load estimate more realistic.

> Per API 521§ 5.10.2,
> "in evaluating relieving requirements due to any cause, any automatic control valves that are not under consideration as causing a relieving requirement and that would tend to relieve the system should be assumed to remain in the position required for minimum normal processing flow. In other words, no credit should be taken for any favourable instrument response." [3]

For example, if the outlets from the system are expected to remain open, the calculation may assume that some of the fluid will flow through the normal path, see Figure 2. This credit is usually at the minimum turndown rate, as common industry practice is to assume that other control system remains in place and does not respond (unless it exasperates the single failure).[3] If the relief scenario is liquid, credit can be taken for the liquid outlet to reduce the fluid from the control valve that needs to be relieved by the pressure relief device by the minimum turndown rate. If the liquid outlet does not provide enough credit, the vapor outlet can be sized to take some relief as long as the vapor line is not lost by passing the liquid. A common example of this would be that a high-level alarm would trip a compressor; in this case, the relief device would have to pass both the liquid and the vapor feed to the vessel as the increased liquid level caused multiple failures. As long as the pressure relief system designer verified the liquid could flow through the vapor line and accumulate in the downstream system, credit can be taken through the vapor line. If the relief scenario is vapor, credit may be taken for flow through the vapor line at the minimum turndown rate (though the potential for credit for vapor flow through the normal liquid outlet is normally not considered). Consideration should be given for the difference in the relief fluid and the normal fluid and to adjust the credit based on the actual volume rate.

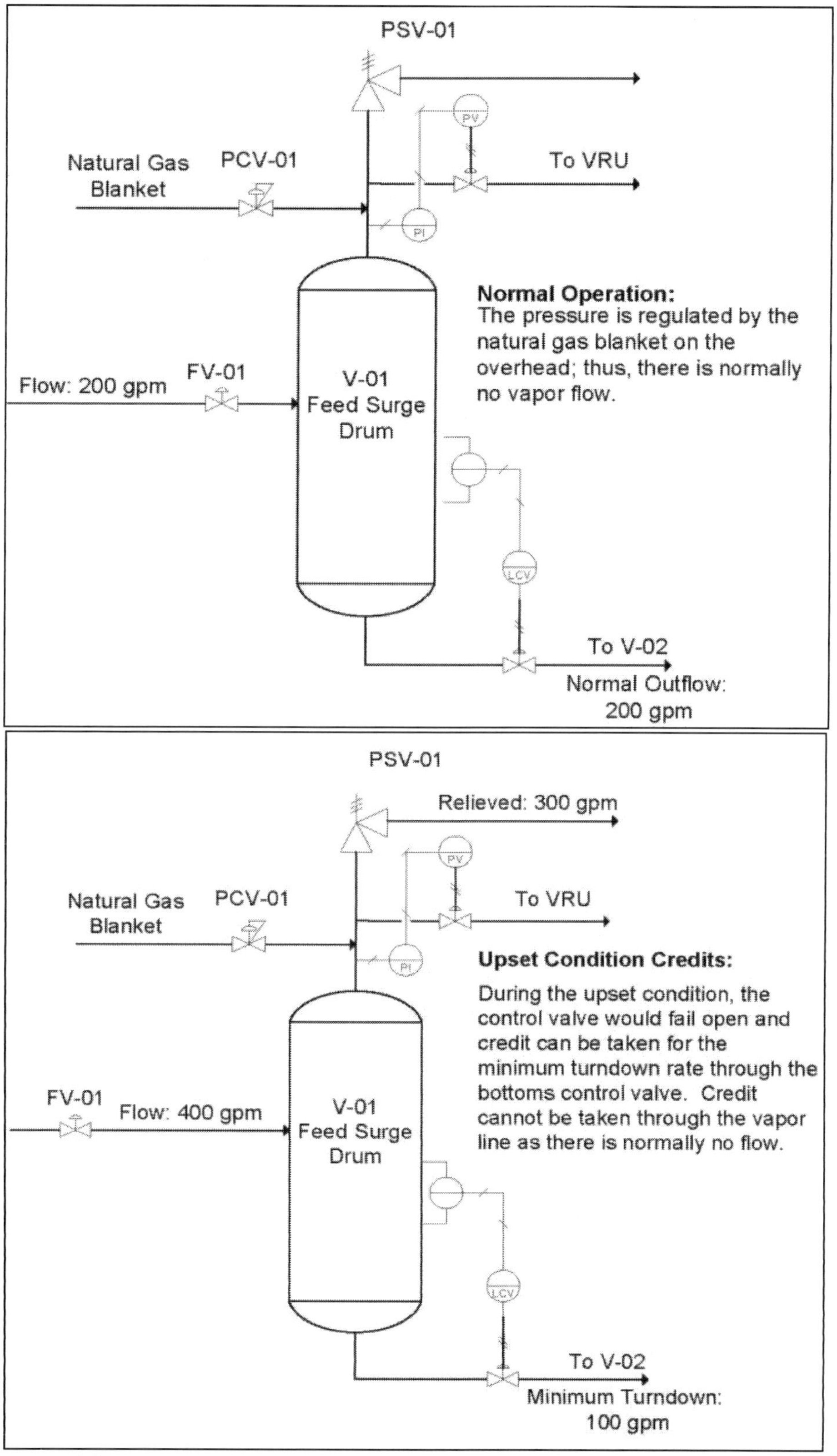

Figure 2: Looking at a feed surge drum during inlet control valve failure can lead to credits to be taken as long as they are reasonable for the system. Credit cannot be taken through the vapor line as there is normally no flow.

There is a special control valve failure case that pressure relief system designers need to be aware of. When a liquid control valve fails open, the vessel may de-inventory and gas may flow through the liquid control valve into the downstream vessel. This scenario, typically called a gas blow-by, should be reviewed for all control valve failures where there is liquid and gas in the upstream vessel. Figure 3 shows a typical system where gas blow-by may occur.

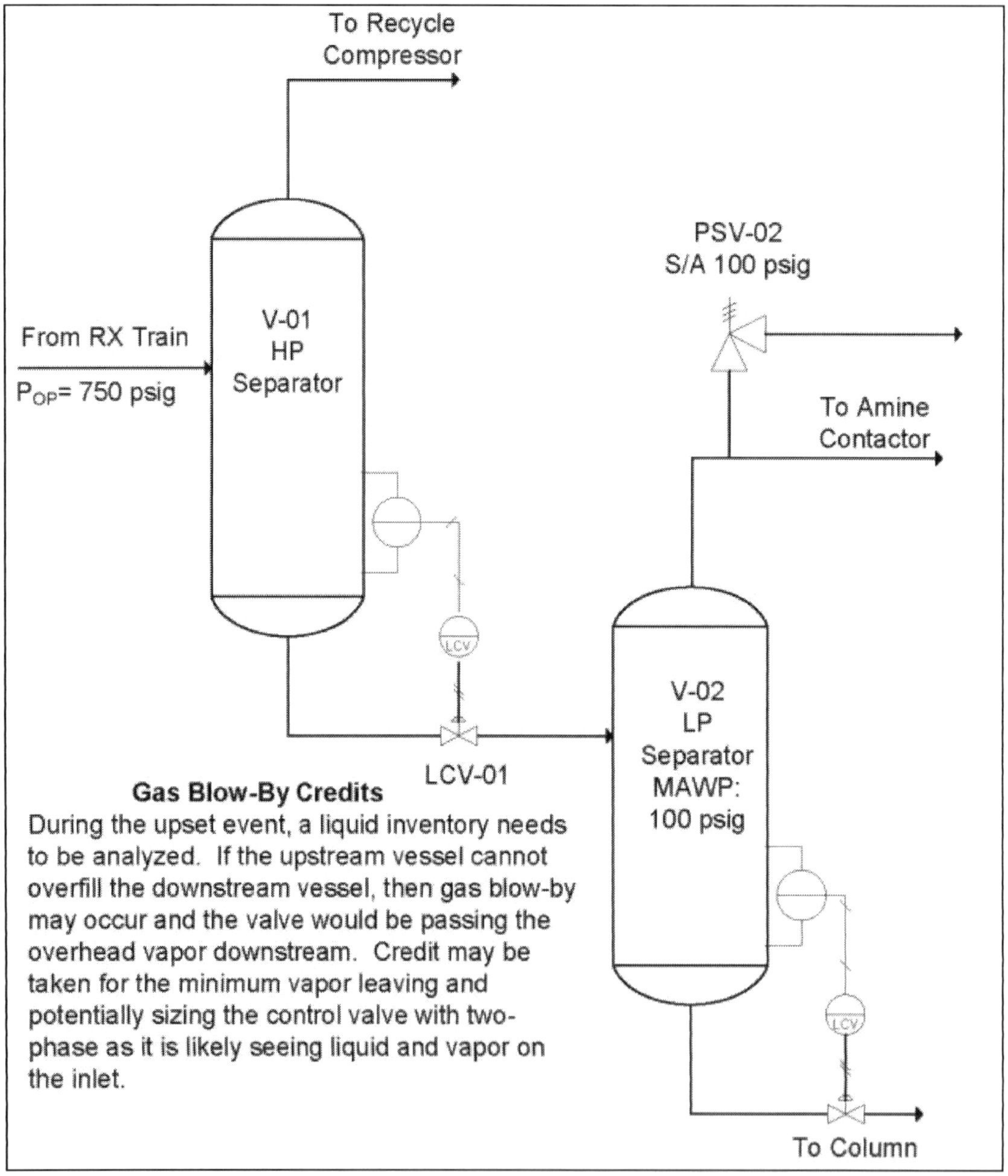

Figure 3: Example of when gas blow-by may occur. When looking at a low pressure separator, the sizing event should be vapor breakthrough the control valve from the upstream high pressure separator.

3. When Realistic Assumptions Make the Estimates Worse

3.1 Compressors Downstream of the System

In some instances, making relief load estimates match closer to reality may lead to a larger relief load estimate. In these cases, the pressure relief system designer must ensure that the relief system is accurately modeled and that the designer uses the larger rates and does not allow designs based on nonsensical assumptions to be used as the basis. There are many different assumptions that could cause this case to be true; the example that we are going to highlight is when the cascading effect of one failure causes the normal plant control system to take an action and increase the relief requirements for a pressure relief device. Oftentimes, a pressure relief system designer may want to take credit for some of the relief effluent to outflow from the system through a compressor. As stated above, this would not be a viable flow path for liquids, as almost all compressors have a high level trip on the inlet drum, but what about vapors? Depending on the scenario, the compressor may pass vapors at the upset conditions or the compressor could trip and result in a blocked outlet. Examples that may cause a compressor to trip under upset conditions include a higher inlet gas temperature trip, a high amp / power requirement trip (due to the increased suction pressure), high temperate trip (due to the potential loss of cooling to the compressor), vibration trips, and etc.[4,5] It is useful for the pressure relief system designer to discuss the scenario with personnel who are familiar with the operation of the compressor (usually either the operators or unit process engineer) to ensure that any assumptions made are consistent with operational history. Figure 4 shows a case where a compressor trip would not result in a larger relief load.

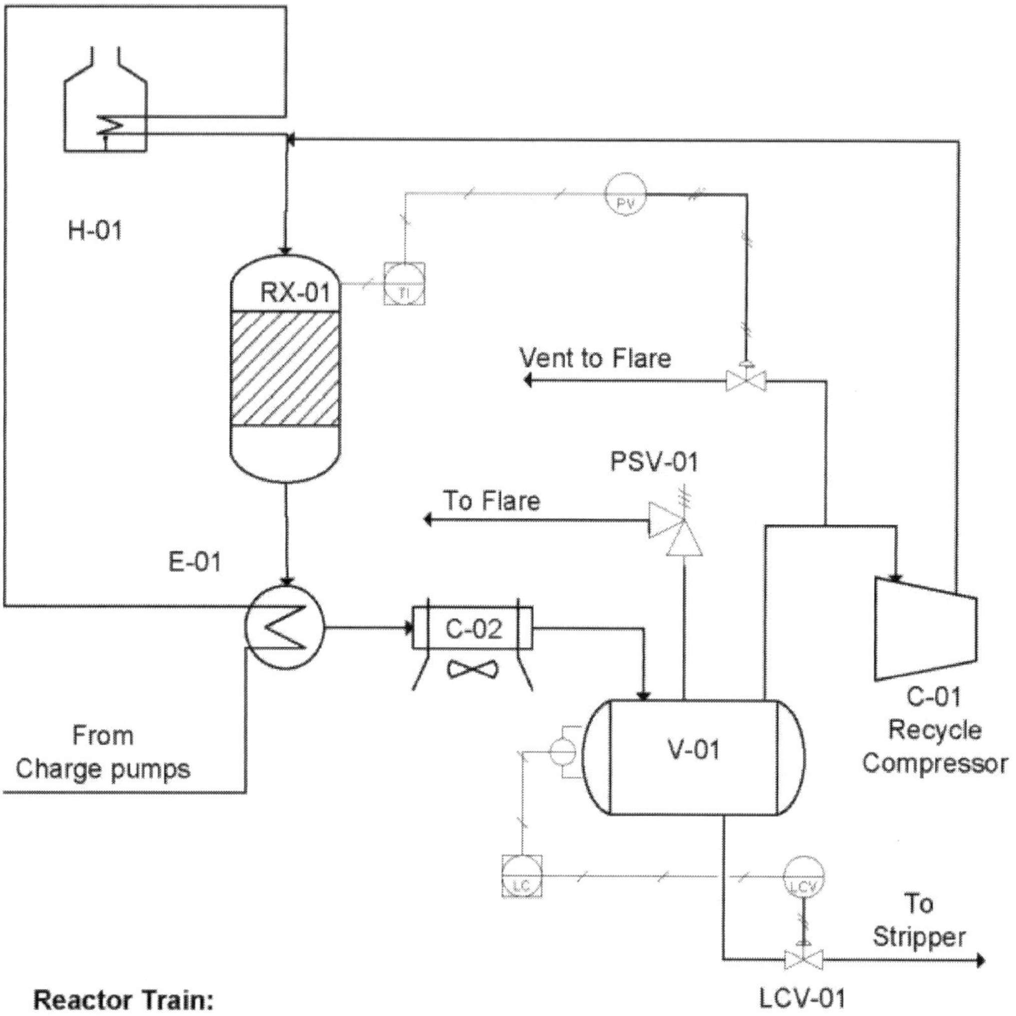

During an upset event due to the loss of the compressor, the relief load is not expected to be any larger as the recycle gas is lost. This source is not a feed source since it never enters or leaves the system. Also, the de-pressuring valve on the suction line to the compressor may open to vent to flare if a high temperature reading occurs.

Figure 4: An example of how compressors can affect the relief load. The loss of the compressor would not result in any excess relief as the recycle gas is lost.

3.2 Example of Compressors Shutdown Analysis

A larger estimated rate may occur in a scenario where there is a partial loss of cooling in a column system that uses a compressor to move the off gas from the overhead accumulator to the downstream system (e.g. a FCC Main Fractionator, Coker Bubble Tower, Atmospheric Tower, and etc.). After the simulation is developed to model the column operation during the partial loss of cooling, the modified characteristics of the off gas (e.g. temperature, density) are consistent with conditions that trip the compressor.

Additionally, the pressure relief system designer must ensure that the scenario itself (e.g. a unit wide power or cooling tower failure) could result in a compressor trip and result in this outlet being blocked. The relief load estimate would then need to include the cascading effects of the compressor shutting down and may be much larger than would be otherwise predicted (see Figure 5).

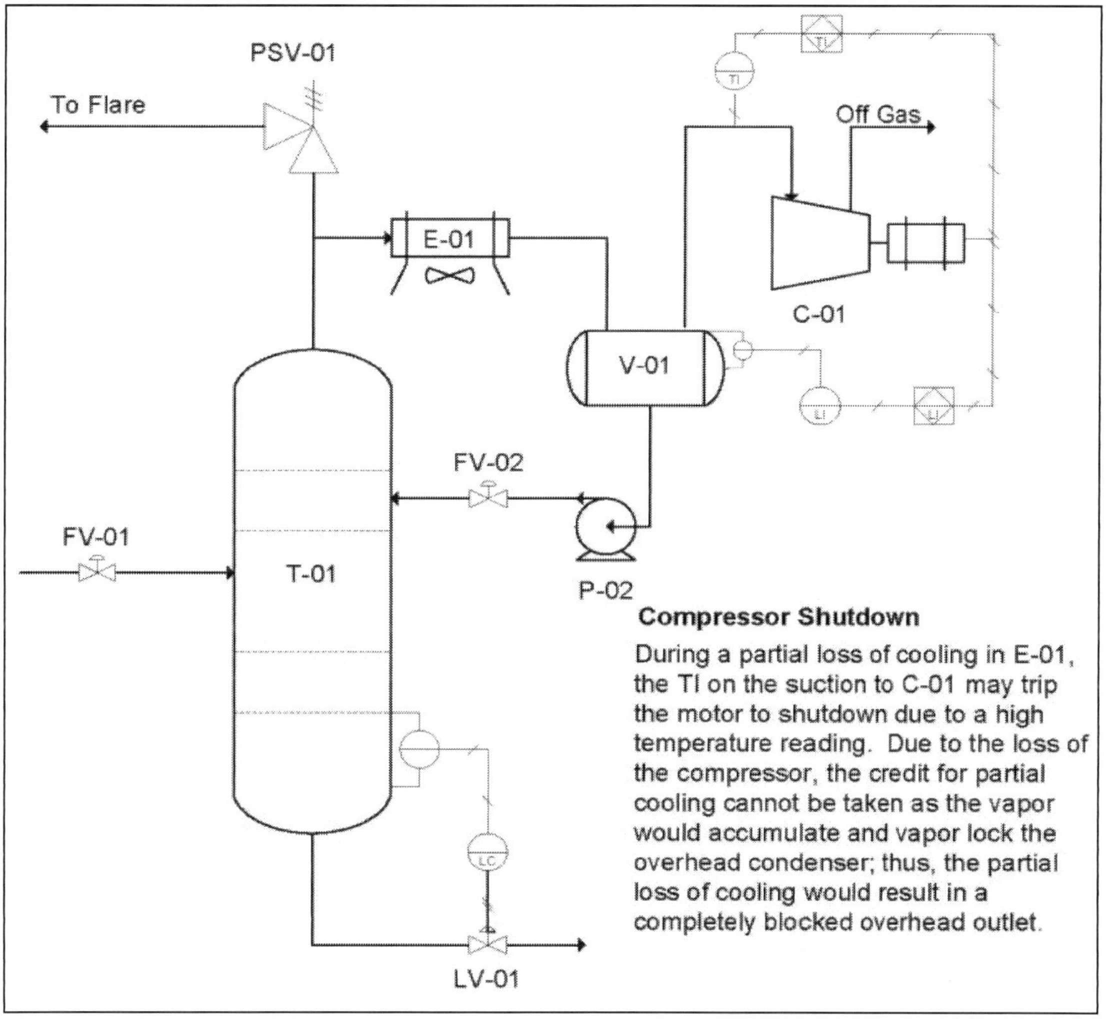

Figure 5: Example of how compressors can affect the relief load. On the right, the loss of the compressor would result in a larger relief as the vapor outlet is completely blocked.

4. Conclusion

To realistically represent the relief load estimate, the pressure relief system designer must understand the system, how it is operated, the limits and shutdowns of the associated equipment, and how these parameters interact during the specific overpressure event being reviewed. The pressure relief system designer must also ensure that all assumptions are consistent with industry, company, and regulatory requirements for the facility. There are many ways to check to see if the relief load estimate is based on realistic assumptions or on conservative/simplifying assumptions. In many cases, conservative/simplifying assumptions are acceptable as long as the assumptions don't result in deviations from standards or practices that result in spending more money. Oftentimes, conservative/simplifying assumptions do result in deviations, and the pressure relief system designer must then use more realistic assumptions to generate the relief load.

The pressure relief system designer must also be vigilant to not use simplifying assumptions that are not conservative. These assumptions tend to result in relief rates that may under-predict the true requirements in an overpressure event. Oftentimes industry, company, and regulatory requirements for the facility are set up to eliminate these scenarios, but the example in Section 3.2 illustrates the pressure relief system designer must check these assumptions (many of which can be implicit).

5. References

[1]. Smith, Dustin and John Burgess, "Relief Valve and flare action items: What plant engineers should know," *Hydrocarbon Processing*, Volume 91, NO.11, 2012

[2]. Middleman, Stanley, *An Introduction to Mass and Heat Transfer: Principles of Analysis and Design*, John Wiley and Sons, New York, NY. 1998

[3] ANSI/API Standard 521, "Pressure-relieving and Depressuring Systems," American Petroleum Institute, 5th ed., 2008.

[4]. GPSA, *Engineering Data Book: Volume I 10th Edition*, Gas Processors Suppliers Association, Tulsa, OK. 1987

[5]. Lieberman, Norman and Elizabeth Lieberman, *A Working Guide to Process Equipment 3rd Edition*, McGraw-Hill, New York, NY. 2008

Explosion incident simulation for Petrochemical plant

Zhao Xiangdi
State Key Laboratory of Safety and Control for Chemicals, SINOPEC Safety Engineering Institute
No.218, Yan'an San Road, Qingdao, P.R.China 266071
cheeryde@gmail.com

Yuan Jiwu
State Key Laboratory of Safety and Control for Chemicals, SINOPEC Safety Engineering Institute
No.218, Yan'an San Road, Qingdao, P.R.China 266071
Yuanjw.qday@sinopec.com

Jiang Chunming
State Key Laboratory of Safety and Control for Chemicals, SINOPEC Safety Engineering Institute
No.218, Yan'an San Road, Qingdao, P.R.China 266071
Jiangcm.qday@sinopec.com

Wang Zheng
State Key Laboratory of Safety and Control for Chemicals, SINOPEC Safety Engineering Institute
No.218, Yan'an San Road, Qingdao, P.R.China 266071
wz.qday@sinopec.com

Prepared for Presentation at
American Institute of Chemical Engineers
2013 Spring Meeting
9th Global Congress on Process Safety
San Antonio, Texas
April 28 – May 1, 2013

UNPUBLISHED

Explosion incident simulation for Petrochemical plant

Keywords: numerical simulation, explosion incident, TNT equivalent, Multi-Energy

Abstract

A new method of explosion simulation which is based on FLACS software was used for simulating and rebuilding an explosion incident which occurred in a coal gasification company. The other two commonly used methods which are based on TNT equivalent method and Multi-Energy method were considered to calculate the explosion overpressure also. Simulation results indicated that this new method can calculate explosion overpressure in three dimensional. It can make up for the shortfall of traditional assessment techniques. The result is more stable and reasonable when compared with the scene of the incident.

1. Introduction

Petrochemical plants may happen explosion incident easily than other facilities since they have many equipments and devices with lots of hazard material. Once the device exploded, it will cause large casualty and property loss. For example, Alon big spring refinery(2007) and Guangwei(2008) occurred fire and explosion incidents and caused many people dead. So it is important for our to calculate plant's capability against explosion. We could use the results to determine whether those plants should be firmed and changed or not. Researchers around the world found a series special assessment methods based on incident consequences or risk. Someone used those methods to design and renovate process safety of petrochemical plant[1][2]. However, researches on those technology were lags behind in China. We used traditional assessment technology which based on empirical formula. Those traditional methods did not consider the actual situation of factory. It couldn't show the those factor's effect on explosion incident such as the layer equipment, wind, fire protection system and other condition [3][4][5].

Nowadays the Computational Fluid Dynamics analysis technique is becoming more sophisticated. Someone in industry and academia studies explosion simulation used CFD. Herrmann, Hansen studied the consequences of gas cloud explosion using FLACS computational fluid dynamics software, then compared results with experiment, and found the results matched each other well[6][7]. Windhors, used AutoReaGas software to simulate the gas explosion and determine safe distance between petrochemical cracker and control room [8]. P.Hoorelbeke studied explosion risk assessment techniques commonly used [9]. However, there were only a little study on numerical simulation model for explosion incidents. Therefore, assessment on explosion incident by using FLACS were performed in the work.

2. Explosion incident and Simulate model

2.1 Explosion incident

23 December 2008 an explosion incident occurred in Hunan province in China. Gas released from one pipe in the carbon monoxide shift.

Then the gas cloud was ignited and caused explosion. The scenes after this incident with the insulation layer damaged was shown in figure 1.

(a)F2012，F2013, F2014

(b) T2101，R2103

(c) Office building with windows and glass damaged

Figure 1: The scene of the explosion incident with the insulation layer damaged

The gas which flowed in the pipeline was hydrogen. The loss rate of contain was 562kNm3/h in the release phase. The wind speed in-situ was about 3m/s with the direction was NE. The key equipment in this unit for simulation were shown in table 1.

Table 1. Information for the key equipment

Number	Name	Size(m)	Description
F2102	Gas-liquid separator	Φ4.2×13.4	40% of insulation damaged
F2103	Gas-liquid separator	Φ3.7×11.6	25% of insulation damaged
F2104	Gas-liquid separator	Φ3.7×15.6	Little effect
T2101	Regulate tower	Φ3.2×26.0	70% of insulation damaged

R2103	Converter	Φ4.9×13.8	50% of insulation damaged
R2104	Converter	Φ4.9×11.7	30% of insulation damaged
Control room		15.0×46.8×4	Some of the windows damaged; Lots of the windows broken
Office building		15×46.8×19.5	Some of the windows damaged; Lots of the windows broken

2.2 Model for simulation

3-D model which was based on the factory's actual layout was shown in figure 1. This model was built using the actual sizes of the carbon monoxide shift unit. The total size for simulation was 100m×80m×40m. Ignition position was in the center of carbon monoxide shift unit nearby leak point. The distance from ignition position to control room and office building was 65m. Shapes of equipment were modified for simulation. Monitor points were set for key equipments, as shown in figure 2.

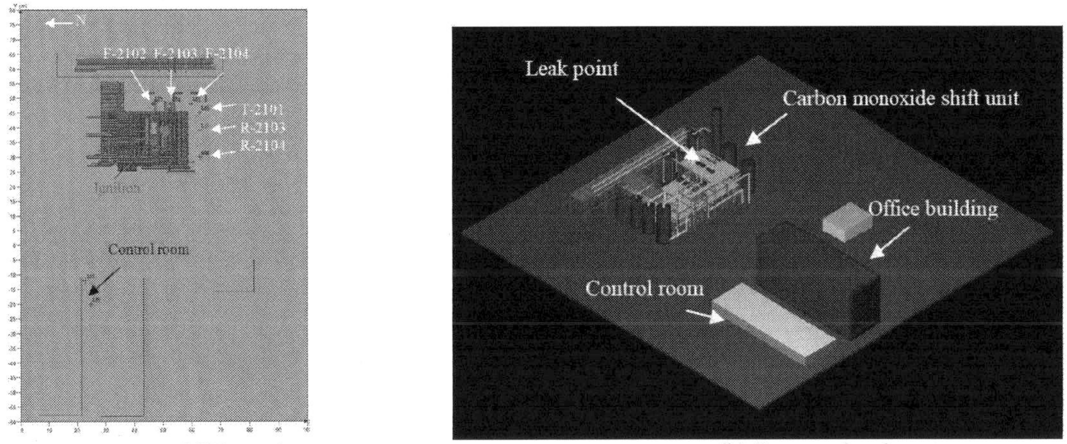

(a)Top view (b) Isometric view

Figure 2: Three dimensional model for simulation

2.3 Results and discussion

2.3.1 Leak and dispersion

The material in pipeline was hydrogen. Loss rate for contain was 562kNm^3/h. The time was 24 second between leak started and gas ignited. Wind speed was about 3m/s with direction was NE when the incident occurred. The new method which was based on FLACS was used to rebuilt dispersion process for hydrogen. Simulation results of fuel size after dispersion were shown in figure 3. The whole volume of the hydrogen was about 10×13×7 m^3 in a mixture concentration with equivalence ratio(ER).

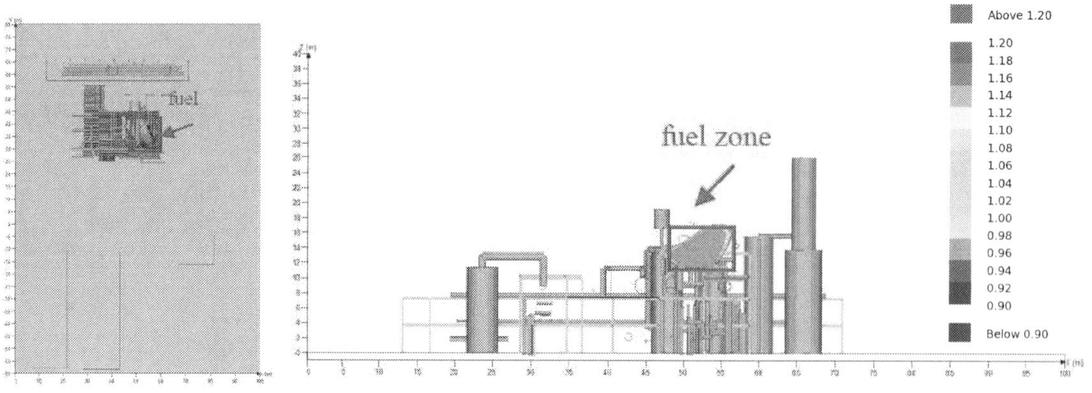

(a) top view (b)side view

Figure 3: The fuel size in this unit after dispersion (0.9~1.2 ER)

2.3.2 Explosion

Simulation assumes that hydrogen and air were in a mixture concentration with equivalence ratio. TNT equivalent method, Multi-Energy method and FLACS method were used to assess the process for gas explosion impact on the equipments and facilities. However, TNT equivalent method was simple and did not consider the effect of actual situation. Although the Multi-Energy method considered the explosion strength as the effect of actual situation, but it also did not consider reflection of complex equipment and did not predict the effect of risk mitigation system such as water spray.

Results of explosion overpressure using those three method were shown in table 2. It can be seen that TNT equivalent method and Multi-Energy method could only take into account explosion overpressure in two-dimensional. For example, overpressure calculated in the position 48m from center of explosion was 5.2 kPa using TNT equivalent method, however, the results increased to 7.9 kPa using Multi-Energy method. But, simulation results using FLACS were different. It could give overpressure in three-dimensional. The overpressure in position 48m from center of explosion was 6.7 kPa with 10m above ground, and 4.7 kPa with 19m above ground , as shows in table 2.

Table 2. Results of explosion overpressure using different methods.

		Overpressure in the different position from the ignition point (kPa)		
		48m (office building)	65m(control room)	74m
TNT equivalent method		5.2	3.48	2.95
Multi-Energy (blast strength 5~10)		7.9 ~ 24.7	5 ~ 16	5.1 ~ 13.4
FLACS method the height above ground(m)	1m	7.8	6.5	3.9
	10m	6.7	-	3.6
	19m	4.7	-	1.9

The overpressure in different monitor points(as shown in figure1) were shown in figure 4. The maximum explosion overpressure in carbon monoxide shift unit was about 62kPa, as well as, the overpressure was about 6.5kPa near the control room and office building.

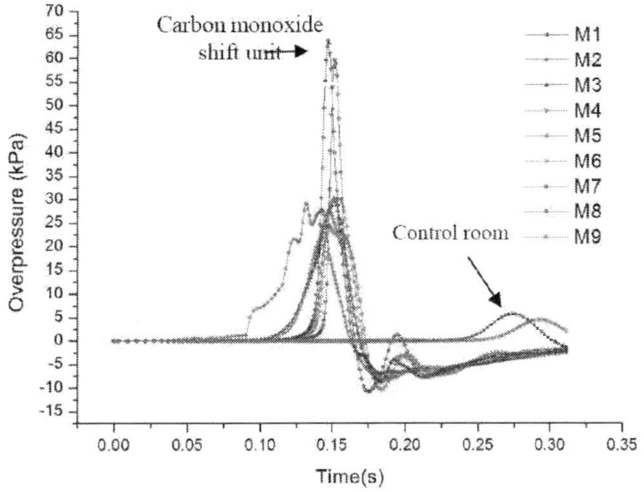

Figure 4: Overpressure of explosion in the different monitor points

The process for explosion overpressure wave affecting on equipments and facilities were shown in figure 5. As soon as explosion starting, a strong shock wave (a), causing major damage to the surrounding equipments, then it would gradually expand and weak (b-c), and subsequently affected office building and control room (d).

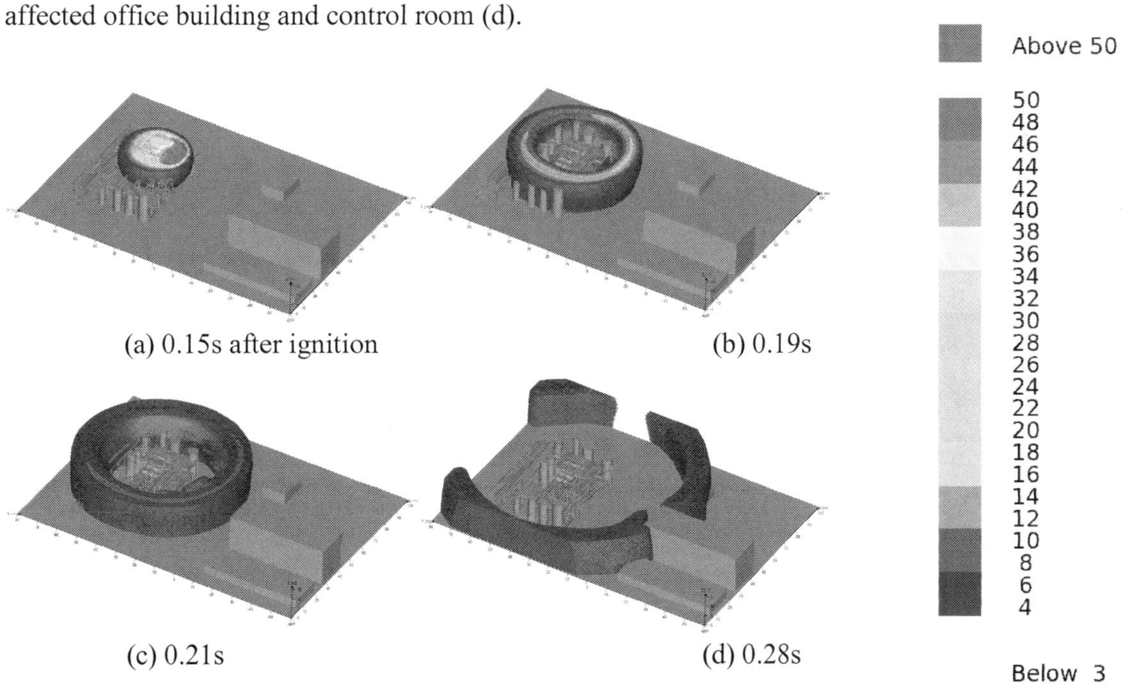

Figure 5: Explosion overpressure wave changed and expanded process (3-50 kPa)

3. Conclusion

The shortfall of traditional assessment techniques is that they can't predict the explosion effects with the layer equipment, wind, fire protection system and other condition's effect on the flame acceleration. A method of explosion simulation which is based on FLACS software for petrochemical plants was introduced in this paper. The results were as follows:

This method can calculate all parts of the space after explosion, and can make up for the shortfall of traditional assessment techniques. The overpressure simulation results using this method are between the results calculated by TNT equivalent method and Multi-Energy method. The result is more stable and reasonable when compared with the scene of the incident.

4. References

[1] D.Bjerketvedt, J.R.Bakke and K. Wingerden(1997), "Gas explosion handbook ,"*Journal of Hazardous Materials*, Volume 52, pp.1-150.

[2] S.B. Dorofeev(2007), "Evaluation of safety distances related to unconfined hydrogen explosions ," *International Journal of Hydrogen Energy*, Volume 32,p.2118-2124.

[3] B.H.Hjertager, I.Moen, J.H.S.Lee, R.k.Eckhoff, k.Fuhre, O. krest,(1981), "The influence of obstacles on Flame propagation and pressure development in a large vented tube ," *CMI Report No. 803403-2,* Chr. Michelsen Institute, Bergen, Norway.

[4] "Guidelines for chemical process quantitative risk analysis,"*American institute of chemical engineers*, NY,2000.

[5] A.C. van den Berg (1985), "The multi-energy method. A framework for vapour cloud explosion blast prediction ,"*Journal of Hazardous Materials*, Vol.12, pp.1-10.

[6] Herrmann, D.D. (1999), " Improved Vapor Cloud Explosion Predictions by Combining CFD Modeling with the Baker-Strehlow Method ," *International Conference and Workshop on Modeling the Consequence of Accidental Releases of Hazardous Materials*. CCPS/AIChE, Sep. 28 ,pp.479-494.

[7] Hansen, O.R., Talberg, O. and Bakke, J.R (1999), "CFD-Based Methodology for Quantitative Gas Explosion Risk Assessment in Congested Process Areas: Examples and Validation Status ," *International Conference and Workshop on Modeling the Consequence of Accidental Releases of Hazardous Materials*. CCPS/AIChE, Sep. 28 ,pp.457-477.

[8] Windhorst, J.C.A (1999), " Deriving Engineering Specifications from Release Dispersion and CFD Explosion," *International Conference and Workshop on Modeling the Consequence of Accidental Releases of Hazardous Materials*. CCPS/AIChE, Sep. 28 ,pp.495-514.

[9] P.Hoorelbeke, C.Izatt, J.R.Bakke, J.Renoult, R.W.Brewerton(2006), "Vapor cloud explosion analysis of onshore petrochemical facilities," *American Society of Safety Engineers Middle East Chapter 7th Professional Development Conference& Exhibition kingdom of Bahrain*. March 18-22,pp.1-15.

Using Reliability Centered Maintenance (RCM) to Reduce Fire Risk in High Temperature Oil Handling

Marc W. Yarlott, P.E.
Veolia Water North America
200 E Randolph St,
Chicago, IL 60601
marc.yarlott@veoliawaterna.com

Ryan Johnson
Veolia Water North America
200 E Randolph St,
Chicago, IL 60601
ryan.johnson@veoliawaterna.com

Steven Waters
Veolia Water North America
200 E Randolph St,
Chicago, IL 60601
steven.waters@veoliawaterna.com

Prepared for Presentation at
American Institute of Chemical Engineers
2013 Spring Meeting
9th Global Congress on Process Safety
San Antonio, Texas
April 28 – May 1, 2013

UNPUBLISHED

Using Reliability Centered Maintenance (RCM) to Reduce Fire Risk in High Temperature Oil Handling

Marc W. Yarlott, P.E.
Veolia Water North America
200 E Randolph St,
Chicago, IL 60601
marc.yarlott@veoliawaterna.com

Ryan Johnson
Veolia Water North America
Steven Waters
Veolia Water North America

Keywords: reliability, maintenance, RCM, critical spares, uptime

Abstract

Veolia Water North America contracted for the technically difficult Biosolids handling process Design, Build, Operate project with the Metropolitan Water Reclamation District of Greater Chicago at the Stickney Water Reclamation Plant. The plant design uses a high temperature oil to operate the Biosolids drying equipment. Several other plants using a similar high temperature oil system have had fires related to leakage of the oil. Having successfully employed the RCM analysis at many of Veolia's projects world wide, Veolia applied the RCM process to gain improvements in the overall reliability of the plant, and particularly the high temperature oil system, by optimizing maintenance strategies and identifying spare parts, procedures, and plans to maximize the uptime of the system. The RCM process resulted in identification of 425 spare parts at a cost of $525,000 which mitigated over $33 million dollars worth of potential risk. For a very technical process with significant risk of failures, the value of a complete RCM study will have a very high return on investment..

1. Introduction

Veolia Water North America (Veolia) operates a 150 dry ton per day biosolids pelletizer operation within the Stickney Water Reclamation plant in Chicago, Illinois. The facility is comprised of 4 independent pelletizer trains. The facility receives thickened solids (22-28%) and the solids are pumped to one of 4 large Seghers pelletizer units. The facility utilizes a gas fired boiler to heat the thermal oil that is used to transfer the heat needed to evaporate the moisture from the biosolids. The systems are very complex in nature and as with most solids processing facilities these systems are a challenge to operate and maintain.

As with any major start-up and commissioning, the operations staff gained a great deal of knowledge during the commissioning and start-up of the system. The process of converting the solids to a pellet is not just about heat transfer. Veolia's contract is a performance based one with plant uptime being the most significant driver for success. Veolia successfully completed its performance testing in late 2009. The management team at Veolia carefully evaluated each part of the performance test data and determined that it would be prudent to focus technical resources to build an uptime focused maintenance strategy.

In the Spring of 2010 Veolia initiated a comprehensive Reliability Centered Maintenance (RCM) review at the facility. This review was facilitated by internal Veolia employees and included the valuable input of the operations and maintenance staff at the facility. Veolia also brought in staff from other projects to help in analyzing specific functional areas. This RCM review incorporated a critical spares analysis that would be used to make cost/risk based storage decisions.

2. METHODOLOGY

2.1 Reliability Centered Maintenance

RCM is "a process used to determine the maintenance requirements of any physical asset in its operating context" [1]

Reliability Centered Maintenance has its roots in the civil aviation industry. In the late 1960s United Airlines invested a great deal of time and money in studying how and why equipment failed. The United Airlines management team understood that safety and reliability were of paramount importance to the operational mission. United Airlines mapped out a plan to better understand equipment failures, and focused the project on improving overall reliability of its aircraft. This analysis resulted in a startling conclusion that "there was no correlation of equipment age and failure" and that "equipment failures could not be prevented." In 1978, United Airlines published its findings in a document titled Reliability Centered Maintenance [2]. This publication was a significant document that challenged maintenance professionals to take a hard look at how and why maintenance tasks were performed.

The study concluded that equipment failure conformed to one of six failure patterns, see Figure 1 [1]. These curves plot the probability of failure on the y-axis and time on the x-axis. The failure curve analysis yielded very surprising results in that the statistics proved conclusively that equipment failures were not primarily dependent upon equipment age. The equipment failure distribution of these patterns is as follows:

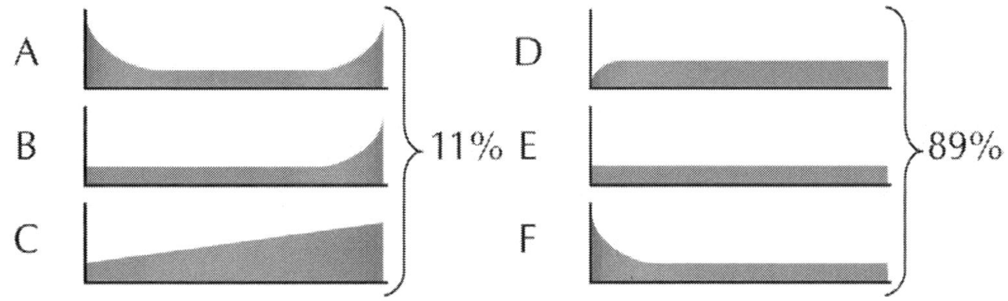

Pattern A	**4%**
Pattern B	**2%**
Pattern C	**5%**
Pattern D	**7%**
Pattern E	**14%**
Pattern F	**68%**

Figure 1, Six Patterns of Failure [1]

Based on this data, government regulators and the airlines decided to revamp the maintenance strategy of the civil aviation industry. In reviewing the failure patterns, one can quickly deduce that patterns A, B, and C have some dependency on equipment age, while patterns D, E, and F are definitely not age related.

It became quite obvious that performing calendar time based overhaul/intrusive maintenance was actually contributing to equipment failures rather than preventing them as planned. This realization resulted in drastic change to the airline maintenance programs and aircraft designs. This was also the beginning of the reliability-centered maintenance movement.

RCM will result in the correct maintenance mix of preventive maintenance (PM), condition assessment, repair (also called reactive maintenance) and proactive maintenance techniques to ensure an increase in the probability that a machine or component will function in the required manner over its design life cycle with a minimum of maintenance. RCM requires that maintenance decisions be based on maintenance requirements supported by sound technical and economic justification. As with any strategies, there are many ways to achieve the end goals.

It is important to note that RCM is not a "silver bullet" that can be applied in a lackadaisical manner. RCM requires discipline and attention to detail in the evaluation process. An RCM analysis is based on a detailed Failure Modes and Effects Analysis (FMEA) and includes probabilities of failure and system reliability calculations. The RCM process ensures that all the critical failure modes are being managed. The analysis is used to determine appropriate maintenance tasks to address each of the identified failure modes and their consequences. RCM is a proven approach that reduces risk and focuses limited maintenance resources on important assets.

In its simplest form the RCM process requires that a group knowledgeable in the operation and maintenance of the facility review systems in a methodical and organized manner. The goal is for the Operations and Maintenance (O&M) staff to learn from one another and freely exchange information related to the process, equipment, reliability, and operation. The RCM process requires that the O&M staff evaluate the following seven questions:

- **Functions:** What does this asset /system do (what do the users want this asset to do)?
- **Functional Failures:** In what ways can the asset/system fail to deliver the functions described above?
- **Failure Modes:** What causes the functional failures to occur?
- **Failure Effects:** What happens when the failure occurs?
- **Consequences:** Does this failure even matter?
- **Proactive Tasks:** Is there anything that can be done to predict or reduce the likelihood of the failure?
- **Default actions:** Otherwise, what other action is available to improve reliability?

Each failure mode is put through a decision tree (Figure 2) to determine its impact on environmental, safety, and financial risk. All of the critical failure modes are subsequently examined and assigned a proactive task (PM, PdM, or failure finding) to manage the risk.

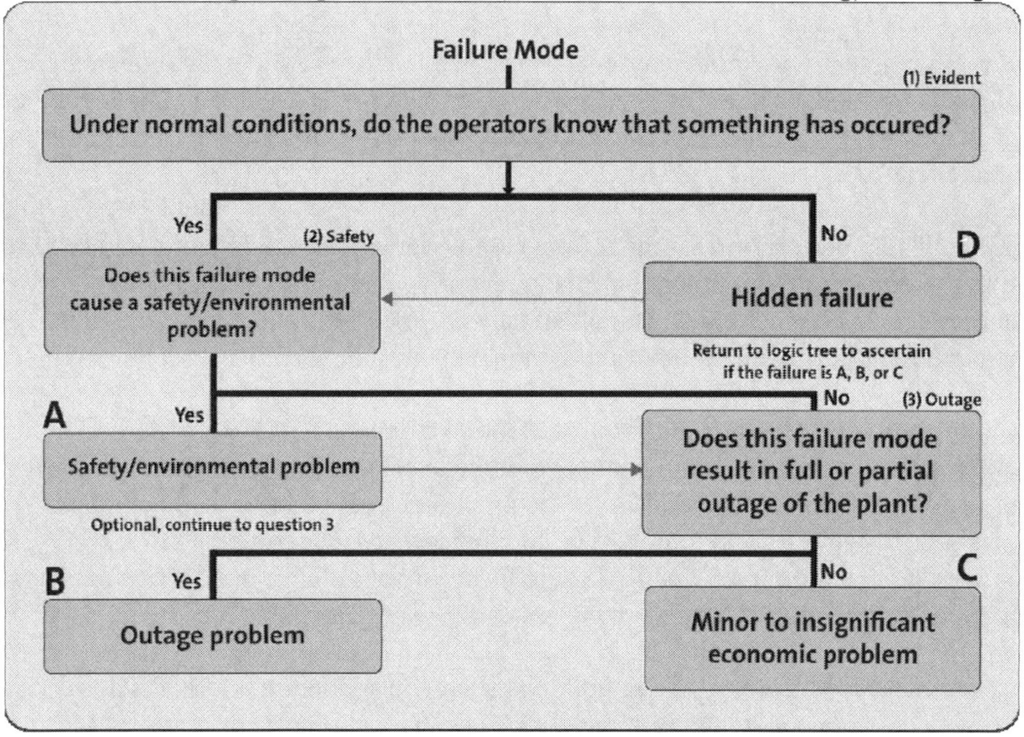

Figure 2, Decision Diagram [3]

3. RESULTS

3.1 Implementation of RCM

Prior to performing the RCM review, the project team was experiencing mechanical failures of system components that were resulting in some periods of downtime. The facility uptime was being negatively impacted by asset failures, and the need for rapid repairs resulted in little to no root cause analysis. The facility uptime had dropped off after performance testing and uptime was running on average of 50% June 2010. Staff worked out a plan that prioritized the RCM reviews based on the principle of analyzing problem systems first.

Initially the RCM project focused on one train of the processing facility. The train was further separated into 13 systems. Over a 16 week period, staff completed 13 classical RCM Reviews. In total over 1784 failure modes were analyzed in total.

One of the benefits of conducting the RCM review was that it also gave the project staff the opportunity to understand in more detail where the system choke points were, and why the failures were occurring. Many useful suggestions on improving uptime and reliability were recorded and noted in the RCM review as "items of interest". In total over 104 items were identified as recommended improvements to the existing installation. For example, one item of interest was related to the biosolids cake pumps. These pumps had some highly unusual failure modes and had become quite problematic from an O&M perspective. During the RCM review there were several failure modes that pointed the review team to questions if the hydraulic unit that was provided with these pumps was adequately sized. The hydraulic unit reservoir was running hot, and there were wide fluctuations and pressure swings in the system. After the RCM review the facility staff worked with a local engineering company to resize the drive unit. This change has significantly improved the uptime of the pumps and eliminated many of the "unusual" failure modes. Staff at the project are now working their way through the "items of interest" and implementing many of the group's recommendations.

A second benefit of the RCM review was a success directly related to the high temperature oil system. During the RCM Review the staff identified Failure Mode #61:

61	***Primary Oil***	Oil heater High Pressure alarm: Staff needs to confirm that there is indeed a High Pressure alarm on the oil heater when the PRV fails closed. This is a safety issue that can be mitigated with existing instrumentation

In following up on this item the plant staff confirmed that there was indeed a high pressure alarm; however, in checking the alarm, they discovered that the Pressure Relief Valves (PRV)'s were not the correct for this service. The Original Equipment Manufacturer (OEM) had supplied air service PRV's instead of liquid service. The process was shut down and immediately new PRVs were ordered and installed on the Heaters. The RCM review, allowed the staff to step through the features and failure modes of the equipment and identify and prevent a failure before it ever happened.

3.2 Critical Spare parts

There are many maintenance management practices that can help to properly manage risk. One area of maintenance management that must be included in a sound asset management program is the selection/identification of "critical spares." To achieve the desired availability of equipment, it is important to give proper consideration to having some spares parts/components available on-site. This is often implicitly considered by maintenance teams as an easy way to mitigate down time risks, thus leading to unnecessary amount of spares. One must be careful when evaluating critical spare parts. It is important that each spare part be properly vetted and justified through a proper risk based method.

Over the past 10 years equipment manufacturers have changed the way that they support their products. The days of having spares readily available for purchase have ended, in favor of the "just in time" manufacturing philosophy. In addition to less available stock, manufacturers have set some limitations on the timeframe that equipment is supported (planned obsolescence). These changes have created a need for operations and maintenance staff to better understand the challenges involved in obtaining needed parts.

Indeed, to be truly cost-effective the proper selection of critical spares has to be developed based on a methodical process. This process must take into consideration a cost-benefit/risk analysis that is aligned with the facility's goals and objectives. This analysis needs to include the following:

- Failure mode identification
- Consequences (safety, environmental, and economic) of not having the spare
- Lead time to procure
- Availability of spare (i.e. obsolete)
- Probability of failure
- Predictability of the failure
- Shelf life
- Storage considerations (space, environmental conditions, etc.)
- Cash flow

Each of these factors will now be considered in more detail.

3.3 Failure Mode Identification

A "failure mode" is an event that is reasonably likely to occur and that would cause the loss of an asset's functionality [1]. When developing a potential list of spare parts for an asset, an understanding of the equipment at the "failure mode" level is necessary.

The Chicago project identified 1784 failure modes in total. During the RCM review, staff decided which parts would be considered for spares. It is important that experienced operations and maintenance staff participate in the development of the failure modes. Special attention should be given to failure modes that are wear-related.

3.4 Consequences

Not all failure modes have the same impacts to the operations at a facility. Great care should be taken in evaluating the consequence of failure, utilizing a criticality review process (Figure 2). One should make an effort to monetize the risk of lacking a spare part when needed. For Chicago a project-specific matrix was developed and used to quantify lack of equipment availability into money. This allowed the staff to factor in items like cost of downtime, contract issues and safety into economic terms. This provided the framework upon which a proper spare part evaluation can be performed.

3.5 Lead Time

It is critically important that we understand the supply chain when it comes to spare parts. The availability and lead time for obtaining spare parts varies significantly based on the manufacturer/supplier and geographical location of the facility. When conducting a critical spares evaluation, one must always fully understand how long it will take to procure an item. Oftentimes excessive lead times may be reason enough to have a spare available at the facility. Initially a decision may be made not to stock a spare because a redundant asset is available. This decision may have to be reconsidered if one has to rely on solely the redundant system for a long period (while one waits for the spare to arrive). Many of the assets at the Chicago plant were custom built and manufactured overseas. Staff made contact with the vendors and determined the lead times on each of the items identified during the review.

3.6 Availability

Oftentimes equipment vendors discontinue product lines and support for products. Some of these changes are the result of market-driven forces, such as mergers and acquisitions within the manufacturing industry itself, while others are purely technologically-driven. Special consideration should be given to electronic components such as Variable Frequency Driver's (VFD) and I/O cards. In the Chicago review we discovered that some of the electronic components were no longer supported by manufacturers.

3.7 Probability of Failure

A proper critical spares evaluation needs to factor the probability of occurrence into the equation. This is usually done by experienced staff familiar with the common equipment in a similar operating context. There are published failure statistics that can also be used to benchmark this type of probability. Veolia has an enterprise Computerized Maintenance Management System (CMMS) that it uses to quantify the Mean Time Between Failures (MTBF). Understanding the failure rates will help in making a sound business decision as it relates to procuring spares.

3.8 Predictability of Failure

Some equipment/components will fail over time due to wear or age/fatigue. It may be possible to properly manage this type of failure by conducting periodic inspections to measure the overall condition of the asset. We identified components like chains, belts, sprockets, liners, and screws can be evaluated for wear and new parts can be ordered prior to loss of functionality.

Properly set up and monitored, condition assessment programs can minimize the amount of spares needed at a facility. Items like fuses, shear pins and electronics fail with little warning and should always be stocked if the asset is critical.

3.9 Shelf Life

Some components deteriorate with time (like probes, sensors, ball bearings, batteries and gaskets). These timing considerations should be factored into the decision-making process.

3.10 Storage Considerations

Prior to procuring any spare parts one must evaluate his/her location to determine if the facility has adequate storage facilities on site.

Consideration should be given to environmental conditions (humidity, temperature, sulfides, etc.) and the available space. Staff also needs to ensure that it has the proper systems in place to track the spares and to perform PM maintenance tasks on the inventory if needed (i.e., a spare pump may require that the shaft be rotated periodically to distribute lubrication and prevent shaft deflection). In Chicago we decided to store some of the spares on site and some in off site storage facilities. We had very good success in working with vendors in some cases to store items for us.

3.11 Critical Spares Methodology

Veolia Water utilized its internally developed critical spares methodology during the RCM review. The following is a small section of the spares analysis that was based on some contract terms and downtime estimates.

The contract is production-based and includes liquidated damages for downtime and loss of availability. What follows is a general description as to how the process was used to determine the need to stock a spare. For presentation purposes, only three scenarios are included.

3.11.1 Step One: "Monetize the Risk"

There are 4 process trains at this facility and three are needed to meet full capacity. The group determined that if all 4 trains were unavailable there would be a $1,250/hour cost of downtime. The group then evaluated loss of capacity and assigned a lower dollar number for some unit operations ($450/hr). After debating the liquidated damages clause of the contract and its impact on the business, the group settled on $225/hr as being the cost of loss of redundancy. This process was extremely rigorous and debated by the review team. The assumptions were sent in a memorandum to the senior management team.

3.11.2 Step Two: Failure Mode Evaluation

	Failure Mode	Local	System	Plant
1	Bearing seizes due to lack of lubrication	Bearing wears, gets hot and eventually fails	Bearing wears, gets hot and eventually fails. Standby pump starts and carries load	Loss of redundancy
2	Temperature transmitter fails	Transmitter fails and thermal oil pumps stop pumping	Oil system shutdown and OLC initiates a system shutdown	Loss of one train
3	Heat trace thermostat fails	Heat trace fails, water line freezes, resulting in loss of flow	Biosolids pumps trip on loss of seal water.	Complete plant shutdown

3.11.3 Step Three: Match Critical Failure Mode to Spare

Failure Mode	Spare
Bearing seizes due to lack of lubrication	Bearing
Temperature transmitter fails	Transmitter
Heat trace thermostat fails	Thermostat

3.11.4 Step Four: Determine Spare Cost and Lead Time and Calculated Downtime Cost

Spare	Lead time (in operating hours)	Downtime cost/hour	Cost of downtime due to lack of spare
Bearing	48	$225	$10,800/event
Transmitter	24	$225	$5,400/event
Thermostat	24	$1250	$30,000/event

3.11.5 Step Five: Adjust Downtime Cost for Failure Probability

Spare	Downtime Cost (DC) due to lack of spare	Failure Probability (FP)**	Downtime Cost adjusted (DC * FP)	Cost of spare and storage
Bearing	$10,800/event	.2	$2,160/year	$1,200
Transmitter	$5,400/event	.1	$540/year	$250
Thermostat	$30,000	.1	$3,000/year	$600

** The Failure Probability (FP) is based on the number of failures anticipated in a ten-year period (e.g. a .1 signifies a 10% chance of at least one failure in a 10-year period). This can be based on past failure data or available MTBF data from other sources. One must consider contractual terms and residual contract years when making spare part decisions. For example, if the contract end date is within a two-year period and the risk of not carrying a spare is tolerable, one should not stock the item.

3.11.6 Step Six: Step Six: Evaluate Cost of Spare versus the Downtime Costs

Spare	Cost of spare and storage	Downtime cost adjusted	Spare required (yes or no)
Bearing	$1,200	$2,100	Yes
Transmitter	$250	$540	Yes
Thermostat	$600	$1250	Yes

As with any type of analysis it is important that the review group uses a bit of common sense when making spare part decisions. One must factor into the decision-making process contractual obligations, contract term, and the practicality of storing the item. Oftentimes the decision may come down to a very small risk (<$1,500 can be mitigated by spending about the same amount of money). One needs to use sound judgment before money is spent on spares.

In total, we analyzed 1784 failure modes for critical spare purposes. In total we decided to stock 425 spares parts at a cost of $525,000. These spares mitigated over $33 million dollars worth of potential risk.

4. CONCLUSIONS

Veolia Water has introduced a number of innovations to the standard framework for delivering RCM that have improved the business case for delivery. These innovations include techniques for prioritizing and simplifying the analysis, as well as more reliable means for estimating maintenance frequencies and estimating the true cost-benefit of critical spares. The use of RCM in the water/wastewater industry has been adapted from the original methodology used by United Airlines. The changes to the process allow for speedier delivery and execution of RCM. Implementation of RCM on 100% of the assets is not practical, nor cost-effective in our industry.

Veolia Water has been able to demonstrate a link between doing RCM well and delivering excellence in asset management. Overall, the benefit to the business and our customers that result from implementing the Veolia Water RCM process is as follows:

- Better identification of risks. Many new, and often significant, operational risks are identified with recommendations to mitigate them.

- Improved maintenance plan. New maintenance plans are developed to specifically manage the risks of asset failure. Frequencies and resource levels are optimized.

- Gains in operational efficiency. Opportunities for improvement are proposed for operating scenarios that offer risk and efficiency benefits.

- List of critical spares. Spares needs are determined, based on cost-benefit assessments, including financial estimates of risk.

- Prioritized capital improvements. Investigations are recommended for a number of potential capital works items designed to enhance contract performance and minimize risk.

- Effective knowledge capture. The RCM captures both documented information as well as the tacit knowledge held by staff about how the systems work.

- Tool for operations training. Participating staff are trained in the methodology and learn in detail how their systems work.

The project staff has been busy implementing the recommendations that came about from the RCM review. The identification of the wrong PRV's on the high temperature oil system was a very important safety related finding, but the benefits of the program continue to add up. The project has already seen drastic improvements in equipment availability and uptime. Each maintenance task, spare part, and equipment check has been developed to manage risk and improve uptime. The staff at the plant continues to learn from equipment failure and has validated that the time spent on the analysis and spares has tremendously improved the overall operation. The following, Figure 3, is a trend of uptime of one train since the RCM reviews were completed.

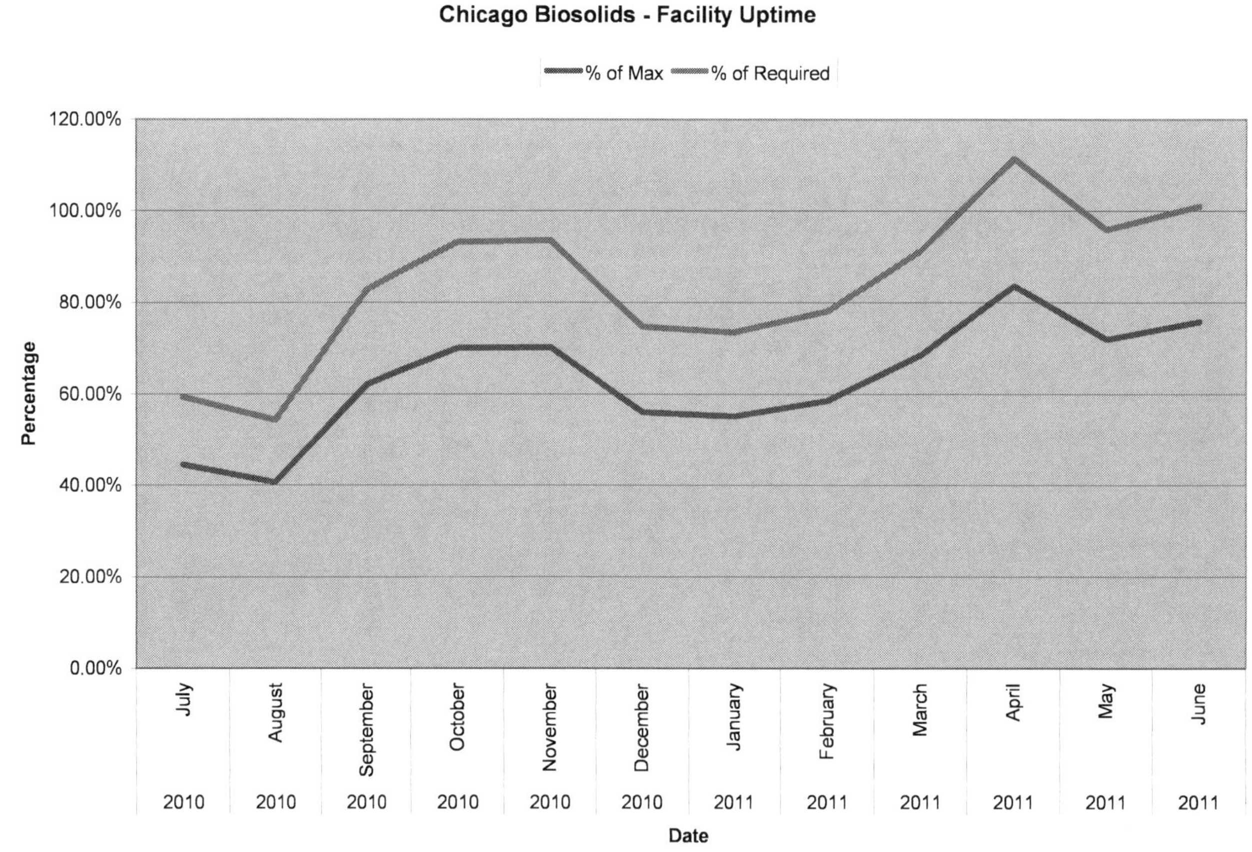

Figure 3, Uptime Performance Through June 2011

5. REFERENCES

[1] Moubray, John, *Reliability-centered Maintenance*, 2nd Edition, Industrial Press Inc., New York, New York, 1997.

[2] Nowlan, F. S.; Heap, H. F., *Reliability-centered Maintenance*, United Airlines and Dolby Access Press, sponsored and published by the office of Assistant Secretary of Defense, San Francisco, California, 1978.

[3] Smith, A.M.; Hinchcliffe, G.R., *RCM Gateway to World Class Maintenance*, Elsevier Butterworth Heinemann., Burlington, Massachusetts, 2004.

(44x) Characterization of the Emissive Power of Hybrid Methane-Air-Dust Flames

Brian Elias and Ali S. Rangwala, Fire Protection Engineering, Worcester Polytechnic Institute, Worcester, MA

Combustible dusts are commonplace in a wide variety of industrial settings ranging from coal mines to manufacturing facilities. The ignition of a dust cloud formed by the mixing of these combustible particles with the surrounding air may result in a deflagration or flash fire. The purpose of this study is to quantify the emissive power of the flame in order to better understand the combustion dynamics and the risk of burn injury associated with a dust flash fire.

Experimental tests are conducted by feeding combustible dust samples into a premixed dust-air burner apparatus to create a stabilized flame for analysis. The two dust types investigated in this study are Pittsburgh seam coal, and iron dust – both with sizes ranging from 70 to 90 um. In addition to the composition of the dust, other controlling parameters including dust concentration and turbulent intensity are also varied and their respective effect on the heat flux is reported. The total emissive power of the flame is calculated from the heat flux measurements in combination with flame surface area and an appropriate view factor.

9TH GLOBAL CONGRESS ON PROCESS SAFETY

Are You in Control of Process Safety?

Basis of Safety Assurance Can Provide the Answer

Graeme Ellis, Principal Lead Consultant
ABB Consulting
Daresbury Park, Warrington, WA4 4BT, UK
graeme.ellis@gb.abb.com

(Acknowledgement: Based on paper written by Phil Eames, ABB Consulting)

Prepared for Presentation at
American Institute of Chemical Engineers
2013 Spring Meeting
9th Global Congress on Process Safety
San Antonio, Texas
April 28 – May 1, 2013

UNPUBLISHED

Are You in Control of Process Safety?

Basis of Safety Assurance Can Provide the Answer

Graeme Ellis, Principal Lead Consultant
ABB Consulting
Daresbury Park, Warrington, WA4 4BT, UK
graeme.ellis@gb.abb.com

(Acknowledgement: Based on paper written by Phil Eames, ABB Consulting)

Keywords: Basis of Safety, Assurance, layers of protection, drill-down, vital signs, audit, safety critical elements

Abstract

Serious process safety accidents continue at an unacceptable frequency in the process industry, and senior managers are right to be concerned about whether all reasonable steps are being taken to control risks. Traditional methods for revalidation of Process Hazards Analysis and auditing of Process Safety Management systems can involve considerable time and resources. This paper will describe a 'Basis of Safety Assurance' approach that has been developed to provide a rapid view on the 'vital signs' related to the state of health of measures to prevent and mitigate major accident hazards. A representative set of scenarios for the site are selected and evidence gathered from documentation and site visits to confirm the robustness of the key layers of protection. Deficiencies related to plant, processes or people are identified resulting in a risk targeted set of improvement recommendations.

1. Introduction

Effective process safety management requires a thorough understanding of process risks so that measures to prevent and control them can be implemented and maintained. Structured techniques such as hazard identification (HAZID) and hazard and operability study (HAZOP) are used to identify major accident scenarios and decide how they will be prevented, controlled and mitigated. The description of major accident scenarios and the agreed layers of protection (or barriers) to reduce the risks to an acceptable level, is often termed the basis of safety.

The purpose of a process safety management system (PSMS) is to provide a framework of high level procedures (or risk control systems), in order to maintain protective equipment and procedures in a healthy state. Therefore many organizations, when they seek to assure themselves that process safety is being managed appropriately, base this assurance on detailed auditing of the PSMS.

However, process safety incidents are usually caused by failures of multiple risk control systems at a detailed level. The potential interactions between risk control systems are often not visible at the level of the PSMS, which is constructed of distinct generic elements (mechanical integrity, incident investigation, management of change etc.). Auditing of distinct elements may not identify the potential for an incident due to such interactions. Moreover, modern PSMS are tending to become more integrated with occupational safety management systems, environmental management systems and quality systems, making their role in the prevention of process safety incidents potentially less visible at system level.

Given that the prime concern in process safety management is to prevent process incidents from occurring, it makes sense to devote some assurance effort directly to the scenarios themselves and to their specific layers of protection. This is the essence of Basis of Safety Assurance. Whereas auditing addresses the component elements of the management system, Basis of Safety Assurance seeks to challenge the effectiveness of the process hazards analysis and robustness of the layers of protection for specific scenarios, thereby providing direct assurance that the risk reduction measures are in place and working.

This is not to suggest that auditing of the PSMS is not a necessary or effective tool. But focusing on an organization's key process incident scenarios can be a very direct and cost-effective way to provide assurance that process safety is being managed appropriately, complementing traditional audit-based assurance processes.

This paper describes the Basis of Safety Assurance technique and the experience of its use across a number of sectors in the process industry. It explains how the technique can be used to enhance an organization's process safety management audit and review program.

2. Limitations of Audit Based Assurance

Traditional process safety assurance or governance approaches are based on audits of the process safety management system (PSMS). However, in many organizations the components of the PSMS form part of a larger safety, health, environmental and possible quality management system, and may not be distinguished from components designed to control personal safety and health risks. This means that focus on process safety may diluted by or obscured by the larger management system.

Even if the PSMS is a stand-alone system, or is visible within a larger system, auditing focuses on the system as a whole and not on the control of specific risks. Auditing is often undertaken on an element-by-element basis, which is likely to miss the potential interactions between elements that characterize complex process safety events, typically represented by the "Swiss cheese" model. Moreover, even within a single element significant audit time is normally spent on understanding how the element is managed rather than on detailed field verification. Field verification on audits typically aims to confirm that the system is in compliance with requirements, for example that trip testing is being carried out to plan. It generally does not test that the system is effective in protecting against the risk of a specific scenario, for example the trip is identified as a layer of protection in HAZOP, there is a written test method, the trip test method is an "end to end", faults identified in previous tests have been corrected, etc.

3. Layers of protection and process safety "vital signs"

In his book on the management of complex process risks, James Reason [1] wrote of the need for proactive checking of the process or plant, or the 'system' as he describes it:

"*In the absence of bad outcomes, the best way - perhaps the only way - to sustain a state of intelligent and respectful wariness is to gather the right kinds of data. This means creating a safety information system that collects, analyses and disseminates information from incidents and near misses as well as from regular proactive checks on the system's vital signs. All of these activities can be said to make up an informed culture - one in which those who manage and operate the system have current knowledge about the human, technical, organisational and environmental factors that determine the safety of the system as a whole. In most important respects, an informed culture is a safety culture*".

The state of health of safety critical layers of protection is very much vital signs for the process in relation to major accident scenarios. If they are specified and designed correctly, maintained and tested, then the process is safe. The proactive checking that layers of protection are in place and healthy provides exactly the current knowledge of the state of the plant and the organization that Reason holds to be so important.

4. The 'Basis of Safety' Event

The testing of the state of health of layers of protection for specific major accident scenarios is the focus for a Basis of Safety Assurance event, a generic agenda for which is shown in Table 1.

Table 1. Basis of Safety Assurance Agenda

Phase	Day	Activity
Preparation	**0**	**Arrange for immediate access to key site personnel and information during time on site**
Introduction	**1**	**Introduction to Process Safety on site.** **Review of major accident hazards** **Site tour** **Examination of Process Hazard Analyses** **Selection of 'Drill Down' scenarios**
Drill Downs	**2-N**	**Detailed drill downs** **Approx 4 scenarios per day including field visits**
Feedback	**Final**	**Preparation and delivery of feedback**

The initial part of the event aims to understand the major accident hazards associated with the process by discussion, field visits and review of the process hazards analysis (HAZID and HAZOP documentation). This would typically take one day, during which time an assessment of the quality of the site's process hazards analysis and documentation of the basis of safety is also made. It culminates in the selection of specific major accident scenarios to be subjected to detailed assessment, sometimes termed a "drill-down" or "deep dive".

It is useful to select a range of different types of scenario where possible (fire, explosion, toxic release, environmental, etc.), so that as many different types of safeguard are subject to assessment. These then include for example; pressure relief systems, safety instrumented systems, passive and active fire protection systems, secondary and tertiary containment systems, etc.

The main part of the Basis of Safety event comprises the scenario assessments. These typically take around 2 hours per scenario, depending on the quality of the basis of safety documentation, the knowledge of the team and the speed of data retrieval, for example how fast plant records can be accessed to verify that maintenance and testing is taking place.

Findings from the scenario assessments provide the required assurance that layers of protection are healthy or lead to the generation of recommendations for improvement. Common findings across a number of scenario assessments can provide a useful indication of systemic issues, for example weaknesses in proof test procedures or compliance with critical equipment testing regimes. This part of the event complements conventional PSMS auditing.

5. The Scenario Assessment Technique

Most value can be gained from the scenario assessment technique if it is conducted as a team exercise. Experience has shown that the best results can be obtained by deploying two process safety specialists as assessors; one with a PSMS and operations background and the other with a plant engineering and asset integrity management background. This approach also helps from a leading and scribing standpoint since the event is recorded in a similar way to a process hazards analysis. In addition to the assessors, the involvement of process engineers, operating managers and maintenance engineers from the site is very important. This provides an understanding of the scenario and the location of information to verify that the identified layers of protection are in place and being maintained.

Since scenarios are normally selected from the site's process hazards analysis or basis of safety documentation, a scenario assessment normally starts with this documentation. The scenario is described and the relevant records are referenced, shown in the upper part of the recording form in Table 2.

It is important to note that the existing documentation may not fully describe the scenario, since the quality of HAZID and HAZOP documentation is heavily influenced by team composition, knowledge available, leader competence, availability of time and recording policy.

Table 2. Scenario Assessment Template (example)

Major Accident Hazard Drill Down Record
Scenario: Storage Tank V7706 (150m3) overfilled during (continuous) run-down and a large volume of solvent released into bund from the breather/emergency vent, subsequent ignition resulting in a flash fire/pool fire and potential for several fatalities.

PHA Reference: No specific scenario was identified for this material as incidents involving acetone were taken as the representative set. The acetone scenario for overfill was estimated to result in two fatalities.			
Safeguard	**Verification**	**Findings**	**Conclusions**
Monitoring of level transmitter by operators and dependent alarm function.	1. Calibration and maintenance of level transmitter and alarm LIA-7706. 2. SOP requiring level monitoring.	1. No regular maintenance is now performed on the level transmitter following the installation of the independent high level switches. Corrective maintenance is performed based on notifications raised by the operators. 2. There is no specific SOP relating to the operation of this solvent storage tank.	Consideration implementing routine level transmitter calibration on solvent storage tanks.
Independent high level switch with trip of run-down line.	1. Proof testing of high level trip LZA-7706H.	1. The trip is proof tested annually 2. The last proof test of this instrument was carried out on the 27/4/10. The proof test is an end- to-end test to a pre-defined test method including checklist. 3. No SIL assessment has been conducted.	Controls appear to be effective. However, conduct a SIL assessment for this scenario to verify that the proof test interval is adequate to provide the required level of protection.

It is therefore important that the assessors work with the team to identify all the relevant safeguards, which are recorded in the left hand column of the template shown in Table 2. Note that the term "safeguard" is used rather than "layer of protection" because the aim is to capture safeguards such as procedures and emergency plans as well as physical independent protection systems that are the focus of formalized Layers of Protection Analysis (LOPA) techniques.

Once safeguards have been listed the assessors ask - for each safeguard in turn - what can we check is in place? This is recorded in the "verification" column. Examples include procedure number and latest version, alarm test methods, maintenance routines and pressure safety valve test records.

Evidence for each verification is then sought and the findings recorded in the "findings" column of the template. It is important that documentary evidence is collected to ensure that the assessment is suitably probing and detailed.

Conclusions are then drawn for each safeguard. This may be a simple "Okay" if everything expected is in place, but may be a recommendation for improvement, a corrective action or a recommendation to consider whether a finding may be evidence of a systemic weakness. In viewing the robustness of the safeguards as a whole the team may determine that additional risk reduction is appropriate and make a recommendation to identify additional risk reduction measures. This would be required in a UK legislative context to demonstrate that risks have been reduced to "as low as reasonably practicable" or ALARP.

It is important to include a field visit for each scenario. This has two purposes, firstly to physically check that safeguards are installed in line with the design intent, for example that relief devices are in place and that passive fire protection and secondary containment are in good condition. Photographs of deficiencies provide high-impact evidence to site management. Secondly to talk to operators and maintenance technicians about their understanding of the process and of the major accident hazard that is being assessed. This gives an insight into the quality and experience of staff and the quality of process-specific training that they have received.

6. Experience with the Technique

There is now a significant body of experience of Basis of Safety Assurance across a number of process industry sectors. Companies have found it effective regardless of sector or scale of process hazard. Major hazards facilities are more familiar with the concepts but it is readily applicable to lower level process safety hazards, including those in the pharmaceutical and power generation sectors.

Some important advantages of using the scenario assessment "drill-down" technique are as follows:

- Drilling down into specific scenarios can identify significant risk exposures and weaknesses at a very detailed level, for example:
    - Trips tested only at commissioning;
    - Absence of detailed test methods and reliance on technicians' experience;
    - Trip testing not from sensor to final element (end-to-end);
    - Reliance on inexperienced 3rd parties for proof testing;
    - Demands on trips more frequent than the design basis.
- Examining multiple scenarios can provide evidence of whether there are systemic weaknesses.
- The technique is useful to identify weaknesses in process hazards analysis and in keeping records and basis of safety documentation up-to-date.
- The technique is a useful vehicle for providing training materials for specific process hazards.

7. Conclusion

Because it focuses on specific risks, Basis of Safety Assurance complements conventional PSM system audit activities. It helps to maintain a focus on major accident hazards and improve the understanding of site personnel and standard of documentation. It aligns with line management responsibilities for specific processes and assets, and it directly exposes risks of major accident hazards, thus providing real time assurance to senior management.

8. References

[1] Reason, J.T (1997), Managing the Risks of Organizational Accidents, Ashgate Publishing, Aldershot, UK.

RAGAGEP 101

Lisa Long
James Lay, P.E.
Michael Marshall, P.E.
Jeffrey Wanko, P.E., C.S.P.
US Occupational Safety and Health Administration
200 Constitution Avenue NW
Washington, DC 20037
long.lisa@dol.gov

Prepared for Presentation at

American Institute of Chemical Engineers

2013 Spring Meeting

9th Global Congress on Process Safety

San Antonio, Texas

April 28 – May 1, 2013

UNPUBLISHED

RAGAGEP 101

Lisa Long
James Lay, P.E.
Michael Marshall, P.E.
Jeffrey Wanko, P.E., C.S.P.
US Occupational Safety and Health Administration
200 Constitution Avenue NW
Washington, DC 20037
long.lisa@dol.gov

Keywords: Recognized and generally accepted good engineering practice, RAGAGEP, OSHA, process safety management, PSM, consensus codes and standards

Abstract

Recognized and generally accepted good engineering practices (RAGAGEPs) play an important role in the Occupational Safety and Health Administration's Process Safety Management standard (29 CFR 1910.119). Since 2007, OSHA issued numerous RAGAGEP-related citations under the Petroleum Refinery Process Safety Management and PSM Covered Chemical Facility National Emphasis Programs. Even though the PSM standard became effective over 20 years ago, noncompliance with PSM's RAGAGEP requirements remains an issue.

This paper reviews RAGAGEP basics from OSHA's perspective, using examples from national emphasis program inspections to illustrate common RAGAGEP compliance problems.

1. Introduction

Good engineering practice (GEP) has always been part of the foundation of safe operations in the chemical industry. GEP is a part of all aspects of chemical processing including the design of equipment, the overall process, and the man-machine interface.

This paper on recognized and generally accepted good engineering practice (RAGAGEP):

- Provides background information;
- Describes how RAGAGEP evolved from early industry descriptions, the PSM promulgation process, Occupational Safety and Health Administration (OSHA) interpretations, and later industry publications;
- Illustrates how other federal agencies use GEP and RAGAGEP;
- Describes OSHA's enforcement of RAGAGEP in the Petroleum Refinery Process Safety Management National Emphasis Program (Refinery NEP) and Process Safety Management Covered Chemical Facility National Emphasis Program (ChemNEP);
- Documents examples of OSHA citations related to RAGAGEP; and

Provides incidents in which RAGAGEP noncompliance was a causal factor.

2. Background

Prior to the enactment of OSHA's Process Safety Management of Highly Hazardous Chemicals Standard (PSM) [1], OSHA primarily enforced GEP through specific consensus standards incorporated by reference into existing OSHA standards. For example, OSHA's 29 CFR 1910.106 standard, *Flammable and Combustible Liquids*, incorporates the American Society of Mechanical Engineers (ASME) *Boiler and Pressure Vessel Code*, Section VIII - 1968.

However, with the promulgation of PSM in 1992, OSHA began enforcing regulations related to GEP differently. Notably, the PSM standard does not incorporate any GEP. Rather, the PSM standard includes nonspecific requirements that employers comply with RAGAGEP in equipment design and in inspection and testing.

A phrase that is common in the industry relating to management system performance is "Say what you do and do what you say." OSHA's PSM standard parallels this philosophy in that it is largely a performance standard. The PSM standard sets forth a management system framework upon which employers build. The design and maintenance of the process is an essential part of the standard in which RAGAGEP plays a critical role.

The PSM standard is a set of requirements for employers with processes containing highly hazardous chemicals intended to prevent releases. The PSM standard is not prescriptive because of the number and types of industries, facilities, chemicals, and technologies covered by the standard. Employers have latitude to develop and initiate the required management systems in a manner that is flexible and takes into account the attributes of their covered processes and equipment while still achieving safe operations.

Specifically for RAGAGEP, the PSM standard:

1. Allows employers to select appropriate RAGAGEP for their processes;
2. Requires employers to document the RAGAGEP to which they comply ("Say what you do"); and

Requires employers perform ("do what you say") to the provisions of the chosen RAGAGEP and document compliance with the chosen RAGAGEP.

3. Early RAGAGEP Recognition

3.1 American Institute of Chemical Engineers

One of the first authoritative references on the use of GEP as a guideline in the chemical processing industry was by the American Institute of Chemical Engineers' (AIChE) Center for Chemical Process Safety (CCPS) in 1989. In the *Guidelines for Technical Management of Chemical Process Safety* [2], CCPS provided guidelines on how to use external and internal GEP to achieve safe facilities. Some of the guidance by CCPS related to GEP included:

- Management of standards, codes, and regulations as a necessary part of chemical process safety activities;
- Application of external codes, standards, and regulations to the layout, design, and operation of chemical plants and the protection of equipment, employees, and the public;
- Means of deciding which codes, standards, and guidelines the facility follows;
- A table listing of some example external codes/regulations including ASME - Boiler and Pressure Vessel Code and Code for Pressure Piping (B31), American Petroleum Institute (API) - RP 520/521 Pressure Relieving Systems, AIChE – Design Institute for Emergency Relief Systems Procedure for Emergency Venting Sizing, OSHA – 29 CFR 1910, etc.;
- Clear documentation of the chosen codes, standards and guidelines so that those involved in design and maintenance know which RAGAGEP to follow.
- Means of documenting the reason for deviating from a code or standard. The facility seeking to use an alternative approach should demonstrate that the approach is at least as safe as the chosen code or standard;
- Development of internal standards to complement external codes and design practices;
- Developing and instituting internal standards to achieve consistency in design, where appropriate, and to safely operate and maintain the facilities; and
- Management commitment to following all chosen RAGAGEP whether internally- or externally-developed.

3.2 American Petroleum Institute

In January 1990, API issued Recommended Practice (RP) 750, *Management of Process Hazards* [3]. RP 750 states the following with respect to GEP:

- The mechanical design information should include design codes employed (Section 2.3.1);

- The mechanical design should be consistent with the applicable consensus codes and standards in effect at the time the design is prepared or, in the absence of such codes and standards, recognized and generally accepted engineering practices (Section 2.3.2);
- When the mechanical design is not consistent with applicable consensus codes and standards, the deviation and its design basis should be documented (Section 2.3.2);
- When a process hazards analysis or other review reveals that existing equipment is designed and constructed in accordance with consensus codes, standards, or practices that are no longer in general use, procedures should be implemented to ensure that the equipment is fit for service (Section 2.3.2);
- Facilities should establish inspection and testing programs for critical equipment. Such programs should include testing and inspection procedures in accordance with common standards and codes, such as API 510 or the API Guide for the Inspection of Refinery Equipment. (Section 8.5(a)).

3.3 OSHA Proposed PSM Rule

In 1990, OSHA issued a Notice of Proposed Rulemaking (NPRM) for the PSM rule [4]. The NPRM proposed regulations including some related to RAGAGEP, however, the NPRM contained no background information on the genesis or intent of the specific RAGAGEP requirements. The NPRM contained the following language on RAGAGEP:

- 1910.119(d)(3)(i)(F) - The employer shall develop and maintain a compilation of written safety information...information pertaining to the equipment in the process shall include....Design codes employed;
- 1910.119(d)(3)(ii) - The employer shall document that equipment complies with applicable codes and standards, such as those published by the ASME, API, AIChE, ANSI, ASTM and NFPA, where they exist; or recognized and generally accepted good engineering practices;
- 1910.119(d)(3)(iii) - For existing equipment designed and constructed in accordance with codes, standards, or practices that are no longer in general use, the employer shall determine and document that the equipment is designed, maintained, inspected, tested, and operating in a safe manner;
- 1910.119(j)(3)(ii) - Inspection and testing procedures shall follow applicable codes and standards, such as those published by ASME, API, AIChE, ANSI, ASTM, and NFPA, where they exist; or recognized and generally accepted good engineering practices; and
- 1910.119(j)(3)(iii) - The frequency of inspections and tests of process equipment shall be consistent with applicable codes and standards, or, more frequently if determined to be necessary by prior operating experience.

4. OSHA Final Rule on PSM and RAGAGEP

In 1992, OSHA promulgated its PSM regulation containing some, but not all, of the RAGAGEP provisions listed in the NPRM. The preamble to the PSM rule explains that the phrase "recognized and generally accepted good engineering practices" included both appropriate internal standards and applicable published codes and standards.

The preamble to the final PSM standard [5] describes how and why RAGAGEP became part of the regulation:

- (57 FR 6375) As part of the process safety information the employer is required to document that the process equipment being used complies with applicable consensus codes and standards, where they exist; or must be consistent with recognized and generally accepted engineering practices.
- OSHA modified the paragraph 1910.119(d)(3)(ii) from the NPRM by eliminating the specific list of codes and standards producing organizations.
- OSHA stated that the discussion in paragraph (j), *Mechanical Integrity*, section of the preamble describes the reasons for the change from the specific list of required codes and standards.
- (57 FR 6390 – 6391) OSHA explained the intent of proposed paragraph (j)(3)(ii) was an effort to assure that inspections and tests are performed properly.

Many commenters to the NPRM, though, objected to OSHA's inclusion of a list of consensus documents in the regulation. While OSHA believed the lists in the proposed (d)(3)(ii) and (j)(3)(ii) were not exhaustive nor attempts at incorporation by reference, OSHA published the final rule without lists.

The following "RAGAGEP rules" show the final text of the existing RAGAGEP standards. Bolded text represents language added to the final standards following the comment period and the struck-through portion represents language removed:

- 1910.119(d)(3)(i)(F) - The employer shall develop and maintain a compilation of written safety information...information pertaining to the equipment in the process shall include....Design codes **and standards** employed;
- 1910.119(d)(3)(ii) - The employer shall document that equipment complies with **recognized and generally accepted good engineering practices.** ~~applicable codes and standards, such as those published by the ASME, API, AIChE, ANSI, ASTM and NFPA, where they exist; or recognized and generally accepted good engineering practices~~;
- 1910.119(d)(3)(iii) - For existing equipment designed and constructed in accordance with codes, standards, or practices that are no longer in general use, the employer shall determine and document that the equipment is designed, maintained, inspected, tested, and operating in a safe manner;
- 1910.119(j~~)(3)~~**(4)**(ii) - Inspection and testing procedures shall follow **recognized and generally accepted good engineering practices.** ~~applicable codes and standards, such as those published by ASME, API, AIChE, ANSI, ASTM, and NFPA, where they exist; or recognized and generally accepted good engineering practices~~; and

- 1910.119(j)~~(3)~~**(4)**(iii) - The frequency of inspections and tests of process equipment shall be consistent with applicable **manufacturers' recommendations and good engineering practices, and** ~~codes and standards, or,~~ more frequently if determined to be necessary by prior operating experience.

4.1 Appendix C to the PSM Final Rule

Appendix C to the PSM Final Rule, *Compliance Guidelines and Recommendations for Process Safety Management (Non-Mandatory),* contains information for the regulated community related to RAGAGEP. Appendix C states the following with respect to the design of equipment:

> The information pertaining to process equipment design must be documented. In other words, what were the codes and standards relied on to establish good engineering practice. These codes and standards are published by such organizations as the American Society of Mechanical Engineers, American Petroleum Institute, American National Standards Institute, National Fire Protection Association, American Society for Testing and Materials, National Board of Boiler and Pressure Vessel Inspectors, National Association of Corrosion Engineers, American Society of Exchange Manufacturers Association, and model building code groups.
>
> In addition, various engineering societies issue technical reports which impact process design. For example, the American Institute of Chemical Engineers has published technical reports on topics such as two phase flow for venting devices. This type of technically recognized report would constitute good engineering practice.

Note, the codes and standards listed above are examples and not mandatory – the employer selects the appropriate RAGAGEP with which to comply.

The following excerpt from Appendix C discusses RAGAGEP as it relates to mechanical integrity inspections and tests:

> Also, applicable codes and standards such as the National Board Inspection Code, or those from the American Society for Testing and Material, American Petroleum Institute, National Fire Protection Association, American National Standards Institute, American Society of Mechanical Engineers, and other groups, provide information to help establish an effective testing and inspection frequency, as well as appropriate methodologies.
>
> The applicable codes and standards provide criteria for external inspections for such items as foundation and supports, anchor bolts, concrete or steel supports, guy wires, nozzles and sprinklers, pipe hangers, grounding connections, protective coatings and insulation, and external metal surfaces of piping and vessels, etc. These codes and standards also provide information on methodologies for internal inspection, and a frequency formula based on the corrosion rate of the materials of construction. Also, erosion both internal and

external needs to be considered along with corrosion effects for piping and valves. Where the corrosion rate is not known, a maximum inspection frequency is recommended, and methods of developing the corrosion rate are available in the codes. Internal inspections need to cover items such as vessel shell, bottom and head; metallic linings; nonmetallic linings; thickness measurements for vessels and piping; inspection for erosion, corrosion, cracking and bulges; internal equipment like trays, baffles, sensors and screens for erosion, corrosion or cracking and other deficiencies. Some of these inspections may be performed by state or local government inspectors under state and local statutes. However, each employer needs to develop procedures to ensure that tests and inspections are conducted properly and that consistency is maintained even where different employees may be involved. Appropriate training is to be provided to maintenance personnel to ensure that they understand the preventive maintenance program procedures, safe practices, and the proper use and application of special equipment or unique tools that may be required. This training is part of the overall training program called for in the standard.

5. Further Guidance on RAGAGEP

5.1 Center for Chemical Process Safety

In 2006, CCPS published, *Guidelines for Mechanical Integrity Systems* [6] which discusses RAGAGEP extensively - what it is, how it is used, application to different classes of equipment, design, fabrication, construction/installation, inspection, testing, and maintenance activities. The text contains the following information on RAGAGEP:

- A description of RAGAGEP with many examples;
- Facility-develop internal standards provided by the manufacturer or licensor of a process and often these are based on RAGAGEP;
- Recognition that to effectively use RAGAGEP, facility management must determine which practices are available and then assess the applicability of each practice to their facility; and
- Several chapters that address the applicability and use of RAGAGEP for detailed operations, e.g., vessel design and fabrications, in-service piping inspections, RAGAGEP for instrumentation and controls, RAGAGEP for prime movers, etc.

5.2 Incorporation of Consensus Standards

In federal regulation, use of the terms recognized and generally accepted good engineering practices and RAGAGEP originated with OSHA. However, the Federal government has long encouraged the use of voluntary consensus standards in regulatory affairs. The National Technology Transfer and Advancement Act of 1995 (NTTA), as interpreted by the Office of Management and Budget's (OMB's) Circular Number A-119 [7], instructs agencies to "use voluntary consensus standards in lieu of government-unique standards except where inconsistent with law or otherwise impractical." In addition, OMB A-119 states that "many voluntary consensus standards are appropriate or adaptable for the government's purposes."

OMB A-119 defines "standards" and "technical standards" as:

1. Common and repeated use of rules, conditions, guidelines or characteristics for products or related processes and production methods, and related management systems practices.

2. The definition of terms; classification of components; delineation of procedures; specification of dimensions, materials, performance, designs, or operations; measurement of quality and quantity in describing materials, processes, products, systems, services, or practices; test methods and sampling procedures; or descriptions of fit and measurements.

OMB further defines "voluntary consensus standards" as standards developed or adopted by voluntary consensus standards bodies, both domestic and international. These bodies must have attributes including openness, balance of interest, due process, an appeals process, and consensus, defined as general agreement, but not necessarily unanimity, with a process for attempting to resolve objections by interested parties. Circular A-119 also defines other types of standards as "non-consensus standards", "industry standards", "Company standards", or "de facto standards" developed in the private sector, but without the full consensus process.

Government agencies regulating highly hazardous chemicals (HHCs) use voluntary consensus standards in a variety of ways. For example, Section 6(a) of the Occupational Safety and Health Act of 1970 [8] gave the nascent OSHA a two-year window in which to "by rule promulgate as an occupational safety or health standard any national consensus standard, and any established Federal standard…" The 6(a) provision enabled OSHA to quickly establish a basic suite of protective standards. National consensus standards incorporated by reference into regulations under this provision of the OSH act are listed in 29 CFR 1910.6 *Incorporation by reference.* Examples include the ASME *Boiler and Pressure Vessel Code*, Section VIII *Unfired Pressure Vessels, Including Addenda* (1969), incorporation by reference approved for OSHA standards §1910.261, 1910. 262, and 1910.263(i)(24)(ii); and API 650 (1966) *Welded Steel Tanks for Oil Storage, 3rd Ed.*, incorporation by reference approved for §1910.106(b)(1)(iii)(a)(2). Since OSHA's incorporation by reference provision expired in 1974, the agency has followed the rule making process referenced in section 6(b) of the OSHA Act, including for updating references to more recent versions of national consensus standards. Incorporation of consensus standards by reference has benefits to both OSHA and the regulated community. However, OSHA must incorporate a specific version of a standard. When the consensus standard is updated, OSHA's incorporated standard becomes dated and can only be updated through the rulemaking process. Commenters on OSHA's ANPRM realized this, and this resulted in a different approach, the introduction of the RAGAGEP concept.

6. OSHA RAGAGEP Enforcement

6.1 OSHA RAGAGEP Policy

In the past, OSHA issued enforcement policy related to RAGAGEP. The first milestone enforcement policy related to RAGAGEP was a 2000 OSHA Letter of Interpretation (LOI) to the Instrument Society of America [9]. In this LOI, OSHA stated the following regarding the agency's RAGAGEP policy:

- 1910.119(d)(3)(i)(F) requires an employer to document which design codes and standards are used for SIS as well as all covered equipment;
- 1910.119(d)(3)(ii) requires an employer to document that SIS comply with RAGAGEP;
- 1910.119(j)(4) requires an employer to conduct inspection and testing on process equipment, including SIS, in accordance with RAGAGEP;
- If an employer documents compliance with ISA S84.01 for SIS in accordance with 1910.119(d)(3)(i)(F) and it meets all S84.01 and other OSHA PSM requirements related to SIS, OSHA considers the employer in compliance with PSM requirements for SIS.

OSHA issued a second important RAGAGEP enforcement policy LOI in 2010 [10]. In this LOI related to blast resistant structures, OSHA reiterated its policy for enforcing the RAGAGEP sections of the PSM standard.

6.2 OSHA RAGAGEP Enforcement

Because Process Safety information and, specifically, compliance with RAGAGEP is such an important part of a strong PSM program, OSHA emphasizes this in both the Refinery and Chem NEPs. Both NEPs rely on focused questions and compliance guidance that Compliance Officers can ask to help assess compliance with the standard. The Refinery NEP contains both static and dynamic (or changing) questions. OSHA published the static questions in the NEP while the dynamic questions changed periodically and limited to distribution within OSHA.

The Refinery NEP static questions focus on the following areas with respect to RAGAGEP:

- Relief Systems / Blowdowns
    - Does the PSI contain information on the RAGAGEP used?
    - Does the PSI contain the RAGAGEP's required design parameters?
    - Do the relief valves comply with the RAGAGEP?
    - Are intervening valves appropriate and adequately controlled?
    - Do the valves / blowdowns relieve to a safe location?
    - Are blowdown alarm systems in accordance with RAGAGEP?
    - Is blowdown quench water system in accordance with RAGAGEP?

    - Does the relief testing program comply with RAGAGEP?
        - Frequency
        - Personnel qualifications
        - Testing procedures
        - Testing blowdown alarms
- Facility Siting
- Pressure vessels/Piping
    - Does the PSI contain information on the RAGAGEP?
    - Are the safeguards properly designed?
    - Does the testing and inspection comply with RAGAGEP
        - Bonded liners
        - Thickness measurement locations
        - Corrosion under insulation
        - Testing frequency
        - Personnel qualifications
        - Testing basis
        - Injection points
        - Appropriate testing locations

It is important to note that OSHA also uses RAGAGEP compliance questions in it dynamic list of inspection priority items in both its Refinery and Chem NEPs.

OSHA analyzed citation data from the Refinery and Chem NEPs to determine the most frequently cited standards and any trends. In both enforcement efforts, the most frequently cited PSM sub-element is 1910.119(d)(3)(ii)—the requirement that employers document that equipment complies with RAGAGEP. Of these citations, most common areas cited involved valves and facility siting.

The following list outlines some of the valve-related issues cited while enforcing the Refinery NEP:

- Valves between pressure vessel and relief device not locked open or controlled.
- Relief device appropriate back pressure not maintained.
- Relief device inlet line pressure drop or chattering problems.
- Relief devices not installed on pressure vessels.
- Relief devices not appropriately sized for the circumstances.
- Valves inaccessible.
- Valves inoperable.

The following list outlines to two most frequent concerns regarding facility siting:

- Facilities not compliant with RAGAGEP chosen by the facility, and
- Structures either unprotected by adequate separation or building construction from consequences of HHC release such as fire, explosion, or toxic releases.

In the case of facility siting, citations often reference multiple on-site structures including control rooms and other buildings. In many cases, siting studies identified issues that the facility had not resolved in a timely manner.

OSHA also commonly cites the failure to document RAGAGEP in Chemical NEP inspections. However, due to the larger variety of facility types inspected, it is difficult to determine trends. The following list is a sample of RAGAGEP violations found during Chemical NEP inspections:

- Safety interlocks do not follow RAGAGEP
- RAGAGEP not documented for equipment (glass lined vessels, piping, rupture disks, chemical hoses, pumps)
- RAGAGEP not documented for ventilation systems design
- Ammonia system labeling not in accordance with RAGAGEP
- Chlorine alarm systems not designed in accordance with RAGAGEP
- Piping supported did not follow RAGAGEP
- Corrosion rates not calculated in accordance with RAGAGEP
- Chlorine did not have shut-off valves in accordance with RAGAGEP
- Relief devices not designed in accordance with RAGAGEP

PSM explicitly requires employers to document that the equipment complies with RAGAGEP at the time the process was originally built. However, employers should stay abreast of current engineering codes, standards, and practices. Changes in codes, standards, and practices often result when industry has discovered new hazards or new means of hazard control. If a the PHA team finds the process design does not adequately protect against a newly discovered or newly recognized hazard, employers should consider compliance with a more recent code or additional safeguards to ensure employees are protected.

7. Chemical Process Safety Incidents and RAGAGEP

Process safety incidents in the chemical and refining industries occur for many reasons. Investigators find multiple broken barriers that result in a process safety incident. Barriers such as engineering and administrative controls often play a role in the prevention of chemical process incidents. Barrier design, operation, and maintenance are frequently based on RAGAGEP.

Highlights of incidents reveal many employer failures regarding RAGAGEP allowing incident trajectories to continue to harmful outcomes.

Bellwood, Illinois - June 14, 2006

The incident occurred when a mixture of flammable liquids boiled in an open-top kettle. Vapors escaped the kettle and flowed into the processing room and into an adjoining warehouse/general purpose room where it ignited. The flash fire and deflagration killed one and injured two.

The facility installed the chemical processing area in direct communication with the general use area not in accordance with NFPA 30. In addition, the ventilation system was designed with only overhead point pick-up and not general ventilation with floor sweeps.

Middleton, Massachusetts - March 13, 2011

The release of acetone vapor into a congested workspace caused a deflagration that nearly destroyed the building. The ventilation system design was not designed to sweep or remove flammable vapors from the room. The facility did not use NFPA 30 in its ventilation system design. The inadequately designed ventilation allowed flammable vapors to accumulate.

Houston, Texas - April 12, 2004

The employer modified a large pressure vessel for its operation. The modification involved cutting and subsequent replacing a large coupon in the vessel wall. The modification did not follow RAGAGEP for performing welding on a pressure vessel. After the facility placed the modified vessel into service, it failed at the welded coupon, ejected its contents, rocketed into a neighboring building, and caused a large fire.

Illiopolis, Illinois - April 23, 2004

An explosion and fire killed five and injured three at this polyvinyl chloride facility. Many factors combined to cause a release of vinyl chloride monomer into the congested workspace. However, the facility had not designed and maintained its safety interlocks in accordance with RAGAGEP as they were easily disabled or bypassed. When needed to prevent the release, the interlock was bypassed and the release occurred.

Two of the most basic backbones of any process safety program are to design equipment in accordance with RAGAGEP and to maintain, inspect, and test equipment in accordance with RAGAGEP. Given these two fundamental tenets of process safety, there is little doubt that contributing to most incidents are failures to adhere to these requirements.

8. Conclusion

The RAGAGEP concept as used by OSHA comprises hundreds of industry-published codes and standards and, possibly, many thousands of internally-produced practices and protocols. When promulgated, the PSM standard made mandatory that facilities adopt code and standards published for their industry or produce internal standards that meet similar requirements. The management system that OSHA envisioned with the PSM standard is no different from any quality cycle (plan-do-check-act) and RAGAGEP plays a critical role.

Because OSHA is not in a position to be expert in any industry, use of RAGAGEP is critical to the success of a performance standard such as PSM. Using the collective wisdom and experience of industry experts, RAGAGEP helps achieve safe operations throughout industries using highly hazardous chemicals.

9. References

[1] "Process Safety Management of Highly Hazardous Chemicals," 29 CFR 1910.119, 57 FR 6356, Feb. 24, 1992

[2] "Guidelines for Technical Management of Chemical Process Safety," (pp 123 – 127), Center for Chemical Process Safety, 1989

[3] "Management of Process Hazards," American Petroleum Institute Recommended Practice 750, 1990

[4] 55 FR 29150 – 29173, Occupational Safety and Health Administration, July 27, 1990

[5] 57 FR 6356, Preamble to OSHA PSM Final Rule, 2/24/1992

[6] "Guidelines for Mechanical Integrity Systems," Center for Chemical Process Safety, 2006

[7] "Federal Participation in the Development and Use of Voluntary Consensus Standards and in Conformity Assessment Activities," OMB Circular A-119, Revised, February 10, 1998, http://www.whitehouse.gov/omb/circulars_a119/ (accessed February 13, 2013)

[8] "Occupational Safety and Health Act of 1970," 29 USC 651 – 678, December 1970

[9] "Compliance with PSM and ANSI/ISA-S84.01 for safety instrumented systems," Occupation Safety and Health Administration Letter of Interpretation, March 23, 2000, http://www.osha.gov/pls/oshaweb/owadisp.show_document?p_table=INTERPRETATIONS&p_id=23722

[10] "Applicability of the PSM Standard's Mechanical Integrity Requirements to Refinery Structures." Occupational Safety and Health Administration Letter of Interpretation, February 1, 2010 http://www.osha.gov/pls/oshaweb/owadisp.show_document?p_table=INTERPRETATIONS&p_id=27443

Disclaimer

This paper represents the views of the authors. Although is the authors based the paper on OSHA programs, it does not represent official OSHA policy.

Summary of Expert Forum on the Use of Performance-Based Regulatory Models in the U.S. Oil and Gas Industry, Offshore and Onshore

M. Sam Mannan*
Mary Kay O'Connor Process Safety Center, Artie McFerrin Department of Chemical Engineering, Texas A & M University, College Station, TX 77843, USA
Email: mannan@tamu.edu

Xiaodan Gao
Mary Kay O'Connor Process Safety Center, Artie McFerrin Department of Chemical Engineering, Texas A & M University, College Station, TX 77843, USA
Email: xiaodan.gao@tamu.edu

Prepared for Presentation at
American Institute of Chemical Engineers
2013 Spring Meeting
9th Global Congress on Process Safety
San Antonio, Texas
April 28 – May 1, 2013

UNPUBLISHED

Summary of Expert Forum on the Use of Performance-Based Regulatory Models in the U.S. Oil and Gas Industry, Offshore and Onshore

M. Sam Mannan, PhD, PE, CSP DHC*
Mary Kay O'Connor Process Safety Center, Artie McFerrin Department of Chemical Engineering, Texas A & M University, College Station, Texas 77843, USA
Email: mannan@tamu.edu

Xiaodan Gao
Mary Kay O'Connor Process Safety Center, Artie McFerrin Department of Chemical Engineering, Texas A & M University, College Station, Texas 77843, USA
Email: xiaodan.gao@tamu.edu

Keywords: Performance-based and prescriptive regulatory regime; Federal agency oversight; Safety case

Abstract

Over the recent decade, catastrophic incidents such as the BP Texas City Explosion and the Deepwater Horizon oil spill have grabbed much public and media attention. It is important now to discuss the effectiveness of the current safety regulatory regime in the US. Recommendations have been made to move forward to performance-based regulation. To gain opinions from different parties and the public, agencies including OSHA, BSEE, USCG and EPA initiated a stakeholder meeting to discuss several important issues of a performance-based regulatory regime. Discussions included: advantages and disadvantages of performance-based and prescriptive regulatory approaches, ways to advance performance-based regulations and standards in the oil and gas industry, and effective oversight of performance-based regulatory models. This paper will summarize the standpoints of different parties such as CCPS, API, Steel Workers and Contra Costa Accidental Release Prevention, and then provide detailed evidence to address the topics based on a comprehensive literature review.

1. Introduction

As presented by May, [1] a regulatory regime must provide a framework with a perfect equilibrium between control, accountability for results, and flexibility to achieve the results. A prescriptive regulatory system [2] "sets specific technical or procedural requirements that must be complied with by regulated entities". A performance-based regulatory (PBR) system [2] which is also referred to as goal-based, "identifies functions or outcomes for regulated entities but allows them considerable flexibility to determine how they will undertake the functions and achieve the outcomes".

After a series of accidents that occurred in the US, such as the 2005 Texas City Explosion and the 2010 Deepwater Horizon oil spill accident, recommendations have been made to adopt thorough performance-based regulatory regimes and develop a proactive and risk-based performance approach, taking the example of Safety Case by UK Health and Safety Executive (UK HSE).

To obtain diverse opinions about this important issue, multiple US agencies including Occupational Safety & Health Administration (OSHA), Bureau of Safety and Environmental Enforcement (BSEE), United States Coast Guard (USCG), and United States Environmental Protection Agency (EPA) invited interested parties to participate in a co-sponsored stakeholder meeting, and to submit comments on the use and implementation of performance-based regulatory models for enhanced safety and environmental performance in the United States oil and gas industry. The agencies proposed specific topics about different aspects of performance-based regulatory regime, which should address: 1) advantages and disadvantages of performance-based and prescriptive regulatory approaches; 2) whether these methods could create synergies between multiple agencies; 3) what types of models or combinations could result in long-term economic benefits; 4) ways to advance performance-based regulations and standards in oil and gas industry; and 5) effective oversight of performance-based regulatory models.

Mary Kay O'Connor Process Safety Center (MKOPSC) was invited to interpret our standpoints as a representative from the academic field. This paper will summarize the standpoints of different parties such as Center for Chemical Process Safety (CCPS), American Petroleum Institute (API), United Steel Workers and Contra Costa Accidental Release Prevention Program, and then provide detailed evidence to address all the topics based on a comprehensive literature review.

2. Standpoints of Federal Agencies and Other Parties

2.1 Federal Agencies

The speeches from the federal agencies explained their expectations and motivations to hold this meeting, including specific examples that came from their experience. Each of their presentations is briefly summarized as followed:

2.1.1 Occupational Safety & Health Administration (OSHA)

The representative from OSHA stated that federal agencies are reorganizing their work in ways that best serve competitive America, which is the spirit to organize this stakeholder meeting. In such a collaboration manner, the differences from agencies could be carefully considered. Much progress will be made through this opportunity to move beyond the tired, old cat-and-mouse relationship between regulators. Given the complexity of the oil and gas industry, it is important in our society to maintain and improve employee and public safety and protect the environment.

2.1.2 Environmental Protection Agency (EPA)

EPA stated that it is a very big challenge for all of the regulators to come up with programs that provide the necessary flexibility for all different facilities to comply with and actually meet the requirement. EPA is trying to develop programs that are with high flexibility for companies to comply with and take their major responsibility. The companies should always try their best to protect the public health, the environment and the workers that they have on their facilities.

2.1.3 White House Office of Information and Regulatory Affairs (OIRA)

President Obama has issued a series of policies, statements, and directives aiming at regulatory reform, and the most principal of which is Executive Order 13563 titled "Improving Regulation and Regulatory Review." This executive order outlines a series of core principles and procedures to direct agency regulatory activity with the aim of promoting predictability and avoiding uncertainty in the marketplace, and in the process, eliminating unnecessary or unjustified burdens on the private sectors. The government does not always have all the answers and we are working with agencies to consider alternatives to regulatory options before they issued those regulations. The representative introduced the six key provisions in this executive order: 1) Predictability and Certainty, 2) The best available science, 3) Public participation, 4) Use of the least burdensome tools, 5) Considering costs and benefits (both quantitative and qualitative), and 6) seeking to improve actual results.

2.1.4 Bureau of Safety and Environmental Enforcement (BSEE)

The responsibility of BSEE is to protect the environment, and conserve offshore resources through regulatory oversight and enforcement. The representative provided a regulatory update of BSEE, including Safety and Environmental Management Systems (SEMS) I&II, and summarized the BSEE way ahead with the following strategies:

- Blend prescriptive and performance-based approaches
- Compel safety, emergency preparedness, environmental responsibility, and appropriate development and conservation
- Build and sustain this capacity within and across BSEE's key functions

2.1.5 United States Coast Guard (USCG)

The USCG representative stated that they have a blend of both prescriptive and performance-based regulations. They utilize international standards, guidance, recommended practices, classification society rules and industry standards. In general, they regulate US vessels that are involved in oil and gas under Outer Continental Shelf Activities (OCSLA) rules.

2.1.6 Pipeline and Hazardous Material Safety Administration (PHMSA)

It was mentioned that PHMSA has been experimenting with risk-based, management-based, and performance-based regulations for about two decades. They made some progress, but still there are a lot of things that they have yet to learn. When there is a major accident, there is a demand for the regulatory agency to fix the problem. Generally people want performance-based regulatory approaches only until we give it to them because of some disadvantages. Federal agencies have limited resources to obtain a perfect regulatory regime. It will be the operator's responsibility to identify the risks for their facilities if a performance-based scheme will be mandated. It will not be easy to conform to performance-based regulations.

2.2 Viewpoints from Other Parties

The American Petroleum Institute (API) supports a performance based regulation to encourage innovation in designs, techniques, processes and materials. API has developed around 600 standards covering oil and natural gas industry, which have been cited in multiple regulations by agencies such as OSHA, EPA, DOT and BSEE. For example, API RP 75 has been cited by BSEE, and API 2610 was cited by four agencies covering regulations in design, construction, operation, inspection, maintenance of a petroleum terminal.

Opinions from the US Steelworkers Union reflect the perspective of workers working in the refining sector. They recommended a combination of prescriptive-based and performance-based regulation. The challenges would come from comprehensive database/lessons learned from industrial accidents. We should also learn lessons from incidents with less severe consequences than catastrophic incidents. They believed that agency duplications should be reduced to increase efficiency.

Dr. Andrew Hopkins from Australia argued that we should move toward a safety case regime. He pointed out that regulators and industry should work in collaboration to prevent accidents in the safety case. The elements of a successful safety case include:

- Hazard management framework, like PSM or SEMS, with PHA as the heart of the safety case
- Case made to the regulator, which forces the operator to choose the best standard for their purposes and justify it to regulators
- Competent and well-resourced regulator
- General duty of care, from which the regulated body must genuinely want to improve safety

The Center for Chemical Process Safety (CCPS) stated that a risk-based process safety regulatory regime is necessary to thoroughly understand potential consequences, apply appropriate systems to achieve desired risk and ensure systems are functioning as expected. CCPS has guidelines that are similar to OSHA PSM, but with several added elements such as process safety culture, measures and metrics, and stakeholder outreach. CCPS stressed the use of metrics and finding root causes to problems, all within a risk-based approach.

The California Accidental Release Prevention Program (CalARP) indicated that they have been mandating an Industrial Safety Ordinance (ISO) since 1999, which is the most stringent process safety regulation in the United States. It is similar to OSHA PSM, but contains additional requirements, such as Human factors, Management of Organizational Changes, Inherently Safer Design, Root Cause Analysis Investigations, and Safety Culture Assessments. The actions to ensure oversight include:

- Conducting unannounced inspections of facilities
- Performing audits of Safety Programs at the Facilities
- Detailed on-site inspections/audits at least every three years (4-5 weeks per audit)
- Screening process before each audit to focus on problem areas
- Audit follow-up on previous issues
- Facilities are required to develop process safety metrics

3. MKOPSC perspective

3.1 Advantages and Limits of Performance-Based Regulatory Approaches

Prescriptive regulatory approaches emphasize control and accountability, and performance-based regulatory approaches promote flexibility with accountability for results. These features must be acknowledged as integral parts of the discussion regarding the advantages and limits for adapting performance-based regulations. The debate must consider two main points: the possibility for a long-term hazard reduction (Health, Safety and Environmental) and the utilization of resources.

3.1.1 Advantages of Performance-Based Regulatory Approaches

Numerous authors have provided discussions concerning the advantages of performance-based regulatory regimes. The key points raised can be summarized as follows:

- *Innovation and adoption of new technologies* [3-7]

  The flexibility that accompanies performance-based regulations allows for a continuous improvement by means of the implementation of new technologies as they become available. This is a desired feature in order to keep up with a complex and continuously changing industry and market, where unique situations often occur. Moreover, contrary to common belief, opportunities for innovation arise not only for the regulated entities, but also for the regulatory agencies. According to EPA [8], regulatory agencies are granted discretion to set performance targets within broad policy areas and to experiment with means to achieve them.

- *Goal-oriented risk management system* [7-8]

  The implementation of a performance-based regulatory system is accompanied by a goal-oriented risk management system. There is a clearer quantification of the performance, and any decision must then be based on the assessment of the hazards implied. Moreover, this management system will be common to all participating companies, thus allowing for unification in the objectives, communication, planning, and response to HSE-related issues.

- *Safety ownership, responsibility and pro-activity* [7-12]

By means of the joint responsibility generated, traditional distinctions between regulated entities, regulatory agencies and regulation beneficiaries become fuzzier [12]. It results in a further understanding of the involvement and support necessary, and a wider use of the industry knowledge [11, 13-14]. The ownership of risk and the generated responsibility switch the focus to company's HSE objectives, abolishing dedication limitations. This provides plenty of opportunities for independent assessment and actions that go beyond minimum performance objectives.

- *Potential for more efficient utilization of resources* [1, 7-8, 11, 14-15]

As a consequence of the flexibility to achieve results, the cost-effectiveness of the compliance strategies is stimulated. Companies are encouraged to innovate with cost-effective solutions [15] and proven successful alternatives that have been applied in different locations of the world [7]. In the long-term, this will not only benefit the companies, but also their workers and the public in general.

3.1.2 Limits of Performance-Based Regulatory Approaches

The Regulatory Policy Program of the Center for Business and Government at Harvard University organized a workshop in May 2002 to discuss performance-based regulations. Several government agencies and researchers from a variety of areas participated in the workshop. The workshop concluded that some of the most important criteria to consider when determining or selecting the regulatory approach are the effectiveness, efficiency, equity, clarity, and the ease and accuracy of enforcement [18]. The discussion provided in this section follows the aforementioned report and incorporates other literature sources.

- *Cost increase pitfalls for companies and regulatory agencies* [11, 13,14,16, 17]

Apart from the obviously overwhelming task that the application of performance-based regulations implies, and the associated economic burden, several possible pitfalls can be identified for both the regulators and the industry. Several authors assure that an increased cost to regulators is unavoidable due to the number of trained personnel that a more active regulator will demand. Some authors even point out monitoring compliance being too costly as a factor for failure [11].

- *Lack of suitable metrics to ensure compliance*

One of the most widely debated issues about performance-based regulations is the lack of performance indicators for both regulators and regulated entities to measure the outcomes. Dave Drummond from the University of Wisconsin-Madison states that as long as the outcome cannot be measured, performance-based regulation should not be used [19]. However, even if the outcome can be measured, the measurements or indicators must be adequately reliable in order to provide appropriate feedback. For this purpose, performance-based regulation may be accompanied by appropriate prescriptive requirements for measuring, reporting, recordkeeping and data management [20].

- *Performance may be difficult to measure directly for rare and catastrophic events*

Performance-based regulations might be difficult to implement when the consequences of poor performance are very serious. In that case, the regulation must be formulated very carefully. Youngblood and Kim [20] recommend the use of an objectives hierarchy for the selection of performance measurements and establishment of performance criteria. These measurements and criteria should focus on levels of performance in which the failure to meet a criterion or a given target will not lead to severe consequences or immediate safety concerns [20].

Performance-based regulations are less desirable when used to address rare catastrophic events, such as big fires, explosions or toxic releases, as it is almost impossible to measure performance for these types of events. Instead, performance must be predicted, which causes uncertainty and difficulties in the implementation [18].

- *Small companies having limited expertise or capacity for innovation*

  With regard to the size of the company, performance-based regulation is usually more difficult to implement in small companies. Medium and large companies have a larger availability of resources to create customized programs to achieve the desired outcomes. They also have the capacity of developing innovative, efficient and cost-effective solutions to meet performance goals [21]. Smaller companies may not have the resources to develop new programs or processes or invest in innovation [21]. Some small companies would rather be told the process they have to follow to achieve the desired outcome instead of creating one on their own [18].

- *There is no detailed understanding of the problem*

  Performance-based regulations can be either tightly or loosely specified, depending on the level of discretion provided to the regulator and the regulated entity [18]. For tightly specified performance-based standards or those where a quantitative threshold is provided, it is necessary to understand in detail the cause-effect or dose-response relationships between the sources and the final objective in order to establish optimal thresholds [18]. Lack of understanding of the problem or loosely specified requirements, including words such as “adequate” or “effective” [22], cause difficulty in the implementation and enforcement of the regulation.

  It is worth pointing out that the workshop at Harvard University [18] was conducted 10 years ago. However, the challenges and difficulties to implement performance-based regulation remain more or less the same. More research is needed to obtain data on the effectiveness of performance-based regulation as compared to prescriptive regulation in order to determine the best approach for each industry/situation.

  Overall, it is desirable to use a combination of performance-based and prescriptive regulations in order to address a specific issue or problem [21].

3.2 Advancement of the Use of Performance-Based Regulation in U.S.

In order to advance the use of performance-based regulations in the US Oil and Gas Industry, several criteria must first be met. The main factors in whether or not the programs are acceptable and successful include the following:

- There must be convincing statistical evidence that the performance-based measures will outperform the prescriptive-based measures they are replacing.
- There must be a fair way to implement these performance-based measures in relation to the prescriptive-based measures (not overregulated, not too much burden on companies to comply too quickly) as well as in relation to the differences in situations that the companies currently have (large vs. small, no one taking advantage of rules).
- There must be an economic benefit to the switch from prescriptive measures to performance-based measures, where the financial sunk costs and operating cost increase is counterbalanced by a tangible economic benefit, whether that benefit is decrease in incidents/fatalities or a decrease in cost through more efficient use of resources and innovation.

These concepts are graphically represented in the following Figure 1.

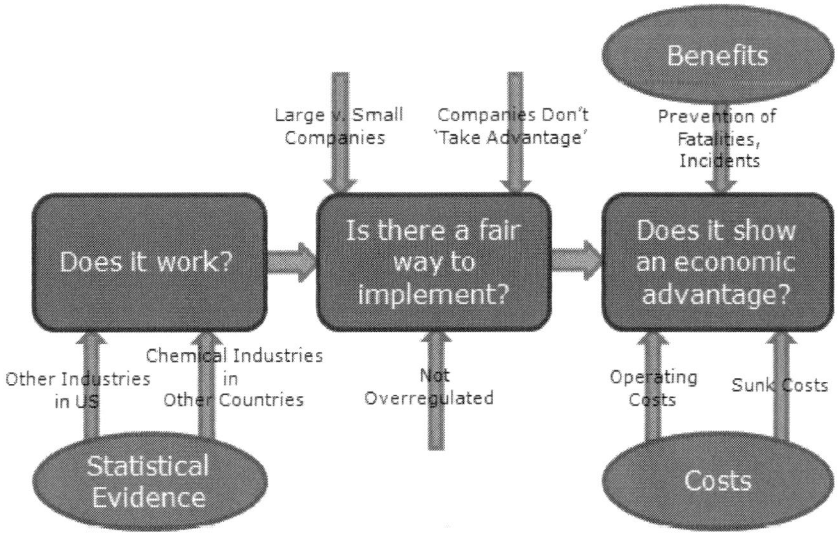

Figure 1. Advancement of performance-based regulation

These points are further discussed in the following paragraphs.

3.2.1 Statistical evidence

It is simple human nature to resist change if the change proposed does not fulfill the need it is intended to satisfy. Therefore, it is natural to expect resistance to change from the current types of regulation if it cannot be proven that the safety performance stands a significant chance of improving. There is precedent for the advancement of performance-based regulation, both in other industries in the US (nuclear industry and their Performance-Based and-Risk Informed regulation), and in the oil and gas industry in Europe (UK and Norway).

The question becomes, then, have these industries seen any tangible increase in safety statistics by either PSM or Safety Case? Unfortunately, the answer to this question is that it is hard to tell for several reasons (Figure 2 and Figure 3). Based upon the statistics from the UK HSE [23] and US API [24], it is difficult to tell whether the statistics show any real upward or downward trend.

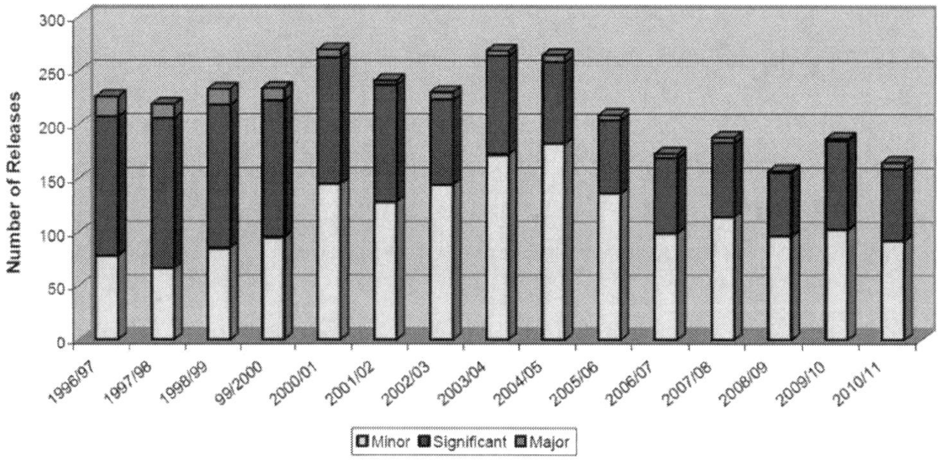

Figure 2: HSE data for offshore hydrocarbon releases in the period from 1996-97 to 2010-11[23]

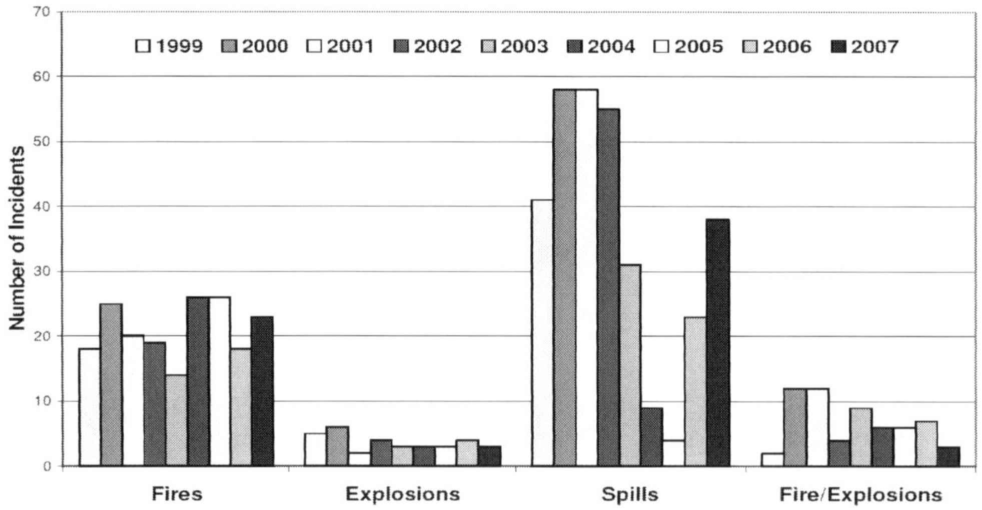

Figure 3: API Process Safety Performance Measurement Report 1999-2007[24]

If there is a trend, either upward or downward, it would also be very difficult to tell whether it was due to the regulatory regime or whether it was a function of some other factor (*e.g.*, an increase in minor accidents due to better reporting or a decrease in major accidents due to more of them being reported as significant). As such, the regulatory bodies would have to prove that these gains that other industries and other countries are making are not just "smoke and mirrors" based on other influencing factors.

Finally, it is difficult to say that any trends can be taken from a mere 15 years of data, especially when the goal of the regulation is to decrease significant events that may happen once every 10, 25, or even 50 years. Furthermore, an extended amount of time may be needed to get the programs fully implemented and working properly before results can be seen. Thus, many years of data collecting may be needed before one can conclusively state that the program is working as intended.

3.2.2 Is there a fair way to implement?

When PSM was first implemented and RMP was on the horizon, industry had several doubts and fears for the two seemingly overlapping regimes with new procedures for compliance. In the two decades since, many of the same fears would apply to the advancement of new regulation.

Not overregulated [25]: When PSM started and RMP was around the corner, one of the main concerns that industry had was whether or not they would be overregulated by the dual regulations. Since PSM and RMP have different aims (occupational vs. environmental), there was concern that each would be quite different although aimed at basically the same end. Thus, Andrews, speaking for industry through his paper, made several key requests:

- EPA chemicals be a subset equal to or less than the chemicals in OSHA PSM
- EPA threshold quantities be greater than or equal to the threshold quantities of PSM, arguing that if they were less, it would effectively extend PSM to operations that OSHA did not deem it necessary for PSM to be extended to
- Exclusions for certain (unmanned low-hazard) facilities
- Recognition that compliance with PSM would constitute compliance with RMP

This shows the level of concern with what industry saw as "fairness" in overregulation. They wanted two mutually inclusive regulations that would not increase their burden, even though the regulations had different goals.

Companies do not take advantage [26]: Companies under a performance-based regime, since it is not as concrete as a prescriptive-based regime, may attempt to take advantage of the added flexibility (not in an innovative way, but in a way such that they would be 'cheating' the system).

May and Koski [26] suggest that extra education could be a helpful way to keep this from happening, basically arguing that it will lead to the owners, operators, and engineers 'owning' safety, thus making it so that they will not be morally capable of cheating. However, they also note that due to differing capabilities and willingness to learn, this may not be the most effective method.

The second method suggested was to hold engineers (and whomever does safety reviews, tests, *etc.*) responsible for the accuracy of the results, saying that if it is found that anything is forged or unreal they can be held legally responsible. They also take it one step further and say that engineers could be held liable for performance (whether they did a good job complying with the regulation or not). They admit that this may cause insurance costs for liability to go too high to afford, causing engineers and consultants to charge more, possibly pricing them out of a job.

The third method mentioned was similar to a peer-review process where documents are reviewed by others for adequacy. The quality of results would depend on the quality of the reviewer and how enthusiastic they are about making suggestions and questioning safety cases presented.

Large vs. Small Companies [18]: Small companies face their own specific set of challenges in the move to performance-based regulation. Coglianese, Nash, and Olmstead [18] posit that small firms may not have the resources to look for a way to achieve a performance-based goal and that it is oftentimes easier for them to follow orders rather than innovate for the simple reason that they may not have the money, manpower, or time to find a better way to reach the moving target. They also point out that it is the job of the government and industry to work together so that the government can understand exactly how difficult it could be for a small firm to implement a performance-based regulation and their possible preference for a prescriptive-based regulation. In such a way, the large companies must help to protect the smaller companies from unduly strenuous regulation by an open communication with those creating the regulation.

3.2.3 Does it show an economic advantage? [25]

It is simple economics that, above all else, a company (whose motive is to maximize economic return, without which they will not stay in business) needs to see an economic benefit in order to accept new regulations. Andrews specifically mentions that many in industry believe that OSHA underestimated the cost of compliance with the new PSM system (OSHA said $810 million per year for the first five years, while Andrews argued it would be five times that). He also argues that though there will obviously be some economic benefit through lower fatality rates and incidents and increased reliability, industry was still concerned about trying to gain a competitive advantage (being able to gain a competitive advantage means being able to maximize economic benefit) in the face of such a large resource investment.

3.3 Multiple Agency Jurisdiction Issues

3.3.1 Duplication of regulations

Reduction of regulatory overlaps is a theoretically attractive way to streamline operations and simplify compliance for industry where there may currently be situations in which similar hazards are managed in seemingly different ways. The following paragraphs explain the origin of some overlaps, advantages that could possibly be gained, and the disadvantages that may be created in the elimination of regulatory overlap.

Origins:

Aagaard [27] has presented several ways in which regulatory overlap could occur. Overlap can occur organically, as a function of the growth of a regulatory body or as an attempt to gain more power, prestige, or oversight for a body, either because they want the extra power or because they believe they can do a better job or fill some gaps in the regulation. It can happen arbitrarily, when two regimes' missions somewhat overlap, giving them both the power to regulate similar areas. It is possible that overlap may arise intentionally, such as with OSHA PSM and EPA RMP, where Congress specifically mandated that both be developed in conjunction with one another; or unintentionally when the delegating bodies are unsure which regulatory body should be responsible for what, or if there is political conflict. It could also arise from overlapping legal fields, which is difficult to avoid and nearly impossible to change in hindsight. For example, OSHA regulates employee health while EPA regulates environmental and public health. Their legal fields necessarily overlap because in many cases what is dangerous to a worker is dangerous to the public or environment, so the legal overlap could not have been avoided, nor can it be eliminated.

In some ways, this overlap could be used to the advantage of the regulators, specifically when the overlap is intentional and is meant to fill gaps in between regulatory bodies that may have similar fields but differing missions. In other ways, the overlap could be destructive, particularly when it is inadvertent and causes conflict in how to deal with compliance in industry. Thus, when handled properly, overlap can be a powerful tool in making sure that regulations cover the largest span possible while causing minimum inconvenience to industry, who must comply, report, and improve.

Disadvantages:

It is obvious that disadvantages would arise from overlap, mostly because of efficiency issues. The extent of these disadvantages depends mainly on the coordination between the regulatory bodies and the degree of the overlap. Four categories were identified by Aagaard [27] as the classical arguments against overlap:

- *Duplication* – a waste of resources if the functions can be consolidated or eliminated from one of the overlapping bodies, though it is not a complete waste if there is benefit to the duplication
- *Conflict* - opportunities for conflicting regulations, whether directly conflicting (cannot satisfy both) or indirectly conflicting (can comply with both with increased effort)
- *Coordination* – necessity of increased coordination between the regulatory bodies
- *Complexity* – difficult to monitor who is in charge of what, or assign blame for failures, *etc.*

Rectanus and Natalicchio [28] further posit that overlap may have harmful effects during regulatory functions such as incident investigation, stating that regulatory bodies may prefer to independently work on investigations so as not to get delayed by other bodies. They state that it may be difficult or time-consuming to have to coordinated investigations and in the end the regulators may not be able to obtain all of the necessary information or work on the timeline of another regulator. Thus, they trade the inconvenience to the operator (having multiple investigations, procuring the same documents multiple times, *etc.*) for the convenience of being able to work on their own terms. Other overlaps exist, such as creation and administration of training materials, development of emergency response procedures, and record-keeping.

Industry itself has also expressed concern over some of the disadvantages of duplicative and inconsistent regulation. For example, Andrews [25] stated in his paper on industry views of regulation in the early days of PSM, that if a regulation (RMP) was implemented after and in conjunction with PSM and included PSM controlled chemicals at lower than PSM threshold values, it would effectively be extending PSM to sites that OSHA did not deem appropriate. This is an example where, with poor coordination, industry could have been severely affected by an overlap in regulation. Thus, care must be taken in the implementation of closely-related regulations so that a regulation is not inappropriately extended into a realm where it is not necessary.

Advantages

Perhaps counter-intuitively, there are also advantages that can be gained by these overlaps, particularly when they are intentional, well-thought out, and allow opportunity for collaboration between regulators that work in similar areas, but have differing missions and expertise.

Rectanus and Natalicchio [28] found that advantages could be gained in precisely these areas through overlap. They state that it is often reasonable to have some degree of overlap in such closely related fields, and that it is not necessarily a bad thing if it allows for competition between agencies, better service delivery, or emergency backup. They do, however, state that overlap must be properly managed in order to retain these advantages. Aagaard [27] further states that overlap can increase reliability by reducing the number of errors in action (regulators do not act when they should have acted) and can foster policy innovation where regulators have a heightened sense of what other bodies are doing and can thus act in a more informed manner, such as making better policy decisions in conjunction with one another, and learning from the successes and mistakes of others.

Should Overlap Be Eliminated?

It is intuitive to want to eliminate as much overlap as possible, especially from the industry standpoint where this elimination could possibly result in less work, greater efficiency, and less confusion over compliance. However, care must be taken to avoid eliminating overlap that was put in place intentionally to gain the possible benefits of competition and collaboration. Thus, as much as one would perform a risk analysis in a hazardous process if they were considering making a change, if overlap is to be eliminated, it must be as a result of careful analysis of the risk of the elimination.

3.3.2 Implement PBR uniformly by multiple agencies

Initially, it is important for every agency to determine how they are going to implement performance-based regulations individually and then uniformly across the industry. Primarily, agencies will have to develop strategic plans that cover the functions and operations of the agencies; then, they will have to identify the strategic goals and objectives. Further, a detailed description of the required resources and processes needed to fulfill the goals and objectives should be completed. Later on, the agencies will have to identify which are the factors that could affect the achievement of the goals and objectives suggested earlier by the agencies. Finally the agencies will have to describe how the performance goals included in annual performance plans will be related to the agencies goals and objectives [1, 29 - 30].

Once all the elements mentioned in the previous paragraph have been addressed, the agencies must receive feedback from the US Congress, the Board of Directors of the companies, the public and the stakeholders (as shown in Figure 4).

Congress will have to review the plan for every agency to achieve the corresponding goals and objectives and analyze the processes and resources required to meet such goals. This is necessary information for Congress to assign the appropriate budget for the agencies to fulfill their objectives. However, for this to be completed a performance plan has to be analyzed and discussed between the agency and Congress. Some points that must be addressed in this plan are:

- Describe the processes and resources required to meet the performance goals;
- Determine which are the performance indicators to be used in measuring or assessing the relevant outputs and outcomes of each activity;
- Provide a comparison basis of the actual program results to the established performance goals [29].

Figure 4. Feedback path for the agencies

Furthermore, before this plan is given to Congress, several perspectives have to be considered. For instance, the expertise of the Board of Directors of the companies, stakeholders and the public will strengthen the performance regulations by making them more achievable and realistic. Wholey *et al.* [29] proposed a criteria methodology to grant the political, senior management and analytical support for the successful implantation of performance-based regulations. These criteria are:

- Agreement on goals and strategies by using the professional judgment of experts.
- Performance measurement systems of sufficient quality. In this respect, the national database system recommended and discussed earlier could be used to develop the performance measurement systems.
- Performance-based management systems to build the communication of the performance information to the relevant agency.
- Accountability, demonstrating effective or improved performance, and supporting policy decision making, establishing communication between the agencies with the stakeholders and the public.

Moreover, it is recommended that a proper inspection methodology for agencies to enforce regulations is established, as well as a uniform terminology across agencies and industries. Finally, a training program has to be implemented. This training should include the political and bureaucratic context for performance-based management and should focus mainly on the following issues [29]:

- Measuring, evaluating, and reporting on performance to analyze the implications of alternative policies and program approaches;
- Using performance information to support resource allocation and other policy decision making;
- Reinforcing performance-based management and building expertise in performance-based management.

4. Conclusions

For all the various combinations of agencies and regulatory programs, it is very difficult to judge which regulatory regime is more effective. To answer this question, performance data should be collected adequately and uniformly under consistent incident reporting system. Otherwise, biased data collection will always lead to incorrect conclusions.

Secondly, it has been discussed that a "one size fits all" regulatory approach is not appropriate due to the diversity of existing industries, processes and company sizes. It is therefore necessary to determine the most adequate regulatory approach to implement according to the situation. Some situations in which prescriptive regulation is preferred are mentioned below:

- High risk situations, particularly, when there are technologies known to work well. For example, in the case of the airline industry where the consequences of an accident are severe, there are prescriptive methods to achieve the required level of safety, which have shown successful results.
- When companies have limited capabilities to develop their own approach to achieve performance targets, such as small companies. These companies would benefit from prescriptive regulations or prescriptive components given as optional guidance to complement performance-based regulations.
- Standard procedures that have little need for innovation.
- In situations where experience exists, or there is little room for innovation, a prescriptive regulation should be preferred. Instead, when the rate of change of technology or innovation is an important consideration, performance-based regulations might be desirable.
- Businesses could benefit from prescriptive components as a complement to performance standards to provide more concrete guidance.

Finally, to answer the question that if we should consider further changes in the regulatory framework, we should figure out how to improve our existing program instead of changing the whole system and allow time for existing regulations (*e.g.,* PSM) to be implemented properly. We could increase training and reinforce the regulation compliance, gradually add regulatory reform and efficiencies to the existing system, and implement overall quantitative risk assessments for plants. We should put in place a well-planned national database system to collect appropriate and statistically significant data on incidents to be able to evaluate the performance of regulatory agencies as well as the regulated entities.

References

[1] May P., "Performance-Based Regulatory Regimes", 13th World Conference on Earthquake Engineering, Vancouver, B.C., Canada August 1-6, 2004 Paper No. 3254.

[2] Jennifer Dagg, Peggy Holroyd, Nathan Lemphers, Randy Lucas and Benjamin Thibault, "Comparing the Offshore Drilling Regulatory Regimes of the Canadian Arctic, the U.S., the U.K., Greenland and Norway", 2011.

[3] C. J. Burke and J. A. Veil., "Potential environmental benefits from regulatory consideration of synthetic drilling muds", Environmental Assessment Division, Argonne National Laboratory 9700 South Cass Avenue, Argonne, Illinois 60439.

[4] Office of Technology Assessment, U.S. Congress. "Environmental Policy Tools: A User's Guide. OTA-EVN-634", Washington DC: U.S. Government Printing Office, 1995.

[5] John, D. "Civic environmentalism: Alternatives to regulation in states and communities", Washington, DC: CQ Press, 1994.

[6] Hopkins Andrew, Professor, Australian National University & consultant to the Chemical Safety Board Deepwater Horizon team.

[7] National Academy of Engineering and National Research Council, "Macondo Well - Deepwater Horizon Blowout: Lessons for Improving Offshore Drilling Safety", The National Academy Press, Washington, D.C., 2011.

[8] EPA, "Reinventing Environmental Regulation".

[9] Coglianese C, Lazar D. "Management-Based Regulation: Prescribing Private Management to Achieve Public Goals." Law and Society Review, 37(4): 691-730, 2003.

[10] Gunningham N, Johnstone R., "Regulating Workplace Safety, Systems and Sanctions", Oxford: Oxford University Press, 1999.

[11] Queensland Goverment, Department of State Development, "Guidelines on alternatives to prescriptive regulations".

[12] Karkkainen, Bradley C., Archon Fung, and Charles F. Sabel, "After backyard environmentalism: Toward a performance-based regime of environmental regulation", American Behavioral Scientist, 44(4): 692-711, 2000.

[13] DNV, "Key Aspects of An effective U.S. offshore Safety regime", Whitepaper.

[14] U.S. Regulatory Council. "Performance Standards, A Practical Guide to the Use of Performance Standards as a Regulatory Alternative. Report of the Project on Regulatory Alternatives", Washington DC: Administrative Conference of the United States, 1981.

[15] Yates Report. "Weathertightness of Buildings in New Zealand. Report of the Government Administration Committee's Inquiry into the Weathertightness of Buildings in New Zealand", Wellington, NZ: House of Representatives, 2003.

[16] Arnold Ken, Senior Technical Advisor, Worley Parsons.

[17] Coglianese C, Nash J, Olmstead T. "Performance-Based Regulation: Prospects and Limitations in Health, Safety, and Environmental Protection." Administrative Law Review, 55: 705-729, 2003.

[18] Coglianese, C., J. Nash, and T. Olmstead, "Performance-Based Regulation: Prospects and Limitations in Health, Safety and Environmental Protection", Regulatory Policy Program Report No. RPP-03. John F. Kennedy School of Government. Harvard University, 2002.

[19] Drummond, D. "Regulatory strategies: Building performance oriented chemical waste rules", Available from: http://www2.fpm.wisc.edu/chemwasteinfo/PerformOrientedRules.htm.

[20] Youngblood, R. and I.S. Kim, "Issues in Formulating Performance Based approaches to regulatory oversight of nuclear poweder plants", Nuclear Engineering and Technology, 37(3): p. 231-244,2005.

[21] Queensland Goverment. Department of State Development, "Guidelines on alternatives to prescriptive regulation", Available from: https://www.wbginvestmentclimate.org/uploads/23.Queensland.pdf.

[22] Regulatory Uncertainty in Performance-Based Regulations, "Analysis of the Uncertainty in STUK Guide YVL 4.3. Fire Protection at Nuclear Facilities in Finland". Available from: http://www.google.com/url?sa=t&rct=j&q=&esrc=s&source=web&cd=2&cad=rja&ved=0CEYQFjAB&url=http%3A%2F%2Fwww.neimagazine.com%2FJournals%2FPower%2FNEI%2FMarch_2004%2Fattachments%2FTrubatchUncertain.doc&ei=xVMxUJ2tE-K02gWlloHACQ&usg=AFQjCNGtK1cILOPvH5ek8Qweg7l9eMKTlQ.

[23] "Offshore Injury, Health, and Incident Statistics 2010/2011," UK Health and Safety Executive, Merseyside 2011.

[24] American Petroleum Institute, "Process Safety Performance Measurement Report", American Petroleum Institute Statistics Department, June 2008.

[25] D. C. Andrews, "Industry views on chemical process safety," Process Safety Progress, vol. 12, pp. 104-105, 1993.

[26] P. J. May and C. Koski, "Performance-Based Regulation and Regulatory Regimes," Pacific Earthquake Engineering Research Center, University of California, Berkeley, Berkeley, CA September 2004.

[27] T. S. Aagaard, "Regulatory Overlap, Overlapping Legal Fields, and Statutory Discontinuities," Virginia Environmental Law Journal, vol. 29, 2011.

[28] L. Rectanus and J. Natalicchio, "Better Coordination Can Improve Safety at Hazardous Material Facilities," United States General Accounting Office, Washington, D.C. 2000.

[29] Wholey, J. “Performance-Based Management: Responding to the Challenges”, M.E. Sharpe, Inc., Vol. 22, No. 3, pp. 288-307, 1999.

[30] Altmayer, C. “Moving to performance-based management”, Government finance review, 2006.

A Guide to the Legal Framework of the PSM Standard for Engineers

Christopher Cunio, Esq.
Cooley Manion Jones LLP
21 Custom House Street
Boston, MA 02110
ccunio@cmjlaw.com

Georges Melhem, Ph.D.
ioMosaic Corporation
93 Stiles Road
Salem, NH 03079
melhem@iomosaic.com

Prepared for Presentation at
American Institute of Chemical Engineers
2013 Spring Meeting
9th Global Congress on Process Safety
San Antonio, Texas
April 28 – May 1, 2013

Keywords: *acceptable limits, audit, citation, compliance, corrective action, coverage, deficiency, fine, flammable, General Duty Clause, highly hazardous chemical, inspection, timely manner, training, mechanical integrity, National Emphasis Program, non-serious, performance based, process, process safety information, process hazards analysis, RAGAGEP, serious, violation, and willful.*

Abstract: *Compliance with the PSM Standard is challenging for even the most sophisticated operators because of the broad scope and highly technical nature of the 14 PSM elements. This article provides guidance on how to comply with the three elements most frequently cited by OSHA—process safety information, process hazards analysis, and mechanical integrity—and the consequences of a failure to do so.*

The Process Safety Management (PSM) Standard is a U.S. regulation that was issued by the Secretary of Labor in 1992. Its enactment was motivated by a series of chemical disasters in the 1980s and early 1990s, including the 1984 Bhopal, India gas leak disaster, which resulted in more than 2000 fatalities, and deadly chemical explosions in the U.S. facilities of Phillips 66, ARCO and BASF.[1]

The primary purpose of the PSM Standard is to prevent, or minimize the consequences of, the release of highly hazardous chemicals into locations that could expose people to serious bodily injury and deadly harm. To achieve its purpose, the Standard requires regulated facilities to implement a "PSM program," which is a systematic approach to proactively review chemical processes and identify, evaluate, and prevent or mitigate chemical releases. PSM programs must contain 14 components (referred to as "elements"): (1) employee participation, (2) process safety information, (3) process hazard analysis, (4) operating procedures, (5) training, (6) contractors, (7) pre-startup safety review, (8) mechanical integrity, (9) hot work permit, (10) management of change, (11) incident investigation, (12) emergency planning and response, (13) compliance audits, and (14) trade secrets.[2]

Compliance with the PSM Standard is challenging for even the most sophisticated operators because of the broad scope and highly technical nature of these 14 elements. This article focuses on the elements of the PSM program—specifically process safety information, process hazard analysis, and mechanical integrity—that have resulted in the majority of OSHA citations under the Petroleum Refinery (>50%) and Chemical (>60%) PSM National Emphasis Programs.[3]

The goal of this article is to provide in-house engineers with guidance on how to comply with these elements and when to consult external resources, such as engineering and/or legal PSM experts. The number of citations issued by OSHA in connection with the National Emphasis Programs demonstrates that there is still considerable misunderstanding of the PSM Standard, which is now over 20 years old. Consultation and collaboration with external PSM experts can facilitate the creation and maintenance of a robust, compliant PSM program. It can also be beneficial when addressing PSM violations identified through audits (1st, 2nd, and 3rd party) and OSHA inspections to ensure efficient and effective corrective action. Finally, it can be beneficial during due diligence performed in connection with mergers and acquisitions. External PSM expertise can help identify PSM violations and quantify corrective action prior to closing, which can substantially impact purchase price and/or the scope of indemnification rights under a purchase agreement.

Performance Based

The PSM Standard is "performance-based," which means that the regulator, in this case the Secretary of Labor, dictates "what to do" (e.g., to implement a PSM program that contains the 14 elements), and the operators determine "how to do it" (e.g. through the adoption of internal policies, external standards, or a hybrid/combination approach) based on their knowledge, industry practice, facility needs and conditions, and economic considerations. A site's PSM program can be a single document or series of documents; the only requirement being that it contains all 14 elements.

Operators should consider consulting PSM experts to determine whether a site's PSM program appropriately incorporates all 14 elements. In addition, they should consider consulting PSM experts to review the content of any existing PSM programs because OSHA can issue citations for a site's failure to follow its own PSM program, even when the site's program is arguably more stringent than industry practice. For example, OSHA recently cited a petroleum refinery for allowing the wall of a flare line to fall below minimum acceptable thickness identified in its internal inspection manual. OSHA v. Wynnewood[4] In short, operators must not only develop compliant PSM programs, but they must be understood, implemented, and followed.

PSM Coverage

The threshold inquiry for PSM compliance is to determine what areas of a facility must be covered by a PSM program. Under the PSM Standard, a PSM program must cover any "process" that involves "highly hazardous chemicals." The Standard broadly defines "process" as any activity involving a highly hazardous chemical, such as the use, storage, manufacturing, or handling of a chemical. This broad definition includes all vessels interconnected within an activity, as well as vessels that are separated from an activity but could be involved in a hazardous release.[5]

The PSM Standard defines "highly hazardous chemicals" in two ways. First, it lists 137 chemicals in Appendix A to the Standard as "high hazardous chemicals" when they are present at a facility in excess of certain threshold quantities (TQ). OSHA has determined that these chemicals have a potential for catastrophic release—*i.e.*, a major uncontrolled emission, fire, or explosion that puts employees in serious danger—when present above their TQs. Second, the Standard defines highly hazardous chemicals as *flammable gases* (as defined by 29 C.F.R. 1910.1200) and *flammable liquids* (with a flashpoint below 100 °F) that are on site in one location in a quantity of 10,000 pounds or more. This definition, however, excludes hydrocarbon fuels used solely for workplace consumption, and flammable liquids stored in atmospheric tanks or transferred that are kept below their normal boiling point without chilling or refrigeration.[6]

The reasonable interpretation of these definitions is that all equipment involved in, or that could have an impact on, any activity related to highly hazardous chemicals must be covered by the PSM program unless it falls under a limited exception.[7] Operators should consider consulting PSM experts in connection with defining the boundaries of PSM coverage. An expert's demarcation of what activities constitute "process" and what chemicals rise to the level of "highly hazardous" will help a facility develop and maintain an effective, compliant PSM program. OSHA has recently issued citations related to the improper delineation of the boundaries of PSM coverage. See, e.g., OSHA v. Delek[8] OSHA cited Delek, a petroleum refinery, for its failure to inspect a ventilation system for a control room on the basis that it was "process equipment" involved in the "handling" of chemicals.

Where a majority of processes at a facility are PSM covered, operators should consider applying PSM requirements to the non-covered processes as well to avoid, among other things, having two different process safety programs in place at the same facility. Indeed, many PSM elements may already be applied facility-wide (e.g., management of change, pre-startup safety review, hot work permits, contractor safety, emergency planning and response, and incident investigation).

Process Safety Information

Process safety information (PSI) is critical data regarding the highly hazardous chemicals, technology, and equipment involved in the chemical process. The PSI element of the PSM Standard requires operators to have the following types of PSI in a complete, accurate, up-to-date and accessible format: (1) information pertaining to chemical hazards, which shall consist of, at a minimum, toxicity information, permissible exposure limits, physical data, reactivity data, corrosivity data, thermal and chemical stability data, and hazardous effects of inadvertent mixing of different materials that could foreseeably occur; [9](2) information pertaining to the technology, which shall consist of, at a minimum, block flow or process flow diagrams similar to exemplar diagrams contained in Appendix B to the Standard, process chemistry, maximum intended inventory, safe upper and lower limits for temperatures, pressures, flows or compositions, and an evaluation of the consequences of deviations, including those affecting the safety and health of employees;[10] and (3) information pertaining to equipment, which shall include materials of construction, piping and instrument diagrams (P&IDs), electrical classification, relief system design and design basis, ventilation system design, design codes and standards employed, material and energy balances for processes built after the Standard was enacted (May 26, 1992), and safety systems (e.g., interlocks, detection, or suppression systems).[11]

The amount of information needed to meet the PSI element can be considerable. There is no requirement that all PSI be compiled in a single document, or that it be located in a single file. Where it is contained in various documents and/or locations, good practice is to compile an index of the PSI and/or locations.

Compliant PSI—including P&IDs, electrical classifications, equipment files, and relief system documentation—is essential to the safe operation of a facility. P&IDs are critical because they depict the connectivity of equipment and the instrumentation used to control the equipment. They are also required to, among other things, conduct effective process hazards analysis (discussed below) and relief system evaluation. Electrical classifications are critical because improperly classified electrical equipment can become a source of ignition and can cause an explosion. Equipment files are critical because operators must be able to document that all equipment complies with current recognized and generally accepted good engineering practices (RAGAGEP), or is designed, maintained, inspected, tested, and operating in a safe manner. Relief system documentation is critical because relief systems are the last line of defense for protecting equipment from scenarios that can cause catastrophic equipment failure and explosions, fires, and toxicity impacts. OSHA has recently issued citations for incomplete and inaccurate PSI. See, e.g., OSHA v. Hunter Panels L.L.C.[12] OSHA cited Hunter, a manufacturer of insulation paneling, because its PSI did not contain specific corrosivity data ("process safety information did not include corrosivity data for resistance of Type 321 Stainless Steel … to n-Pentane").

Operators should consider consulting PSM experts to determine whether a site's PSI is compliant and, if not, how to close the compliance gap. Common errors include assuming that all chemical hazard information is located on material safety data sheets; however, many MSDSs do not contain complete information regarding reactivity, corrosivity, stability and the hazardous effects of inadvertent mixing. This gap can be addressed by creating hazardous chemical interaction matrices.

Process Hazard Analysis

Process hazard analysis (PHA) is an organized set of assessments designed to identify and analyze the significance of potential hazards associated with the processing or handling of highly hazardous chemicals. PHAs help facilities determine how to control such hazards. The PHA element of the PSM Standard requires operators to have conducted all initial PHAs by May 26, 1997, and to conduct follow-up PHAs at regular intervals not to exceed five years.[13] In order to conduct an effective PHA, all PSI should be complete, accurate, and up-to-date. Any deficiencies in PSI must be identified in the PHA and listed as a recommendation.

In addition, PHAs must be conducted by a team that includes individuals with expertise in engineering and process operations using one of the following methodologies: What-If, Checklist, What-If/Checklist, Hazard and Operability Study, Failure Mode and Effects Analysis, Fault Tree Analysis, or an appropriate equivalent method.[14] A popular new methodology being used is the Layer of Protection Analysis (LOPA). LOPA is used to identify safety integrity level requirements for safety instrumented systems as well as to provide better quantification of the frequencies of scenarios identified in PHAs.

PHAs must address the hazards of the process, any previous incidents with potentially catastrophic consequences, engineering and administrative controls, and the consequences of failure of the controls. PHAs must also address facility siting, human factors and failure of controls.[15] Although many companies use checklists to address facility siting issues during PHAs, good practice is to also conduct separate site-wide facility studies focused on occupied buildings according to industry standards such as API 752.

Upon their completion, PHAs may contain recommendations, which must be resolved (i.e., accepted, rejected, or withdrawn) in "a timely manner." Corrective action pursuant to recommendations must be completed "as soon as possible."[16] Although the PSM Standard does not define these terms, government publications indicate, for example, that OSHA expects employers to complete corrective action within a one or two year timeframe absent unusual circumstances.[17] In order to obtain consistent results from PHAs across a facility or an entire company, it is important to establish a preferred PHA methodology and other supporting tools such as risk ranking matrices, and recommendation tracking software.

Operators should consider consulting PSM experts to help determine whether PHAs are being conducted in compliance with the Standard. OSHA has recently issued citations for failure to properly conduct PHAs. See, e.g., OSHA v. Westlake Vinyls Co.[18] OSHA cited Westlake, a manufacturer of chemicals and plastic products, for failing to address hazards of the process, administrative controls, facility siting and human factors in its PHAs. PSM experts can also elucidate unfixed compliance terms like "a timely manner" and "as soon as possible" to assist a facility's compliance effort. OSHA has recently issued citations for failure to address recommendations within an appropriate time-scale. See, e.g., OSHA v. Heritage-WTI, Inc.[19] OSHA cited Heritage-WTI, an environmental services company that incinerates hazardous wastes and other materials, for failing to complete certain PHA recommendations for over 5 years. OSHA has also recently cited companies for failure to complete corrective actions in shorter time periods. See, e.g., OSHA v. Cargill Meat Solutions Co. (recommendations from a 2 year old PHA revalidation were not documented as closed)[20] and OSHA v. Hunter Panels L.L.C. (operating company cited in August 2012 for failure to "document the resolution of action items from the 2012 Process Hazard Analysis" related to providing a spill kit for small releases at a tank area).

Mechanical Integrity

Mechanical Integrity (MI) is the system of assuring that process equipment is in satisfactory condition to safely and reliably perform its intended design function and operate properly within the limits of established PSI. Process equipment is equipment that can control, prevent, or mitigate catastrophic chemical releases. It includes pressure vessels, storage tanks, piping systems (including piping components such as valves), relief and vent systems and devices, emergency shutdown systems, controls (including monitoring devices, sensors, alarms, and interlocks), pumps, and utility systems.[21]

The MI element of the PSM Standard requires operators to establish and implement an MI program to maintain the ongoing integrity of process equipment.[22] A critical first step in developing an MI program is to identify the process equipment and instrumentation that needs to be included. A good practice is to look at the PHAs to determine if failure of equipment could result in the release of hazardous chemicals. Also, any safeguards listed for PHA scenarios should be included.

Additionally, MI requires written procedures by which an operator shall: (1) conduct MI training; (2) perform inspections and testing of process equipment following RAGAGEP; and (3) correct deficiencies in process equipment that are outside of acceptable limits (defined by the PSI, or RAGAGEP) before any further use, or in a timely manner while necessary means are taken to assure the safe operation of the equipment during an interim period.[23]

The term RAGAGEP includes "national consensus" codes and standards like those published by National Fire Protection Association, American Society for Testing and Materials, American National Standards Institute, and American Petroleum Institute.[24] Examples of deficiencies outside of "acceptable limits" include equipment that was not designed, fabricated, constructed or installed per RAGAGEP; equipment with mechanical defects that interfere with its capability to function as intended; a degradation in the equipment exceeding acceptable limits (including retirement thickness); operating equipment outside its normal limits; equipment leaks; and by-passed equipment.[25] For example, equipment that is operated below retirement thickness is outside acceptable limits until it is repaired or replaced to meet PSI requirements or it is determined fit-for-service "as is" under API 579, which establishes a new acceptable limit.

Individuals who perform inspections and testing may need to have certifications, such as for inspecting pressure vessels and welding. Also, the testing and inspections can be costly, particularly if equipment must be taken out of service to perform the test or inspection. Implementing risk-based inspection programs such as outlined in API 580/581 may allow resources to be more effectively utilized.

Operators should consider consulting PSM experts concerning MI to help determine whether the scope of an existing MI program is adequate, whether inspections are being properly conducted, and whether deficiencies are being addressed in compliance with the Standard. OSHA has recently issued citations for an MI program's failure to correct process equipment deficiencies in a safe and timely manner. See, e.g., OSHA v. Arkema Inc.[26] OSHA cited Arkema, a chemical company, for operating with compromised rupture discs in relief systems.

PSM Violations

Deficient PSI, PHAs and/or MI violate the PSM Standard, which could result in an OSHA citation. There are three categories of PSM violations: (1) serious; (2) non-serious; and (3) willful. A violation is "serious" if death or serious physical harm (i.e., a substantial impairment to bodily function) could result from the violation. Conversely, a violation is "non-serious" if death or serious physical harm could not result from the violation. OSHA assesses civil penalties of up to $7,000 for every serious and non-serious violation. A "willful violation" is one that is committed with either intentional disregard or plain indifference to the requirements of the PSM Standard. OSHA assesses civil penalties between $5,000 and $70,000 for every willful violation. Willful violations that cause death to an employee are subject to criminal sanctions, including imprisonment of up to 1 year.[27]

The severity of each OSHA penalty is determined by the gravity of the violation. Two factors largely determine the gravity of a violation: (1) the severity of the injury that could occur from the violation (i.e., high, medium, or low); and (2) the probability that the injury could result from the violation (i.e., greater probability and lesser probability). Other factors OSHA may consider include the size of the operator, the operator's good faith, and the prior history of violations at the site. These factors, however, are not defenses to the underlying violations.[28]

When confronted with a potential PSM violation (e.g., inadequate PSI, PHAs and/or MI), OSHA may issue a citation under the PSM Standard or, in the alternative, under the OSHA General Duty Clause in the event that the PSM Standard does not apply. The General Duty Clause is a "catchall provision" that imposes an independent duty on operators to provide a safe work environment. Specifically, it requires an operator to provide a place of employment that is free from "recognized hazards" that are currently "causing, or are likely to cause [should they occur] death or serious physical harm to . . . employees." A hazard is "recognized" where: (a) the employer has identified it; (b) it is known in the industry; or (c) it is blatantly obvious. Penalties under the General Duty Clause are the same as those issued under the PSM Standard.[29]

Conclusion

Personnel who collaborate with PSM experts will understand the complexities of today's PSM standard. This understanding will make their facilities better equipped to maintain PSM compliance, help avoid costly OSHA citations, better control cost of corrective action, and foster a safe and reliable work environment for all employees.

References

[1] 29 CFR 1910.119 (OSHA PSM Standard); 57 FR 6356-01 (Preamble and Final PSM Rule); Public Law 101-549 §304 (Clean Air Act Amendments 1990).
[2] 29 CFR 1910.119(c-p).
[3] See CPL 03-00-010 (Petroleum Refinery NEP 2009); CPL 03-00-014 (Chemical NEP 2011).
[4] OSHRC Docket No. 07-0609 (8/4/08); 340 Fed. Appx. 462 (10th Cir. 2009).
[5] 29 CFR 1910.119(a)(1)(i)-(ii) and 29 CFR 1910.119(b).
[6] 29 CFR 1910.119(a)(1)(i)(Appendix A) and 29 CFR 1910.119(a)(1)(ii).
[7] The PSM Standard lists a few, limited exceptions where the PSM program will not apply: at retail facilities, oil or gas well drilling or servicing operations, or normally unoccupied remote facilities that are geographically remote and have no permanent employees (29 CFR 1910.119(a)(2)(i)-(iii)).
[8] OSHRC Docket No. 08-1386 (4/27/11).
[9] 29 CFR 1910.119(c)(3), (d), (l), Appendix C, par. 3(complete, accurate, up-to-date and accessible requirements). Material Safety Data Sheets that meet the requirements of 29 C.F.R. 1910.1200(g) can be used to comply so long as they contain the requisite information.
[10] Where original technical information no longer exists, operators may develop such information in conjunction with process hazard analysis.
[11] 29 CFR 1910.119(d)(1) - (3).
[12] Inspection #544919 (citation issued: 1/25/13).
[13] 29 CFR 1910.119(e).
[14] Id.
[15] Id.
[16] 29 CFR 1910.119(e)(5).
[17] 57 FR 6356-01, p. 6379.
[18] Inspection #329260 (citation issued: 9/18/12).
[19] Inspection #189909 (citation issued: 8/6/12).
[20] Inspection #315 121 848 (citation issued: 12/6/11).
[21] 29 CFR 1910.119(j).
[22] 20 CFR 1910.119(j)(2).
[23] 29 CFR 1910.119(j)(3) – (5).
[24] 57 FR 6356-01, p. 6390.
[25] CPL 03-00-010, pp. 6-7.
[26] Inspection No. 316 217 835 (citation issued: 7/10/12).
[27] 29 USC §666 (Penalties); Trinity Indus., Inc., 504 F.3d 397 (3d Cir. 2007) (serious); OSHA Field Operations Manual Ch. 4-II-C-3(non-serious); J.A. Jones Constr. Co., 15 OSH Cas. (BNA) par. 2201(willful). In 2009, a bill known as the "Protecting America's Workers Act," was proposed in both the U.S. Senate (S.1580) and House of Representatives (H.R.2067) to increase OSHA civil penalties to $250,000 (maximum) per violation, and criminal penalties to up to 20 years for repeat deaths.
[28] 29 USC §666(j) (criteria); OSHA Field Operations Manual Ch. 6-III-A (gravity).
[29] 29 USC §654 (OSHA General Duty Clause); Occupational Health & Safety Law; Compliance and Practice §§7.24-7.25.

New CCPS Guideline Book: Guidelines for Enabling Conditions and Conditional Modifiers in LOPA

J. Wayne Chastain
Plant Protection Technical Services, Eastman Chemical Company
200 South Wilcox Drive
Kingsport, TN 37660
chastain@eastman.com

Stanley A. Urbanik
DuPont Corporation
Wilmington, DE
Stanley.A.Urbanik@dupont.com

Robert W. Johnson
Unwin Company
Columbus, OH
rjohnson@unwin-co.com

John F. Murphy
Process Safety Services
Punta Gorda, FL
hamjfm@embarqmail.com

Prepared for Presentation at
American Institute of Chemical Engineers
2013 Spring Meeting
9th Global Congress on Process Safety
San Antonio, Texas
April 28 – May 1, 2013

UNPUBLISHED

New CCPS Guideline Book: Guidelines for Enabling Conditions and Conditional Modifiers in LOPA

J. Wayne Chastain
Plant Protection Technical Services, Eastman Chemical Company
200 South Wilcox Drive
Kingsport, TN 37660
chastain@eastman.com

Stanley A. Urbanik
DuPont Corporation

Robert W. Johnson
Unwin Company

John F. Murphy
Process Safety Services

Keywords: Layers of Protection Analysis, LOPA, conditional modifier, enabling condition

Abstract

This paper outlines the content of a new CCPS book entitled *Guidelines for Enabling Conditions and Conditional Modifiers in LOPA*. An enabling condition is a condition that is not a failure, error or a protection layer but makes it possible for an incident sequence to proceed to a consequence of concern. It consists of an operating phase or condition that does not directly cause the scenario, but that must be present or active in order for the scenario to proceed to a loss event; expressed as a dimensionless probability. A conditional modifier is one of several probabilities included in scenario risk calculations, generally when risk criteria endpoints are expressed in impact terms (e.g., fatalities) instead of in primary loss event (e.g., release, vessel rupture). Conditional modifiers include, but are not limited to: probability of a hazardous atmosphere, probability of ignition, probability of explosion, probability of personnel presence, probability of injury or fatality, and probability of equipment damage or other financial impact. These *Guidelines for Enabling Conditions and Conditional Modifiers in LOPA* contain information useful to the experienced analyst and accomplished practitioner. In this book, various kinds of enabling conditions and conditional modifiers are defined and illustrated as they might be used in Layer of Protection Analyses (LOPAs). There is a discussion on how enabling conditions and conditional modifiers can be used in other risk analysis applications. The appendices discuss "double jeopardy" and simultaneous failures, peak risk, and example rules for enabling conditions and conditional modifiers.

1. Introduction

Enabling conditions and conditional modifiers are two types of factors that have long been used in quantitative risk analysis. Both of these factors were introduced for use in LOPA in *Layer of Protection Analysis: Simplified Process Risk Assessment [1]*. Enabling conditions are conditions that generally influence the initiating events for scenarios. They are conditions that do not directly cause the scenario, but which must be active in order for the scenario to proceed to a loss event. Enabling conditions are generally included in a LOPA independent of the type of endpoint used in the analysis. Conditional modifiers are probabilities included in LOPA when the risk criteria endpoints are expressed in terms of impact as opposed to primary loss event terms. Neither enabling conditions nor conditional modifiers are independent protection layers and do not typically meet all of the criteria used to qualify an independent protection layer (IPL). The development of enabling conditions and conditional modifiers in *Layer of Protection Analysis: Simplified Process Risk Assessment* was sparse given that the major foci of the book were the overall methodology, initiating events, and independent protection layers [1].

Guidelines for Enabling Conditions and Conditional Modifiers in LOPA provides a more in-depth treatment of these two important factors used in LOPA. Although these factors are used in other types of risk analysis, the focus of this text is on their use in LOPA. The text provides suggested values or means of obtaining appropriate values for these factors, and also provides guidance on when the values should and should not be used. Misuse of these factors can result in an analysis that can significantly under-predict or over-predict the actual risk of a scenario.

2. LOPA Overview

Layer of Protection Analysis is a simplified form of quantitative risk analysis that uses order-of-magnitude categories for initiating event frequency, consequence severity and probability of failure of independent protection layers (IPLs) to analyze and assess the risk of one or more incident scenarios. LOPA is a risk analysis method that was developed to answer questions such as “what layers of protection are needed to meet our risk goals?” and “how much risk reduction does each layer provide or need to provide?” LOPA can help to answer these questions with less time and effort than more complex methods such as quantitative risk analysis.

LOPAs are typically conducted after a qualitative hazard evaluation has been used to identify the hazard scenarios of the process. Often only a subset of the identified hazard scenarios from a qualitative study is evaluated using LOPA. Likewise, LOPA can be used as an endpoint for risk analysis or can be used as a screening tool to determine scenarios that should be evaluated using more detailed quantitative risk analysis techniques.

LOPA is applied to a selected incident scenario. For the scenario being evaluated the initiating event and the frequency associated with that initiating event must be determined. The safeguards that are present to prevent the scenario are then evaluated to determine if they meet the requirements to be used as an independent protection layer. Each safeguard should meet seven key attributes in order to qualify as an IPL. These attributes are:

- Independence
- Functionality
- Integrity
- Reliability
- Validation, maintenance, and auditing
- Access security
- Management of change

For each IPL, the probability of failure on demand must be determined.

By combining the initiating event frequency with the probability of failure on demand for each IPL, the scenario frequency can be determined. This predicted frequency can be compared to an established risk criterion.

Enabling conditions and conditional modifiers are additional factors that are included by some organizations in their LOPAs. Enabling conditions can be required for the evaluation of some scenarios in order to obtain an accurate evaluation of the risk of a scenario. Their exclusion from a LOPA can result in overestimation of the risk of an event.

Conditional modifiers are used when the risk criteria associated with the LOPA process have been selected based on their use in the analysis. When risk criteria used in the analysis are based on releases from primary containment, conditional modifiers are typically not applicable. However, if the risk criteria selected are based on the potential ultimate consequences associated with an event, the use of conditional modifiers may be warranted.

Enabling conditions and conditional modifiers have associated probabilities, similar to IPLs.

3. Enabling Conditions

Enabling conditions make the initiation of a scenario possible. Enabling conditions are not initiating events, but for a given scenario, if the enabling condition is not true, the scenario will not progress from the initiating event to the hazardous outcome under consideration. Mitigating factors such as probability of personnel presence and probability of ignition, which come into play after the primary loss event, are not enabling conditions but conditional modifiers. Enabling conditions generally fall into one of several categories.

3.1 Time At Risk

Time-at-risk enabling conditions come into play when an incident sequence may only be realized a certain fraction of time when the conditions are right for the event sequence to proceed to a loss event. One of the most important considerations about time-at-risk enabling conditions is that they only apply to revealed failures. A revealed failure is one that becomes apparent and will be corrected after its occurrence. In contrast, an unrevealed failure can occur at any time, but only becomes apparent when a specific event occurs that reveals it or a test reveals that the failure has occurred.

3.1.1 Seasonal Effects

One group of common time-at-risk enabling conditions are various seasonal effects. For example, low ambient temperatures that can cause freezing of process or utility lines may only be present for a certain amount of time during the year. Likewise, sufficiently high ambient temperatures to effect cooling capacity or cause excessive vapor pressure of some process components may only be present during a portion of the year. Seasonal enabling conditions can typically be characterized by climatological data available for the site. When the seasonal effect itself is the initiating event, then it is not also an enabling condition.

3.1.2 Process State Risks

Another type of time-at-risk enabling condition is process state risks that express the probability that the process is in a certain state. These enabling conditions apply when an initiating failure only has the undesired impact when the process is in the particular state. Process state time-at-risk can apply to batch processes where the risk being evaluated is only present during certain periods of the batch cycle, or to scenarios that are only applicable during the startup, shutdown, or clean-out of the process.

3.1.3 Situations Where Time-at-Risk Should Not be Used as an Enabling Conditions

Time-at-risk enabling conditions are not appropriate in all situations. Some examples where time-at-risk enabling conditions are not appropriate include:

- Continuous operations where hazards are constant over time
- Infrequent short-duration operating modes with high potential severity
- Procedure-based operations involving human error in the execution of a stepwise procedure
- Continuous or procedure-based operations involving any combination of factors that are failure or fault conditions, regardless of whether these conditions are revealed or latent
- Incident sequences where time-at-risk applies to the severity of consequences of the loss event rather than its likelihood of occurrence

3.1.4 Campaign Enabling Conditions

Campaign enabling conditions reflect different conditions that are present due to changes in processes from batch to batch or from time to time. In a production facility where some batches pose certain hazards while other batches do not because of the presence of certain raw materials, final products, catalysts, or operating conditions, campaign enabling conditions may be appropriate in a LOPA. Campaign enabling conditions may also apply to operations with certain modes that are present for limited periods of time such as recycle or a final control element being down in an effluent handling system or a process that only operates a limited portion of the year due to low demand.

Campaign enabling conditions may potentially provide high levels of risk reduction. When evaluating the use of campaign enabling conditions several factors must be considered.

Campaign enabling conditions may change over time as the production needs of a facility change. Campaign enabling conditions are not appropriate to be used in situations where there is the potential for an unrevealed failure that only becomes apparent with the initiation of the loss event sequence when the condition is true.

An example of an unrevealed failure that would prevent the use of a campaign enabling condition is a level control loop on a weigh tank that is only used during a particular campaign. If the level control loop failure could occur at any point in time but would only be revealed during the hazardous campaign, the campaign enabling condition could not be applied. If the level transmitter is checked prior to the campaign being started and the failure can only occur during the campaign, then the campaign enabling condition could be properly applied.

Because of the high levels of risk reduction that may be claimed with a campaign enabling condition, caution must be used. In certain cases, the application of an enabling condition can result in a risk that meets the criteria being evaluated, but for brief periods of time exposes personnel to very high risks. This concept is called "peak risk" and is evaluated in the text as a special case.

4. Conditional Modifiers

Conditional modifiers are probabilities included in a LOPA or other risk analysis when risk criteria are expressed in impact terms (e.g. fatalities) as opposed to loss of primary containment. These modifiers include:

- Probability of a hazardous atmosphere
- Probability of ignition or initiation
- Probability of explosion
- Probability of personnel presence
- Probability of injury or fatality
- Probability of equipment damage or other financial impact.

Although conditional modifiers have the same effect on the predicted frequency for an event in a LOPA as independent protection layers, they differ significantly since IPLs are engineered or administrative controls that are capable of responding to a process deviation to either avoid or mitigate a loss event. Conditional modifiers, by contrast, are not response actions but relate to the condition that the facility is likely to be in at a particular point or period of time in the incident sequence. The conditional modifier indicates that probability that the condition is true at any time. Even though conditional modifiers are not engineered and administrative controls, the probability that a conditional modifier is true can be impacted by the way that the facility is managed. As an example, access control procedures and training can have an impact on the probability of personnel presence.

Conditional modifiers, like IPLs, need to meet the basic requirement of independence; i.e. independent of other considerations and factors in the scenario risk calculations. Particular care needs to be taken to ensure that conditional modifiers are independent of the severity of the event.

The improper use of conditional modifiers in a LOPA can result in an underestimation of the risk of an event. Because of this, their use needs to be intentional, disciplined, and consistent. To ensure appropriate application, a good approach is to develop a standardized baseline set of conditional modifiers and the rules for their use. Common pitfalls that exist in the use of conditional modifiers are:

- Using more conditional modifiers than an incident scenario warrants
- Being overly optimistic in estimating conditional modifier probabilities
- Including more additional modifiers until a desired risk level is reached
- Using conditional modifiers to justify avoiding a standard practice
- Using conditional modifiers in situations where the risk criteria are based solely on loss of primary containment or implicitly assume conditional modifiers

When selecting a conditional modifier for use in a LOPA, the value ought to be expected to remain relatively constant over time. A change to the process that results in a change to probability of a conditional modifier can invalidate the risk decision made using LOPA unless such a change is properly accounted. A company's management of change process may need to be expanded to capture the potential impacts to conditional modifiers that a process change may entail.

4.1 Probability of a Hazardous Atmosphere

Probability of a hazardous atmosphere is a conditional modifier that may be applied to loss of primary containment events with the potential to cause harm, but may not depending on the specific situation. A loss-of-containment event of an ignitable material may form a flammable atmosphere or not depending on the ambient temperature. When evaluating the use of a probability of hazardous atmosphere, care needs to be taken to ensure that the conditional modifier is independent of other aspects of the LOPA that may reflect a similar effect such as time-at-risk or procedural controls that may be counted as IPLs.

4.2 Probability of Ignition or Initiation

When a flammable atmosphere is present, there is a potential that it will be ignited. The probability that ignition will occur is a affected by a broad range of factors that have made detailed treatment of this topic beyond the scope of *Guidelines for Enabling Conditions and Conditional Modifiers in LOPA*; however, several pages are devoted to providing initial guidance for the LOPA practitioner in dealing with this important conditional modifier. A closely related event is the initiation of an uncontrolled reaction in a reactive material mixture.

There are three basic scenarios in which probability of ignition needs to be determined if it is going to be included in a LOPA.

- Ignition of a flammable atmosphere inside process equipment
- Ignition of a flammable vapor, gas, or dust cloud external to process equipment
- Initiation of uncontrolled reaction in a reactive material or mixture.

Each of these is given some level of attention in the text.

Probability of ignition inside process equipment is applicable whenever a flammable or combustible atmosphere exists normally or can be formed as the result of a process upset condition in process equipment. Each situation needs to be evaluated carefully, taking into account a broad range of process parameters, details about the equipment and its operating and upset conditions, the properties of the materials being processed, historical data, and other applicable data in reaching the determination of the value to be used for conditional modifier.

Many of the same considerations that are pertinent to the probability of ignition inside process equipment also apply when determining an appropriate value for the probability of ignition of a flammable or combustible atmosphere outside process equipment. Other ignition mechanisms that are not applicable to flammable and combustible atmospheres inside process equipment must also need to be considered such as traffic, hot work, personnel with cell phones or other non-classified equipment, personnel smoking, etc. The size of the flammable cloud outside equipment and the duration of release are also significant factors in determining the probability of ignition. The potential for immediate vs. delayed ignition is a factor to be considered, since delayed ignition can potentially pose a more hazardous situation than immediate ignition.

The probability of initiation of an uncontrolled reaction is a function of the different mechanisms that are available for imparting the necessary energy to begin the reaction. These mechanisms may include shock or impact, friction, electrostatic discharge, and heat. The effect and probability of contamination of the process material needs to be accounted for, since contamination can result in significantly lower levels of energy required to cause the reaction to begin.

4.3 Probability of Explosion

Several different conditional modifiers can be applied to situations in which an explosion could potentially occur. Some of the conditional modifiers related to probability of explosions that are evaluated in the text are the probability of dust explosion, probability of a vessel rupture, probability of a vapor cloud explosion, and probability of a deflagration-to-detonation transition.

4.4 Probability of Personnel Presence

Probability of personnel presence is one of the most commonly used conditional modifiers within LOPA. This conditional modifier reflects the probability that one or more people are in the affected area when a loss event occurs. Several key considerations for using (or not using) probability of personnel presence as a conditional modifier include:

- When using the conditional modifier in conjunction with the probability of injury, the LOPA practitioner needs to ensure that the factors are not "double-counted"
- Using probability of personnel presence may require a more detailed evaluation of the incident sequence which may require source term calculations and dispersion modeling

- Probability of personnel presence needs to take into account all potentially impacted personnel including not only operations personnel, but also maintenance workers, and any other personnel that may be in the area
- The probability of personnel presence needs to be independent of the scenario being evaluated which will often not be the case when operator response to an alarm will take personnel into the field
- Probability of personnel presence should not be credited when the initiating event is an operator error that immediately results in the loss of containment
- When personnel presence is embedded in the determination of the severity of consequences, a conditional modifier should not be used
- Probability of personnel presence should not be credited when the corporate risk criteria are not based on personnel presence

An alternate approach to using probability of personnel presence as a conditional modifier is to use access control as independent protection layer. Administrative access control should only be used as an IPL when it will effectively limit access and the control meets all of the requirements of an IPL. When administrative controls are used as an IPL to limit the access of personnel to an area, it would be incorrect to also use probability of personnel presence as a conditional modifier.

The effect area determination is an important part of the probability of personnel presence. Effect areas can be determined from dispersion modeling, the determination of toxic release effect areas, fire and explosion effect areas. Once the effect area has been determined, the sum of the time that personnel are going to be in the effect area can be determined. In the event that more than one person can be present in the effect area, corporate risk criteria may require that a different severity category be used in the analysis.

4.5 Probability of Injury or Fatality

The probability of injury or fatality is a conditional modifier that reflects the probability that personnel that are present in the effect area will suffer the consequence being evaluated. Four approaches are presented in the text for addressing the probability of injury or fatality.

- Approach #1: Use P=1 for the probability of injury or fatality conditional modifier, because of how the effect boundary is defined.
- Approach #2: Apply the same probability of injury or fatality conditional modifier (P<1) to everyone in the effect area.
- Approach #3: Estimate the probability of serious injury or fatality conditional modifier for individual(s) inside the effect area.
- Approach #4: Use of a combined approach of the first three in the evaluation.

There are several notes and cautions when using the probability of injury or fatality.

- More than one consequence may need to be evaluated.
- The conditional modifier needs to apply to the entire effect area, not just at the outside boundary of the effect area.

- Controls that limit personnel from entering a dangerous area that has been created can be addressed as independent protection layers.
- The characteristics of the particular chemical involved in the release may impact the probability of injury or fatality.

Probability of equipment damage or other financial impact as a possible conditional modifier is also discussed in the text.

5. Application to Other Methods

Guidelines for Enabling Conditions and Conditional Modifiers in LOPA briefly addresses the use of these factors in other types of risk analyses. The applicability and use of these factors in the various types of quantitative risk analysis is discussed included their use in fault tree analysis, event tree analysis, and quantitative impact analysis. The use of the factors is also discussed as they might apply to scenario identification methods such as Hazard and Operability Studies and What-If Analysis. Finally the text includes a brief discussion of the applicability to barrier analysis and diagrams.

6. Appendices

The first two appendices of the text address two common issues that arise in LOPA when using enabling conditions or conditional modifiers; namely, simultaneous failures and the concept of "double jeopardy" and peak risk concepts. The last appendix in the text provides an example rule set for LOPA enabling conditions.

6.1 Simultaneous Failure and "Double Jeopardy"

The first appendix describes what is meant by "double jeopardy" in the scope of risk analysis. Specifically "double jeopardy" is the concurrent incidence of two independent initiating events or other revealed failures. Per this definition, the text points out that the true incidence of "double jeopardy" events is very small. It is important to note that the text draws a firm distinction between revealed and unrevealed failures. If any of the failures in the "double jeopardy" case being evaluated are unrevealed failures, then the scenario is not a case of double jeopardy.

6.2 Peak Risk

The second appendix deals with the concept of peak risk. Peak risk was developed to help address situations in which a relatively high risk is present in a facility for a very short period of time. When this type of risk is annualized, as would be done when an enabling condition is added to the scenario LOPA indicating the amount of time the process is in the state where the event can occur, the risk can be shown to be low enough. Peak risk is a means of setting a limit on the risk that personnel may be exposed to without consideration being given to the duration of the event.

6.3 Example Rule Set

The final appendix in the text provides an example rule set for LOPA enabling conditions. This example is not meant to provide recommended guidance, but simply a possible path forward for what guidance on the topic might look like.

7. Conclusions

Guidelines for Enabling Conditions and Conditional Modifiers in LOPA is a text that will move the understanding and consistent application of these important LOPA factors forward in the industry. The warnings and caveats given in the text will help to prevent the misuse of these factors that can result in under-prediction of the risk being evaluated in LOPA. The text gives an overview of the major types of enabling conditions and conditional modifiers being used by LOPA practitioners.

8. References

[1] *Layer of Protection Analysis: Simplified Process Risk Assessment*; American Institute of Chemical Engineers, Center for Chemical Process Safety: New York, New York, 2001.

Implementing LOPA Recommendations into Design of Instrumented Protective Systems

Rajeev Limaye PE, CFSE
Director - Control Systems & Instrumentation
Praxair, Inc.
1585 Sawdust Road
The Woodlands Texas 77380
rajeev_limaye@praxair.com

Prepared for Presentation at
American Institute of Chemical Engineers
2013 Spring Meeting
9th Global Congress on Process Safety
San Antonio, Texas
April 28 – May 1, 2013

UNPUBLISHED

Implementing LOPA Recommendations into Design of Instrumented Protective Systems

Rajeev Limaye PE, CFSE
Director - Control Systems & Instrumentation
Praxair, Inc.
1585 Sawdust Road
The Woodlands, Texas 77380
rajeev_limaye@praxair.com

Keywords: ISA84, SIS, Instrumented Protective System, IPS, SIL, IEC61511, IEC61508, OSHA, PSM, LOPA, SIF, ISS, ISF, Process Safety Time, Risk Reduction, PHA, HAZOP

Abstract

When Independent Protection Layers (IPLs) are identified during Layer of Protection Analysis (LOPA), there may not be enough time to verify the validity of each IPL. In functional safety lifecycle, the next steps after LOPA are preparation of Safety Requirement Specification (SRS) and conceptual design of the Instrumented Protective System (IPS) which are typically performed by control system engineer(s). If certain protection layers are found to be inadequate, an iterative approach to revisit the LOPA is required to ensure the required risk reduction is achieved by the IPLs.
In most cases, multiple Instrumented Protective Functions (IPFs) and control functions require the same process measurement. For example, an alarm, a trip and a control loop may require the same process value. Adequate instrumentation must be provided to meet the independency criteria of IPL. Various scenarios are discussed on how and when to share the process signals between an IPS and the Basic Process Control System (BPCS). Good engineering practices to achieve safety as well as reliability of the system by means of different fault tolerant configurations are discussed. Typical Piping & Instrumentation Diagram (P&ID) representation of some of the common scenarios is also presented.
If operator response to an alarm is one of the IPLs, then some additional requirements need to be taken into consideration such as operator response time, human factors, etc.

1. Introduction

Inherent process risk is reduced to a tolerable level by implementing protective functions. Each organization has to define the tolerable risk level for safety, environmental and commercial hazards. Each protective function reduces the risk by a certain order of magnitude. LOPA is one of the most widely used semi-quantitative methods of analyzing and documenting protective functions. An important outcome of LOPA is identification of IPLs essential for required risk reduction. The required Safety Integrity

Level (SIL) of each IPL is also determined during LOPA. SIL defines the target performance level of an IPL in terms of a range of average probability of failure on demand (PFD_{avg}).
Safety Instrumented System (SIS) design, implementation, maintenance and operation is covered by ISA84 standard as functional safety lifecycle. ISA84 is endorsed by the Occupational Safety and Health Administration (OSHA) as a Recognized And Generally Accepted Good Engineering Practice (RAGAGEP). If an employer documents that it will comply with ISA84 and meets all ISA84 requirements, the employer will be considered in compliance with OSHA Process Safety Management (PSM) requirements for the SIS.

2. IPS classification

Over the last decade, several terms and definitions were introduced through ISA84, IEC61508 and IEC61511 standards as well as CCPS books and other publications. It is important to get clear understanding of these terms and acronyms to clearly define the scope of this discussion.

A safeguard is any device, system or action that would likely prevent an undesirable process incident triggered by an initiating event [1]. The safeguards could be non-instrumented (e.g. pressure safety valve) or instrumented (e.g. trip). In this paper, the focus is only on the instrumented safeguards.

A safeguard can be classified using following three attributes

Purpose	• Personnel Safety • Environmental Protection (releases) • Asset Protection (commercial)
Performance (Integrity Level)	The amount of risk reduction offered by the safeguard is measured in terms of Risk Reduction Factor (RRF). The RRF is also represented in terms of average Probability of Failure on Demand (PFD_{avg}). $RRF = 1/PFD_{avg}$
System	System in which the safeguard is implemented (BPCS, PLC, SIS, BMS, HIPPS etc.)

Fig. 1 shows the visual representation of the classification of safeguards and IPSs.

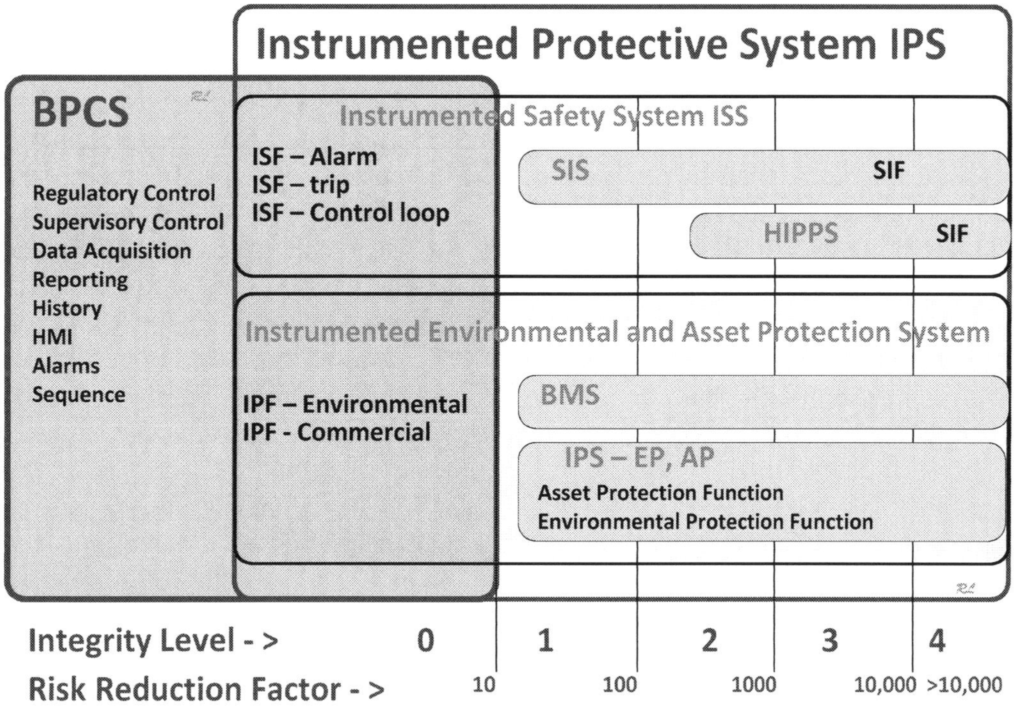

Fig. 1. Classification of safeguards and instrumented protective systems

Instrumented Protective System (IPS) refers to all systems that implement instrumented protective functions (IPFs) for the safety, environmental protection and asset protection. BPCS could be used to implement some of the IPFs.

Instrumented Safety Function (ISF) is a type of IPF for safety.

Instrumented Safety System (ISS) is a type of IPS intended for implementing ISF [2]. Some of the ISFs could be implemented in BPCS.

Safety Instrumented System (SIS) is an ISS that requires compliance with ISA84/IEC61511 standard for all safety lifecycle steps [3].

Basic Process Control System (BPCS) can be used to implement the IPF including ISFs. As per the ISA84, risk reduction credit cannot be more than 10 for the ISF implemented in BPCS.

Safety Instrumented Function (SIF) is a type of ISF that is implemented in SIS and must follow safety lifecycle as per ISA84/IEC61511 standard

High Integrity Pressure Protection System (HIPPS) is a type of SIS that offers higher risk reduction factor and must comply with ISA84/IEC61511 standard. Standards like API 521 and ASME Section VIII, Division 1 & 2 allows the use of HIPPS in lieu of a pressure relief device, as long as the HIPPS meets or exceeds the protection provided by the pressure relief device [4].

Burner Management System (BMS) implements furnace burner startup / shutdown logic as per the standards like NFPA 85.

Independent Protection Layer (IPL) is a type of IPF that is credited in LOPA for reducing the risk of undesired event and meets the criteria of being independent, specific, auditable and dependable defined in ISA84/IEC61511 Part 3 [5].

3. IPS design

While the BPCS is independent from the rest of the IPSs, often it is not practical to physically separate the BMS and other IPSs from the SIS. There is no requirement that these protective functions need to be implemented in separate systems. It may not be economical to do so. Whenever possible, SIFs should be separated from non-safety related IPFs. Most likely the IPFs with RRF of up to 10 are implemented in BPCS. The IPFs requiring higher level of risk reduction are implemented in SIS. ISA84 clause 11.2.2 states that "Where the SIS is to implement both safety and non-safety instrumented functions then all the hardware and software that can negatively affect any SIF under normal and fault conditions shall be treated as part of SIS and comply with the requirements of the highest SIL" [3].

In most discussions and literature, the focus of LOPA is identifying the layers of protection for personnel safety. However, safety cannot be addressed in isolation. It is also important to design the systems for asset protection and to prevent the environmental releases. Similar to the risk tolerance for safety incidents, each organization should develop the risk tolerance for commercial and environmental risks. The same LOPA approach should be used for analyzing and identifying the protection layers for safety, environmental and commercial risks. The scenarios should be properly classified during LOPA to indicate the type of risk it is addressing. The IPLs should be classified for protection against safety, environmental or commercial risk. An IPL may offer protection against combination of safety, environmental and commercial risks. Depending on the tolerable risk of consequence in each category, the integrity level requirement of the IPL may be different for safety, environmental and commercial risks. The highest integrity level of all three categories should be selected for the IPL.

Usually, the control systems engineer is responsible for allocating the IPFs to various IPSs and determining the number and type of systems needed. A comprehensive list of safeguards that are credited for risk reduction in LOPA with the classification using three attributes explained above is very handy. It is usually called an "IPL List". Each IPL should be given a unique ID which can be referred to on the P&IDs and the rest of the safety lifecycle documentation and planning including the SRS.

If most IPLs are SIL1 with an exception of just a few that require higher SIL rating, it is worth evaluating the cost of implementing just those IPLs in a separate small High Integrity Protection System, while selecting relatively less expensive SIS hardware for the rest of the IPLs that constitute the majority of the inputs and outputs.

4. LOPA issued for design

One of the main outcomes of the LOPA is identification of IPLs to mitigate the frequency of an undesired process event below the tolerable level. There have been discussions and theories on how to effectively conduct LOPA. Some recommend that it be done by a smaller group of people after the HAZOP for the scenarios with risk ranking above certain predetermined level. With this approach, a focused attention can be placed on the

scenarios identified for LOPA. This approach adds another activity on the project schedule and is perceived to take a longer time. There is another concept of performing LOPA during the HAZOP while the captive audience is present in the room. Usually HAZOP is performed by a team of much larger size compared to the LOPA team. There are pros and cons of each method. There is no one method better than the other. The setting in which LOPA can be effectively done depends on each organization. The quality of LOPA outcome depends on the team members' familiarity with LOPA concepts, nature and amount of experience, familiarity with other ISA84 safety lifecycle steps as well as the enthusiasm. If the team performing HAZOP is tired due to its long duration and schedule pressure, there is a tendency to cut corners and wrap up the LOPA scenarios quickly. Usually in such cases, the documentation of IPLs is inadequate and poorly done. There is not enough time during LOPA to analyze in detail the validity of each IPL.
Once the initial LOPA report is published, the next step is the SRS development and conceptual design of the IPS which is typically performed by control system engineer/s. It is often necessary to revisit the LOPA during the conceptual design of IPS. Some of the most commonly found reasons to revisit the LOPA are as follows:

- Improve the description of the IPLs, consequence and initiating event for better clarity.
- There is not enough time during LOPA to verify details such as process safety time, and the safety function response times.
- Availability of instruments and/or the process taps to meet the independency criteria. If identified ISFs do not meet the criteria of IPL, additional means of mitigating risk need to be identified.
- IPL listed is also part of the initiating event.
- Consideration of ISFs in BPCS (either excessive or ignored).
- During HAZOP the items that need further investigation are documented as HAZOP action items. These items are resolved after the HAZOP and proposed modifications may impact the ISFs.

Once the LOPA is revisited to incorporate the changes or improvements described above, it is typically referred to as LOPA issued for design (IFD). It is important to go through this iteration at least once to ensure the correct documentation of LOPA – which is the basis of any IPS design. During the verification step of each milestone in safety lifecycle, the functionality and performance of each ISF is verified against the LOPA documentation.

5. Process example

To illustrate these concepts, consider a process example consisting of high pressure knock out drum as shown in Fig 2.

The product gas at high pressure and temperature exiting the reformer is cooled in a process gas cooler, which results in condensation of water in the gas. The condensate water is removed in the knockout drum V-100. The level of the liquid in the knockout drum is controlled by a level control loop LC-100. The normal operating range of 35-

50% has been established to maintain a liquid level blanket in the knockout drum. If the level goes high, the separation will not occur as the liquid will get carried over in the product stream. If the level drops too low, there is a risk of the high pressure gas entering low pressure system downstream of the control valve LV-100. Maintaining the liquid level blanket to avoid this hazardous event is very important.

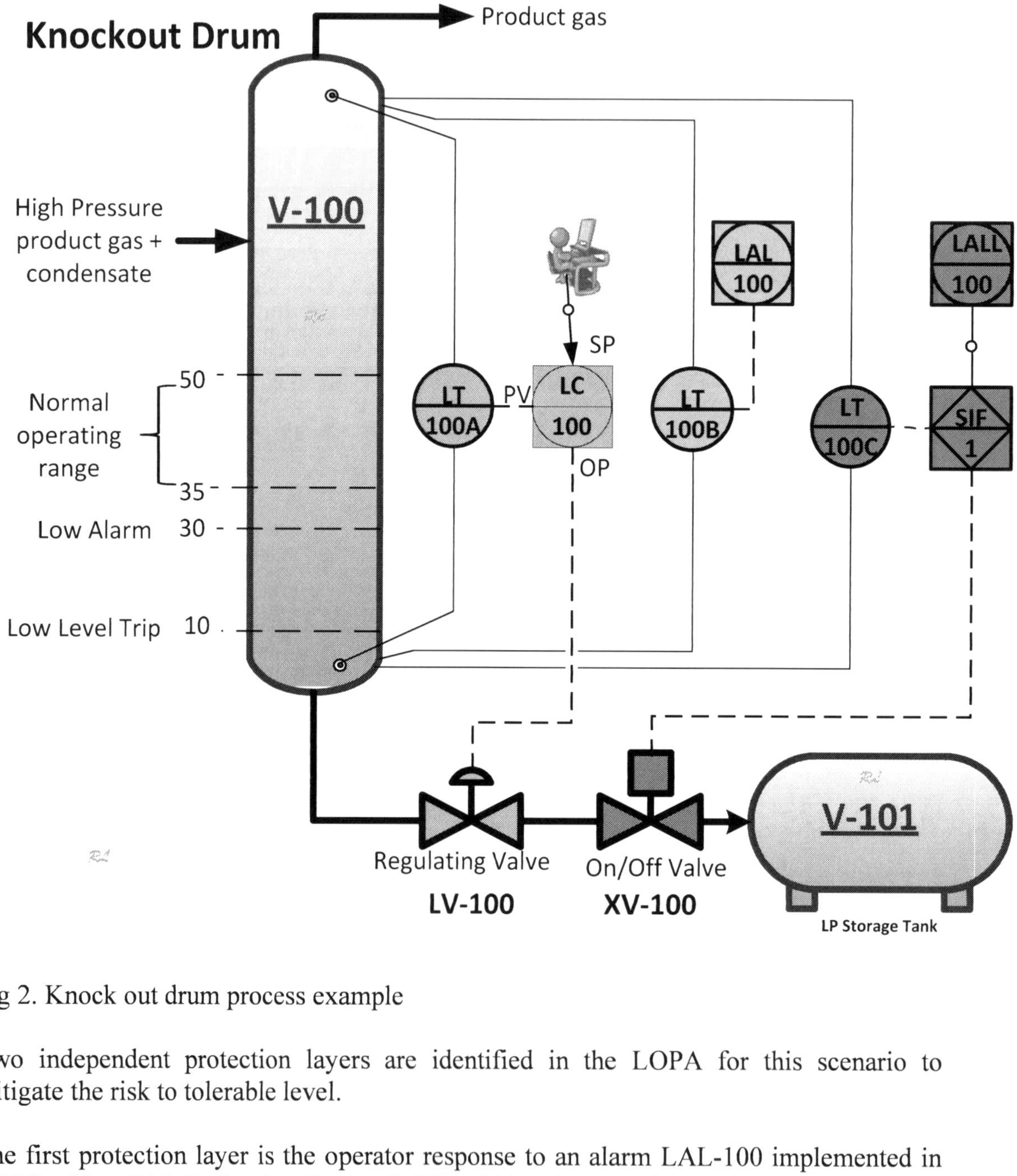

Fig 2. Knock out drum process example

Two independent protection layers are identified in the LOPA for this scenario to mitigate the risk to tolerable level.

The first protection layer is the operator response to an alarm LAL-100 implemented in BPCS with a Risk Reduction Factor (RRF) of 10. The low alarm limit for this pre-trip Safety Related Alarm is set to 30%.
$PFD_{avg} = 1/RRF = 1/10 = 1 \times 10^{-1}$

As per ISA84, the maximum risk reduction than can be assigned to operator response to an alarm implemented in BPCS is 10. It is commonly referred to as SIL0. An alarm is called as Safety Related Alarm (SRA) when operator response to the alarm is used as a protection layer and the risk reduction credit of one order of magnitude is claimed in LOPA.

The second protection layer to prevent the high pressure gas from entering low pressure system is SIF-1 with an RRF greater than 100. This SIL2 SIF is implemented in the SIS. If the level falls below the trip limit of 10%, then the on-off valve XV-100 closes. SIF is a combination of sensor, logic solver and final control element.

6. Time considerations

The response time of ISF must be less than the Process Safety Time.

6. 1 Process Safety Time (PST)

Process Safety Time is the difference between the time at which the unacceptable condition occurs ($T_{CONDITION}$) and the time where unwanted event occurs (T_{EVENT}) [1].

Process Safety Time = $T_{EVENT} - T_{CONDITION}$

In the above example, there are two protection layers. The first ISF is operator response to an alarm.
The process safety time for an alarm is the time when level reaches 30% till it goes to the trip set point of 10%. The time can be calculated by dividing the volume of the knockout drum for the 20% of instrument range (difference between the alarm set point and trip set point) by the worst case flowrate of condensate when the level control valve LV-100 stays wide open.

The process safety time for the SIF can be calculated in a similar manner. It will be the time when the level reaches trip point of 10% till it goes to 0% at maximum flowrate.
Typically the Process Safety Time calculation is done by process engineers. Such calculations become easy when the process model is available.

6.2 Alarm Response Time

Alarm Response Time is the difference between the time at which the alarm condition occurs and the time when process starts responding in the direction to correct the alarm condition. It includes the sensor lag, BPCS lag, operator response time and any process lag. Process deadtime is the amount of time it takes for the process to begin reacting after corrective action.

Alarm Response Time = Sensor delay + BPCS delay + Operator response time + process deadtime

The Process Safety Time for alarm has to be greater than the Alarm Response Time. These different time elements are shown in Fig. 3.

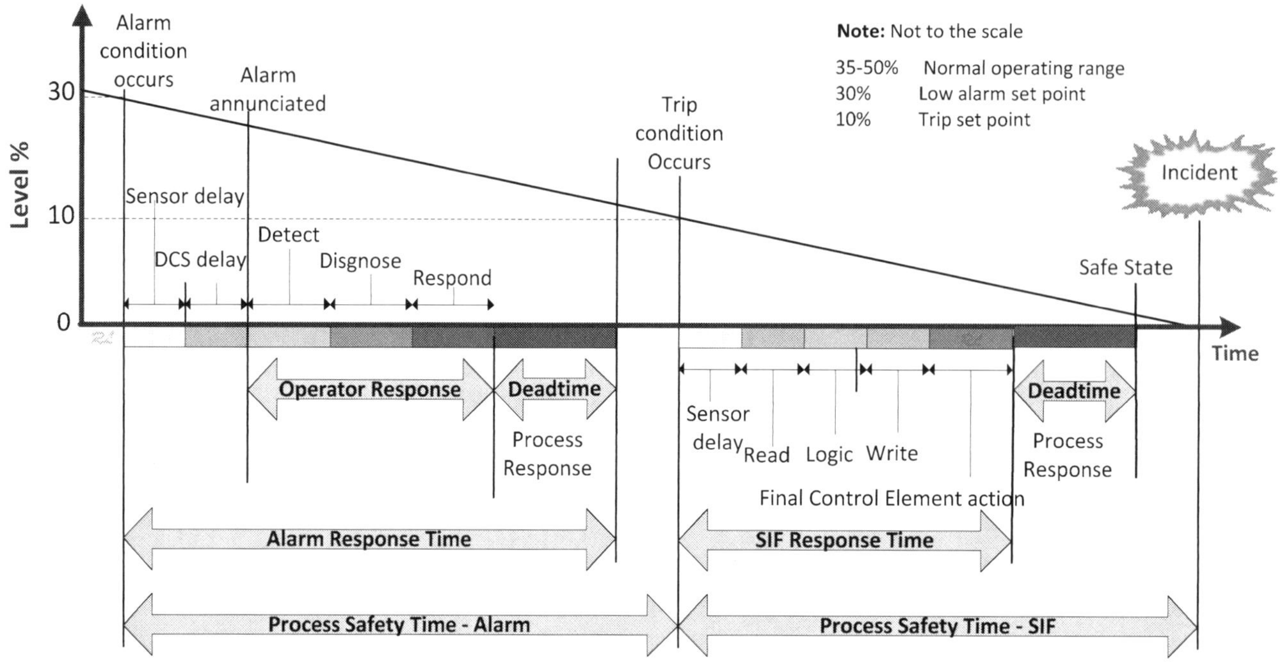

Fig. 3. Various time elements in relation to Process Safety Time

6.3 Operator Response Time

Operator response time is impacted by human factors, ergonomics, training, etc. collectively called as performance shaping factors. As per ISA18.2 feedback model of operator process interaction, the operator response time constitutes the following human interactions.

Detect: The operator becomes aware of the deviation from the desired condition. The design of the alarm system and the operator interface impact detection of deviation.

Diagnose: The operator uses knowledge and skills to interpret the information and diagnose the situation and determine the corrective action to take in response.

Respond: The operator takes corrective action in response to the deviation.

6.4 Minimum Time To Respond

This is defined in the alarm philosophy document. Each organization or plant site should develop an alarm philosophy document as defined in ISA18.2 [6]. Minimum Time To Respond is the quickest possible time to allow operator to go through detect – diagnose – respond steps. It is not physically practical to take necessary corrective actions in less than this time. Three to ten (3 - 10) minutes is most commonly used value as a minimum time to respond.

If the required operator response based on the PST for alarm is less than the Minimum Time To Respond, then no credit can be taken for the operator response to alarm as protection layer. This requirement is applicable to not just a Safety Related Alarm, but any alarm configured in the system. In such situations, various options should be reviewed to allow sufficient time of operator response. In the above process example, the simplest option is to check if the low alarm set point (LAL-100) can be increased to get more Process Safety Time.

If the alarm set point cannot be increased, using a restriction orifice to limit the maximum flow could be another option.

6.5 SIF Response Time

SIF response time includes the sensor delay, analog input card scan time, logic execution, writing output to the final control element and the time it takes for the valve to close. Typically valve closing time is most significant in these time elements. The logic may have ON or OFF delays. All these delays need to be included while calculating the SIF response time. SIF response time should be less than half of the Process Safety Time for SIF. If that is not the case, various options should be considered. The easiest option is increasing the trip limit if it is possible. Often times, adjusting trip settings does not have much impact as much faster SIF response is needed. The first step is to understand the most significant contributor to the SIF response time. Are there any logic elements with excessive delays? Can those be safely reduced? If the valve is the biggest component, are there options to install quick exhaust and/or volume boosters to improve the response? If none of the solutions are possible, selecting altogether a different valve with fast response may be necessary.

7. Instrument selection

The independency requirement of IPL greatly influences the selection of instruments and the process vessel design as well. The brownfield vs. greenfield project may also have an impact on deciding the instrumentation.

7.1 Level measurement

If there is a requirement of multiple level measurements on the vessel, and if the process taps are not enough, it is easier to influence the vessel design to add the required number of taps on a greenfield project as most likely, the vessel design is still not issued for fabrication when the initial LOPA is performed.

On a brownfield project, if LOPA determines the need for additional independent level measurements, it is unlikely that the new taps will be drilled into the vessel that is already in service. In such situation, different measurement technologies and options need to be explored. For example, if the vessel has a flange on the top, a radar level gauge or Guided Wave Radar (GWR) could be selected. Another option that requires no changes to the vessel integrity is a nuclear level gauge. Although it is expensive and prone to high maintenance and may require permits, it might be the best alternative of all the available options.

If the same process tap is to be shared by two level instruments, then common mode failure analysis using Quantitative Risk Analysis (QRA) techniques need to be performed. For clean service, if the common mode failure analysis determines that it may be possible to share the process taps, it is highly advisable to choose transmitters with advanced diagnostic features such as detection of plugged impulse lines. The type of process fluids involved and the process conditions play a big role in deciding whether a pair of common taps can be shared by two instruments. The process taps should not be shared if there is a history of impulse line plugging or possibility of deposit formation or the impulse line freezing.

7.2 Flow measurement

Orifice plate is the most common flow element used to measure the flow using differential pressure transmitters. Most often LOPA identifies multiple IPLs requiring the same flow measurement. A triple tap flange is most commonly used to provide three independent taps when independent flow measurements are required for the same flow. On a brownfield project, where only one process tap for the flow measurement is available, replacing the orifice flange is an option. Typically the whole pipe spool upstream and downstream of orifice plate is replaced. The new pipe spools with triple tap flanges could be fabricated and pressure tested in advance. It is relatively easy to replace the pipe spools during turn around.

If replacing orifice flanges is not an option, there are various other ways of additional independent flow measurement such as sonic flow meters, magnetic meters, anemometers etc. Depending on the process, an appropriate method should be selected.

7.3 Temperature measurement

When independent temperature measurements are required, it is often possible to use the measurements upstream or downstream of the vessel that are indicative of the problem. When no other measurements are available, it is possible to drill thermowells on process piping or use a new pipe spool with required number of thermowells. Using a bundle of thermocouples in the same thermowell and connecting the thermocouples to different temperature transmitters introduces a common cause failure of measurement from the same thermowell and needs to be taken into account during PFD_{avg} calculation of the SIF.

7.4 Pressure measurement

Independent pressure measurement of a vessel is not as complicated as independent level measurement as, adequate pressure taps usually exist on most vessels. It is also possible to use pressure taps on the downstream pipeline as long as there are no isolation valves between the pressure tap and the vessel.

7.5 Final control element

In most of the cases, final control element is the most significant contributor to the PFD_{avg}. SIL verification calculations will confirm if the selected device and test frequency can meet the target SIL of the SIF.

Increasing the proof test frequency and/or partial stroke testing options should be considered to improve the PFD_{avg} if required.

If the final control element involves solenoid valves to cut the instrument air to the regulatory control valve, it must be verified that the regulatory loop is not part of the initiating event. Other considerations include the tight shutoff requirements, time taken to complete the fail safe action and metallurgy.

7.6 Failure rate data requirements

The components used in SIF should be suitable for safety application with appropriate SIL rating. It is important to verify the Failure Modes, Effects, and Diagnostic Analysis (FMEDA) data sheets from the manufacturers while selecting and purchasing the instrumentation for SIF. This information is required during SIL verification calculations.

8. IPL being part of initiating event

Many times the IPL listed is part of the initiating event. If the LOPA team is not careful or lacks experience, one may find such IPLs listed in the LOPA. In the above process example, the initiating event is failure of regulatory loop to control the level. This failure may be due to a valve malfunction, frozen transmitter, impulse line plugging or human error. The operator response to alarm is used as the first IPL. If the transmitter used to generate the alarm is the same as the one used in the regulatory loop (LT-100A), then it is not a valid IPL. It is important that the alarm is generated from an independent transmitter.

9. Sharing of devices between BPCS and SIS

ISA84 clause 11.2.10 states that "A device used to perform part of SIF shall not be used for BPCS where the failure of that device results in failure of basic control function which causes a demand on SIF".
Therefore, the same sensor used for generating alarm cannot be shared with the control function in BPCS and cannot be shared with the SIF implemented in SIS. When independent sensors are available for each function, they can be configured in fault tolerant mode to achieve higher reliability and to increase the diagnostic coverage. Some of the most commonly encountered scenarios are discussed below. These examples are for illustration purpose only. Each owner operator company is responsible for doing analysis to ensure their configurations are valid and can satisfy the independency criteria of ISA84.

9.1 Standalone SRA

When the operator response to an alarm is identified in LOPA as a protection layer, and there are no other ISFs or control functions associated with the measurement, the transmitter is wired to BPCS as shown in Fig. 4.

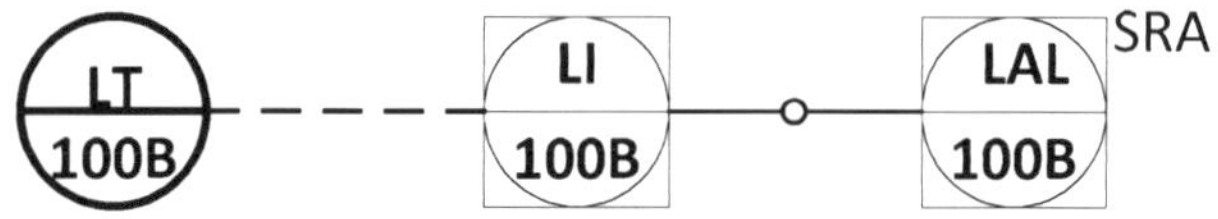

Fig 4. Standalone SRA P&ID representation

9.2 SRA and control function

Fig 5a shows a scenario where SRA and control function (such as PID loop) is implemented in BPCS. When the same process measurement is required for both the functions, separate transmitters should be used for alarm and control as shown in the Fig 5a. Each transmitter should be wired to separate cards in the controller or preferably separate controllers of BPCS.

The two transmitters could be configured in a BPCS as shown in Fig. 5b to improve availability, facilitate maintenance and improve diagnostic coverage. As shown in Fig. 5b, a software switch HS-100x is provided for the operator to manually change the source of input for the alarm as well as for the control function. This allows taking one of the transmitters out of service for maintenance or proof testing. Depending on the functionality in BPCS, a deviation alarm should be configured for the maintenance technician. If the difference between the readings of two transmitters is more than a pre-set threshold, a low priority deviation alarm is generated. The operator action for this alarm is typically to generate the maintenance work order for instrument technician to check the transmitters and correct the situation.

When the hand switch is used to switch the input to another source, a timer KS-100x should be started with alarm KAH-100x. If the time in switched input mode exceeds preconfigured limit, an alarm should be generated. The preconfigured timeout limit to generate warning alarm should be less than the Mean Time to Repair (MTTR) of the transmitter. If the time in switched state exceeds MTTR, then the SIF should initiate the action to put the process in safe state.

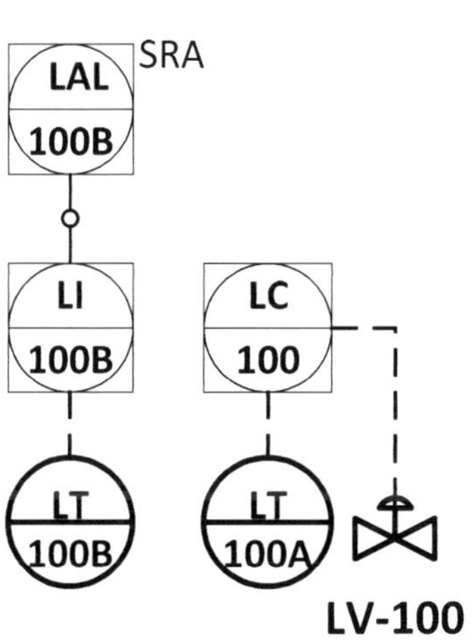

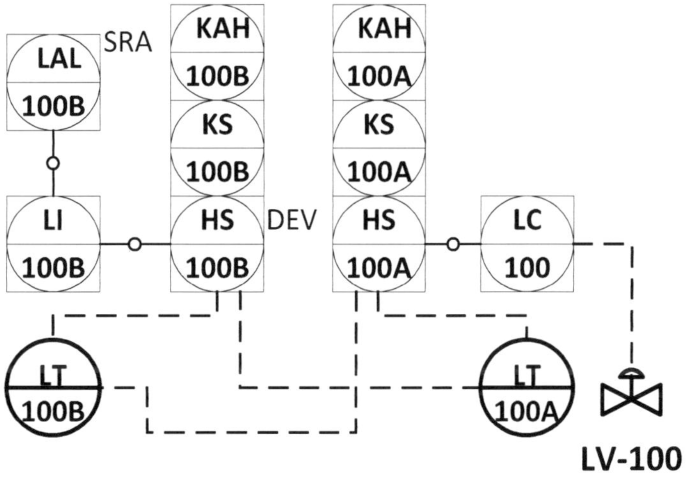

Fig 5a SRA & Control

Fig 5b Arrangement for easy maintenance and higher diagnostic coverage

Fig 5: Safety Related Alarm and control function both implemented in BPCS

9.3 SRA, control function and SIF

Fig. 6 shows a scenario where the SRA, control and SIF are all using the same process measurement. The SRA and control function are implemented in the BPCS, while the SIF is implemented in the SIS. When the available instrumentation is adequate to meet the independency criteria of each function, it is beneficial to wire it in fault tolerant configuration to improve reliability.

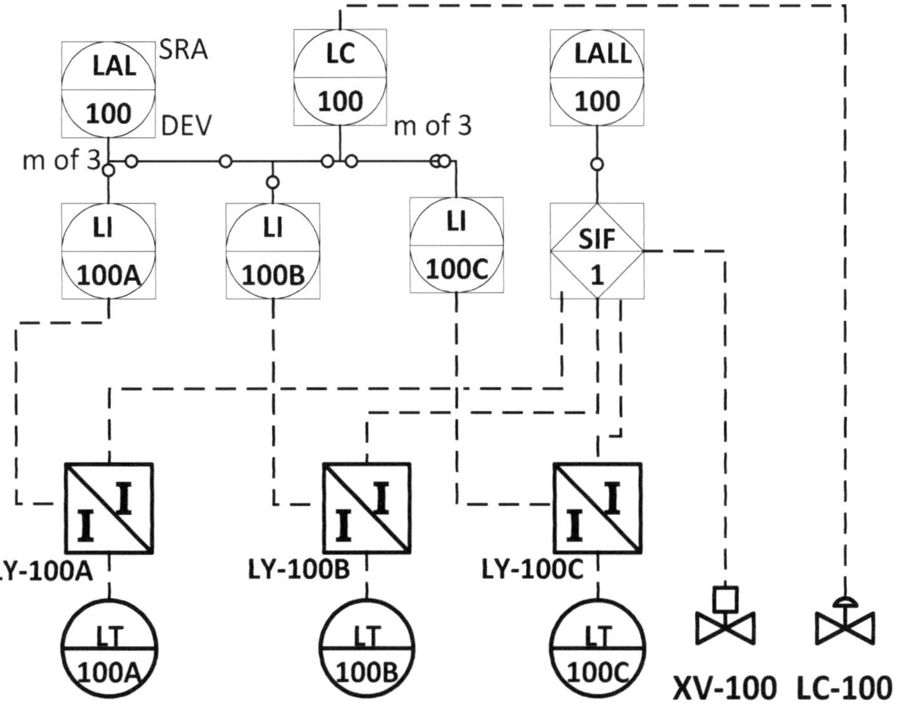

Fig 6. SRA, SIF and control function implementation example to achieve high reliability

In this example the three level transmitters using independent taps for process connection are wired to SIS through the safety certified current loop isolator and repeater. The transmitters using Highway Addressable Remote Transducer (HART) communication protocol are powered by SIS and the isolators have capability to pass through the HART signals on each channel.

SIL 1 or SIL2 SIF is implemented in SIS with 2oo3 voting logic for the level input signals. Depending on the BPCS, the actual implementation may differ. Fig 4 shows a generic representation where a middle of 3 selector block is used in BPCS which has SRA configured. Some BPCS have a standard 2oo3 block which can be used as well. Output of another middle of 3 selector block is used as a PV for the control function.
When sharing the transmitters between BPCS and SIS, the following considerations should be taken into account [7].

- The failure of any hardware or software outside the SIS should not prevent any SIF from operating correctly.
- The failure of a BPCS component does not result in the initiating cause for the process hazard and the failure (or defeat/bypass) of the SIF that protects against the specific scenario under evaluation.
- The probability of common mode, common cause or dependent failures, such as plugged impulse lines, maintenance activity including bypasses, incorrectly operated line isolation valves, etc., has been adequately evaluated and determined to be sufficiently low. It is often recommended to use diverse measurement technology to reduce the common cause failure problems. A combination of differential pressure and GWR transmitter is an example of using diverse technologies to measure the same process value.
- The shared components are managed according to ISA84, including proof testing, access security and management of change.
- The sensor (e.g., transmitter, analyzer, switch) is powered by the SIS. The signal is transmitted to the BPCS by an optical isolator or other means to ensure that no failure of the BPCS affects the functionality of the SIS.

10. ISFs in BPCS

BPCS is normally the first line of defense against process excursions. Typically, three types of ISFs are implemented in BPCS:

- Regulatory control
- Alarm
- Trip (or BPCS interlock)

Risk reduction credit can be taken in LOPA for the functions implanted in BPCS as long as following guidelines are observed:

- ISF is independent of the initiating event.
- Credit for each ISF in BPCS is no more than RRF of 10 (PFD_{avg} >= 1 X 10^{-1}).
- No more than two ISFs are credited for the same scenario (combined RRF of two BPCS functions is no more than 100). A detailed analysis recommended in CCPS

LOPA book should be performed if credit is taken for two ISFs implemented in BPCS [8].

- Separate measurement devices and final control elements are used for each ISF. For example, the transmitters and valves used for regulatory control should be separate from those used for the trip action. Alarm should be generated from transmitter that is not used in regulatory control or trip. (They could be wired in fault tolerant configuration as shown in Fig. 5 and 6).
- The devices used in ISF for same scenario are wired to different IO cards of BPCS.
- Procedures for BPCS maintenance exist and are followed. Good Management of Change (MOC) procedures are followed for changes to BPCS logic and parameters, all changes are auditable and well documented, appropriate access control to BPCS is in place.

11. Instrument index and P&ID updates

Each IPL is evaluated in detail for its validity as part of the conceptual IPS design. It may result in modifications such as addition or deletion of instruments and sensor elements, tag name changes, logic changes and re-assignment of signals to BPCS and other IPSs. It is important to maintain and update the IO database and with appropriate comments that could prove very useful in later stages of the lifecycle. For example, tag assignment to IO cards of IPS is done at the beginning of the detail design phase of IPS. A comment to allocate the transmitters used in the same scenario to different IO cards in the system is very handy during IO allocation step. Maintaining and updating the IO index is an essential step for project change control. IO count of each IPS is an important key quantity in any project. Reporting the changes to this key quantity to project management on a routine basis is an absolute must to keep all stakeholders informed of the scope changes which may impact the project cost and schedule.

Any updates to P&ID as a result of changes to IPL should be marked on the master P&ID set and be reviewed and approved per the P&ID MOC procedures of the organization.

12. Conclusion

LOPA should address the mitigation of risks in all three categories – safety, environmental and asset protection. An IPL list should be maintained with at least the three attributes to classify the safeguards - integrity level, system and the purpose. An IPL list is very useful in allocating them to appropriate IPS.

Once the initial LOPA report is published, the next step in the functional safety lifecycle involves preparation of SRS and conceptual design of IPS. During the conceptual design, some IPLs may be found invalid. Some of the most common reasons for invalid IPLs are: process safety time, independency criteria, common mode failures and poor documentation. It is important to revisit the LOPA during the conceptual design to correct all discrepancies and publish the LOPA as issued for design. Usually it requires at least one revision to LOPA to update the findings from the conceptual design phase.

The greenfield vs. brownfield project could influence the selection of instrumentation. When independent sensors are available for each IPF, they can be configured in fault tolerant mode to achieve higher reliability and to increase the diagnostic coverage.

13. References

[1] Guidelines for Safe and Reliable Instrumented Protective Systems; Center for Chemical Process Safety, John Wiley & Sons 2007 ISBN 978-0-471-97940-1

[2] ANSI/ISA-84.91.01-2011. Identification and Mechanical Integrity of Instrumented Safety Functions in the Process Industry; International Society of Automation, Research Triangle Park, NC

[3] ANSI/ISA-84.00.01-2004 Part 1 (IEC 61511-1 Mod). Functional Safety: Safety Instrumented Systems for the Process Industry Sector - Part 1: Framework, Definitions, System, Hardware and Software Requirements

[4] Angela E. Summers, "Consider an instrumented system for overpressure protection"; Chemical Engineering Progress, November 2000

[5] ISA-84.00.01-2004 Part 3 (IEC 61511-1 Mod) Functional Safety: Safety Instrumented Systems for the Process Industry Sector - Part 3: Guidance for the determination of the required Safety Integrity Levels

[6] ANSI/ISA–18.2–2009. Management of Alarm Systems for the Process Industries

[7] Technical Report ISA-TR84.00.04-2005 Part 1 Guidelines for the Implementation of ANSI/ISA-84.00.01-2004 (IEC 61511 Mod)

[8] Layer of Protection Analysis; Center for Chemical Process Safety; American Institute of Chemical Engineers 2001 ISBN 0-8169-0811-7

Biography

Rajeev Limaye (Rajeev_Limaye@praxair.com) is Director of Control Systems & Instrumentation at Praxair, Inc. in their Global Hydrogen business unit in Houston, Texas. He has a masters degree in Chemical Engineering from IIT Bombay India and MBA from University of Houston. Rajeev has worked in process automation industry for over 25 years and holds a PE license in Texas. He is an advisor to the University of Houston Downtown for their degree program in Control Systems Engineering. Rajeev is a Certified Functional Safety Expert (CFSE) and a member of ISA84 standards committee.

9TH GLOBAL CONGRESS ON PROCESS SAFETY

Safety controls, alarms, and interlocks as IPLs

Angela E. Summers, Ph.D., P.E.
SIS-TECH Solutions
12621 Featherwood Dr. Suite 120, Houston, TX 77034
asummers@sis-tech.com

Prepared for Presentation at
American Institute of Chemical Engineers
2013 Spring Meeting
9th Global Congress on Process Safety
San Antonio, Texas
April 28 – May 1, 2013

UNPUBLISHED

Safety controls, alarms, and interlocks as IPLs

Angela E. Summers, Ph.D., P.E.
SIS-TECH Solutions
12621 Featherwood Dr. Suite 120, Houston, TX 77034
asummers@sis-tech.com

Keywords: safety controls, alarms, interlocks, SIS, BPCS, process control, layers of protection analysis

Abstract

Layers of Protection Analysis (LOPA) evaluates the sequence of events that first initiate and then propagate to a hazardous event. This semi-quantitative risk assessment technique can expose the role that automation plays in causing initiating events and in responding to the resulting abnormal operation. Automation that is specifically designed to achieve or maintain a safe state of a process in response to a hazardous event is now referred to as safety controls, alarms, and interlocks (SCAI).

Guidelines for Initiating Events and Independent Protection Layers addresses four basic types of SCAI: safety controls, safety alarms, safety interlocks, and safety instrumented systems (SIS). This paper discusses the design, operation, maintenance, and testing practices necessary for SCAI to be considered as independent protection layers (IPL). It also provides guidance on claiming multiple layers of protection in the basic process control system.

1. What are SCAI?

SCAI or *Safety Controls, Alarms, and Interlocks* are process safety safeguards implemented with instrumentation and controls, used to achieve or maintain a safe state for a process, and required to provide risk reduction with respect to a specific hazardous event (ISA 84.91.01-2012). They are the safety subset of instrumented protective systems (IPS), which are discussed in *Guidelines for Safe and Reliable Instrumented Protective Systems* (CCPS, 2007). An IPS is any instrumented system that addresses risk related to health and safety effects, environmental impacts, loss of property and business interruption costs. As shown in Figure 1, there are many terms that can be used to further classify SCAI.

Figure 1. Classification of SCAI (adapted from ISA 84.91.01-2012)

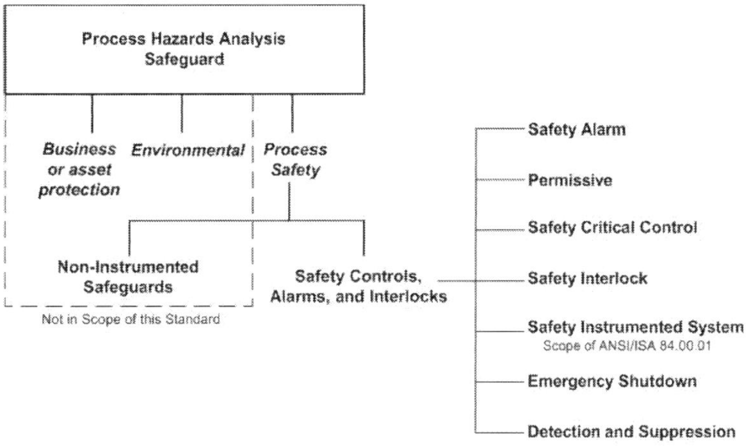

As with any IPL, SCAI should be independent of the initiating cause and any other IPL used for risk reduction for a particular hazardous event. To understand its capability in stopping the event, the SCAI function must be clearly identified, e.g., what does it detect, what is the setpoint, how does it affect the process, how fast does it need to act, and how does the process respond to its action. Typically, SCAI have sensors to detect the process condition, a logic solver to determine what action to take based on the process condition, and final elements to take action on the process (Figure 2). The decision and action can be manual (e.g., operator response to a safety alarm) or automatic (e.g., via continuous control, interlock, or SIS). SCAI also includes other interconnected equipment, such as human interfaces, wiring, process connections, communications, and utilities, which are necessary in order to achieve the desired functionality and to operate and maintain the SCAI.

Figure 2. Example elements required for SCAI successful operation

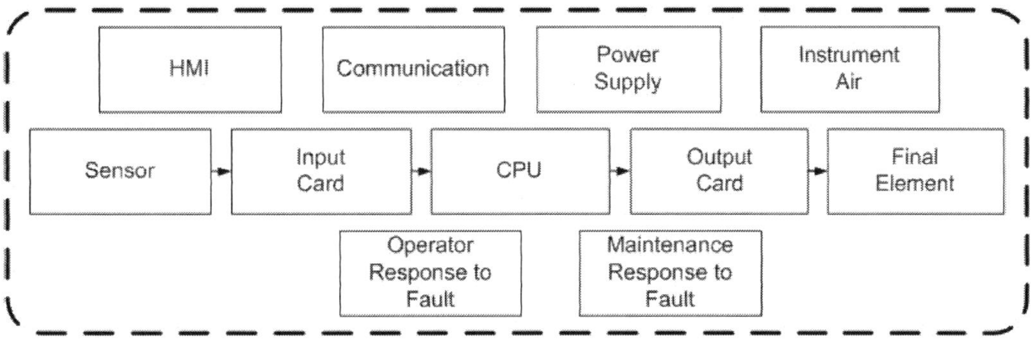

SCAI are complex systems requiring many different devices to operate successfully in order to stop the hazardous event propagation. The logic solver technology is typically implemented using electrical, electronic, or programmable electronic equipment. Older installations may use mechanical switches, which change state based on manipulating pneumatic (or hydraulic) flow and pressure. Field device technology ranges the gamut from simple switches to complex analyzers. While there is no limit on the designer's choice, the technology must provide the desired level of risk reduction in the operating environment and under the user's management system. As technology moves from the drawing board to sustained operation, the proof of integrity must also move from estimated performance to historical operation as evidenced through site mechanical integrity records. Historical demonstration of successful prior use is essential for assuring that equipment is fit for service.

The ability of SCAI to achieve the desired functionality and the claimed risk reduction is limited by the performance achieved by the installed equipment. Each SCAI device has distinct failure modes and has a failure rate that contributes to the overall SCAI performance. For this reason, equipment is selected based on its expected capability in the operating environment, but the SCAI is judged by the cumulative capability of the system. The actual level of risk reduction achieved should be substantiated by mechanical integrity and human reliability data. The records associated with any SCAI should confirm that the equipment operates as specified during all intended operating modes. Failure tracking and analysis is important to verify hazards and risk analysis assumptions and to support continuous improvement.

2. SCAI practices

According to *Guidelines for Initiating Events and Independent Protection Layers*, SCAI specifications should include:

- Functional requirements.
- Configuration, installation, and maintenance requirements to achieve and sustain the claimed performance.
- Failure modes, means used to detect these failure modes, and expected system and operator response to detected failure.
- Compensating measures required to continue safe operations with faulted SCAI
- Conditions required to safely bypass the SCAI (includes override or manual operation) and any compensating measures to be in place during bypass.
- Conditions required for safe reset of the SCAI.

SCAI equipment should be sufficiently robust to withstand the actual stresses of the operating environment and provide the required integrity and reliability. Some technologies used in control applications may not be acceptable for SCAI due to inadequate installation history, poor in-service reliability/integrity, or unpredictable/erratic behavior. For example, wireless technology as defined by ISA 100.11a is not generally considered an acceptable technology for executing SCAI, but may be used for monitoring, status or diagnostic communication (CCPS, 2013).

An equipment list should be maintained which identifies SCAI equipment by a unique designation, such as a tag number or equipment ID that is traceable to the mechanical integrity requirements necessary to maintain the required integrity and reliability throughout its life. Mechanical integrity includes a variety of activities, such as inspection, calibration, preventive maintenance, repair/replacement, and proof testing.

Inspections and proof tests are conducted as necessary to identify and correct equipment deficiencies. Visual inspection conducted during plant rounds may need to be conducted more frequently based upon process and environmental severity. Inspection and proof test are performed according to a written procedure that ensures that the required operation is demonstrated to the degree feasible under safe conditions. Typical information recorded includes as-found/as-left conditions, the tester, when tested, the procedure and equipment used, and calibration records.

It is inherently safer to implement SCAI equipment such that it takes the safe state action on failure rather than fails to an alarm (or dangerous) state. Examples of failures to consider are loss of signal, out of range values, loss of communications, power failure, instrument air failure, and loss of other utilities. If it is intended to continue operation of the process with out-of-service SCAI equipment, the risk should be assessed and compensating measures should be provided to address increased risk. Out-of-service periods should be tracked and minimized to the degree possible through equipment design and maintenance practices.

Documented procedures, training, and auditing ensure effective management of change control for the hardware and software. As with any IPL, SCAI are only bypassed (e.g., suppressed or setpoint changed) with administrative approval and after deployment of any necessary compensating measure. Administrative approval may involve formal MOC, bypass control procedures, or special operating procedures.

Information about SCAI should be included in operating procedures, since the operation of SCAI affects the process or requires that the operator take action. When programmable electronic systems are used, management of change and access security procedures are needed to ensure that changes to the embedded software and application program are reviewed and approved. Any change potentially affecting SCAI should be examined to determine the requirements for validating the functionality prior to the SCAI being placed in service. Access to engineering interfaces should also be controlled to reduce human error and prevent compromise of the system.

3. BPCS versus SIS

The type of equipment and the design and management practices used to implement SCAI limits achievable performance. SCAI can be implemented using basic process control system (BPCS) or safety instrumented system (SIS) equipment. To be counted as IPL, SCAI are designed and managed to achieve an order-of-magnitude risk reduction (PFD $\approx$ 0.1). Because of the typical design and management of BPCS equipment, SCAI implemented in a single BPCS is limited to a single order-of-magnitude risk reduction. Higher claims (e.g., PFD $\leq$ 0.01) can be supported by a SCAI designed and managed in compliance with IEC 61511.

SCAI claimed as IPL are limited to PFD = 0.1, unless they are designed and managed per IEC 61511. IEC 61511 describes the conditions necessary to allow two BPCS loops to be credited for the same hazardous event with an overall claim of PFD = 0.01. IEC 61511 identifies independence and separation of non-SIS (including control, diagnostics, and SCAI) and SIS as a critical requirement; otherwise the implementation must meet the full requirements of IEC 61511.

3.1 BPCS

3.1.1 Type

BPCSs are most commonly implemented in pneumatic control loops, programmable logic controllers (PLCs), distributed control systems (DCS), discrete control systems (e.g., on/off, relay), and single loop controllers (SLCs).

3.1.2 Practices

The design and management of the BPCS is usually focused on ensuring that it can reliably perform the control functions that maintain normal operation, such as PID (proportional-integral-derivative) and batch or sequential controls. The response to detected faults and momentary loss of communication is generally continued process operation using the last 'good' value. Failure of the BPCS in executing normal control functions is one of the leading causes of hazardous events. A BPCS is also capable of executing other functions such as alarming and process shutdown when special design and management practices are followed.

3.1.3 Performance capability

A BPCS's performance is limited by the hardware and software design. Most BPCS have little online redundancy in the input/output cards (I/O) or in the CPU and have limited diagnostic coverage of these components. When redundancy is available, it is generally a backup processor that does not operate until internal diagnostics (typically not themselves redundant) detect a problem. Redundant BPCS are generally comprised of hot standby or hot swappable processors, which significantly limits the potential performance claim.

With multiple devices required to operate correctly, individual devices within the SCAI usually need to achieve an installed performance an order of magnitude better than the required system performance. For example, if it is desired to claim that a BPCS loop achieves a PFD = 0.1, the BPCS controller would need to have a PFD ≈ 0.01. Industry data (ISA TR84.00.04 and PDS Data Handbook, 2010) does not support a PFD ≈ 0.01 claim for controllers that are not designed and managed in compliance with IEC 61511.

3.2 SIS

3.2.1 Type

SIS are most commonly implemented using discrete control systems (e.g., on/off relay) and programmable electronic systems.

3.2.2 Practices

The design and management of the SIS ensures that it can reliably take action to achieve or maintain a safe state of the process to stop propagation of the hazardous event. In the process industrial sector, the term SIS applies to instrumentation and controls designed and managed in accordance with international standard IEC 61511 (IEC, 2003).

Use of programmable electronic systems (PES) requires diagnostics on input and output cards to detect stuck-on or stuck-off conditions and on the main processor to detect a stuck or locked-up program execution. These diagnostics are necessary for the programmable electronic system to meet the minimum requirements for SIL 1 (PFD<0.1).

A PES relies on software - embedded and application – to execute the internal and user functions of the system. This software is patched, upgraded, and updated under a revision control system that ensures adequate testing and validation. Changes to embedded software require full system validation (ISA TR84.00.03), while changes to the application program must be tested based on their traceable impact to the system (IEC 6511).

Data communication and engineering interface to PES should be separated from other credited SCAI. Any interconnections that can compromise performance should be protected with firewalls and strict access security provisions.

Diagnostics are not required for hardwired logic solvers, such as relays or trip amplifiers, which have low failure rates and well-defined failure modes. Hardwired logic solvers are separate and diverse from other control systems and are not vulnerable to cyber-security risks or data highway corruption.

3.2.3 Performance capability

While it has many prescriptive requirements, IEC 61511 is fundamentally a performance-based standard that requires the establishment the safety integrity level (SIL). The target SIL is normally determined using some type of risk assessment. For example, a LOPA determines the PFD that the SIS needs to meet in order for the event risk to be reduced below the risk criteria. The required PFD establishes the target SIL. Table 1 shows the relationship between SIL and the PFD ranges.

Table 1. Safety Integrity Level Relationship to PFD and Risk Reduction (CCPS, 2013)

SIL Level	PFD is greater than:	But less than:	Risk Reduction
SIL 1	0.01	0.1	1 order of magnitude
SIL 2	0.001	0.01	2 orders of magnitude
SIL 3	0.0001	0.001	3 orders of magnitude

When claiming SIL 2 and 3, the standard has minimum redundancy requirements to protect against systematic errors. Redundant devices and signal paths using multiple sensors, controllers, or final elements to provide the same function may be necessary to ensure SIL 2 and are required to ensure no single points of failure for SIL 3.

4. The 4 SCAI and LOPA

SCAI are instrumented safeguards identified during the hazard identification and risk assessment that are necessary to achieve risk reduction related to a potential process safety event. As discussed earlier, SCAI can be referred to by many other names – the key to the identification is the function's purpose with regard to the event. This paper divides SCAI into four types of systems: control, alarms, interlocks, and SIS (CCPS, 2013).

IEC 61511 restricts the PFD claim that can be made for a BPCS loop to ≥ 0.1, while the PFD claim for SIS can be < 0.1. For the purposes of LOPA, both are given the rounded value of 0.1. Regardless of the PFD performance claim, SCAI must be under an appropriate mechanical integrity program that includes documentation, procedures, and administrative controls (ANSI/ISA 84.91.01).

4.1 Safety Controls

These normally operate to support process (or regulatory) control and during a process upset acts to stop the hazardous event progression. These controls are not typically designed or managed in accordance with IEC 61511, so they may also be referred to as BPCS IPL. The risk reduction claimed for a safety control (which does not conform to IEC 61511) must be below 10 (IEC 61511 Clause 9.4.2). In LOPA, a risk reduction factor of 10 or PFD = 0.1 is used.

4.2 Safety Alarms

These initiate an alarm that requires that the operator take action in accordance with a response procedure to stop the hazardous event propagation. The risk reduction claimed for a safety alarm (which does not conform to IEC 61511) must be below 10 (IEC 61511 Clause 9.4.2). The safety alarm includes not only the alarm, but also the interfaces and final control elements used by the operator to take the required action. Refer to ISA 18.2 and IEC 61511 for additional guidance related to the instrumentation and control design. Operator reliability in responding to the alarm should also be considered in making claims for safety alarms. Operator knowledge of required actions and the capability to take timely action should be validated using tests, drills, or simulated environments. When these conditions are met, a risk reduction factor of 10 or PFD = 0.1 is used in LOPA.

4.3 Safety Interlocks

These are instrumentation and controls that take automatic action to achieve or maintain a safe state of the process when a process variable reaches a defined limit. When an interlock is not designed and managed in accordance with IEC 61511, it may also be referred to as a BPCS IPL. The risk reduction claimed for a safety interlock (which does not conform to IEC 61511) must be below 10 (IEC 61511 Clause 9.4.2). With proper design and management, a risk reduction factor of 10 or PFD = 0.1 may be used. When the safety interlock is designed and managed per IEC 61511, it is an SIS.

4.4 SIS

Safety instrumented systems consist of instrumentation and controls designed and managed in accordance with IEC 61511 (IEC, 2003) that take action to achieve or maintain a safe state of the process in response to process demand. Unless the BPCS is an SIS, the SIS must be independent and separate from the BPCS to the extent that the safety integrity of the SIS is not compromised (IEC 61511 Clause 11.2.4). Safety instrumented systems are typically designed to operate in low demand mode but may operate in high demand/continuous mode.

The ISA84 committee has developed a series of complementary technical reports to provide guidance and practical examples related to various SIS topics and applications. Three of these technical reports, ISA-TR84.00.02, ISA-TR84.00.03, and ISA-TR84.00.04, provide a comprehensive overview of the SIS lifecycle and associated requirements.

5. LOPA - Approach A versus Approach B

In the initial LOPA guideline (2001, CCPS), two approaches were presented for considering the BPCS in LOPA. Approach A allowed the identification of the BPCS as providing an IE or an IPL, while Approach B provided the basis for additional analysis of the BPCS design and management system.

Whether choosing Approach A or B, special consideration should be given to situations where low-demand mode functions are implemented in a BPCS. It is inherently safer to implement low-demand mode functions in a separate and independent SIS that is specifically designed and managed for safety interlocks. Further, some application-specific practices require that safety interlocks be implemented in an independent SIS.

5.1 Approach A

The key criteria are (1) the initiating cause is not related to the BPCS and (2) the BPCS IPL is properly designed and managed to claim one order-of-magnitude risk reduction. This approach takes the position that a single BPCS loop failure invalidates all other functions that could be implemented in BPCS. It is used for LOPA because its rules are clear and conservative. It provides a high level of protection against common cause failures between the BPCS and SCAI.

5.2 Approach B

This approach potentially allows a maximum of two loops within BPCS to be identified for the same scenario. Approach B assumes that each BPCS is designed and managed such that it is capable of supporting the specified function and is sufficiently independent from the other such that the likelihood (probability) of simultaneous failure is low enough to support another order-of-magnitude risk reduction. Approach B requires that the analyst be knowledgeable in the BPCS design architecture, has adequate data available on the actual performance of the BPCS, and understands how to identify and account for common cause failures between the different automation layers.

Manufacturer claims regarding an individual device are not sufficient to justify sharing the device between two BPCS IPL or between BPCS IPL and BPCS IE for the same hazardous event.

5.2.1 Analytical Approach

LOPA is not the best tool for detailed evaluation of whether the BPCS architecture can support two loops. Analysis and test data should demonstrate that the two particular BPCS are designed and managed in a manner such that the two BPCS loops, in combination, can achieve the overall performance claim of $\approx$ 0.01/year. Consider use of advanced quantitative analysis techniques, such as fault tree analysis, to ensure that common cause (including systematic errors) is assessed fully. Since a BPCS failure could result in the simultaneous loss of control and safety protection, carefully examine the independence and separation of the control system and the SCAI. When strict independence is not maintained, the analysis can become quite complex necessitating higher expertise and knowledge of automation design.

If the risk reduction claimed for a BPCS is greater than 10, IEC 61511 Clause 9.3.3 requires that the BPCS be designed and managed as SIS.

At a minimum, the system analysis should include (CCPS, 2013):

- Assessment of potential common cause, common mode, and systematic failures between the BPCS IE and IPL or the two BPCS IPLs to determine that their impact is sufficiently low as compared to the performance claim.
- Written specification covering all functions within the system, e.g., instrument diagrams, P&IDs, loop diagrams, functional specifications.
- For the Generic Data approach to validation, the assessment of previous historical performance of the BPCS logic solver, input/output cards, sensors, final elements, human response, etc. (*NOTE: manufacturer information must be examined critically to ensure that it applies to situations similar to a particular installation*).
- For the Site-Specific Data approach to validation, evaluation of inspection, maintenance, and test data over a significant period to demonstrate that the system achieves the performance claim.
- Assessment of access security measures for hardware and software.
- Management of change and revision control for the hardware and software, including setpoints, fail configuration, and operator overrides.

5.2.2 Approach B – Example Configurations

Two independent and separate BPCS controllers can be implemented such that there is no communication (or interconnection) between the two BPCS loops (e.g., IEC 61511-2). This architecture is an inherently safer design, as it is physically separate as well as independent, so it provides a high degree of protection against data corruption and inadvertent software changes. This architecture can theoretically achieve the 0.01 PFD claim for the overall performance of the two BPCS loops if the two systems are designed and managed according to good engineering practices.

A maximum of two BPCS controllers (when designed and managed in accordance with good engineering practices) can be credited in LOPA for the same hazardous event as an initiating cause and IPL or as two IPL (CCPS, expected 2013).

Figure 3. Illustration of two BPCS functions credited for the same scenario per IEC 61511 (Adapted from Figure 2, IEC 61511, 2003)

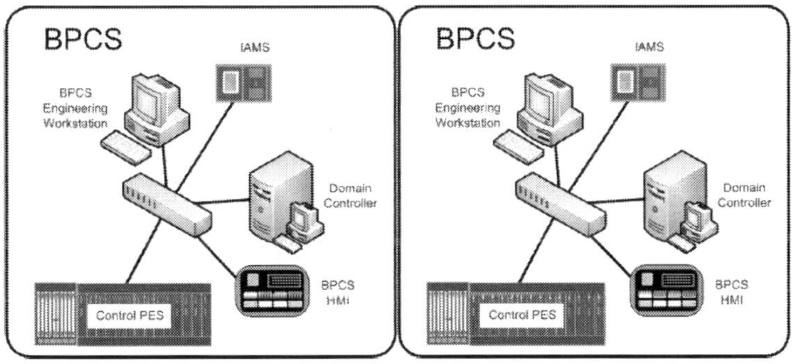

CCPS *IPS* (CCPS, 2007) recognized that the controllers could be functionally separate while safely sharing information. An example of this architecture (Figure 4) consists of separate controllers that communicate in a secure manner using firewalls and communication controls. When the controllers share data or interfaces, integrity of the two BPCS loops must be assured with rigorous management of change and access control. Ensuring cyber security and software/data integrity necessitate special design and management practices (dISA TR84.00.09 and ANSI/ISA 99). This architecture can also theoretically achieve the 0.01 PFD claim as long as the two systems are designed and managed according to good engineering practices.

Figure 4. Illustration of two BPCS functions credited for the same scenario per CCPS IPL/IE (Adapted from Figure 5.5, CCPS, expected 2013)

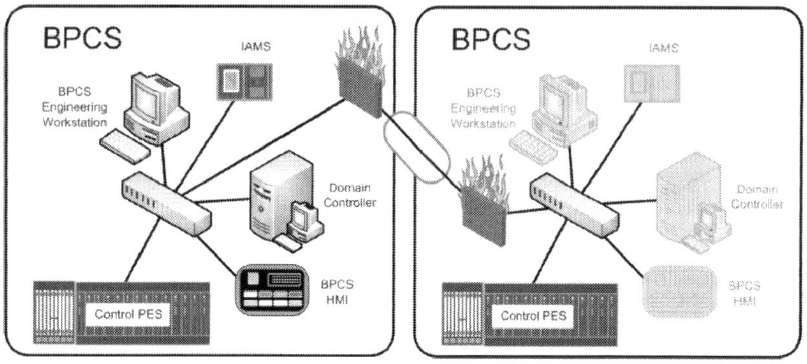

When both loops are implemented in the same BPCS (Figure 5), the shared equipment is generally insufficient for crediting the two BPCS loops for the same hazardous event. The CCPS LOPA book (CCPS, 2001) indicated that the BPCS CPU had at least two orders of magnitude better performance than the field devices. More recent field data demonstrates that the failure rate of a typical BPCS logic solver is greater than 0.01/year (PDS Data Handbook, 2010). In fact, industry and manufacturer data indicates a random hardware failure rate for typical BPCS equipment (such as DCS/PLCs) in the range of 10^{-5}/hr to 10^{-6}/hr (ISA TR84.00.04-2005 Appendix F.2.1), so the BPCS performance is equivalent to many other electrical, mechanical, or programmable electronic devices. Therefore, it is difficult to substantiate that a single BPCS qualifies for Approach B with a claimed failure rate of ≈ 0.01/year without advanced quantitative techniques.

When the two loops share equipment or support systems that can cause the dangerous failure of the integrated BPCSs, the resulting integrated system should be analyzed as a single BPCS. A BPCS architecture that uses a primary and hot backup controller is not sufficient for crediting as two independent controllers due to potential common cause failure. For example, a hot backup controller shares common equipment with the primary controller, such as the backplane, firmware, diagnostics, and transfer mechanisms. A hot backup controller improves operational reliability by reducing the likelihood of spurious process shutdown by a factor of perhaps 2 to 3, but it is not sufficient for claiming a dangerous failure rate ≈ 0.01/year.

Thus, designing a PE controller to achieve a PFD less than 0.01 requires advanced diagnostics, fault tolerant architectures, and rigorous software controls. IEC 61511 Clause 11.5 provides criteria and considerations for achieving this level of performance. Normal and customary BPCS practices will not meet these requirements.

Figure 5. Illustration indicating what is necessary to achieve ≈ 0.01/yr in a single controller

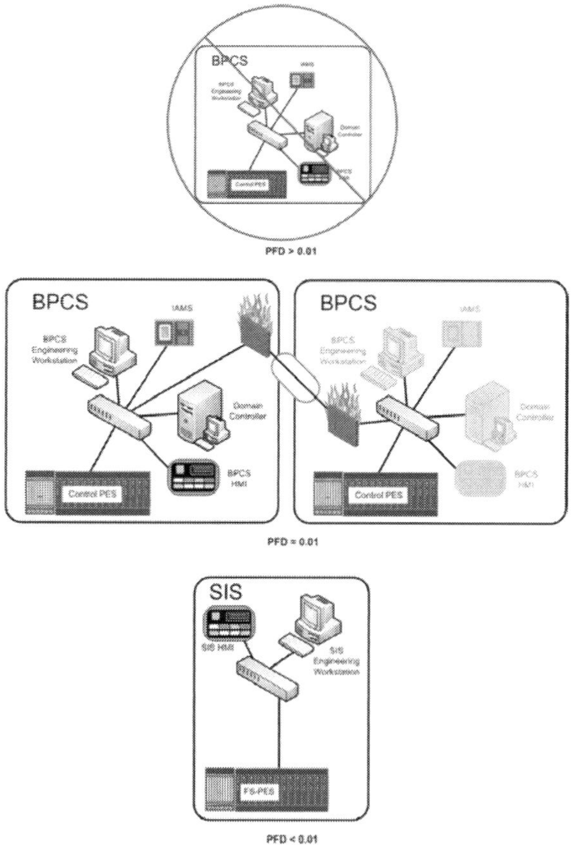

6. Summary

SCAI are an important subset of the process safety safeguards used to reduce the risk of hazardous events. The instrumentation and control systems need rigorous design and management practices to ensure that they are available when rare hazardous events begin to propagate. There are many words that can be used to identify, describe, define, or otherwise classify SCAI. As a rose is a rose, if it is an instrument or control, is process-safety related, and is required for risk reduction, it is SCAI.

When considering use of BPCS for SCAI, the first thing to remember is that risk reduction is not free. It requires effort. Getting an order of magnitude risk reduction from the BPCS is hard. Independence and separation require special design and management practices, focusing on eliminating single points of failure and human error. Claiming two orders of magnitude from two independent BPCS controllers must be supported with quantitative analysis (such as fault tree) that accounts for real world performance and human errors. Claiming two orders of magnitude from a single BPCS controller requires that the BPCS be designed and managed according to IEC 61511.

7. References

[1] CCPS (2007), Guidelines for Safe and Reliable Instrumented Protective Systems, Hoboken, NJ, John Wiley & Sons.

[2] CCPS (expected 2013), Guidelines for Initiating Events and Independent Protection Layers, Hoboken, NJ, John Wiley & Sons.

[3] ANSI/ISA 84.91.01 (2012), Identification and Mechanical Integrity of Safety Controls, Alarms, and Interlocks in the Process Industry. Research Triangle Park, NC, Print.

[4] ANSI/ISA 100.11a (2011), Wireless system for industrial automation: Process control and related applications, Research Triangle Park, NC, Print.

[5] IEC 61511 (2003), Functional safety - Safety instrumented systems for the process industry sector, Geneva, Switzerland, Print.

[6] ANSI/ISA 18.2 (2009), Management of Alarm Systems for the Process Industries, Research Triangle Park, NC, Print.

[7] ISA TR84.00.02 (2002), Safety Instrumented Functions (SIF) Safety Integrity Level (SIL) Evaluation Techniques, Research Triangle Park, NC, Print.

[8] ISA TR84.00.03 (2012), Guidance for Testing of Process Sector Safety Instrumented Functions (SIF) Implemented as or Within Safety Instrumented Systems (SIS), Research Triangle Park, NC, Print.

[9] ISA TR84.00.04 (2012), Guidelines for the Implementation of ANSI/ISA-84.00.01-2004 (IEC 61511 Mod), Research Triangle Park, NC, Print.

[10] CCPS (2001), Layer of Protection Analysis: Simplified Process Risk Assessment, Hoboken, NJ, John Wiley & Sons.

[11] ISA TR84.00.09 (expected 2014), Electrical/Electronic/Programmable Electronic Systems (E/E/PES) for Use in Process Safety Applications Security Protection Layers and Considerations Related to SIS, Draft 7.4, Research Triangle Park, NC, Print.

[12] ANSI/ISA 99.00.01 (2007), Security for Industrial Automation and Control Systems, Research Triangle Park, NC, Print.

[13] SINTEF (2010), Reliability Data for Safety Instrumented Systems Data Handbook, Trondheim, Norway, Gruppen.

Journey to Process Safety Excellence

Shanahan Mondal
Marathon Petroleum Company LP
Texas Refining Division
502 10th Street South
Texas City, TX 77590
smondal@marathonpetroleum.com

Brian McPherson
Marathon Petroleum Company LP
Texas Refining Division
bmmcpherson@marathonpetroleum.com

Prepared for Presentation at
American Institute of Chemical Engineers
2013 Spring Meeting
9th Global Congress on Process Safety
San Antonio, Texas
April 28 – May 1, 2013

UNPUBLISHED

Journey to Process Safety Excellence

Shanahan Mondal
Marathon Petroleum Company LP
Texas Refining Division
502 10th Street South
Texas City, TX 77590
smondal@marathonpetroleum.com

Brian McPherson
Marathon Petroleum Company LP
Texas Refining Division

Keywords: process safety leadership, process safety excellence

Abstract

This is a perfect time to call for action and focus on the reduction of process safety events. An organization that wants to achieve process safety excellence must look beyond compliance. This paper details how Marathon Petroleum Company's (MPC) Texas City refinery took on several initiatives to move the needle on process safety in the right direction.

1. Introduction

The refining and petrochemical industries have come a long way since the OSHA PSM standard was enacted in 1992. However, we are still seeing process safety events take the toll on the lives of our loved ones and property. There is still progress that we can make and therefore now is a good time for a call for action to further reduce process safety events. What are some ways we can proactively take steps towards achieving "process safety excellence"? Strong performing process safety programs are on the right path to achieving excellence.

This paper illustrates how a small Gulf Coast refinery embarked on the journey toward process safety excellence. Marathon Petroleum Company's Texas City refinery has a 76,000 bpd throughput and is one of seven refineries within the company. In order to be a sustainable business, the Texas City refinery must make efficient use of the people and resources. This story of the MPC Texas City refinery takes us several steps beyond process safety compliance. Folks throughout the refinery were able to embrace several new initiatives in process safety with that "can do" attitude Texans are known for. As a team this small plant has helped move the needle on process safety within our industry.

The main sections of this paper include:

- Everyone is a Process Safety Leader!
- Employee engagement
    - LOPC reduction efforts

    - Process Safety Bulletins
    - Process Safety Roundup
    - Celebrating our wins
- Industry outreach
    - API site assessment
    - AFPM Regional networks

2. Everyone is a Process Safety Leader!

2.1 Why is this important?

Many organizations have set aggressive goals to have ZERO process safety events. For these goals to develop momentum and come to fruition, they must reach the folks in the company that "have the boots on the ground" and are actively engaged in regular process safety activities. These activities include:

- Operating process units within safe operating limits
- Performing tasks to support the mechanical integrity of process equipment
- Identifying and mitigating hazards through process hazard analysis
- Auditing process safety programs for compliance with OSHA regulations
- Performing management of change (MOC)

An organization in which the leadership team empowers the entire workforce to be process safety leaders is one that is on the path to excellence.

2.2 Background

MPC partnered with Stars International to develop a "Leadership Skills for Extraordinary Safety Performance" course several years ago. Marathon employees and contractors participated in this "Safety 1" course and learned new ways to promote personal safety. One of the fundamental concepts taught in this class was that effective communication skills play a key role in achieving world class personal safety programs. The course focused on participants being able to identify which of the four personality quadrants they tended to define as their comfort zone: Controller, Thinker, Expressive, or Team Builder. The participants were then taught how to communicate effectively with others that were in different personality quadrants.

This course taught MPC employees and contractors various leadership skills to promote personal safety. Some of these include the following:

- Permission and pledge : "You have my permission to stop me if I am doing something unsafe and I pledge to give you the same feedback"
- Be in the moment : Safety outcomes are a function of choices we make in the moment
- Coaching others to be safe

The Safety 1 basic skills and refresher courses have been the foundation of Marathon's personal safety programs for the past five years. Marathon saw how this course was able to shift the personal safety culture in a positive direction.

2.3 What was done?

MPC partnered again with Stars International to build on this foundation but this time shifted the focus to process safety. The basic tenets of effective communication among different personalities still hold true to improve process safety. The key learning from the new course was that "Everyone is a Process Safety Leader! "

MPC Texas City hosted a pilot of this Process Safety Leadership course. A diverse group of employees were chosen as course participants. The refinery manager, department managers, refinery operators, lab personnel, training specialists, and support personnel were examples of those represented in the student demographics. The intent of inviting a diverse group of participants was so they could accept that process safety was everyone's responsibility.

The course highlighted a need to maintain a sense of vulnerability. Without a sense of vulnerability, complacency can set in. To illustrate how vulnerable we were, the course used a large scale Swiss cheese model to visually demonstrate how holes in the layers of protection can lead to catastrophic process safety events. Examples of real industry process safety events were used to make these personal to the course participants.

Once we had instilled a sense of vulnerability in the participants, they were given the task of figuring out how they play a role in blocking the holes in the Swiss cheese process safety model. We distilled down employee functions within the organization to 15 different categories. Each employee was given a placard with their category and on it there were various tasks they did that impacted the 14 elements of process safety. The "A-ha!" moment came in the course when participants realized what specifically they could do to prevent process safety events. A majority of the course participants found that they could block several holes in the process safety model through different tasks they were associated with. To make this even more personal, each participant was given a cardboard circle with a picture of their face on it. Through different interactive sessions during the class, they put their "hole-blocker" in the Swiss cheese model to cover up the holes. By the end of the class, the diverse group of participants had covered all the holes in the model and prevented catastrophic process safety events!

The take home message for course participants was that everyone can be a process safety leader by realizing that if they are not vigilant, holes can form in the Swiss cheese process safety model.

.

3. Employee Engagement

3.1 Loss of Primary Containment (LOPC) reduction initiative

3.1.1 Why is this important?

A site culture where small leaks are allowed to go unaddressed can lead to more catastrophic process safety events. A sense of vulnerability helps us realize that small leaks must be mitigated expeditiously in order to prevent larger incidents. An organization that focuses on a reduction in loss of primary containment (LOPC) events is on the right track to process safety excellence.

The reporting of LOPC events and subsequent incident investigations to address root causes is helpful in preventing process safety events. Facilities in our industry operate continuously 24 hrs a day 7 days a week. In order to capture the LOPC events that can occur at any given time, it is important to empower the entire shift workforce with the tools to identify and report LOPC events.

3.1.2 Background

Marathon Petroleum has adopted a suite of process safety metrics in accordance with API Recommended Practice 754 "Process Safety Performance Indicators for the Refining and Petrochemical Industries". This practice allows process safety incidents to be classified into different categories as shown in Heinrich's triangle below:

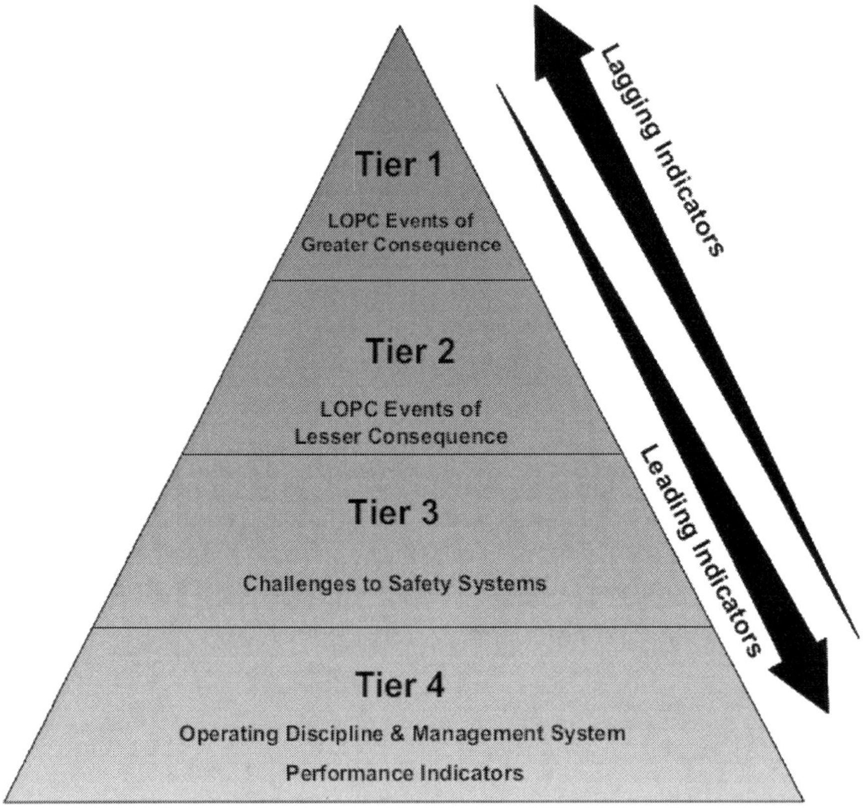

The PSE Tier 1 and Tier 2 events are lagging indicators of process safety while the Tier 3 and Tier 4 incidents are leading indicators. Small LOPC events typically are categorized in the Tier 3 bucket. Tier 3 events are leading indicators that could have led to more severe PSE Tier 1 and 2 events if conditions were different. If we keep highly hazardous materials in the pipes, there is less of a chance they will lead to the catastrophic process safety events.

MPC's personal safety culture is such that first aids and near misses are promptly reported. We have attributed the reporting of first aid injuries to preventing more severe personal injuries. Process safety is similar to personal safety in this sense. One can think of the LOPC Tier 3 incidents as precursors to Tier 1 and Tier 2 events. Focusing on keeping material within its proper designated process boundaries (eg. piping and vessels) can help prevent the catastrophic process safety events.

3.1.3 What was done?

MPC Texas City adopted a slogan "Let's Keep it in the Pipes!" to keep LOPC reduction in the forefront of the minds of employees. We developed a hardhat sticker contest for employees and contractors to participate in. They were to design a catchy logo to match the "Let's Keep it in the Pipes!" theme. We received several original submissions from all levels of the organization. Here is the logo Marathon Texas City ended up selecting:

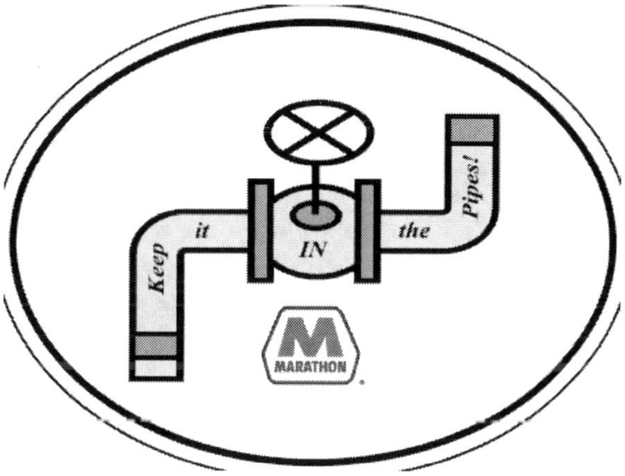

MPC Texas City employees and contractors see this logo and slogan on process safety communication and understand our focus is on reducing LOPC events. We have explained to our Marathon Texas City family that prompt reporting of small leaks and spills is the first way to prevent larger more severe LOPC events. We have seen the culture shift in the reporting of these small leaks and spills. Folks understand that by investigating the small leaks and spills we can help prevent the Tier 1 and Tier 2 process safety events from occurring.

3.2 Process Safety Bulletins

3.2.1 What does this mean?

One of the key aspects of a good Process Safety Management program is effective communication. The lessons learned from incident investigations can be turned into teachable moments with the workforce. Repeat events in our industry are a good justification for a process safety communication tool.

3.2.2 Background

MPC Texas City focused on communicating process safety Tier 3 incidents with the workforce. The intent was to develop a catchy medium through which we could grab the attention of the casual reader. A second challenge was finding a vehicle to deploy the process safety initiative through. Efforts were made to tie into our current Health, Environmental, Safety, & Security communication initiatives. A stand alone process safety initiative is often received as "flavor of the month" and may not make the desired impact on the workforce.

3.2.3 What was done?

In order to strengthen our communication of PSM, the decision was made to generate an "Event Sharing" tool at MPC Texas City. This would become known as the Process Safety Bulletin. Creating a monthly bulletin to highlight a short article on an element of PSM helped give a face to each element team. Additionally, the Process Safety Bulletin captured the refinery's PSE Tier 3 metrics to identify weaknesses which were likely contributing to loss of primary containment (LOPC) events. Finally, it conveyed a few process safety related incident investigations and gave the reader a visual picture of the incident, short description of what happened, and a take home "What can YOU do" message.

The Process Safety Bulletin was incorporated into the site's monthly Responsible Care Safety Meeting which allowed all personnel at the site to review within their workgroups. The impact of the bulletin was notable at Marathon Texas City. It was very well received by all; each month the bulletin was new, fresh, and exciting. As the development progressed and word got out, the bulletin became a company-wide PSM Best Practice and was shared with PSM groups throughout the company. This initiative has more than created a spark to our Process Safety Culture and has advanced our site further towards achieving process safety excellence.

3.3 Process Safety Roundup

3.3.1 What does this mean?

The journey toward process safety excellence is not just the journey of a single individual. All employees and contractors within the organization are on this journey together. Excellence can be achieved when folks are working together on process safety initiatives in unison – much like an engine firing on all cylinders. Just like any journey in life, it's important to pull over periodically and make sure you are headed in the right direction. Marathon Texas city's Process Safety Roundup activity provided such a venue for the reflection on process safety.

3.3.2 Background

MPC Texas City has implemented the concept of process safety element teams. An element team has an "Element Champion" which is often a department manager. The element champion leads a team of 4-5 folks that consist of subject matter experts, process safety engineers, and hourly workforce members. The purpose of these element teams included support of the following activities:

- Development of element specific goals

- Development of element specific leading and lagging indicators
- Identification of gaps between Marathon corporate policies and Texas City site policies
- Performing audits at the refinery to ensure we are doing what we say we will do in our PSM policies

As you can see these are teams with aggressive agendas that must work outside their traditional silos when it comes to process safety activities. Often a site's goals are shared across several element teams. Due to these shared goals, a forum through which PSM element teams could learn from one another was in high demand.

3.3.3 What was done?

For the past two years, MPC held an annual "Texas City PSM Round-Up." This meeting was held off-site in order to have the PSM element teams meet face to face without day to day distractions. In November 2012, approximately fifty element team members attended. The event was designed to facilitate communication styles, to discuss past accomplishments, and to establish future goals. The event also helped team members understand the goals of other PSM element teams and to gain a better perspective on MPC corporate and Texas City specific goals. As an extension of this effort, each team will be featured in the Texas City Process Safety Bulletin to make sure that the workforce understands who the teams are and what they are designed to achieve.

In 2012 we carved out a dedicated slot of time on the PSM Round-Up agenda to focus on one specific element. Mechanical Integrity (MI) was the spotlight last year. During this discussion we passed around a piece of piping that had failed due to corrosion in the presence of residual elements. This visual prop really engaged the audience and gave them a better understanding of the site MI programs.

3.4 Celebrating our Wins with our Family

3.4.1 What does this mean?

At MPC Texas City we have developed a sense of family among our Marathon and contractor employees. When we accomplished big goals, we have done them as a team. In order to have continued success in future initiatives, everyone in the organization should feel that they are valued for their contributions.

As mentioned earlier, occasional glances in the rearview mirror help us understand where we came from and give us purpose to head in the right direction. The same is true on the journey toward process safety excellence. The solution to the process safety puzzle is not just those two guys in the corner offices in the PSM department – it involves all employees and contractors. We all play a critical piece in the puzzle and as an organization we should thank them for their contributions.

3.4.2 What was done?

In the fall of 2012 MPC Texas City went one entire year with ZERO PSE Tier 1 events. This was a big success for the site and we wanted to celebrate it. It took all of the employees and contractors doing their process safety roles right each time to reach this milestone.

To acknowledge everyone's involvement in getting us to this milestone we did the following:

- Held a BBQ lunch for all MPC Texas City employees and contractors to say thanks
- Distributed the "Let's Keep It In the Pipes ! " stickers during the event
- Distributed a tri-fold pamphlet to further educate folks on the various tiers of process safety events
- At the end of the event, both employees and contractors signed commitment posters for the process safety elements that they felt related to their roles in the refinery.

4. Industry Outreach

4.1 Defining what "good" looks like

4.1.1 What does this mean?

Industry wide process safety statistics can be used to benchmark a site's process safety performance. A more meaningful tool is a thorough audit of a site's process safety management systems by a team with industry experience. In addition, a field audit to verify that the site is executing what is in the policies and procedures can help identify gaps in process safety. Both these activities can identify opportunities for improvement. When selecting an external process safety audit program, focus should be made to help ensure the external team has the right industry experience level.

4.1.2 Background

In 2012 the American Fuel & Petrochemical Manufacturers (AFPM) and the American Petroleum Institute (API) rolled out a suite of products to advance process safety in the industry. These included the following:

- Event sharing
- Process safety metrics
- Hazard Identification
- Regional Networks
- Training Certification
- Process Safety Site Assessment

The intent of the Process Safety Site Assessment (PSSA) was to have a team of assessors with significant industry experience assess the following areas of process safety:

- Process safety leadership
- Management of Change / Pre-Startup Safety Review
- Mechanical Integrity
- Safe Work practices
- Facility siting

- Process Hazard Analysis
- Operating Practices
- HF Alky RP 751

4.1.3 What was done?

Marathon Texas City volunteered in June 2012 to pilot the Process Safety Site Assessment. The motivation behind this was to see how we compared to how industry had defined what "good looks like". Industry through the API trade association had developed a set of protocol questions for each of the eight process elements being assessed. MPC Texas City compiled sets of answers on how the site would answer those questions. In addition, Marathon Texas City created "evidence binders" to support how they answered the protocol questions. These binders were very helpful during the assessment since the assessors were able to make more efficient use of their week long time at the facility.

The term "assessor" was used rather than the term "auditor" since the idea here was to instill a sense of learning during the assessment. During the week, Marathon subject matter experts were able to rub elbows with approximately 300 years of industry experience. We learned where our gaps were in our management systems and tactical execution of our policies and procedures. By working closely with the experienced assessment team, we received guidance on how these issues could be resolved in industry. The overall learning from this site assessment for Marathon Texas City was immense. Marathon Texas City now had a tool to identify what industry had defined as "best in class" and had an action plan in place to get to that level.

4.2 Sharing with Industry

4.2.1 What does this mean?

In order to move toward process safety excellence, an organization should learn from their peers in industry. Small companies with limited resources could seek out these networking opportunities. Through this networking, information can be shared regarding practices to prevent process safety events.

4.2.2 What was done?

MPC Texas City actively participated in the AFPM's Texas Gulf Coast Regional Network meetings. This meeting was an effective forum to share process safety topics among industry stakeholders. Texas City was able to share the following with their peer group:

- Value of participating in the Process Safety Site Assessment
- Marathon Petroleum Company's "Process Safety Advisory" as a means to communicate industry and company process safety events
- Interactive sharing of Texas City specific events

5.0 Conclusion

In order to move the needle on process safety in the right direction, an organization should continue to look for creative ways to go beyond traditional compliance activities. This paper has highlighted some of the activities that a small Gulf Coast refiner like Marathon Petroleum Company's Texas City has participated in to improve process safety. We value the sharing of what has worked for us and encourage our peers in industry to continue this exchange of information.

Management of Safety Critical Elements as a Base for Risk Management of Major Accident Hazards

Mariana Bahadian Bardy
Det Norske Veritas
Rua Sete de Setembro 111/12th floor
mariana.bardy@dnv.com

Flávio Luiz Barros Diniz
Det Norske Veritas
flavio.diniz@dnv.com

Paula Silveira
Det Norske Veritas
paula.silveira@dnv.com

Prepared for Presentation at
American Institute of Chemical Engineers
2013 Spring Meeting
9th Global Congress on Process Safety
San Antonio, Texas
April 28 – May 1, 2013

UNPUBLISHED

Management of Safety Critical Elements as a Base for Risk Management of Major Accident Hazards

Mariana Bahadian Bardy
Det Norske Veritas
Rua Sete de Setembro 111/12th floor
mariana.bardy@dnv.com

Flávio Luiz Barros Diniz
Det Norske Veritas
flavio.diniz@dnv.com
Paula Silveira
Det Norske Veritas
paula.silveira@dnv.com

Keywords: safety critical element, barrier, major accident hazard

Abstract

Considering the already established relevance of barriers to avoid Major Accidents, the objective of this paper is to present a methodology for management of Safety Critical Elements (SCE), from the identification of them, definition of relevant importance to each activity performed by the installation and establishing alternatives and contingencies for the failure or absence of the SCE. The proposed methodology, adapted from common use methodologies from Offshore Industry to Process Industries, is developed in 5 steps, being Step 1 the use of a Hazard Identification technique and indication of Major Accident Hazards. Following on Step 2 bowtie diagrams are developed for the MAH and SCE are identified. The SOOB – Summary of Operational Boundaries on Step 3 identifies the activities that may or may not proceed or caution is applied in the case the SCE is defeat and on Step 4 a Contingency Plan is develop to maintain operation for the cases indicated on the SOOB that operation may not proceed or proceed with caution. Finally, on Step 5, definition of prioritization of maintenance and inspection activities shall be in place for each SCE, including preventive maintenance routines, inspections routines and definition of spares, where applicable. This methodology can help on the identification of gaps and management of critical elements and consequently improving the performance of safety systems increasing their availability.

1. Introduction and background

Recent accidents have indicated the importance of safety barriers on management of major accidents, reducing its likelihood or minimizing consequence. Buncefield, Texas City and Macondo, just to indicate some, have stated in their accident investigation reports the failure of safety barriers or non-existence of adequate ones as potential causes for the major accident occurrence.

This paper presents a methodology for management of Safety Critical Elements (SCE), adapted from common use methodologies from Offshore Industry to Process Industries, from the identification of them, definition of relevant importance to each activity performed by the installation and establishing alternatives and contingencies for the failure or absence of the SCE.

Several reference have definitions of SCE and how they must be managed, as NORSOK[1], that indicated that Safety Critical Equipment is equipment that shall be in operation to ensure escape, evacuation and /or to prevent escalation.

According to HSE UK[2] "any structure, plant, equipment, system (including computer software) or component part whose failure could cause or contribute substantially to a major accident is safety critical, as is any which is intended to prevent or limit the effect of a major accident".

For this paper, SCE is defined as indicated by HSE UK, as being the Barriers that can avoid or mitigate Major Accident Hazards.

2. Description of Methodology

For the objective of systematic management of Safety Critical Elements, the methodology outlined in Figure 1 is proposed, covering the 5 steps described below.

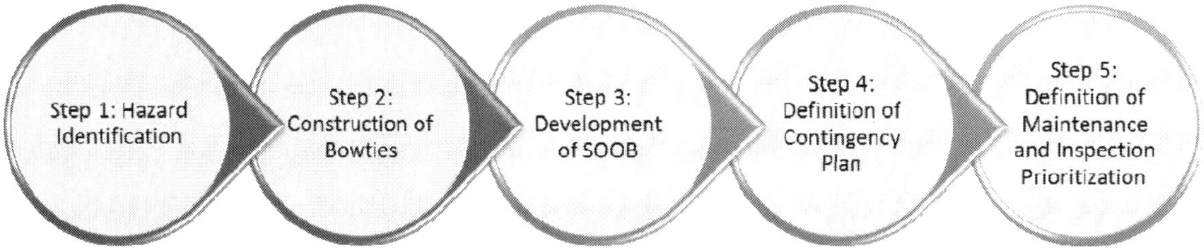

Figure 1 – Methodology for SCE Management

2.1 Step 1 – Hazard Identification

The first step is to identify the accidental scenarios from the specific process under analysis. For that purpose, it is proposed to perform a Process Hazard Analysis (PHA) for identification of accidental scenarios and classification according to a Risk Matrix, defined by each company according to its risk management process. Figure 2 represents an example of spreadsheet to be applied for the PHA.

Process Hazard Analysis (PHA)											
System:				Hazard/Event Group:							
1.Hazard	2.Causes	3.Effects	4.Freq	5.Sev	6.Risk	7.Safeguards	8.Final Freq	9.Final Sev	10.Final Risk	11.Recommendations	12.#
											1
											2
											3

Figure 2 – Example of PHA Spreadsheet

The spreadsheet has 12 columns, and two classification of the risk for each scenario. Columns 4, 5 and 6 have the classification without considering existing safety barriers for the scenario. The barriers are listed on Column 7 and Columns 8, 9 and 10 indicate the classification of the risk, considering the existence of the barriers and these are operating or ready to operate when needed.

For the classification of severity, likelihood/frequency and risk, a risk matrix shall be used, representing the risk tolerability of the company. An example of risk matrix is on Figure 3, extract from ISO 17776:2000 [3].

Consequence					Increasing probability			
Severity rating	People	Assets	Environ-ment	Reputation	A	B	C	D
					Has occurred in E&P industry	Has occurred in operating company	Occurred several times a year in operating company	Occurred several times a year in location
0	Zero injury	Zero damage	Zero effect	Zero impact				
1	Slight injury	Slight damage	Slight effect	Slight impact	Manage for continued improvement			
2	Minor injury	Minor damage	Minor effect	Limited impact				
3	Major injury	Local damage	Local effect	Considerable impact				
4	Single fatality	Major damage	Major effect	Major national impact	Incorporate risk-reducing measures		Fail to meet screening criteria	
5	Multiple fatalities	Extensive damage	Massive effect	Major international impact				

Figure 3 – Example of Risk Matrix

Note that this matrix has 5 different severity ratings and analyze four different effects: people, assets, environment and reputation. A common approach to define Major Accidental Hazards (MAH) is to consider whose with highest consequence classification, as the ones classified with Severity Category 5, from the matrix indicated on Figure 3, that represents multiple fatalities as impact on people, extensive damage for environment, massive effect to assets and major international impact on reputation.

The main advantage to select the MAH to go to Bowtie, as described above, is that the barriers related to those events can be clearly identified and consequently be managed properly and in a focused way. On the other hand, when there is no distinction between MAH and other scenarios, with lower damage potential, the number of barriers to be managed increase reducing the focus on the major impact scenarios MAH.

Note that some safety barriers are normally identified on this PHA and shall be reviewed and detailed on the next steps.

2.1 Step 2 – Development of Bowties

The following step of this methodology is to develop bowtie diagrams for each of the MAH, or combination of MAH, if applicable, as exemplified in Figure 4. The BowTie methodology is designed to give a picture of the risks, to help people understand the relationship between the risks and organizational events and to identify where barriers in place can act, on the prevention or on the mitigation and consequently give a better overview if those are enough to mitigate the risks related to the MAH.

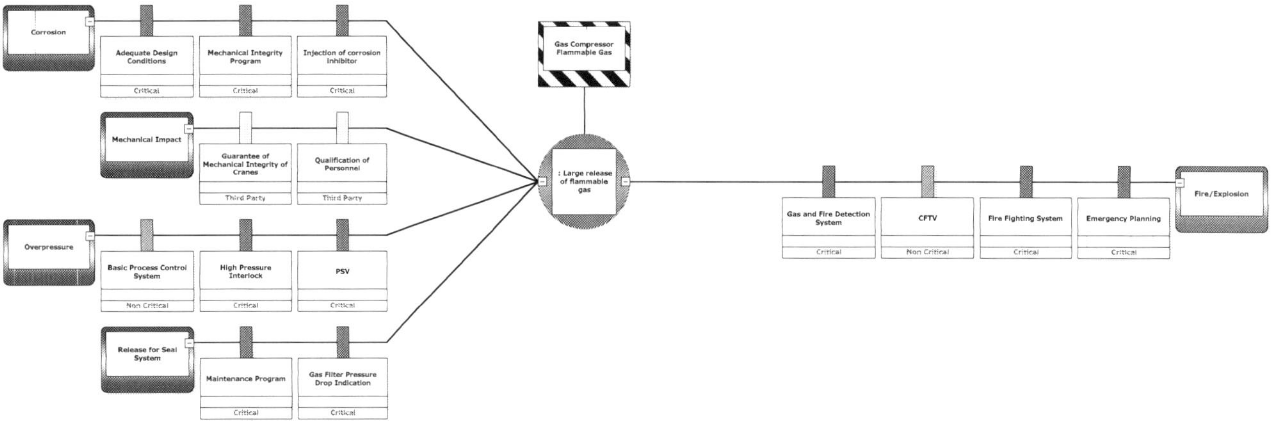

Figure 4 – Example of Bowtie

With the use of multidisciplinary team from the company, starting from a Top Event, localized in the center of the diagram, causes, preventive barriers, consequences and mitigating barriers are identified. Following, each barrier, preventive or mitigating, is classified as:

- Critical: essential barrier to avoid the causes or associated consequences.
- Non-critical: barrier that reduces likelihood or minimize consequence, but does not avoid the occurrence of the top event or associated effects.

- Third Party: barriers, critical or not, that are not under the company responsibility for management.

Also responsible person or function can be indicated on the bowtie for each barrier. The list of SCE are composed by those barriers classified as critical for each bowtie.

The Safety Critical Elements (SCE) can be an Equipment, System or Procedure. In the example presented in Figure 4, for Top Event *Large Release of Flammable Gas, from Compression System*, the following barriers were classified as Safety Critical Equipment or System:

- Safety interlocks
- PSVs
- Filter Pressure Drop Indication
- Injection of Corrosion Inhibitor
- Gas and Fire Detection System
- Fire Fighting System
- CFTV

The other critical barriers such as Mechanical Integrity Program and Emergency Planning are considered as Safety Critical Procedure.

All of those critical elements, equipment, systems and procedures, shall be managed but specially for the equipment and systems contingency procedures shall apply considering they are operating under degraded conditions or out of operation. As part of this scope a Summary of Operations Boundaries (SOOB) analysis is carried out as stated below.

2.3 Step 3 – Development of SOOB

The Step 3 of this methodology consists in developing the Summary of Operations Boundaries - SOOB analysis. This is based on a matrix which crosses main operations and activities with the Operational Risk Factors. Operational Risk Factors includes controls, identified in the BowTie analysis, under reduced effectiveness and risk factors such as severe weather/sea conditions. The matrix is completed row by row by reviewing all combinations.

The main objective is to examine if operations can be permitted or prohibited when certain controls have been defeated or running under reduced effectiveness and examine if operations can proceed in the case of occurrence of external factors that can potentially influence the risk of doing these operations e.g. severe weather conditions.

This will distinguish when a "stop work" is applied or if that shall be a "proceed with caution" condition, as indicated by IADC[4]. A "traffic light" system may be applied, indicating:

- Red: stop the work or do not proceed;
- Yellow: evaluate conditions, perform risk analysis or implement additional protection;
- Green: continuous normal operation.

Note that the activities will vary depending on the type of installation. Some examples are:

- loading or unloading of trucks or railcars;
- operation above normal conditions;
- increase of capacity;
- confined space entry;
- working at height.

An example of analysis is for Gas Detection System failure, it is allowed to proceed with works at heights and confined space entry; with caution, requiring additional evaluation normal operation and loading/unloading; and not permitted operation above normal condition, increase of capacity and hot work.

Operations vs. Operational Risk Factors	Normal Production	Operation Above Normal Conditions	Increase of Capacity	Loading/Unloading Truck	Loading/Unloading Railcar	Confined Space Entry	Hot Work	Working at Heights
· Safety interlocks	RA	X	X	RA	RA	P	RA	RA
· PSVs	RA	X	X	RA	RA	P	P	RA
· Filter Pressure Drop Indication	RA	RA	RA	NA	NA	NA	NA	NA
· Injection of Corrosion Inhibitor	RA	RA	RA	NA	NA	NA	NA	NA
· Gas and Fire Detection System	RA	X	X	RA	RA	P	X	P
· CFTV	P	RA	RA	RA	RA	P	P	P

P- Permitted
RA – Perform Risk Analysis
X- Do not Proceed

Figure 5 – Example of SOOB Matrix

2.4 Step 4 – Definition of Contingency Plan

The Step 4 of this methodology consists on the definition of Contingency Plan for each SCE. The immediate response action that will normally be applied are:

- to stop or limit operations to within the limits of remaining barriers; or

- identify and assess any temporary substituted safety system barrier that may be implemented to support continued operation.

The company shall establish and document contingency procedures and a system of approval and control of SCE to be used when those are under degraded conditions or out of operation.

The following items shall be considered:

- Implementation of alternative controls equivalent;
- Limitation and reduction of production;
- Isolation and stopping of equipment, systems, installations;
- Deadline for the temporary procedure to be allowed until corrective measures are taken.

A specific contingency plan is then developed for each SCE, using, for instance, the example indicated in Figure 6.

<table>
<tr><th>SCE</th><th>Permitted Acitivities</th><th>Activities with Restriction</th><th>Prohibited Activities</th></tr>
<tr><td>Gas Detection System</td><td>Confined Space Entry
Working at heights</td><td>Normal Production
Loading/Unloading</td><td>Hot work
Operation above normal conditions
Increase of capacity</td></tr>
<tr><td>Permitted Activities:</td><td colspan="3">No limitation for the development or continuation of activity, event with loss of the SCE</td></tr>
<tr><td>Activities with Restriction:</td><td colspan="3">No limitation for the development or continuation of activity, event with loss of the SCE</td></tr>
<tr><td>Prohibited Activities:</td><td colspan="3">Not allowed to perform the activity and must be interrupted, even with</td></tr>
<tr><th colspan="4">Alternative Procedures for Activities with Resctriction</th></tr>
<tr><th colspan="2">Activity</th><th>Deadline</th><th>Responsible</th></tr>
<tr><td colspan="2">Normal production to continuous with one extra Operation Supervisor per shift, with focus on Control Room supervision</td><td>One month</td><td>Operation Manager</td></tr>
<tr><td colspan="2">If SCE not returned to full operation after first deadline, reduce production and safe stop production</td><td>-</td><td>Operation Manager</td></tr>
<tr><td colspan="2">Performed loading/unloading activities with one extra fiel operator</td><td>One month</td><td>Operation Manager</td></tr>
<tr><th>SCE Resposible</th><td colspan="3">Maintenance Manager</td></tr>
</table>

Figure 6 – Example of Contingency Plan for SCE

2.4 *Step 5 – Definition of Maintenance and Inspection Prioritization*

The final step for implementation of this methodology of Management of SCE is to incorporate on maintenance and inspection routines and procedures prioritizations that will consider the findings of the analysis of the SCE. Some important points shall be considered:

- Guarantee that all SCE are classified as high priority for maintenance routines;
- Guarantee no delays for inspection routines for the elements associated with MAH and classified as SCE;
- Evaluate the need of spares of SCE, where applicable.

3. Conclusion

As initially indicated, this paper presents a 5-step methodology for management of SCEs, being those defined here as safety barriers that can avoid or mitigate Major Accident Hazards. The objective of each step as well as practical approach and examples are presented, adapting common use methodologies from Offshore Industry to Process Industries.

As extension of this work, considering all aspects for the presented, some improvements can be implemented. The inclusion of procedures as part of the analysis, after the identification of the critical procedures, with guarantee of correct training or certification of operators, is one of these points. One additional relevant aspect is to incorporate a 6th step on the above methodology of management of SCE, with the audit of process of management of the critical barriers.

Finally, it is importance to note that, this methodology was developed with the intention of giving support for companies to systemic manage Safety Critical Elements and comply with relevant regulation and best practices.

4. References

[1] Norsok S-001, Edition 4, February 2008, item 3.1.11. Norway. 2008

[2] Health and Safety Executive, A guide to the Offshore Installations (Safety Case) Regulations 2005, item 83. London. 3rd Edition. 2006.

[3] ISO 17776:2000, Petroleum and natural gas industries — Offshore production Installations —Guidelines on tools and techniques for hazard identification and risk assessment, Table A.1. Geneva. 2006.

[4] IADC, HSE Case Guidelines for Mobile Offshore Drilling Units, Issue 3.2.1, 2009.

Promoting Commitment to Process Safety

Russell A. Ogle,
Andrew R. Carpenter and Sean J. Dee
Exponent, Inc.
4580 Weaver Parkway, Warrenville, IL 60555
rogle@exponent.com, acarpenter@exponent.com, and sdee@exponent.com

Prepared for Presentation at
American Institute of Chemical Engineers
2013 Spring Meeting
9th Global Congress on Process Safety
San Antonio, Texas
April 28 – May 1, 2013

UNPUBLISHED

Promoting Commitment to Process Safety

Russell A. Ogle,
Andrew R. Carpenter and Sean J. Dee
Exponent, Inc.
4580 Weaver Parkway, Warrenville, IL 60555

Keywords: process safety management, safety culture, game theory

Abstract

Successful implementation of process safety management (PSM) typically requires the use of a combination of physical and procedural safeguards. Ideally, these safeguards must be maintained in a constant state of readiness throughout the life of the process facility. The management system provides the framework for maintaining and sustaining these safeguards, but the critical ingredient that ensures that the work gets done is the individual employee's commitment to PSM.

Game theory provides a useful framework for understanding how an organization can foster cooperation between management and its employees towards the common goal of safety. Using a specific game-theoretical model known as the repeated Prisoner's Dilemma, it is shown how cooperative behavior is the natural consequence of a balanced system of rewards and penalties. Here the term "balanced" means that both employees and their managers are subject to this system of rewards and penalties. By implementing a balanced framework for feedback, an organization can promote a commitment to process safety by fostering cooperation between both the employees and the managers.

1. Introduction

Successful implementation of process safety management (PSM) typically requires the use of a combination of physical and procedural safeguards. Ideally, these safeguards must be maintained in a constant state of readiness throughout the life of the process facility. The management system provides the framework for maintaining and sustaining these safeguards, but the critical ingredient that ensures that the work gets done is the individual employee's commitment to PSM. The social sciences, especially psychology and organizational behavior, can provide valuable insight towards understanding the management of employee commitment to safety. However, other fields of study can make useful contributions to our understanding. Game theory is one of these tools.

Game theory is the scientific study of interactive decision-making, and is especially useful in understanding the interplay of conflict and cooperation between different stakeholders. Consider the interaction between the individual employee and the management organization. The objective of both management and the individual employee is to secure a strong commitment to process safety. But there are many time and attention demands placed on the individual employee, and numerous financial pressures on management. There can be a strong temptation to cut corners and sacrifice operational discipline in favor of simplifying a task and getting it done faster or cheaper. In game theory, this situation is known as the Prisoner's Dilemma. The Prisoner's Dilemma is one of the simplest models of the conflicts that can arise between the goals of an individual and an organization.

Game theory provides the tools which can help an organization to understand how to foster cooperation between management and its employees to achieve the common goal of safety. Using a specific game-theoretical model known as a repeated Prisoner's Dilemma, it is shown how cooperative behavior is the natural consequence of a balanced system of rewards and penalties. Here the term "balanced" means that both employees and their managers are subject to this system of rewards and penalties. By implementing a balanced framework for feedback, an organization can promote a commitment to process safety by fostering cooperation between both the employees and the managers.

2. Seeking Commitment to Process Safety

The effective control of the risks associated with a chemical process requires more than just physical and procedural safeguards. It requires people who demonstrate a continuous effort to sustain these safeguards. In short, it requires commitment to process safety at all levels of the organization: employees, supervisors, and managers. The fundamental role of commitment to successful process safety management is discussed at length in the book, *Guidelines for Risk Based Process Safety*, by the Center for Chemical Process Safety (CCPS), a directorate of the American Institute of Chemical Engineers (AIChE) [1]. The CCPS risk based process safety paradigm is based on four pillars:

- Commit to process safety
- Understand hazards and risk
- Manage risk
- Learn from experience

The first component, the commitment of the organization to process safety, is perhaps the most fundamental. It is our thesis that a useful measure of this commitment is through the individual's operational discipline in executing their assigned process safety responsibilities.

Why is commitment to process safety important? Commitment is important because it is the individual's personal pledge to perform their assigned responsibilities. There are always going to be unanticipated or non-routine aspects to any work assignment. Individual commitment to process safety transcends simple compliance with a written work order or procedure. When confronted with the unexpected, it is the individual's commitment to process safety that spurs them to ask the important questions: 1) What can go wrong? 2) How bad is it? 3) How likely is it? 4) What can we do to manage the risk?

A true commitment to process safety introduces a difficult challenge for the employee and her supervisor. Commitment to process safety requires extra effort beyond simply getting the work assignment done. Commitment to process safety requires additional planning and thinking before the work is started. It requires a collaborative work approach with colleagues who are busy with their own work assignments. And, to be perfectly frank, it requires paperwork.

How is commitment achieved with employees, supervisors, and managers? The behavior-based safety literature tells us that what gets rewarded is what gets done [2, 3]. Individuals at all levels of the organization need to believe that they will be rewarded for their commitment to process safety, and furthermore, that they will be penalized if they fail in that commitment. This requires a sustained effort by upper management to observe, evaluate, and administer feedback—rewards or penalties—to the supervisors of the employees. The supervisors, in turn, must also observe, evaluate, and administer feedback. To successfully create and maintain a commitment to process safety from the front-line employees, the employees must see that this commitment will be rewarded, and that a failure in commitment will be penalized. Game theory is a tool that can inform management on how to develop a balanced system of rewards and penalties to achieve this commitment at all levels of the organization.

3. The Repeated Prisoner's Dilemma

Game theory applies mathematical modeling to understand the payoffs and costs associated with two or more individuals making interactive decisions [4]. By understanding the net cost benefit of an individual's actions given the available information regarding the potential decisions of the other players, one can develop strategies and predict behavior based on the mathematical results, and basic human instincts. In particular, the application of game theory has been useful in studying interactions between parties who are in conflict, but would benefit from employing a cooperative strategy. While the mathematics can become very complicated, valuable insights can be gained without heavy reliance on the more complex mathematical modeling components. The exercise of identifying players, considering each player's potential decisions, and determining the payoffs and costs associated with the various outcomes, can provide significant insight in solving problems involving conflict, and structuring optimized solutions.

When considering commitment to process safety, individual employees are left to optimize a multivariable problem. From an organizational level, multiple constraints are imposed on individual employees with respect to their job. First and foremost, employees are expected to perform their jobs tasks adequately, and when posed with the additional requirement of performing the job safely, an employee must sacrifice labor, attention, and thought that could otherwise be spent on their other responsibilities. It is then tempting for an employee to ignore, or shirk, their commitment to process safety. The benefit from shirking is twofold. First, a shirking employee does not need to invest the effort into the process safety management program. But additionally, if the other employees continue to put in the effort associated with maintaining a superior process safety program, the shirking employee also gains the additional benefits that all the employees realize with having a safe workplace. Since commitment to process safety is often an investment to prevent future loss associated with system failures, the benefits of a safe work environment can seem less tangible that an individual employee's day-to-day time and effort. However, if all employees shirk their responsibility, then the process safety management program becomes ineffective, and all the employees only reap the benefit of the additional time and effort that is not spent on their commitment to process safety. In the long run, this makes the employees more susceptible to accidents, and the high costs associated with not having an adequate safety management program. Management has often turned to incentive, reward, and penalty programs to reinforce individual commitment to process safety. The field of psychology is often used to develop these programs [5, 6]. This type of problem has also been extensively discussed in the game theory literature relating to economics and social sciences. Using game theory, management can better understand strategic programs to reaffirm and promote commitment to process safety.

This type of problem, where an individual can benefit by defecting from the beneficial collective action of a group but still reap the group's collective benefits is known in Game Theory as the free rider problem. From the perspective of this work, the free rider problem involves an individual (free rider), who reaps the benefit of a safe work environment (communal resource) without putting in any effort to generate that benefit. A free rider problem can be constructed as a two player game referred to as the prisoner's dilemma. On one axis, the individual employee has the opportunity to Commit to, or Defect from the process safety culture, while the rest of the collective group is represented as a single player on the second axis [7]. The payoffs for committing or defecting for the individual player are represented by the numbers in the two-by-two matrix below. There associated payoffs for the collective group represented by the second player have been omitted for simplicity.

Individual

	Enough employees Commit	**Not enough employees Commit**
Commit	Reward = 3	Sucker = 0
Defect	Temptation = 5	Penalty = 1

The four possible outcomes for the individual employee in this game construction are Reward, Temptation, Sucker, and Penalty. Reward corresponds to both the individual employee and enough other employees cooperating to maintain a superior safety management program. The Temptation provides the highest payout for the individual employee because in addition to reaping the safety benefits of, there are no costs (time, effort) to the defecting employee as others pick up the slack. But, if an insufficient number of other employees carry out their organizational safety responsibilities, then no one, including the individual employee, receives the benefits associated with good operational discipline. Instead, there is a smaller payout (penalty) associated with shirking that represents the employee not investing time or effort into organizational safety. Additionally, if an employee cooperates and fulfills their responsibilities but few others cooperate, then the individual receives the lowest payout and can be termed as a Sucker.

The payoff values shown in the matrix are for illustration only. The rankings are intentionally ordered such that: Temptation > Reward > Penalty > Sucker.[1] The evaluation of payoffs in terms of expected utility is an important subject, but it is quite involved. We refer the reader to decision analysis texts for a discussion of practical strategies for the evaluation of utilities [9]. For our

[1] For technical reasons, it is typically assumed that the payoffs are structured such that Reward > ½ (Sucker + Temptation). This relation ensures that continued cooperation results in a higher total payout than alternation of cooperation and shirking: which simply means that the dilemma cannot be beaten by players taking turns receiving the temptation payout. See Appendix B in the book by Axelrod [8].

purposes, the ranking of the payoffs is sufficient. In addition to the payout rankings, another key concept associated with the prisoner's dilemma is the appearance of a dominant strategy. Regardless of whether the collective group chooses to commit, or defect, the individual employee will also have a higher payout if they defect (5 > 3, 1 > 0). Hence, the strategy of Defect strictly dominates the strategy of Commit.

Repeated interaction has a strong influence on the ability to coordinate the actions of multiple players. In the previously discussed matrix, one would predict the individual employee to defect. However, if there are repeated iterations of the prisoner's dilemma, it stands to reason that the defect strategy would soon dominate the entire population, resulting in the penalty payout. In this situation, one can use the promise of prolonged reward payouts in the future, to outweigh the incentive to defect in the present. This idea, of fostering cooperation and coordination to maximize the benefit for all employees is effective as long as the promise of future interaction is credible. If the game is played a finite number of times, the employee can benefit from shirking when the promise of future interactions has dwindled. If, however, the employee believes that the game will be played repeatedly without end, then the reward of continued cooperation minus the threat of future penalties outweighs the benefit of shirking. It makes more sense for the employee to cooperate.

When the reward for continued cooperation outweighs the incentives to defect, the coordination of action becomes self-enforcing. This is not only true mathematically, but it has also been demonstrated in laboratory and field work associated with the social sciences [8]. Furthermore, as discussed in Schneier's book, the importance tangible, meaningful rewards and penalties to foster coordination out of dominant strategies is critical [10]. Commitment to process safety is more than just a mission statement, or a policy, it requires active participation from players at all levels (employee and management) to reinforce the message, provide credibility, and rewards and penalties directly tied to performance, utilizing leading indicators or process safety.

4. Promoting Commitment through Cooperation

At facilities that handle highly hazardous chemicals, most employees have process safety responsibilities in their job. It is essential that those responsibilities are defined and measures of success are communicated and documented. Supervisors should then conduct observations, both scheduled and random, to verify that these responsibilities are being implemented. Studies have shown that the conduct of supervisory observations leads to higher levels of voluntary employee participation [11].

Process safety observations gathered by the supervisor should be classified into compliant and non-compliant categories. Employees should be rewarded for their compliance with their process safety responsibilities. Rewards should be consistent with the organizations value system and do not need to be expensive. Simple, frequent rewards like public recognition, a free lunch, or a gift certificate can contribute significantly to the employee's motivation for continued success [12, 13]. The key point is that regardless of the type of reward used, it must be delivered in a genuine manner with public recognition. Taken to its ultimate conclusion, employees who are leaders in demonstrating their commitment to process safety should receive special consideration for promotion over those who fail in this commitment. And serious consideration should be given to rewarding the employee's supervisor.

For those who fail in their commitment to process safety must feel the sting of penalties. The penalties should be consistent with the organizations values and should be delivered privately. Only in exceptional circumstances should a penalty be delivered publicly. The idea of the penalty is to encourage the employee to change their behavior. Penalties demonstrate the organization's commitment to process safety as an organizational value [14]. And just as with rewards, depending on the seriousness of the situation, consideration should be given to exacting a penalty on a supervisor with employees who fail to demonstrate the desired level of commitment to process safety.

5. Conclusions

Individual commitment is a cornerstone to a successful process safety management program. This commitment must be maintained at the employee, supervisor, and manager levels to ensure the safety and well-being of all those involved with hazardous processes. Additionally, this commitment requires continual reaffirmation and improvement. However, at times, the commitment to safety can be in direct conflict with financial constraints, deadlines, or production goals. This conflict can lead to individuals to shirk their commitment and make decisions based on factors that are not consistent with the safety management program. In these instances, the use of reward and penalties can reinforce the importance of process safety, and change behavior in impactful ways. Particularly, from a game theory perspective, it is possible to create scenarios where the incentives to ignore one's commitment are discounted to levels that make them insignificant in the day to day operations of a chemical process. By reinforcing positive safety habits, and correcting unsafe behavior, it is possible to improve plant wide process safety.

6. References

1. Center for Chemical Process Safety, Guidelines for Risk Based Process Safety. American Institute of Chemical Engineers, New York, 2007.
2. R.E. Medina, et al. "Behavioral Safety in a Refinery," *Professional Safety*, pp. 36-40, August 2009.
3. E.S. Geller. "Social Influence Principles," *Professional Safety*, pp. 25-31, October 2002.
4. J. Watson. Strategy: An Introduction to Game Theory, Second Edition. W.W. Norton & company, New York, p. 1, 2008.
5. E.S. Geller. The Psychology of Safety. CRC Press, Boca Raton Florida, 1998.
6. T.E. McSween. The Values-Based Safety Process. Wiley, New York 2003.
7. L. Fisher. Rock, Paper, Scissors: Game Theory in Everyday Life. Basic Books, New York, pp. 61-68, 2008.
8. R. Axelrod. The Evolution of Cooperation, Revised Edition. Basic Books, New York, 2006.

9. R.T. Clemen. Making Hard Decisions: An Introduction to Decision Analysis, Second Edition. Duxbury Press, Belmont California, Chapters 13 and 14, 1996.

10. B. Schneier. Liars and Outliers: Enabling the Trust that Society Needs to Thrive. John Wiley & Sons, New York, pp. 1-14, 2012.

11. S. Cook and T.E. McSween. "The Role of Supervisors in Behavioral Safety Observations," *Professional Safety*, pp. 33-36, October 2000.

12. T.R. Krause. "Motivating Employees for Safety Success," *Professional Safety*, pp.22-25, March 2000.

13. P.M. Gendinning. "Employee Safety Incentives: a Best Practices Survey of Human Resource Practitioners," *Professional Safety*, pp. 22-24, February 2001.

14. L.I. Perkinson. "Discipline in the Extremes," *Professional Safety*, pp.31-35, April 2005.

Operational Discipline: Panacea for Performance?

James A. Klein
North America Operations SHE, DuPont

Brian D. Rains
DuPont Sustainable Solutions, DuPont

Prepared for Presentation at
American Institute of Chemical Engineers
2013 Spring Meeting
9th Global Congress on Process Safety
San Antonio, Texas
April 28 – May 1, 2013

UNPUBLISHED

Operational Discipline: Panacea for Performance?

James A. Klein
North America Operations SHE, DuPont

Brian D. Rains
DuPont Sustainable Solutions, DuPont

Abstract

Over the last few years, many companies have had an increased focus on implementing or upgrading programs to improve Operational Discipline (OD) in pursuit of excellent PSM performance, leading to reduced process-related injury and incident rates. OD describes human behavior in following required systems, procedures, and practices to achieve reliable, high quality, and safe operations. DuPont defines OD as a deeply rooted dedication and commitment by each member of the organization to carry out each task, the right way, each time. But is a focus on OD enough? This paper will discuss the elements and precursors of a strong OD program, how leadership can help ensure successful improvement programs, and, ultimately, how a focus on OD and PSM performance must be integrated with operations and business processes throughout the life cycle of the facility.

Keywords: Operational Discipline, Safety Leadership, Process Safety

Introduction

DuPont has for many years defined Operational Discipline (OD) as the deeply rooted dedication and commitment by every member of an organization to carry out each task, the right way, each time [Klein 2005]. The corporate PSM standard requires that:

> *Sites shall implement and maintain programs and systems to achieve and maintain a high degree of OD for all PSM elements in a manner that supports business and operating objectives.*

Specific accountability for improving OD is assigned to line supervision, and sites are required to appoint OD Leadership Champions to help identify and support OD improvement activities. Ideally this person should meet the following criteria:

- A member of site line leadership (not SHE/PSM)
- Have a view of site operations and SHE, not just PSM
- Be able to impact site resources, priorities, and activities

The intent is that the Site OD Champion is a line leader who is able to provide significant leadership and influence site activities in support of improved OD. Successful OD efforts, of course, must involve all employees to achieve desired results.

Relationship of OD to PSM Performance

Once strong PSM systems have been implemented, the greatest opportunity to maintain and improve performance in avoidance of serious injuries and incidents often lies with a focus on improving OD. If people are not following system requirements, at all levels of the organization, then poor performance must result and the risk of serious injury and incidents necessarily increases. Safety programs are only as effective as the day-to-day commitment of everyone in an organization to rigorously follow system requirements correctly every time, changing safety systems from concept to reality in order to achieve desired performance. The failure of just one person in completing a job task correctly, just one time, can initiate or contribute to serious injuries and potentially catastrophic incidents. The design, implementation, and daily execution of effective PSM management systems are dependent on workers at all levels in the organization doing their job tasks correctly every time. Performance is ultimately a function of strong PSM systems and engaged workers rigorously following system requirements.

Are OD Programs Sufficient to Achieve Improved PSM Performance?

If PSM systems have been established for many years, such as for sites covered by the OSHA PSM regulation in the United States, is a new or renewed focus on OD sufficient? Most companies now have or can obtain the technical knowledge and capabilities to safely identify and manage process hazards and risks. As improved guidance and tools for implementing PSM have been developed, the focus on underlying causes of major incidents has shifted in many cases to understanding and improving organizational and cultural causes that relate to OD and other performance factors [Klein and Dharmavaram 2012].

What we've found, of course, is that strong PSM systems, culture, and OD excellence are closely related. Ensuring that many PSM systems are well-designed and implemented as planned is essential to achieving excellent OD. Excellent OD, in turn, is required to develop, implement, and maintain good PSM systems and to daily reflect the safety culture. The safety culture that underlies these efforts ultimately can support strong performance or may need to be improved to help address existing performance issues. This interdependence means simply that a focus on OD as a new or strengthened focus will probably not by itself be successful in achieving desired performance improvements. For example, good OD is certainly dependent on PSM systems and safety culture for development of comprehensive and useful operating procedures, training programs, and overall organizational capability, including fitness-for-duty programs. OD efforts that don't include a focus on these and other precursors as discussed in this paper are likely not going to achieve program goals.

What's required for strong OD efforts?

Strong Safety Culture

PSM/OD programs and performance will only be as effective as the underlying safety culture supports. If an organization has a weak safety culture, PSM efforts will probably be harmed by conflicting priorities, such as production needs or lowered cost. A strong safety culture, based on a commitment to SHE as a core value, and as evidenced by organizational policies, goals, metrics, and day-to-day decision-making, supports establishing strong PSM systems and achieving high levels of performance. Safety priorities are recognized, not as conflicting with other priorities, but rather as inherently necessary for completing any task the right way.

Focused Leadership

Management leadership and commitment which is visible throughout the organization is required for continued excellence in PSM. Beginning with the formation and fostering of a strong safety culture, continuing with policies that set high expectations for excellent safety performance, and implemented in day-to-day decisions that provide resources and establish high priority for safety-related activities, the actions of all levels of management must support and reinforce strong PSM programs and accountability.

Good OD will not happen on its own, but rather is something that requires continued management attention to implement appropriate systems, to monitor and evaluate actual performance, to demonstrate personal attention and dedication, to provide appropriate resources as needed, and to develop and support effective processes to facilitate employee understanding, involvement, and follow-through. The many benefits of OD in helping achieve world class manufacturing operations, reduce risk, and prevent injuries provide a basis for engaging leadership to support and maintain effective OD programs.

DuPont has begun implementing the Flame Model, which was developed to identify best practices for leadership processes in leading improvement of OD. Five key areas have been identified:

- **Future Direction** – An evergreen future-oriented vision process is in place that includes inter-dependent behavior that supports strong OD, has been well communicated and explained to engage site personnel, and is supported daily through leadership focus and behaviors.
- **Leadership Processes** – The leader is a role model for the site and actively works to build a strong leadership group committed to SHE and OD improvement. Coaching and other operations excellence practices are used to structure leadership activities as well as immerse site personnel in the desired culture.
- **Accountability** – Site and personal goals that include a focus on OD have been established with clear expectations on performance. People are challenged to contribute to improved OD, and commitments are periodically reviewed and reinforced. Recognition is provided as appropriate.
- **Managing Processes** – Metrics, audits, and other activities are defined to continually assess site progress and focus around OD improvement. Situational leadership practices are used to support site leaders and personnel as needed.
- **Engendering Trust** – Leaders focus on supporting a trustful work environment to engage and support site personnel through consistent leadership principles and behaviors. Programs are personalized in terms of impact to all employees, feedback is sought, and

inter-dependent behavior is encouraged. Site personnel know they are important to success and that their contribution is valued.

Effective Risk Management

Rigorous hazard identification, process hazard assessment, and risk evaluation and management are necessary for ensuring appropriate PSM systems are implemented to help achieve excellent PSM performance. If poor OD impacts risk management practices, then the necessary safeguards for risk management are not identified, implemented, or maintained, leading to performance problems. Risk management is essential in establishing the required foundations of effective PSM and OD programs. If this is not done well, a focus on OD will be like climbing a ladder placed on the wrong wall; it will be impossible to get to where you want to go.

Organizational Capability and Training

With increasing complexity associated with manufacturing processes, continuous addition of new PSM requirements, continued development of new RAGAGEPs and regulations, and potentially increased staffing turnover expected in the future, a focus on organizational capability and training is essential to achieving good OD. A well-defined training strategy should be developed and implemented, with refresher training provided at appropriate intervals. Effective management of change procedures associated with personnel turnover to ensure appropriate understanding of PSM systems and OD practices must be in place to maintain capability. In addition, fitness-for-duty programs based on alcohol, drug, and fatigue problems require continuous attention.

Well Designed Procedures

In many ways, the heart of a strong OD Culture is the Procedures. Within the procedures, the collective knowledge and know-how of an organization should be codified. Repeatable, excellent results are ideally delivered by the organization by robustly following the procedures. The discipline and organizational confidence to comply with procedural requirements comes foundational from the belief that the procedures are helpful, accurate, and ultimately, the best way to achieve the desired results.

Eleven steps for strengthening OD through High Quality Procedures have been developed. Details are summarized below:

1. A Procedures “Standard” that details requirements
2. Single Person Owners for every procedure
3. Well written, easily followed and visual
4. Consistent, comprehensive and user friendly content
5. Importance and Risk Classified
6. Routine Reviews by Users and Subject Matter Experts
7. Appropriate Verification
8. Change Management Process
9. Initial & Refresher Training plus Assessments

10. Revisions Controls while maintain Accessibility
11. Metrics and Audits

What else is needed to achieve the desired PSM Performance?

Clearly defined Roles and Responsibilities starting at the top

At DuPont, like most companies, Safety is a key performance measure. DuPont considers Safety a Core Value. Other companies express the relative importance of safety in other ways. All responsible companies are committed to keeping their employees safe. But Process Safety, as recent experiences have indicated, cannot be assessed with a high degree of accuracy with safety statistics only. Process Safety results will also be reflected in other key performance indicators including environmental performance, company reputation, asset productivity, business/supply chain continuity, and even employee morale. Because Process Safety performance is so critical, leadership responsibilities from the top of company must be delineated.

The attached describes the two key organizational functions. On the right side is Corporate Competency and how the DuPont Company is organized to ensure we are utilizing the best available knowledge and experience. The left side indicates the responsibilities of Line Management beginning with the CEO and the company's Board of Directors. Fundamentally, the Corporate Competency Team owns "What" the company program will be. Line Management is responsible for the "How, When, Where, and Who" of execution. In the DuPont PSM Standard, 14 pages of text are used to describe these duties for organizations within the Company, including Global, Regional and Site Operations, Corporate HSE and Engineering, Manufacturing Technology, Sourcing, Integrated Health and Business Leadership.

Said succinctly as written in the DuPont PSM Standard:

"Line management has the responsibility to implement this standard."

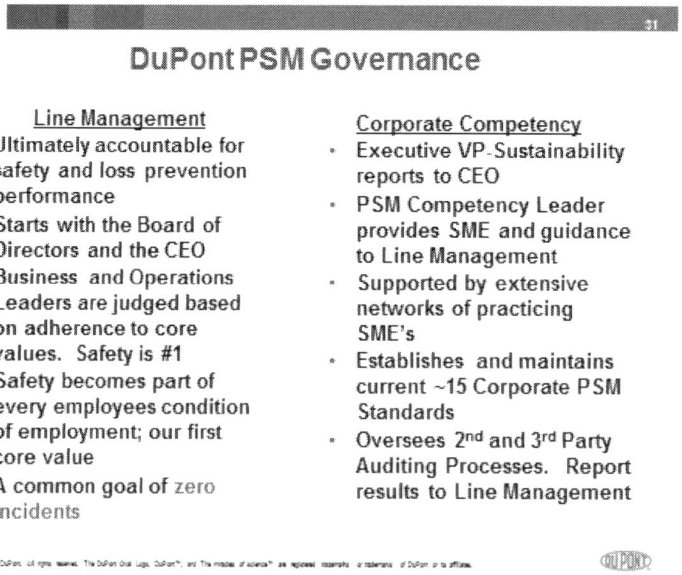

Operations & Business Integration

In recent years, the PSM technical requirements have been further incorporated into broader operations and production systems, including Operational Excellence (i.e. OEMS.) This degree of integration is critical as Process Safety Management must not be viewed by an organization as a Safety initiative, but rather how "we do business" within operations. The degree to which the PSM Managing System is adopted and integrated and to which all employees are involved in those elements that touch their day to day work will speak volumes about the quality of the overall PSM Managing System deployment and implementation. And will ultimately determine the organizations success in reducing the frequency and severity of Process Safety Incidents.

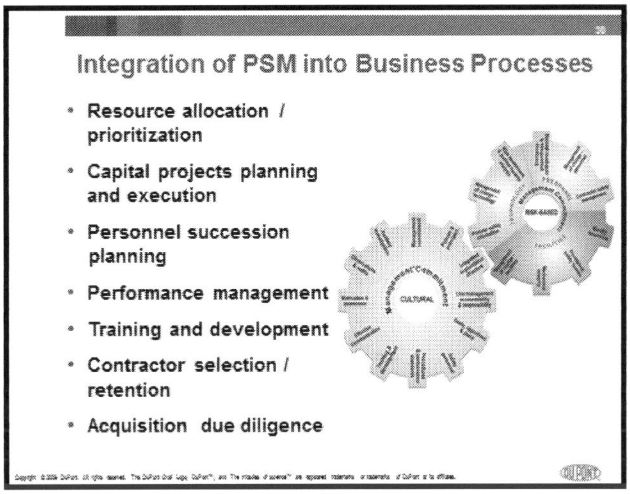

Strong Managing Processes

What brings together the Technical System Requirements with the Cultural requirements for successful implementation of Process Safety Management, including OD, is the PSM Managing System. Recommendation #2 from the Baker Panel Report into the BP-Texas City Incident described the recommended approach with the following words:

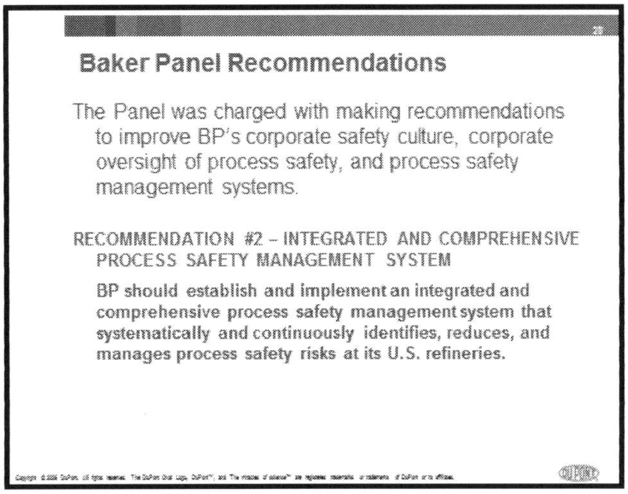

DuPont has graphically displayed the interdependent nature between leadership, culture and technical requirements of PSM Implementation for more than 20 years using the "PSM Wheel" also shown below.

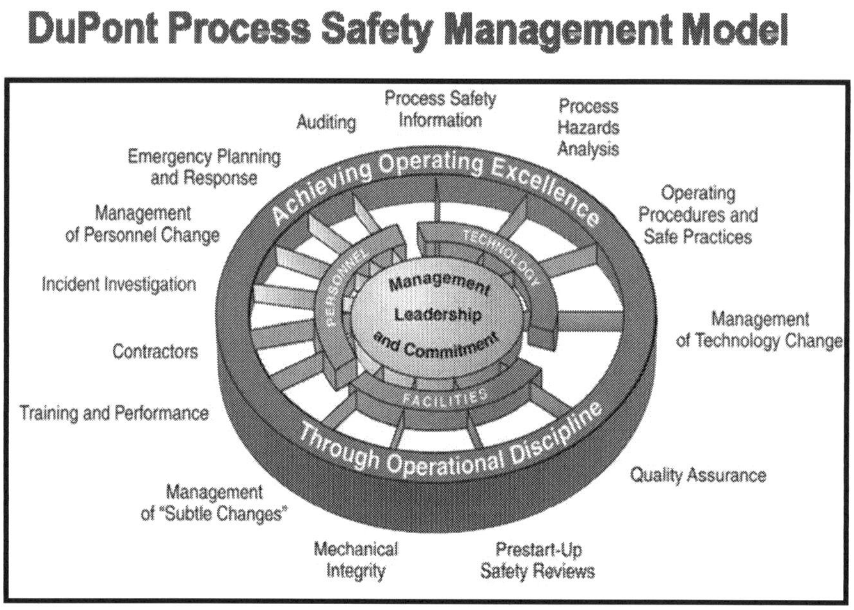

Management's Leadership and Commitment is in the center of the wheel where energy, drive and power and provided. From the center flows the resolve to make the wheel function as a smooth and single unit. Strength is delivered from the center by the 14 individual spokes or technical elements. Each spoke is important and weakness in any spoke will cause the wheel's "ride" to be uneven and sub-optimum. But where the "rubber meets the road" and where true risk mitigation occurs is in the outer rim that comes in contact with daily operations. Operational Excellence is achieved through rigorous adherence to the managing system requirements via a strong culture of OD. OD transforms an "on paper" managing system into an effective "in action" managing system.

PSM Standards incorporated into the Design/Build of new Facilities

An all too frequent conversation with companies around the world has been that the standard to which operating facilities are managing their risks (via a documented PSM System) is not the same standard that was used by the organizations that designed and built this same facility. The observation is usually expressed in the attitude of a complaint, as it is an enormous challenge to "catch up" the facility's process safety systems when those who delivered it to them "didn't do their job."

One area of particular weakness seems to be the management of Process Safety Information at the interface between the Project Teams and the Operations Teams. Too frequently there has not been an acceptable plan with clear ownership for such key project documents as Process Hazards Analyses, Quality Assurance inspections, Process Flow Diagrams, and Process & Instrument Diagrams. Once this information is lost, either literally or figuratively, it is extremely difficult (expensive and painful) to re-construct. Millions of dollars can easily be spent to re-create even a basic level of Process Safety Information.

"Beginning with the end in mind" is essential in the Design and Construction Phases of any capital project. Including as members of the capital project team the key, future operations, technical and maintenance leaders is a very effective way of ensuring operational continuity. It will also contribute to a common approach of PSM standards implementation in the design and construction of a new facility.

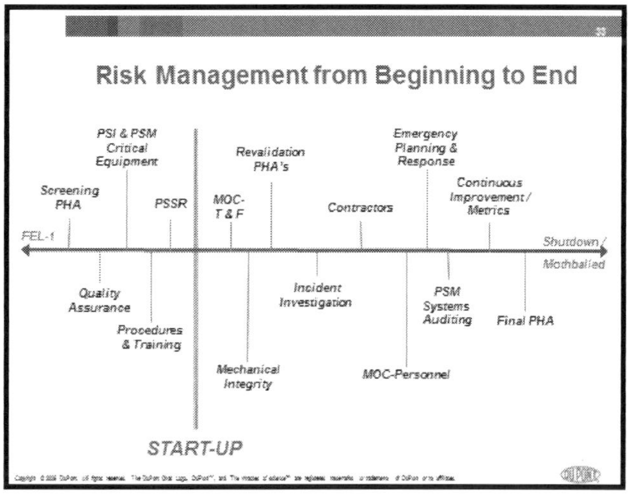

Active Employee Ownership and Involvement

Employee Participation is a required element of PSM Programs at USA sites that are covered by the PSM Regulation 29 CFR 1910.119. Key language in the regulation includes:

- *Employers shall develop a written plan of action regarding the implementation of the employee participation required by this paragraph.*
- *Employers shall consult with employees and their representatives on the conduct and development of process hazards analyses and on the development of the other elements of process safety management in this standard.*
- *Employers shall provide to employees and their representatives access to process hazard analyses and to all other information required to be developed under this standard.*

Compliance with these requirements is not optional but employee participation should not end here. True employee involvement and ownership of the PSM system should be a critical leadership objective and an essential part of the organization's safety and PSM culture.

From DuPont's, Process Safety Management Standard:

> *"Site leaders shall play a key role and be accountable for establishing and sustaining site safety culture, PSM expectations, accountabilities and objectives; resourcing program elements; providing reward and recognition; and measuring ongoing*

performance on a routine basis. Site management shall be responsible for demonstrating ongoing management leadership, commitment, and personal involvement in site PSM programs plus providing for and encouraging employee involvement."

Specific Employee Involvement requirements are detailed below:

Providing for and Encourage Employee Involvement
Effective PSM relies on both the individual and collective efforts of everyone involved in managing process safety. Recognizing that employees are uniquely knowledgeable about the process and are "key managers" of process safety (often the only "managers" immediately available), it is important that site management provide for and encourage a broad spectrum of employee involvement in the design, implementation, and ongoing operation of the site's PSM program within all elements.

Sites shall maintain programs to achieve employee involvement in process safety activities, including the following:

- *Membership on site PSM committees, subcommittees, or special project teams*
- *Participation in PHA teams*
- *Participation in incident investigation teams*
- *Participation in pre-start-up reviews*
- *Conducting equipment tests and inspections*
- *Writing and/or reviewing operating and maintenance procedures*
- *Participation in development of training procedures and programs*
- *Participation in process safety audits*
- *Participation in emergency response planning, drills, and critiques*
- *Participation in MOC-technology (MOC-T) and MOC-subtle facility (MOC-SF) change reviews and implementation*
- *Participation in preparations and implementation of MOC-Personnel changes*

Within each element of PSM and as part of the site PSM policy, sites shall solicit, on an ongoing basis, participation and input from employees regarding ideas for improving and strengthening site PSM programs and systems. Sites shall also establish, document, and communicate local mechanisms and accountabilities to collect, track, and respond to employee PSM concerns and suggestions.

Continuous Improvement driven by Metrics, Audits and ever-strengthening Standards

The goal of all Process Safety Management systems is the mitigation and control of operational risks such that significant process safety incidents are eliminated. Although great progress has been made over the decades that PSM has been practiced, all can agree that the desired "Goal of Zero" has not yet been consistently achieved and that more work is needed. Continuous Improvement of the robustness of the managing systems and rigor in their execution is needed.

Three tools that organizations have to drive PSM Continuous Improvement are metrics (lagging and leading), audits (first, second and third party) and ever strengthening PSM system requirements and standards.

Performance metrics have received considerable attention as we also seek to focus in on those vital few indicators that most accurately describe overall system health and which most consistently and accurately predict future process safety incident performance. Collaborative efforts sponsored by CCPS and API plus the Responsible Care Initiative have brought much new thinking to this field. DuPont has elected to collect corporately on a quarterly basis 10 fundamental metrics, as described below. Additional leading metrics are kept locally for periodic review by site and regional leaders. Finally, additional near-miss and other learning event metrics are recommended but remain optional for local leaders to consider in the journey via continuous improvement towards the Goal of Zero.

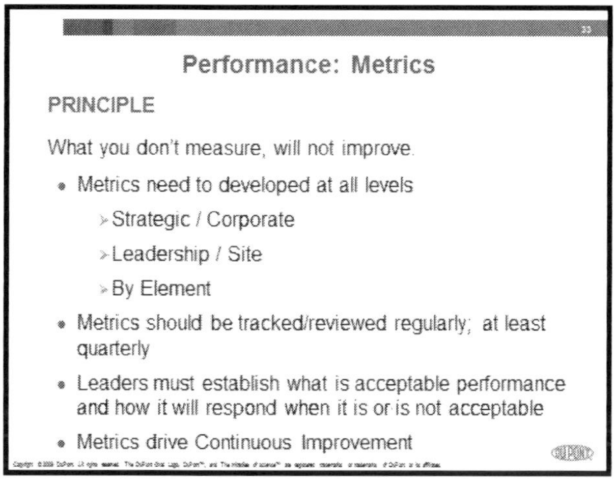

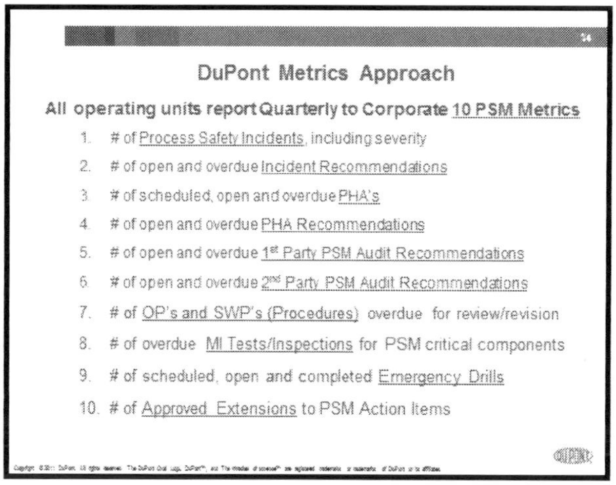

Proactive auditing of both the cultural and technical elements of PSM can provide leaders and the organization at-large valuable insight as to the relative strength of each. Insightful leaders understand the importance of seeking and obtaining unfiltered data that provides a clear picture of reality. Only then can leaders be confident that their decisions will achieve the desired outcomes.

First Party Auditing should be the most effective as long as the organization approaches the audit opportunity with brutal honesty and in the spirit of improvement at all costs. Second Party Auditing can provide valuable insight from outside the unit while keeping the learning's and auditor development within the company. Third Party Auditing should provide the most independent and potentially most valuable audit recommendations. Each should be incorporated into a comprehensive PSM Audit Program.

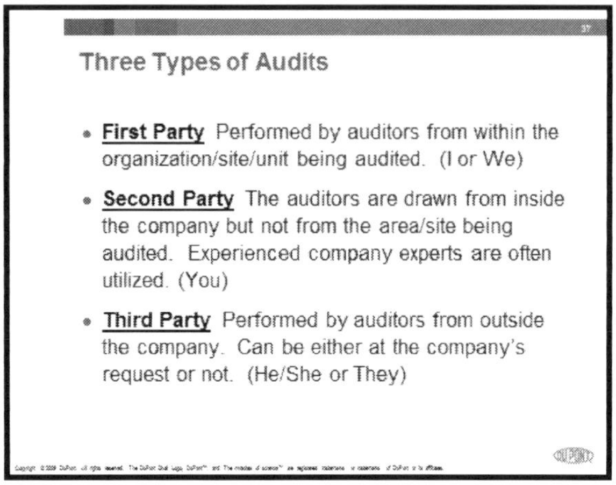

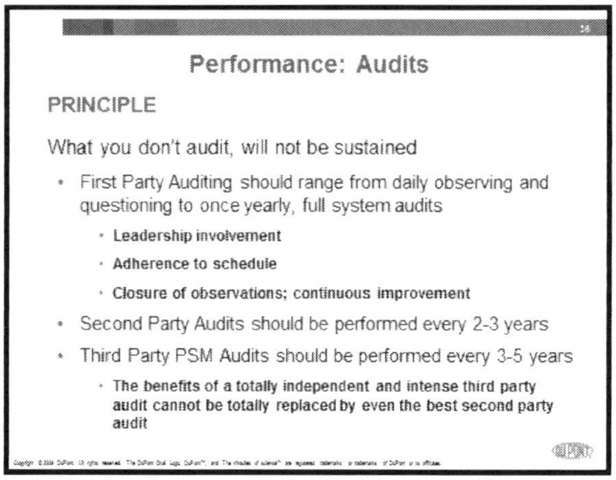

Many, if not most medium to large sized companies, especially in geographies that are not covered by prescriptive government regulations have developed comprehensive PSM Standards. The standard is intended to produce consistent implementation of company best practices and the highest level of performance. As mentioned earlier, PSM requirements are increasingly being integrated and incorporated into broader Operations Excellence or Asset Integrity Managing Systems.

One of challenges of establishing a PSM Standard is keeping it current as company and industry knowledge advances. Assuming that PSM Audits are based upon and conducted against this PSM Standard, PSM Audit questions must also be kept current. This requires effort and commitment.

The DuPont Company issued its first comprehensive Process Safety Management Standard in 1979. It was one of the approaches or systems to achieve PSM excellence that was considered when OSHA promulgated 29 CFR 1910.119. One might think that with such a rich history, the standard would be achieving a level of relative maturity and that changes would be minimal. In actual fact, the opposite has occurred. In the last 5 years, three revisions to the DuPont PSM Standard have been finalized and issued. And as detailed below, quite a number fairly significant changes have been introduced and incorporated into these three revisions.

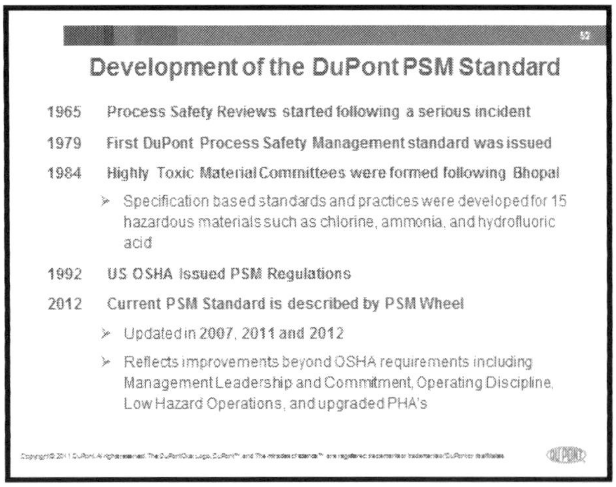

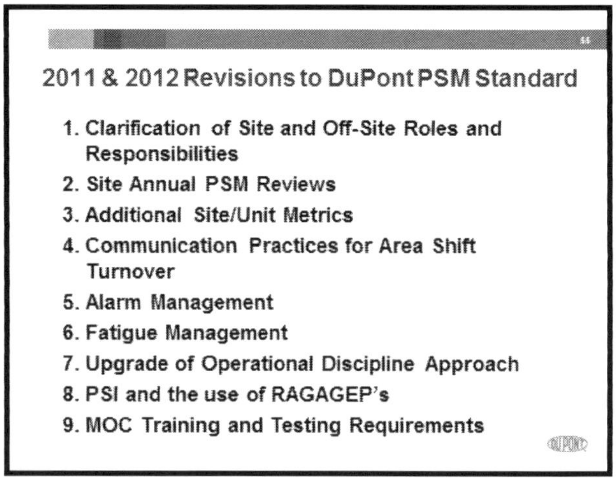

Continuous improvement in the standards can quickly elevate the entire company's PSM managing systems and its performance. Metrics, Audits and Standards are three of the most effective tools for driving and achieving PSM excellence via Continuous Improvement.

Conclusion

Process Safety Management has as its overall objective to mitigate operational risks such that process related incidents and injuries are prevented; DuPont's commitment is to zero. Organizations that embrace Operational Discipline are striving for excellence by having every member carrying out each task the right way, every time. As important as Operational Discipline is to achieving Process Safety Excellence, it should not be viewed as a panacea. It is an essential ingredient but in and of itself, is not sufficient.

What is needed to progress are robust PSM Systems, a strong Safety Culture and OD Excellence. These three legs of the PSM stool are highly interdependent and are mutually supportive. OD excellence needs focused leadership, effective risk management, organizational capability and well-designed procedures. A strong safety culture demands clearly defined roles, employee ownership/involvement and the integration of safety into operations and business processes. A robust PSM system requires an integrated and comprehensive approach, incorporation into the design and installation of new facilities, and continuous improvement as demonstrated by metrics, audits and ever-strengthening standards.

Knowing what to do is critical to preventing process safety incidents. Doing what we know saves lives.

References:

1 James A. Klein, Operational Discipline in the Workplace, Process Safety Progress, Vol. 24, No. 4, p. 228-235, December, 2005

2 James A. Klein and S. Dharmavaram, Improving the Performance of Established PSM Programs, Process Safety Progress, Vol. 31, No. 3, p. 261-65, Sept., 2012

(43m) Process Safety Information Management - It Is More Than You Probably Think

Mason Martin, Process Safety Management, Phillips 66, Borger, TX and Patrick Nonhof, Provenance Consulting, Borger, TX

Many businesses continue to struggle with managing the expectations of the Process Safety Management standard--even twenty years after the rule was first promulgated. In particular, the sheer enormity of Process Safety Information (PSI) continues to overwhelm many organizations. PSI is a foundational element of PSM upon which other elements such as Process Hazards Analysis are directly dependent. Relationships to the other elements may be less obvious, but PSI is ultimately the foundation upon which all risk-based decisions are based. PSI has to be based on the accuracy of core documents; tools must be created to identify and correct any discrepancies among the various systems that generate, store, and use PSI; and the same tools can be used to drive alignment outward into the other elements. By relying on these fundamentals, this paper details a system that can be put in place by management to improve a site's process safety culture through facilitating a better understanding of PSI and the inherent connections to other elements of the PSM standard.

Connecting the Dots – Managing the Little Things That Lead to Big Accidents

Dave Jones
Chevron Energy Technology Company
6001 Bollinger Canyon Road, San Ramon, CA 94583
dwjones@chevron.com

Anne O'Neal
Chevron Oronite
6001 Bollinger Canyon Road, San Ramon, CA 94583
anneoneal@chevron.com

Prepared for Presentation at
American Institute of Chemical Engineers
2013 Spring Meeting
9th Global Congress on Process Safety
San Antonio, Texas
April 28 – May 1, 2013

UNPUBLISHED

Connecting the Dots – Managing the Little Things That Lead to Big Accidents

Dave Jones
Chevron Energy Technology Company
6001 Bollinger Canyon Road, San Ramon, CA 94583
dwjones@chevron.com

Anne O'Neal
Chevron Oronite
6001 Bollinger Canyon Road, San Ramon, CA 94583
anneoneal@chevron.com

Keywords: Major accidents, details, dots, culture

Abstract

Major accidents often result from a failure to attend to seemingly little details. Examples might include an improperly designed injection point; a procedure that is mostly, but not completely followed; a small part with the wrong metallurgy; or other design, maintenance or operating discrepancies. In many cases, system flaws can exist for extended periods of time, detected or undetected, with little or no consequence. However, on occasion, and with the right initiating event, they can combine with other system flaws, allowing an unforeseen chain of events to propagate into a major accident.

This paper will review aggregated information from a range of process safety incidents, including "subtle" accidents resulting from a series of "little" failures. It will highlight the management system breakdowns that contributed to the incidents, and examine the complex interactions among people, process safety systems, and equipment as orchestrated by the overall organizational culture. The paper will reinforce the concept that an effective process safety culture is key to "connecting the dots" that, if unchecked, can result in a major accident.

1. Introduction

Following the Piper Alpha tragedy in the North Sea, in which 167 men lost their lives, Dr. Tony Barrell of the United Kingdom Health and Safety Executive said, "I wouldn't put it above or below other disasters. There is, actually, an awful sameness about these incidents."[1]

The "awful sameness" to which Dr. Barrell refers may reflect an observation that, with rare exception, all process safety accidents can be described within a common, recurring group of management system failures. For example, a failure to manage change was a key contributor to process safety events at Flixborough and Chernobyl, among others.[2] Similarly, the accidents at

Longford and Grangemouth were attributed to inadequate hazard identification/risk assessment.[2] Permit-to-work systems/safe work practice failures were identified as contributing to the Piper Alpha and Pasadena Texas (1989) fires and explosions.[2]

Clearly, major accidents have multiple management system contributors. The Bhopal tragedy involved, among other things, breakdowns in hazard identification/risk management, management of change, asset integrity, incident investigation, process safety competency, and emergency response.[2] Other events have different combinations of contributing factors.

However, when we evaluate a range of process safety incidents, we find that the management system failures leading to these events begin to repeat themselves into well defined categories. The CCPS *Guidelines for Risk Based Process Safety*[3] represent, in the authors' view, the current best compilation of these categories, often called elements (CCPS elements). The CCPS book discusses 20 elements altogether, ranging from process safety culture to management review and continuous improvement – and including key areas such as risk analysis, asset integrity and auditing. For a complete list and comprehensive treatment of these elements, see reference 3.

In this paper we will attempt to show that while the process safety management systems themselves are relatively simple, their actual application requires attention to innumerable "little things," any one of which has the potential to initiate or propagate a major accident. Identifying and managing these "little things" necessitates a combination of "big picture" thinking and rigorous attention to detail across all elements of process safety management and all relevant organizational functions.

2. "Little Things" That need to be Managed

In this section we will illustrate the dynamic nature and seemingly limitless numbers of details at work within each of the process safety management systems and highlight some of the complexities involved in managing them.

2.1 Accident Models

Many accident models have been developed. These include variations on the "Swiss Cheese" model, first proposed by James Reason[4] and the "Bow Tie" model emerging out of European and UK risk assessment applications. More recently, Nancy Leveson from the Massachusetts Institute of Technology has proposed the System-Theoretic Accident Model and Process (STAMP)[5] which attempts to include human error and culture into the accident equation. In the authors' view, all of these models have merit, yet none is perfect. For internal educational purposes Chevron uses a hybrid model as shown in Figure 1.

It is worth noting that each of the above models characterizes large-scale accidents as a coming together of multiple system failures. This theme is directly counter to the traditional concept of "double jeopardy" which holds that the combined failure of two independent systems is so unlikely as to not require special consideration. Industry accidents over the last 30 years have clearly demonstrated the fallacy of a "double jeopardy" approach to process safety.

2.2 Dynamic Nature of Process Safety

The "Swiss cheese" shown in Figure 1 represents basic prevention equipment, engineering controls, software, human actions, etc. These include facility-specific procedures, training, protective devices, inspection programs, and design integrity, among others and are the real-world manifestations of process safety management systems. They are the mechanisms that actually intercede to stop an accident from starting or propagating.

Figure 1 – Hybrid Accident Model

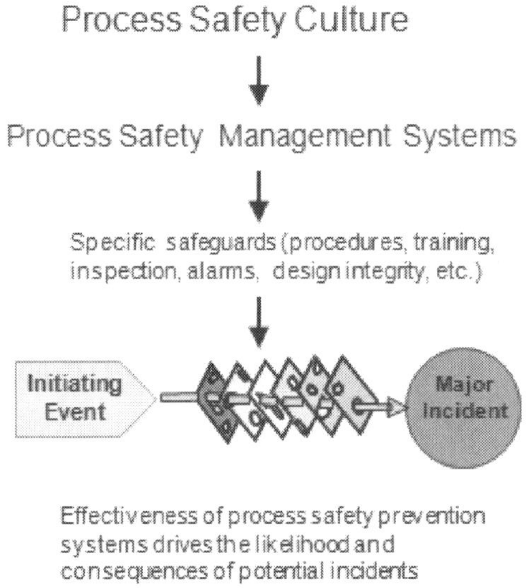

Driving the effectiveness of the "Swiss cheese" are the process safety management systems themselves - the CCPS Elements. Ultimately, the organizational culture as applied to process safety determines how well the management systems perform. In real time, each component of this and any other accident model is in constant flux. For example:

- The effectiveness of safeguards can change;
  - Safety systems can degrade, go untested, be compromised, or improve in functionality
  - Corrosion mechanisms can increase or decrease, or be introduced to previously unaffected areas
  - Procedures and training can vary in effectiveness, depending upon facility changes
- Process safety management systems change;
  - Critical controls such as management of change can get overwhelmed, allowing higher risk changes to "sneak" through
  - Personnel movements can lead to lost understanding of risks, or bring new, beneficial approaches
  - Process safety information can become quickly out of date
- Overriding process safety culture can degrade or strengthen with retirements and reassignments of key personnel, acquisitions or external pressures.

Given the constant change experienced by each of the above components in the accident chain, it is clear that process safety can never be viewed as a static condition. There is no milestone that can be achieved and final "victory" declared over major accidents. Effective process safety management approaches all aspects of the accident model with a sense of vulnerability, aware that one or multiple components may be compromised at any given time. Ongoing diligent monitoring and continuous improvement are the best defense.

2.3 Management Systems vs. Facility Complexity

The distillation of process safety into 20 key elements can mask the magnitude of the effort required to prevent major accidents. Given facility complexities, large equipment counts, variations in administrative controls, and human performance, the potential for flawed execution within each element is seemingly endless.

As an example, a process safety management system for Asset Integrity and Reliability requires that safety-critical shutdown systems be identified and tested at prescribed frequencies and repaired as necessary. Some of the factors that could influence how well this fundamental requirement is implemented for a specific shutdown system are suggested in Table 1.

Table 1 – Factors Potentially Compromising A Specific Shutdown System Integrity
• Failure to identify the system as safety critical • System identified but not included in testing program • System not designed to be testable on line • System designed to be mostly testable, except for final control element • System designed to be mostly testable, except for the process connection valves subject to plugging • Testing poses a potential to shut down or damage the facility • Procedure for testing is complex • Modifications to system design not included in testing procedures • Personnel not trained in testing the system • Shutdown system bypassed during startup and not reactivated • Shutdown system bypassed because of threat of spurious trips • Importance of shutdown system not understood • Software modification altering shutdown system set point • Internal control system failure • Testing poorly documented • Identified non-conformance not acted upon • ....ad infinitum

The list in Table 1, which is by no means exhaustive, helps illustrate the challenge in implementing just one aspect of one process safety management system for one piece of equipment. Nor does the above list include the human performance or cultural forces that can affect the viability of the system tests, such as operating distractions or lack of management emphasis and monitoring. More importantly, any one of the above shortcomings in execution could become part of a sequence of events ultimately leading to an accident.

If we began to construct similar facility-specific lists for all the CCPS Elements, applied to the entire inventory of equipment and activities, a picture would start to emerge of the enormous variety and complexity of the "little things" that can degrade process safety and ultimately contribute to a major accident. We don't have to look too far to find verification that it is these little things that present the challenge. In July 2000, the crash of the Air France Concorde jet was attributed to a small strip of titanium that had fallen off another airplane minutes earlier.[6] In 1980 the floating dormitory for North Sea oil workers *Alexander Keilland* sank, killing 123 people, when a cracked strut failed. A shipyard painter had painted over the crack during maintenance so it went undetected. [6] Similarly, one link in the accident chain that caused the Ocean Ranger drill ship to go down in 1982 off the coast of Newfoundland, killing 54 people on board, was the failure of small glass porthole window allowing seawater to enter the ballast control room.[6]

2.4 Picture the "Dots"

If we were to graph the relationships among the management systems and the "little things" ultimately precipitating an accident, they might look like Figure 2.

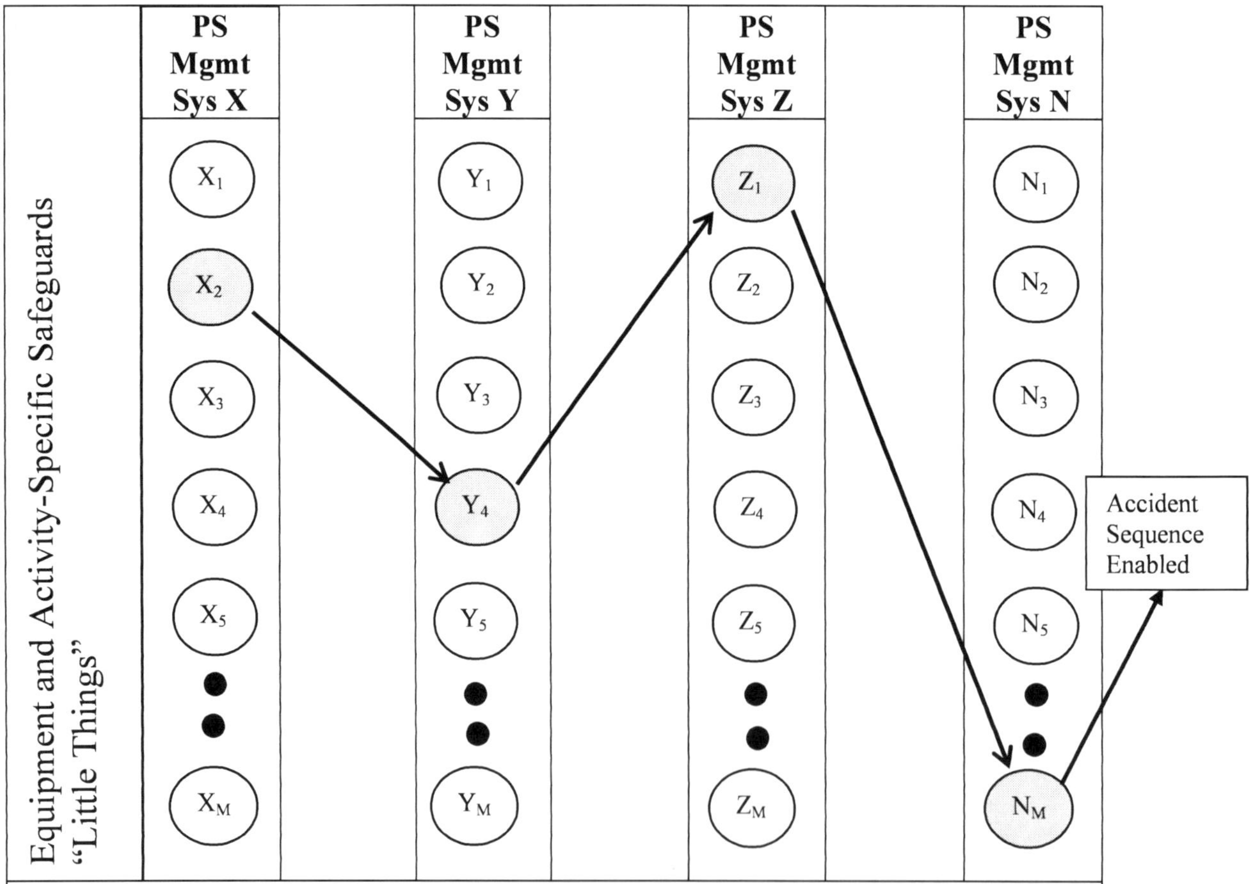

Figure 2 – Graphical Representation of Little Details"

3. Case Study[7]

This case study helps to illustrate the complex nature of the "little things" associated with just a couple of management systems – Operating Procedures, Training and Performance Assurance, and Asset Integrity and Reliability.

Figure 3 – LPG Treater Flow Diagram

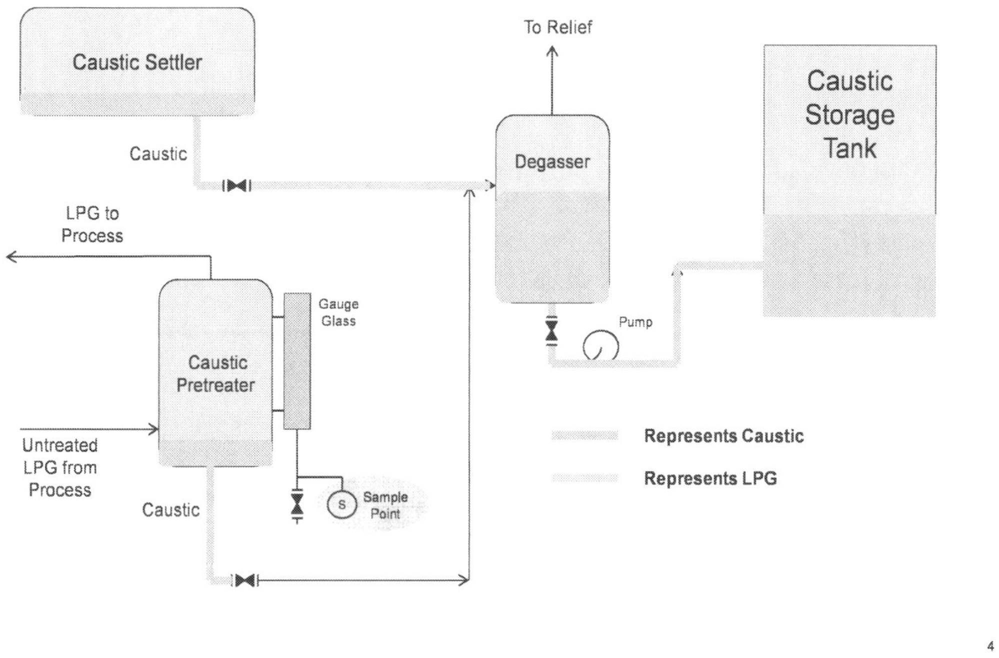

Figure 3 shows the flow diagram for a fairly standard process for treating liquefied petroleum gas (LPG) with caustic (sodium hydroxide) to remove certain impurities prior to further processing. The green represent caustic inventories, while the remaining volumes in the Caustic Pretreater are LPG. The Caustic Settler has an LPG inventory as well a vapor space. While performing a routine duty to drain caustic from the Caustic Pretreater, the draining operation was continued to the point where all the caustic material was removed and LPG was directed to the atmospheric Caustic Storage Tank resulting in a loss of containment at the tank. There were no injuries related to this incident.

3.1 Sequence of Events

The operator's routine duties included periodically transferring caustic from both the Pretreater and Settler to the Degasser and then to the Storage Tank. After sampling the Pretreater for caustic, the operator opened the manual valve on the outlet of the Pretreater to start the caustic transfer. The operator also started the Degasser pump to transfer caustic from the Degasser to the Storage Tank. The gauge glass on the Pretreater was dirty and difficult to read, but it was assumed that the caustic level was above the gauge glass. The actual level was most likely below the gauge glass but above the sample point.

The Pretreater level was then depleted by the operator, with LPG going to the Degasser and with flaring through the relief system increasing. The valve from the Pretreater was then closed. Caustic transfer was then started from the Settler, and the valve from the Pretreater was reopened, sending pure LPG to the Degasser from the Pretreater. The operator saw increased flaring as confirmation that procedures were being done properly, since two vessels were being fed to the Degasser, with increased vapor release from the entrained LPG in caustic. Some of the increased flaring was from the pure LPG being sent from the Pretreater.

With the Degasser pump continuing to run, the Degasser caustic level was depleted and LPG was transferred to the Storage Tank. The low level alarm on the Degasser was not functioning. As the level of caustic and LPG increased in the Storage Tank the operator stopped flow from the Settler and Pretreater. Attempts were then made to bring the level of caustic back up in the Degasser by opening the Pretreater valve, which sent more LPG to the Degasser since it remained depleted of caustic. Shortly afterwards, the Storage Tank high level alarm sounded, and material was released from the top of storage tank.

3.2 Analysis

The above incident description suggests that, as with many other accidents, there were "a lot of things going on" increasing the potential for incomplete diagnosis and overlooking of details. Focusing initially on just three of the management systems (CCPS Elements) involved we can see firsthand some of the "little things" that need to be managed to prevent incidents of this type.

3.2.1 Operating Procedures

The procedure for transferring caustic from the Pretreater did not specify whether the Degasser pump should be used simultaneously to the transfer. The purpose of such a restriction is to avoid inadvertently sending LPG to the Storage Tank, and also to allow sufficient retention time in the Degasser for disengagement of any entrained LPG from the caustic. Had either one of these process-related details been included in the operating procedures and training, it would have provided operating personnel with a broader view of the potential consequences related to this transfer. It might have helped them "connect the dots."

The investigation also concluded that the operator was monitoring too many items at once and did not notice that the caustic level in the Degasser had been depleted. Since the Degasser level alarm was dysfunctional, this would have meant that the operator would have to verify the level manually. This would be another "little thing" that might have compensated for the level alarm failure.

3.2.2 Training and Performance Assurance

The gauge glass on the Pretreater was difficult to read. Although included in field training the technique for blowing down and clearing the gauge glass was not fully understood by the operator. Improved training learning objectives might have provided the operator with a better understanding of the Pretreater level.

Risks and precautionary steps associated with LPG transfer during caustic transfer were not fully understood after classroom and field training. Improved training and learning objectives might have aided the operator in detecting potential LPG carryover to the Storage Tank.

3.2.3 Asset Integrity and Reliability

The Degasser low level alarm did not function correctly. It had been falsely indicating a low level condition for several months prior to the incident. A previous alarm replacement did not address the root cause of the malfunction.

3.3 Process Safety is all About Details

As the "Swiss Cheese" model suggests, had any one of the "Swiss cheese" layers been effective, this incident might not have progressed. It is also worth noting that the actual incident investigation went beyond the three process safety systems indicated above, in alignment with other areas identified in Figure 1.

Even the above limited review illustrates the many "little things" that contribute to an incident, and the many seemingly small opportunities for preventing it. Slightly modified operating procedures, better training on level gauges, clearer understanding of risks in transferring LPG, and others are all minor improvements on systems and activities already in place that might have stopped the accident from progressing It is this little bit extra that often means a difference between a non-event and an accident. For example, this can mean going one step further in training, adding one additional precaution in an operating procedure and, of course, ensuring that all alarms, even those perceived as less important, are functioning properly.

This incident also illustrates an important process safety opportunity. How do we know when one "little thing" can interact with another, progressing the accident sequence? How do we "connect the dots?" Put another way, how do we broaden our thinking so that when something abnormal is occurring in one area of an operation we are aware that other areas can potentially be impacted? For example, as the operator was having difficulty reading the Pretreater level gauge, the additional knowledge that the Degasser level alarm was not functional and that LPG could be sent to the Storage Tank might have driven different actions.

4. Connecting and Managing the "Little Things"

If we extend the lessons from the Case Study to a facility or even company-wide scale, several key focus areas are suggested that might help to prevent similar incidents.

4.1 Understand that Details Drive Accidents

Major accidents almost always come down to "little things" done, not done, or performed incompletely. This extends beyond the operating and maintenance roles into engineering and management. For example, a major fire occurred on an offshore platform because the compressor seal oil pumps had no back-up power supply – an engineering detail overlooked. On the management side, the Space Shuttle Challenger represents a clear example of where management ignored the recommendations of engineers and failed to fully appreciate the catastrophic consequences that could result from an O-ring failure if a launch were to occur in cold weather.[2]

4.2 Get the Big Picture

Process safety elements are often viewed in silos, rather than as holistically interactive management systems that depend upon each other to function properly. When one system has a breakdown assumptions may be made that other prevention devices will be working properly. Statements like, "the shutdown system will handle it" or "it will be picked up on operator rounds" reflect this tendency to rely on other mechanisms to stop the event. These other safeguards may or may not be working well enough to prevent the scenario from developing.

Before determining that a serious potential consequence is unlikely to occur after an initial failure, management, engineering and operating personnel should look across the spectrum of affected process safety systems, and verify that "backup" safeguards are functional and sufficiently effective. In some cases, simple questions might help broaden the identification process and identify seemingly small, but important breakdowns in the protection layers:

- What else can be affected? (Texas City)[2]
- How cold can we go? (Challenger Space Shuttle, Longford Gas plant)[2]
- Where is the feed going? (Three Mile Island)[2]
- Was the job finished and ready to return to service? (Piper Alpha)[2]
- In what order do I open these valves? (Feyzin)[2]
- What happens if we turn this up? (Chernobyl)[2]
- What does my Technical Authority say about this? (Flixborough)[2]

For the accidents indicated in the parentheses above, the answers to the preceding questions are obvious in hindsight. The challenge is to ask the right questions ahead of time and encourage personnel to explore consequences and safeguards beyond the obvious.

Understanding the big picture can be complicated by the natural segregation of functional activities. Performing separate but interrelated tasks in silos can make it difficult to fully comprehend how one effort might create "dots" that negatively combine with the "dots" of another effort. A review of the sinking of the *Herald of Free Enterprise* at the Zeebrugge, Belgium port might help demonstrate this point.

Reference 2 discusses the process safety system failures involved in the Zeebrugge accident. The ferry departed port with the bow doors open, and capsized about a mile offshore. 193 lives were lost. The "little things" contributing this tragedy included: (1) nobody checked to verify the doors were closed; (2) there were no clear procedures on who exactly was responsible for checking the doors; and (3) the Assistant Bosun, who had the task of closing the doors, was asleep in his cabin and did not hear the ship intercom call to attend to stations. Other contributing factors included basic design flaws, incompatibility with berth designs and overloading.[2]

Specific process safety management systems involved included hazard identification and risk analysis, operating procedures and incident investigation (there were five previous incidents of sailing with the doors open). In addition, the ferry designs were modeled after World War II beach landing craft, which were intended for a radically different purpose. The application of these designs to the safe transport of people should have been (but were not) subjected to a rigorous management of change procedure.[2]

Figure 4 – The Complex Pattern of the Zeebrugge Accident – Taken From Rasmussen[11]

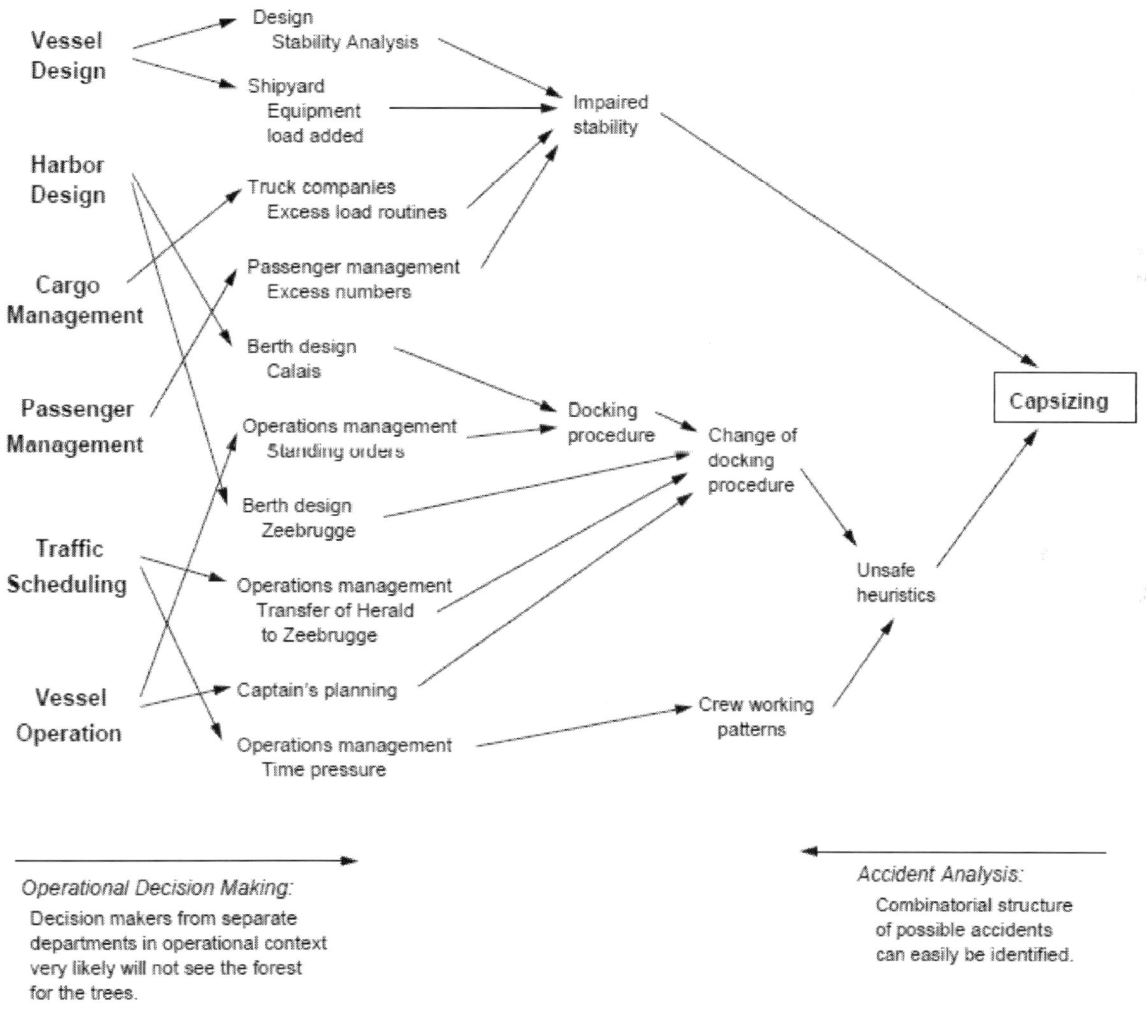

Figure 4 presents a causal tree for the Zeebrugge incident taken from Rasmussen.[11] He contends that major accidents are not a "coincidence of independent failures and human errors" but rather a coming together of different activities often working towards different objectives. The interactions among them can enhance the potential for accident.

As illustrated in Figure 4, each of the functions necessary for ferry operations (Vessel Design, Harbor Design, Cargo Management, Passenger Management, Traffic Scheduling and Vessel Operation) were conducted as relatively separate activities. Each also tried to optimize performance within their scope of work. This resulted in negative impacts to other functional areas. For example, optimizing cargo and passenger management meant adding additional weight to the ferry, directly compromising vessel stability.

Similarly, since the ferry was not designed for the berth at Zeebrugge, docking procedures had to be modified to allow passenger vehicle loading and unloading. These procedures negatively impacted requirements for traffic scheduling and standing orders for vessel operation.

Within the processing industries, engineering, operations, maintenance and other departments have the potential to operate in silos, each focused on achieving their own metrics. This brings attention to the importance of applying process safety management systems across all relevant parts of the organization to help identify and address interrelated impacts.

As a final example, a process plant upgrade included significant piping and equipment modifications, and new additional equipment. Parts of the piping were subject to a well-known corrosion mechanism. Piping and valves were specified appropriately to minimize corrosion. After installation and commissioning, one of the valves failed within 48 hours, requiring a widespread facility shutdown. The investigation revealed that while originally specified correctly, the Purchasing Department had substituted a different valve material, helping to meet their goal of lower cost and faster delivery. This valve did not undergo a positive material identification upon receipt or prior to installation.

The above example helps illustrate the breadth of application required for process safety to be effective across organizational boundaries. In this case, extending process safety to the purchasing department and developing appropriate procedures that ensure alignment with engineering specifications would have prevented the incident. It was also the "little thing" of not verifying materials of construction upon receipt of this valve that finally initiated the accident.

4.3 Perform Thorough Hazard Identification and Risk Assessments

Hazard and operability studies (HAZOPs) and other qualitative risk assessments represent the best formal opportunity for understanding the interactions among "little things." Fully "connecting the dots" requires the most experienced team members be made available. Also, the HAZOP team leader must be highly qualified and able to encourage the team to look beyond their normal, day-to-day experiences. Team qualification requirements can be difficult to sustain, study after study. However, it is essential risk assessments be done with the right personnel.

For a risk assessment to be effective, the team must be able to connect the dots in two distinct ways. Firstly, the team must have the ability to imagine a wide range of scenarios, including the unlikely. If the consequences are not fully developed and include low-frequency events, only a limited set of safeguards will be identified. These may not include the ones necessary to prevent or mitigate the full scenario.

Secondly, safeguards must be verified. This means confirming that "little things" are in place and working effectively: Do equipment or activity-specific procedures really exist and are they included in training? Is the alarm or shutdown tested? Is it testable? How often is it tested? What is done with the test results? Does the corrosion monitoring program include the piping system of concern? What are the inspection results? Is the relief valve sized for this case? When was the relief valve tested? Has the discharge piping been designed for the anticipated Joule-Thompson cooling?

All of above are indicative of the kinds of questions a qualified leader would guide the HAZOP team into exploring. Moreover, the team must have the expertise and documentation to be able to answer the questions.

4.4 Apply Discipline in Performing Basic Process Safety

The hunt for "little things" does not preclude the need to continuously manage all process safety systems. Alarms need to be tested, piping needs to be inspected, equipment needs to be maintained, changes need to be controlled, and engineering designs need to be appropriately executed consistent with relevant technical codes and standards, as examples. These overriding management systems, if effective, set the framework for finding the "little things" and allowing the interconnections among safeguards to be understood.

4.5 Stay Vigilant to Signals

Many major accidents were preceded by "dry runs." For example, a control valve slammed shut in a hot crude oil feed line, causing a pressure surge that damaged the feed pump seal. A year later, the same valve abruptly closed with the subsequent pressure surge causing a flange to open. Hot oil was released and a fire burned for several hours, requiring significant reconstruction. By thoroughly investigating these kinds of earlier incidents and taking corrective actions, the potential for later more serious events to recur is reduced. Reference 2 devotes an entire chapter to this subject, titled "Not Learning from Near Misses."

Since major accidents or even major near misses are infrequent occurrences, relying on them as indicators for corrective actions may be too late. When process safety management systems start to degrade, there are usually many small indicators. For example, important aspects of process safety information might be out of date, a few changes might be slipping through the system without proper management of change analysis, or there may be a handful of overdue inspections or corrective actions. Overall, the process safety systems appear to be functioning well, but with some gaps. Gaps can become "dots" in the accident chain. These leading indicators of system degradation represent the earliest opportunities to improve effectiveness. Special efforts should be made, through self auditing, to identify them.

Appendix A of the CCPS publication, *Recognizing Catastrophic Incident Warning Signs* (Reference 8) provides a useful checklist of "warning signs" as indicators that some degradation in process safety effectiveness is occurring. Of relevance, many of these checklist questions address “little details” ranging from “loose bolts” to “instruments bypassed.”

Recently there has been renewed interest in the concept of “High Reliability Organizations (HROs)” Much of the early work on this subject emerged from research efforts on aircraft carrier and air traffic control operations.[9] Subsequent work attempted to identify key aspects of excellence exhibited by HROs. These include a “preoccupation with failure” and “reluctance to simplify interpretations” among others.[10] While a clear mapping between those organizations giving birth the HRO concept (highly repetitive yet variable activities) and complex processing facility operations has yet to be established, the “reluctance to simplify interpretations” is, in the authors’ opinion, directly transferrable. By searching for a deeper meaning behind even the smallest of failures, organizations will be “connecting the dots” and evaluating those other components that, had circumstances been different, would have allowed that small failure to progress.

4.6 The Role of Culture

A wealth of literature is available reinforcing the principle that the overall organizational culture drives the effectiveness of the process safety management systems. Figure 1 conveys this concept. Leadership, starting with top management, sets the culture.

In Chevron, process safety is deployed within the context of our overall Operational Excellence (OE) efforts. As stated in Company literature:

> “Leadership is the single largest factor for success in OE. Leaders establish the vision and set objectives that challenge the organization to achieve world-class results. ….
>
> “Leaders visibly demonstrate their commitment through personal engagement with the workforce and by showing concern for the health and safety of every individual. They demonstrate the same commitment to protecting the environment and process safety risk mitigation.”

The second sentence holds the key to success in process safety and OE. In order to affect the culture, leaders must do more than pronounce support. They must demonstrate it. Part of that demonstration is reinforcing the need for attention to detail and to the “little things” that lead to big accidents.

5. Conclusion

In this paper we’ve attempted to show that “the devil is in the details.” Major accidents are often the result of small things that are easily overlooked or tolerated. Identifying the little details that can enable an accident to develop requires a “big picture” approach to understand the full breadth of the potential accident chain. It also requires a renewed rigor in such basics as qualitative risk assessments and incident investigation in order to ferret out those “little things.” There is neither a Holy Grail nor a quick solution to process safety. Accident prevention comes down to implementing the CCPS elements consistently, thoroughly, and across all affected functions.

6 References

[1] Barrell, Tony, PhD, Former Head of UK HSE Offshore Safety Division (1998), as quoted in the BBC Video "*Spiral to Disaster*."

[2] CCPS, *Incidents that Define Process Safety,* BP; Center for Chemical Process Safety, New York, NY. (2008)

[3] CCPS *Guidelines for Risk Based Process Safety,* Center for Chemical Process Safety, New York, NY. (2007)

[4] Reason, James, "The Contribution of Latent Human Failures to the Breakdown of Complex Systems," *Philosophical Transactions of the Royal Society of London.* (1990)

[5] Leveson, Nancy, *Engineering a Safer World, Systems Thinking Applied to Safety,* Unpublished Draft. 2009)

[6] Chiles, James *Inviting Disaster, Lessons from the Edge of Technology* HarperCollins Publishers. (2002)

[7] *LPG Release from Caustic System,* API AFPM Operating Practices Symposium, Los Angeles, CA, Nov 13, 2012.

[8] CCPS, *Recognizing Catastrophic Incident Warning Signs,* Center for Chemical Process Safety, New York, NY. (2012)

[9] Roberts, K. H., *Some Characteristics of High-Reliability Organizations.* Organization Science, 1, 160-177. (1990)

[10] Weick, K. E., Sutcliffe, K. M., & Obstfeld, D., "Organizing for High Reliability: Processes of Collective Mindfulness." In B. M. Staw & L. L. Cummings (Eds.), *Research in Organizational Behavior* (Vol. 21, pp. 81-123). Greenwich, CT: JAI Press, Inc. (1999).

[11] Rasmussen, J., "Risk Management in a Dynamic Society: A Modeling Problem," *Safety Science,* Vol 27 No. 2/3 Pp 187-213 Elsevier Science Limited. (1997)

Practical Implementation of Safety Management Systems at Unregulated Upstream Oil & Gas Facilities

Kristin D. Norton, PE CFSE
Risk Management Professionals, Inc.
300 Goddard, Suite 200
Irvine, California 92691
Kristin. Norton@RMPCorp.com

Michael B. Saura
Colin R. Scholtz
Risk Management Professionals, Inc.
Michael.Saura@RMPCorp.com
Colin.Scholtz@RMPCorp.com

Prepared for Presentation at
American Institute of Chemical Engineers
2013 Spring Meeting
9th Global Congress on Process Safety
San Antonio, Texas
April 28 – May 1, 2013

UNPUBLISHED

Practical Implementation of Safety Management Systems at Unregulated Upstream Oil & Gas Facilities

Kristin D. Norton, PE CFSE
Risk Management Professionals, Inc.
300 Goddard, Suite 200
Irvine, California 92691
Kristin. Norton@RMPCorp.com

Michael B. Saura
Colin R. Scholtz
Risk Management Professionals, Inc.

Keywords: Process Safety Management, PSM, Upstream, Exempt

Abstract

Upstream oil and gas facilities are typically exempt from the OSHA Process Safety Management (PSM) standard; however, in light of recent regulatory scrutiny on the safety of upstream operations, many operating companies have extended their safety management programs to their upstream facilities creating challenges for implementation. Many of these facilities have been operating without significant incidents for many years without proper Process Safety Information, Process Hazard Analyses, Management of Change programs, etc. making the challenge of accepting and implementing these programs difficult to embrace. This paper is focused on providing insights and practical tips for the implementation of safety programs at upstream facilities to enhance the commitment to implementing process safety with limited resources.

1. Introduction

Numerous facilities process and store highly hazardous materials, both toxic and flammable, below the PSM regulatory threshold. While these facilities are not required to comply with the rigorous requirements of PSM, because these facilities process highly hazardous materials they still have a responsibility to identify hazards associated with the process and ensure the facilities are designed, operated, and maintained to minimize hazards. In order to meet this objective, many facilities have elected to implement PSM elements at the unregulated facilities.

For example, upstream oil and gas facilities with over ten thousand pounds of hydrocarbons meeting the definition of flammable liquids or gases (e.g., with a flash point below 100 F) are subject to the OSHA PSM standard, with several exemptions; however, many upstream oil and gas facilities do not meet this requirement and are therefore not covered under the standard. These upstream facilities are typically geographically dispersed and include facilities within each area for collecting oil from one or more oil well and separating produced oil, gas, and water.

Additional processing may be completed within the area or the products may be exported for further treatment and/or processing. This results in numerous nearly identical facilities throughout the geographical area that have the same functional design. Figure 1 below provides a simplified overview of upstream oil production.

Figure 1. Upstream Processing Overview

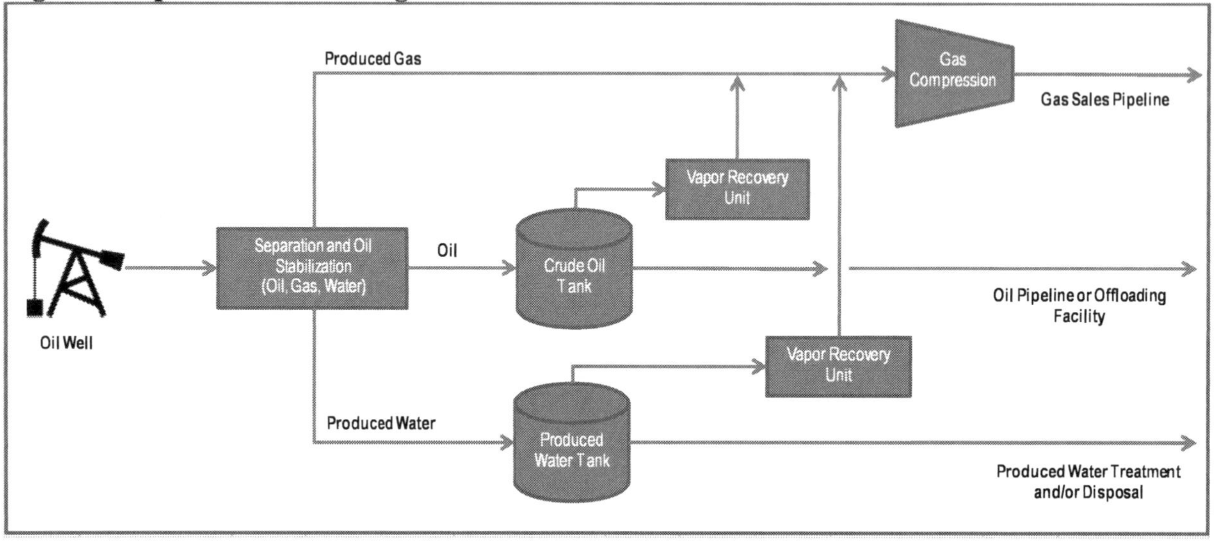

This paper gathers insights from working with companies with upstream operations of various sizes and complexity to provide guidance for organizing and managing the safety management system (both organization and recommendation tracking), as well as specific tips for the following PSM elements:

- Process Safety Information
- Process Hazard Analysis
- Operating Procedures

2. Safety Management System Management

2.1 Organization

PSM program responsibilities extend to various disciplines at the facility, including operations, maintenance, engineering, instrumentation, safety, etc. With numerous stakeholders managing their respective portion of the program it can become difficult to ensure the program is properly implemented and requirements can slip through the cracks. Designating a specific individual or group to be responsible for the overall management of the PSM program provides centralized focus; however, it is also important to have coordination between all disciplines to support the implementation of the program. Some of the most successful PSM programs have support from management at the top and are implemented from the bottom-up, with process safety being embraced as part of the daily duties of all personnel. For example, the Operations Department utilizes the Operating Procedures on a daily basis and as Operations identifies opportunities for improvement and implements changes, the documentation is then updated and coordinated with the PSM Manager. The coordination between the specific departments and the PSM Manager is key to ensuring effective implementation of the program. Thus, it is imperative to have a clear structure in place with defined responsibilities and systems in place to foster coordination among different department.

2.2 Recommendation Tracking

Recommendations are the result of various safety management system studies, including Process Hazard Analyses, Incident Investigations, and Compliance Audits. However, due to the volume of data to process from numerous facilities and studies, recommendation tracking and closure presents one of the most difficult tasks for companies with limited manpower resources. The mechanism to track recommendation closure is left to the discretion of the company and varies a wide spectrum - from simple spreadsheets to customized software solutions. While the specific tool utilized to track the data can streamline the overall management process, the most significant means to streamline data management is to develop and prioritize recommendations based upon risk (e.g., limiting the recommendations to those that will close a risk gap).

A simple way to reduce the number of PSM recommendations precipitating from the studies is to develop specific criteria to drive the team to develop a study recommendation – typically based upon risk and regulatory requirements (including codes and standards). As a result, all suggestions that the team develops that are not risk or requirement driven, but are "nice-to-have", can be tracked separately from the study recommendations, thus requiring less documentation when implemented. Having a separate mechanism to document the catch-all recommendations, such as a formalized "parking lot", allows the team to identify potential operational improvements without having these suggested improvements fall through the cracks and also fosters creative solutions. Additionally, another benefit of designating implementation criteria is to supply management with a tool to prioritize the implementation of recommendations based upon the amount of risk reduction achieved by the respective recommendation.

3. Process Safety Information

3.1 Piping & instrumentation Diagrams (P&IDs)

The successful implementation of PSM, even if not required, relies on accurate and complete Process Safety Information (PSI). An integral part of the PSI documentation is the P&IDs, which represents the process plant in a level such that it illustrates the process equipment, associated piping, basic process controls, and possibly safety instrumented systems. The P&IDs serve as a robust tool of understanding the facility. For legacy upstream facilities that may have switched owners several times, there is a significant gap in the availability of accurate P&IDs. Typically, either the P&IDs are very old and outdated or do not exist.

For facilities that do not have detailed P&IDs available it can be a significant undertaking to develop the P&IDs from scratch, requiring support from engineering and operations personnel via site walk-downs to work with the drafters to develop accurate P&IDs. This can be extremely time consuming and cost prohibitive for companies with limited engineering and operations support that are spread thin managing and operating numerous facilities. However, as mentioned previously, one characteristic of upstream facilities is that there are numerous facilities with the same functional design. While P&IDs are an efficient mechanism to document PSI and are considered best practices, there are other mechanisms a facility can utilize for documenting the technology of the process and the associated equipment. For simplified processes, such as those utilized for upstream operations, a process flow diagram can provide an overview of the technology of the process. Additionally, print-outs from the operations human machine interface (HMI) screen provide information regarding the normal pressures, temperatures, flow rates, shutdowns in place, etc. as well as the functional layout of the process. Figure 2 provides an example of a HMI screen shot for a Lease Automatic Custody Transfer (LACT) Unit that is utilized to accurately measure the amount of crude that is sold. These screen shots (if available) can be utilized as supporting PSI when conducting the Process Hazard Analysis, developing Operating Procedures, and implementing training programs.

Figure 2. HMI Screen Shot Example

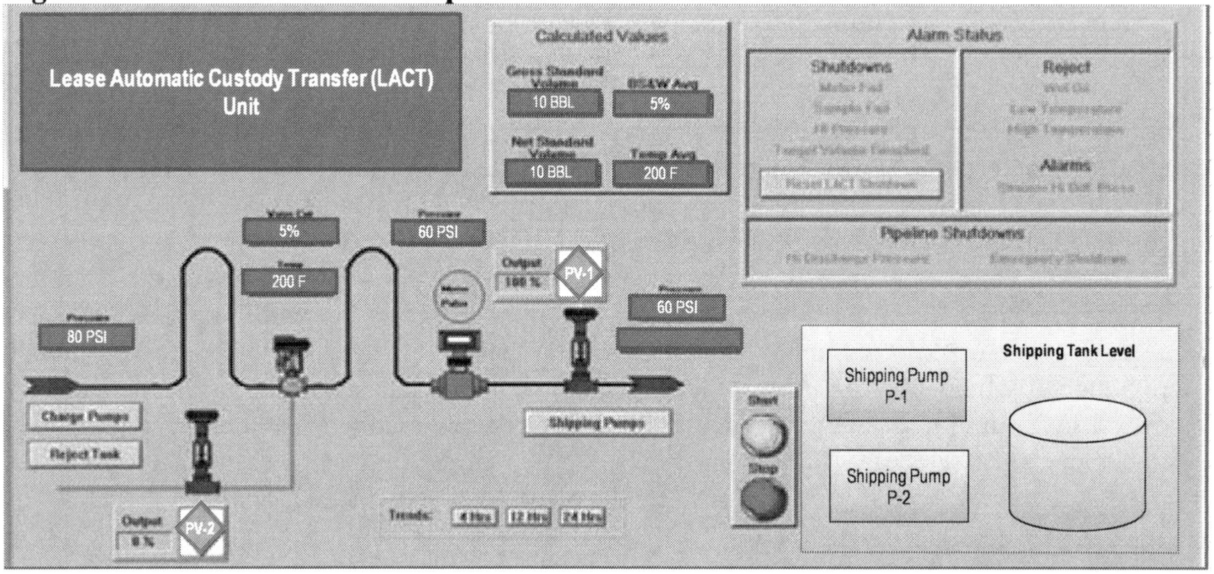

Additionally, for facilities with developed P&IDs, without a drafting department or document control program there is a potential for facilities to postpone or sometimes even neglect P&ID changes. One of the best mechanisms for a P&ID review is the Process Hazard Analysis. During the study, the team systematically reviews the P&IDs in detail and typically finds errors and improvements during the study (and in some cases specific systems are field verified). It is in the facility's best interest to ensure that the PHA sessions document these P&ID errors and changes on a "red-line" copy, typically assigned to one process engineer in the study. This copy can serve as the master red-lined copy until the changes can be implemented by a drafter. This provides a direct means of controlling the required changes to the P&IDs and the responsibility for implementation of changes lies with the assigned process engineer.

4. Process Hazard Analysis

The PHA is another element of the PSM program that requires extensive resources and commitment from the company. Additionally, since upstream facilities are typically dispersed geographically and may be under different operational management, there can be additional challenges ensuring consistency among operating groups. The following sections provide tips to streamline the PHA study process to ensure consistency of the results and make efficient use of available resources.

4.1 Characterizing Hazards of the Process

When managing numerous facilities it can be difficult to ensure consistency in risk ranking for the PHAs, since the ranking is based upon the collective risk perception of each specific team. In order to provide a mechanism to calibrate the risk ranking, management can provide supporting information for the hazards of the process and what level of risk this presents to personnel safety, health and the environment. For example, providing dispersion modeling for various H_2S concentrations, with associated release sizes (e.g., leak vs. rupture) and heights (e.g., process failures vs. relief valves to atmosphere) helps to illustrate the potential for personnel exposure and define whether the associated event can result in a fatality or injury. This can be applied at each facility PHA to assist in ensuring that the risk ranking is applied consistently.

4.2 Checklist Development

There are several PHA techniques that can be applied with various levels of detail, including What-If?, Checklists, and Hazard and Operability (HAZOP) studies. The HAZOP Study technique provides a comprehensive evaluation of the hazards of the process using a team-based, cause-by-cause methodology. While this provides a thorough evaluation, it also requires significant manpower resources when applied at numerous facilities. However, a typical characteristic of upstream operations is that while there are numerous widespread facilities, the facilities also tend to be patterned. By grouping similar facilities, such as tank batteries, compressor stations, vapor recovery units, gauge settings, automatic well test units, etc., we can evaluate each facility group systematically.

One effective approach to evaluate the facility groups is to conduct a baseline HAZOP Study for each facility type, then based upon the results of the study develop a checklist to evaluate the remaining facilities individually, while noting any differences in the evaluated facility from the baseline. These checklists are typically more detailed than the typical PHA checklists in that they are specific for each facility type, with a focus on mitigating specific consequences and safeguards.

4.2.1 Vapor Recovery Unit Checklist Development Example

The following section provides an example illustrating the checklist development for a Vapor Recovery Unit (VRU) Compressor (see Figure 3) based upon specific HAZOP scenarios. This example is not comprehensive, but provides an overview of the technique to transform HAZOP scenarios into checklist questions.

Figure 3. Block Flow Diagram Example –VRU Compressor

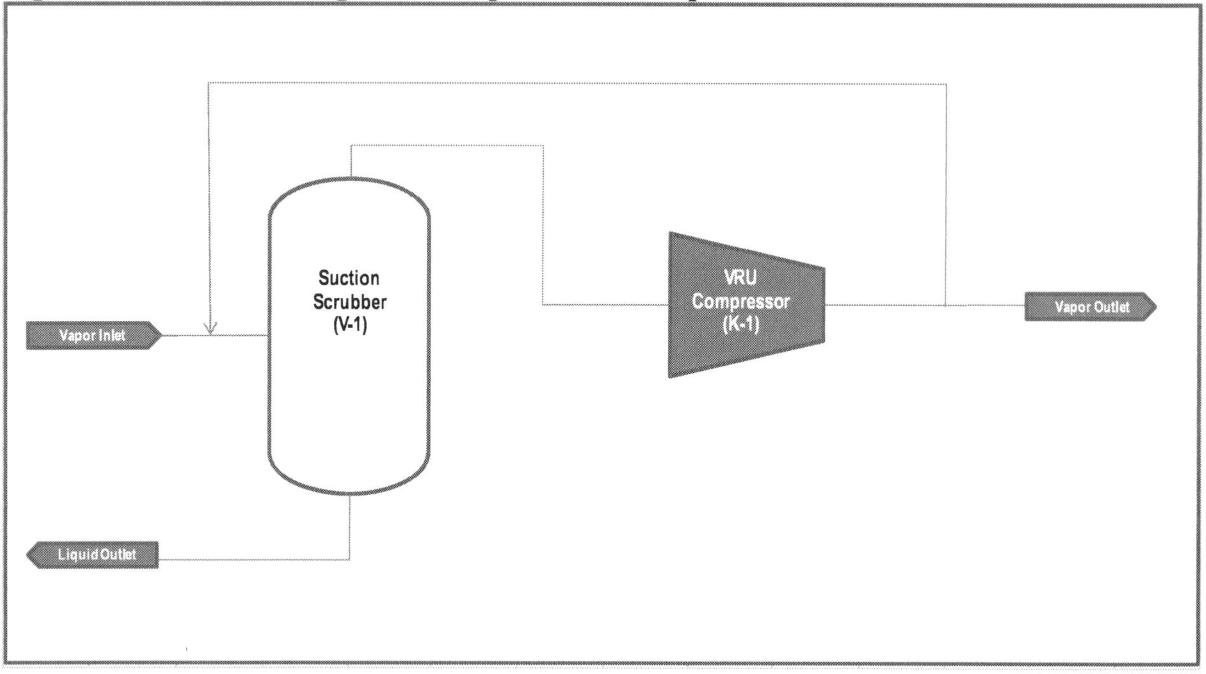

The following table provides an overview of several "No Flow" scenarios from the VRU Compressor baseline HAZOP Study:

Table 1. HAZOP Scenario Example – VRU Compressor: No Flow

HAZOP Scenarios					
Cause	**Consequence**	**Safeguards**	**S**	**L**	**RR**
1.1. Block valve on the suction of the VRU Compressor (K-1) inadvertently closed.	Potential loss of suction to the VRU Compressor resulting in increased temperature and bearing damage. This is an asset damage issue.	Operator Training and Procedures for valve operation. PALL-001 trips the VRU Compressor on low low suction pressure. TAH-001 alarms on high VRU Compressor bearing temperature. TAHH-001 trips the VRU Compressor on high high discharge temperature.	4	4	L
1.2. Block valve on the discharge of the VRU Compressor (K-1) inadvertently closed.	Potential increased VRU Compressor discharge pressure resulting in overpressurization of the discharge piping. Potential loss of containment with fire/explosion and personnel exposure.	Operator Training and Procedures for valve operation. PV-001 opens on low VRU Compressor suction pressure to provide recycle. PAHH-001 trips the VRU Compressor on high high discharge pressure. PSV-001 on the discharge of the VRU Compressor is designed to relieve pressure for the blocked discharge case.	1	4	M

In reviewing these examples, the following questions can be deduced from the consequences and safeguards:

Design Questions

- What are the operating suction and discharge pressures?
- What are the maximum suction and discharge design pressures?
- What are the operating suction and discharge temperatures?
- What are the maximum suction and discharge design temperatures?

Blocked Suction Questions

- Will a loss of suction flow result in increased temperature?
- If the temperature will increase, is there a potential for bearing damage?
- Is there any potential for loss of containment due to bearing damage?
- Are there alarms configured for high gas temperature? If so, will the Operator have sufficient time to respond?
- Are there alarms configured for high bearing temperature? If so, will the Operator have sufficient time to respond?
- Will the compressor trip on a loss of suction pressure?
- Is the suction piping or equipment designed for the maximum settle-out pressure for the compressor?

Blocked Discharge Questions

- Will a blocked discharge exceed the piping or equipment design pressure rating?
- If so, is overpressure protection (e.g., relief valve) provided to protect the piping and equipment?
- Will the compressor trip on high discharge pressure?
- Is there a minimum flow recycle to protect the compressor on a blocked discharge?
- Is the take-off for the minimum flow recycle upstream of the first block valve?
- Will a complete recycle of the flow result in increased temperature that could cause compressor damage?
- If so, are safeguards provided to mitigate increased gas temperature (e.g., cooler, high temperature alarm, high temperature trip)?

General Compressor Questions

- Area lock-out / tag-out procedures in place to minimize the potential for valve mis-operation?
- Is the compressor material selection appropriate for the expected vapor properties to minimize potential corrosion?
- For vibrating services, are pulsation dampeners included in the design?

- Is the discharge piping adequately supported for vibration?
- Is vibration monitoring provided for the compressor?
- Are emergency isolation valves provided in the design to safely isolate the compressor in an emergency?
- Are the emergency isolation valves (manual, automatic, remotely operated) appropriate for the specified service (location, fireproofing, fail position)?
- Are procedures in place for routine testing of the emergency isolation valves?
- Is fire water provided in the vicinity of the compressor?
- Can the fire water be safely operated in the event of a fire in the vicinity of the compressor?

After brainstorming potential questions for the scenarios, a checklist can be developed for these scenarios for the VRU Compressor. A sample abbreviated checklist format is provided on the following page:

Table 2. Checklist Example –VRU Compressor

VRU Compressor		
VRU Location:		
Checklist Team:		
Date of Completion:		
Design Parameters Design Pressure: Operating Pressure: Design Temperature: Operating Temperature:		
Checklist Questions	**Yes/No/NA**	**Comments / Recommendations**
Will a loss of suction flow result in increased temperature? • If the temperature will increase, is there a potential for bearing damage? • Is there any potential for loss of containment due to bearing damage?		
Are there alarms configured for high gas temperature? • If so, will the Operator have sufficient time to respond?		
Are there alarms configured for high bearing temperature? • If so, will the Operator have sufficient time to respond?		
Will the compressor trip on a loss of suction pressure?		
Is the suction piping or equipment designed for the maximum settle-out pressure for the compressor?		

VRU Compressor		
Will a blocked discharge exceed the piping or equipment design pressure rating? • If so, is overpressure protection (e.g., relief valve) provided to protect the piping and equipment?		
Will the compressor trip on high discharge pressure?		
Is there a minimum flow recycle to protect the compressor on a blocked discharge? • Is the take-off for the minimum flow recycle upstream of the first block valve?		
Will a complete recycle of the flow result in increased temperature that could cause compressor damage? • If so, are safeguards provided to mitigate increased gas temperature (e.g., cooler, high temperature alarm, high temperature trip)?		
Are lock-out / tag-out procedures in place to minimize the potential for valve mis-operation?		
Is the compressor material selection appropriate for the expected vapor properties to minimize potential corrosion?		
For vibrating services, are pulsation dampeners included in the design? • Is the discharge piping adequately supported for vibration? • Is vibration monitoring provided for the compressor?		
Are emergency isolation valves provided in the design to safely isolate the compressor in an emergency? • Are the emergency isolation valves (manual, automatic, remotely operated) appropriate for the specified service (location, fireproofing, fail position)? • Are procedures in place for routine testing of the emergency isolation valves?		
Is fire water provided in the vicinity of the compressor? • Can the fire water be safely operated in the event of a fire in the vicinity of the compressor?		

5. Operating Procedures

Though not required by PSM, it is considered an industry best practice to develop and implement rigorous Operating Procedures. A well-developed Operating Procedures program can provide significant benefits to both the facility and its operators, including minimizing risk, documenting safe operating limits, capturing the knowledge of experienced operators, and providing valuable training tools.

Upstream facilities deal with a number of hazards, so from a safety perspective, strong Operating Procedures are an effective management tool to minimize the risk of potentially disastrous incidents from human error. They are most commonly applied to frequent, low-risk tasks though they are often applied to infrequent but high-risk tasks as well. Outside of process controls, administrative controls such as Operating Procedures are a crucial element in minimizing the occurrence of hazardous events.

In addition to reducing the risk of process safety incidents, Operating Procedures also provide operational benefits. By the Center for Chemical Process Safety's (CCPS) "Guidelines for Writing Effective Operating and Maintenance Procedures" [2] definition, Operating Procedures are designed to keep a process within its operating limits. Operating Procedures capture these operating limits and ensure the process is operating within its design intent. This ensures consistency of operations, which is also highly beneficial to facilities with multiple operators operating the same process. [1]

Additionally, one of the most significant challenges posed to upstream facilities is knowledge transfer. Many facilities have been operating for years without robust Operating Procedures, relying primarily on the knowledge of experienced Operators to safely operate the facility. However, as operators retire or switch positions, the knowledge of that particular facility or system is no longer available to those left behind to operate the facility. Developing Operating Procedures from scratch is a significant time investment; however, in the long-run developing a knowledge transfer program with comprehensive Operating Procedures will not only help to support operations, but will also provide a valuable training tool. An effective way of implementing a knowledge transfer program is to develop detailed Operating Procedures utilizing the operator shadowing method (i.e., documenting actions for specific procedures based upon observing the actions of the experienced operator). When operating a facility becomes second nature, it is easy to overlook all of the implicit actions associated with a simple task. For example, when discussing starting a pump with an experienced operator is typically documented as a single task (e.g., Start Feed Pump #1). However, this could be expanded to include the specific valve line-up (including recycle), monitoring of specific pressure or flow indicators, expected suction and discharge pressures, troubleshooting advice based upon operating experience, etc. One of the best ways to extract this information is by coupling less experienced operators or engineers with the experienced operators for the shadowing. This method also provides a valuable training tool for the less experienced operators.

5.2 Keeping the Operating Procedures Current

The implementation of successful Operating Procedures is reliant on the accuracy of the procedure. A key method of ensuring accuracy is to conduct an annual review of the operating procedures by key stakeholders from management and operations. Additionally, Operating Procedures should be reviewed following the completion or revalidation of the PHA to ensure the consequences of deviations from the normal operating limits are documented accurately. The resultant updated consequences of deviation can be used in the O&M procedures. Both the annual review and the PHA ensure that the O&M Procedures are current, thus continually minimizing risk and ensuring that the process is kept within its operating limits.

6. Conclusion

Effectively applying PSM at upstream facilities presents numerous challenges due to the geographic extent of facilities and limited available manpower resources. However, as evidenced in this paper there are systems available to streamline the implementation which requires pre-planning and setting up the systems as part of the overall management structure, including the following:

- Defining a management system and clearly assigning PSM responsibilities.
- Developing and maintaining PSI, including P&IDs, and implementing document control policies.
- Standardizing the PHA approach, including characterizing hazards to consistently apply risk ranking, setting risk criteria to define and prioritize recommendations, and organizing facilities such that resources can be systematically applied to baseline HAZOP Studies and subsequent checklists.
- Implementing a knowledge transfer program to develop robust Operating Procedures.

7. References

[1] Scholtz, Colin R. and Maher, Steven T. "Tips for the Creation and Application of Effective Operating Procedures"

[2] "Guidelines for Writing Effective Operating and Maintenance Procedures", Center for Chemical Process Safety, American Institute of Chemical Engineers, 1996.

Overcoming PSM Challenges for Small Chemical Companies: Observations, Insights and Guidance from the Trenches

Philip M. Myers
Advantage Risk Solutions, Inc.
Sunbury, OH, USA
pmyers@arisksolution.com

Prepared for Presentation at
American Institute of Chemical Engineers
2013 Spring Meeting
9th Global Congress on Process Safety
San Antonio, Texas
April 28 – May 1, 2013

UNPUBLISHED

Overcoming PSM Challenges for Small Chemical Companies: Observations, Insights and Guidance from the Trenches

Philip M. Myers
Advantage Risk Solutions, Inc.
Sunbury, OH, USA
pmyers@arisksolution.com

Keywords: PSM, process safety, small company, limited resources, guidance, process safety management, process safety culture, chemical facilities

Abstract

Complete implementation of the OSHA Process Safety Management regulation can be a daunting task for small chemical companies. Small companies typically are much more resource limited in terms of finance and man-power. There may be no process safety professionals on staff, leaving the company without the associated current process safety expert knowledge, skills, abilities and attendant tools. So how can a smaller chemical company navigate the Process Safety Management (PSM) landscape and make steady progress in development and implementation of a practical, functional program that can be maintained and improved? From program development to the technical details of process safety information and mechanical integrity, management of ongoing changes, to the all important "buy-in" of personnel from the president to operators and technicians in creation of a process safety culture – how can it be achieved? This paper presents observations and insights from years of experience working "in the trenches" with small companies - regarding common pitfalls and limitations at small sites and guidance for successful methods to move forward and overcome obstacles encountered.

1. Introduction

Small companies are unique organizations, and there is a wide range of what constitutes a "small" company. Small chemical companies may have revenues in the millions of dollars, rather than the billions of dollars for many Fortune 500 companies. This also means that operating and capital budgets may be limited. Small companies are often privately held, may be family owned and managed, and/or have a single or few plant sites. Staffing is generally "lean" with the total number of employees often less than one hundred. Staff tend to "wear multiple hats," with responsibilities for more than one area that may be seemingly unrelated. It is not uncommon for a human resources manager to also be responsible for process safety at a site with a single process covered by the Occupational Safety and Health Administration (OSHA) Process Safety Management (PSM) standard [1] and/or the Environmental Protection Agency (EPA) Risk Management Program Rule (and Prevention Program - Level 3 provisions) [2]. Many small companies covered by the PSM regulation do not consider themselves to be "chemical companies," yet they use regulated chemicals above the threshold quantities in processes covered by the PSM standard in the conduct of business. While the PSM standard is not applicable to certain limited types of facilities and conditions, there is no general exemption or modification for small chemical companies. And while the EPA Prevention Program is tiered, it is based upon hazard and/or covered processes not on the size of the company. This leaves small chemical companies with the requirement to meet all aspects of the PSM standard. Observations and insights are provided herein, as well as guidance to small companies working to achieve PSM compliance and develop a positive process safety culture within the constraints of a small company environment.

2. Why is PSM important for small companies?

Small chemical companies, and other small companies that handle Highly Hazardous Chemicals (HHCs), may have significant hazards. Significant risks may be posed to employees and to the company without hazard identification, control, and risk management.

Small companies, that do not even consider themselves to be chemical companies, may not:

- Process toxic materials...but do have a large ammonia refrigeration unit, or use chlorine cylinders
- Produce explosive materials...but do use propane as a raw material to the process
- Produce corrosives...but use nitric acid (also toxic) in making a product
- Process at high temperatures...except for kettles, boilers, or heat exchangers
- Process at high pressures...but do have high pressure steam

and, yet, may be covered by the PSM standard due to the hazards and risks associated with those materials [3].

The Chemical Safety Board (CSB) is an independent United States (U.S.) federal agency charged with investigating industrial chemical incidents. From a review of CSB investigations, it is clear that many of those investigated are not Fortune 500 companies – most of the investigations are for accidents occurring at smaller companies and facilities.

Given small company constraints, management and staff may not be aware of the hazards or of regulatory requirements. Limited knowledge of the hazards, and in particular chemical reactivity, on the part of personnel at small companies has been a common theme and key factor in many of the CSB incident investigations.

At the onset, it was acknowledged that small chemical companies (and smaller companies generally) are unique. Process safety is not defined by the size of the company. There are small companies with excellent process safety programs, and much larger ones with programs that are lacking. In fact, much larger, but highly decentralized companies may effectively have problems similar to those at small companies.

The focus of this work is to highlight and provide insights regarding key factors and trends from observations while supporting small companies over a number of years - as well as apparent from CSB incident investigations. This is a solution-oriented approach to look at what makes small companies different, and to provide guidance regarding how to effectively improve process safety in a small company setting. This guidance may be equally applicable and valuable to small facilities within larger organizations as well.

3. The challenges

There are many challenges associated (rightly or wrongly) with small chemical companies and other small companies using Highly Hazardous Chemicals (HHCs). A number of these factors are apparent in specific consulting assignments, are readily recognized by staff at small companies, and are specifically identified in CSB investigations.

Resource limitations are well understood (at least superficially) – both in terms of finance and staffing. Smaller organizations with revenues in the millions of dollars have much more limited operating budgets as well as smaller and more focused capital budgets. Process safety activities need to be included in the budgeting process. This can be the first "hurdle" for a small company, to simply develop and include a process safety budget or line item. For example, process safety activities that will involve changes in equipment, non-destructive examination or testing, facility siting, or otherwise incur capital expenditure need to be communicated to management and included in the budget planning process. Smaller companies often do not have process safety professionals on staff, and operate as very "lean" organizations. PSM responsibilities are delegated from management to an environmental manager or engineer, engineering manager or engineer, human resources manager, or any of the many other possibilities.

Process safety knowledge as well as a complete understanding of the hazards and process safety risks is a difficulty for small companies. With limited staff that already have multiple responsibilities, and often with no process safety experience readily available onsite, adequately addressing process safety is certainly challenging. Even for those staff tasked with process safety and diligently attempting to do what is needed, it can be a matter of "you don't know…what you don't know."

Small chemical companies typically are very flexible and innovative organizations. This enables them to respond quickly to customer requests and needs. Many small chemical companies serve as suppliers or toll manufacturers for major chemical companies. Having to respond quickly, and given the potential for somewhat limited process safety knowledge, may lead small companies to be more prone to fall back on experience-based (i.e. "seat of the pants") solutions. Certainly, this can be a problem at larger organizations, but in those cases there are more people to ask questions, additional viewpoints, access to more professional process safety personnel and tools, and larger staffs and budgets to address the questions that arise. A person with process safety responsibility at a small company or site may feel he/she "has to do it all." At a small company, budget pressures may also drive someone assigned process safety to bypass the analyses and studies perceived to be "expensive" and, again, rely on experience or judgment. Experience certainly can play an important role, but it does not replace process safety expertise, tools, and analysis when they are needed.

Staff at small chemical companies may also have very limited budgets (if any) for training, conferences, and even reference materials. Many small companies do not have the resources to send persons with process safety responsibilities (as one of many) to specialized training courses or to national process safety events such as the American Institute of Chemical Engineers (AIChE) Global Congress on Process Safety. If financial resources are budgeted, it is typically for local industry association or trade group meetings. In addition, with a small staff, management finds it harder to justify sending someone away for several days of training, and providing the necessary "coverage" for that person's duties while they are away.

A significant problem that arises all too often is "normalization of deviation". "Normalization of deviation" is a state when a deviation (perceived or known by the company or not…and often it is not) continues for a long enough time period that it becomes the norm or expectation. Normalization of deviation can involve physical aspects of the plant equipment, site practices, procedures, use of methods or techniques, and others. This problem is sometimes exaggerated in small companies due to some of the factors already noted – small staffs where it is harder to obtain other views or opinions, limited process safety knowledge, and others. Small companies often have key personnel that have been with the company for years. However, if these deviations become in-grained in the key senior staff, they get passed along to new hires and the deviations are propagated. One example of "normalization of deviation" occurred at a small chemical company in the conduct of Process Hazard Analysis (PHA), with the technique misapplied so many times that it became "the way we do it." At another chemical company located in the United States (US), the first engineering and construction company hired was a Japanese firm. Throughout the plant flexible expansion joints could be seen - intended (in Japan) for earthquake design, but unnecessary at this plant location. The flexible expansion joints became unintentionally designed-in weak links in this plant, and were continued in later designs as the site norm.

Small chemical companies tend to use manual operations to a higher degree than larger counterparts (who are often customers). As small companies have to be responsive and flexible, sometimes it is not possible to hard-pipe or automate the process. Many more processes at small companies are batch, used to supply relatively small quantities when they are needed by the customer. In contrast, at major chemical facilities, a large proportion of the processes are continuous (or at least semi-continuous) as the processes are kept on-stream to meet larger, more regular product demands. At smaller companies, where processes may need to be routinely started up, shut down, and/or used to produce multiple products, more manual activities often result. All of these changes (often at a rapid pace) put pressure on the process safety program generally, and specifically on the Management-of-Change (MOC) process.

4. Overcoming challenges – a view and guidance from the "trenches"

4.1 Program leadership and the core "team" - first who…then what

The first and most important step a small company can take in developing or improving a process safety program is to focus on the individual to lead the process safety effort, and then on key person(s) for the core "team". This step will likely determine whether the process safety program will be successful, functional and compliant or an ineffective "drag" on the company. After leadership and the core "team" is established, the company can then focus on what needs to be accomplished. A defining characteristic of great companies presented in *Good to Great: Why*

Some Companies Make the Leap and Others Do Not by Jim Collins [4] is having the right program leadership – the "process safety leader." "Process safety leader" and core "team" are in quotations here because (1) at a small company process safety may be a functional role (not in title) with only a fraction of the responsible person's time spent on process safety, and (2) the "team" might be a "team" of one in some cases. The program is far more likely to be successful with a "champion" who is able to focus attention on the program and is committed to its successful implementation. After finding the right "process safety leader", next is identification of the core "team." This core team is key to a successful transformation. Jim Collins states that "People are not your most important asset. The right people are." [4] It is important to include persons that are highly respected and valued by the organization, and may include long-time technical managers, engineers, or other technical staff members. For small companies, this small core team will likely need support by outside expert resources to bring process safety knowledge, skills, abilities, and an industry-wide view and experience. The process safety consultant should be someone trusted (or in whom trust is built) and fully integrated as part of the core team from the beginning. This person can help guide the core team; take on specific, more difficult tasks requested; and/or serve as an expert advisor. It is critical that this core team works together well as a team. The team could consist of as few as two members (process safety leader and outside support) for a very small, single covered process site, or include two to three more for somewhat larger organizations with multiple covered processes.

Figure 1. Select the Right People First

4.2 Confront the brutal facts – initial assessment

Building a solid process safety program (or initiating a focused effort to improve one) begins with an accurate, honest, critical evaluation and appraisal of the "starting point" for the organization. The primary regulatory references are the PSM standard [1], OSHA compliance directive [5], and may also include (if applicable) the EPA Risk Management Program Rule [2], and EPA guidance for auditors [6] and inspectors [7]. Even in development of new process safety management programs, there likely is information from existing business operations that can be used. It is important to confront the brutal facts upfront – the whole, unvarnished truth. This can sometimes be a somewhat "uncomfortable" review process, with vested personnel, interests, long-standing procedures, and ways of doing the business. If a program is not working as intended, is inefficient, mistakenly "glosses over" serious risks, or if significant personnel issues exist, it can be tenuous to honestly assess the situation and develop a plan of action. Even within "good" programs or operations, there may be problematic areas – or simply complacency may have set in with certain personnel with important roles. To effectively move forward, the company needs to confront reality and document the "starting point" from the initial assessment – it will become the foundation for development and action. The initial review and assessment can be accomplished by leading with questions (not answers), engaging in open dialog, and conducting a solutions-oriented review geared toward the desired development or improvement. When a thorough formal review of the program or starting situation is conducted and the facts are gathered, the path forward often becomes apparent. A positive process safety culture is important at plant sites of any size. Smaller companies, however, are often focused on a limited process safety program that is compliant and works within the existing company structure and constraints. Most small companies are not interested in the "latest and greatest" and don't want to be on the "leading edge," rather, they want a program that is compliant, functional, and "good enough."

4.3 Decide "where to begin" – on what basis?

Once the upfront evaluation or appraisal of the organization's "starting point" is complete, the next step is to decide where or how to begin the work of developing or improving the program. For a new process safety program, the best place to start is with development of a written program that is documented in a PSM program manual. This is also a good starting point for any company without a written PSM program. It is important to have an overall written program in place. In developing the written program, many issues tend to get sorted out that would require resolution later anyway – when working on individual PSM elements. Development of an overall written PSM Program helps with assignment of responsibilities and to resolve other site issues. Through development of a written program, the core process safety team can begin to build the linkages between company or site management, the core team, and the program. It is important for the core process safety "team" and management to be "on the same page" before launching into development or improvement of the various individual technical elements and supporting site programs and activities.

Once a written PSM program (manual) is prepared, and the key issues have been sorted out with management, the team can then decide how to begin the detailed work for the program or individual elements. There are a number of ways that this can be approached, taking into account the results of the initial assessment of the company's "starting point." The first step is to determine if there are actions the site can take to exempt the process(es) from the OSHA PSM standard. This should be done even if the company will still develop a process safety program. In that case, the process safety program becomes driven by the site needs, without the regulatory requirements. For instance, if the maximum process inventory onsite or the chemical concentration can be reduced to below the regulatory thresholds, the OSHA PSM standard may no longer be applicable - though the OSHA (and/or EPA) General Duty Clause still applies.

It is important to prioritize work that will need to be done in development or improvement of a process safety program. For small companies, with limited staffing, this becomes even more important. There are a number of ways that that the tasks can be prioritized. Following are example approaches that can be used [8]:

- Begin with areas that have an existing basis and materials that only need to be improved or enhanced to include items specific to PSM
- Start with elements that can be implemented short-term and with little or no capital requirements
- Start with elements that present the highest risks (or losses) for the company
- Initiate work on elements with the longest anticipated time to completion (the most difficult problems to "tackle")
- "Divide and Conquer" – Implement several elements at the same time by assigning elements to different groups (e.g. engineering, environmental, research and development) and/or people

Figure 2. Choosing the right direction for a small company process safety program

However the core team decides to proceed, it is important to make sure that the program or element developed will work for a small company – given the type of facility, size, resources, management, and culture (unless the culture simply needs to be changed...a long term proposition). The program and its components need to be set up so that they can be routinely used as part of the mainstream business processes, and maintained on an ongoing basis once the initial development effort is over. If the program isn't developed in this manner, it may soon be worthless or effectively "shelved". At this point in the program development, it is important for the core process safety team to communicate with site management about resources required (financial and human), and to assist with provision of rough (conservative) estimates for budgeting and planning.

4.4 Plan for discrete, manageable, step-wise progress

As the core process safety team considers plans for further development of the process safety program, it is important to divide the work into discrete, achievable steps. A trusted process safety consultant can be invaluable in this process, as the company considers its "starting point," "where the program needs to go" (a vision of an optimal program for the company), and how the team can incrementally "get there". This involves a comparison to the regulatory requirements, as well as to current industry practices and norms, since the OSHA PSM standard is performance-based (as is the EPA Prevention Program – Level 3). There may be existing materials for certain elements of process safety that are "close" to what is needed for PSM, and simply need to be better or more specifically documented. Other elements may require development from little or nothing. The plan developed should:

- Begin with existing management systems (to the extent possible) that are used on a daily or regular basis. Small companies generally cannot afford to create another, entirely separate management system for process safety [9].
- "Begin with the end in mind." [10] Though the program will be developed in steps, it is good to think about the desired goal, and then build in "hooks" or "placeholders" as may be needed in the interim. For most small chemical companies, the goal is not to be "on the leading edge" of process safety. More often, the goal is to be in compliance, and to have a program that is functional and "good enough."
- Include business growth plans so that the PSM program isn't soon outstripped by the business activity.
- Clearly assign responsibilities and milestones – make sure that you know…and they know… (who is responsible, what is to be done, and when it is to be completed).
- Set reasonable targets – this work will take longer than the process safety leader or core team thinks – this is especially true for small companies where human resources often get diverted.
- Track work and completion progress and dates – what is measured and tracked gets done. Monthly status/progress charts can be used for safety/PSM meetings, and to communicate progress to management.
- Be adjusted/revised periodically – give the company time to make progress, but expect and understand that the plan will need to be revised and adjustments made as the work progresses. Some of these changes are likely to be significant if a new process safety management program is being developed.
- Include periodic checks and audits by others – the checks and "audits" can be formal or informal, as long as they are effective and keep the overall efforts and progress on track
- Include regular communication with company senior and executive management – communicate actions, what is being achieved, what is ahead/behind plan, and most importantly what will be needed from them – be "realistic" and don't understate resource needs, but also be credible…it's a balance.

4.5 People – the "wider" team - assign responsibilities and priorities

In the development or improvement of a process safety program or individual elements, personnel should be assigned specific responsibilities. This includes assignment of "owners" for the overall PSM program and for individual program elements. These individuals can be "point persons" for the development or improvement process, as they ultimately will be the primary persons responsible for that area or element. The "owners" can be supported by others with secondary responsibility for the element – potentially serving an important role both in development and in later function of the program. Still others may be expected to participate in the development of the program element through provision of inputs, though their overall level of involvement in development is lower.

At small companies, it is important to involve all technical staff and management to the degree possible – and "not let anyone off the hook." That includes senior and executive management. This is a chance for the core process safety team to communicate to management the important role management needs to fulfill (including visible commitment and leadership, provision of resources, authorizations and approvals), and that process safety is not "just" a compliance program. With a limited staff, it is helpful to identify how most company human resources can be assigned some process safety responsibilities that can be accomplished as part of day-to-day activities. An example is given in Table 1 (following the references) for assignment of process safety development and implementation responsibilities for the overall PSM program. The example is for a facility that is also covered by the EPA Risk Management Program Rule and Prevention Program requirements. [It should be noted that in the example table provided, the CEO is not involved in the day-to-day management or operations of the company. Rather, the CEO represents the company owners/shareholders, with a focus on overall company and financial performance.] It is recommended that a similar table be developed specifically for the mechanical integrity element due its importance to the PSM program and the level of effort typically required.

Assignment of responsibilities is very specific to the individual company and the knowledge, skills and abilities (KSA) of individual team members. Further, for smaller companies, a number of the roles listed in the example table may not exist (at any given organization or single location), or the roles may be functional, and without the title of "manager." While it is generally advisable to have one "owner' for the various program elements, there may be exceptions. One notable exception is for "visible support and actions" where management and leaders throughout the organization should consider themselves to have "primary" responsibility. Other cases may arise depending upon organizational preferences, personnel expertise, and/or a further functional split of responsibility – such as the engineering manager may have responsibility for mechanical integrity of new designs and engineering projects while the maintenance manager has responsibility for existing, in-plant equipment. The division of responsibilities does not necessarily appear to be logical at times, as with limited staff, sometimes it is simply necessary to review and approved each other's work. Assignment of responsibilities also depends upon reporting relationships, how "hands-on" an individual may be in a given area, and peculiarities of specific organizations. The table also highlights the need and expectation to get inputs to the program and individual elements from plant hourly personnel. In the division of process safety responsibilities, what makes sense for one organization may not work for another. Having the right people involved is important in all organizations – but is even more critical for smaller companies. The example roles and responsibilities matrix given (in Table 1) is provided to demonstrate the range of personnel that can be assigned various process safety responsibilities in a small chemical company.

4.6 Process safety knowledge and tools – identify areas of weakness and supplement

People at small companies are "motivated, hard-working, genuinely concerned people without the time, confidence and training" [3] for process safety. Often, they are missing the desired reference materials and tools as well. For a small company to develop an effective process safety program the "process safety leader" and core "team" should know and be focused on what is *required* and what is *needed* by the site. There is no "ignorance is bliss" application in process safety. And, even strictly in terms of compliance, a lack of knowledge or understanding is not a legal defense for non-compliance, but can potentially expose the company for gross negligence if an accident occurs. This is another reason that while PSM certainly is a compliance program, it is never "just" a compliance program. Personnel with process safety responsibilities at small companies want to "do the right things," but often have a more challenging situation given the difficulties associated (potentially) with insufficient process safety knowledge – unless local knowledge is supplemented with outside process safety expertise.

A difficulty for small companies comes in the area of process safety knowledge generally, as well as in a complete understanding of the hazards and process safety risks. Adequately addressing process safety is certainly challenging - with limited staff that already have multiple responsibilities, and often with no process safety experience readily available onsite. The often repeated "you don't know…what you don't know" is a real concern in process safety - even for personnel with process safety responsibilities and diligently attempting to do what is needed. Simply stated, staff may not know enough to be able to ask the right questions. Further, some process safety questions are difficult to address unless the person is immersed in process safety. For instance, how is a small company to know that the OSHA regulated concentration for formaldehyde is 37%? It was not published in the original list of HHCs, the list has not been updated with this information, and it only exists in an OSHA letter of interpretation. And, some process safety issues can "get into the weeds" even for process safety professionals, such as in defining upstream and downstream covered process limits (depending on the material, concentration, interconnections, co-location, etc.), inclusion or exclusion of various plant support systems and structures, understanding facility siting and human factors requirements, and so forth.

A more dangerous process safety knowledge issue occurs when "you don't know…what you think you know" – and have limited readily available resources to check into the issue or problem with others. For example, a chemical site had a large release of a toxic liquid, that covered a storage containment area, and nothing really "bad" happened. A level of "comfort" can develop that even if the worst event imagined goes wrong, it's still not a problem. Only…the release experienced was for a material with a boiling point near atmospheric temperatures, and the event happened on a cool spring day. What, then, happens on a hot summer day? The hazard zone or area affected could be significantly larger and have much different impacts. On that hot day, the plant "experience" with the previous release may be worth little. At a large company, when such process safety questions arise, the local site personnel may be able to access corporate Subject Matter Experts (SMEs) in process safety. At a small company, the "process safety leader" or team member may be left to find answers on their own…or with resources on the internet (and, hopefully, supplemented by a trusted outside advisor). A further complication can arise when checking with peers at other small, local chemical companies. While this is generally a helpful approach, the feedback should be carefully considered, as it sometimes may be misleading. Checking with and remaining comparable with peers doesn't help if they are also wrong. This situation arose for one small chemical company when comparing how a number of small local companies defined the limits of regulatory applicability. This situation involved bulk storage of a listed chemical (above the threshold quantity), and transfer of the material via process piping to downstream batch reactor processes using the material in small quantities (significantly below the threshold quantity in the reactor). In this case, the small chemical company had defined the limit of the covered process as the bulk storage and process piping – not including downstream batch reactors and equipment. Its local peers defined similar processes in the same way. The only problem was that they were all wrong. Working with the outside process safety expert, who also consulted (for further verification) with an individual at the regulatory agency who had primary responsibility for setting up the regulatory program, the batch reactor processes were then included in the PSM program. It is worth remembering that while benchmarking with or seeking the advice of other small chemical companies in the areas may be helpful, they also may "be in the same boat," with limited process safety knowledge.

The tools used to manage and implement the process safety program can also help set the stage for success…or frustration. Migration from paper-based systems to electronic ones – at least in part - is a positive, and generally advisable step for many small chemical companies. It is not uncommon, at least initially, for small company process safety programs to be paper-based systems. Transitioning to even simple, electronic (MS-Word®, MS-Excel®, MS-Access®) systems for a PSM program helps both in development and in the longer-term maintenance of the program. An electronic system helps makes it easier to manage and track various aspects and elements of the program, enables document control, provides easier access to information, prevents information from getting lost, and can be backed up with other business systems and data. For small companies with multiple covered processes use of a Computerized Maintenance Management System (CMMS) can become a worthwhile investment and a significant aid in handling all of the voluminous data, actions, and records associated with a mechanical integrity program. It does require effort to input information and build the database, but can yield significant benefits in the mid- to long-term. However, one word of caution…a number of small companies have found that trying to build a PSM program into an enterprise resource planning (ERP) system (designed for other business processes) can be a difficult and sometimes ineffective approach. And, it may be "overkill" for a small company. ERP systems often have maintenance functions, for example. However, use of a specialized CMMS – designed for and used in the process industries – may be a better choice. For companies with a single or few covered processes, simpler electronic systems may be sufficient.

4.7 Process equipment – limitations and finding a path forward

Small chemical companies often utilize used equipment that is in long-term storage or purchase used equipment on the open market, due to capital budget limitations, and with direct impacts on process safety. Often, the focus is short-term with the key driver being functionality at the lowest possible cost [11]. Compliance with codes and standards, facility siting, related process safety information, and other long-term issues that may require additional studies and incur additional costs may be "put off" to be dealt with at a later time.

Assembling a process with used equipment often results in a process with varying safe operating limits for each major equipment item, and equipment that may not have been designed for the currently intended service. Equipment may not be properly sized, including design for pressure control devices. A small company may change process parameters and the process design to enable use of the much cheaper used equipment. All of this leads to a process that may be designed to different codes and standards, with differing safe operating limits, different materials of construction, and a number of unknowns. Sometimes small companies will even forgo purchase of the vessel code documents (e.g. U-1 forms) that, if available, may be provided at an added expense in the equipment purchase price. Further, the used equipment may be physically modified to meet requirements of the project. In the case of physical modifications to equipment, if they are not made by a "code shop" the equipment is no longer a "code" vessel.

Documentation of the necessary information for process safety for used vessels purchased on the open market, in terms of the process safety information and information for the mechanical integrity program, can become a real challenge. This may then impact control of the process, management of change, as well as inspection, testing, and preventative maintenance for the equipment and entire process.

When purchasing used equipment, small companies should make every effort to obtain construction drawings, code information (e.g. U-1 forms), inspection records and certifications, and other information that will be needed for the process safety information and mechanical integrity program. This includes payment of additional fees in the equipment purchase price, if necessary, for vessel code documents and other information. It is well worth the small additional charge to have the equipment information, and will same time, money, and frustration later on.

If used equipment is incorporated in process design, the engineering team will need to carefully address issues regarding materials of construction and wall thickness; provision of properly sized pressure relief; and inspection, testing and preventative maintenance, for example. The OSHA PSM standard requires (in 29 CFR Part 1919.119(d)(3)(H)(iii)) that the company "…determine and document that the equipment is designed, maintained, inspected, tested, and operating in a safe manner." In addition, the safe operating limits for the process likely will be defined by the "weakest link" - the equipment or portion of the process with the lowest rating (e.g. pressure rating).

4.8 Build upon what already exists and develop what is needed

Small chemical companies have to be extremely effective in the use of human resources. It is vital that the process safety program is built upon information, methods, and systems actively used in the business, to the extent practicable. This should start with use and/or tie-in to an existing, functioning management system at the site. Small companies often use quality management systems or may have environmental management systems, for example, that may be used for process safety management as well.

Existing information that can be used as part of the process safety program (identified in the initial review and assessment) should be assembled for inclusion. Examples of materials that are often already in use at small companies and can be incorporated into the process safety program include:

- Material Safety Data Sheets (MSDSs)/Globally Harmonized System of Classification and Labeling Safety Data Sheets (SDSs)
- Procedures
- Batch tickets or protocols
- Safe work practices (e.g. hot work, line break)
- Maintenance program (in some form) and work order system

Plant sources of this information may include operations, research and development, the site safety program, environmental programs, operations, and others. The program development process can make use of this existing information so that not everything has to be built from the "ground-up." Some areas will need to be enhanced or improved, and others may require new development.

Key technical aspects or elements of process safety programs that are particularly important, and will need to be rigorously developed (and likely improved even if they already exist) include:

- Process Hazard Analysis – the primary means to identify the hazards.
- Management of Change – as it "touches" personnel throughout the business on a daily basis, is used to manage the risk associated with changes, and can be used to drive and build a positive process safety culture.
- Audit – formal/regulatory and more importantly informally as a means to evaluate progress and what is needed – feeds back into planning and may initiate "course correction."

4.9 Use available resources

There are a number of resources that small companies can use to develop process safety programs. The most important resource is a trusted process safety consultant. Working together with, and as an advisor to the "process safety leader" and core team, this individual should be ideally qualified and able to help point the core "team" to additional resources. Examples of available resources follow:

- Regulator websites (e.g. OSHA, EPA)
- Federal agencies (e.g. CSB)

- Reference materials, such as:
  - Center for Chemical Process Safety (CCPS) publications, including *Guidelines for Risk Based Process Safety* [12]
  - Standards from a number of organizations (e.g. American Petroleum Institute, National Fire Protection Association)
  - Journals (e.g. Process Safety Progress – provided with membership to the AIChE Safety and Health Division, Journal of Loss Prevention in the Process Industries)
  - Electronic and online services and libraries (e.g. Knovel® - provided as a benefit of AIChE membership)
- Networking sources:
  - Local networks (e.g. chemical companies)
  - Staff contacts at previous employers or vendors
  - Emergency services (e.g. fire department)
  - Professional contacts (e.g. trusted peers, mentors)
  - Linked-In
- Process safety websites, blogs, newsletters
- Electronic mailing lists and accident alerts/databases
- Professional associations (e.g. American Institute of Chemical Engineers – local and national level, American Chemical Society, International Society for Automation)

In many instances, quality information can be accessed for free on the internet. Due care should be taken, however, to ensure that information obtained is from a credible source – otherwise, the materials may be worth exactly what they cost…nothing. Most current standards are not available on the internet for free, as the standards organizations use income from sales to fund the organization's activities and further standards development.

4.10 Budget for process safety needs - operational and capital budgets

As early as possible in the process, the company planning and budgeting cycle should include process safety management resource requirements. Initially, at many small companies, process safety expenditures may be "ad-hoc" or as approved by management. It is important to develop a budget for process safety and include it within the normal business cycle. It is best to incorporate process safety into existing business systems and processes when possible. Inclusion of process safety as a budget item facilitates better planning and allows for steady progress. Two further recommended actions are to:

- Include process safety costs for specific customer-driven projects or changes in customer contracts – often the incremental cost to the customer is small and the impact in providing resources for a small company's process safety budget is quite large.
- Consider process safety projects or tasks as engineering projects that are planned, budgeted, and tracked.

4.11 Re-evaluate and adjust the process safety development or improvement plan

No matter how effectively and diligently the "process safety leader" and core "team" plan, changes to initial plans will be needed. Every small company is unique. The "starting point" for small companies can vary significantly, including the amount of information that can be used from existing business activities. And, the process and work likely *will take longer than initially planned.* This is especially true for small companies and teams – as personnel needed for other day-to-day tasks and projects in operations may become unavailable for process safety work planned. The "process safety leader" at a small company needs to be flexible and reasonable, while consciously working to hold staff accountable and ensure the process safety work progresses.

4.12 When an orderly process is not feasible – make progress when and where possible

All of the previous discussion assumes that a reasonable level of support is given by management, and that at least "pockets" of support can be found within the staff to develop a process safety program. If it appears that support is so limited that such a planned, orderly approach is not initially possible, a humorous alternative is presented by Berger in "The Process Safety Guerilla: Giving Your Company A Future, Even if it Doesn't Want One." [13] Berger explicitly states regarding terms used and the approach presented that "....their use is only in the metaphorical sense...not to advocate illegal activity of any kind." Berger's paper outlines how an employee concerned about process safety at a facility can (as a "process safety guerilla") accomplish positive change. The approach described by Berger [13] is to:

1. "Arouse and organize the people" – identifying supporters and involving people with similar concerns and interests.
2. "Achieve internal unification" – creating an identity and mission, and holding information discussions.
3. "Establish bases" – building a group of persons willing to promote process safety and networking with others.
4. "Equip forces" – gathering and obtaining materials and tools, and learning how to use them.
5. "Gain strength" – by becoming an authority and taking a stand as the opportunity arises.
6. "Gain control" – by converting senior or executive management into supporters, or by using a near-miss or an accident to draw attention to the importance of process safety.

Typically, support for process safety can be found somewhere within the company, so a planned, orderly path can be taken. However, at one point or another, the "process safety leader" at a small company may never-the-less feel like a "process safety guerilla."

5. Commitment and perseverance

Building, maintaining, and sustaining an effective process safety program within a small chemical company setting requires a high level of professional and personal commitment and sheer perseverance. At a small company there are many demands on each staff member, and most "wear more than one hat." Process safety can be compared to a heavy flywheel on an engine, using the analogy from Jim Collins [4]. It takes a lot of energy to get the flywheel going. Once it is going at the desired speed, it smoothes out the running of the engine, and continues the motion of cylinders.

Figure 3. The "Flywheel" – Through Steady Efforts Develop Momentum

The primary concept from the "flywheel" analogy that applies to process safety is to *build and maintain momentum.* Anyone that has been tasked with starting a new process safety program can express the frustration caused by a lack of momentum during initial stages of the program. It requires steady, hard work to bring the flywheel into constant motion and progress. The transformation of process safety programs works much like a flywheel – and, there is no secret or magical solution – it takes steady, hard work. There are many sound fundamentals that lead to a "good" process safety program, but no single answer. Process safety requires constant work, pushing in the same direction, more or less according to plan. To build, improve, maintain, and sustain a good process safety program – "work hard" and "work smart" are familiar and appropriate advice.

6. Summary

In spite of many challenges small chemical companies face the requirement, for a complete process safety management program and its full implementation, is the same as required for much larger organizations. Small companies have to be creative and highly effective in development of practical, functional programs that utilize existing resources and reasonably delegate responsibility for various program elements across the organization.

Key technical programs for process safety include Process Hazard Analysis, Management of Change, and Audit, as they are critical to identifying hazards, managing the ongoing changes affecting process safety and risk, and providing the means to identify gaps and areas needing further attention. The management of change program also can be effectively used to "drive" a change in process safety culture over time - with regular interactions of personnel throughout the organization and by focusing attention on changes, decisions and actions taken and the implications on process and company risk.

There are many challenges that small companies must overcome. It starts with good process safety leadership and development of a core team that is professionally and personally committed to process safety. The brutal facts regarding the company "starting point' should be identified early on, as well as identification of existing materials and processes that can be incorporated. This provides a basis for deciding "where to begin" with the work and enables planning of discrete, manageable steps in the process safety development or improvement process. At small companies, the process safety responsibilities generally have to be delegated across the small staff, with logical assignments made that recognize and incorporate other normal staff assignments and daily activities when possible. While there are a number of resources available to small companies, the best resource is likely to be a trusted process safety professional/consultant that becomes an integral part of, and advisor to, the core team. For small companies, one of the most difficult aspects of process safety to address is *insufficient process safety knowledge*. You don't know...what you don't know. This challenge can be overcome largely through judicious use of a process safety expert and additional resources he or she is able to provide. Throughout the entire process, it is important to integrate process safety into existing business systems and operations to the extent possible. And, this includes the company planning and budgeting cycle. Flexibility, commitment, and perseverance are all qualities needed in the "process safety leader" and core "team" for development, implementation, and maintenance of effective process safety programs at small companies – that when applied, can be used to creatively overcome the process safety challenges encountered.

7. References

[1] Occupational Safety and Health Administration (OSHA), 29 CFR Part 1910.119, Process Safety Management of Highly Hazardous Chemicals; Explosives and Blasting Agents; Final Rule, Washington, DC, February 24, 1992.

[2] Environmental Protection Agency (USEPA), 40 CFR Part 68, Accidental Release Prevention Requirements: Risk Management Programs Under Clean Air Act Section 112(r)(7); Final Rule, June 21, 1996.

[3] Howat, Colin S, "Simplified PSM for Smaller non-Regulated Sites: Development History, Program Overview & Recommendations," 8th Global Congress on Process Safety and 14th Process Plant Safety Symposium, Houston, Texas. April 2012.

[4] Collins, JC, *Good to Great: Why Some Companies Make the Leap...and Others Don't*, New York, NY: HarperCollins Publishers Inc., 2001.

[5] Occupational Safety and Health Administration (OSHA), Instruction CPL 02-02045, Process Safety Management of Highly Hazardous Chemicals - Compliance Guidelines and Enforcement Procedures, September 13, 1994.

[6] Environmental Protection Agency (USEPA), EPA 550-B99-008, Guidance for Auditing Risk Management Plans/Programs under Clean Air Act Section 112(r), August, 1999.

[7] Environmental Protection Agency (USEPA), EPA-550-K-11-001, Guidance for Conducting Risk Management Program Inspections under Clean Air Act Section 112(r), January, 2011.

[8] Goddard, Sheryl, "A Pre-Plan for Process Safety," 8th Global Congress on Process Safety and 14th Process Plant Safety Symposium, Houston, Texas. April 2012.

[9] Myers, Philip M, "Retool Your Process Safety Program for Efficiency and to Meet Multiple Management System Requirements," 4th Global Congress on Process Safety and 10th Process Plant Safety Symposium, New Orleans, Louisiana. April, 2008.

[10] Covey, Stephen R, *The 7 Habits of Highly Effective People: Powerful Lessons in Personal Change*, New York, NY: Simon & Schuster, 1989.

[11] Herber, John M, "Why Smaller Companies Struggle with Process Safety," 8th Global Congress on Process Safety and 14th Process Plant Safety Symposium, Houston, Texas. April 2012.

[12] Center for Chemical Process Safety, *Guidelines for Risk Based Process Safety*, Hoboken, NJ: John Wiley and Sons, 2007.

[13] Berger, Scott, "The Process Safety Guerilla: Giving Your Company a Future, Even if it Doesn't Want One," 5th Global Congress on Process Safety and 11th Process Plant Safety Symposium, Tampa, Florida. April, 2009.

Table 1. Overall PSM and Risk Management Program - Roles and Responsibilities Matrix

Roles/Positions	Vision, Goals, Objectives	Program Oversight & Direction	Provide Resources - People, Equipment, & Financial	Establish Organizational Roles, Responsibilities & Accountability	Visible Support & Actions	Overall Program & Reporting	Overall PSM Program & Reporting	Overall Risk Management Program & RMP & Reporting	Employee Participation	Process Safety Information (PSI)	Process Hazard Analysis (PHA)	Operating Procedures	Training	Contractors	Management of Change (MOC) & Pre-Startup Safety Review (PSSR) / Process Change Request and Review (PCRR)	Mechanical Integrity (MI)	Safe Work Practices (SWPs) - incl Hot Work	Incident Investigation	Emergency Management	Compliance Audits	Trade Secrets
CEO	A				*																
President	A	A	A/P	A	P	A	A	A	*									A	A	A	
Plant Manager/Site Leader	A	A	P	P	P	A	A	A	*				I		A	A		A	A	A	
Business Managers (Business Development)					*				*						I						S
Research & Development Manager	I				*	I	I	I	*	S	S	S	S		S	I		A		I	P
Controller/Purchasing Manager			S		*				*					S	I	A					
Human Resource Manager	I		S		*	I	I	I	*				S/A	I	S					I	
Engineering Manager	S	S	I	I	P	S	S	I	I	A	A	A	P	A/S	A	A/P	I	A/S	A	A	S
Environment, Health, Safety & Security Manager	P	P	I	S	P	P	A	P	A/S	I	I	I	A/S	I	I	I	A	P	P	P	A
PSM Manager/Coordinator	S	S	I	I	P	S	P	S	P	P	P	S	S	S	P	S	I	S	I	S	S
Maintenance Buyer			S		*				*	I				S	S	S				I	
Environmental Manager	I				*	I	I	S	*		I		S		I	I	I	I	S	S	I
Safety Manager	I				*				*	S	I	I	P	P	I	I	P	S	I	I	I
Process/Plant Engineer					*	I	I	I	*	S	S	P	I	I	S	S	I	I		I	S
Production Manager			S		P				*		S	S	I	I	S	I	I	I	I	I	
Systems Administrator					*				*			I			I					I	I
Instrumentation & Controls Engineer					*	I	I	I	*	S	I	I	I	I	S	S	I	I		I	
Maintenance Manager					P	I	I	I	*	I	I	I	I	S	S	P	S	I		I	
Chemist					*				*	S	I	I			I	I	I	I			I
Maintenance - I&C Technician(s)					*		I	I	*	I	I	I	I	I	I	I	I	I	I	I	I
Maintenance - Mechanical Technician(s)					*		I	I	*	I	I	I	I	I	I	I	I	I	I	I	I
Supervisors & Operators					*		I	I	*	I	I	I	I	I	I	I	I	I	I	I	I

A = Approve Activity or Awareness
I = Input and Participation Expected
P = Primary Responsibility
S = Secondary Responsibility or Support
* = Important Activity for All

9TH GLOBAL CONGRESS ON PROCESS SAFETY

Small Site MOC: Big Company Challenges without Big Company Resources

Rainer Hoff, Ph.D., P.Eng.
Gateway Consulting Group, Inc,
8610 Transit Road, Suite 100, East Amherst, NY, 14051
rhoff@gatewaygroup.com

Gordon C. Blake, P.Eng.
Gateway Consulting Group, Inc,
8610 Transit Road, Suite 100, East Amherst, NY, 14051
gblake@gatewaygroup.com

Prepared for Presentation at
American Institute of Chemical Engineers
2013 Spring Meeting
9th Global Congress on Process Safety
San Antonio, Texas
April 28 – May 1, 2013

UNPUBLISHED

Small Site MOC: Big Company Challenges without Big Company Resources

Rainer Hoff, Ph.D., P.Eng.
Gateway Consulting Group, Inc,
8610 Transit Road, Suite 100, East Amherst, NY, 14051
rhoff@gatewaygroup.com

Gordon C. Blake, P.Eng.
Gateway Consulting Group, Inc,
8610 Transit Road, Suite 100, East Amherst, NY, 14051
gblake@gatewaygroup.com

Keywords: Management of Change, MOC, business process improvement,

Abstract

Many have argued that the PSM regulations impact small sites particularly hard. While small sites must comply with the same regulations as much larger sites, small sites are hamstrung by a lack of resources. This is particularly true for Management of Change due to the inherent complexity of MOC.

This paper focuses on the challenges of improving the MOC business process at small sites. An MOC improvement initiative generally involves improving the MOC business process, and then assisting with some form of technological automation.

The authors interviewed representatives from small sites to obtain their first-hand lived experiences, in order to properly ground the study.

1. Introduction

Managing change effectively has always been important at chemical plants and refineries. Lack of a proper MOC program the prime cause in a number of well-known accidents, including the Flixborough explosion [1], and a contributing factor in other well-known accidents [2, 3]. Certainly small sites are not immune to accidents where improper MOC was a factor.

Since the promulgation of OSHA 1910.119 [4]—the Process Safety Management regulations—proper Management of Change is a regulatory requirement, for Covered Processes (as defined in the regulation).

The Process Safety Management regulations cover 14 elements, much more than just Management of Change. As a result, when the regulations were first promulgated in 1992,

substantial effort was needed to bring most plants into compliance. In this environment, with deadlines and potential penalties, Management of Change business processes were either hastily created, if they did not previously exist at the site, or quickly reworked and formalized from an existing business process.
Poorly or inadequately applied MOC processes have contributed to many incidents [2, 5] , and have been identified as issues in Refinery National Emphasis Program [6] and Chemical Plant National Emphasis Programs [7]. Based on the authors' experience, personnel who are required to use the MOC process, the "users", often have a negative view of the MOC process. User dissatisfaction may be due to a poorly-designed or poorly-implemented process.

Herber [8] notes that 85% of the Chemical Safety Board investigations occur at small companies, implying that small companies are disproportionally impacted by PSM issues, including MOC. Both Herber [8] and Goddard [9] elaborate on the reasons why small sites find PSM compliance so challenging.

Many small sites, whether triggered by an incident or simply a desire for enhanced process safety, attempt to improve their MOC systems. Improving MOC normally requires improving the MOC business process itself (the policies, procedures, forms, engineering aids, etc.). Efficiency improvements, and often better compliance, can be achieved by using electronic technology to implement MOC.

This paper summarizes the results of a survey of 64 sites who have attempted to improve their MOC processes. Some were independent small sites, while others belonged to a larger corporate entity.

This paper does not attempt to define MOC best practice, since that is well-covered by other authors [10-17]. This paper focuses on the effort, challenges and enablers that allow a small site to move on the path to best practice.

1.1 MOC Overview

A high-level model of MOC, referred to as a "lifecycle" model, is shown in Figure 1. The diagram depicts a permanent, non-emergency MOC process. A lifecycle is composed of "states": Initiation, Scoping, Change Design, etc. While the MOC is active, it could be considered to be "in" a state: e.g. the MOC is in the Approvals state.

Further details of lifecycle models can be found in the references [10].

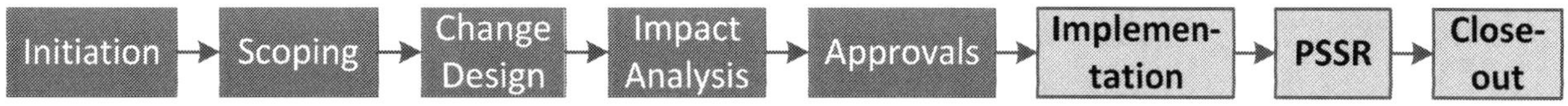

Figure 1. Typical permanent, non-emergency MOC lifecycle.

2. Characterizing Small Sites

2.1 What is a Small Site?

There are various metrics which can be used to characterize the size of a site:

- number of units or processes
- number of employees
- physical size
- proximity of coworkers
- revenue

At a small site, there may only be a few or a dozen units or processes, unlike a large refinery, where there may be up to 100 units.

A small site often has fewer than 100 people on site (without regard to whether these are employees or contractors).

Also, a site may be physically small so that it is convenient to walk, say, from the administration building to the control room(s); in a large site, it may be more practical to drive.

At a small site, technical resources are usually in the same building or even the same room. In contrast, a large site is bigger and has more formal communications, often respecting organizational boundaries and hierarchies.

2.2 How Do Small Sites Behave?

The small size engenders a number of behaviors, which promote efficiency:

- informal communication
- rapid decision-making
- easier consensus-building

Communication tends to be less formal at a small site, since coworkers see each other on a routine basis.

Decision-making is rapid, because the people that need to be involved, are close at hand.

Consensus-building is often easier, since it's often based on personal relationships, which may be stronger in a smaller organization.

On the other hand, a small site has:

- fewer total resources
- fewer opportunities for specialization

With fewer individuals, a given person may have many roles and responsibilities, so it is more difficult to become an expert in one domain—staff at small sites are required to be generalists [9].

2.3 The Best/Worst of Both Worlds: Small Sites as Part of a Larger Corporate Entity

The prior discussion contains an implicit assumption, being that the small site is completely independent of any other site or corporate entity. There are comparatively few small sites that enjoy that level of independence. More common is for a small site to be part of a larger corporate entity. This provides opportunities for collaborating with other sites and leveraging corporate resources, but also has a concomitant obligation to conform to corporate policies, procedures and technology standards.

The position of a site as an independent entity versus part of a larger corporation has an important impact on site autonomy and how the site may proceed in any MOC improvement initiative.

2.4 Research Methodology

Small sites may not have the same opportunity to improve their MOC systems as large sites have, due to resource constraints. This paper reports the results of a qualitative research study with input from a variety of sites.

A total of 64 sites were surveyed for this research. However, the sample was not random. The sample was biased in favor of companies and sites that had expressed an interest in improving their MOC systems. Consequently, 0% of the survey participants believed that their MOC systems were best practice, and 100% believed that improvements were possible and desirable.

Small, independent sites were surveyed as well as small sites that belong to a larger corporate entity, the latter being more prevalent.

The results presented in the following chapters, are be ordered by the feedback that was most prevalent: that is, the most common issues are stated first, the second most common next, and so on. Since this is a non-randomized, qualitative study, presenting quantitative results (e.g. x% of respondents had characteristic Y), is not statistically valid. As is common for qualitative research, the results could be used to identify relevant questions for a future quantitative study.

3. MOC Business Process Improvement

MOC improvement initiatives are generally conducted in two phases:

1. MOC Business Process Improvement: This is where the MOC process itself (policies, procedures, guidelines, forms, engineering aids, etc.) are updated.
2. MOC Technology Improvement: This is where the MOC process is implemented (or updated) using some form of technology.

3.1 Independent Site MOC Business Process Improvement Challenges

Based on the survey results the most common challenges for small independent sites are:

1. Uncertainty over who "owns" the business process
2. Lack of user engagement at the proper time.
3. Lack of compliance with existing MOC process

There is often overlap between the Quality organization and the Process Safety organization at a site. Both the ISO9000 quality systems program [18] and OSHA1910.119 process safety management regulations [4] govern how procedures are modified, approved and released. This creates situations where both organizations (rare) claim ownership of the MOC process, or neither organization (more common) claims ownership of the MOC process. While there may be a perceived need for MOC improvements, no one organization feels that it has the mandate to advance this kind of initiative.

An MOC Improvement Team ("Team") may work diligently at improving the MOC process. This team will no doubt solicit input from interested parties at the site, in an attempt to sincerely address MOC requirements contributed by several disciplines. Often, the site response is lackluster, and meaningful feedback isn't received until after the Team has completed its work, has updated the MOC procedure, and is now syndicating the new process throughout the facility. Of course, at that point, it is too late to incorporate the feedback without redoing much of the Team's work.

Sometimes site personnel do not execute MOCs in accordance with their existing MOC procedures: checklists are ignored, forms are not completed, and signatures are overlooked. User reactions while being trained in the new MOC process may be that it is "a lot more complicated". In fact, the new process may be much simpler than the former process, but if the former process was ignored, then the mistaken perception is one of increased complexity.

3.2 Independent Site MOC Business Process Improvement Enablers

Based on the survey results the most common success enablers for small independent sites are:

1. MOC redesign can be funded from local (often engineering) budgets
2. Someone is willing to "champion" the initiative
3. A streamlined process, without unnecessary steps, leads to higher compliance

MOC process improvement initiatives are often done with a combination of local resources and outside expertise. For a small site MOC redesign, the cost of the outside resources is normally within the approval limits of local site management. Funds are often available in budgets for engineering studies.

Good leadership of an MOC improvement initiative is important. A project champion, like a good project manager, is likely to keep the initiative moving, and is able to overcome obstacles.

Needlessly complex processes discourage compliance. Even a well-intentioned person may find it difficult or impossible to comply with a confusing process or one with unnecessary steps. A simpler process, streamlined to the essentials should encourage better compliance with MOC requirements.

3.3 Corporate Site MOC Business Process Improvement Challenges

This covers the case where a small site desires or is required to participate in an MOC process improvement initiative in conjunction with other sites or in compliance with a corporate initiative. Based on the survey results the most common challenges for small sites that are part of a larger corporation are:

1. Lack of a regulatory mandate
2. Difficulty in arriving at a consensus on the path forward
3. Impacts of acquisitions and divestitures

Some companies have dozens of sites, but only some of them have any PSM-covered processes. Nonetheless, the corporation mandates that business processes will be consistent across the organization, and therefore the non-PSM sites must implement an MOC process. A non-PSM site has less impetus for deploying MOC than a regulated site, so the effort and cooperation are necessarily less at the non-PSM sites.

Many companies pride themselves on their adoption of consensus management. Consensus management normally implies that most or all parties must agree to a new initiative or approach. This becomes problematic when a consistent MOC process is sought to address multiple sites whose needs are inconsistent. Sites with simpler requirements require considerable coaxing to accept approaches that address the needs of other, more complex, sites.

The current business climate is very dynamic with acquisitions and divestitures of assets (e.g. entire plants or facilities) being quite common. Unfortunately these may not be rapid events, but multi-year processes as a site is made available for sale, groomed for the sale (by cutting all expenses beyond the bare essentials), becomes the subject of bidding and speculation, and is finally sold. Once a potential sale is announced, MOC improvement initiatives stop almost immediately, since the participants await direction from the new owners.

3.4 Corporate Site MOC Business Process Improvement Enablers

Based on the survey results the most common success enablers for small sites, which are part of a larger corporate entity, are:

1. Effective championing
2. Effective facilitation
3. Effective buy-in

As mentioned in §3.2, having a person willing to champion the MOC process improvement initiative is a big contributor to its overall success.

One of the difficulties of redesigning the MOC process, on a corporate or multi-site basis, is that none of the participants may be experts in business process design, and each of them may simply understand his/her local process and not fully appreciate the processes at other sites. A facilitator, experienced with business process redesign and having MOC experience, is a vital member of the MOC Redesign Team. Without effective facilitation, the Team may degenerate into cycles of the form: first “A” is seen as the solution, then “B”, then “C”, then back to “A”, etc. The facilitator should avoid that.

An improved MOC process, with enthusiastic multi-site support, will experience better compliance than if such support were lacking.

4. MOC Technology Improvement

The second phase of MOC improvement initiatives is often to realize business process efficiencies by deploying an electronic MOC system.

Often, the MOC technology improvement is performed concurrently with the MOC business process improvement. That is a reasonable approach, since the available technology options put constraints on what is possible or practical during the redesign.

4.1 Independent Site MOC Technology Improvement Challenges

Based on the survey results the most common technology challenges for small independent sites are:

1. Budget limits
2. Positioning of MOC as a “tag-along” to another initiative
3. Lack of leadership

There are very real budgetary limits on what a small site can spend on a technology solution for what is perceived as an administrative process (in contrast with process control systems which are viewed differently). If the proposed technology fits into the site’s spending limits, then the project can often proceed. If the project is more costly, then the justification process is so onerous, that site personnel either don’t have the time or desire to pursue it.

There may be other technology upgrade initiatives taking place at the site, and the participants feel that they may be able to take advantage of economies of scale if MOC were upgraded at the same time. Often, there is a team with a mandate to implement better incident investigation and tracking; that team may be well-resourced. The person(s) responsible for MOC is invited to participate in the technology evaluation, again, with the hopes of obtaining a lower unit cost. The difficulty with this is that incident investigation (when measured by the number of tasks) is a much less complex process than MOC. So, the evaluation effort is actually inverted, whereby the simple process (incidents) receives a lot of attention, and the complex process (MOC) "tags along" and receives comparatively little attention. This will usually result in a sub-optimal MOC deployment, even if the technology selected is acceptable.

Similarly, there may be a team investigating electronic document management ("EDM") with the goal of managing process safety information and other important facility documents and drawings. Again, the person(s) responsible for MOC is invited to participate in the technology evaluation, again, with the hopes of obtaining a lower unit cost. This arrangement has a different problem. From a project management perspective, MOC is well-bounded. MOC lends itself to fixed-price, short-duration implementation because the MOC process requirements are similar from site-to-site—only the details differ. EDM, in contrast, is comparatively open-ended with many opportunities to expand the scope of the initiative by simply addressing more document types (PSI, environmental, maintenance, inspection, ...). An EDM initiative is not typically fixed-price, and has a longer implementation duration than MOC. Combining MOC and EDM can be done. But the participants must understand that these different kinds of projects.

Lack of leadership is an oft-cited issue when implementing improved MOC technology at small sites. The difficulty lies in the fact that there are many people who want a better MOC solution—PSM Managers, MOC Coordinators, environmental representatives, the user community—but there may be no one who feels qualified to lead what is seen as an "IT project".

4.2 Independent Site MOC Technology Improvement Enablers

Based on the survey results the most common success enablers for small independent sites, undertaking an MOC technology improvement initiative, are:

1. Experienced partners
2. Effective championing
3. Supportive IT resource

Specifying and implementing an electronic MOC solution requires a number of specialized skills. It is unlikely that these skills exist at a small site, nor is it cost-effective to develop those skills. A technology partner with the requisite skills and MOC experience is an invaluable asset.

As stated previously, a project champion is a significant asset to an MOC technology improvement initiative. Technology changes often bring unexpected results, and the change the way people are accustomed to working. A champion is needed to reassure the doubters by explaining the merits of the "new" approach.

A small site electronic MOC implementation does not require much (measured in man-hours) support from IT. But it is usually difficult to accomplish with no IT resources. So a supportive IT person can significant streamline the implementation process.

4.3 Corporate Site MOC Technology Improvement Challenges

This covers the case where a small site desires or is required to participate in an MOC technology improvement initiative in conjunction with other sites or in compliance with a corporate initiative. The end-user survey had far more responses in this section than any of the others. Based on the survey results the most common challenges for small sites that are part of a larger corporation are:

1. An IT organization that wants to create a home-grown solution
2. An IT organization that wants to leverage existing vendor relationships, even when the existing vendor has little to offer
3. An IT organization that views MOC as an extension of the work order process
4. An IT organization that refuses to accept any solution a Team proposes, but offers no solution of its own

Many IT organizations have a mandate to provide technological solutions. IT may conduct "make versus buy" decisions, and decide that MOC is an application that can be accomplished in-house. Sometimes the decision is driven by the existence of programming resources that don't have any current assignments, so it appears that these resources are "free". The success rate of in-house MOC initiatives is poor. Oftentimes, the development team grossly underestimates the complexity of the MOC process and delivers solutions that are inadequate (by trimming the requirements to fit the budget) or consuming far more resources than planned (actual resource consumption of 10 times the planned resource consumption, is not uncommon). In-house development creates other forms of end-user dissatisfaction: as time goes on, the original developers move on to other projects, and the support for the in-house software wanes.

A common IT organization goal is to minimize the "technology footprint". In one sense, this can be restated as reducing the number of major application vendors that IT deals with. This is a reasonable goal, since each new vendor relationship has an associated cost. The problem arises when IT attempts to engage existing vendors that have no MOC solution or an MOC solution in name only (that is, something called "MOC", but lacking important functionality). The end-user community is quickly frustrated by this, because they see themselves teaching the vendor about MOC, when the expectation is that the vendor should already be proficient in this area, and, in fact, be able to share experiences and best practices with the end users.

Almost all sites have an electronic mechanism for generating work orders. Some even believe that MOC could be an offshoot of the work order system, illustrated at the top of Figure 2, and evidenced by checkboxes on the work orders entitled "MOC Needed ? ☑". The creates a situation where MOC is a reactive process. By the time the work order exists, the *final* solution has already been accepted and is ready for implementation. Even if an MOC is triggered at this point, it causes all the essential MOC tasks— risk ranking, task scoping, design proposal, impact analysis, and approvals—to be conducted in a hurried manner, after the fact. These are really the activities that minimize risk and ensure that the MOC has been safety designed, so these are not the tasks that should be rushed. In contrast, the MOC best practice begins with a *proposed* solution, and the MOC concept is developed and risk mitigated throughout its lifecycle. The latter approach, illustrated at the bottom of Figure 2, is the recommended approach. Work order systems are used to implement the MOC activities in the physical plant—so work order systems are important to the overall process. But a work order system on its own is too little, too late, to be a viable automation tool for MOC.

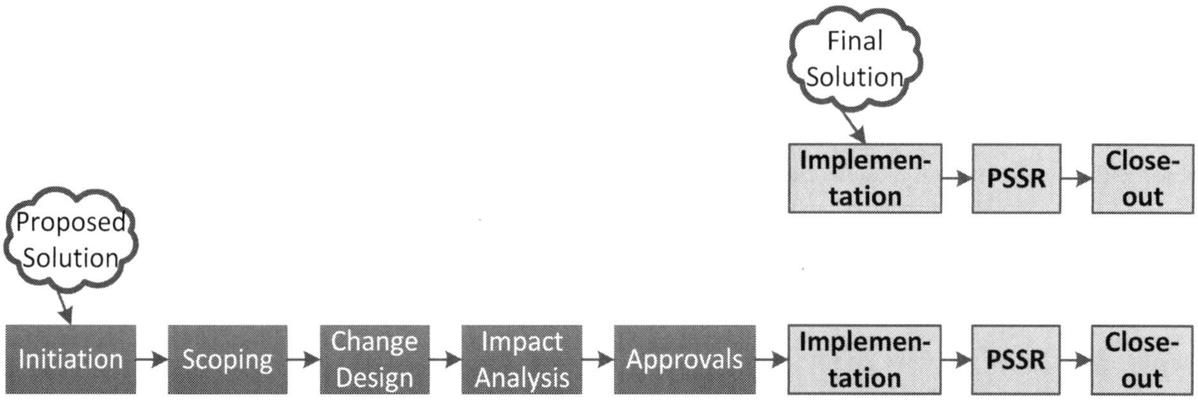

Figure 2. MOCs triggered by work orders (top) versus MOCs independently triggered by a need for change (bottom).

Finally, some IT organizations simply refuse to assist with an MOC improvement initiative. They are unwilling to accept the technology recommendations from the MOC Improvement Team, yet are unwilling to recommend an alternate solution. There may be legitimate reasons—manpower, budget, other priorities but these reasons are often not exposed to the MOC Team.

4.4 Corporate Site MOC Business Process Improvement Enablers

Based on the survey results the most common success enablers for small sites, which are part of a larger corporate entity, are:

1. Supportive IT organization
2. Experienced partners who have experience with enterprise MOC rollouts
3. Executive commitment

Section 4.3 identified 4 barriers to success of an initiative to improve MOC technology, on a corporate basis. All of the barriers dealt with IT issues. The counterpart to that, and one mentioned by all the successful MOC Team, was the support of a capable IT organization. Much of the work is performed by third-party vendors, so the work load on IT is minimal, but the support of this organization must still be there.

Corporate MOC deployments are complex. Not only is the MOC process itself complex, but there are legitimate variations between business units (usually because the processes or hazards are different). An experienced partner will understand and highlight the trade-offs of various levels of standardization and site-specific deployment.

In previous sections, the importance of having someone champion the initiative was stressed. That is no less important for a corporate MOC technology initiative. But also, due to the cost of these larger initiatives and the impact on the organization, it is necessary to have executive support as well.

5. Conclusions

Independent smaller sites have some advantages over larger sites, since communication is easier and decision-making more rapid. They also have some disadvantage due to a lack of specialized resources and resources in general. Small sites that are part of a larger corporate entity may be able to leverage corporate resources, but may also be constrained by corporate policies.

An MOC improvement initiative generally has two phases: a business process improvement phase and a technology improvement phase. The two phases may be concurrent.

A survey of 64 companies identified a number of challenges experienced during MOC process improvement initiatives, primarily:

- uncertainty over who "owns" the MOC process
- lack of a regulatory mandate (at some sites).

The success enablers experienced during MOC process improvement initiatives were primarily:

- redesign efforts are comparatively inexpensive and can be funded from local engineering budgets
- effective championing of the effort

The challenges experienced during MOC technology improvement initiatives are, primarily:

- budget limits
- IT organizations who want to create their own MOC solutions, but may not be able to complete this project successfully

The success enablers experienced during MOC technology improvement initiatives are, primarily:

- experienced partners
- supportive IT organizations

6. References

[1] Parker, R. J., et al., 1975, "Formal Investigation into Accident on 1st June 1974 at the Nypro Factory in Flixborough," London.

[2] Baker III, J. A., Bowman, F. L., Erwin, G., Gorton, S., Hendershot, D. C., Leveson, N., Priest, S., Rosenthal, I., Tebo, P. V., Wiegmann, D. A., and Wilson, L. D., 2007, "The Report of the BP U.S. Refineries Independent Safety Review Panel," Report, Washington.

[3] Travers, I., 2010, "Learning the lessons from Buncefield," AIChE National Spring Meeting, AIChE, San Antonio, TX, p. 10pp.

[4] OSHA, 1992, "Process safety management of highly hazardous chemicals," 29CFR1910.119, OSHA, Washington.

[5] Board, U. S. C. S. a. H. I., 1998, "Investigation Report, Explosives Manufacturing Incident, Sierra Chemical Company," Chemical Safety Board, Washington, DC.

[6] OSHA, 2007, "Directive CPL 03-00-004 Petroleum Refinery Process Safety Management National Emphasis Program," OSHA, ed.Washington, p. 104.

[7] OSHA, 2010, "Directives -10-05 (CPL 02) - PSM Covered Chemical Facilities National Emphasis Program," OSHA, Washington.

[8] Herber, J. W., "Why Smaller Companies Struggle with Process Safety," Proc. 8th Global Congress on Process Safety, AIChE.

[9] Goddard, S., "A Pre-Plan for Process Safety Management," Proc. 8th Global Congress on Process Safety, AIChE.

[10] Hoff, R., 2007, "Lifecycles," MOC Best Practices, 1(2).

[11] Hoff, R., 2011, "Quantitative Models for Assessing MOC Performance," AIChE National Spring Meeting, 7th Global Congress on Process Safety, AIChE, Chicago, p. 15pp.

[12] Sanders, R. E., 2005, Chemical process safety : learning from case histories, Elsevier Butterworth Heinemann, Amsterdam ; Boston.

[13] Chosnek, J., and Clifton, R., 2008, "Improved Process Safety Management and Simple Metrics," Process Safety Progress, 27(4), pp. 284-288.

[14] Keren, N., West, H. H., and Mannan, M. S., 2002, "Benchmarking MOC Practices in the Process Industries," Process Safety Progress, 21(2), pp. 103-112.

[15] American Institute of Chemical Engineers. Center for Chemical Process Safety., 2008, Guidelines for the Management of Change for Process Safety, Wiley-Interscience, Hoboken, N.J.

[16] Chosneck, J., 2010, "Managing management of change," AIChE National Spring Meeting, AIChE, San Antonio, TX, p. 12pp.

[17] Garland, R. W., 2004, "Electronic Management of Change Process," Process Safety Progress, 23(4), pp. 244-251.

[18] International Organization for Standardization, 2000, "Quality management systems--Fundamentals and vocabulary," International Organization for Standardization, Switzerland, p. 36pp.

Novel Development of Asset Integrity Management for Refinery Equipment: Theory, Application and Effect of Setting Critical Operational TAN and Sulfur Level for Distillation Units

Dingrong Qu
SINOPEC Research Institute of Safety Engineering
No. 218, Yan'an 3rd Road, Qingdao, P.R. China, 266071
Qudr.qday@sinopec.com

Shanjun Mu
SINOPEC Research Institute of Safety Engineering
No. 218, Yan'an 3rd Road, Qingdao, P.R. China, 266071
mushanjun@vip.sina.com

Xiaohui Liu
SINOPEC Research Institute of Safety Engineering
No. 218, Yan'an 3rd Road, Qingdao, P.R. China, 266071
Liuxh.qday@sinopec.com

Xiu Jiang
SINOPEC Research Institute of Safety Engineering
No. 218, Yan'an 3rd Road, Qingdao, P.R. China, 266071
Jiangx.qday@sinopec.com

Yanyuan Li
SINOPEC Research Institute of Safety Engineering
No. 218, Yan'an 3rd Road, Qingdao, P.R. China, 266071
Liyy.qday@sinopec.com

Prepared for Presentation at
American Institute oF Chemical Engineers
2013 Spring Meeting
9th Global Congress on Pr℃ess Safety
San Antonio, Texas
April 28 – May 1, 2013

UNPUBLISHED

Novel Development of Asset Integrity Management for Refinery Equipment: Theory, Application and Effect of Setting Critical Operational TAN and Sulfur Level for Distillation Units

Dingrong Qu
SINOPEC Research Institute of Safety Engineering
No. 218, Yan'an 3rd Road, Qingdao, P.R. China, 266071
Qudr.qday@sinopec.com

Shanjun Mu, Xiaohui Liu, Xiu Jiang, Yanyuan Li
SINOPEC Research Institute of Safety Engineering

Keywords: naphthenic acid corrosion, sulfidation corrosion, crude distillation units, and mechanical integrity

Abstract

In today's market, refineries are continuously challenged by the low-grade crudes with increased concentration of sulfur and naphthenic acid. Operators need to know the limits of Total Acid Number (TAN) and sulfur content for their distillation units to avoid unexpected accelerated corrosion, which may lead to fires, explosions and toxic releases. New concepts and theory of critical TAN (Total Acid Number) and sulfur level will be discussed, and the procedure of such MI project is reviewed by using a successful case. Critical operational TAN and sulfur level are determined on the basis of comprehensive evaluation of high temperature corrosion risk for every important part of distillation units. A corrosion model, which included main factors such as TAN / sulfur content, materials of construction, and temperature, has been developed for this assessment. Historical corrosion inspection data is used to validate the accuracy of the model. Experience of setting TAN and sulfur limits for 11 refineries over last 10 years is summarized. Comments and recommendations to improve this technology were made. The benefit of such MI project has been widely acknowledged for the increased reliability of crude distillation units where it is implemented.

1. Introduction

A hierarchical asset integrity management (AIM) programme (Figure 1) is suggested and applied for a major refining and petrochemical company, which runs 30+ refineries with a total crude refining capacity approximately 1.6 billion Bbls/year. In this AIM programme, the operation system of AIM covers principle guidelines for many management aspects, such as organization and responsibility of AIM, resource and training, management of change, inspection, data management, material management, turnaround management, and preventative maintenance. The operation system is documented through standards and guidelines of different levels, and is embodied in AIM Software. AIM tools include some well-known tools for risk management and inspection planning, such as RBI (Risk-based Inspection), IOW (Integrity Operational Window) and SIL (Safety Integrity Level). Setting critical operational TAN and sulfur level is a recently developed AIM tool for distillation units. This AIM system also includes some necessary engineering techniques to ensure asset integrity.

Figure 1. Hierarchical structure of an asset integrity management progamme

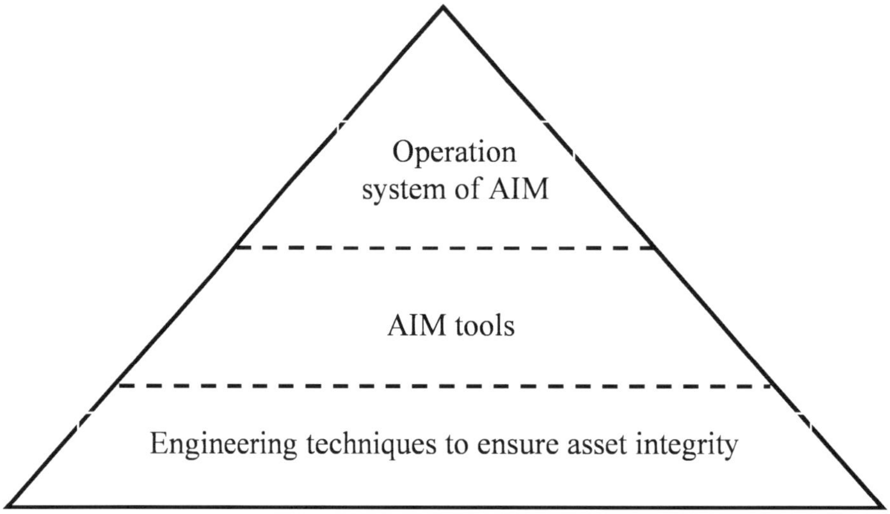

2. Challenge of Processing Corrosive Crudes

One of the worldwide trends in crude oil refining industry is purchasing and processing of corrosive crudes with increasingly higher TAN and sulfur content. Such "opportunity" crudes are intentionally purchased with great discounts. Since the cost of crude oil accounts for approximately 90% of the economics, the cost saved on crude purchases can potentially provide a huge economic advantage over industry competitors. On the other hand, inherently corrosive crude oil could negatively impact the refineries' profit margin by significantly increasing the corrosion rate, and thereby reduce the reliability and integrity of the equipment. Processing corrosive crudes and maintaining equipment integrity is a great challenge for refineries. Data shows that the average TAN and sulfur content of crudes for the company has increased markedly over the last decade. This change in crude qualities has resulted in equipment failures in both the high temperature equipment as well as the distillation overhead systems. As a result of these unpredicted failures, additional attention has been focused on these crudes that contain increased levels of naphthenic acid and sulfur.

Another great challenge for the company is the combination of uncertainty of crude resource and capability of distillation units bearing corrosive crude. Due to availability, refineries have to process any available crude oil. Normally, the crude slate is changed monthly. However, some refineries have had to adapt to crude slates that change weekly or even daily. As a result, processing crudes with various TAN and sulfur content could present a potential and unpredictable risk due to unpredicted corrosion. Moreover, most distillation units were designed and built decades ago for "benign" crudes with limited corrosiveness, such as organic acid and sulfur. During the life cycle of refining units, critical equipment and/or key pipes might be partially changed or upgraded. Therefore, components of distillation units have different remaining life and different capability of bearing corrosive crude. The company would like run its distillation units at a corrosion rate lower than the bearable corrosion rate, which is the maximum limit of corrosion rate in terms of safety and economic concerns. These facts make it of paramount importance for the company to know the bearable corrosiveness limit of its distillation units. And therefore the company can distribute crude slates appropriately among its refineries on the basis of their bearable corrosiveness limits.

3. Concept and Theory

The major problems induced by "opportunity" crudes are Naphthenic acid (NAC) and sulfidation corrosion (SC). Refinery equipment is attacked in the high temperature range of 425 °F (218 °C) up to about 750°F (400 °C). Over the last 60 years many efforts have been made to understand and predict NAC and SC[1-8]. Below are some rules of thumb and useful facts regarding NAC and SC:

- There are two different corrosion prediction tools for SC. The modified McConomy curves are used for SC in H2-free streams (which is the default type SC discussed in this paper) and the Couper-Gorman curves that are used for SC in H_2/H_2S-containing streams[9,10].
- SC in H_2-free streams is thought to occur at temperatures above 446 °F (230 °C). High temperature SC is typically a general corrosion. The protective iron sulfide scale that forms from SC will partially inhibit NAC[7].
- Limited plant experience shows that the modified McConomy curves generally overestimate SC rate. However, the modified McConomy curves could underestimate SC rate at some specific side cuts or locations.
- Fractions may vary significantly in their potential to cause SC due to the type of sulfur compounds. Some sulfur contents can be grouped as reactive or non-reactive, and therefore, their corrosivity could deviate from that predicted by the modified McConomy curves[11].
- Crudes with a TAN value higher than 0.5 mg KOH/g and crude fractions with TAN greater than 1.5 mg KOH/g are typically considered to be corrosive.
- NAC usually starts from 425 °F (218 °C) and severity increases with temperature up to about 750 °F (400 °C) according to API 571. However, some field evidence and laboratory tests show that the increment trend may end at 650 °F (343 °C) due to either thermo-decomposition of the acid or coking. Products of coking at high temperature may act as a physical barrier for NAC and SC, and decrease their rates.
- Naphthenic acids can also be grouped as reactives ("bad naphthenic acid") and inactives ("good naphthenic acid") by both their corrosivity and molecular weights. Some efforts are made to distinguish them by comparing their corrosivity and chemical structures[12]; however, engineering applications are not seen yet.
- Turbulent flow increases the NAC corrosion rate.
- Since SC and NAC can co-exist at the same temperature range, it is difficult and usually unnecessary to distinguish their respective contributions to the total corrosion rate. Occasionally, ''naphthenic acid corrosion index'' (NACI), a simple ratio of the corrosion rate to the weight of corrosion scale, could be used to compare the relative contributes of SC and NAC[13].

- API RP 581 provides a detailed database for estimating corrosion rates of concurrent NAC and SC of normal materials in refineries, including carbon steel, low alloy steel, austenitic stainless steel without Mo, and austenitic stainless steel with Mo.

However, precisely predicting NAC and SC rates based on common knowledge is still a challenging task. Compromise and assumption are needed to make an experiential corrosion predicting model simple and technically feasible:

- Sulfur compounds and naphthenic acids are similarly corrosive, supposing the source of crudes is disregarded. This assumption contradicts the knowledge of SC and NAC for specific crude; however, it is statistically true for most crude.
- Distribution of sulfur compounds and naphthenic acids over boiling points are the same, supposing the source of crudes is disregarded. Again, it is statistically true for most crude.

The corrosion model is used to calculate the bearable TAN and sulfur of distillation cut for every specific location. And then, the bearable TAN and sulfur of crude slate of every location can be calculated proportionally. Critical operational TAN (COTAN) and Sulfur Level (COSL) are then determined for the lowest bearable TAN and sulfur of crude slate of all locations in the distillation unit. The theory detail can be found in the reference[14].

4. Case Study

A small crude distillation with capacity 3 million Bbls/year suffers from heavy corrosion. The plant was originally made of carbon steel and partially upgraded recently. It was forced to shut down for inspection and reparation regularly even its major product paraffin wax was still profitable. Figure 2 through 5 shows that high temperature naphthenic acid corrosion and sulfidation corrosion extensively exist in high temperature part of unit, such as heat-exchanger, pipes, tower, and column internal.

Historical data shows that corrosives have been kept relatively constant over the monitoring period (Figure 6). The TAN of slate is around 1 mg KOH/g oil and sulfur content is about 0.25%. Concentrations of both acid and sulfur are predicted to increase slowly, but consistently in the future. Corrosives distribution is also measured for atmospheric and vacuum distillation as shown in Figure 7, which is used to proportionally calculate the bearable corrosives of distillation cut and those of crude slate for every specific location. Table 1 shows some selected corrosion model calculation results. Vacuum tower furnace piping is the weakest part of whole unit. It is recommended to upgrade the vacuum tower furnace piping with 316 stainless steel. Once the replacement is done, the COTAN and COSL of unit can be set as 1 mg KOH/g and 0.6%.

Figure 2. Corrosion pits on a heat-exchanger

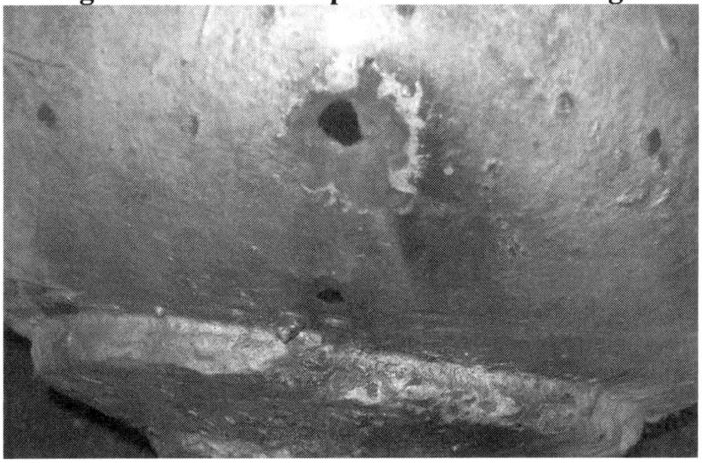

Figure 3. Corrosion of a tongue-like tray in an atmospheric distillation tower

Figure 4. Corrosion of a supporting beam in a vacuum distillation tower

Figure 5. Corrosion of a column tray in a vacuum tower with float valves missing

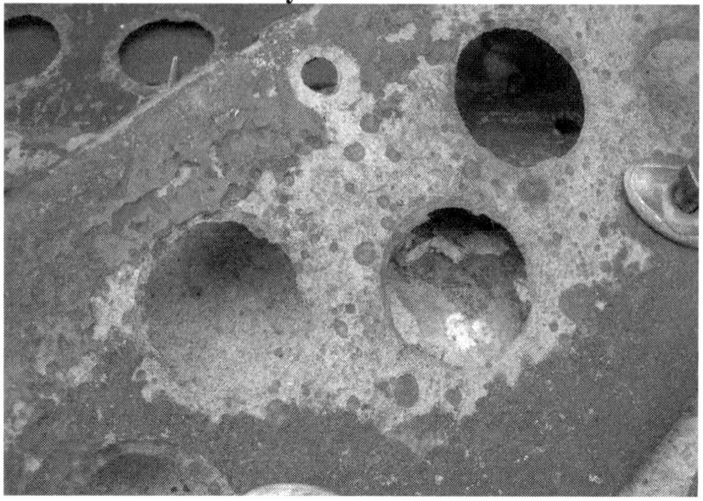

Figure 6. Historical date of acids and sulfur content for the crude slate

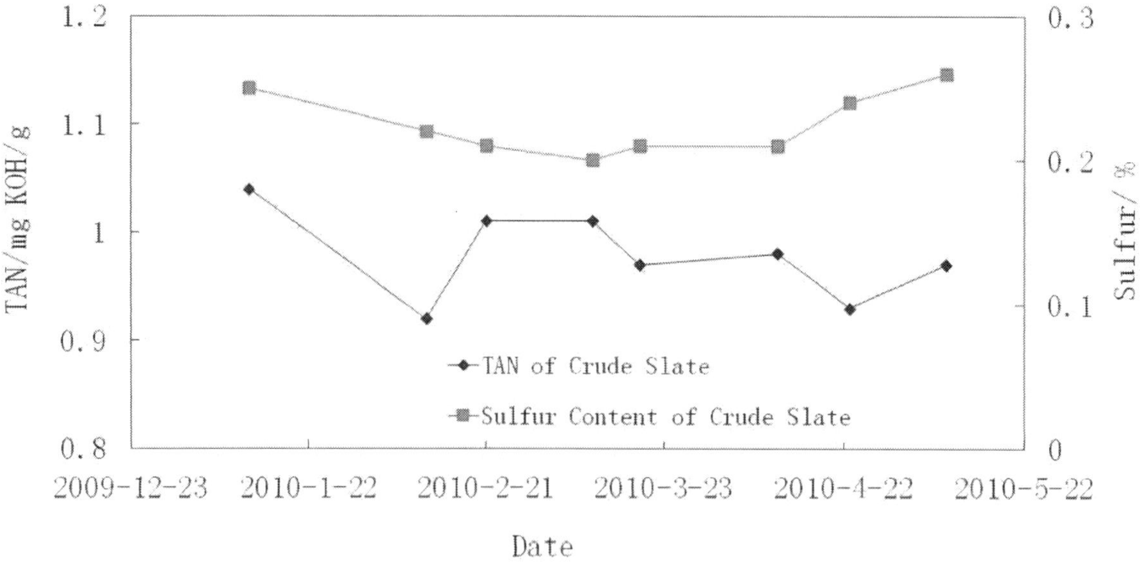

Figure 7. Corrosives distribution over atmospheric and vacuum distillation.

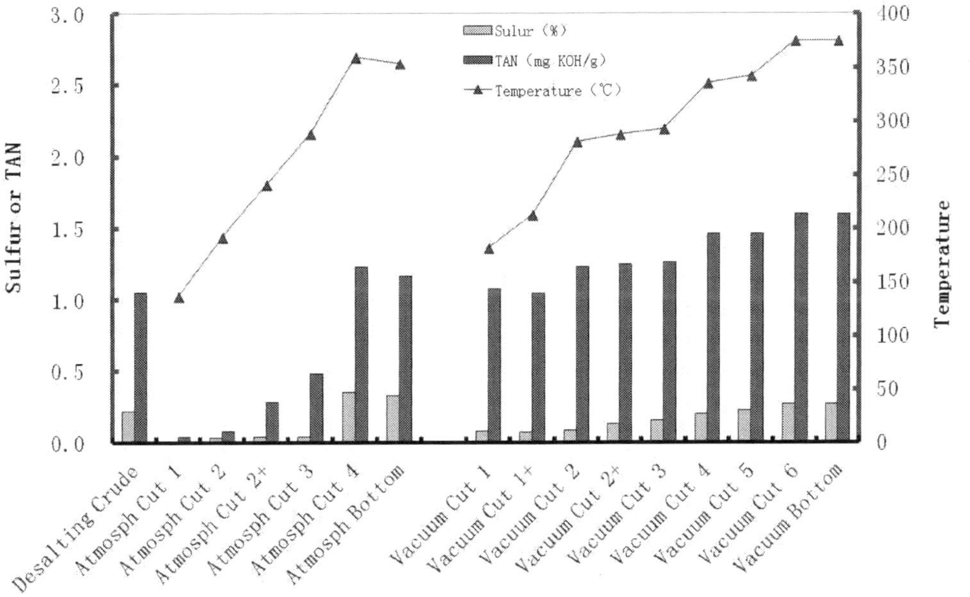

Table 1. Corrosion model calculation results.

Parts	Distillation Cut	Material	Bearable sulfur of crude (%)	Bearable TAN of crude (mg KOH/g)
Atmospheric tower furnace pipe	Desalting crude	1Cr5Mo	0.6	1.5
Vacuum tower furnace pipe	Atoms. bottom	1Cr5Mo	0.2	0.7
Pipe for Atoms. cut 3	Atoms. cut 3	Carbon steel	1	N/A
Pipe for Atoms. cut 4	Atoms. cut 3	304	N/A	N/A
Heat Exchanger E1009	Desalting oil/ Atoms. cut 2	Carbon steel	3	1
Heat Exchanger E1020B	Desalting oil/ Vacuum bottom	Carbon steel	0.6	2

N/A: not applicable, concentration of corrosives unlikely exceeds the bearable value according to the model.

5. Applications and Effects

The theory of COTAN and COSL originated years ago when the company was suffering a great challenge with "opportunity" crudes. The application of COTAN and COSL not only set sulfur and TAN limits for units, but also identified many units' "weak points". Some suggestions to execute necessary remedies were made for many units after applying the knowledge of COTAN and COSL, which had precluded many "would-be" accidents. The theory was gradually developed and the first application started from an old unit on 2007, and then gradually extended to others. So far, it has been implemented to 30+ units, and has contributed to the company's reputation of safety excellence.

6. Conclusion

A corrosion predict model plays a vital role in determining COTAN and COSL. Its accuracy is still controversial. Refinery experience shows that the corrosion rate predicted by the modified McConomy curves is normally conservative, and usually higher than the actual corrosion measured. However, more severe corrosion than the McConomy curves predicted is also observed. This diversity could be attributed to many factors, such as flow condition and source of crude slate.

Critical operational TAN (COTAN) and Sulfur Level (COSL) are still new concepts for refining industry. It is a successful attempt to apply current knowledge of NAC and SC on the basis of refineries experience.:

7. References

[1] W. A. Derungs, "Naphthenic Acid Corrosion – An Old Enemy of the Petroleum Industry", *Corrosion* 12, (1956): p.617.

[2] S Tebbal, "Critical Review on Naphthenic Acid Corrosion", *CORROSION/99*, paper no.380 (Houston, TX: NACE, 1999).

[3] K.R. Lewis, M.L. Daane, and R. Schelling, "Processing Corrosive Crude Oils", *CORROSION/99*, paper no.377 (Houston, TX: NACE, 1999).

[4] D. Johnson, G. McAteer, and H. Zuk, "The Safe Processing of High Naphthenic Acid Content Crude Oils Refinery Experience and Mitigations Studies", *CORROSION/2003*, paper no.03645 (Houston, TX: NACE, 2003).

[5] D. Qu, Y. G. Zheng, H. M. Jing, X. Jiang, and W. Ke, "Erosion-corrosion of Q235 and 5Cr1/2Mo Steels in Oil with Naphthenic Acid and/or Sulfur Compound at High Temperature", *Materials and Corrosion* 56, 8(2005):p.533.

[6] D. Qu, Y.G. Zheng, H.M. Jing, Z.M. Yao, and W. Ke, "High Temperature Naphthenic Acid Corrosion and Sulphidic Corrosion of Q235 and 5Cr1/2Mo Steels in Synthetic Refining Media", *Corrosion Science* 48, (2006): p.1960.

[7] G. Bota, D. Qu, S. Nesic, and H.A. Wolf, "Naphthenic Acid Corrosion of Mild Steel in the Presence of Sulfide Scales Formed in Crude Oil Fractions at High Temperature", *CORROSION/2010*, paper no.10353 (Houston, TX: NACE, 2010).

[8] H.D. Dettman, N. Li, and J. Luo, "Refinery Corrosion, Organic Acid Structure and Athabasca Bitumen", *CORROSION/2009*, paper no.09336 (Houston, TX: NACE, 2009).

[9] API RECOMMENDED PRACTICE 939-C (FIRST EDITION, MAY 2009), "Guidelines for Avoiding Sulfidation (Sulfidic) Corrosion Failures in Oil Refineries", (Washington, D.C: API).

[10] *NACE International Publication 34103* (2004), "Overview of Sulfidic Corrosion In Petroleum Refining", (Houston, TX:NACE, 2004).

[11] V. Kanukuntla, D. Qu, S. Nesic, and A. Wolf "Experimental Study of Concurrent Naphthenic Acid and Sulfidation Corrosion", *17th ICC* (2008), paper. 2764.

[12] D. Qu, Y. Zheng, X. Jiang and W. Ke, "Correlation between the Corrosivity of Naphthenic Acids and Their Chemical Structures", *Anti-Corrosion Methods and Materials* 54,4 (2007): p211.

[13] J.H. Lee Craig, "Naphthenic Acid Corrosion in the Refinery", *CORROSION/95*, paper no. 333 (Houston, TX: NACE, 1995).

[14] Dingrong Qu, Xiaohui Liu, Xiu Jiang, Zhenggui Lan, Guangbin Shan, "Setting Critical Operational TAN and Sulfur Level for Crude Distillation Units" *CORROSION/2011*, paper no. 11362 (Houston, TX: NACE, 2011).

Large Eddy Simulation – Based Mixing Tee Analysis and Comparison with In-Service Failure Data

Seetha Ramudu Kummari, Ph.D.
The Equity Engineering Group Inc., Shaker Heights, Ohio
srkummari@equityeng.com

Michael F.P. Bifano, Ph.D.
The Equity Engineering Group Inc., Shaker Heights, Ohio
mfpbifano@equityeng.com

David J. Dewees, P.E.
The Equity Engineering Group Inc., Shaker Heights, Ohio
djdewees@equityeng.com

Prepared for Presentation at
American Institute of Chemical Engineers
2013 Spring Meeting
9th Global Congress on Process Safety
San Antonio, Texas
April 28 – May 1, 2013

UNPUBLISHED

Large Eddy Simulation – Based Mixing Tee Analysis and Comparison with In-Service Failure Data

Seetha Ramudu Kummari, Ph.D.
Michael F.P. Bifano, Ph.D.
David J. Dewees, P.E.
The Equity Engineering Group Inc., Shaker Heights, Ohio
srkummari@equityeng.com

Keywords: Mixing tee, turbulent thermal mixing, Large Eddy Simulations (LES), Fatigue Crack Growth, Computational Fluid Dynamics (CFD)

Abstract

High frequency temperature fluctuations resulting from turbulent thermal mixing of hot and cold streams in a tee junction are known to cause surface cracks on inner surfaces of the tee junctions, a phenomenon called thermal stripping in the petrochemical industry. While temperature fluctuations due to turbulent thermal mixing have been linked to surface cracks in the mixing tees, to date, how such surface cracks grow and cause leakages has not been assessed in the literature. The current study evaluates, not only surface crack initiation due to turbulent thermal mixing but also the effects of operating fluctuations (global transients) on crack initiation and growth. An objective of the study is to develop a comprehensive methodology to address mixing tee failures utilizing Large Eddy Simulations (LES) and standard two parameter Computational Fluid Dynamics (CFD) turbulence models, finite element stress analysis, fatigue life estimation and fatigue crack growth approaches. The developed methodology is used to assess an industrial mixing tee that experienced cracks after 15 years of service. The results obtained from the assessment are benchmarked against in-service failure data. The study suggests that a combination of advanced analyses such as LES based CFD, FEA, fatigue life estimation, fatigue crack growth approaches are necessary to analyze the effect of turbulent thermal mixing and operating fluctuations for an evaluation of a mixing tee failure.

1. Introduction

Tee junctions are common in petrochemical industries and are used to mix two fluid streams at different temperatures, and often consisting of different compositions. The mixing process of hot and cold fluids in the tee junctions is chaotic (turbulent) in nature and can result in high cycle temperature fluctuations on the inner surfaces of the tee junction and/or downstream piping. Such high cycle temperature fluctuations can initiate surface cracks; a phenomenon well known as thermal stripping in the petrochemical industry. Though thermal stripping alone may not be sufficient to propagate initiated cracks, operating fluctuations such as substantial variations in flow rates, pressure, and temperature, and plant shutdown and startup could potentially propagate cracks. This could lead to leakage and structural integrity and process safety concerns. The high frequency temperature fluctuations resulting from turbulent thermal mixing cannot be sufficiently measured by modern thermocouples [1]. This makes it very difficult to identify and remediate locations where thermal stripping occurs. In the present study, advanced numerical methods such as Large Eddy Simulations (LES) are used to identifying mixing point locations prone to thermal stripping.

Standard Computational Fluid Dynamics (CFD) using two parameter turbulence models (Reynolds Averaged Navier Stokes Models, RANS) or unsteady RANS (URANS) cannot adequately predict the turbulent thermal mixing phenomenon [2], and advanced analysis, specifically LES, is required to obtain meaningful results [1, 3–7]. LES resolves detailed characteristics of the flow, such as the "large" eddies, and their propagation in time and hence the resulting temperature fluctuations. This paper presents a comprehensive analysis of a petrochemical industry mixing point. The analysis utilizes a combination of LES and RANS CFD turbulence models, finite element stress analysis, fatigue life estimations and fatigue crack growth approaches. The investigation is driven by extensive analysis of operational data, and is rigorously benchmarked against the observed in-service failure of a mixing tee. Specifically, the tee junction under consideration introduces cold hydrogen at variable flow rates into the primary hot hydrocarbon stream. After approximately 15 years of service, multiple cracks were identified downstream of the mixing point on the inside surface of the run pipe.

2. Methods

2.1. Severity of Mixing – Momentum Ratio Calculations

Severity of mixing depends upon the relative momentums of the two flows (branch and run pipe flows) and can be assessed by a parameter called, momentum ratio which is defined as:

$$MR = \frac{\rho_r U_r^2 (D_r \times D_b)}{\rho_b U_b^2 \times \pi \left(\frac{D_b}{2}\right)^2} \qquad \text{(Eq. 1)}$$

Where:

ρ, U and D represent density, velocity and inner diameter of the pipes and the subscripts, *r* and *b* represent run pipe and branch pipe, respectively. In the mixing tee analysis literature [4, 5, 8, 9], flow regimes are classified based upon the value of the of the momentum ratio as follows:

Momentum Ratio	:	**Flow Regime**
< 0.35	:	Impinging jet
0.35 – 1.35	:	Turn jet
1.35 – 4.0	:	Re-attaching jet
> 4.0	:	Wall jet

A lower momentum ratio lead to a higher expectation of mixing, i.e., impinging jet flows result in heavier mixing compared to that of wall jets in which the branch fluid is forced against the upper portion of the run pipe. Though this type of flow regime classification can be used for comparing relative severities of mixing, the classification does not provide much insight into thermal stripping phenomenon and hence surface crack initiations. It is shown in the current study that even the wall type flow can result in severe local turbulent thermal mixing and lead to surface crack initiations. This suggests that the flow regime classification is not very predictive for typical petrochemical mixing points. Operating conditions that result in momentum ratios significantly higher than 4.0 still demonstrate enough branch fluid energy for random and turbulent penetrations into the run pipe fluid, followed by the forcing of these fluid structures back against the pipe wall. The result is alternating hot and cold shocks randomly distributed along the pipe that can be characterized in a statistical sense.

Momentum ratios are calculated using (Eq.1) for six months of actual operating history of the mixing tee under consideration. Based on the calculated momentum ratios, four momentum ratios are chosen to represent "heavy", "moderate", "less" and "very less" mixing conditions in the six months of operating conditions. Corresponding momentum ratios are as shown below. Note that classification of mixing conditions is relative to the operating history and applies only to the mixing tee under consideration.

Momentum Ratio	:	**Mixing Condition**
22	:	Heavy mixing
106	:	Moderate mixing
1038	:	Less mixing
7712	:	Very less mixing

2.2. Large Eddy Simulations – Simulation of Turbulent Thermal Mixing

A commercial CFD software, STAR-CCM+ version 7.02.008 is used for the CFD analysis of the mixing tee. Only one half of the mixing tee model is considered because of symmetries in the model geometry – the restriction this enforces on the turbulent flow is considered reasonable. The run pipe is extended by 3.0 times the diameter upstream and 9.0 times the diameter downstream of the tee to minimize the effect of the assumed inlet velocity distribution and backflow at the region of interest. Similarly, the branch pipe is extended by 5.0 times its diameter upstream of the tee junction. All the regions, in the model, except boundary layers, are discretized using polyhedral cells in STAR-CCM+. A focused mesh is used in the regions of interest and near the fluid/solid interface and where heavy mixing is expected. The model contains approximately half a million polyhedral cells. Figure 1 shows the discretized model of the mixing tee.

A Volume of Fluids (VOF) multiphase model with surface tension effects is chosen to represent the fluid interaction between the branch and run pipe streams. Velocity inlet boundary conditions are applied to both the run and branch pipe inlets and pressure outlet boundary conditions are defined at the outlet. The following power law (n = 7) velocity profiles are defined at the inlets to better represent fully developed inlet flows,

$$V(r) = \frac{(n+1)(2n+1)}{2n^2} V_{avg}\left(1 - \frac{r}{R}\right)^{\frac{1}{n}} \qquad \text{(Eq.2)}$$

$$\text{for } n = 7, \qquad V(r) = 1.2245\, V_{avg}\left(1 - \frac{r}{R}\right)^{\frac{1}{7}} \qquad \text{(Eq.3)}$$

Where:
r is radial coordinate of the fluid volume
R is radius of the pipe
V_{avg} is average velocity of the fluid in the pipe

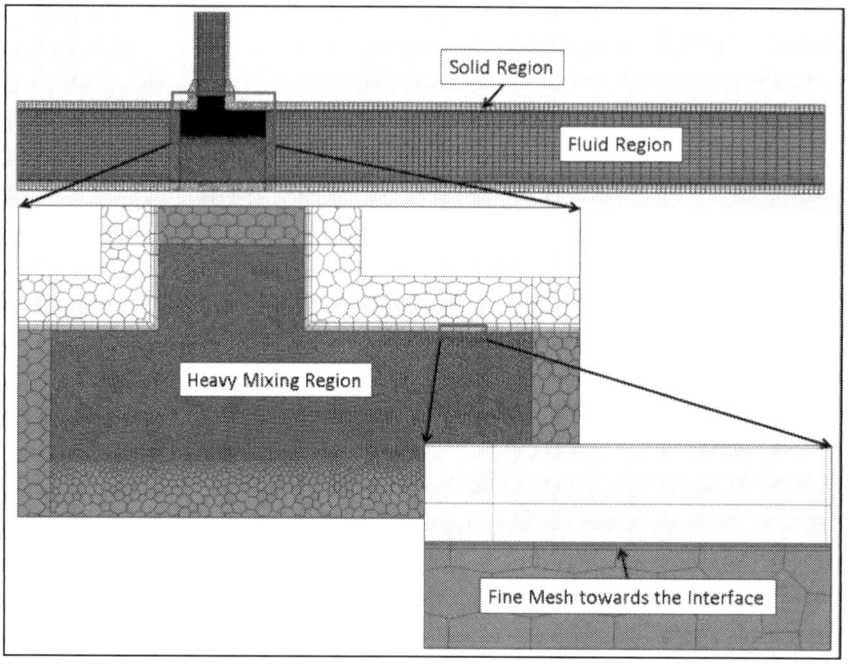

Figure 1: Different regions of the mixing tee CFD model are shown. The model is discretized using polyhedral cells in STAR-CCM+. Fine mesh is included in the regions of heavy mixing and near the fluid/solid interface.

The liquid hydrocarbon mixture in the mixing tee is modeled as an incompressible fluid, whereas the hydrogen gas is modeled as a compressible ideal gas. No slip conditions are defined at the fluid/solid interface. The effect of pipe insulation is simulated by defining an effective heat transfer coefficient of 0.568 $W/m^2.K$ and a sink temperature of 21°C on the outer surfaces of the pipes (Figure 2), which includes the combined resistance of the external insulation and free external convection in still ambient air. Initially, the CFD model is run with a conventional (RANS) k-ε turbulence model. To increase speed and stability in the iterative LES solver, the converged RANS solution is used as an initial condition. A time step of 5.0E-5 seconds is chosen for LES simulations. To remove all artifacts of the RANS solution, the LES simulation is run for 1.5 seconds prior to any data collection. Following the development of a stable LES solution, data is collected at the rate of 200 Hz. Fluid/solid interface temperatures are recorded at multiple locations along the length of the run pipe; at the center of the tee, weldolet, 6 inches away, 12 inches away and 18 inches away from the center of the tee (Figure 3). The simulation is run at the moderate mixing condition for about 2.0 seconds and then the flow is switched to the heavy mixing condition, and is run for an additional 2.0 seconds, in an attempt to capture any transient effects of unsteady operating conditions.

2.3. *RANS Simulation – Simulation of Operating Fluctuations*

Conventional RANS simulations are performed for the four mixing conditions; heavy, moderate, less and very less mixing conditions. The intent of simulating different mixing conditions is that the effect of bulk operating fluctuations is isolated. The CFD model and boundary conditions are identical to those described in the previous section except that only steady state RANS simulations are performed for each mixing condition.

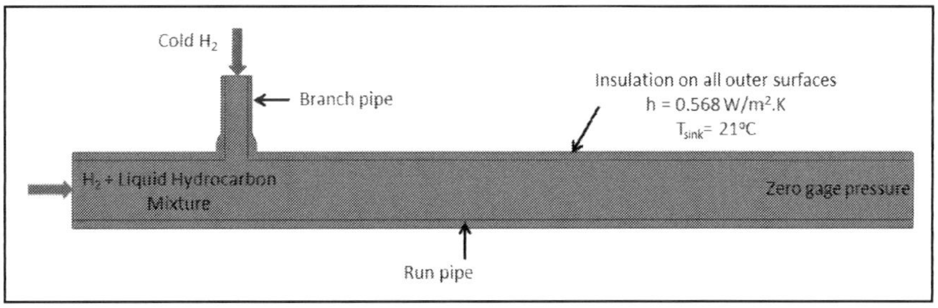

Figure 2: Velocity inlet and pressure outlet boundary conditions are applied. An effective heat transfer coefficient of 0.568 W/m^2.K is applied to the outer walls of the pipe to account for the effect of insulation.

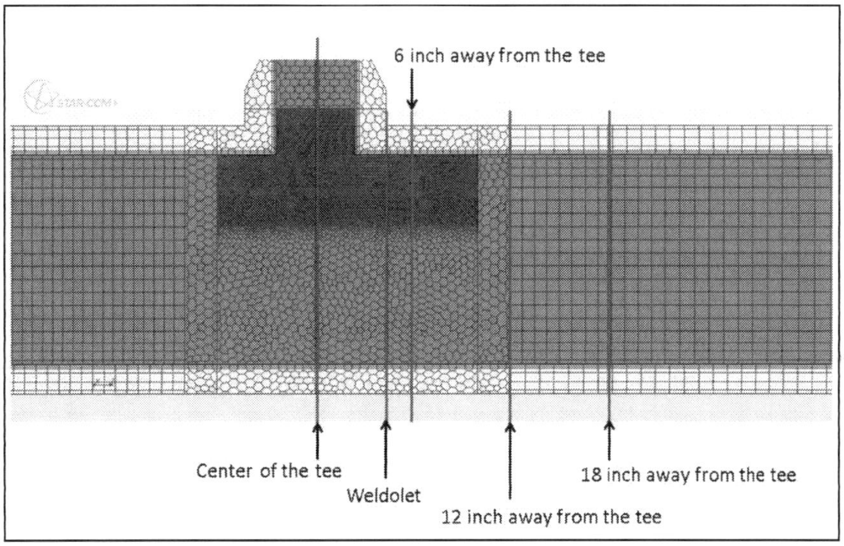

Figure 3: Five different planes are chosen downstream of the mixing tee for temperature data collection at fluid/solid interface.

2.4. *Finite Element Stress Analysis*

Stresses due to the temperature fluctuations are evaluated using a simplified axisymmetric finite element representation. The commercial finite element software, Abaqus Version 6.11 is used for the stress evaluations. An axisymmetric representation of the run pipe is discretized using quadratic coupled temperature–displacement axisymmetric elements (Abaqus element, CAX8RT). The finite element model consists of only one element along the length direction of the pipe, but contains 48 elements in the thickness direction with a fine mesh towards the inner diameter of the pipe. To evaluate stresses due to turbulent thermal mixing, the fluid/solid interface temperature fluctuations at various points in the LES CFD model are applied at the inner surface of the pipe as time dependent (transient) temperature boundary conditions. An effective heat transfer coefficient of 0.568 W/m^2.K with a sink temperature 21°C is defined on the outer surface of the pipe to simulate the effect of insulation and external convection. Adiabatic boundary conditions (no thermal gradients) are assumed on the transverse cut-planes of the pipe (Figure 4).

Additionally, the transient effect of switching mixing conditions (operating fluctuations) is simulated by changing the sink (fluid) temperature (steady state temperatures obtained from the RANS simulations) instantly at the inner surface of the pipe and specifying a nominal heat transfer coefficient of 1704 W/m^2.K. This value is calculated using standard forced convection pipe flow correlations and is typical for the relevant fluid properties and flow conditions applicable to the tee. The boundary conditions are shown in Figure 5.

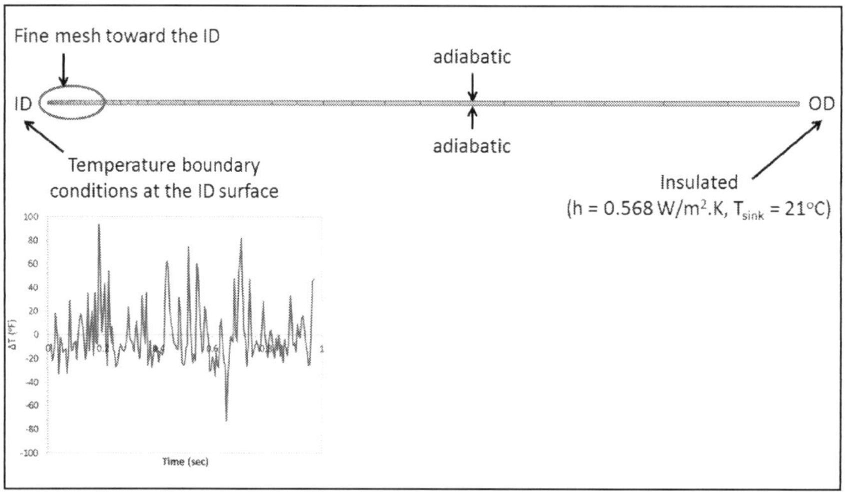

Figure 4: The boundary conditions on the axisymmetric finite element model to evaluate the stresses due to turbulent thermal mixing are shown. Temperature fluctuations from the LES are applied on the inner diameter as transient temperature boundary conditions.

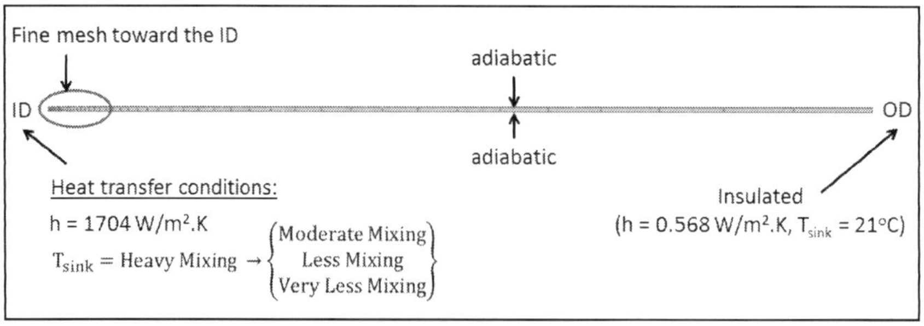

Figure 5: The boundary conditions on the axisymmetric finite element model to evaluate the stresses due to operating fluctuations are shown. Steady state temperatures obtained from the RANS simulations are applied as sink (fluid) temperatures.

2.5. Fatigue Life for Crack Initiation

Stress histories obtained from the axisymmetric finite element analysis are rain flow cycle counted to obtain the number of stress cycles that a particular location experiences. Then the number cycles to failure are obtained using the ASME design fatigue curve for stainless steel. The time to initiate a crack is determined by applying Miner's rule for all the cycles with stress amplitudes larger than the endurance limit of the pipe material (stainless steel, 13.6 ksi). A usage factor is calculated by dividing operating duration (15 years) by time to failure.

2.6. Fatigue Crack Growth Analysis

Following fatigue crack initiation, final crack depths are estimated with knowledge of the operating duration using API 579-1/ASME FFS-1 2007 Edition (API 579). In the present study, crack depths are estimated for 15 years of typical service. A semi-elliptical crack of size 0.04 inches in depth (a), 0.24 inches in length (2c) is assumed to be initiated at the time of crack initiation. The initial crack sizes are chosen based on engineering observations and fracture mechanics literature that indicate typical crack sizes between 0.5 mm (0.02 inches) and 3.0 mm (0.12 inches) at the time of crack initiation (i.e. at the end of a fatigue test). In this analysis a crack size of 1.0 mm (0.04 inches) is assumed to initiate at the time of fatigue crack initiation.

The resulting stress histories from the FEA model are rain flow cycle counted to determine the most common stress amplitude for the crack growth analysis. Stress intensities (K_{max} and K_{min}), corresponding to the maximum and minimum stresses and for the given crack size, are calculated according to Annex C of API 579. The crack growth rate is estimated using a Paris law model for stainless steel, given in Annex F of API 579. The Paris law used for the estimation of crack growth rate in both depth and length direction is given by (Eq.4) and (Eq.5). The final crack depth and length (a and 2c) are estimated by assuming that the crack growth is constant for 10 cycles. The validity of this assumption is verified by a sensitivity analysis.

$$\frac{da}{dN} = 8.61 \times 10^{-10} \left(\frac{E_{amb} \times \Delta K}{E_{at}}\right)^{3.0} \quad \text{for } \Delta K \geq \Delta K_{th} \qquad \text{(Eq.4)}$$

$$\frac{da}{dN} = 0 \quad \text{for } \Delta K < \Delta K_{th} \qquad \text{(Eq.5)}$$

Where:

$\frac{da}{dN}$ is crack growth rate per cycle

ΔK is stress intensity range ($K_{max} - K_{min}$)

ΔK_{th} is threshold stress intensity range below which $\frac{da}{dN}$ is assumed to be zero

E_{amb} is elastic modulus at ambient temperature

E_{at} is elastic modulus at assessment temperature

3. Results

3.1. Turbulent Thermal Mixing Results (LES Simulations)

Figure 6 show the distinction between the temperature distributions obtained using the RANS k-e turbulence model and the LES turbulence model. As can be seen from the figure, the conventional RANS turbulence model does not resolve eddies whereas LES models do. Temperature fluctuations of 50°F to 100°F at 5.0 Hz are common near the tee junction. Figure 7 shows temperature fluctuations due to turbulent thermal mixing at 27° (from the branch pipe), center of the tee and 30°, 6 inches away from the tee. Maximum temperature fluctuations around the circumference of the run pipe point to the locations of maximum thermal fatigue (Figure 8).

The results indicate that stresses corresponding to the simulated temperature fluctuations (Figure 9) may be as high ± 40 ksi (i.e. both in tension and compression). Figure 10 shows stress amplitudes verses mean stresses of the cycles obtained from the rain flow cycle counting algorithm. Fatigue analysis suggests that the time to initiate a crack varies from location to location. The allowable cycle limit can be readily determined as a function of the location.

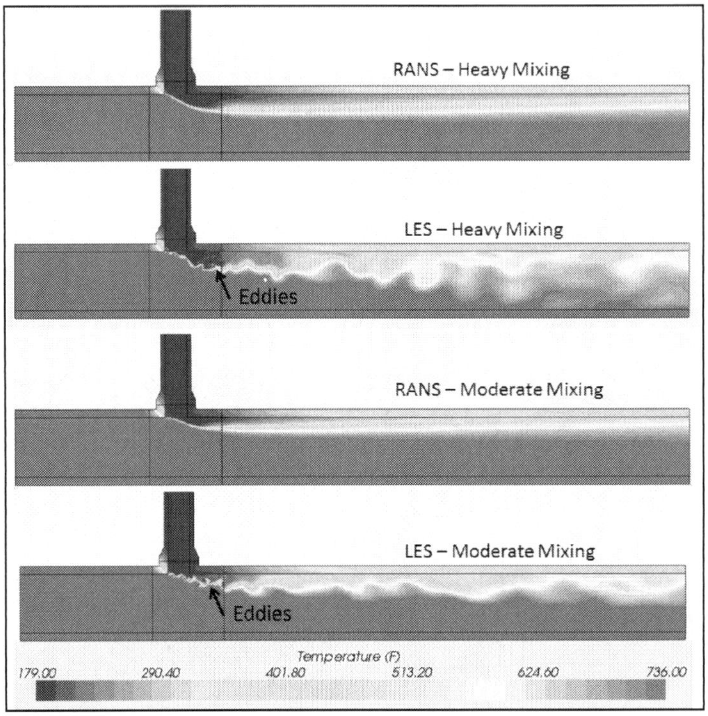

Figure 6: The distinction between RANS and LES simulations is illustrated for both heavy and moderate mixing conditions. LES resolve large eddies of the turbulent thermal mixing and hence the resulting high frequency temperature fluctuations.

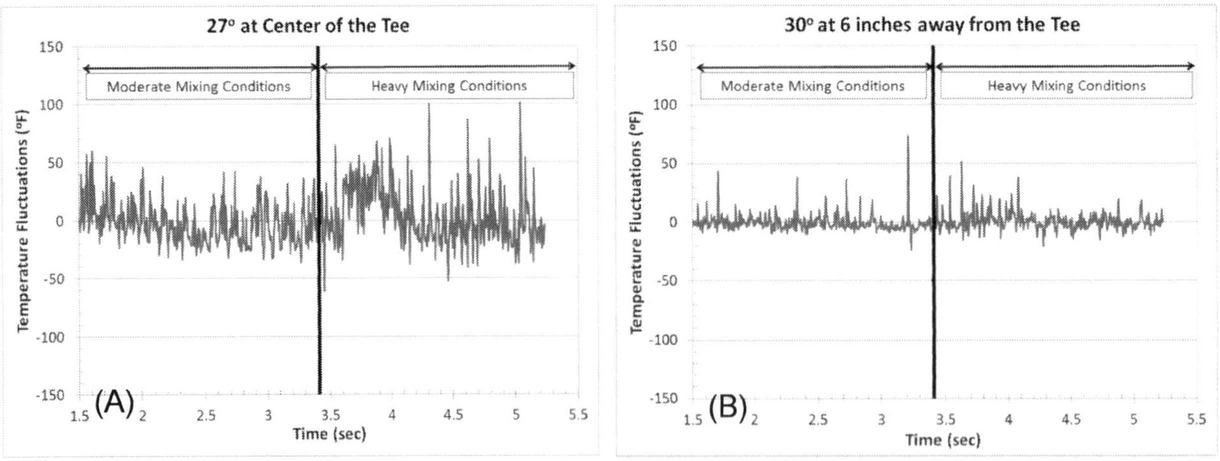

Figure 7: Temperature fluctuations due to turbulent thermal mixing at (A) 27°, center of the tee and (B) 30°, 6 inches away from the tee

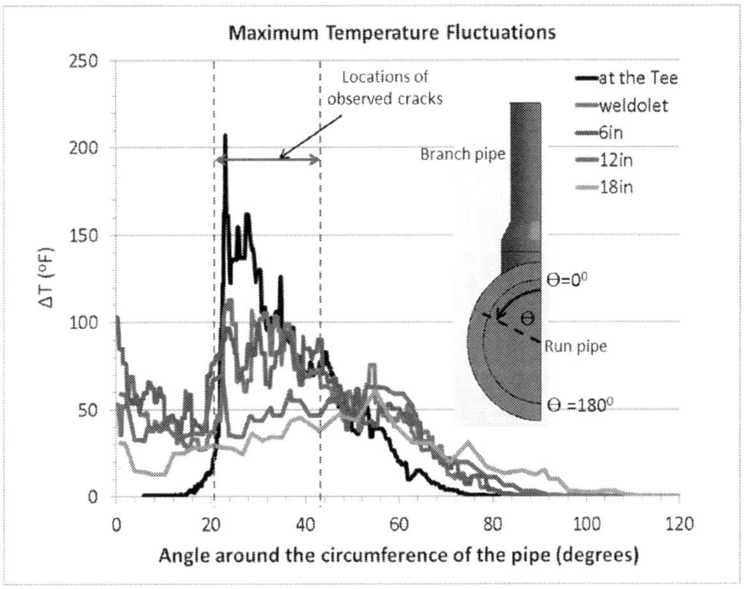

Figure 8: Maximum temperature fluctuations around the circumference of the pipe are plotted. Maximum temperature fluctuations location correlates well with the location of the observed cracking

The results of the crack growth analysis indicate that the stress intensity ranges (ΔK) for the maximum and minimum stress distributions caused by thermal fluctuations are below the threshold stress intensity range, ΔK_{th}. Therefore, while the temperature fluctuations are sufficient to initiate cracking, the thermal fluctuations due to thermal mixing do not penetrate far enough into the wall to create a sufficient through wall stress distribution capable of driving crack growth.

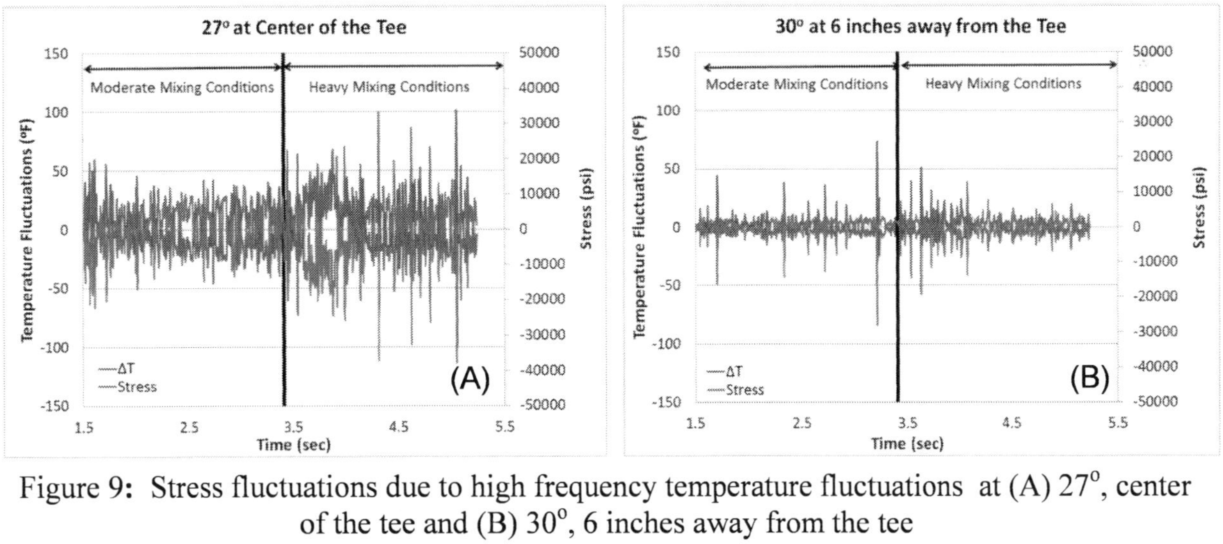

Figure 9: Stress fluctuations due to high frequency temperature fluctuations at (A) 27°, center of the tee and (B) 30°, 6 inches away from the tee

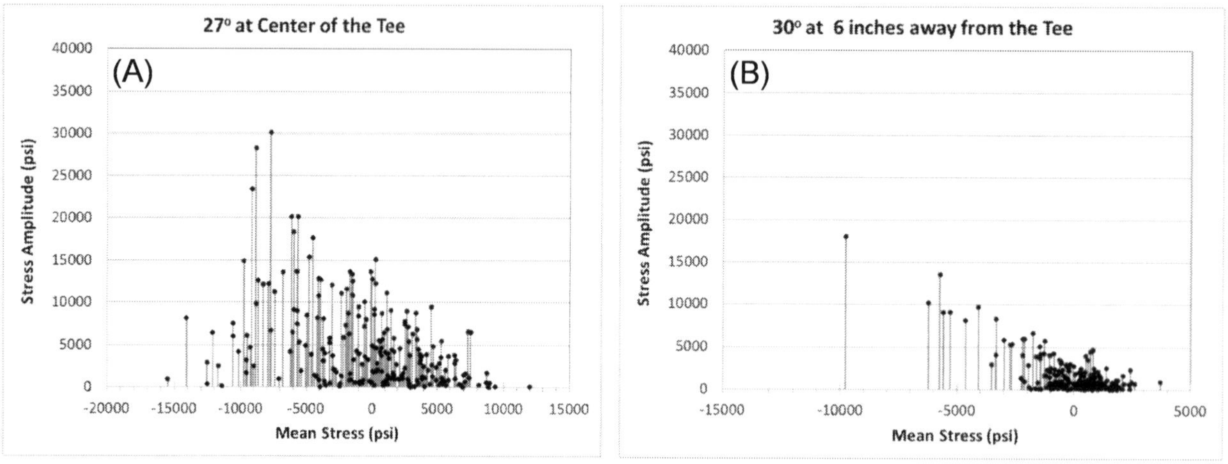

Figure 10: Stress amplitude verses mean stress from Rain Flow cycle counting algorithm for (A) 27°, center of the tee and (B) 30°, 6 inches away from the tee

3.2. Variations in Operating Conditions (RANS Simulations)

Figure 11 shows steady state temperature distributions for different mixing conditions simulated. Figure 12 illustrates the temperature variation that occurs around the pipe circumference when the mixing condition is changed from one steady value to another (operating fluctuations). The temperature fluctuations due to these operating fluctuations are as high as 450°F (Figure 12). The results indicate that the maximum temperature fluctuations occur at 6 inches away from the tee and near 40° from the centerline of the branch pipe. Remarkably, the results show that the locations of the maximum temperature fluctuations (30° to 40°, Figure 12) correspond to the locations of the observed cracks.

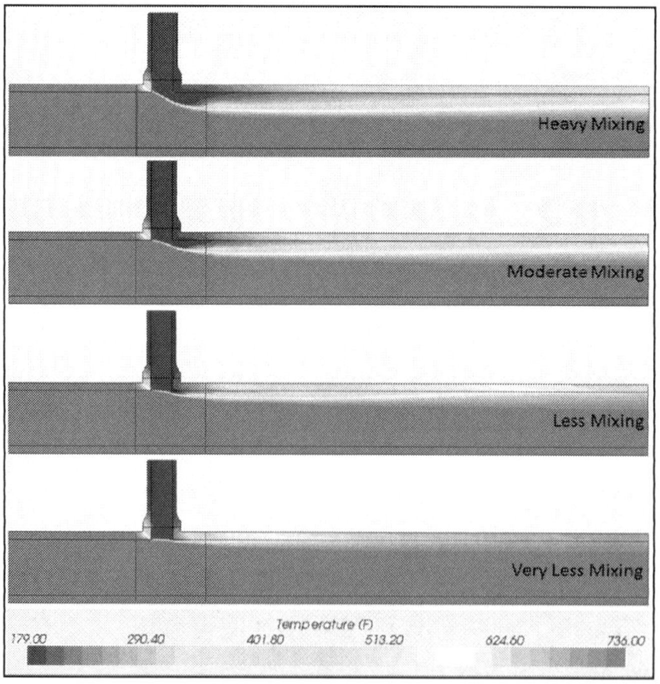

Figure 11: Steady state temperature distributions for different mixing conditions.

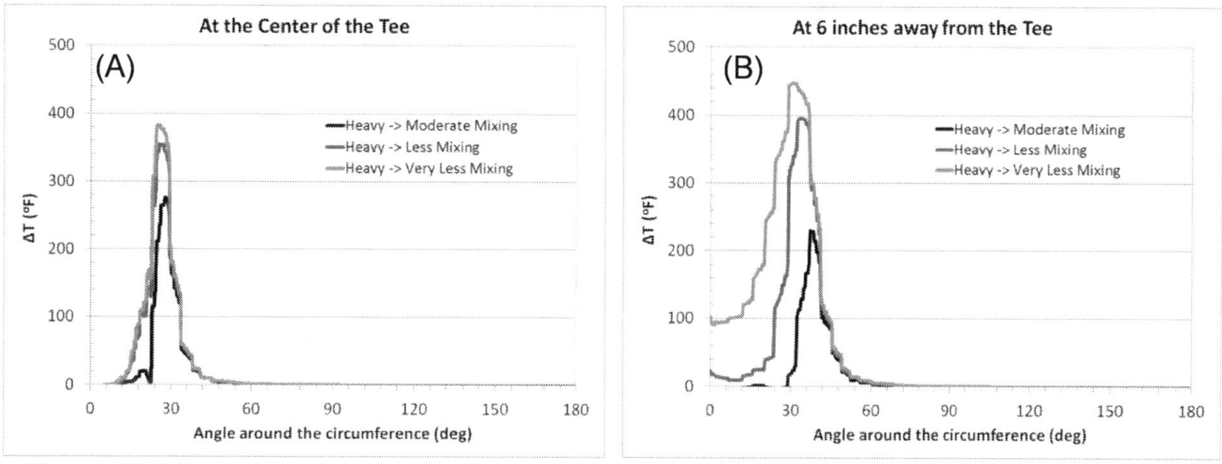

Figure 12: The effect of operating fluctuations is shown. The temperature fluctuations due to the operating fluctuations could be as high as 450°F and maximum temperature fluctuations occur between 30° and 40° from the branch pipe.

Stress analysis reveals that membrane plus bending stresses peak at a different time than that of the total stresses at the inner diameter of the pipe. The total stresses peak near 10 seconds after the change of mixing conditions whereas the magnitude of the membrane plus bending stresses peak near 22.5 seconds. The total stresses are used for fatigue initiation life estimations whereas through wall thickness stresses at which membrane plus bending stresses are maximum are used for crack propagation analyses (Figure 13). Fatigue life calculations for the operating fluctuations predict a usage factor of 0.25 suggesting that operating fluctuations exhaust ~25% of the fatigue life.

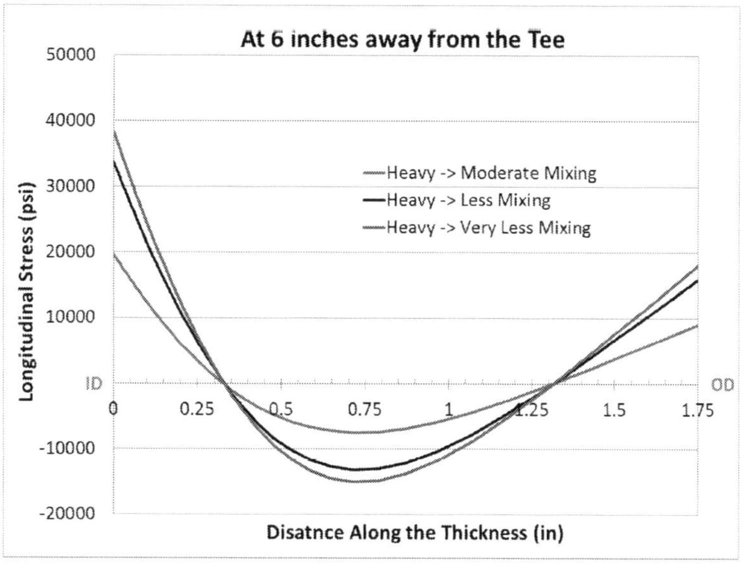

Figure 13: Through wall stress distributions due to operating fluctuations.

The results from the crack growth analysis are shown in Figure 14. The analysis predicts that an initial semi-elliptical crack of size, 0.04 inches in depth and 0.24 inches in length, grows by 0.0123 inches and 0.0113 inches in the depth and length directions, respectively, resulting in a final predicted crack size of 0.052 inches in depth and 0.251 inches in length. These results indicate that the operating fluctuations alone are unlikely to cause fatigue crack initiation but significantly contribute to fatigue crack growth. The stress fluctuations extend deep enough into the pipe walls to drive fatigue crack growth.

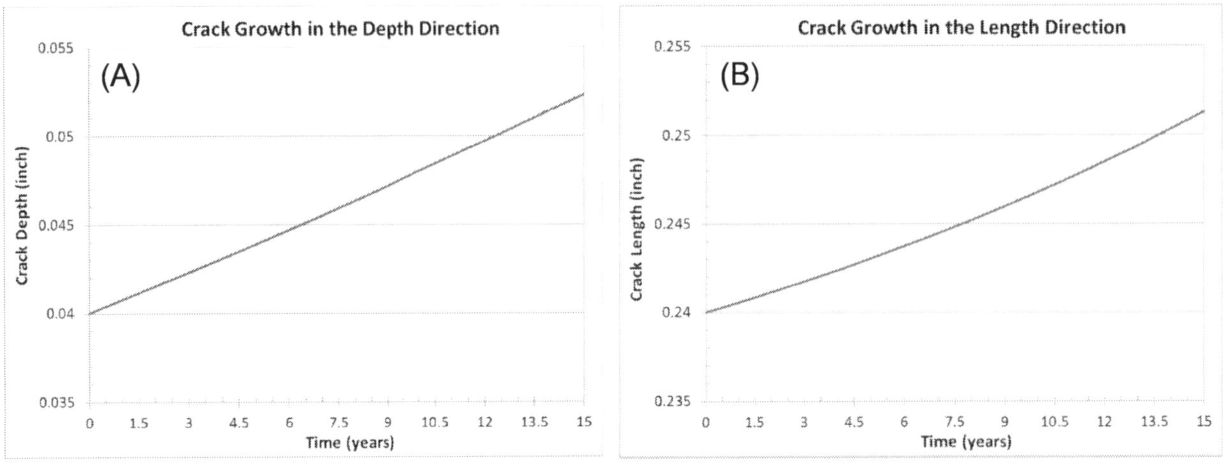

Figure 14: Fatigue crack growth in the operating duration of 15 years is shown. The simulated crack of initial size 0.04 inches in depth and 0.24 inches in length grows to a final size of 0.052 inches in depth and 0.251 inches in length.

4. Discussion

In conclusion, it can be stated that the comprehensive analysis including LES based CFD, FEA, fatigue crack initiation, and crack growth calculations, largely predict the cracking observed in the field. However it's important to understand the strengths and limitations of the methodologies used and how they can influence the results.

The results from the CFD analysis show clear distinctions between LES and conventional RANS turbulence models. The study shows that LES can be used to resolve eddies caused from the thermal mixing (Figure 6) and it is capable of obtaining the resulting thermal fluctuations at the fluid/solid interface (Figure 7). The current study also shows that the maximum temperature fluctuation locations correlate well with the locations of the actual observed cracks (Figure 8). This suggests an ability of the adopted methodology to locate the areas of possible thermal fatigue concerns.

Fatigue life estimations in this study are based on the ASME Section VIII, Div. 2 mean fatigue curve for industrial environments. There may be variations in the life estimation depending upon the fatigue curve used (for example, mean fatigue curve vs. -2σ fatigue curve). The results are also based on only two flow conditions (heavy and moderate mixing conditions). Practical time requirements and computational efforts limit LES analysis to only a few flow conditions, i.e., performing LES with actual transient inlet conditions or even a large number of inlet conditions is not a practically viable option.

The Paris law used for crack growth analysis is based on conservative estimates for crack growth rates. The final crack sizes may vary depending upon the constants in the crack growth model. The current analysis indicates that stress distributions due to turbulent thermal mixing do not penetrate deep enough into the metal to drive crack growth. This suggests that, though the high frequency thermal fluctuations are significantly capable of initiating cracking, the stress distributions are comprised of mainly peak stress at the surface of the metal and do not penetrate deep enough to propagate a crack front. Therefore variations in operating conditions such as pressure, temperature, and flow rates (as analyzed in the present study) are necessary to explain crack growth. The methodology adopted accurately predicts the most susceptible crack prone locations and a significant degree of fatigue crack growth. However, the method is not fully predictive since the projected crack depths are less than those found in service. Variations in operating temperatures and pressures likely account for these differences, and are logically the next element of complexity to add to the methodology adopted.

The major contribution of the present study is the finding that fluctuations in operating conditions, normally assumed to be steady in conventional LES studies, are found to be significant in driving a crack that is readily initiated due to turbulent thermal mixing. A multipart investigation that utilizes actual operating histories, LES CFD, conventional RANS CFD, finite element analysis, and fatigue crack initiation and growth models are necessary in identifying the vulnerability of industrial mixing points to fatigue damage.

5. References

1. M.H.C. Hannink, A.K. Kuczaj, F.J. Blom, J.M. Church, E.M.J. Komen. 2008, A coupled CFD-FEM strategy to predict thermal fatigue in mixing tees of nuclear reactors, EUROSAFE 2008, Paris.
2. C. Peniguel, 1998, Heat transfer simulation for industrial applications: needs, limitations, expectations, Int. J. Heat Fluid Flow, 19, 102-114.
3. J. Westin, P. Veber, L. Anderson, C. Mannetje, U. Andersson, J. Eriksson, M. Henriksson, F. Alavyoon, C. Andersson, 2008, High-cycle thermal fatigue in mixing tees. Large Eddy Simulations compared to a new validation experiment, Proc. 16th Int. Conf. On. Nuclear Engineering ICONE16-48731, Orlando, Florida, May 11-15.
4. L.-W. Hu and M. S. Kazimi, 2006, LES benchmark study of high cycle temperature fluctuations caused by thermal striping in a mixing tee, Int. J. Heat Fluid Flow, 27, 54.
5. M. Igarashi, N. Kimura, N. Tanaka and M. Kamide, 2003, LES analysis of fluid temperature fluctuations in a mixing tee pipe with the same diameters, Proc. 11th Int. Conf. On. Nuclear Engineering ICONE-11, Tokyo, Japan, April 20-23.
6. A.K. Kuczaj, E.M.J. Komen, M.S. Loginov, 2010, Large-Eddy Simulation study of turbulent mixing in a T-junction, Nuclear Engineering and Design, 240(9), pp. 2116-2122.
7. G. Grotzbach, M. Worner, 1999, Direct numerical and larege eddy simulations in nmuclear applications, Int. J. Heat Fluid Flow, 20, 222-240.
8. M. Igarashi, M. Tanaka, S. Kawashima, H. Kamide, 2002, Experimental study on fluid mixing for evaluation of thermal stripping in T-pipe junction, Proc. 10th Int. Conf. On. Nuclear Engineering ICONE10-22255 pp. 383-390, Arlington, Virginia, April 14-18.
9. S.M. Hosseini, K. Yuki, H. Hashizume, 2008, Classification of turbulent jets in a T-junction area with a 90-deg bend upstream, Int. J. Heat Fluid Flow, 51, 2444-2454.

9TH GLOBAL CONGRESS ON PROCESS SAFETY

Improving Reliability & Safety Performance of Solenoid Valves by Stroke Testing

Loren L. Stewart, BSME
exida
64 N. Main St., Sellersville, Pa 18960
lstewart@exida.com

Julia V. Bukowski, Ph.D.
Dept. of Electrical & Computer Engineering
Villanova University, Villanova, PA 19085
Julia.Bukowski@villanova.edu

William M. Goble, Ph.D., CFSE
exida
64 N. Main St., Sellersville, Pa 18960
wgoble@exida.com

Prepared for Presentation at
American Institute of Chemical Engineers
2013 Spring Meeting
9th Global Congress on Process Safety
San Antonio, Texas
April 28 – May 1, 2013

Improving Reliability & Safety Performance of Solenoid Valves by Stroke Testing

Loren Stewart, BSME
exida
64 N. Main St., Sellersville, Pa 18960
lstewart@exida.com

Julia V. Bukowski, Ph.D.
Villanova University

William M. Goble, Ph.D., CFSE
exida

Keywords: Stiction, static friction, binding, solenoid valve, emergency shutdown valve, valve stroke testing, proof testing, PFDavg, SIL rating

Abstract

Solenoid valves integrated into the design of emergency shutdown (ESD) valves used in industrial process systems, can tend to bind, i.e., to become stuck in one position, when not moved for long periods of time. This binding, also known as failure due to excessive stiction, has significant negative impacts on the valve's reliability and safety performance. It is a serious and costly problem normally addressed by expensive and time-consuming manual proof tests which typically require a process shutdown to perform testing. This paper describes an effective, alternative in-service testing protocol, known as valve stroke testing, which verifies whether or not the solenoid valve is stuck in position. It recommends a best practice procedure for implementing the valve stroke test. It provides a quantitative example of how valve stroke testing significantly improves safety performance when performed frequently (at intervals of one week or less) or even infrequently (at intervals of three to six months).

1. Introduction

In an industrial setting such as a chemical plant, oil refinery or a deep sea oil rig, there are safety instrumented systems in place which will promptly bring about a safe shutdown of a process if certain hazardous conditions are detected. Such safe shutdowns frequently involve isolating a process fluid flow, often accomplished by the use of an ESD valve.

Two of the ways an ESD valve can fail are it can function spuriously (also known as a spurious trip) or it can fail to function. If the ESD valve functions spuriously, it unnecessarily shuts down a properly working process in a safe manner; this type of failure is classified as a safe failure or fail safe. If the ESD valve fails to function, it loses the ability to shut down the process should a shutdown be required; this type of failure is categorized as a dangerous failure or "fail danger." If the process does not require shutdown while the ESD valve is in a state of dangerous failure then no harm is done provided the dangerous failure is discovered and repaired before a hazardous condition necessitating a shutdown occurs. If, however, the ESD valve is in a state of dangerous failure when a hazardous condition necessitating a shutdown occurs and the ESD valve does not perform its function when required, the worst case consequences of this "failure on demand" can be catastrophic including loss of life, significant destruction of or damage to equipment and property, and severe environmental damage.

Some ESD valve failures can be detected by various diagnostic procedures while the ESD valve is in service. If these detected failures are promptly repaired, safety performance is improved. Some ESD valve failures cannot be detected while the ESD valve is in service and consequently, safety standards, e.g. [1, 2] require periodic "proof testing" to verify that the ESD valve will operate properly if called upon to initiate a process shutdown. Such testing usually requires an interruption of process operations and in some instances may require physical removal of various ESD elements from the process for testing. Hence, proof testing is relatively time-consuming and expensive compared to in-service diagnostic testing.

ESD valves often include a solenoid valve of some type in the ESD valve design with the solenoid valve remaining in an energized state while the process operates normally and with the solenoid valve being de-energized, causing its seat to move to a different position, to initiate a safe process shutdown. Solenoid valves can fail to function on demand due to a number of different conditions called dangerous failure modes. In particular, based on a validated database of failure rates and failure modes for mechanical components [3, 4], one finds that a significant failure mode contributing to the dangerous failure rate of the solenoid valve is that of sticking or adhesion. This dangerous failure mode can lead to a failure to shut down when needed, i.e., a failure on demand.

Because the solenoid valve is one element of an ESD valve, it is often "at rest," i.e., stationary, during normal plant operation. The failure mode where it has become "stuck" in one position cannot be observed, i.e., the dangerous failure is undetectable in normal operation. Traditionally, this failure mode is discovered during proof testing and, for this particular ESD element, proof testing usually necessitates a process shutdown. As proof testing is a time-consuming and expensive proposition, the proof test interval is scheduled as infrequently as possible consistent with the reliability and safety objectives of the end-user.

Safety standards rate safety performance by safety integrity level (SIL) which is an order of magnitude rating based on the computed value of average probability of failure on demand (PFDavg) over the time interval [0, T_P] as defined in (Eq. 1). T_P is the time interval between proof tests and PFD(t) is the time varying probability of failure on demand.

$$PFDavg(t) = \frac{1}{T_P}\int_0^{T_P} PFD(t)dt \qquad \text{(Eq.1)}$$

For a fixed value of T_P (Eq. 1) results in a particular value for PFDavg. Table 1 shows SIL levels assigned to different ranges of PFDavg.

Table 1. Relationship Between SIL Level and PFDavg.

SIL Level	PFDavg
1	$[10^{-2}, 10^{-1}]$
2	$[10^{-3}, 10^{-2}]$
3	$[10^{-4}, 10^{-3}]$
4	$[10^{-5}, 10^{-4}]$

But suppose we do not have a definite value of T_P in mind and want to set T_P as high as possible while still attaining a particular SIL rating. Then we are interested in PFDavg as a function of time as given in (Eq 2).

$$PFDavg(t) = \frac{1}{t}\int_0^t PFD(s)ds \qquad \text{(Eq. 2)}$$

Assuming that the rate of dangerous failures can be represented by the constant λ_D and that the device for which PFDavg(t) is being computed operates correctly when it is installed, then one major portion of PFDavg(t) can be reasonably approximated by

$$PFDavg(t) \cong \frac{1}{2}\lambda_D \times t \qquad 0 \leq t \leq T_P \qquad \text{(Eq.2a)}$$

Thus, PFDavg(t) increases approximately linearly as t increases up to T_P. The SIL rating that the device achieves can be altered by adjusting T_P. In general, mechanical devices do not undergo in-service diagnostic testing and it is not until the time of proof testing that dangerous failures are discovered and corrected. Just after the time of proof testing, assuming that the proof test was complete and perfect so that all dangerous failures were discovered and corrected, PFDavg(t) will return to zero and begin to increase linearly until the next proof test. Thus, over all t, (Eq. 2a) is properly written as

$$PFDavg(t) \cong \frac{1}{2}\lambda_D t - (j-1)\frac{1}{2}\lambda_D T_P \qquad (j-1)\,T_P < t \leq j\,T_P \quad j = 1, 2, 3\ldots \quad \text{(Eq. 2b)}$$

Thus, the behavior of PFDavg(t) over repeated proof test cycles results in a saw-tooth plot as shown in Figure 1 where T_P has been arbitrarily set at 2 years. Note that the purple line in Figure 1 is the maximally allowed PFDavg to attain a SIL 3 rating. So with T_P set at 2 years, the example device would receive a SIL 2 rating. T_P would have to be decreased to approximately 1.25 years (the value of time when the PFDavg(t) plot and the horizontal purple line first intersect) in order to receive a SIL 3 rating.

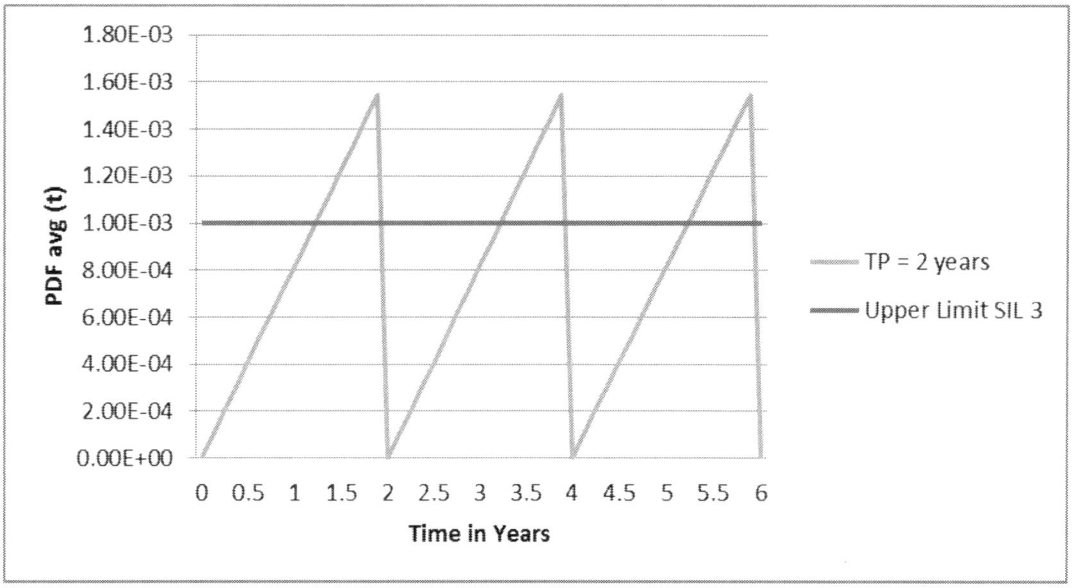

Figure 1. Plot illustrating the behavior of PFDavg(t) over time three inspection cycles when T_P equals 2 years.

Solenoid valve failures due to sticking or adhesion contribute significantly to reducing safety performance because they increase λ_D and result in reductions in T_P, i.e., they result in more frequent process downtime to facilitate proof testing, in order to meet safety objectives. Thus, finding less costly ways to address this adhesion failure mode has value.

The remainder of this paper

- Describes the basic operation of a solenoid valve to the extent required to understand the remainder of the paper
- Explains the concept of "stiction"
- Describes the effects of stiction on solenoid valve operation
- Presents evidence that stiction increases to some maximum level over relatively short periods of time
- Proposes a means to effectively reduce or eliminate the increased stiction through appropriately timed valve stroke tests
- Describes how to assess the impact of valve stroke testing on PFDavg(t)
- Details "best practices" guidelines for implementing solenoid valve stroke testing
- Provides a quantitative example illustrating improvement of reliability and safety performance when stroke testing is implemented

2. Solenoid Valve Basics

A solenoid coil generates an electromagnetic field which can be used to change the position of a metallic plunger. In a solenoid valve, this action can then be used either directly or indirectly to control process fluid flow. Figure 2 illustrates direct control of process fluid flow. In Figure 2a, when an electric current passes through the solenoid coil the resulting magnetic field attracts the plunger and holds it within the coil allowing the process fluid to flow from left to right. In Figure 2b, when the electric current is removed from the coil, the electromagnetic field vanishes and the plunger is no longer held within the coil but falls blocking the path of the process fluid flow.

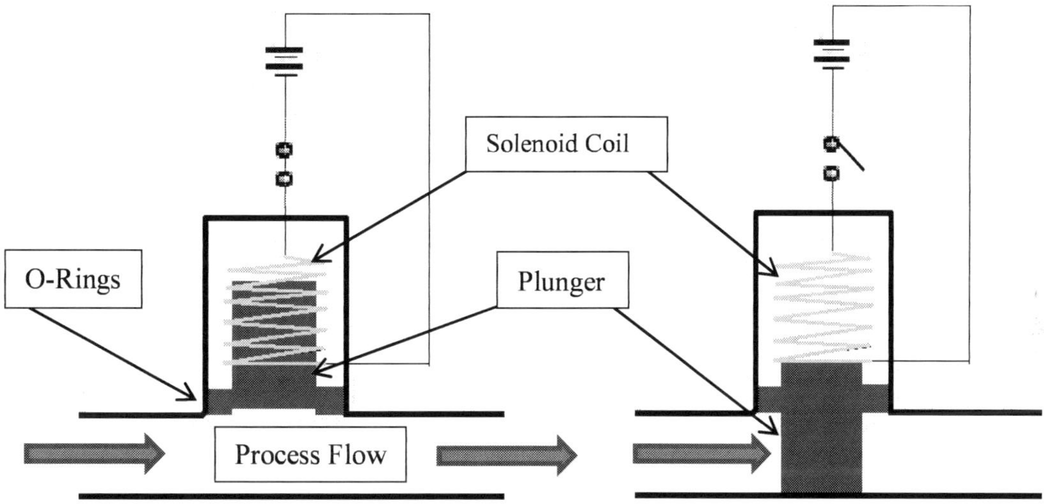

Figure 2a. Open solenoid valve

Figure 2b. Closed solenoid valve

Figure 2. Cross section of an open and closed soleniod valve, showing how plunger movement directly controls process fluid flow.

An example of using a solenoid coil for indirect *process* fluid control is a solenoid actuated pneumatic spool valve (see Figure 3). The solenoid valve typically controls compressed air which, in turn, positions another valve (not shown) which directly controls the process fluid. Thus the solenoid valve *indirectly* controls the process fluid.

Figure 3a shows the normal position of the valve for a particular application. In this coil design (which is different from that of Figure 2), when the solenoid coil is energized the magnetic field repels the plunger which is of a more complex design known as a spool. The spool, repelled by the magnetic field from the solenoid, is moved from left to right, changing the pathways allowing the flow of compressed air fluid. Consequently, when the solenoid coil is energized, the spool compresses the spring at the right end of the spool valve. In this spool position, ports x and y of the valve provide a pathway allowing compressed air to flow through the spool valve.

Figure 3b shows the emergency position of the valve in the event that an emergency shutdown of the process fluid is required. Now the solenoid coil is de-energized; the removal of the magnetic field means that the spool is no longer being repelled by the magnetic field. Without this magnetic force to push the spool to the right, the spring decompresses pushing the spool to the left, back inside the solenoid coil. In this spool position, port x is blocked which prevents compressed air from flowing through the spool valve to pressurize a valve actuator while ports y and z are open to provide a vent for pressure in the actuator.

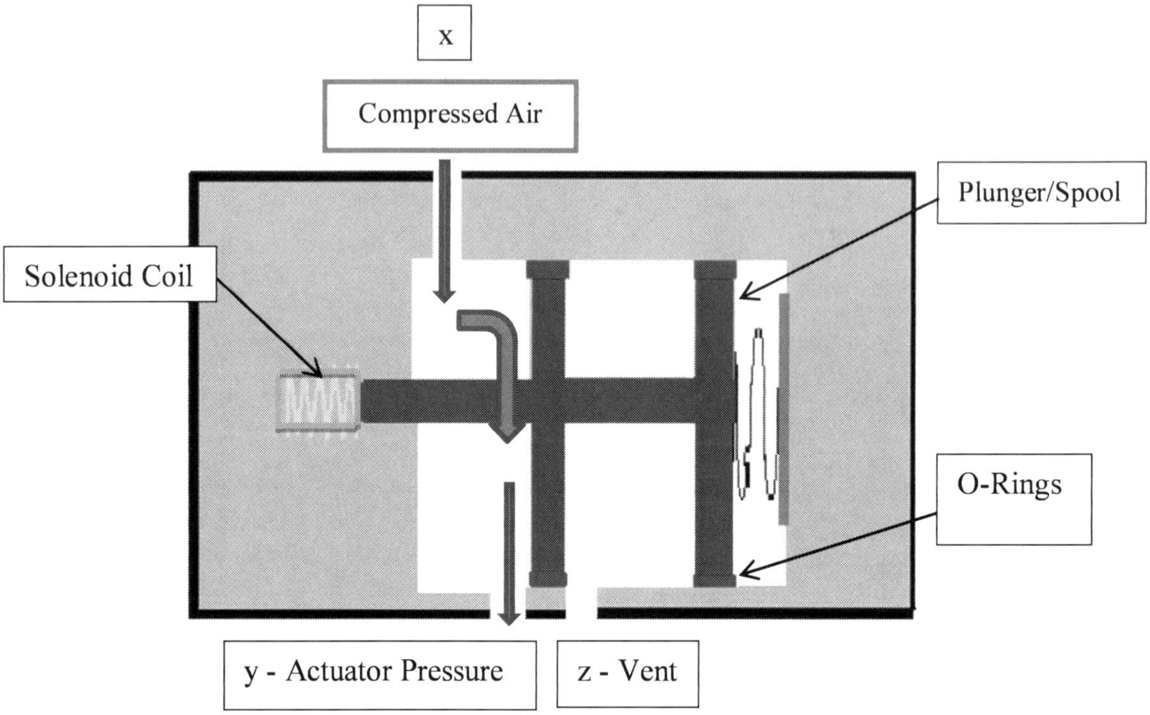

Figure 3a. Spool valve showing an open pathway from ports x to y for compressed air flow due to the soleniod being energized and compressing the spring.

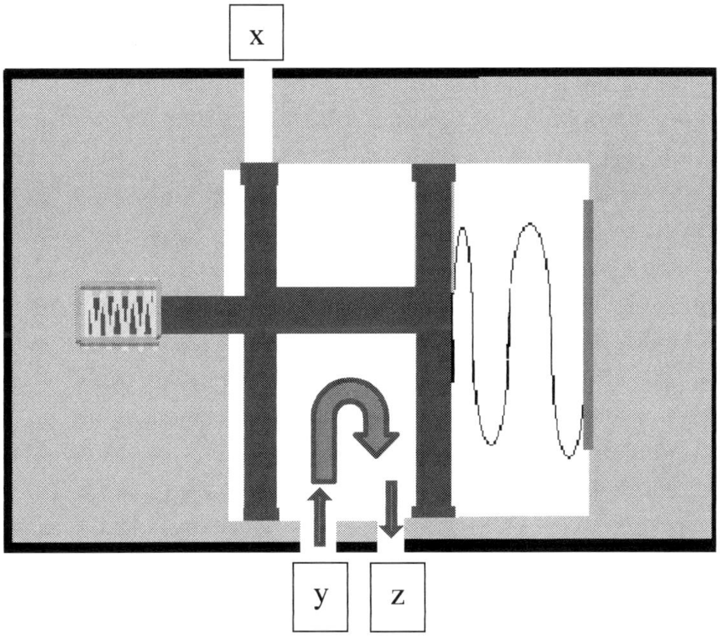

Figure 3b. Spool valve showing a closed pathway from ports x to y, blocking the flow of compressed air and venting air pressure by opening ports y to z. The soleniod is de-energized and the spring is not compressed.

Note the presence of O-ring seals (shown in blue) in both the simple solenoid valve (Figure 1) and the spool valve (Figure 2). They are in place to ensure that the fluid does not escape the valve chambers while also ensuring that no foreign elements can enter the fluid's environment. The O-rings surround both the plunger (Figure 1) and the spool (Figure 2) on the surfaces where the plunger or spool are in very close proximity (but not actually touching) the valve chamber walls.

While this introduction captures the main ideas needed to understand the remainder of this paper, the reader should be aware that actual valve designs are more complicated than presented here. Some use fluid pressures rather than springs to move the spool. Many other design differences exist. But all contain O-rings which are involved in a particular failure mode known to significantly affect solenoid valve reliability and safety performance. This failure mode is caused by excessive "stiction".

3. Stiction

3.1 What is Stiction?

Many studies and organizations have defined stiction, i.e., static friction, in different ways. A few examples of the various definitions are as follows:

- According to Olsson [5], stiction is defined as *"short for static friction as opposed to dynamic friction. It describes the friction force at rest. Static friction counteracts external forces below a certain level and thus keeps an object from moving."*

- According to Entech [6], *"stiction is a tendency to stick-slip due to high static friction. The phenomenon causes a limited resolution of the resulting control valve motion. ISA terminology has not settled on a suitable term yet. Stick-slip is the tendency of a control valve to stick while at rest and to suddenly slip after force has been applied."*

- According to Horch [7], *"The control valve is stuck in a certain position due to high static friction. The (integrating) controller then increases the set point to the valve until the static friction can be overcome. Then the valve breaks off and moves to a new position (slip phase) where it sticks again. The new position is usually on the other side of the desired set point such that the process starts in the opposite direction again."* This is an extreme case of stiction. On the contrary, once the valve overcomes stiction, it might travel smoothly for some time and then stick again when the velocity of the valve is close to zero.

- According to Ruel [8], the Instrument Society of America (ISA)(ANSI/ISA-S51.1-1979) described *"stiction as the resistance to the start of motion, usually measured as the difference between the driving values required to overcome static friction upscale and downscale."* The definition was first proposed in 1963 in American National Standard C85.1-1963.

- According to Ruel [8], *"stiction is a combination of the words stick and friction, created to emphasize the difference between static and dynamic friction. Stiction exists when the static (starting) friction exceeds the dynamic (moving) friction inside the valve. Stiction describes the valve's stem (or shaft) sticking when small changes are attempted."*

In the definitions above, all agree that stiction is the act of being "stuck" by static friction which prevents one surface from moving against another. Further, if the external force becomes greater than the static friction, the stiction between two surfaces will be overcome and the object will begin to move again [9].

Stiction between two surfaces can be the result of many factors including: corrosion, cold welding, i.e., metal on metal adhesion, breakdown of the lubrication boundary layer, change in the viscosity of the lubrication, build-up of deposits, contamination or foreign material on the contact surfaces, chemical reactions between the metal or lubrication, deterioration and breakdown of the valve's sealing components, etc.

This paper defines stiction as the resistance to the start of motion usually measured as the difference between the external force being applied in order to overcome the static friction and the force to maintain movement between the two contacting or working surfaces.

3.2 Stiction in the Context of a Solenoid Valve

In normal operation, the O-rings in the solenoid valve create a smooth transition while the plunger or spool is in motion. Because the O-rings are in direct contact with the valve chamber walls, in order for the plunger/spool to begin to move, it must first overcome the stiction between the O-rings and the chamber walls. Once in motion, the plunger/spool must overcome the sliding or dynamic friction between the O-rings and walls but, since the dynamic friction is usually significantly less than the nominal level of stiction, it is generally not of concern. By design the magnetic and spring forces in the solenoid valve are sufficient to overcome the *nominal* level of static friction and, depending on the design specifics of the solenoid valve, generally have excess force sufficient to overcome about 2 to 2.5 times nominal stiction. Once the stiction is overcome, the plunger/spool will continue to move. However, a common and troublesome problem can occur when the valve is at rest, i.e., stationary, for a length of time because the stiction between the O-ring and the valve chamber walls increases over time from its nominal level until it reaches some maximum value. It is possible that the increases in stiction may reach a level where the forces generated by the solenoid coil and/or spring (in the case of a spool valve) are no longer sufficient to overcome the increased stiction. Consequently, when the valve is called upon to close, it is unable to do so. However, if the excessive stiction is overcome, stiction reverts to its original nominal level though it does begin to increase again once the valve stops moving.

3.3 Evidence that Stiction Increases with the Length of Time the Contact Surfaces are Stationary

3.3.1 Expert Knowledge

Mechanical engineers and technicians who routinely work with solenoid valves are familiar with the experience of trying to "stroke", i.e., move through its range of motion, a valve that has been stationary for an extended time. They report that it is not uncommon for extra force to be required to move the valve after it has been stationary for a month or more. However, they are unable to say when within the month the increased stiction reaches a critical level that causes the valve to stick because it cannot generate sufficient force to overcome the increased stiction.

3.3.2 O-Ring Manufacturer's Findings

In the Parker O-ring Handbook [10] it is noted that the "coefficient of starting friction", i.e., stiction, increases when the O-ring has been stationary for between 1 week and 1 month after which time the stiction plateaus. One specific graph (Figure 5-7 in [10]) of "break out friction", i.e., stiction, vs. "delay between cycles" shows the stiction force plateauing at approximately 300 hours. "The theory has been proposed and generally accepted that the increase in friction on standing is caused by the rubber O-ring flowing into the micro-fine grooves or surface irregularities of the mating part"[10].

3.3.3 Experimental Study to Determine Equilibrium Rest Time

Zhao and Bhushan [11] conducted an experimental study on the effects of lubrication thickness, lubrication viscosity, and the length of rest time, i.e., the length of time the contact surfaces remain stationary, on the increases in stiction in magnetic thin-filmed disks. Based on their results they introduced the concept of equilibrium rest time, i.e., the time by which the maximum stiction is reached due to lack of motion. After conducting experiments with three different lubricants and holding the disks in a stationary position over extended periods of time, they found that a maximum stiction is reached by approximately 275 hours or about 11.6 days. The equilibrium rest time for the different lubricants are reached at slightly different points in time, but all are reached before about 275 hours [11]. Even though these experiments do not involve O-rings, it is interesting that a maximum level of stiction is reached and results regarding the time required to reach the stiction plateau are consistent with the information available in [10].

3.3.4 Reliability Declaration Per ISO 13849

In a declaration [12] regarding the reliability indicators and information for use of a specific solenoid actuated pneumatic valve with respect to the safety standard EN ISO 13849-1 [13], it is specifically stated that the "valve must be operated at least once per week or once per shift to insure the intended function." The data supporting this declaration was obtained by cycle testing during which the solenoid valve was never at rest for any significant time. This statement recognizes that the results of cycle testing are valid only if periodic valve movement is maintained. In the absence of such periodic movement, the failure rates derived from the results of cycle testing [14] cannot be considered valid when applied to components in applications where these components spend significant times at rest.

3.3.5 Section Summary

From this multi-source evidence we can conclude that, as a rule of thumb, O-ring stiction increases with valve inactivity and reaches a maximum value at approximately 275 hours. Further, stiction levels can be returned to nominal or near nominal levels if the increased stiction can be overcome by valve motion. This suggests that increased stiction can be dealt with in a more cost effective way than proof testing and that solenoid valve reliability and safety performance can be significantly improved.

4. Solenoid Valve Stroke Testing

In a stroke test, the solenoid coil is de-energized for a short period of time to see if the spring can fully decompress moving the spool in one direction through its full range of motion. Then when the coil is re-energized, the spool must again move, in the opposite direction, through its full range of motion. The reader may be familiar with the term "partial stroke testing" which refers to moving a valve through only a small portion of its range of motion to establish that it can be moved. In the case of solenoid valves, a partial stroke test is not possible. All stroke testing of solenoid valves will involve a full stroke.

If the solenoid valve is *directly* controlling process fluid, the stroke testing could lead to an unacceptable disruption of process fluid, so in the case of *direct* process fluid control, in-service solenoid valve stroke testing is impractical and without special redundant architectures, must be done when the process is shut down. However, if a solenoid valve is *indirectly* controlling process fluid, a full stroke of the solenoid valve (provided that the full stroke takes place over a short enough period of time) will generally not cause a significant disruption of the process fluid flow but will briefly alter the position of the actuator/valve assembly that the solenoid is controlling. This may be negligible or may result in a small perturbation to the process fluid flow that in most cases will be acceptable. At the completion of the valve stroke test, which is performed with the valve in service, either the solenoid valve has been verified to be operable or a dangerous failure due to stiction has been detected.

The success or failure of the solenoid valve stroke test can be determined indirectly by observing the actuator pressure and/or actuator valve stem movement because a full stroke test of the solenoid valve also performs a partial stroke test of the ESD. A small decrease in actuator pressure with corresponding actuator valve stem movement indicates a successful solenoid valve test *as well as* a successful ESD test. No decrease in pressure indicates a failure somewhere in the ESD but not necessarily of the solenoid valve itself. Some automatic stroke test devices measure the actuator pressure and actuator stem movement then automatically indicate pass/fail via a discrete output. Stroke testing the solenoid valve provides reliability and safety improvements not only to that device but also to the entire ESD as well.

4.1 Assessing the Reliability & Safety Improvements Due to Stroke Testing

This concept of solenoid valve stroke testing offers an opportunity to eliminate or at least address the failures of solenoid valves due to stiction as well as to address detectable non-stiction failures and, consequently, to improve solenoid valve reliability and safety performance. To aid in clearly understanding the beneficial reliability and safety effects of stroke testing, we note that the dangerous failures will be due either to the stiction failure mode or some non-stiction failure modes. The failures due to stiction are all detectable via valve stroke testing. The non-stiction failures are further divided into those failure modes which are detectable and those which are undetectable by valve stroke testing. We define the failure rates associated with these three cases as $\lambda_{Dstiction}$, $\lambda_{Dnon\text{-}stictionDetectable}$, and $\lambda_{Dnon\text{-}stictionUndetectable}$, respectively.

4.1.1 Case 1: No Valve Stroke Testing

In the case where no valve stroke testing is performed, the failure rates of all dangerous failure modes are included in λ_D, i.e.,

$$\lambda_D = \lambda_{Dstiction} + \lambda_{Dnon\text{-}stictionDetectable} + \lambda_{Dnon\text{-}stictionUndetectable} \quad \text{(Eq. 3)}$$

and (Eq. 3) is substituted into (Eq. 2a) to compute PFDavg(t) at $t = T_P$. Provided all dangerous failures are discovered and repaired as a result of the manual proof test, the resulting plot of PFDavg(t) will be a saw-tooth plot where PFDavg(t) returns to 0 after each proof test.

4.1.2 Case 2: Infrequent Valve Stroke Testing

By "infrequent valve stroke testing" we mean testing that occurs less frequently than once per week. Typically infrequent testing occurs once every three to six months. As with Case 1, the failure rates of all dangerous failure modes are included in λ_D so that (Eq. 3) again applies. However, there are some noteworthy differences between Case 2 and Case 1.

In Case 2, we assume that any dangerous failures detected as a result of the valve stroke test are promptly and completely repaired. Indeed, if prompt repair (within 168 hours) is not made, the value in performing the stroke test is reduced. Thus, Case 2 has nearly the same effect as a manual proof test, albeit a slightly incomplete one. It will detect all but the failures associated with the failure rate $\lambda_{Dnon\text{-}stictionUndetectable}$. Thus, PFDavg(t) is governed by (Eq. 2a) expect that t now equals T_{VST}, the interval of the valve stroke test, and after each interval of length T_{VST} PFDavg(t) does not return to zero but is reduced to a small residual value (due to $\lambda_{Dnon\text{-}stictionUndetectable}$) which will not be eliminated until the time of a complete manual proof test. Thus, the plot of PFDavg(t) for this case will resemble a saw-tooth plot as is shown in the quantitative example in Section 6 but one whose peaks increase slightly each cycle until the time of manual proof test. [Note: for the example parameter values used in Section 6, this increase in peak values is too small to be seen in the plots.]. So, infrequent valve stroke testing with prompt and complete repair of detected failures will decrease PFDavg. As a result, it may be possible to increase T_P without negatively impacting the safety performance of the solenoid valve.

4.1.3 Case 3: Frequent Valve Stroke Testing

By "frequent valve stroke testing" we mean in-service testing that occurs at least once per week and includes complete and correct repairs of all detected failures within 168 hours. Since the stroke testing is performed before the increase in stiction reaches its maximum, there is a significant chance that the stroke test will succeed in moving the valve **and** disrupt the buildup of stiction returning it to its nominal level. Thus, while the valve stroke test is designed to detect stiction-related failures there is a **significant chance that frequent valve stroke testing will prevent the failures** due to this failure mode from occurring effectively setting $\lambda_{Dstiction}$ to 0. Therefore, for Case 3 only the failure rates associated with non-stiction failure modes are included in λ_D, i.e.,

$$\lambda_D = \lambda_{Dnon\text{-}stictionDetectable} + \lambda_{Dnon\text{-}stictionUndetectable} \qquad \text{(Eq. 4)}$$

and (Eq. 4) is substituted into (Eq. 2a) to compute PFDavg(t) at $t = T_{VST}$ which in this case is 168 hours or less. Thus frequent valve stroke testing clearly decreases PFDavg. As a result, it may be possible to increase T_P without negatively impacting the safety performance of the solenoid valve.

4.2 Setting the Frequency of the Valve Stroke Test

While valve stroke testing will have beneficial results on PFDavg even if performed infrequently, **ideally we would like not simply to detect failures due to stiction but also to actually prevent them**. So, is stroke testing once per week sufficiently frequent to insure that stiction build up will be disrupted before it exceeds the valve's available excess force? Unlike random failures truly represented by a constant failure rate (meaning that failures occur with equal probability at any time during the interval between diagnostic tests), failures due to stiction are unlikely to occur very early in the time interval between tests when stiction levels have not increased very much and more likely to occur once stiction levels have increased beyond the level that the excess force of the valve can overcome. Thus, while we know that, as a rule of thumb, the maximum level of stiction is reached at about 275 hours, we also need to consider how *fast* the stiction builds towards its maximum value in order to determine the most appropriate interval for valve stroke testing.

In [11], plots of $\log_{10}$ stiction force vs. $\log_{10}$ rest time (time of inactivity) are given for the three lubricants studied. These *log-log plots* show an approximately linear increase in stiction from a nominal stiction level until a plateau is reached at about 275 hours. This indicates that the stiction increases as a power function meaning the increase is proportional to t^n. Thus the increase in stiction force over time would be modeled as

$$\text{Increase in stiction force} = \alpha\, t^n \qquad 0 \leq t \leq t_{max} \approx 275 \text{ hrs} \qquad \text{(Eq.5)}$$

where α is a constant of proportionality. Now, for the purposes of this example let us assume that the maximum stiction is three times nominal stiction. This implies that the maximum *increase* in stiction force at 275 hours would be twice nominal stiction which when added to the initial nominal stiction would give a total stiction force of about three times nominal stiction. For the lubricants studied, $0.07 \leq n \leq 0.17$. Thus stiction can increase *very* rapidly to its plateau value. For example, if $n = 0.17$, then at one week (168 hours) the stiction would increase by almost 92% of its maximum possible increase making the actual stiction force equal to nominal + (0.92 * 2 * nominal) for a total stiction of 2.85 nominal value.

If these example values applied to the solenoid valve, would it have enough excess force to overcome this amount of stiction? Possibly, but maybe not. However, our problem is that the experiments described in [11] were not on solenoid valves and we do not know what values of *n* might apply to any given solenoid valve or even it the solenoid valve O-ring stiction rises as a power function.

4.3 Considering an Alternative to Valve Stroke Testing

In some end-user circles there persists the common belief that valve stroke testing risks a plant shutdown due to a false trip and that a better safety solution is to provide a much higher safety margin (greater excess force) to overcome increases in stiction. However, those who are well versed in solenoid design realize that a higher safety margin to overcome stiction actually results in a *greater* false trip rate. This occurs because a higher safety margin to overcome increases in stiction requires a bigger spring. A bigger spring requires a bigger coil which, in turn, requires more energy to power the coil. These increased energy requirements result in a higher likelihood that the coil burns out which results in a higher false trip rate. Therefore, if we seek to optimize the false trip rate while improving reliability and safety, then valve stroke testing provides a good solution.

5. Recommended "Best Practices" for Implementing Solenoid Valve Stroke Testing Based on Current Knowledge (March 2013)

Based on currently available information consistent over multiple sources, i.e., that stiction reaches a maximum value in about 275 hours, we recommend as best practice, the following approach to solenoid valve stroke testing:

- Begin solenoid valve stroke testing on a weekly basis recording whether each test verified valve movement or indicated the valve was stuck (in which case, initiate prompt repairs)
- After a few months of testing, if the number of times the valve is found stuck seems excessive (by engineering judgment) then halve the test interval (to 3-4 days)
- Repeat the above procedures with this new test interval until an optimum test interval is identified such that stroke testing generally verifies valve movement.

In some instances, the initial one week test interval may not be adequate. Because low power (less than 2 Watts) solenoid valves generally are designed with less excess force available, they may not develop adequate force to overcome increases in stiction that may build up even before the one week test interval. In these cases, a shorter initial stroke test interval may be warranted.

To avoid false trips, the partial stroke testing device which initiates the full stroke of the solenoid valve must be set up properly:

- the pressure drop in the actuator should not be allowed to go to zero during the stroke test. Many designs limit the pressure drop to a minimum value while still fully exercising the solenoid valve.
- the minimum actuator pressure should be established to overcome deadband yet provide minimum movement of the valve controlling the process fluid. Many valves allow a small movement without imposing a flow restriction. A small movement is all that is necessary.
- "quick exhaust valves" should not be used as these devices are non-linear and often lock in the pressure exhaust causing a false trip. Instead, linear pneumatic boosters should be used.

6. Assessing the Reliability & Safety Improvements Due to Stroke Testing

In order to present a realistic quantitative example of the reliability and safety performance improvements that can result from valve stroke testing, a failure mode effects and diagnostic analysis [15] was performed on a solenoid valve from manufacturer X, model Y. The predicted values for the various constant failure rates are given in Table 2.

Table 2. Values of failure rates from the solenoid valve (manufacturer X, model Y) used in computing PFDavg.

Parameter	Failures/10^9 hrs operation
$\lambda_{Dstiction}$	103.7
$\lambda_{Dnon\text{-}stictionDetectable}$	84.3
$\lambda_{Dnon\text{-}stictionUndetectable}$	1.9

Finally, consider the behavior of PFDavg(t) due to the solenoid valve for each of the three cases. For our example parameter values, PFDavg(t) was computed for Case 1 with proof testing performed every 2 years (T_P = 2 years), for Case 2 with valve stroke testing performed once every six months (T_{VST} = 6 months) and for Case 3 with valve stroke testing performed once per week (T_{VST} = 168 hours). The results are plotted in Figure 4. Figure 5 illustrates the plot of PFDavg(t) for Case 3 magnified to show its detail. Note that in Figure 4 the purple horizontal line represents the maximally allowed value of PFDavg for which the valve could receive a SIL 3 rating. Without valve stroke testing, the valve can only be rated at SIL 2. With valve stroke testing every six months or once a week the PFDavg(t) remains below the SIL 3 rating level without proof testing. Note that this does not mean that proof testing can be completely eliminated as other dangerous undetected failures can exist that need to be discovered and corrected through proof testing. It does mean that it may be possible to increase the proof test interval significantly while maintaining a SIL 3 rating.

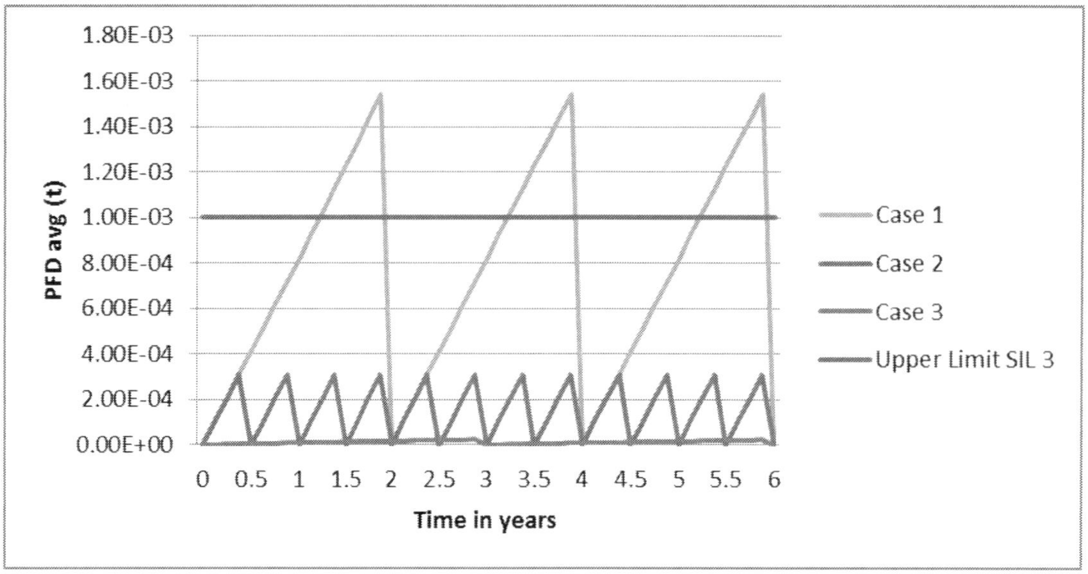

Figure 4. Plots of PFDavg(t) for Case 1 with T_P = 3 years, Case 2 with T_{VST} = 6 months and Case 3 with T_{VST} = 1 week.

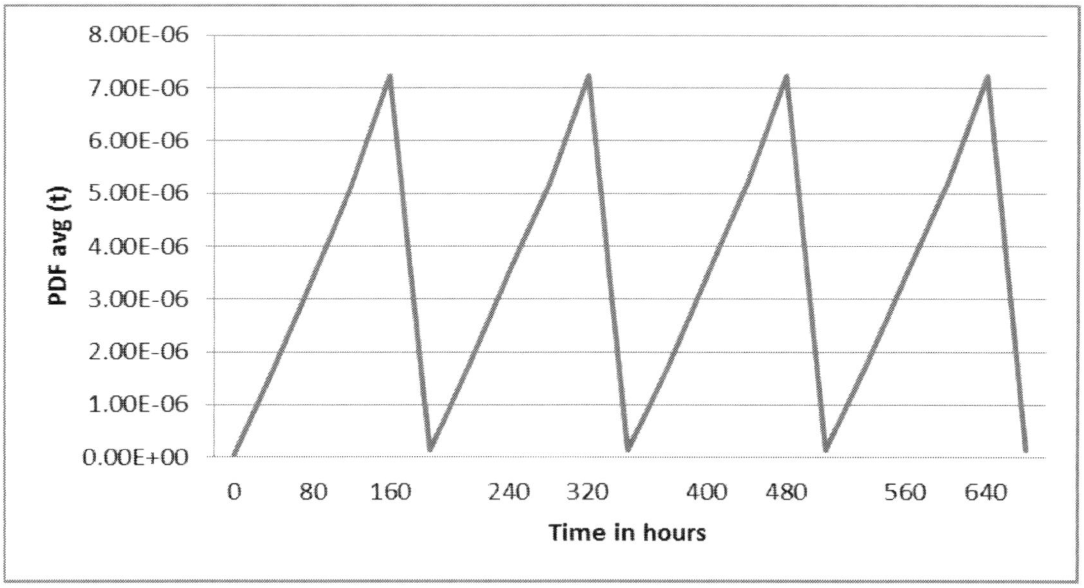

Figure 5. PDF_{avg} for the example solenoid valve as a function of time, in hours showing a weekly stroke test (every 168 hours).

7. Conclusions and Recommendations

Even infrequent solenoid valve stroke testing can have significant beneficial impacts with respect to improvements in reliability and safety performance. However, if a solenoid valve stroke test at a frequency of once per week (or more) is feasible within a given application, it should be implemented per the best practices strategy given above.

Clearly, the timing of valve stroke tests could be optimized if we had better models of how stiction increases with valve inactivity. Manufacturers may wish to undertake controlled tests and data collection on different valve designs within their product lines to better understand this stiction phenomenon. This would permit them to make recommendations regarding valve stroke testing that would improve reliability and safety performance and possibly increase the time intervals between required proof testing.

8. Acknowledgement

The authors would like to acknowledge the helpful discussions with and comments from Steven Close and Hal Thomas both of whom are with exida.

9. References

[1] IEC 61508 (1997), *Functional safety of electrical/electronic/programmable electronic safety-related systems,* Geneva: International Electrotechnical Commission.

[2] IEC 61511 (2003), *Functional safety – safety instrumented systems for the process industry,* Geneva: International Electrotechnical Commission.

[3] W. M. Goble and J. V. Bukowski, "Validation of a Mechanical Component Failure Database," *Proceedings of the 2009 Annual Reliability and Maintainability Symposium*, Fort Worth, TX, January 2009.

[4] *Electrical and Mechanical Component Reliability Handbook*, 3rd Ed, exida, Sellersville, PA, 2012.

[5] H. Olsson(1996), Control systems with friction, PhD thesis, Lund Institute of Technology, Sweden.

[6] EnTech (1998), *EnTech Control Valve Dynamic Specification (version 3.0).*

[7] A. Horch (2000), Condition Monitoring of Control Loops, PhD thesis, Royal Institute of Technology. Stockholm, Sweden.

[8] M. Ruel (2000), *Stiction:* The hidden menace, *Control Magazine.* http://www.expertune.com/articles/RuelNov2000/stiction.html.

[9] S. Choudhury, N. F. Thornhill, S. L. Shah, A Data-Driven Model for Valve Stiction, *ADCHEM 2003*, Jan. 11-14, 2004, Hong Kong.

[10] Parker Hannifin Corporation, *Parker O-ring Handbook: ORD 5700,* 2007.

[11] Z. Zhao and B. Bhushan, Effect of Lubrication Thickness and Viscosity and Rest Time on Long-Term Stiction in Magnetic Thin-Film Rigid Disks, *IEEE Transactions on Magnetics*, Vol 34. No 4, July 1998.

[12] Rexroth Bosch Group, *Declaration: Reliability indicators and informations for use with respect to the utilization of EN ISO 13849-1:2008-12,* 2009.

[13] ISO 13849-1 (2009), *Safety-related parts of control systems -- Part 1: General principles for design,* Geneva: International Standards Organization.

[14] TÜV Rheinland Report No. V372 2010 S1, TÜV Rheinland Energie and Umwelt GmbH, 2011

[15] W. M. Goble and A. C. Brombacher, "Using a Failure Modes, Effects and Diagnostic Analysis (FMEDA) to Measure Diagnostic Coverage in Programmable Elecronic Systems," *Reliability Engineering and System Safety*, Vol. 66, No. 2, November 1999.

10. Abbreviations and Terms

ESD – Emergency Shutdown

PFD – Probability of Failure on Demand

PFDavg – Average of the Probability of Failure on Demand on the time interval [0, t]

PFDavg(t) – Average of the Probability of Failure on Demand at time t

SIL – Safety Integrity Level

T_P – Time interval between proof tests

T_{VST} – Time interval between valve stroke tests

λ_D – Constant failure rate for dangerous failures

$\lambda_{Dstiction}$ – Constant failure rate for dangerous failures due only to stiction

$\lambda_{Dnon\text{-}stictionDetectable}$ - Constant failure rate for dangerous detectable failures due to causes other than stiction

$\lambda_{Dnon\text{-}stictionUndetectable}$ – Constant failure rate for dangerous undetectable failures due to causes other than stiction

9TH GLOBAL CONGRESS ON PROCESS SAFETY

Lessons Learned from Fire in Air Supply Ducts

Al Waller
FMC Corporation
Philadelphia, PA
alvin.waller@fmc.com

Michelle Brown, CSP
FMC Corporation
Philadelphia, PA
michelle.brown@fmc.com

Prepared for Presentation at
American Institute of Chemical Engineers
2013 Spring Meeting
9th Global Congress on Process Safety
San Antonio, Texas
April 28 – May 1, 2013

UNPUBLISHED

Lessons Learned from Fire in Air Supply Ducts

Al Waller
FMC Corporation
Philadelphia, PA
alvin.waller@fmc.com

Michelle Brown, CSP
FMC Corporation
Philadelphia, PA
michelle.brown@fmc.com

Keywords: Lessons learned, fire, HEPA filter, air filter, thermal inertia

Abstract

In 2012, loss of air flow caused activation of interlocks to shut off steam and electrical heat inputs to an air heating system that was part of a drying/milling operation. Shortly thereafter, upstream filters and air ducts ignited causing more than US$500K in damage and business interruption. Investigations concluded that the interlocks functioned as designed on loss-of-airflow but were not sufficient to prevent overheating the inlet air filters (including HEPA filters) by the residual heat in the air heaters. The two main lessons learned from this incident apply to a broad range of operations in the CPI:

1) Mundane unit operations such as air filters that have combustible components must be evaluated as thoroughly as the rest of the chemical process.
2) Designers and PHA leaders must recognize the potential of thermal inertia of heaters (and coolers) to heat (or cool) a stagnant volume of fluid when flow is interrupted. Gas streams are particularly susceptible due to the low heat capacity relative to the heat capacity of a metallic heat exchanger. Interlocking the supplies (to steam and electrical heaters in this incident) is often necessary but may not be sufficient.

Other lessons "re-learned" include the importance of learning from a previous near miss, the importance of limiting escalation aspects of this fire, and the importance of preventative maintenance.

1. Introduction

A drying and milling operation experienced a fire inside its air supply ducts in 2012. Total losses exceeded US$500k, including replacement of air duct components and business interruption. An internal investigation was conducted with assistance from a consulting company by first focusing on the source of the fuel and then on the ignition mechanism.

Figure 1 below is a simplified diagram of a portion of the system; instrumentation and controls are omitted for clarity. Ambient air is pulled into the system by negative draft where it is filtered through pre-filters then HEPA filters. This filtered air is then heated via a steam heat exchanger then an electrical resistance heater to approximately 200C. This hot air then contacts damp combustible powder to convey it through a drying zone and into a cyclone. The distance between the HEPA filters and the steam heater is less than 2 meters while the distance from the electrical heater to the damp solids entry point is approximately 25 meters. Section 3.2 will more completely describe the air duct system and the process mills.

FIGURE 1 – Sketch of Air Duct for Dryer System

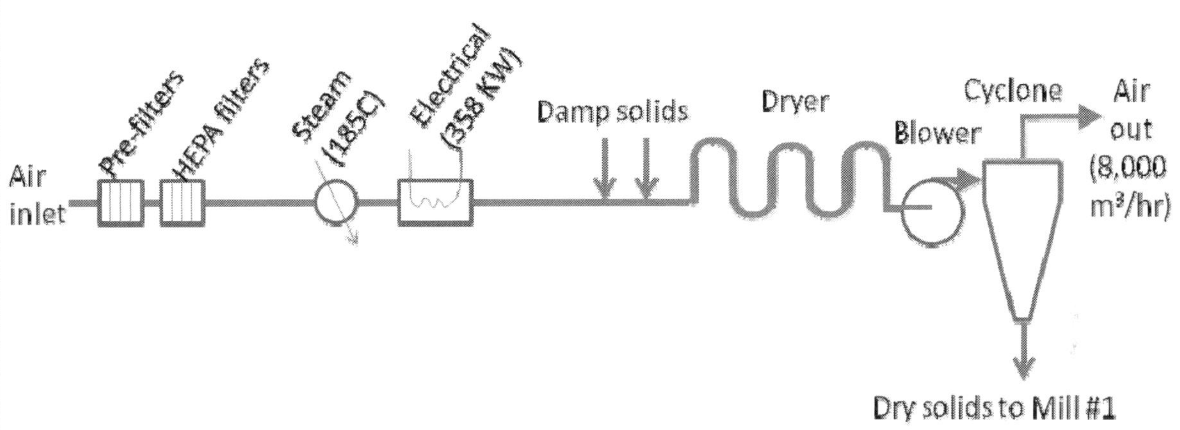

FIGURE 2 – Fire In Progress

The blower drive belts failed in September 2012 resulting in rapid loss of airflow so interlocks shut off steam and power to the heaters. Approximately 17 minutes later a portion of the air duct internally ignited.

2. "New" Lessons Learned

Forensic evidence pointed to the air filters as the source of fuel for the fire. Damage was limited near the air filters in Figure 1 plus there were no process solids or process ash upstream of the entry points for damp solids. The steam heater and electric heater were confirmed as tripped off. Calculations indicate that the residual heat (thermal inertia) was sufficient to ignite the air filters.

2.1 Filters As Fuel

The specific pre-filters and HEPA filters used at the incident site were rated for maximum continuous operation at 70C and 80C, respectively. Tables 1 and 2 list the key aspects of combustion testing of new pre-filters and HEPA filters. All components were susceptible to decomposition at 200C or less with the glass fiber and polyester component of the HEPA filter supporting a self-propagating flame in the oven test. Estimated energy release from combustion of the air filters was consistent with the damage observed from the incident.

Table 1. Stability of Air Filters

Unit	Component	Stability in 200C Oven Test	Stability in 500C Oven Test
Pre-filter	Glass fiber & polyester	Exothermic self-heating at 200C (to 600C). 47% weight loss	Exothermic self-heating (to 600C), 56% weight loss
HEPA Filter	Glass fiber & polyester	Exothermic self-heating at 160C (to 600C), self-propagating flame. 23% weight loss	Exothermic self-heating (to 600C), self-propagating flame. 39% weight loss
	Polyurethane sealant	38% weight loss	97% weight loss

Table 2. Estimated Air Filters Combustion Contributions

Unit	Component	Mass (kg)	Combustible Mass, from 500C Oven Test (kg)	Available Energy Release	
				(MJ)	(% of Total)
Pre-filter	Glass fiber & polyester	6.0	3.3	88	5%
HEPA Filter	Glass fiber & polyester	52.8	20.5	539	31%
	Polyurethane sealant	52.8	51.5	1089	63%

2.2 Residual Heat As Ignition Source

Recorded temperatures from the incident were not helpful in quantifying the residual heat remaining in the heaters after heating sources were tripped off. Recorded temperatures from a prior normal shutdown with the blower operating at 8000 cubic meters per hour were useful to calculate residual heats in the heaters as summarized in Figure 3 and Figure 4. The steam heater retains approximately 48 MJ while the electrical heater retains approximately 15 MJ.

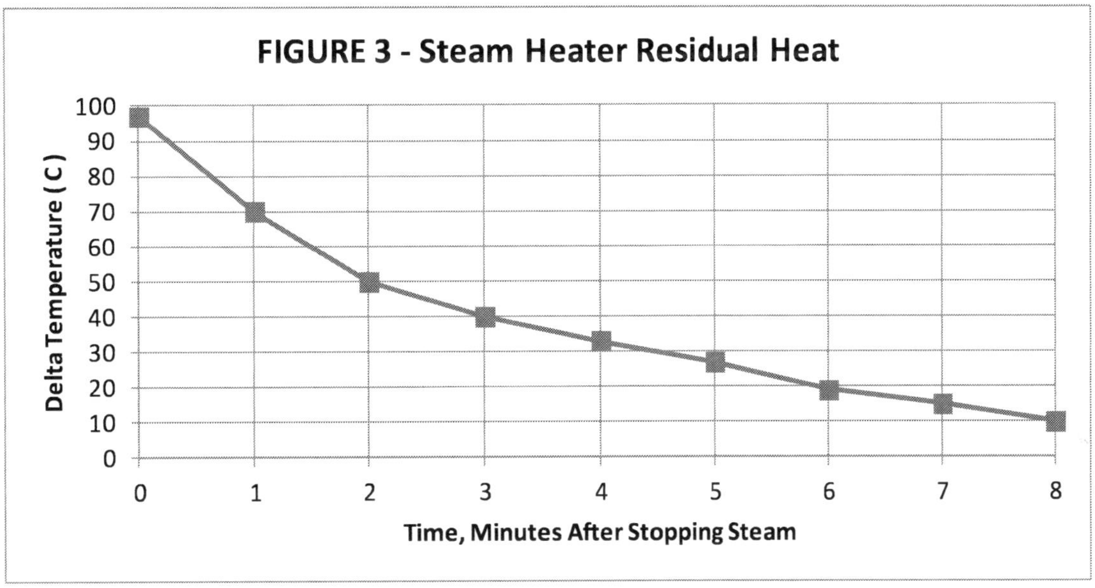

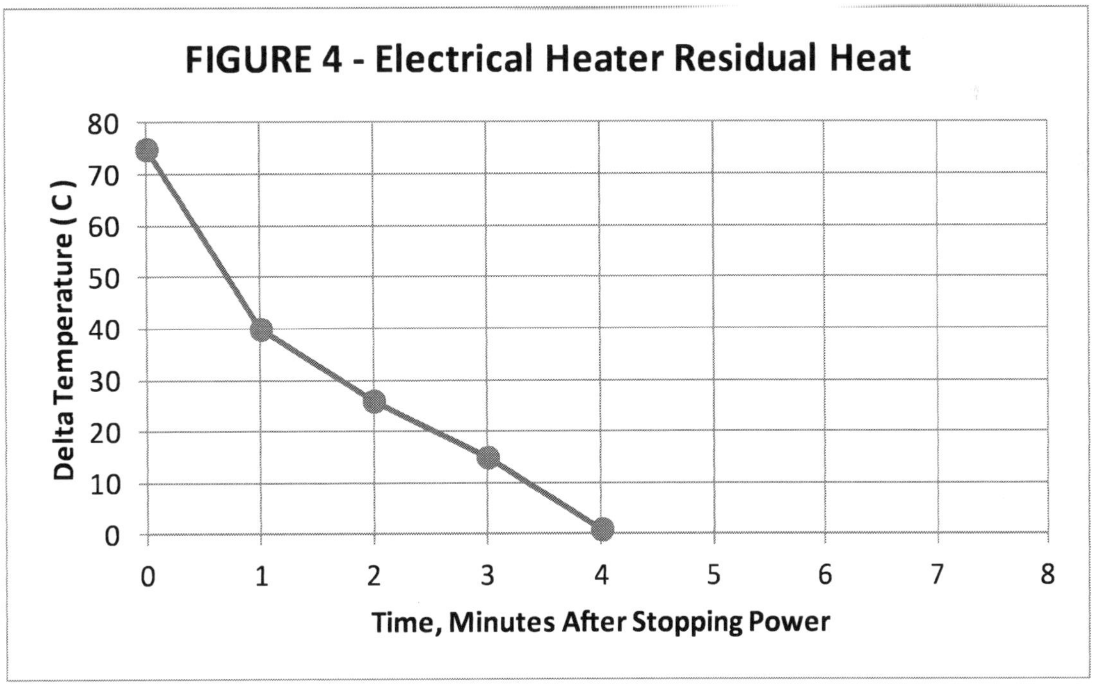

For the incident initiated by blower failure, this combined residual heat of 63 MJ can increase the temperature of stagnant air in the heater portion of the duct to well over 200C resulting in expansion driving the hot air backwards into the air filters. Alternatively, for this incident this combined residual heat could establish a convection current of dense cool air flowing forward at the bottom of the air filter duct, rising in the hot heaters zone, and hot air flowing backwards at the top of the air filter duct to the air inlet. Either incident mechanism results in excessively hot air in air filters that are rated for 70C or 80C maximum temperatures. The Standard Operating Procedures (SOP) for startup recognized the need to start the blower before heating and the SOP for normal shutdown recognized the need to keep the blower running until the air cooled down. Interlock designs recognized the necessity of tripping heating sources on loss-of-airflow. However tripping heating sources on loss-of-airflow is not sufficient to prevent the hazard of hot air reverse flow; therefore, an additional safeguard is needed. For this system, a new vertical vent duct with actuated valve was added between the heaters to open on loss-of-airflow as the primary additional safeguard to provide chimney-effect cooling of the heaters.

3. "Old" Lessons Re-Learned

In addition, lessons from this air duct fire have been refreshed multiple times and are widely known in the CPI.

3.1 Need to Investigate Near Misses and Minor Incidents

Prior to the September 2012 incident, a separate minor incident occurred during a normal shutdown. Some charring of the HEPA filters was attributed to a leaking steam heater isolation valve. The response was to repair the leaking steam valve; thus, the opportunity to investigate this precursor incident was missed.

3.2 Need to Limit Potential Escalation

The fire damage from the September 2012 incident extended beyond the Dryer air duct system (shown in Figure 1 above). The damage was much more severe to the other air filters (pre-filters and HEPA filters), other heat exchangers, and other ductwork interior surfaces as shown in Figure 5. The 3 other blowers in Figure 5 stayed on-line following the Dryer airflow failure and trip of the Dryer air heaters. Hot fumes of pyrolysis/combustion products from the Dryer air filters were pulled into the connected ductwork by these other blowers resulting in direct escalation of the fire or perhaps the Mill #2 electrical heater ignited fumes which backflashed to the air filters. While the Dryer air filters likely experienced restricted oxygen availability, the other air filters had ample oxygen to support combustion. The common air intake duct reduced the initial capital cost by a very small amount compared to separate intake ducts yet added unnecessary complications that led to major escalation. The system was rebuilt with separate air intakes.

FIGURE 5 – Sketch of Air Ducts for Complete System

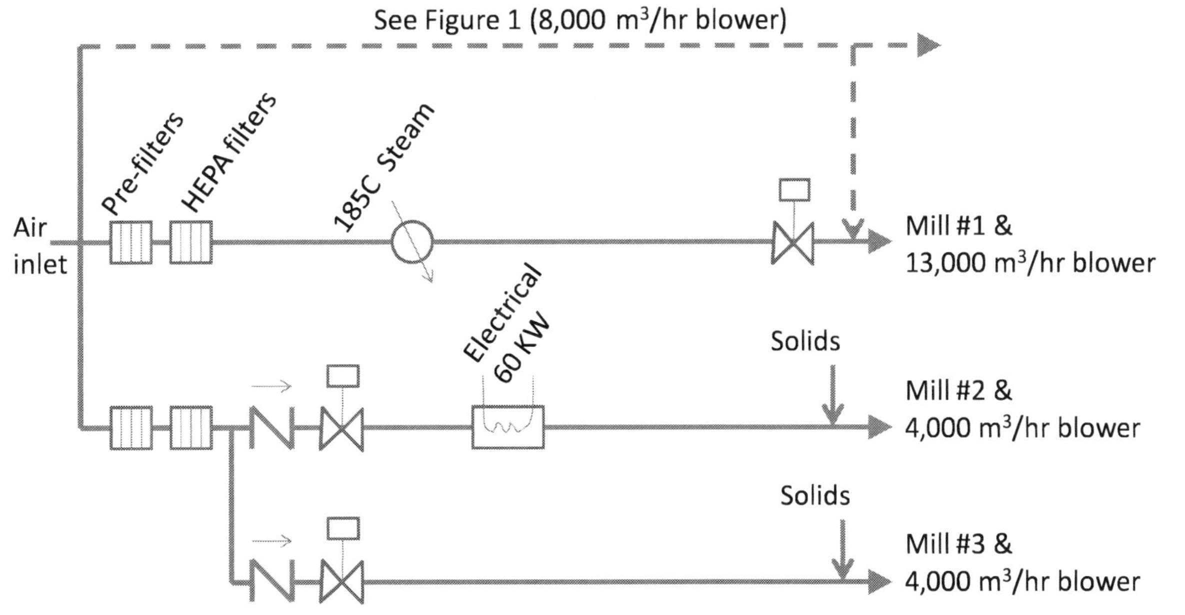

Modest external damage was also visible from smoldering combustible dust on the surfaces of some ducts due to accumulation. Housekeeping practices need to limit combustible dust accumulations below 3 mm (1/8-inch).

3.3 Need to Maintain Equipment Properly

Finally, the blowers did not have a specific installation procedure to ensure belt alignment plus they were not on a preventative maintenance program.

4. Conclusion

Those who forget the past are doomed to repeat. The main lessons learned from this incident apply to a broad range of operations in the CPI:

1) Mundane unit operations such as air filters that have combustible components must be evaluated as thoroughly as the rest of the chemical process.
2) Designers and HAZOP leaders must recognize the potential of thermal inertia of heaters (and coolers) to heat (or cool) a stagnant volume of fluid when flow is interrupted. Gas streams are particularly susceptible due to the low heat capacity relative to the heat capacity of a metallic heat exchanger. Interlocking the supplies (to steam and electrical heaters in this incident) is often necessary but may not be sufficient.
3) Near misses and minor incidents should be thoroughly investigated; an opportunity was missed at this site to prevent the major incident.

4) Keep it simple and keep areas clean. Significant escalation occurred due to a combined air duct that added unnecessary interactions between the systems. Minor escalation occurred from accumulated combustible dusts on exterior surfaces.
5) Maintain equipment properly. Rigorous installation procedures (such as belt alignment) and preventative maintenance improve both safety performance and on-stream reliability.

5. References

Additional references not cited:

[1] Kletz, Trevor. "What went Wrong? - Case Histories of Process Plant Disasters and How They Could Have Been Avoided" (5th Edition, 2009), Elsevier.

[2] DOE (U.S. Department of Energy), "Nuclear Air Cleaning Handbook" Chapter 10 – Fire Protection, DOE-HDBK-1169-2003 (2003).

9TH GLOBAL CONGRESS ON PROCESS SAFETY

Hydrogen Sulfide Exposure from Molten Sulfur – A Forgotten Hazard?

Mark J. Viz,
Sean J. Dee, Russell A. Ogle, and Brenton L. Cox
Exponent, Inc.
4580 Weaver Parkway, Warrenville, Illinois 60555
mviz@exponent.com, sdee@exponent.com, rogle@exponent.com,
and bcox@exponent.com

Prepared for Presentation at
American Institute of Chemical Engineers
2013 Spring Meeting
9th Global Congress on Process Safety
San Antonio, Texas
April 28 – May 1, 2013

UNPUBLISHED

Hydrogen Sulfide Exposure from Molten Sulfur – A Forgotten Hazard?

Mark J. Viz,
Sean J. Dee, Russell A. Ogle, and Brenton L. Cox
Exponent, Inc.
4580 Weaver Parkway, Warrenville, Illinois 60555

Keywords: molten sulfur; hydrogen sulfide; railcar; tank car; tank trailer; process safety management; PSM; hazardous materials; hazmat; PHMSA; OSHA; NA2448

Abstract

Molten sulfur is one of the most frequently transported hazardous materials (hazmat) in North America. Molten sulfur is typically transported at an elevated temperature by rail tank cars or over-the-road tank trailers as a Class 9 "Miscellaneous Hazardous Material" under the identification number NA2448. Perhaps the most prominent hazard of molten sulfur is that it is shipped at an elevated temperature, which can cause thermal burn injuries. A less obvious hazard is the potential release of hydrogen sulfide (H_2S), a highly toxic gas.

In this paper we first summarize the key features of the H_2S-molten sulfur equilibrium that are relevant to this hazard. Then we address the relative risk of H_2S exposure from an analysis of molten sulfur incident release statistics. Next, we discuss a few case studies in which vented H_2S gas from molten sulfur led to unexpected consequences for handling personnel. We close our discussion with suggestions for how to possibly prevent H_2S exposure through a combination of engineering controls, operating procedures for both loading and unloading operations, air monitoring, and personal protection equipment.

1. Introduction

Molten sulfur is one of the most widely used global feed stocks. In 2010, the global demand for sulfur reached 78.3 million tons, with elemental sulfur accounting for 66% of global production.[1] With growing demand in the Middle East and Asia, continued use of sulfuric acid, and a growing need for nickel, sulfur production is projected to grow steadily over the next decade.[2] In 2011, elemental sulfur was a $1.6 billion dollar industry in the United States, with 8.1 million tons of production.[3] Additionally, over the last thirty years there has been a significant shift in the production of molten sulfur, moving from mining of geologic sulfur deposits (Frasch process) to environmental byproduct sulfur (Claus process). This technology shift has increased the H_2S

[1] Messick, D. L., World Sulfur Outlook, The Sulfur Institute, 2010.

[2] Kitto, M., Sulfur Supply Demand Balance: The Outlook to 2015, British Sulfur Consultants, 2005.

[3] Apodaca, L. E., Sulfur, U.S. Geological Survey, Mineral Commodity Summaries, January 2012.

hazard in molten sulfur. Given the livelihood of the global sulfur market and the change in the source of the material, the need to communicate and control the hazards associated with the handling of molten sulfur is more critical than ever.

The hazard of dissolved H_2S in molten sulfur became widely recognized in 1963 with the disappearance of the molten sulfur tanker, the Marine Sulfur Queen.[4] This tanker disappeared without a distress signal off the coast of Florida with all 39 crew members. The U.S. Coast Guard searched for the vessel but was only able to recover a few life preservers and some floating debris. Reports came to light of frequent fires on the sulfur tanker and the frequent odor of rotten eggs.[5,6] Investigators from various countries performed headspace measurements in molten sulfur transportation vessels and found H_2S concentrations approaching the minimum flammable level. Recommendations were published that the dissolved H_2S content of molten sulfur should be maintained at 10 parts per million (ppm).[7]

As early as 1970, the U.S. Department of Transportation (USDOT) issued a rule under docket HM-70, that even though it was intended to address requirements for rail tank cars and tank trailers used to ship H_2S as its own commodity, clearly emphasized the hazards associated with exposure to H_2S. The comments to the federal rule state that H_2S is a "severe irritant and is highly toxic"; that exposure leads to "olfactory fatigue"; that H_2S is "heavier than air" and "tends to 'pool' rather than to dissipate into the atmosphere."[8]

Later, in the mid-1980s, the USDOT issued a Final Rule under docket HM-198 that clearly identified two significant hazards associated with the transportation of molten sulfur: the presence of unknown but potentially significant levels of H_2S and that the elevated temperature of molten sulfur could cause burns.[9] As reported in the Final Rule: "Several incidents involving the transport of molten sulfur, including three in the state of California since 1985, have emphasized the need for regulation of this material. Accidents involving spills near Barstow County and Culver City in 1986, and another in Benecia in 1985, demonstrated a lack of hazard communication. In each of these incidents, lack of swift and accurate identification of the material led to delays in handling the incidents and, in at least two instances, contributed to injuries sustained by persons on the scene."

A review of available Material Safety Data Sheets (MSDSs) for molten sulfur suggests that the range of H_2S concentrations and corresponding safe loading, unloading, and handling procedures for molten sulfur vary considerably among producers and shippers. As a consequence, the handling processes may be performed in a manner ignorant of the possible high concentrations of H_2S. According to several MSDSs, the primary hazards associated with molten sulfur are

4 King, F.W., "Taking H_2S from liquid sulfur-why and how," Oil and Gas Journal, pp. 164-165, June 1974.

5 Anon., "Investigations: The Queen with the Weak Back," TIME, March 8, 1963.

6 U.S. Coast Guard, Marine Board of Investigations, "Disapearance of the SS Marine Sulphur Queen at sea on or about 4 February 1963 with the presumed loss of all persons on board," March, 1964.

7 Lagas, J.A., "Stop emissions from liquid sulfur," Hydrocarbon Processing, pp. 85-89, October 1982.

8 35 FR 18919 et seq., December 12, 1970, docket no. HM-70, "Hydrogen Sulfide Gas in Cargo Tank Trucks and Tank Cars."

9 53 FR 17158 et seq., October 21, 1986, docket no. HM-198, "Molten Sulfur and Molten Materials Generally."

thermal burns, sulfur-fueled fires, dust explosions, exposure to H_2S, and, in the event of a sulfur fire, inhalation of sulfur dioxide.[10,11,12,13,14] The hazard of H_2S is not prominently communicated in many of these current MSDSs.

Elemental sulfur is typically transported in a molten state in railcars or tank trailers. Commonly, the H_2S hazard is reported as a potential risk, because the concentration of H_2S within the vapor space of the railcar, tank trailer or stationary storage or process vessel is specified as a range or threshold value. The actual concentration of H_2S present in the vessel headspace will likely change due to temperature variations and agitation (as occurs in railcars and tank trailers during movement), and the potential exposure will be dependent on the producer, shipper or consignee's handling procedures and best practices. The authors' own field experiences have also shown that particular to the unloading of molten sulfur railcars, unloading personnel can have a false sense of safety when opening the top fittings or manway cover to prepare a tank car for steaming. Unloading personnel occasionally believe that the solidified "crust" of molten sulfur that is usually present at the surface, especially in cold weather, will somehow prevent any significant release of H_2S. Our own investigatory experience has shown that this is not always true.

Occupational exposure to H_2S during handling can generally be categorized as an example of a low probability, high consequence risk. This is supported by a review of the Pipeline and Hazardous Materials Safety Administration (PHMSA) and Occupational Safety and Health Administration (OSHA) incident report databases, which confirms that over the last 20 years, there have been six OSHA-reported fatalities from H_2S exposure associated with molten sulfur containers and over the last 24 years, there have been no PHMSA-reported fatalities from molten sulfur releases (while the commodity is still under the jurisdiction of the federal hazmat transportation regulations).[15,16] Our collective experience based on our investigations of numerous hazmat releases is that the tendency is for successful management of low probability, high consequence hazards to lead to complacency and the occasional elimination of seemingly superfluous safeguards.

[10] Hovensa, Sulfur MSDS No. 1794, 2006.

[11] Tesoro, Safety Data Sheet: Sulfur, Molten, 2012.

[12] Keyera, Sulfur, Molten MSDS, 2011.

[13] Devon, Material Safety Data Sheet: Sulfur (Liquid), 2010.

[14] Marathon, Material Safety Data Sheet: Marathon Sulfur, 2010.

[15] Note that there can be overlap between the release incident reporting required by OSHA and PHMSA based on where the incident occurred and what function was being performed when the release occurred. So, for example, an accidental release of molten sulfur involving an injury or fatality that occurred while a tank car was connected to plant apparatus during unloading might be reported to OSHA and not to PHMSA even though releases during unloading can be covered by PHMSA regulations as well. Efforts to harmonize the regulations and reporting requirements promulgated by a variety of federal agencies has been the subject of federal rulemaking in dockets such as HM-223, "Applicability of the Hazardous Materials Regulations to Loading, Unloading and Storage," Docket No. PHMSA-98-4952 (HM-223), under which PHMSA issued a Final Rule on June 1, 2005.

[16] Note that at times, exposure to H_2S can result in a prolonged hospitalization that results in fatality at a date after the original paperwork related to regulatory agencies is submitted.

According to the USDOT PHMSA "Incident Reports Database," there have been 428 reported releases of molten sulfur in transportation between the years 1989 to 2012, inclusive. For all 428 reported releases, no fatalities and 19 injuries were reported. A portion of these injuries were related to exposure to fumes and odors that resembled "rotten eggs", a description commonly used for H_2S. Additionally, incident reports indicated that H_2S gas was released at various times while the commodity was in transit, being loaded or unloaded, or while being stored in transit. Although there may be some overlap between exposure incidents reported in PHMSA's Incident Reports Database and the analogous Accident Investigation Summaries from OSHA (OSHA Form 170), which are maintained in OHSA's Integrated Management Information System (IMIS), the OSHA database includes six reported fatalities from H_2S exposure associated with the handling of molten sulfur over the last twenty years.

Our review of H_2S exposure incidents is intended to act as a reminder of the importance of employing a hierarchy of safeguards when managing high consequence hazards like H_2S.

2. Technology Background

A major technological shift in the production of sulfur has resulted in an increased potential for exposure to H_2S. For most of the 20th century, the Frasch process was used to recover sulfur from underground, elemental deposits.[17] In the Frasch process, superheated water is used to melt solid deposits of sulfur which are driven to the surface with compressed air for recovery. However, with the growth of the refining industry, the supply of elemental sulfur from the desulfurizing of sour crude oil and natural gas overtook production from the Frasch process in 1970. The Claus process, which can produce elemental sulfur from feed stocks containing up to 20% H_2S, is now the preeminent production process for producing sulfur.

To convert H_2S to elemental sulfur, the Claus process uses a two-step process, first utilizing a high temperature thermal step, followed by a lower temperature catalytic step. In the thermal section, air is added to H_2S at temperatures in excess of 1800°F, to covert one-third of the H_2S to sulfur dioxide (SO_2) as shown in Equation 1.

$$H_2S + 3/2\ O_2 \rightarrow SO_2 + H_2O \qquad \text{[Eq. 1]}$$

The resulting SO_2 then reacts with the remaining H_2S in an endothermic equilibrium reaction to form elemental sulfur vapor as shown in Equation 2.

$$2H_2S + SO_2 \leftrightarrow 3/2\ S_2 + 2\ H_2O \qquad \text{[Eq. 2]}$$

To drive conversion in the endothermic equilibrium reaction shown in Equation 2, S_2 vapor is removed through condensation. As the vapor cools, the S_2 equilibrates with the S_6, and S_8 forms through multiple equilibrium reactions.

If one-third of the sulfur is consumed in Reaction 1, then the remaining two-thirds is consumed in Reaction 2 resulting in stoichiometric consumption of H_2S. However, several side reactions usually lower the conversion to SO_2 in Reaction 1, subsequently lowering the overall H_2S

[17] Greenwood, N. N. and Earnshaw, A., Chemistry of the Elements, 2nd ed., 1997.

conversion. This highlights both the importance of the air to H_2S feed ratio as a process control variable, and the need for a secondary catalytic reactor to further convert the H_2S that is not consumed in the thermal section.

The catalytic section of the Claus process is used to convert H_2S as well as side products including COS and CS_2 that are formed from side reaction in the thermal section. The catalytic section runs at significantly lower temperatures than the thermal section (inlet 400°F - 450°F, outlet 600°F - 630°F), but uses a Clause catalyst (usually aluminum (III) or titanium (IV) oxide) to covert H_2S to elemental sulfur as shown in Equations 3 and 4.[18,19]

$$2\,H_2S + SO_2 \rightarrow 3/6\,S_6 + 2\,H_2O \qquad \text{[Eq. 3]}$$

$$2\,H_2S + SO_2 \rightarrow 3/8\,S_8 + 2\,H_2O \qquad \text{[Eq. 4]}$$

The effluent of the catalytic section can be cycled through additional catalytic beds, or condensed to a final product stream at temperatures between 265°F - 350°F.

The one common feature to all of these chemical reactions is the constituent H_2S. While the Claus process can achieve 96% - 97.5% conversion with one thermal reactor, and three catalytic reactors in series, the solubility of H_2S in molten sulfur adds an additional processing constraint. The solubility of H_2S in molten sulfur actually increases with temperature up to approximately 425°F. Solubility remains approximately constant from 425°F until 630°F, where it begins to drop with further increases in temperature.[20] Thus, in the catalytic reactors and condensers, there is still an appreciable amount of H_2S dissolved in the molten sulfur. If the hot sulfur is then stored in a tank, as the tank cools, the dissolved H_2S will release into the headspace, increasing the concentration to potentially dangerous levels. If the sulfur continues to cool, then a crust will form as the surface of the product reaches temperatures below 240°F. This solid crust can serve as a physical barrier protecting personnel from potential H_2S exposure. In cases where a crust is observed, special consideration should be noted to avoid agitation that could disrupt the solid interface, resulting in a potential H_2S vapor exposure. However, it should be noted, that formation of a protective crust should not be regarded as a reliable safety indicator, since crust formation is not correlated to the concentration of H_2S in the vapor space.

The headspace concentration of H_2S that can be achieved with molten sulfur in a transportation or storage container is influenced by four primary factors: the dissolved H_2S concentration, the liquid temperature, agitation of the liquid, and the presence of any organic contaminants. To maintain an H_2S concentration below its lower flammable limit (LFL), typical specifications for the dissolved H_2S concentration range from 5 to 15 ppm. At the temperature range at which molten sulfur is typically handled, the LFL corresponds to a headspace concentration of 3.4%, or

[18] Goar, B. G. and Fenderson, S., Fundamentals of Sulfur Recovery by the Claus Process, presented at The Laurence Reid Gas Conditioning Conference, Norman, Oklahoma, March 3, 1996.

[19] Sulfur and Hydrogen Sulfide Recovery, Kirk-Othmer Encyclopedia of Chemical Technology, John Wiley & Sons.

[20] Fanelli, R., Solubility of Hydrogen Sulfide in Sulfur, Industrial and Engineering Chemistry, pp. 2031-2033, September 1949.

34,000 ppm (at room temperature the LFL for H_2S is 4.3%).[21] From the viewpoint of respiratory exposure, this is a potentially instantaneous lethal dose.

3. Hazard Communication

The MSDSs for molten sulfur identify several hazards. Common features were noted in the 19 MSDSs reviewed for this study. First, nearly all of the surveyed MSDSs noted the potential dangers associated with H_2S and listed H_2S as a material present in molten sulfur. The amount of H_2S present was usually noted as a small amount; descriptions included "trace", "< 0.1%", "< 1%", "< 500 ppm", and "< 1% (variable)". It was also common for MSDSs to include additional information about the potential hazards of H_2S exposure, the acceptable exposure limits (which even today are still a subject of debate), and to identify the need for proper ventilation during storage and handling.

In the MSDS review, there is significant confusion and room for improvement regarding the description of PPE and unloading procedures. In addition to H_2S hazards, respiratory protection also included consideration for dust hazards since molten sulfur can solidify to a yellow solid when cooled. In most MSDSs, it is noted that respiratory protection should be used, but only if the concentrations of H_2S are unknown/above regulated levels, or other engineering controls are not sufficient to control the exposure. One MSDS stated that respiratory protection was not commonly required when handling molten sulfur. While several MSDSs noted that unloading/opening of tanks should be done in well ventilated areas, it was not noted if proper ventilation is a sufficient engineering control to protect workers from potential exposure to H_2S. One MSDS included a picture of a dust mask in the recommended PPE, which is appropriate for the potential dust hazard, but not for H_2S exposure.

In addition to not specifying when a respirator should be worn, very few MSDSs had detailed recommendations for unloading procedures related to the H_2S hazard in molten sulfur. As previously discussed, most of the MSDSs recognized several key components to safe unloading: proper ventilation, respiratory protection if the H_2S concentration was unknown or above acceptable limits, and confined space warnings. Air monitoring was not commonly listed as a recommended engineering control in the MSDSs. Additionally, recommendations regarding venting of tank headspace were not usually recommended before sampling or unloading. This omission, however, could be the result of the inclusion of opening the tank car or tank trailer before steaming as part of the unloading procedure. Details such as these regarding unloading might not typically be specifically addressed in the MSDS. Other MSDSs specified venting, only if the product was excessively heated or exposed to elevated temperatures. The potential evolution of H_2S into the vapor space can be driven both by temperature increases in the presence of water or by temperature decreases which drive dissolved H_2S out of the molten sulfur. Including specific directions for when respiratory protection is required and directions for proper unloading safety controls before handling/unloading could improve the hazard communication for H_2S in molten sulfur.

[21] NFPA 655, Standard for Prevention of Sulfur Fires and Explosions, National Fire Protection Association, Quincy, MA, Section A.5.1, p. 17, 2007.

4. Review of PHMSA and OSHA Incident Data

Two meaningful sources of incident data regarding H_2S exposure from molten sulfur handling are the PHMSA Incident Reports Database and the Accident Investigation Summaries from OHSA Form 170. Although there may be some overlap in incident reporting between these two databases, in general the PHMSA database contains mandatory reports that involve the release of a regulated hazardous commodity while in transit, while being loaded or unloaded, or while in storage during transit. The OSHA database contains mandatory reports that involve incidents or accidents that have resulted in the injury or fatality of an employee. The differences between the type of incidents reported in each database and the corresponding limitations in the reports should be understood. For example, personnel may have been briefly exposed to H_2S shortly after a tank car has been opened for steaming or unloading, but if there was no release of molten sulfur, the incident would not necessarily need to be reported in the PHMSA database. In a similar manner, if a release of molten sulfur occurred after a transport vessel had been connected to plant apparatus but it did not involve the unloading equipment, this incident might only be reported in the OSHA database, and then only if a reportable injury or fatality occurred as a result. In this manner, then, the review and subsequent use of the information provided in these reports should be considered with care and properly qualified when attempting to estimate occurrence probabilities and severities.

Bulk shipments of molten sulfur in North America are typically identified by the number "NA2448." Using the PHMSA Incident Reports Database, we performed a search of reportable incidents involving NA2448 by all modes of transportation (air, water, highway and rail) over the entire range of dates for which electronic records are available. From 1989 to 2012, the database contained 428 reports of unintended releases of molten sulfur, all of which occurred either by rail or highway. In these 428 reports, no fatalities were reported and 19 injuries were reported. An interesting finding is that of the 428 reports, 43 of them were described (based on the responder's description provided in Part VI of the Form DOT F 5800.1) as being initially discovered based on the detection of a strong odor. In addition, the 43 reports that explicitly identified "odor" in the description all involved rail tank cars. Of these 43, three of them further described the odor as being that of "rotten eggs," which is the typical way that people refer to the unique smell of H_2S.

It is compelling that roughly 10% of the molten sulfur releases reported from 1989 to 2012 indicate that they were initially detected by the "odor" of sulfur, and in a few cases, the more telling odor of "rotten eggs." Many of these instances occurred when a rail tank car was technically "in transit" but was not "in movement," such as when a railcar is in a classification yard. The irony involved in the discovery of these situations is that while the inhalation hazard of the H_2S is a high consequence hazard, had it not been for personnel noticing the odor coming from these tank cars, detection of the lading loss might not have occurred.

For releases that occurred in rail transit, the most common causes were a corroded or otherwise breached rupture disk, an improperly secured manway cover (usually involving a number of loose or missing bolts) or an improperly seated gasket. These causes create concern for railcar handling personnel because unless they can detect a potential exposure by smell or see the physical indicators of lost lading (such as crusted sulfur on the outside of the tank car), they would likely approach the railcar assuming that it was completely sealed. This might lead to personnel being placed in danger if they approach an unsealed railcar that is venting high concentrations of H_2S that are not detected quickly enough for personnel to detect and identify the hazard and approach the railcar using more appropriate PPE.

The majority of releases from tank trailers involved human error on the part of the driver who failed to properly monitor the loading operation and allowed the tank trailer to be overfilled. Reports involving overfilling do not refer to sulfur or H_2S odors, likely because the loss of lading is immediately apparent by visual observation.

A review of the Accident Investigation Summaries from OSHA revealed five cases that noted an H_2S release from molten sulfur tanks during unloading.[22] A total of 16 employees were involved in the five incidents, with six total fatalities. There are several common features related to each of these incidents. First, all five incidents occurred in the fall or winter season in locations where the ambient temperatures are low. Second, none of the five incident reports makes mention of the exposed employees wearing proper respiratory protection, and two investigations noted that employees were not wearing respirators. In two of the incidents, the exposure to H_2S resulted in loss of consciousness which caused a fall for the employees who were not wearing fall protection.

The most recent OSHA report of an accident related to H_2S exposure from molten sulfur occurred in 1995. An employee was taking samples from a molten sulfur tank truck. After opening the manway cover, he reached approximately 15 inches into the tank to collect the sample. A respirator was not provided, and even though he was on a nearly nine-foot tall platform, he was not wearing fall protection. The employee was found unconscious on the gravel covered ground approximately 10 minutes after taking the sample. An autopsy listed blunt force injuries from the fall as the cause of death, but testing revealed H_2S levels up to 800 ppm using similar unloading procedures on site. OSHA maintains that H_2S was a causal factor related to the fall.

In addition to potential fatalities due to falls, H_2S can also cause poisoning or asphyxia. In 1993, three employees were unloading a molten sulfur railcar. Two employees were on a platform on top of the railcar, while the third employee was on the ground hooking up steam lines. The third employee found his two coworkers unconscious on the platform shortly after they had opened the manway cover on the railcar. The employee who opened the manway cover was found dead on arrival, the other employee on the platform was hospitalized for several days, while the third employee was examined and released.

5. Safeguards to Prevent or Control H_2S Exposures

H_2S exposure during the handling of molten sulfur can be generally categorized as a low probability, high consequence event based on an assessment of the data available from PHMSA and OSHA. However, these incidents were of high consequence due to the fatalities that have

[22] Incident reports from www.ohsa.gov. A total of 172 reports were reviewed with "hydrogen sulfide" as a key word. Five reports were related to molten sulfur tanks. Those incident reports are numbers 109097600, 109713388, 17713199, 106387822, and 1383249.

been reported. Managing this or any low probability, high consequence risk may benefit from the use of multiple safeguards for a variety of reasons.

Suggested safeguards to consider, most of which can be implemented independently or together, are:

- Closed-loop venting and treatment of H_2S vapor (e.g., adsorption systems, flares);
- Administrative controls (e.g., standard procedures and training for unloading, buddy system);
- External ventilation of unloading areas;
- Air monitoring systems with alarms;
- Respiratory protection (PPE) sufficient to the hazard level or that ensures the safety of unloading personnel when the hazard level is unknown.

In fact, some states have mandated safety procedures associated with the handling of H_2S, including specifications for personal protective equipment, venting requirements, and equipment specifications.[23] As always, the best system of safeguards will be site-specific and tailored to the capabilities of site personnel. It is also important to consider the consequences of any changes prior to their implementation, even the addition of safeguards, lest any new hazards be introduced.

6. Recommendations

The safe handling of molten sulfur requires understanding and planning regarding the potential release of H_2S. Herein, we have discussed the technical reasons for H_2S hazards in molten sulfur, reviewed accident statistics, and relevant safety documentation. The following recommendations might be considered when handling molten sulfur.

- Sulfur supply contracts, which typically include requirements for a dry weight analysis of the sulfur based on minimum or maximum amounts of ash, carbon, arsenic, selenium and possibly other elements, should also address maximum levels of H_2S. In addition, these requirements should address when and how the product sample should be taken and then analyzed, especially if it is determined that the H_2S content is strongly dependent on sample temperature and storage time.
- Documentation regarding the shipping, receipt, and unloading of molten sulfur should be maintained per regulatory and "best practices" requirements. These documents, as well as sulfur supply contracts, might include warnings regarding the potential increase in H_2S concentration during shipment.

[23] For example, California Title 8, Section 4414 and Washington Administrative Code 296-79-250.

- Development of site-specific loading and unloading procedures should be written and followed given due consideration to the assessment and management of the hazards associated with H_2S in molten sulfur. These procedures might include:
  - Forced exhaust ventilation requirements;
  - Proper temperature, humidity, and air monitoring regarding H_2S concentrations at tank inlets, outlets, and bulk concentrations within the unloading area;
  - Proper PPE for burn protection and H_2S exposure, possibly including a requirement for a SCBA respirator and an escape bottle;
  - Controlled tank headspace venting before sampling or unloading;
  - Consider fall protection requirements knowing that non-lethal exposure to H_2S can induce a loss of consciousness and potential injuries related to falls.

Finally, it is emphasized that there is typically no fixed set of safeguards that will work at every facility. A well-designed system of safeguards to protect workers from H_2S exposure should likely consider the personnel, equipment, and safety management system at that particular facility.

9TH GLOBAL CONGRESS ON PROCESS SAFETY

Calcium Stearate Deflagrations – A series of unfortunate events

Albert Ness
ABS Consulting
10301 Technology Drive
Knoxville, TN 37932
aness@absconsulting.com

Konanur Manjunath
Dow Chemical Co.
Freeport, Tx
KManjunath@Dow.com

Prepared for Presentation at
American Institute of Chemical Engineers
2013 Spring Meeting
9th Global Congress on Process Safety
San Antonio, Texas
April 28 – May 1, 2013

UNPUBLISHED

Calcium Stearate Deflagrations – A series of unfortunate events

Albert Ness
ABS Consulting
10301 Technology Drive
Knoxville, TN 37932
aness@absconsulting.com

Konanur Manjunath
Dow Chemical Co.
Freeport, Tx
KManjunath@Dow.com

Keywords: Calcium Stearate Solids, Combustible Dust, Bulking Brush Discharge, Particle Size Distribution, Resistivity, Bonding, FIBC, Inertion

Abstract

A series of deflagrations occurred in a plant unloading Calcium Stearate from bulk bags. Each one had a different root cause and illustrates the importance of management systems for ignition control and the need to evaluate a system for many types of ignition sources. The first event was caused by static discharge from an ungrounded tote. When upgrading the system Type C bags were specified for the product and a second deflagration occurred due to static discharge from and ungrounded bag. In the third deflagrations cause was most likely a conical pile discharge. The presence of fines and coarse material plus the use of a larger cubic vessel probably enhanced the likelihood of the brush discharge.

.

1. Introduction

This paper describes three deflagrations that occurred in a bulk bag unloading system that had previously run for over six years without incident.

1.1 Product data for Calcium stearate

Table 1 lists data for the Calcium Stearate that was being used in the facility.

Table 1 Calcium Stearate Properties

Parameter	Value
D_{50}	500 microns
Minimum Ignition Energy (MIE) (< 75 μ fraction)	8 mJ
Volumetric resistivity	$14\ \ 10^{14}$ Ω-m

2. Unloading station description

The unloading station consisted of a bulk bag of Calcium Stearate (CaSt) hung u above a feed hopper. An Iris valve was used to control the flow rate of the CaSt into the hopper. A flexible feeding chute was attached from the hopper outlet to the lid of a 2500 L tote. Local exhaust ventilation was provided at the feed hopper and tote.

Figure 1: Unloading station for calcium stearate bulk bags

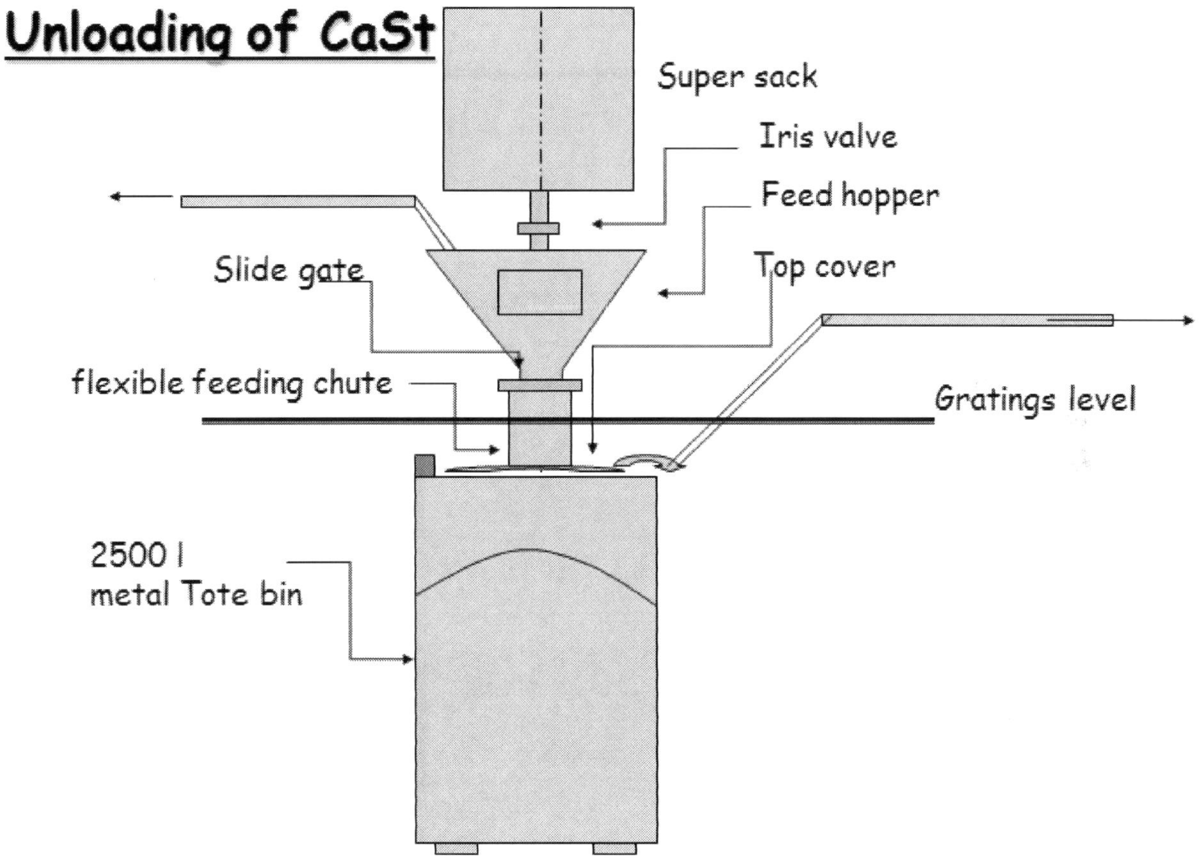

2.1 *Event #1 while unloading*

At the end of the loading cycle a deflagration occurred in the head space of the tote. The fire was quickly extinguished and little actual damage occurred. Investigation showed that the Feed hopper and the lid of the tote were not grounded and that the flexible chute was not static dissipative. Additionally, it was discovered that the tote vacuum system was not used. Figure 2 shows where the damage occurred.

Figure2: Unloading station at the time of event 1

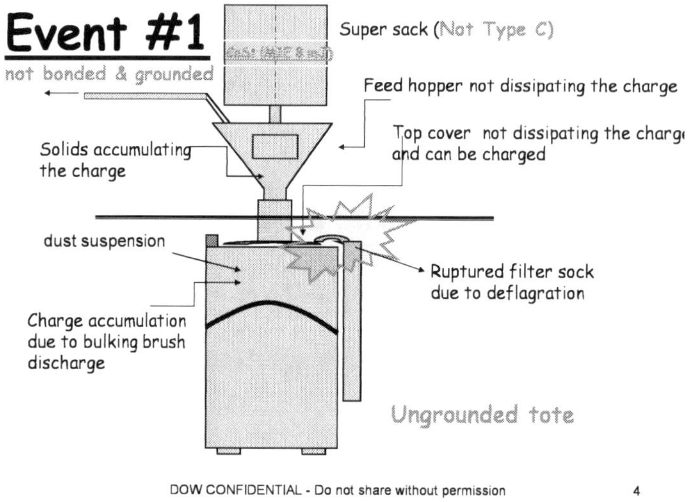

With the vacuum system inoperative, fines were able to stay suspended in the head space of the tote. A calculation of the capacitance of the lid of the lid showed that it had a capacitance of 180 pico Farads. Figure 3 shows that at that level, a static discharge energy of about 30-40 mJ for a charge of 20 EV, well above the 8 mJ MIE of the fraction under 75 microns, was possible. Therefore it was concluded that the most likely cause of the deflagration was a spark between the lid and tote ignition suspended fines in the tote.

Figure 3: Calculated electrostatic build up on process equipment

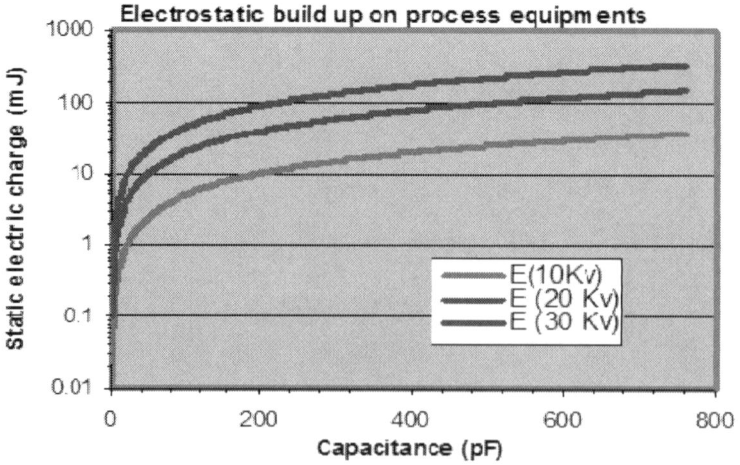

2.2 Event #2

After the first event, the metal equipment was grounded and the tote vacuum system was put back into operation. Also, it was decided to specify use of Type C bags for the bulk bags. Previously, the bulks bags were type B, that is, an all plastic bag that cannot be grounded, but has a breakdown voltage of 4kV. This allows for discharge of energy to the air. A Type C Bag has continuous conductive threads are sown into these bags, hence continuity is provided if the bag is grounded.

After starting up the improved system, a second deflagration occurred. Upon investigation, it was found that; 1) the bulk bag had not been grounded and 2) the Iris valve was not grounded due to its Teflon coating and the powder buildup on it. The failure to ground the new bulk bag is a reminder of the expression "No good deed goes unpunished". It is also a reminder of the need for a good Management of Change System. In the case of this change, the operator's training should include instruction on the hazard of NOT grounding the Type C bag.

2.3 Event #3 while unloading

Prior to restarting the unit, a ground verification system was installed to ensure the bulk bag was grounded before charging. Figure 4 shows where the damage occurred.

Figure 4: Unloading station at the time of event 3

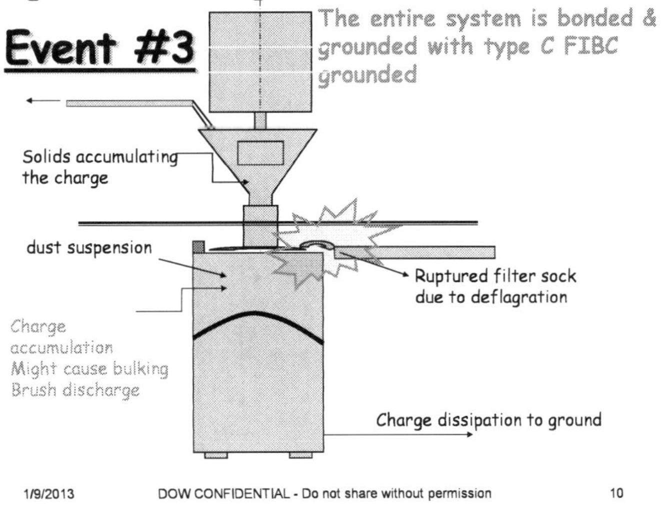

At this point, there was no obvious source of ignition. Several experts were called in and a more thorough investigation was done. Findings were:

1. The FIBC was examined for continuity of grounding. Except for the area of the bag that was damaged by the fire, the continuity was acceptable. Testing of FIBCs at the supplier found all the bags were conductive.
2. The CaSt in the FIBC from the last event was somewhat smaller than in the initial events. Table 2 shows the difference in size between the CaSt particles from the first and third events. The material from the third event was not out of spec, it was considered

acceptable material by the plant. The key point here is that normal variation in the raw material can affect the degree of hazard.

Table 2: Size Comparison

Cumulative Vol. %	Event 1	Event 3
90	800	800
50	500	425
10	200	150
% < 75 μ	2-3	~5

3. Fines collected from the top of a different tote had a D50 of approximately 10 μ and were found to have an MIE of 3 mJ.
4. The dust extraction system was again found to have been off.

2.4 Bulking or Conical Brush Discharges

Whenever solids are filled into containers, bulk bags and so on, bulking brush discharges are developed. It is the type of discharge observed on the conical pile of a heap of powder; thus it is often called conical pile discharge. Bulking brush discharges have an apparent maximum effective energy of 10-25 mJ (with respect to dust ignition), and are believed to be responsible for dust explosions in grounded silos [5, 17, 18].

Britton [17] states that bulking brush discharges cannot always be practically eliminated in equipment above about 1 meter diameter when non-conductive powders are transferred.

The following scenarios can lead to bulking brush discharges:

1. Fines in the suspension that coat the conductive wall rendering it less conductive or nonconductive. Now we have a situation of a system that is not grounded in practice. As an example the fines may be all fraction < 10 microns and the MIE for such a fraction will be much lower than the tested sample (< 74 microns). Even though this is a hypothetical situation, one needs to consider if the attrition is a factor in product processing.
2. Fines are in suspension in the free head room of the vessel and the filling continues with the mixture containing coarser particles. As the particle size increases, the equivalent energy increases. See Figure 3. This combination with fines to super fines and coarser particles is a concern as this can result in an incident.

Scenario 2 perfectly describes the situation that occurred in the third event.

2.4 Causes

The root cause of the third events is the high fines content of the CaSt in the FIBCs. The lack of dust extraction may have led to a suspension of the fines in the head space of the metal tote bin. Ignition could have been caused by either a defective Type C FIBC, or a conical brush discharge.

Testing of the existing bag was inconclusive due to the damage it sustained. Experts in static electricity and combustible solids believed that the conical pile discharge could not be ruled out.

Potential mitigation for conical brush discharges can include:

- Flow rate reduction
- Use of active neutralizers for relatively small silos.
- To minimize the probability of ignition, the first consideration
- Minimize the sub-200 mesh fines concentration to increase the MIE to greater than 20 mJ. Inertion, especially if process design or powder properties make deflagrating venting and suppression undesirable alternatives. Inerting is often considered where the powder MIE falls below the 3-10 mJ range.

In this application, inerting was chosen as the fire and explosion prevention technique

3. *Concluding remarks*

This series of events illustrates many points.

Not fully understanding the source of hazards can lead to jumping to obvious, but not comprehensive, conclusions in an investigation. It pays to consult the appropriate subject matter experts even for seemingly simple events.

No good deed goes unpunished. Changes made to improve the safety of a system can introduce new hazards. In this case the switch to Type C bags introduced a new potential ignition source. Training of operating personnel should include not only what the change is and how the procedures may change, but also on new hazards introduced.

Even on a simple system, it pays to consult with subject matter experts on potential hazards. The effect of variations of particle size on MIE and the concept of conical pile discharges, may very well not be known to process designers.

8. REFERENCES

1. Bartknecht, W., **Dust Explosions**, Berlin: Springer-Verlag, 1989.
2. Glor, M., **Electrostatic Hazards in Powder Handling**, Letchworth UK: Research Studies Press Ltd., 1988.
3. "**Recommended Practice on Static Electricity**", NFPA 77, 2003.
4. "**Recommended Practice for the Classification of Flammable Liquids, Gases, or Vapors and of Hazardous (Classified) Locations for Electrical Installations in Chemical Process Areas**" NFPA 497, 2004
5. "Recommended Practice for the Classification of Combustible Dusts and of Hazardous (Classified) Locations for Electrical Installations in Chemical Process Areas" NFPA 499, 2004
6. "Standard for the Prevention of Fire and Dust Explosions from the Manufacturing, Processing, and Handling of Combustible Particulate Solids, NFPA 654, 2006

7. Eckhoff, R.K., **Dust Explosions in the Process Industries**, London UK: Butterworth-Heinemann, 2003.
8. Britton, L.G., **Avoiding Static Ignitions Chemical Operations**, AIChE, CCPS Concept Book, 1999, Publication G-67
9. Jaeger, Norbert and Siwek, Richard; "**Prevent Explosions of Combustible Dusts**", Chemical Engineering Progress, June 1999
10. T.H. Pratt **Electrostatics ignitions of fires and explosions**. Center for Chemical Process safety. AIchE Industry technology Alliance, USA, 1997.
11. L. Britton Private communications, November , 2003
12. M. Glor **Ignition hazard due to static electricity in particulate processes**, Powder Technology 135–136 (2003) 223– 233

Effectiveness of improvised gas absorption techniques for emergency responders at releases of toxic gases

Hannes Kern
Montanuniversität Leoben
Franz-Josef Strasse 18, 8700 Leoben
hannes.kern@unileoben.ac.at

Matthias Taferner, Harald Raupenstrauch
Montanuniversität Leoben
Franz-Josef Strasse 18, 8700 Leoben
matthias.taferner@stud.unileoben.ac.at, harald.raupenstrauch@unileoben.ac.at

Prepared for Presentation at
American Institute of Chemical Engineers
2013 Spring Meeting
9th Global Congress on Process Safety
San Antonio, Texas
April 28 – May 1, 2013

UNPUBLISHED

Effectiveness of improvised gas absorption techniques for emergency responders at releases of toxic gases

Hannes Kern
Montanuniversität Leoben
Franz-Josef Strasse 18, 8700 Leoben
hannes.kern@unileoben.ac.at

Matthias Taferner, Harald Raupenstrauch
Montanuniversität Leoben
Franz-Josef Strasse 18, 8700 Leoben
matthias.taferner@stud.unileoben.ac.at, harald.raupenstrauch@unileoben.ac.at

Keywords: toxic gas release, emergency response, absorption techniques

Abstract

In September 2011 a spill of nitric acid at a galvanic company in Trieben, Austria caused the release of large amounts of nitrous fumes. An IBC tank with 1000 liter of nitric acid (53%) was flipped over by a forklift and cracked open in the top area. The content was fully released and lead to serious corrosion effects on storage racks combined with the production of critical amounts of nitrous fumes. 27 workers had to undergo medical treatment and a storage and chemical handling area of several hundred square meters was filled by nitrous fumes with concentrations of up to 100 ppm.

Emergency responders used an improvised gas absorption system to clear the area of toxic fumes. Outside the sealed area concentrations of nitrous fumes well below the AEGL-2 (4h) value for nitrous dioxide where measured. In the course of the analyses of the response the effectiveness of improvised gas absorption techniques was investigated.

For the investigations chlorine and ammonia in combination with a mobile exhauster, (10.000 m^3/h) commonly used by emergency responders in Austria were used. Different methods of gas absorption where investigated and tested upon their practical relevance.

1. Introduction

Toxic gases are commonly used in industrial and public areas, both as base products for chemical production and as technical gases for various purposes. In Europe gases like chlorine or ammonia are often used as disinfecting agents or coolant in cold storage warehouses. In February 2012 a release of several hundred kilograms of ammonia from a cold storage warehouse in Vienna (Austria) injured 62 Persons, 32 firefighters and 4 emergency medical technicians right in the busy city center of Vienna. Firefighters were able to contain the released ammonia within the warehouse and tried to absorb and disperse it with water by using water curtains and water shields. Half a year earlier, on September 5th 2011 the spillage of 1000 liter nitric acid 53% in a metallurgical production site Trieben (Austria) caused the occurrence of significant amounts of nitrous fumes in a large storage area. The nitrous fumes were produced due to the reaction of the acid with the material of storage racks and metal materials in the surrounding area. The nitrous fumes were contained in the storage area, but ventilation procedures were very critical due to the close surrounding neighborhood. At the incident in Trieben 27 workers were injured. Due to a lack of knowledge regarding improvised gas absorption techniques experiments with ammonia, chlorine and different absorption agents were carried out. The experiments were supported by the Hazmat Team of the Graz Fire Brigade.

2. Application and testing of absorption techniques

2.1 General overview on techniques for gas defense

In the field of gas defense and control different techniques are applied by emergency responders. Depending on the strategy followed by the responding crews and the situational demands there are three main categories of technical and strategic measures that can be used during a response situation dealing with toxic or hazardous gases. These methods are described in the following sections.

2.1.1 Active methods

As "active methods" all response techniques can be considered that aim to directly fight and disperse gas clouds or change their direction of spread. Most of the active response methods are based on technical devices that are able to generate large volume flows of water or/and air that are directed against the spreading gas cloud. Depending on the size of the gas release water nozzles, water monitors or different sized ventilators are used for the active response approach. Appliances like the LUF 60 (firefighting support vehicle) combine a high volume rate ventilator (90.000 m^3/h) with a circular water dispersion system and provide therefore an aerosol cloud that shows better kinetic effects than the sole air stream. For water soluble gases also absorption effects can be achieved by this method.

2.1.1 Passive methods

"Passive methods" aim to contain the released gas cloud or try to protect areas from gas by a restriction of spread. An example for such systems are, water shields normally used to protect from radiative heat during firefighting operations. Also known are water curtains mainly consisting of fire hoses with different numbers of upward directed spray nozzles. Water curtains are more flexible in application. Water shields and water curtains are available with adjustable flow rates and for different operating pressures. The effectiveness largely depends on the spilled gas type and the nozzle and flow configuration of the chosen appliance.

2.1.1 Absorption techniques

Absorption techniques for emergency responders are in general not well established. For different purposes the effects of water on water soluble gases is known and well understood. The application of knowledge from industrial processes and their applicability for emergency responders is scientifically not well reviewed. Commercial portable gas absorbers are established in some industrial emergency response crews but mostly have very low volume rates. For water soluble gases some absorption techniques concerning the use of water aerosols or spray streams are known but often improvised. For chlorine spills the effects of sodium thiosulfate were investigated and have proved to be very effective [1,2], Other chemicals for absorption purposes regarding emergency response have not been studied very well. Effluents from absorption processes could cause hazardous situations too, but absorption in general offers the chance to reduce the three dimensional problem of a hazardous gas release to a two dimensional problem of a hazardous liquid.

2.2 Response at Trieben (Austria)

On September 5th 2011 a forklift at a metallurgical production site in Trieben (Austria) by accident dropped an IBC with 1000 liter 53% of nitric acid in the chemical storage area. The local fire brigade was called at about 11:00 am and arrived about five minutes later. The incident commander noticed during size up operations orange/brown fumes inside the cellar of the building. Most of the cellar was affected, about 200m^2 were directly contaminated with nitric acid. It is assumed that before the fire brigade was called, workers tried to dilute the nitric acid with up to 3000 liter of water, which caused severe a deterioration of the situation. About 30 minutes later the hazmat team arrived and started measurements with detector tubes. Values of around 100 ppm for nitrous fumes were measured in all areas of the cellar. The nitrous fumes occurred from the reaction of nitric acid with the material of the storage racks in the storage area. The lower parts of the storage racks were principally protected by a coating material but at some points the coating was damaged. Emergency responders over nearly 24 hours tried to contain the spill with chemical binding agents and chemical vacuum cleaners. Due to the contamination with nitrous fumes gastight suits were used which lead to huge logistical demands. The normal working time for the cleaning teams was around 45minutes to an hour, they were supplied by air hoses from a mobile compressor for respiratory air. Due to the close vicinity of the surrounding neighborhood it was not possible to vent the nitrous fumes directly in the atmosphere. A mobile exhauster was placed a spray nozzle was fixed near the exhaust opening of the ventilator. The flow rate of the water stream was around 200 l/min at 6 bar. The exhauster had an output of around 20.000 m^3/h. During the exhaust process the concentration of nitrous fumes was constantly monitored and well below 8,2 ppm (AEGL 2, 4h for NO_2) at the direct vicinity of the exhauster. The effect might have not only occurred through absorption effects but also because of a dilution effect of excess air at the input side of the exhauster.

2.3 Absorption tests with the MSA Auer COMVEX Ventilator

To investigate the potential of improvised gas absorption techniques the applicability of the MSA Auer COMVEX Ventilator was chosen as an example. The COMVEX Ventilator is used by many Austrian fire departments and offers the possibility to mount a water spray system on the ventilator that is normally used for the production of high expansion foam.

2.3.1 MSA Auer COMVEX [3]

The MSA Auer COMVEX Ventilator mainly consists of an air fan and tubing for suction and exhaust purposes. The fan has an air consumption of 10.000 m³/h and a power of 2,6 kW. Additional to the fan, a top piece for the production of high expansion foam is available (Pic.1). The foam generating kit consists of short tube with five spraying nozzles and a mixer to be able to ad foaming agents. At an operating pressure of 4-8 bar the water consumption of the foam generating kit is around 100 l/min. Mixing rates of 1-3% are possible.

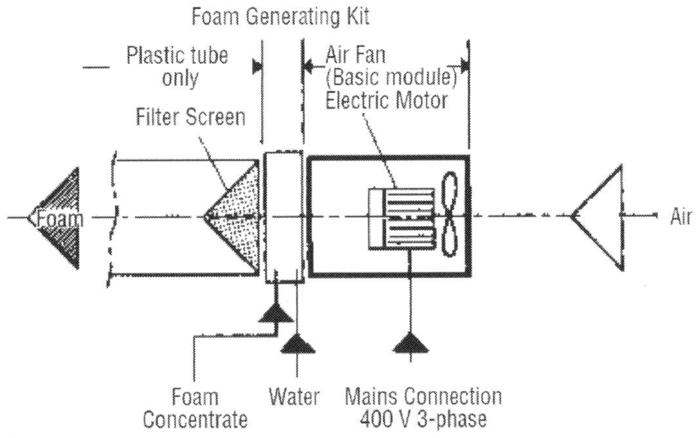

Pic. 1: Principal layout of the COMVEX Ventilator system [3]

2.3.2 Sodium thiosulfate as absorption material

Many absorption materials for technical purposes are well known. For emergency response purposes it is crucial to have absorption materials that present a low hazardous potential and have a good availability in large quantities. Logistically it is not possible to store large amounts of different absorption chemicals for specific tasks. For high risk chemicals (concerning toxicity and used quantities) it may be reasonable to foresee special measures. In case of chlorine that has a high toxic potential and is commonly used for chemical production processes and disinfection purposes, sodium thiosulfate gives good results as absorption agent in combination with water [2].
Sodium thiosulfate in an aqueous solution reduces elementary chlorine to chlorine anions. The reaction is shown in Equation 1:

$$2Na^{+} + S_2O_3^{2-} + 4Cl_2 + 13H_2O \rightarrow 2Na^{+} + 2HSO_4^{-} + 8\ H_3O^{+} + 8Cl^{-} \quad \text{[Eq. 1]}$$

One mol of sodium thiosulfate is able to reduce 4 mol of elementary chlorine and has therefore a high potential to be used as absorption agent. Sodium thiosulfate pentahydrate has a solubility of around 680 g/l at 20 °C which equals a concentration of about 2.7 mol/l.

2.3.2 The MSA Auer COMVEX as improvised absorber

The focus of the investigations was the absorption of ammonia and chlorine at concentrations around 200-300 ppm. The ventilator was equipped with a suction tube (10 m) and an equally built exhaust tube. Tests with ammonia where carried out with inlet concentrations of 250 and 290 ppm. The ammonia feed stream was provided by a bottle of pressurized ammonia, the concentration was monitored by detector tubes. At the start of the experiment a constant ammonia stream was adjusted and measured at the exhaust tube while the ventilator was operating. When constant concentration conditions were reached, the injection of water started and the gas concentration was measured again. At the first test the ammonia concentration dropped from 250 ppm to 100 ppm (33.3%). The second run lead to a concentration of 150 ppm at a starting concentration of 290 ppm (48.3%).

Pic. 2: Improvised absorption system during experiments with chlorine

Due to the good solubility of ammonia in water, the tests were only carried out with water. The second series of experiments was carried out with chlorine in the same way. The first test with chlorine was started at a concentration of 490 ppm, the injection of water caused no measurable reduction of the concentration. In the second experiment with chlorine, a 2,7 mol/l solution of sodium thiosulfate was added to the water stream by the mixer of the foam generating kit (Pic.2). A mixing ratio of 3% was chosen. The use of sodium thiosulfate resulted in a reduction of the chlorine concentration from 260 ppm to 100 ppm (61.5%).

3. Conclusion

The effectiveness of gas absorption techniques is not very well known by emergency responders in Austria. Examples from real incidents show that the availability of gas absorption techniques may improve the effectiveness of emergency response at toxic gas releases. It is clearly not possible to have ready gas absorption appliances and agents for a wide range of specific problems. Scientific research on available improvised techniques and the development of technical improvements may enhance their effectiveness. Experiments with sodium thiosulfate in cases of chlorine spills show an effectiveness of over 60% at low chlorine concentrations. Keeping in mind that the investigated ventilator/nozzle system was not designed for gas absorption in general, the results may still leave a lot of room for improvements. Future research will focus on the improvement of the effectiveness of different methods for the control of toxic gas releases.

4. References

[1] Engelhardt, F., "Chlorabsorption an fallenden Tropfen und Übertragung auf Wasservorhänge," *Bergische Universität –Gesamthochschule Wuppertal*, Dissertation Juni 2002

[2] Bauch, M., Peine, T., "Versuche zur Erhöhung der Effizienz beim Niederschlagen von Chlor," *Brandschutz - Deutsche Feuerwehr Zeitung*, NO. 1, 2010

[3] "MSA AUER COMVEX Portable Ventilator and High Expansion Foam Generator - Instruction Manual," MSA AUER

Case Studies in Process Safety: Lessons Learned from Software-Related Accidents

Terry L. Hardy
Director, Safety & Risk Management
Great Circle Analytics, LLC
1238 Race Street
Denver, CO, 80206
thardy@gcirc.com

Prepared for Presentation at
American Institute of Chemical Engineers
2013 Spring Meeting
9th Global Congress on Process Safety
San Antonio, Texas
April 28 – May 1, 2013

Case Studies in Process Safety: Lessons Learned from Software-Related Accidents

Terry L. Hardy
Director, Safety & Risk Management
Great Circle Analytics, LLC
1238 Race Street
Denver, CO, 80206
thardy@gcirc.com

Keywords: Safety Analysis, Software Safety, Computing Systems, Process Safety Management

Abstract

Software and automation can play a significant role with respect to safety in the chemical process and energy production industries. Because computing systems are increasingly used to control critical functions, software may directly contribute to an accident. On the other hand, software can also be used as part of the hazard control strategy to reduce risks, and computing systems can provide valuable information to help make safety decisions. The importance of including software as part of an organization's efforts to analyze and manage hazards and risks seems clear, but for many organizations software is not effectively incorporated into process safety efforts. This paper reviews lessons learned from accidents and incidents to illustrate the potential for a software-related accident even when process safety management tools and techniques are used. This discussion is intended to provide insights to help improve process safety and software safety efforts.

1. Introduction

Process Safety Management generally refers to the application of management principles and systems to the identification, understanding, and control of process hazards to protect employees, facility assets, and the environment. Process Safety Management is focused on prevention of, preparedness for, mitigation of, response to, and restoration from catastrophic releases of chemicals or energy from a process associated with a facility. In Process Safety Management, a process is defined as any use, storage, manufacturing, handling, or the on-site movement of highly hazardous chemicals, or combination of these activities. Process Safety Management is regulated by OSHA through the Process Safety Management (PSM) Standard.

Software and computing systems are becoming essential to the operation of most industries, often through the automation of tasks, and as such can have a major impact on process safety. Software and computing systems offer tremendous advantages that encourage the use of automation. Software can allow complex operations to be performed, and can perform complicated calculations. Computing systems are reliable in the sense that there are a small number of components to perform a large number of actions. Computing systems are also flexible and can be modified to fit the application. From a safety perspective, automation can reduce the risks to operators by separating them from dangerous environments such as those in hazardous chemical processing facilities. In addition, automation can help by performing tasks that might pose a threat to humans, such as lifting large objects or operating high-speed cutting tools.

Software, computing systems, and automation have disadvantages as well. A major disadvantage to computing systems is that they often introduce increased complexity to systems, making them harder to understand, operate, maintain, and test. This complexity can also introduce new failure modes and interactions. Complex systems have a greater number of requirements, more interfaces, more subsystems, and more logical operations, increasing the potential for error. Computing systems can create a situation where no single operator can immediately foresee the consequences of a given action in the system. Computing systems can also increase coupling. Tight coupling occurs when processes are highly integrated and intrinsically time-dependent; for example, once a process has been set in motion, it must be completed within a certain period of time. Tightly coupled systems allow little time for operators to take action to prevent adverse effects. In addition, automation may require that operators think about their tasks in new ways, and may change the fundamental role of the operator from one who physically performed tasks to someone who monitors and communicates information. Finally, software will only do what a programmer has told it to do. Therefore, unlike humans, software cannot react to an abnormal situation unless specifically told to.

In spite of the fact that software is such an important part of complex systems, the analysis of hazards and risks from software has been inconsistent in the process industries. Safety analyses have historically been hardware-focused. Therefore, many analysts may not understand how to incorporate software into their system hazard analyses, and evaluators of those analyses may not understand what should be included and assessed. Organizations may be focused on compliance to regulations, which often do not address software, and as a result organizations may not properly assess or mitigate software risks. Organizations need to increase the attention given to addressing and analyzing the potential for hazards related to software and computing systems.

Software includes computer programs, procedures, scripts, rules, and associated documentation and data pertaining to the development and operation of a computer. Software can be developed by the organization implementing the system, by an outside software developer, or may be purchased as Commercial Off-The-Shelf (COTS) software. Software safety encompasses not just the software but also the computing system. A computing system includes the software and supporting hardware, sensors, effectors, humans who interact with the system, and data necessary for successful operation. Examples of computing systems include Programmable Logic Controllers (PLC) and Supervisory Control and Data Acquisition (SCADA) systems.

Through the review of hundreds of software-related accidents and incidents, and the author's personal experience in this discipline, common themes and lessons have been found.[1] This paper will use case studies to illustrate those themes and lessons learned. The discussion will center on software safety as part of a broader process safety effort and will include recommendations that can be used to improve the safety of software-driven systems.

2. Lessons Learned: Software and Computing System Related Accidents

This section discusses a number of lessons learned related to accidents and incidents where software and computing systems were contributors. These accidents are taken from detailed reports and investigative summaries from multiple industries and organizations, and are included to broadly illustrate hazards and risks in software and computing systems. Note that in discussing these accidents, this paper does not intend to oversimplify the events and conditions that led to the accidents, or place blame on any individuals or organizations. There is rarely a single identifiable cause leading to an accident. Accidents are usually the result of complex factors that include hardware, software, human interactions, environments, and procedures. The descriptions are meant to provide examples of where the analysis process failed in some way. Readers are encouraged to review the complete accident and mishap investigation reports referenced here to understand the often complex conditions and chain of events that led to each accident discussed here.

2.1 Decisions made in the acquisition and planning phases of development can profoundly affect safety

Planning typically involves trade-offs between many different facets of the program, including cost, schedule, performance, and safety. Poor planning can lead to unexpected safety consequences, and many safety decisions are actually made in the planning and acquisition phase. For example, inadequate resources may be allocated to the software safety effort. This can result in a failure to perform hazard analyses and identify safety requirements early in the program when these activities have the most impact. Therefore, safety personnel should be included in early phases of a program, especially when developing contractual requirements, and software safety considerations should be included in schedule and resource discussions.

On August 11, 1985, an Institute, West Virginia chemical production facility leaked methylene chloride and aldicarb oxime, toxic chemicals used to manufacture the pesticide Temik. Six workers were injured, and more than a hundred residents living near the facility were sent to the hospital. The toxic chemical leak occurred when a tank overheated after a steam valve failed. Due to the failure, steam leaked into the heating jacket of a storage tank containing a mixture of methylene chloride and aldicarb oxime. The methylene chloride vaporized from the heating, leaving concentrated aldicarb oxime to settle on the bottom of the tank. Operators later pumped the material from the storage tank for processing. They thought they had emptied the tank, but concentrated aldicarb oxime remained in the storage tank because a level gauge was broken and the operators performed no additional checks. The concentrated aldicarb oxime continued to be heated over several days by the steam in the heating jacket, ultimately resulting in a runaway chemical reaction days later. The high pressure from the reaction led to release of aldicarb oxime through a flare, and large amounts of the chemical were also released to the atmosphere when a burst disc failed.

The root causes of the leak were found to be the use of a tank that was not designed to hold aldicarb oxime, faulty level indicators, defective safety valves, mistaken transmission of steam to a tank, and failure of control room operators to notice critical pressure and temperature changes. In addition, a computerized monitoring system contributed to the severity of the accident. Earlier in the year, the company had installed a new air-monitoring system that could automatically identify and detect releases of hazardous gases. That computerized system could also predict whether the gases would be contained within the release area or would migrate and potentially harm the public. This information was important because it could be used to mitigate risks through evacuation or other measures if a vapor cloud appeared to be spreading. Company officials initially stated that the computer system had failed and therefore they were not informed of the lack of containment of the aldicarb oxime gas cloud. However, they later admitted that they had not purchased a version of the software programmed to detect aldicarb oxime. Had the company invested in the more expensive version of the software, they could have predicted that a cloud of aldicarb oxime would not have been contained in the production area. However, the company had only ordered the basic software model that did not contain aldicarb oxime properties.[2, 3]

2.2 Hazard causes may not include software and computing systems

Many organizations are improving their safety analyses with respect to software. However, software safety still lacks sufficient numbers of qualified practitioners, and the methods for performing comprehensive analyses are not universally known. In many cases, software causes may not be included in hazard analyses. Or if they are, the software causes may be stated in general terms in hazard analyses, such as including a generic hazard cause called "software error," instead of defining specifically what software functionality can lead to an undesirable outcome. In addition, the software hazard analyses may not pay enough attention to those cases where the software works exactly as intended, but the implemented functionality was unsafe.

On October 18, 2005, an explosion occurred at a natural gas decompression/recompression facility near Empress, Alberta. Extensive damage occurred to the facility, but no injuries were reported. The facility decompressed and conditioned high-pressure gas from a pipeline to achieve pressures and temperatures consistent with what was needed in the Natural Gas Liquids (NGL) stripping facilities. At the time of the accident the facility's "A" PLC lost communication with the NGL facility control room. Transportation Safety Board of Canada (TSB) investigators found that a processor card had failed, shutting down the PLC. When the PLC failed its outputs were de-energized. However, some of the microprocessor functions required an energized signal to initiate physical shutdown of some functions. These functions included main and auxiliary lube oil pumps, purge/damper valves, and suction and discharge valves. As a result of this condition, the pumps did not shut down and valves did not close in a safe position when the PLC failed. Operators in the NGL control room received an alarm indicating that the communications were lost. They investigated the problem and believed that the failure in communication had triggered an emergency shutdown of the facility. The operators believed that the emergency shutdown meant that all components were in a safe state. However, because the recompressor motor did not shut down, a surge event occurred in the pipe. This surge caused a pipe to break, leading to a natural gas leak. The leaking natural gas ignited, resulting in an explosion. The TSB noted that the loss of a PLC had not been identified as a safety concern in the hazard analyses prior to the accident.[4]

2.3 Risks may be underestimated or optimistically evaluated

The risk assessment process helps decision makers in their risk decisions (e.g., accept the risk without change, reduce the risk to an acceptable level through hazard elimination and mitigation, transfer the risk, or forego the activity), assists in justifying the acceptance of residual risk, and allows communication of risks throughout the organization. A risk assessment can either be qualitative or quantitative, although the emphasis in process safety is often on qualitative risk assessment. Evaluating software and computing system risk can be difficult, especially if an extensive history does not exist with the automated equipment. Risk must be carefully evaluated based on known technical factors, such as design complexity and maturity, degree of system testing, use of unproven technologies, potential for unexpected human intervention, and so on. Accidents have illustrated unfounded optimism in the use of software to assure safety.

On April 7, 2000, a pipeline failure occurred at the Chalk Point Generating Station in Maryland. The pipeline released approximately 140,400 gallons of fuel oil into surrounding wetlands, Swanson Creek, and the Patuxent River. The cost of the environmental response and cleanup exceeded $71 million. The U.S. National Transportation Safety Board (NTSB) determined that the probable cause of the accident was a fracture in a buckle in the pipe. The buckle went undiscovered because data from an in-line inspection tool were interpreted incorrectly. The NTSB stated that a contributor to the severity of the accident was inadequate operating procedures and practices for monitoring the magnitude of flow through the pipeline. When the leak occurred the 12-inch-diameter pipeline was being cleaned. An automated pipeline monitoring system existed, but this automated system could not adequately monitor pipeline conditions during this operation because the locations of the meters, pressure-sensing points, and temperature-sensing points had not been configured to be in the direct liquid flow path. Therefore, the automated system did not provide alarms or other information to operations personnel. Field personnel manually recorded tank level measurements at the upstream and downstream points, but did not use this information to evaluate whether any fuel oil had been lost. The manual monitoring procedures and lack of alarms from the automated system led to a 7-hour delay in responding to the leak. The field personnel finally realized there was a problem when the fuel pump began to cavitate. Using hand calculations the crew then realized that they had not received over 3000 barrels of fuel oil and they shut the pipeline down. Following the accident the company installed a SCADA system with software-based leak detection and radar tank gauges. In this case the operators underestimated the risks associated with a leak and overestimated their ability to detect a failure.[5]

2.4 Hazard controls may rely only on good software processes and testing

Hazard controls are devices and approaches to mitigate risk presented by a hazard by reducing either the severity of the hazard or the probability of its occurrence. Strong hazard control design follows the design order of precedence, where the first approach is to try to design out the hazard or minimize the risks through design selection. Many hazard reports will focus on the implementation of good software processes, or extensive unit testing, to prove that the design is safe. However, the software processes and testing will not help if the software design is flawed with respect to safe system operations (including hardware, software, processes, humans, and environments). In these cases the software may operate exactly as designed, but the operation may be unsafe.

On February 11, 2003, an employee of a manufacturer in Gonzales, Texas was fatally injured while performing maintenance on a reaction tank. The U.S. Mine Safety and Health Administration (MSHA) determined that the cause of the accident was a failure to close and secure a manual gate valve for a steam line and a failure to place the batch PLC in the stop mode. The company was a surface clay mill that purchased clay and blended, refined, milled and processed the material into products used in paints, inks, and grease. On the day of the accident the employee had been informed that there had been a product change in one of the batch processing systems. The employee was assigned to perform cleanup duties on a reactor tank. Two valves controlled steam entry into the tank: a manual gate valve and a butterfly valve with an automatic pneumatic actuator. The PLC controlled the functioning of the batch system based on sensors that monitored material flow. At the time of the accident the PLC was in "slurry hold" mode. In this mode the system was programmed to actuate the steam valve when the clay slurry level reached 5.5 feet. An aluminum extension ladder used by the employee caused the level sensor to falsely sense that slurry was in the reactor, which resulted in the PLC sending a command to open the steam valve. Because the manual valve had been left open, steam at 350° F then entered the tank, fatally burning the employee.[6]

2.5 Organizations may fail to consider the possibility of hazard controls not working

Hazard controls are never 100 percent reliable. Therefore, it is important for the analyst to consider what could happen if the controls fail. Consideration should be given to unexpected hardware, software, and human interactions, and a "defense in depth" approach should be used where multiple, independent methods are used to prevent a hazard from becoming an accident. Controls should be carefully evaluated to identify what could happen if they do not work as expected, especially if those controls rely heavily on software and personnel actions.

On February 18, 2009, an employee was fatally injured at a coal preparation plant in the Hunter Valley region of New South Wales, Australia. The accident occurred when 10 tons of waste rock were inadvertently released from the reject bin and fell onto the cabin of the employee's truck. At the plant, raw coal was extracted from the mine and usable coal was separated from waste rock. The waste rock was transferred approximately 2 kilometers on conveyers to the reject bin. The waste rock was then loaded from the reject bin onto trucks and hauled away. The process of loading the trucks with waste rock was controlled by a PLC system. The PLC system included truck detection sensors, traffic lights, bin capacity sensing, and remote control, hand-held transmitters used by the truck drivers. On the day of the accident the truck driver drove his truck under the reject bin delivery chute. A signal was sent from the handheld remote control to command the chute to open. The accident report stated that it was not clear whether the signal was sent inadvertently or intentionally. Opening the chute required that two of three lines of truck detection sensors be blocked in addition to a command from the remote control to assure that the truck was in the correct location. Each sensor line contained three sensors, and all three sensors had to be blocked for the entire line to be considered as blocked. At the time of the accident the truck was obscuring one line of sensors, and a second line of sensors was obscured by dirt on the lenses and therefore was not working correctly. Because two of the sensor lines were blocked and the remote control signal had been sent, the PLC automatically opened the reject bin chute door and dropped 10 tons of material on the truck cab before the driver had safely cleared the chute, resulting in the fatal injury.[7]

2.6 Testing may not provide information on subsystem and component interactions or hazard control operations

The safety verification and validation process is intended to determine that the design solution has met all the safety requirements (verification) and that the correct system is built (validation). The verification and validation process, if performed correctly, will provide evidence that risk has been reduced. Testing is an important part of the verification and validation process, and comprehensive software testing should be conducted throughout the development cycle. Testing should include not only nominal conditions based on requirements but also abnormal conditions such as improper inputs, inputs outside expected conditions, inputs exactly at and near the boundary conditions, and inputs stuck at some value. Safety-critical software must include full system integration testing of end-to-end events. That testing must include stressing of the software, and should include interactions of the software with hardware, humans, and environments. A number of accidents have occurred when no component failed in the conventional sense, but the interaction of components caused a system failure. Testing must include verifications of hazard controls, and it should assure that redundancy works when needed.

On October 24, 2002, a grinder exploded at a limestone processing plant in Foreman, Arkansas. An operator was killed when flammable waste fuel covered him and ignited. The operator had started the pump for solid waste fuel processing when the accident occurred. The MSHA stated in its report that the cause of the accident was that the safety monitoring system designed to shut off the waste fuel system pump had not been maintained so that it functioned properly. Kilns were used in processing activities at the plant, and these kilns were heated by burning coal, natural gas, and liquid waste fuel. The liquid waste fuel was delivered by truck or railcar and pumped into large storage tanks. From the storage area it was pumped through a grinder to reduce the particle size of the solids in the fuel. Two independent systems monitored and controlled the waste fuel delivery. A Foxboro Intelligent Automation Distribution Control System monitored and recorded normal operating parameters. The Foxboro also issued audible and visual alarms that were available at the plant control room. A PLC provided basic startup and shutdown of the system and responded to commands from the Foxboro. On the day of the accident the Foxboro sensed that the fuel delivery pressure was low, apparently due to blockage in the line. As designed, the Foxboro sent a command to the PLC to shut down the pumps. However, the PLC failed to respond and the pumps kept running. Three months prior to the accident this PLC had been installed; this was supposed to be a simple replacement of an older PLC of similar capability. However, the Foxboro had not been connected to the newer PLC, and the connections remained to the older non-functioning PLC. The complete system had never been tested with the new PLC. A system test had been scheduled three days prior to the accident but had been aborted when a pump failed during the test; the test had never been rescheduled. The accident report stated that the blockage may have broken free just prior to the accident. With the pumps running, the pressure elevated significantly and a "water hammer" effect caused overpressurization in the system at the grinder. The grinder was torn loose from its base, spraying fuel and pulling loose a 480-volt cable that ultimately served as an ignition source.[8]

2.7 *Software change management and hazard analyses processes may not be integrated*

Engineering by its very nature is an activity that requires change. Changes can come about for a variety of reasons, including the discovery of problems during development, changes in requirements, or routine upgrades of software or hardware. As the development cycle proceeds and as an organization gains more operational experience, new hazards are often uncovered and some hazards may no longer be relevant. If the hazard analysis is not updated to reflect these changes then resources may be expended on previously identified hazards that may no longer be relevant, and new hazards may not be discovered as the design matures. In addition, routine changes and modifications to software and computing systems must be analyzed for potential hazards as part of the broader hazard analysis process.

On June 10, 1999, a 16-inch-diameter pipe carrying gasoline ruptured in Bellingham, Washington. The ruptured pipeline released approximately 237,000 gallons of gasoline into a nearby creek, according to the NTSB. That gasoline then ignited, burning approximately 1½ miles along the creek. Three people died in the fire. One home and the Bellingham water treatment plant were also damaged in the accident.

The NTSB investigated the accident and determined that the probable cause of the rupture was damage to the pipe during a modification project performed in 1994. This damage weakened the pipeline, making it susceptible to rupture under increased pressure in the pipe. The NTSB also stated that inspections of the pipeline during the project were inadequate and the company did not identify and repair damage. The report noted that in-line pipeline inspection data should have prompted the company to excavate and examine that section of pipeline, but the company failed to perform such work after reported anomalies. The SCADA system computers also played a role in the accident. The SCADA system was used for operation of the pipeline, for example to open and close valves remotely as required or to operate pumps as needed. Just prior to the accident the operator was preparing to initiate delivery of gasoline to a terminal in Seattle, diverting delivery from another facility. During the process of switching delivery destinations, the pressure in the pipeline began to increase, which is a normal condition but one that required the operator to start a pump to reduce pressure. However, when the operator tried to start that pump, the SCADA system failed to execute the start command issued by the operator. The operator soon found that the SCADA system was unresponsive to any commands, something that had never happened before. The report stated that, "Had the controller been able to start the pump at Woodinville, it is probable that the pressure backup would have been alleviated and the pipeline operated routinely for the balance of the fuel delivery." Instead, the pressure in the pipe increased, and the increased pressure likely caused the damaged pipe to rupture.

The cause of computer system failure was likely a change made to the system database just prior to the accident. The NTSB accident report stated that the SCADA system administrator entered new records into the live database at the time of the accident. The system administrator however did not check the records or test the system software to see if those changes introduced any problems. The computing system problem could not be replicated after the accident and therefore the cause of the anomaly could not be definitively identified. The report stated, "The Safety Board concludes that, had the SCADA database revisions that were performed shortly before the accident been performed and thoroughly tested on an off-line system instead of the primary on-line SCADA system, errors resulting from those revisions may have been identified and repaired before they could affect the operation of the pipeline."[9]

2.8 Human-software interactions have significant safety implications that are often underestimated

Humans interact with hardware and software in a variety of ways. A number of accidents have occurred where these interactions have been a significant contributing cause. Of special concern is when a new computing system is implemented that changes what is expected of the operator in normal and emergency situations.

On August 28, 2008, an explosion occurred at a pesticide manufacturing plant in Institute, West Virginia. Two workers were killed in the explosion and eight others were injured. The U.S. Chemical Safety and Hazard Investigation Board (CSB) found that the explosion was the result of a runaway chemical reaction. The company was starting up a methomyl unit for the first time after several months of down time for maintenance to install a new computer control system and reactor vessel. Normal operations called for dissolved methomyl and waste chemicals to be fed into a preheated residue treater vessel partially filled with solvent. The treater allowed the methomyl to decompose safely, after which it was mixed with other waste chemicals and used to fuel facility boilers. On the day of the accident a methomyl-solvent mixture was added prematurely to the residue treater vessel before solvent had been added to the tank. This event occurred when operators were troubleshooting equipment problems during the startup. This mixture was supposed to be added to the vessel only after that vessel was filled with clean solvent and heated to a minimum safe temperature. An interlock system existed as part of the automatic feed control system to prevent inadvertent introduction of methomyl. This interlock was password-protected to prevent inadvertent override, but operators intentionally overrode the interlock. Bypassing the interlocks had apparently been a practice condoned by management in the past as a workaround for operational problems. Once the methomyl decomposition reaction started it could not be stopped, and the pressure rapidly rose in the vessel due to gas from the reaction, leading to the explosion. Uncrystallized methomyl existed in the tank due to equipment problems, and this material greatly increased the methomyl concentration in the residue treater, contributing to the runaway reaction.

The CSB stated that the company initiated this process startup before the company had completed critical checks, including valve lineups, equipment checkouts, and computer calibrations. The CSB said that the operating procedure for the startup was complex, but this procedure had not been reviewed or approved. In addition, the company had not performed training on the new computer system put in place as part of the maintenance. This new computer system offered significant improvements to help automatically control the operation. For example, the control system included graphical display screens that simulated the process flow. But the system was also complex and the modifications changed the way operators interacted with the system. The operators thought the screens were difficult to navigate, and responding to troubleshooting alarms was difficult. The accident report stated that, had the operators received adequate training on the new computer system, they may have been able to recognize problems in operation before the explosion.

The report faulted the company for not performing an adequate pre-startup safety review. System operators did not participate in the safety review, and review checklists showed items as completed when they were not. The CSB stated that the company had also failed to perform a thorough Process Hazard Analysis. The report stated that the Process Hazard Analysis was performed quickly because management had not allotted sufficient time for analysis. In addition, the CSB stated that the Process Hazard Analysis included invalid assumptions, and said that the team did not apply the tools properly, resulting in unmitigated accident scenarios.[10]

2.9 Support software, including models and simulations, may be as critical to safety as control software

While the focus of software safety efforts is usually on software directly controlling an application, support software may contribute to an accident or to the effectiveness of the emergency response. Examples of support software include databases used for maintenance activities, software to estimate load stability, models that provide design calculations or assurance information, and so on. A failure to analyze the hazards associated with this support software can result in unforeseen system failures.

On November 12, 2008, a 2 million gallon liquid fertilizer tank at in Chesapeake, Virginia collapsed. Two workers performing welding operations at the site were seriously injured and an adjacent neighborhood was partially flooded as a result of the accident. The CSB found that the company had not assured that welds to replace vertical joints met accepted industry standards, and the CSB faulted the company for its failure to perform inspections of the welds. The company was also faulted for not having proper procedures in place for filling the tanks following major facility modifications. In its report, the CSB also noted that the contractor hired by the company to calculate the maximum fill height had used some faulty assumptions. The maximum liquid level was supposed to be calculated in part based on the minimum measured shell thicknesses and the extent of the weld inspection (full, spot, or no radiography). The contractor used the maximum (not minimum) measured thickness, and assumed full inspection of the welds. These assumptions led to an overestimation of the allowable liquid level. The tank failed at a fill level of 26.74 feet, below the calculated maximum of 27.01 feet. The CSB also noted a number of previous overfilling accidents. The CSB found 16 other tank failures at nine facilities in other states between 1995 and 2008. These 16 failures resulted in one death, four hospitalizations, one community evacuation, and two releases to waterways. Eleven occurred due to defective welding.[11]

3. Recommendations for the Software System Safety Process

There is a great deal of information in the literature on how to improve software development processes. However, the above themes and lessons learned indicate that improving software and computing system safety is more than just providing additional training or improved software development techniques or tools. Improving software safety requires a change in the way process safety practitioners think. Software safety must be fully integrated into the process safety analysis process. And process safety practitioners must begin to ask critical questions about their software safety efforts. Some examples of these questions include the following:

- Do plans reflect how business is really done? Are plans reviewed? Do plans have unrealistic schedules or resource allocations? Are software and computing systems considering in the planning and acquisition processes? Poor or unrealistic plans may reflect an organization that does not truly place a priority on safety activities.
- Is there a convincing story that the safety analysis is complete? Is there a sufficient description of the system to support an analysis, including the software and computing system? Are computing systems treated as "black boxes"? Have software causes been thoroughly evaluated using a systematic approach, or are the software causes listed as "software error" with no further explanation? Do the hazard analyses include support software, such as design models or databases? Failure to show that the problem is being looked at systematically could indicate that there are holes in the analysis, with potentially significant problems overlooked.
- Are the hazard reports detailed enough? Are causes descriptive? Are the recommendations clear and sufficiently detailed? Does the logic make sense and is it complete? Do controls match up with the causes, showing a one-to-one or many-to-one relation? Lack of detail could be an indication of insufficient knowledge of the system, or lack of information on the system.
- Do the risk assessments have a logical basis? Have the risk assessments considered software complexity, maturity, testing, use of unproven technologies, etc.? Are assessments overly optimistic based on what is known about the computing system? A failure to provide a basis for your risk assessments may result in unrealistic assessments.
- Are the hazard controls primarily procedural rather than design changes, safety features, or devices? Is there an overreliance on alarms and signage? Do software controls only rely on good software processes? Is there an overreliance on humans and software to "save the day"? Overreliance on operational controls may indicate a weak safety design.
- Are hazard control recommendations being implemented? Can the selected hazard control strategy actually be implemented and verified? Is the control strategy so complex that it will be impossible to determine whether it will work when needed? Complex controls, overlapping control strategies, or inadequate implementation may be an indication of a weak safety design.
- Has the risk assessment truly considered the worst case? What is the basis for the likelihood levels? Is the risk analyzed only for steady-state operation, or have startups and shutdowns also been considered? Failure to provide good answers to these questions indicates a potential misunderstanding of the risk.
- Are problems found in test and design included in the hazard reports and factored into the design? Have problems and incidents been fully investigated? Does a process exist for tracking action items, including whether recommendations have been implemented? Are changes to software and computing systems properly factored into the hazard analysis? Failure to incorporate problems, changes, and corrective actions is an indication of the potential to miss serious design flaws.

These questions help to identify whether the hazard analysis process is robust. However, we must also ask questions specifically related to the use of software and computing systems in complex systems. The best questions come from real-world examples of accidents where software has been a contributor. Some examples of questions are as follows, and others can be found in the literature.[12]

- Have safety-critical software, commands, and data been identified?
- Do hazard controls for software-related causes combine good practices and specific safeguards?
- Is software and system testing adequate, and do tests include sufficient off-nominal conditions?
- Is the computing system design overly complex?
- Is the design based on unproven technologies?
- What happens if the software locks up?
- Are the sensors used for software decisions fault tolerant?
- Has software mode transition been considered?
- Has consideration been given to the order of commands and potential out of sequence inputs?
- Will the software and system start up and shut down in a known, safe state?
- Are checks performed before initiating hazardous operations?
- Will the software properly handle spurious signals and power outages?

These are by no means all the questions a decision maker should ask, and positive answers to these questions provide no assurance that an accident will be prevented. These questions should encourage critical thinking about safety and generate additional questions to provide further insight on system risk. A failure to ask these questions could mean that the potential for an accident is higher than we had assumed.

It is up to all stakeholders to look for those conditions that could lead to an accident and to recognize that the worst can happen. This means we should all express concerns about software safety management and engineering when necessary based on our knowledge, experience, and judgment, including lessons learned from accidents. We must ask questions to understand the potential for harm, to understand the steps taken to assure that the risks have been reduced, and to assure that there is proof that hazard controls are effective. We do not have to be software experts to ask questions, and in fact a lack of expertise could be an advantage when trying to understand how a system works. Most importantly, when thinking of software and computing system safety, we must think of the system and its interactions between software, hardware, humans, processes, and environments.

4. Conclusion

Process Safety Management can provide immense benefits in reducing operational risks in the chemical process and energy production industries. By proactively identifying hazards, assessing and characterizing risks, and taking actions to reduce those risks, organizations can prevent accidents and reduce the potential for death, injury, property damage, and environmental impacts. Given the importance of software and computing systems in the design and operation of many systems, software must be included as part of the broader process safety effort. However, the accidents provided in this paper, along with many others, illustrate that software is often not systematically considered as part of process safety activities. For example, software may not be identified as a hazard cause, risks may be underestimated, software-related hazard controls may be oversimplified, testing may not include sufficient safety scenarios, and system changes may not be factored into the design or hazard analysis. Process safety practitioners must take advantage of available software and computing system lessons learned, going beyond examples presented here, to improve their own safety efforts and, most importantly, to prevent accidents.

5. References

[1] Hardy, T.L., *Software and System Safety: Accidents, Incidents, and Lessons Learned,* AuthorHouse, 2012.

[2] Pool, B., "System Wrongfully Blamed in Union Carbide Leak: SAFER Fights Off Dark Cloud of Bad Publicity," *Los Angeles Times,* August 20, 1985.

[3] Schlager, N., *Breakdown: Deadly Technological Disasters,* Visible Ink Press, 1995.

[4] Transportation Safety Board of Canada, "Programmable Logic Controller Failure, Foothills Pipe Lines Ltd. Decompression/Recompression Facility, BP Canada Energy Company Empress Natural Gas Liquids Facility Near Empress, Alberta 18 October 2005," Report Number P05H0061, July 12, 2006.

[5] U.S. National Transportation Safety Board, "Rupture of Piney Point Oil Pipeline and Release of Fuel Oil Near Chalk Point, Maryland, April 7, 2000," Pipeline Accident Report NTSB/PAR-02/01, July 23, 2002.

[6] U.S. Mine Safety and Health Administration, "Report of Investigation: Fatal Other Accident (Steam Burns), February 11, 2003, Southern Clay Plants & Pits Southern Clay Prod. Inc., Gonzales, Gonzales County, Texas," Mine I.D. No. 41-00298, 2003.

[7] State of New South Wales (Australia), Department of Industry and Investment, "Fatality involving David Hurst Oldknow Ravensworth Underground Mine Coal Preparation Plant Reject bin 802 18 February 2009," May 2010.

[8] U.S. Mine Safety and Health Administration, "Report of Accident : Exploding Vessels Under Pressure Accident, October 24, 2002, Foreman Quarry and Plant, Ash Grove Cement Company, Foreman, Little River County, Arkansas," Mine I.D. No. 03-00256, 2003.

[9] U.S. National Transportation Safety Board, "Pipeline Rupture and Subsequent Fire in Bellingham, Washington, June 10, 1999," Pipeline Accident Report NTSB/PAR-02/02, October 8, 2002.

[10] U.S. Chemical Safety and Hazard Investigation Board, "Pesticide Chemical Runaway Reaction Pressure Vessel Explosion, Bayer CropScience, Institute, West Virginia, August 28, 2008," Report No. 2008-08-I-WV, January, 2011.

[11] U.S. Chemical Safety and Hazard Investigation Board, "Investigation Report: Allied Terminals, Inc. – Catastrophic Tank Collapse, Allied Terminals, Inc., Chesapeake, Virginia, November 12, 2008," Report No. 2009-03-I-VA, May 2009.

[12] Hardy, T.L., *Essential Questions in System Safety: A Guide for Safety Decision Makers,* AuthorHouse, 2011.

Technical Report on Chevron 2012 Pipe Rupture and Fire

Steve Cutchen, Lauren Grim
United States Chemical Safety and Hazard Investigation Board
1 Denver Federal Center Way
Building 20, Room A1419
Denver, CO 80225

Roger Evans, Don Holmstrom, Amanda Johnson, Dan Tillema, Kelly Wilson, Mark Wingard
United States Chemical Safety and Hazard Investigation Board

The CSB is an independent federal agency charged with investigating serious chemical incidents. The agency's board members are appointed by the President and confirmed by the Senate. CSB investigations look into all aspects of chemical incidents, including physical causes such as equipment failure as well as inadequacies in regulations, industry standards, and safety management systems. Please visit the CSB's website at www.csb.gov. For further information contact Communications Director Hillary Cohen at (202) 446-8094 or Sandy Gilmour at (202) 251-5496

Prepared for Presentation at
American Institute of Chemical Engineers
2013 Spring Meeting
9th Global Congress on Process Safety
San Antonio, Texas
April 28 – May 1, 2013

UNPUBLISHED

Technical Report on Chevron 2012 Pipe Rupture and Fire

Steve Cutchen, Lauren Grim
United States Chemical Safety and Hazard Investigation Board
1 Denver Federal Center Way
Building 20, Room A1419
Denver, CO 80225

Roger Evans, Don Holmstrom, Amanda Johnson, Dan Tillema, Kelly Wilson, Mark Wingard
United States Chemical Safety and Hazard Investigation Board

Keywords: corrosion, sulfidation corrosion

Abstract

The U.S. Chemical Safety and Hazard Investigation Board (CSB) and the California Division of Occupational Safety and Health (Cal/OSHA) released a technical evaluation report on piping samples taken from the Chevron Refinery in Richmond, California, where a hydrocarbon release and massive fire occurred on August 6, 2012. The report, prepared by Anamet, Inc., concluded the piping ruptured due to severe sulfidation corrosion, and that tested pipe samples showed a very low concentration of corrosion-inhibiting silicon.

1. Background

On August 6, 2012, a hydrocarbon release and resulting fire occurred at the Chevron U.S.A. Inc. Richmond Refinery in Richmond, California The incident occurred when a combustible hydrocarbon liquid known as "gas oil" leaked from an 8-inch pipe connected to an atmospheric crude oil distillation column in the refinery's crude unit. Workers initially noted the leak and were in the process of attempting to diagnose the source of the leak in the still-operating crude unit when the pipe ruptured catastrophically. Due to the high temperature, in excess of 600 degrees Fahrenheit, and physical properties of the material in the equipment, the gas-oil immediately formed a large hydrocarbon vapor cloud.

The CSB determined that nineteen Chevron employees were engulfed in a vapor cloud formed by the hydrocarbon release. Eighteen employees escaped before the fire started and one employee escaped without injury after the fire ensued. The fire burned for approximately 5 hours, and water was applied to the rupture site for 23 days. The incident resulted in six minor injuries. Production at the crude unit was suspended for over six months after the incident. More than 15,000 residents in the surrounding area sought treatment at area medical facilities as a result of the release and fire. Post incident on-site visual examination by Chevron, Cal/OSHA, and the CSB revealed extensive thinning of the pipe wall at the location of the rupture.

2. Sample Testing and Report

The evaluation, performed by Anamet, Inc. of Hayward, California was designed and overseen by the CSB, the California Division of Occupational Safety and Health Administration (Cal/OSHA), the United Steelworkers union, and Chevron. The evaluation was funded by the CSB. The report concludes that the 8-inch steel pipe, from a section designated as 4-sidecut which was installed in 1976, ruptured due to severe sulfidation corrosion, and that tested pipe samples showed a very low concentration of corrosion-inhibiting silicon.

The report cites wall thinning due to sulfidation corrosion as the cause of the piping failure. In crude oil distillation, the report notes, naturally occurring sulfur and sulfur compounds are available to react with steel components, particularly plain carbon steels. Corrosion rates vary according to the sulfur content of the oil being processed, temperature, and other factors, including silicon content and other materials in steel pipe.

The report concludes, "The reported service conditions of the #4 Crude Unit 4-sidecut piping, the composition of the carbon steel, presence of thick sulfide scale on the inside surfaces of the pipe, and generally uniform wall thinning indicate that sulfidation corrosion was the cause of the 8-inch 4-sidecut rupture."

The report noted that the failed pipe section was subjected to a higher corrosion rate than neighboring components due to low silicon content in the pipe. The report states that "Experience has shown that silicon in carbon steel is known to inhibit sulfidation" when the concentration is above a threshold value. Chemical analysis of the Chevron pipe showed the silicon concentration of the ruptured section to be far below this value, resulting in a significantly higher rate of sulfidation corrosion.

The report, which contains numerous photographs of the corrosion and pipe rupture, also notes the presence of an inward deformation toward the inside surface of the pipe. The CSB has determined this deformation was likely caused by a fire pike used during the plant's emergency response. A photograph depicts a tight fit between the fire pike tip and this inward deformation of the failed pipe.

3. Report Findings

The Anamet report provides the following conclusions:

1. Rupture of the 8-inch 4-sidecut resulted from wall thinning caused by sulfidation corrosion.
2. Post rupture corrosion destroyed fracture morphology that could have indicated the rupture initiation site. However, rupture likely originated in the thinnest region and initially followed a longitudinal path driven by the hoop stress. Consequently, an area of likely rupture initiation was identified.

3. The wall thickness of the ruptured section was less than all the other sections of the 8-inch 4-sidecut that were evaluated.

4. Chemical analysis showed the silicon concentration of the ruptured section was 0.01-wt%.

5. Six of twelve specimens (50%) from unique sections of the 8-inch 4-sidecut, 12-inch 4-sidecut, and 2011 12-inch samples were found by chemical analysis to have a silicon concentration of less than 0.1-wt%.

6. Deformation of a small region of the 8-inch 4-sidecut rupture edge was consistent with mechanical force applied from the outside surface of the pipe, possibly from a fire pike, such as samples E-099 and E-082-2, or other pointed object. This deformed region was located within the area of likely rupture initiation.

4. CSB Future Plans

The CSB investigation to determine the root causes of the incident is ongoing. The CSB will release a report detailing its findings and recommendations to key stakeholders later this year. The CSB, Cal/OSHA, the USW, and Chevron are cooperating under an agreement to test 4-sidecut carbon steel piping from Chevron's refinery in El Segundo, California, south of Los Angeles.

AUTHOR INDEX

AUTHOR INDEX